AF804171

Methods in Molecular Biology

Series Editor
John M. Walker
School of Life and Medical Sciences
University of Hertfordshire
Hatfield, Hertfordshire, AL10 9AB, UK

For further volumes:
http://www.springer.com/series/7651

Virus-Derived Nanoparticles for Advanced Technologies

Methods and Protocols

Edited by

Christina Wege

Institute of Biomaterials and Biomolecular Systems, University of Stuttgart, Stuttgart, Germany

George P. Lomonossoff

Department of Biological Chemistry, John Innes Centre, Norwich, UK

Humana Press

Editors
Christina Wege
Institute of Biomaterials and Biomolecular Systems
University of Stuttgart
Stuttgart, Germany

George P. Lomonossoff
Department of Biological Chemistry
John Innes Centre
Norwich, UK

ISSN 1064-3745 ISSN 1940-6029 (electronic)
Methods in Molecular Biology
ISBN 978-1-4939-7806-9 ISBN 978-1-4939-7808-3 (eBook)
https://doi.org/10.1007/978-1-4939-7808-3

Library of Congress Control Number: 2018940647

© Springer Science+Business Media, LLC, part of Springer Nature 2018
This work is subject to copyright. All rights are reserved by the Publisher, whether the whole or part of the material is concerned, specifically the rights of translation, reprinting, reuse of illustrations, recitation, broadcasting, reproduction on microfilms or in any other physical way, and transmission or information storage and retrieval, electronic adaptation, computer software, or by similar or dissimilar methodology now known or hereafter developed.
The use of general descriptive names, registered names, trademarks, service marks, etc. in this publication does not imply, even in the absence of a specific statement, that such names are exempt from the relevant protective laws and regulations and therefore free for general use.
The publisher, the authors and the editors are safe to assume that the advice and information in this book are believed to be true and accurate at the date of publication. Neither the publisher nor the authors or the editors give a warranty, express or implied, with respect to the material contained herein or for any errors or omissions that may have been made. The publisher remains neutral with regard to jurisdictional claims in published maps and institutional affiliations.

Printed on acid-free paper

This Humana Press imprint is published by the registered company Springer Science+Business Media, LLC part of Springer Nature.
The registered company address is: 233 Spring Street, New York, NY 10013, U.S.A.

Preface

Nanotechnology is a subject that continues to attract considerable public interest, not all of it favorable. Indeed, nanoparticles themselves can, in certain cases, such as those associated with diesel emissions, be considered harmful. However, there are an increasing number of examples of the beneficial deployment of nanoparticles for a variety of different applications ranging from their use in medicine to create novel diagnostics or therapies through to the design and fabrication of nano-scale electronic devices.

One prerequisite for any application involving nanoparticles is the availability of a source of the particles with consistent size and properties; it is also highly desirable that the properties should be modifiable in a controlled manner. Biologically produced nanoparticles, such as viruses and virus-like particles (VLPs), have these desirable features. In addition, virus particles and VLPs, often collectively known as viral nanoparticles (VNPs), are in many cases capable of self-assembly, are generally biocompatible, and may be modified genetically as well as chemically. Due to these features, it is therefore unsurprising that they have attracted considerable attention for use as nanoparticles for a number of applications and are now being introduced into "real-world" fabrication processes.

In this volume, we have assembled protocols for the use of VNPs for a number of different applications. The protocols have been divided into three parts: Part I concerns the production of a variety of VNPs derived from plant, animal, and bacterial viruses using both prokaryotic and eukaryotic expression systems; it also includes protocols for the incorporation of the VNPs into supramolecular structures. Part II includes protocols for the encapsulation of heterologous materials within VNPs, essentially using them as nano-containers. Part III describes the modification of the outer surface of VNPs, combined approaches, and how such modified VNPs can be developed into functional entities. Inevitably, there is a certain degree of arbitrariness in the assignment of a given chapter into a particular part of this volume but we feel that it is, nonetheless, useful. We anticipate that those interested in using VNPs will be able to "mix and match" the technologies described to achieve the particular result they require.

Finally, we would like to offer our sincerest thanks to all the authors of the various chapters that have made this volume possible. We commend you for your attention to detail in the preparation of the protocols and, particularly, for your forbearance regarding the length of time that it has taken to produce the final version of this volume. We also thank Anke Liedek and Kerstin Ruoff for their excellent help with the editing of the contributions, John M. Walker for his continuous supportive advice, and all those at Springer involved in the publication process.

Stuttgart, Germany — *Christina Wege*
Norwich, UK — *George P. Lomonossoff*

Contents

Part II Functional Compounds in Virus-Based Containers: From Preparation to Applications of Hybrid Nanoobjects

Part III Functions Exposed on Virus Backbones and Combined Approaches: From Fabrication to Technical Uses of Hybrid Particles, Films, and Materials

Contributors

IOANA L. AANEI · *Department of Chemistry, University of California, Berkeley, CA, USA; Materials Sciences Division, Lawrence Berkeley National Laboratories, Berkeley, CA, USA*

RIGOBERTO C. ADVINCULA · *Department of Biomedical Engineering, Case Western Reserve University, Cleveland, OH, USA; Department of Macromolecular Science and Engineering, Case Western Reserve University, Cleveland, OH, USA*

ALAA A. A. ALJABALI · *Faculty of Pharmacy, Yarmouk University, Irbid, Jordan*

KLARA ALTINTOPRAK · *Department of Molecular Biology and Plant Virology, Institute of Biomaterials and Biomolecular Systems, University of Stuttgart, Stuttgart, Germany*

AGNÈS ANNE · *Laboratory of Molecular Electrochemistry, CNRS-Université Paris Diderot, Paris, France*

PETIA ATANASOVA · *Institute for Materials Science, University of Stuttgart, Stuttgart, Germany*

INA BALKE · *Latvian Biomedical Research and Study Centre, Riga, Latvia*

LUCY BEALES · *Mologic, Bedfordshire, UK*

ALEXANDER M. BITTNER · *CIC nanoGUNE, Donostia - San Sebastián, Spain; Ikerbasque, Bilbao, Spain*

ADAM D. BROWN · *Institute for Bioscience and Biotechnology Research, University of Maryland, College Park, MD, USA; Fischell Department of Bioengineering, University of Maryland, College Park, MD, USA*

MICHAEL A. BRUCKMAN · *Department of Biomedical Engineering, Case Western Reserve University, Cleveland, OH, USA; NanoBio Systems, Elyria, OH, USA*

BINRUI CAO · *Department of Chemistry and Biochemistry, Stephenson Life Sciences Center, University of Oklahoma, Norman, OK, USA*

STACY L. CAPEHART · *Department of Chemistry, University of California, Berkeley, CA, USA*

ROGER CASTELLS-GRAELLS · *Department of Biological Chemistry, John Innes Centre, Norwich Research Park, Norwich, UK*

ARNAUD CHOVIN · *Laboratory of Molecular Electrochemistry, CNRS-Université Paris Diderot, Paris, France*

ULRICH COMMANDEUR · *Institute for Molecular Biotechnology, RWTH Aachen University, Aachen, Germany*

JEROEN J. L. M. CORNELISSEN · *Laboratory of Biomolecular Nanotechnology, MESA+ Institute for Nanotechnology, University of Twente, Enschede, The Netherlands*

JAMES N. CULVER · *Institute for Bioscience and Biotechnology Research, University of Maryland, College Park, MD, USA; Department of Plant Science and Landscape Architecture, University of Maryland, College Park, MD, USA*

ANNA E. CZAPAR · *Department of Pathology, Case Western Reserve University, Cleveland, OH, USA*

CHRISTOPHE DEMAILLE · *Laboratory of Molecular Electrochemistry, CNRS-Université Paris Diderot, Paris, France*

CHRISTINA DICKMEIS · *Institute for Molecular Biotechnology, RWTH Aachen University, Aachen, Germany*
TREVOR DOUGLAS · *Department of Chemistry, Indiana University, Bloomington, IN, USA*
BOGDAN G. DRAGNEA · *Department of Chemistry, Indiana University, Bloomington, IN, USA*
WALTER J. DRESSICK · *Center for Bio/Molecular Science and Engineering, US Naval Research Laboratory, Washington, DC, USA*
FABIAN J. EBER · *Department of Molecular Biology and Plant Virology, Institute of Biomaterials and Biomolecular Systems, University of Stuttgart, Stuttgart, Germany*
SABINE EIBEN · *Department of Molecular Biology and Plant Virology, Institute of Biomaterials and Biomolecular Systems, University of Stuttgart, Stuttgart, Germany*
DAVID J. EVANS · *John Innes Centre, Norwich, UK*
RAINER FISCHER · *Institute for Molecular Biotechnology, RWTH Aachen University, Aachen, Germany*
MATTHEW B. FRANCIS · *Department of Chemistry, University of California, Berkeley, CA, USA; Materials Sciences Division, Lawrence Berkeley National Laboratories, Berkeley, CA, USA*
REES F. GARMANN · *Harvard John A. Paulson School of Engineering and Applied Sciences, Harvard University, Cambridge, MA, USA*
FANIA GEIGER · *Department of Cellular Biophysics, Max Planck Institute for Medical Research, Heidelberg, Germany*
WILLIAM M. GELBART · *Department of Chemistry and Biochemistry, UCLA, Los Angeles, CA, USA; California NanoSystems Institute, UCLA, Los Angeles, CA, USA; Molecular Biology Institute, UCLA, Los Angeles, CA, USA*
JEFF E. GLASGOW · *Department of Chemistry, University of California, Berkeley, CA, USA*
RADHIKA GOPAL · *Department of Cell and Molecular Biology, The Scripps Research Institute, La Jolla, CA, USA*
RICHARD H. GUENTHER · *Department of Entomology and Plant Pathology, NC State University, Raleigh, NC, USA*
JULIA HEGER-STEVIC · *Internal Medicine II/Molecular Biology, University Hospital Freiburg, Freiburg, Germany*
CHUNXI HOU · *State Key Laboratory of Supramolecular Structure and Materials, College of Chemistry, Jilin University, Changchun, China*
SLATER A. JAMESON · *Department of Biomedical Engineering, Case Western Reserve University, Cleveland, OH, USA*
HYO-EON JIN · *Department of Bioengineering, University of California, Berkeley, Berkeley, CA, USA; Biological Systems and Engineering, Lawrence Berkeley National Laboratory, Berkeley, CA, USA; College of Pharmacy, Ajou University, Suwon, Korea*
SUKWON JUNG · *Department of Chemical and Biological Engineering, Tufts University, Medford, MA, USA*
EUNAE KANG · *Department of Chemical and Biological Engineering, Tufts University, Medford, MA, USA*
CHARLES M. KNOBLER · *Department of Chemistry and Biochemistry, UCLA, Los Angeles, CA, USA*
CLAUDIA KOCH · *Department of Molecular Biology and Plant Virology, Institute of Biomaterials and Biomolecular Systems, University of Stuttgart, Stuttgart, Germany*
PHILIPP KOLB · *Department of Medical Microbiology and Hygiene, Institute of Virology, University Hospital Freiburg, Freiburg, Germany*

MAURI A. KOSTIAINEN · *Biohybrid Materials, Department of Bioproducts and Biosystems, Aalto University, Espoo, Finland*
EMILY J. LARKIN · *Institute for Bioscience and Biotechnology Research, University of Maryland, College Park, MD, USA*
SEUNG-WUK LEE · *Department of Bioengineering, University of California, Berkeley, Berkeley, CA, USA; Biological Systems and Engineering, Lawrence Berkeley National Laboratory, Berkeley, CA, USA; Tshinghua Berkeley Shenzhen Institute, Berkeley, CA, USA*
CHRISTINA L. LEWIS · *U.S. Army Natick Soldier Research, Development and Engineering Center, Natick, MA, USA*
VEIKKO LINKO · *Biohybrid Materials, Department of Bioproducts and Biosystems, Aalto University, Espoo, Finland*
JUNQIU LIU · *State Key Laboratory of Supramolecular Structure and Materials, College of Chemistry, Jilin University, Changchun, China*
STEVEN A. LOMMEL · *Department of Entomology and Plant Pathology, NC State University, Raleigh, NC, USA*
GEORGE P. LOMONOSSOFF · *Department of Biological Chemistry, John Innes Centre, Norwich Research Park, Norwich, UK*
ANDREW J. LOVE · *The James Hutton Institute, Dundee, SCT, UK*
CHUANBIN MAO · *Department of Chemistry and Biochemistry, Stephenson Life Sciences Research Center, Institute for Biomedical Engineering, Science and Technology, University of Oklahoma, Norman, OK, USA; School of Materials Science and Engineering, Zhejiang University, Hangzhou, Zhejiang, China*
KIMBERLY MCCOY · *Department of Chemistry, Indiana University, Bloomington, IN, USA*
KAMOLRAT METAVARAYUTH · *Department of Chemistry and Biochemistry, University of South Carolina, Columbia, SC, USA*
THIERRY MICHON · *Fruit Biology and Pathology, INRA-Université Bordeaux, Villenave d'Ornon, France*
JOONA MIKKILÄ · *Biohybrid Materials, Department of Bioproducts and Biosystems, Aalto University, Espoo, Finland*
MICHAEL NASSAL · *Internal Medicine II/Molecular Biology, University Hospital Freiburg, Freiburg, Germany*
HUONG GIANG NGUYEN · *Department of Chemistry and Biochemistry, University of South Carolina, Columbia, SC, USA; National Institute of Standards and Technology, Gaithersburg, MD, USA*
ZHONGWEI NIU · *Key Laboratory of Photochemical Conversion and Optoelectronic Materials, Technical Institute of Physics and Chemistry, Chinese Academy of Sciences, Beijing, China; University of Chinese Academy of Sciences, Beijing, China*
EMANUELA NORIS · *Institute for Sustainable Plant Protection, National Research Council of Italy (IPSP-CNR), Torino, Italy*
CHARLES H. OPPERMAN · *Department of Entomology and Plant Pathology, NC State University, Raleigh, NC, USA*
HADRIEN PEYRET · *Department of Biological Chemistry, John Innes Centre, Norwich Research Park, Norwich, UK*
ANDRZEJ S. PITEK · *Department of Biomedical Engineering, Case Western Reserve University, Cleveland, OH, USA*
ARSHAK POGHOSSIAN · *Institute of Nano- and Biotechnologies, Aachen University of Applied Sciences, Jülich, Germany*

FERNANDO PONZ · *Centro de Biotecnología y Genómica de Plantas (CBGP, UPM-INIA), Madrid, Spain*

RINDIA M. PUTRI · *Laboratory of Biomolecular Nanotechnology, MESA+ Institute for Nanotechnology, University of Twente, Enschede, The Netherlands*

YUPENG REN · *Department of Biological Sciences, National University of Singapore, Singapore, Singapore*

GUNTA RESEVIČA · *Latvian Biomedical Research and Study Centre, Riga, Latvia*

PATRICK VAN RIJN · *Faculty of Medical Sciences, University of Groningen, Groningen, The Netherlands*

AMY ROE · *School of Molecular and Cellular Biology, Faculty of Biological Sciences, University of Leeds, Leeds, UK*

DAVID J. ROWLANDS · *School of Molecular and Cellular Biology, Faculty of Biological Sciences, University of Leeds, Leeds, UK*

MARK V. DE RUITER · *Laboratory of Biomolecular Nanotechnology, MESA+ Institute for Nanotechnology, University of Twente, Enschede, The Netherlands*

FLORA SÁNCHEZ · *Centro de Biotecnología y Genómica de Plantas (CBGP, UPM-INIA), Madrid, Spain*

KEITH SAUNDERS · *Department of Biological Chemistry, John Innes Centre, Norwich Research Park, Norwich, UK*

MICHAEL J. SCHÖNING · *Institute of Nano- and Biotechnologies, Aachen University of Applied Sciences, Jülich, Germany*

ANETTE SCHNEEMANN · *Department of Cell and Molecular Biology, The Scripps Research Institute, La Jolla, CA, USA*

SOURABH SHUKLA · *Department of Biomedical Engineering, Case Western Reserve University, Cleveland, OH, USA; Case Comprehensive Cancer Center, Case Western Reserve University, Cleveland, OH, USA*

TIM L. SIT · *Department of Entomology and Plant Pathology, NC State University, Raleigh, NC, USA*

CARISSA M. SOTO · *Center for Bio/Molecular Science and Engineering, US Naval Research Laboratory, Washington, DC, USA; Materials and Applications Division, US Bureau of Engraving and Printing, Washington, DC, USA*

NICOLE F. STEINMETZ · *Department of Biomedical Engineering, Case Western Reserve University, Cleveland, OH, USA; Case Comprehensive Cancer Center, Case Western Reserve University, Cleveland, OH, USA; Department of Radiology, Case Western Reserve University, Cleveland, OH, USA; Department of Materials Science and Engineering, Case Western Reserve University, Cleveland, OH, USA; Department of Macromolecular Science and Engineering, Case Western Reserve University, Cleveland, OH, USA*

SAM L. STEPHEN · *School of Molecular and Cellular Biology, Faculty of Biological Sciences, University of Leeds, Leeds, UK; CPI, National Biologics Manufacturing Centre, Darlington, UK*

NICOLA J. STONEHOUSE · *School of Molecular and Cellular Biology, Faculty of Biological Sciences, University of Leeds, Leeds, UK*

MICHAEL E. TALIANSKY · *The James Hutton Institute, Dundee, SCT, UK*

EVA C. THUENEMANN · *Department of Biological Chemistry, John Innes Centre, Norwich Research Park, Norwich, UK*

Ye Tian · *Key Laboratory of Photochemical Conversion and Optoelectronic Materials, Technical Institute of Physics and Chemistry, Chinese Academy of Sciences, Beijing, China; University of Chinese Academy of Sciences, Beijing, China*
Brylee David B. Tiu · *Department of Biomedical Engineering, Case Western Reserve University, Cleveland, OH, USA; Department of Macromolecular Science and Engineering, Case Western Reserve University, Cleveland, OH, USA*
Irina B. Tsvetkova · *Department of Chemistry, Indiana University, Bloomington, IN, USA*
Frank A. Veliz · *Department of Biomedical Engineering, Case Western Reserve University, Cleveland, OH, USA*
Stella E. Vieweger · *Department of Chemistry, Indiana University, Bloomington, IN, USA*
Andreas Walker · *Institute of Virology, University Hospital Düsseldorf, Düsseldorf, Germany*
Qian Wang · *Department of Chemistry and Biochemistry, University of South Carolina, Columbia, SC, USA*
Qiangbin Wang · *Division of Nanobiomedicine and i-Lab, Suzhou Institute of Nano-Tech and Nano-Bionics, Key Laboratory of Nano-Bio Interface, Chinese Academy of Sciences, Suzhou, China*
Christina Wege · *Department of Molecular Biology and Plant Virology, Institute of Biomaterials and Biomolecular Systems, University of Stuttgart, Stuttgart, Germany*
Sek-Man Wong · *Department of Biological Sciences, National University of Singapore, Singapore, Singapore*
Hong Xu · *Department of Chemistry and Biochemistry, Stephenson Life Sciences Center, University of Oklahoma, Norman, OK, USA*
Cuixian Yang · *Department of Chemical and Biological Engineering, Tufts University, Medford, MA, USA*
Mingying Yang · *Institute of Applied Bioresource Research, College of Animal Science, Zhejiang University, Hangzhou, China*
Hyunmin Yi · *Department of Chemical and Biological Engineering, Tufts University, Medford, MA, USA*
Andris Zeltins · *Latvian Biomedical Research and Study Centre, Riga, Latvia*
Kun Zhou · *Division of Nanobiomedicine and i-Lab, Suzhou Institute of Nano-Tech and Nano-Bionics, Key Laboratory of Nano-Bio Interface, Chinese Academy of Sciences, Suzhou, China*

Part I

Preparing Virus and Virus-Derived Nanoarchitectures as Containers, Carrier Scaffolds, and Structure-Directing Agents: From Particles to Functional Super-Assemblies

Chapter 1

Production of Mosaic Turnip Crinkle Virus-Like Particles Derived by Coinfiltration of Wild-Type and Modified Forms of Virus Coat Protein in Plants

Roger Castells-Graells, George P. Lomonossoff, and Keith Saunders

Abstract

When the coat protein reading frame of turnip crinkle virus (TCV) is transiently expressed in leaves, virus-like particles (VLPs) are readily formed. However, after introducing genetic modifications to the full-length coat protein sequence, such as the introduction of an epitope-specific sequence within the coat protein sequence or the in-frame carboxyl terminal fusion of GFP, the formation of such modified VLPs is poor. However, by coexpression of one of these modified forms with wild-type TCV coat protein by the coinfiltration of appropriate *Agrobacterium* suspensions, VLP generation is enhanced through the formation of "mosaics," that is, individual VLPs consisting of both modified and wild-type subunits (also known as phenotypically mixed VLPs). Here we describe methods for the introduction of genetic modifications into the TCV coat protein sequence, the production of mosaic TCV VLPs and their characterization.

Key words Turnip crinkle virus, Plant infiltrations, Virus-like particles, Mosaic, *Agrobacterium*, Bionanotechnology, Transient expression, Epitope display

1 Introduction

TCV is composed of 180 copies of a single type of coat protein molecule that can adopt one of three possible structural conformations in the mature virus structure. The crystal structures of the mature particle [1] and its swollen form [2] have been elucidated. During virus assembly pairs of TCV coat protein molecules associate with each other via the coat protein Projecting (P) and Shell (S) domains (*see* Fig. 1). Coat protein pairs are then assembled into mature capsid structures by the interaction of one S domain of a coat protein pair with an adjacent S domain of another coat protein pair. The S domains of structurally related coat proteins are held together by calcium ions and it is the removal of these ions that results in the swollen form of the virus structure. In addition and as a consequence of capsid formation, each mature

Christina Wege and George P. Lomonossoff (eds.), *Virus-Derived Nanoparticles for Advanced Technologies: Methods and Protocols*, Methods in Molecular Biology, vol. 1776, https://doi.org/10.1007/978-1-4939-7808-3_1,
© Springer Science+Business Media, LLC, part of Springer Nature 2018

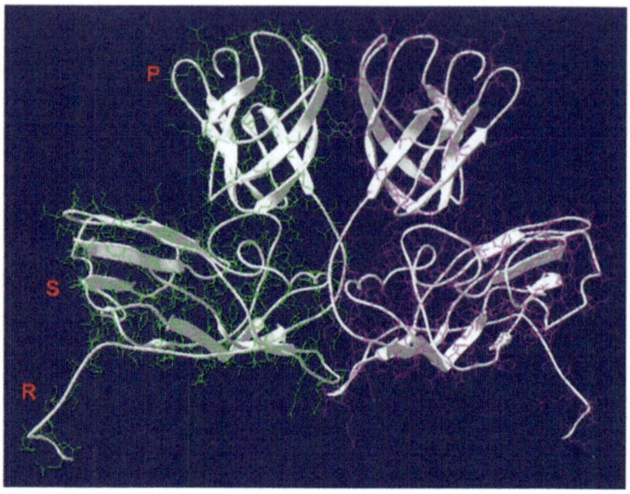

Fig. 1 The position of R (random), S (shell), and P (projecting) structural domains in an associated pair of coat protein molecules of TCV (adapted from [1])

virus particle possesses a single pair of coat protein molecules that become permanently fused together [3]. This stable dimer has been proposed to play a role in the uncoating of the mature virus particle [2].

Virus-like particles (VLPs) composed of turnip crinkle virus (TCV) coat protein can be readily generated through *Agrobacterium*-mediated transient expression using pEAQ-*HT* constructs [4, 5] containing the sequence encoding the TCV coat protein (TCVCP) [6]. Virus-sized, stable, RNA containing $T = 3$ particles result from the expression of full-length TCV coat protein; smaller, empty RNA-free $T = 1$ particles can be formed by the expression of coat protein which lacks the region that interacts with the viral genomic RNA (the random or R domain) [6]. Modified TCV VLPs that display either GFP or a hepatitis B virus epitope on the particle outer surface via the projecting (P) domain can be formed by the expression of modified forms (*see* Fig. 2) of the TCV coat protein. However, the yield of these modified VLPs is greatly diminished compared to the yield obtained with unmodified coat protein [6]. It is likely that the formation of dimers of the modified subunits is adversely affected by the additional amino acid sequences engineered into the P domain. As a result, subsequent VLP formation is not accomplished at the same rate as found with wild-type coat protein. A detailed analysis of the properties of these peptide-displaying VLPs and their potential use in bionanotechnological applications [7–9] such as vaccines [10] and as cancer targeting imaging agents [11] has been hindered as a consequence of the low yield. However, we rationalized that by performing coinfiltrations with plasmids expressing either wild-type or modified coat protein sequences it should be possible to enhance the production of peptide-displaying VLPs through the

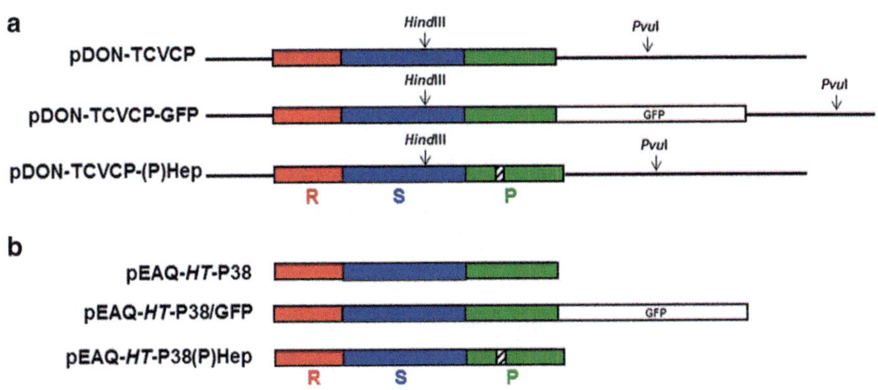

Fig. 2 Schematic representation of the gene constructions in this study. (**a**) Entry clones showing the positions of pertinent restriction sites. (**b**) Destination clones derived from the entry clones that were subsequently used to transform *Agrobacterium* LBA4404. R = random; S = shell and P = projecting structural domains, GFP = green fluorescent protein. Striped box = the location of the inserted hepatitis B epitope MIDIDPYKEFG amino acid sequence [6]

creation of mosaic particles. Using the methods described below (sucrose gradients and affinity chromatography) we have shown that the production of such mosaics, containing at least one modified subunit, significantly enhances VLP yield and is thus a route to the production of TCV VLPs displaying heterologous sequences.

2 Materials

2.1 Enzymes and Cloning Reagents

1. BP and LR clonase enzymes for Gateway cloning (Invitrogen).
2. *Hind*III, *Pvu*I restriction enzymes, Phusion polymerase, and T4 DNA ligase.
3. QIAprep spin miniprep kit (Qiagen), QIAquick gel extraction kit (Qiagen), MinElute reaction cleanup kit (Qiagen) (*see* **Note 1**).
4. Primers KS 35 and KS 36 (*see* Table 2).

2.2 Plants, Plasmids, and Bacterial Strains

1. *Nicotiana benthamiana* plants are grown in glasshouses with supplemental lighting for 16 h at a constant temperature of 24 °C. Infiltrations are performed on plants ranging from 3 to 4 weeks old (*see* **Note 2**).
2. One Shot ® TOP10 chemically competent *E. coli* (Invitrogen) is used for propagation of recombinant plasmids.
3. *Agrobacterium tumefaciens* strain LBA4404 is used for plant-based transient expression.
4. Plasmids for recombination cloning, subcloning, and *Agrobacterium* transformation are described in Table 1.
5. Bacterial glycerol stock of *A. tumefaciens* strain LBA4404 containing pBin61-TCV DNA [2, 12].

Table 1
Plasmids for TCV-like particle cloning and expression

Name	Description	Resistance
pDONR-207	Empty entry vector (Invitrogen)	Gentamycin
pDON-TCVCP	Entry vector possessing the TCV coat protein	Gentamycin
*pDON-TCVCP-GFP	Entry vector of the TCV coat protein with the GFP sequence cloned at its carboxyl terminus	Gentamycin
*pDON-TCVCP-(P)Hep	Entry vector of the TCV coat protein with the hepatitis B amino acid sequence MDIDPYKEFG cloned into the P domain of the coat protein	Gentamycin
pBin61-TCV	Infectious clone of the TCV-M strain cloned in pBin-61	Kanamycin
pEAQ-*HT*-DEST1	Empty binary vector	Kanamycin
pEAQ-*HT*-P38	Binary vector for transient TCV coat protein expression	Kanamycin
*pEAQ-*HT*-P38/GFP	Binary vector for transient TCV coat protein expression with a carboxyl terminal GFP	Kanamycin
*pEAQ-*HT*-P38(P)Hep	Binary vector for transient TCV coat protein expression with an amino acid sequence within the P domain specific for a monoclonal antibody to a hepatitis B virus amino acid sequence	Kanamycin

*Plasmids possessing modifications, addition of GFP and the insertion of the hepatitis B monoclonal antibody peptide sequence, were constructed during this study

Table 2
Oligonucleotides for cloning the TCV coat protein gene

Oligonucleotide	Sequence
KS 35	5′-GGGGACAAGTTTGTACAAAAAAGCAGG CTTAATGGAAAATGATCCTAGAGTC-3′
KS 36	5′-GGGGACCACTTTGTACAAGAAAGCTGG GTTTACTAAATTCTGAGTGCTTGC-3′

2.3 Buffers and Solutions

1. MilliQ water.
2. 2× YT medium: 16 g/l tryptone, 10 g/l yeast extract, 5 g/l, NaCl, pH 7.4.
3. SOC medium: 20 g/l tryptone, 5.0 g/l yeast extract, 0.58 g/ml NaCl, 0.19 g/l KCl, 2.03 g/l $MgCl_2$, 2.46 g/l $MgSO_4$ $7H_2O$, 3.6 g/l glucose, pH 7.5.
4. LB agar: 10 g/l Bacto tryptone, 10 g/l NaCl, 5 g/l yeast extract, pH 7.0, 10 g/l agar. Add antibiotics from stock solutions for a final concentration as indicated: LB + Kan (50 μg/ml), LB + Gen (7 μg/ml), LB + Kan (50 μg/ml) + Rif (50 μg/ml).
5. 7 mg/ml gentamycin sulfate (stock solution) in water, stored at −20 °C.
6. 50 mg/ml kanamycin sulfate (stock solution) in water, stored at −20 °C.
7. 10 mg/ml rifampicin (stock solution) in methanol, stored at −20 °C.
8. EB: 10 mM Tris–HCl, pH 8.5.
9. MMA: 10 mM 2-(*N*-morpholino)ethanesulfonic acid (MES), pH 5.6, 10 mM $MgCl_2$, 100 μM acetosyringone (3′, 5′-dimethoxy-4′-hydroxyacetophenone). A 100 mM stock solution of acetosyringone is prepared in ethanol and stored at −20 °C.
10. Extraction buffer: 1 mM $MgSO_4$, 1 mM sodium phosphate, pH 7.4.
11. Elution buffer: 200 μM glycine, pH 2.5, adjust pH with HCl.
12. 1 M Tris, pH 10.4, adjust pH with NaOH.
13. MOPS buffer: 50 mM MOPS, 50 mM Tris base, 0.1% SDS, 1 mM EDTA, pH 7.7.
14. SeeBlue® Plus2 prestained protein marker (Invitrogen).
15. Instant Blue for staining NuPage gels (Expedeon).
16. Western blot transfer buffer: 25 mM Tris, 190 mM glycine, pH 8.3, 20% methanol.
17. PBS-T: 80 g NaCl, 2 g KCl, 14.4 g Na_2HPO_4, 2.4 g KH_2PO_4, pH 7.4 (adjust pH with HCl) in 1 l for a 10× stock solution, 0.05% (v/v) Tween 20.
18. 5% (w/v) skim milk powder in PBS-T.
19. Immobilon western chemiluminescent HRP substrate.
20. 25% and 70% (w/v) sucrose in 1 mM $MgSO_4$, 1 mM sodium phosphate, pH 7.4.
21. 2% (w/v) uranyl acetate.
22. NuPAGE LDS sample buffer (Invitrogen).

23. 1× TBE buffer: 10.8 g/l Tris–HCl, 5.5 g/l boric acid, 2 mM EDTA.
24. Agarose.
25. 100% (w/v) glycerol.
26. Liquid nitrogen.

2.4 Antibodies

1. Monoclonal primary antibody against the HBcAg protein epitope (ref 10E11, Abcam Ltd.).
2. Goat anti-mouse secondary antibody conjugated to horseradish peroxidase (ref W4021, Promega).
3. GFP Tag antibody conjugated to horseradish peroxidase (A10260 Invitrogen).

2.5 Consumable Materials and Devices

1. Hypodermic needles.
2. 1 ml syringes (without needle).
3. Syringe with a long needle.
4. Miracloth (Merck Chemicals Ltd.).
5. Filter paper.
6. GFP affinity chromatography columns consisting of GFP-Nano-Trap A beads (Chromotek).
7. 4–12% (w/v gradient) NuPAGE gels in NuPAGE® gel system (Invitrogen).
8. Gel transfer apparatus, wet chamber (Mini Trans-Blot®, Bio-Rad).
9. Nitrocellulose membrane (for Western blot analysis).
10. ImageQuant LAS 500 detection equipment.
11. Plastic (pyroxylin) and carbon-coated copper grids with one side flash coated with palladium (400 mesh, Agar Scientific Ltd.).
12. FEI Tecnai20 TEM microscope.
13. Eppendorf electroporator 2510.
14. Spectra-Por Float_A_Lyzer G2 dialysis devices (Spectrum Labs).

3 Methods

3.1 Genetic Modifications

For all expression studies, PCR-derived TCV cDNA from plasmid pBin61-TCV [12] was cloned into the entry vector pDONR-207 utilizing the Gateway BP recombinational ligation reaction giving rise to pDONR-TCVCP (*see* **Note 3**). This was subsequently transferred, by the Gateway LR reaction, to the pEAQ-*HT*-DEST1 expression vector [5] for plant infiltration and expression studies.

Thus, TCV VLPs are generated by expression of the full-length coat protein open reading frame from the plasmid construct pEAQ-*HT*-P38. Changes to the coat protein, by the fusion of GFP to the coat protein carboxyl terminus or the introduction of a specific sequence within the P domain of the coat protein have been accomplished with modified forms of pDONR-TCVCP. Here DNA corresponding to the necessary sequences was designed and synthesized commercially. By utilizing a unique *Hind*III site within the TCV coat protein sequence and a *Pvu*I recognition sequence flanking the coat protein sequence in the plasmid backbone, the synthetic DNA can then be ligated into pDONR-TCVCP. Modified forms of the TCV sequence in pDONR-TCVCP are subsequently transferred into pEAQ-HT-DEST1 for expression (*see* Fig. 2).

3.1.1 Cloning TCV Coat Protein Reading Frame Sequence into pDONR-207

1. Prepare plasmid pBin61-TCV DNA [2] from a −80 °C bacterial glycerol stock by inoculation of 3 ml of 2× YT medium supplemented with 50 μg/ml kanamycin. Grow at 37 °C overnight with constant agitation. Isolate plasmid DNA (Qiagen kit).
2. Set up PCR with pBin61-TCV as the DNA template, and appropriate primer pairs (KS 35 and KS 36, *see* Table 2, with reactions of 98 °C, 2 min and for 30 cycles of 98 °C for 10 s, then 65 °C for 20 s, and 72 °C for 45 s). The primer pairs are complementary to the TCV coat protein nucleotide sequence and incorporate sequences necessary for subsequent Gateway cloning (*see* Table 2). Perform PCR with Phusion polymerase.
3. Verify DNA synthesis by agarose gel electrophoresis under standard conditions on a 1.1% (w/v) agarose gel in 1× TBE buffer and recover PCR product of 1050 bp by Qiagen QIA quick gel extraction kit according to the manufacturer's instructions, eluting the DNA into EB buffer.
4. Perform the Gateway clonase BP reaction (follow manufacturer's instructions) with the PCR derived DNA and the entry vector pDONR-207. After transformation of TOP10 *E. coli*, (follow manufacturer's instructions) select for transformed *E. coli* on LB agar plates supplemented with gentamycin. Verify clones by DNA sequencing and set up the Gateway clonase LR reaction (follow manufacturer's instructions) with the destination vector pEAQ-*HT*-DEST1. Again after transformation of TOP10 *E. coli*, isolate positive transformed *E. coli* on LB agar supplemented with kanamycin. Verify positive clones by DNA sequencing.

3.1.2 Modifications to the TCV Sequence in pDONR-TCVCP

1. Design the TCV coat protein sequence to possess the GFP or other desired nucleotide sequence fused in frame to the carboxyl terminus by consulting and downloading (online) the appropriate sequence file from GenBank (accession code for

TCV: HQ589261). Assemble, in silico, such sequences to achieve the desired amino acid sequence in the expressed gene construct, *see* Fig. 2 for the location of the unique restriction sites, *Hind*III and *Pvu*I employed in this cloning procedure. Subsequently order designed synthetic sequences from a commercial synthesizer company. Similarly design the required nucleotide sequence that encodes the hepatitis B epitope sequence (MIDIDPYKEFG), or any other amino acid sequence, so that it can be inserted into the TCV coat protein P domain [6]. For both synthetic sequences, the 5′ terminal sequence and the 3′ terminal sequence will be the unique *Hind*III recognition sequence located within the TCV coat protein S domain and the unique *Pvu*I recognition sequence in the plasmid backbone respectively, Fig. 2.

2. Perform standard restriction digestions of the plasmid that contains the synthetic DNA sequence and pDONR-TCVCP with *Hind*III and *Pvu*I. Recover and purify the necessary DNAs after agarose gel electrophoresis by the Qiagen QIA quick gel extraction kit.
3. Set up ligation reactions with these DNAs with T4 DNA ligase. Transform TOP10 *E. coli*, recover and verify positive clones as before. The insert of the resulting plasmids can then be transferred into pEAQ-*HT*-Dest1 by the Gateway LR clonase reaction.

3.2 Transient Plant Expression

Agrobacterium tumefaciens is first transformed individually with the selected pEAQ-*HT*-DEST1 plasmids containing the desired nucleotide sequences, and cultures are then prepared and infiltrated into plant leaves for protein expression. Mosaic VLP generation is achieved following the coinfiltration of wild-type and the modified TCV coat protein gene constructions. By adjusting the ratio of the wild-type to the modified TCV coat protein construct in the infiltration solution, the ratio of the two forms of the protein in the mosaic VLPs can be controlled (*see* Fig. 4).

1. Store stock cultures of exponentially grown untransformed *Agrobacterium* in 25% (v/v) glycerol at −80 °C. Briefly, *Agrobacteria* cultures are collected at log phase by gentle centrifugation, washed three times, and resuspended with 100-fold concentration in 25% (v/v) glycerol. Cultures are snap frozen in liquid nitrogen prior to storage at −80 °C.
2. Transformation of *Agrobacterium* with pEAQ-based expression plasmids is achieved by electroporation [13]. Approximately 50 ng of plasmid DNA is sufficient for electroporation of 40 μl of cells at 2.5 kV. Check that a time constant between 5.8 and 6.5 μs was achieved before proceeding with the next step. If not, repeat with a fresh batch of electro competent cells. After adding 0.8 ml of SOC and incubating at 28 °C for 1 h with

shaking at 200 rpm, 10% (v/v) of the transformation mixture is spread onto LB agar containing kanamycin and rifampicin. Incubate plates at 28 °C.

3. Prepare 2× YT liquid medium with appropriate antibiotics for the *Agrobacterium* strain (rifampicin 50 μg/ml for LBA4404) and expression plasmid (kanamycin 50 μg/ml for pEAQ-based plasmids; *see* **Note 4**), to grow *Agrobacteria* for preparation of an adequate infiltration solution (*see* **Note 2**).
4. Inoculate 2× YT liquid culture, 5 ml, by picking a single colony from a plate. Grow at 28 °C in a shaking incubator until the OD at 600 nm is ≥2.
5. Centrifuge the cultures at 4000 × *g* for 10 min at room temperature to pellet the cells and discard the supernatant.
6. Resuspend the cells gently in the required volume of MMA (*see* **Note 2**) to make a solution of final OD (600 nm) = 0.4. For coexpression of two gene constructs, prepare solutions of individual OD (600 nm) = 0.8 which, when mixed 1:1, will result in a final OD (600 nm) = 0.4 for each gene construct. Ratios of 3:1 and 9:1 of wild-type TCV coat protein to modified TCV coat protein are similarly obtained by appropriate dilution of the relevant cultures (*see* **Note 5**).
7. Leave the infiltration solution at room temperature for 0.5–3 h to allow the bacterial culture to adapt to the buffer conditions.
8. Gently scratch the leaf surface with a hypodermic needle. Syringe the infiltration solution, at the damaged point on the leaf, into the leaf ensuring that the entire leaf takes up the infiltration solution (*see* **Note 6**).
9. Harvesting is typically done between 5 and 9 days post infiltration.

3.3 Extraction and Purification

1. Harvest infiltrated leaves, weigh and homogenize the leaf tissue with three volumes (e.g., for 1 g tissue, use 3 ml) of extraction buffer (*see* **Note 7**) using a blender in the cold room, at 4 °C.
2. Squeeze the homogenate through two layers of Miracloth and centrifuge at 13,000 × *g* for 20 min at 4 °C to remove cell debris.
3. To prepare the double sucrose cushion, pour the plant clarified extract into a suitable ultracentrifuge tube, then add 2 ml of 25% (w/v) sucrose solution underneath the extract by using a syringe with a long needle passing through the supernatant layer. Next add 250 μl of the 70% (w/v) sucrose underneath the previous sucrose solution by similar means. Balance the tubes and centrifuge at 274,000 × *g* for 2.5 h at 4 °C. Depending on the volume of the supernatant, use either a TH641 or Surespin rotor (or similar) [14].

4. After centrifugation, puncture the bottom of the tube with a needle (*see* **Note 8**). Recover the first 500 μl, i.e., bottom fraction (B) of the tube. Discard the next 1.5 ml and collect the next 500 μl, the middle fraction (M). Collect the supernatant fraction (S; 500 μl) directly from the top of the tube with a pipette.
5. Store samples at 4 °C for gel electrophoresis and western blot analysis (*see* Subheading 3.5).
6. For GFP affinity chromatography, ultracentrifugation fractions are dialyzed against extraction buffer overnight at 4 °C using a dialysis membrane with a molecular weight cutoff of 100 kDa. Dialyzed fractions are stored at 4 °C.

3.4 Characterization and Purification of Mosaic VLPs by Affinity Chromatography

To confirm the formation of mosaic VLPs formed of both wild-type and GFP-displaying subunits and to separate these from VLPs containing just wild-type coat protein, GFP affinity chromatography can be performed. For mosaics displaying other sequences, alternative methods specific for the inserted sequence will be required.

1. Equilibrate beads according to manufacturer's instructions (commercial product protocol). This is to ensure that the preservative in the solution in which the beads are dispatched to the customer is removed.
2. Add the dialyzed VLPs fraction (0.5–1 ml) to the equilibrated Nano-Trap®_A beads. Mix by repetitive inverting for 1 h at 4 °C on a rotating wheel.
3. Remove the bottom cap from the spin column and place it in a new 2 ml tube. Centrifuge at 100 × *g* for 10 s. Retain this fraction for subsequent gel analysis.
4. Wash the beads by resuspending them in 500 μl ice-cold dilution buffer. Place the spin column in a new 2 ml tube and centrifuge at 100 × *g* for 10 s. Collect the flow-through for immunoblot analysis and wash two more times. These fractions will consist of VLPs entirely formed of wild-type coat protein. Close the column with the bottom plug.
5. Add 100 μl of elution buffer to the nano-Trap® A beads. Pipette the beads up and down for 30 s. Remove the bottom plug of the spin column and place it in a new 2 ml tube containing 10 μl 1 M Tris pH 8.5 to neutralize the eluate. This fraction will contain mosaic VLPs (*see* **Note 9**).
6. Store samples at 4 °C for gel electrophoresis and western blot analysis (*see* Subheading 3.5).

3.5 Gel Electrophoresis and Western Blot Analysis

Protein extracts, ultracentrifugation fractions, washed and eluted chromatography fractions are analyzed by electrophoresis in 4–12% (w/v gradient) NuPAGE Bis-Tris gels resolved with MOPS buffer. In order to visualize the protein bands the gels are stained with

Instant Blue. TCV coat protein is resolved as a protein band at an apparent molecular weight of 38 kDa. Similarly, TCV coat protein fusion products are resolved at approximately 40 kDa for coat protein/Hep and at 65 kDa for coat protein/GFP respectively.

3.5.1 Instant Blue Staining

1. Set up 4–12% (w/v gradient) NuPAGE Bis-Tris gel according to manufacturer's instructions. Add 5 μl of SeeBlue protein marker in the first lane of the gel.
2. Gel samples, derived from the affinity chromatography procedure, are denatured by boiling for 5 min after the addition of 4× NuPAGE LDS sample buffer. Load the samples (up to 20 μl) in the remaining lanes and run the gel with MOPS buffer at 200 V for about 50 min.
3. Place the gel in Instant Blue staining solution for 1–4 h (*see* Figs. 3a and 4a). Destaining is not necessary.

3.5.2 Western Blot Analysis

For the western blot analysis repeat **steps 1** and **2**.

1. After electrophoresis, transfer the proteins from the gel to the nitrocellulose membrane in a gel transfer apparatus set at 100 V for 1 h using the western blot wet chamber transfer system.
2. Block the membrane overnight with 5% (w/v) milk in PBS-T with constant agitation in a cold room.

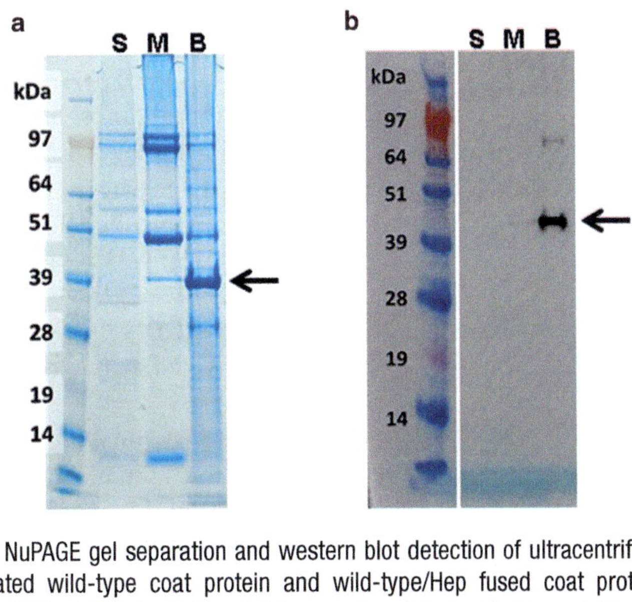

Fig. 3 NuPAGE gel separation and western blot detection of ultracentrifugation separated wild-type coat protein and wild-type/Hep fused coat proteins in mosaic VLPs. (**a**) Instant Blue staining; coat protein (molecular weight 38 kDa) arrowed. (**b**) Western blot detection of coat protein/Hep fusion product (40 kDa) arrowed. S = supernatant; M = middle and B = bottom fractions

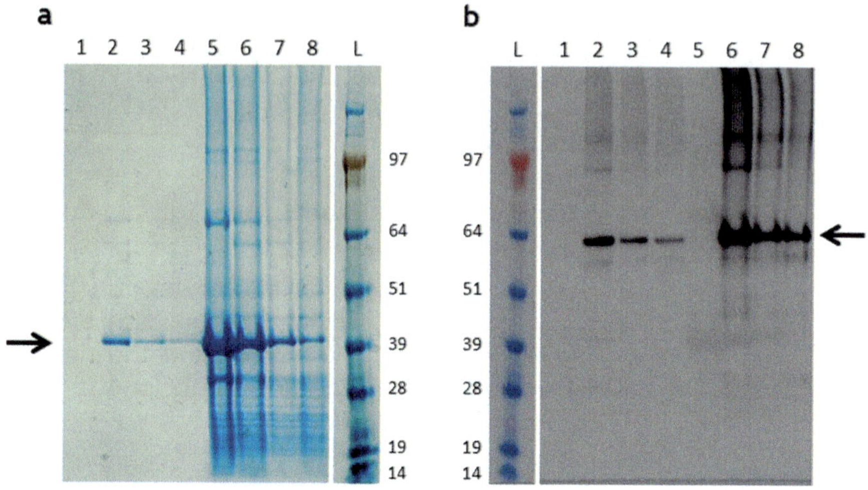

Fig. 4 NuPAGE gel separation and western blot detection of wild-type coat protein and wild-type/GFP fused coat proteins in mosaic VLPs. (**a**) Proteins revealed following NuPAGE gel electrophoresis subsequently stained with Instant Blue. Wild-type coat protein (38 kDa, arrowed) is only present in the elution fractions of the mosaic VLPs (lanes 2–4) and in plant leaf extracts (lanes 5–8) (*see* **Note 9**). (**b**) Western blot detection of GFP following NuPAGE gel electrophoresis. The presence of coat protein fused with GFP (65 kDa, arrowed) in the elution fractions of the mosaic VLPs (lanes 2–4) and in plant leaf extracts (lanes 6–8). Samples in lanes 1 and 5 are derived from infiltrations with pEAQ-*HT*-P38. Samples in lanes 2 and 6 are from infiltrations with a 1:1 (v:v) mixture of pEAQ-*HT*-P38 and pEAQ-*HT*/GFP. Samples in lanes 3 and 7 are from infiltrations with a 3:1 (v:v) mixture of pEAQ-*HT*-P38/GFP and pEAQ-*HT*-P38. Samples in lanes 4 and 8 are from infiltrations with a 9:1 (v:v) mixture of pEAQ-*HT*-P38/GFP and pEAQ-*HT*-P38 (*see* **Note 10**)

3. Incubate with antibody solutions diluted 1:10,000 in PBS-T is for 1 h at room temperature. Remove unbound antibody by washing three times in PBS-T for 20 min at room temperature.
   (a) To detect the hepatitis B-specific amino acid sequence, the membrane is incubated with a monoclonal primary antibody against the HBcAg protein epitope followed by its detection with a goat anti-mouse secondary antibody conjugated to HRP (*see* Fig. 3b).
   (b) GFP is detected by the use of an anti GFP-HRP conjugate antibody (*see* Fig. 4b).
4. HRP is detected with the chemiluminescent substrate ECL plus following 2 min incubation. Protein bands are subsequently visualized in an ImageQuant LAS 500 detection equipment (*see* **Note 10**).

3.6 Negative-Stained Transmission Electron Microscopy

Transmission electron microscopy can be used as a tool to confirm the assembly of virus-like particles and to study their shape and size (*see* Fig. 5).

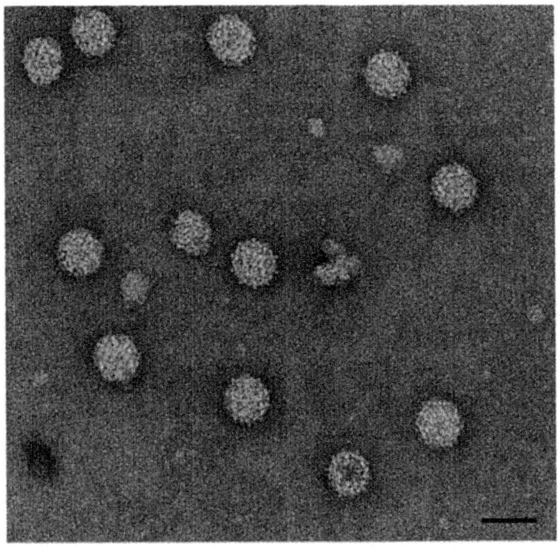

Fig. 5 TEM image of uranyl acetate stained TCV VLPs, bar = 40 nm

1. Place 5–10 μl of VLP preparation onto carbon-coated copper–palladium grids (400 mesh) and allow the particles to settle (20 s) before blotting dry with filter paper.
2. Wash the grids with five drops of MilliQ water and blot dry.
3. Negative stain the VLPs with 2% (w/v) uranyl acetate solution for 15 s then dry the grid through by blotting its edge with filter paper.
4. View grids in a transmission electron microscope such as the FEI Tecnai20 TEM microscope and obtain pictures with a bottom-mounted digital camera.

4 Notes

1. Other commercial kits for plasmid preparation, gel extraction, and enzymatic reaction cleanup can also be used.
2. The volume of the culture depends on the scale of your experiment. Generally, 5 ml of infiltration solution is enough to infiltrate three leaves of one *N. benthamiana* plant (approximately 10 g of fresh-weight tissue). If possible, prepare infiltration solution in excess to avoid being short of it during the infiltration process in the glasshouse. Typically, inoculate the culture in the afternoon and grow overnight. Resuspending the infiltration solution to an OD 600 of 0.4 usually requires 4–5 volumes of MMA to that of the starting culture.
3. For convenience, the plant expression vector pEAQ-*HT*-Dest1, a Gateway compatible vector was used throughout the current study. Restriction enzyme cloning of the TCV coat protein nucleotide sequence directly into the *Age*I and *Xho*I

restriction sites of pEAQ-*HT* [6] would have resulted in similar expression vectors.

4. Although each culture tends to grow at a different rate, allowing cultures to grow to stationary phase generally ensures that all cultures have similar densities. Other highly efficient *Agrobacterium* strains such as AGL1 may also be used, although the commonly used GV3101 (or related "nopaline" strains) are not recommended due to low-level transient expression.
5. The highest amount of fused coat protein/GFP hybrid protein was evident in preparations isolated from leaves that had been infiltrated with equal volumes of pEAQ-*HT*-P38 and pEAQ-*HT*-P38/GFP. The pH is neutralized to preserve the VLP structure.
6. The expression level is generally higher before the plant starts flowering. Plants that are 3 weeks post potting-on from the seedling stage are ideal. For small-scale experiments (useful for checking clones) smaller plants may be used for syringe infiltration of small leaf patches that can be extracted on a small scale using a bead beater or bead mill. To infiltrate leaves, nick the leaf surface with a sterile needle. Aspirate infiltration solution into a sterile 1 ml plastic syringe (take care to avoid bubbles), place the syringe over the leaf wound while keeping a finger behind the leaf for support. Gently press the solution into the intercellular space.
7. It is important not to employ buffers that contain EDTA thereby avoiding the loss of calcium from the VLPs that would result in the loss of viral structures.
8. The major concentration of VLPs is found in the interface between the 70% and 25% (w/v) sucrose solutions. Therefore, it is recommendable to take the first 0.5 ml from the bottom when using 13 ml ultracentrifuge tubes. In the middle and supernatant fractions, the presence of VLPs is low or undetectable.
9. TCV coat protein does not bind to GFP affinity beads. However TCV coat protein that has been formed into mosaic VLPs with TCV coat protein/GFP will be eluted from GFP affinity beads due to the interaction between TCV coat protein fused to GFP and the beads. Therefore in lane 1 Fig. 4a, there is no TCV coat protein observed because these VLPs are entirely formed of wild-type coat protein.
10. We have demonstrated the formation of mosaic VLPs through the coexpression of both wild-type and modified coat proteins. This approach has clear benefits, compared to infiltrations with just the modified coat protein [6]. The formation of mosaic VLPs by other plant and animal viruses, whose capsid is formed of repeated single coat protein molecules, like TCV, is worthy of investigation.

Acknowledgments

This work was supported by the UK Biotechnological and Biological Sciences Research Council (BBSRC) Institute Strategic Programme Grant "Understanding and Exploiting Plant and Microbial Secondary Metabolism" (BB/J004596/1) and the John Innes Foundation. We thank Elaine Barclay for TEM assistance and Andrew Davis for photography.

References

1. Hogle JM, Maeda A, Harrison SC (1986) Structure and assembly of turnip crinkle virus. I. X-ray crystallographic structure analysis at 3.2 Å resolution. J Mol Biol 191:625–638
2. Bakker SE, Ford RJ, Barker AM, Robottom J, Saunders K, Pearson AP, Ranson NA, Stockley PG (2012) Isolation of an asymmetric RNA uncoating intermediate for a single-stranded RNA plant virus. J Mol Biol 417:65–78
3. Stockley PG, Kirsch AL, Chow EP, Smart JE, Harrison SC (1986) Structure of turnip crinkle virus: III. Identification of a unique coat protein dimer. J Mol Biol 191:721–725
4. Sainsbury F, Lomonossoff GP (2008) Extremely high-level and rapid transient protein production in plants without the use of viral replication. Plant Physiol 148:1212–1218
5. Sainsbury F, Thuenemann EC, Lomonossoff GP (2009) pEAQ: versatile expression vectors for easy and quick transient expression of heterologous proteins in plants. Plant Biotechnol J 7:682–693
6. Saunders K, Lomonossoff GP (2015) The generation of turnip crinkle virus-like particles in plants by the transient expression of wild-type and modified forms of its coat protein. Front Plant Sci 6:1138. https://doi.org/10.3389/fpls.2015.01138
7. Lomonossoff GP, Evans DJ (2011) Applications of plant viruses in bionanotechnology. Curr Top Microbiol Immunol 375:61–87
8. Saunders K, Lomonossoff GP (2013) Exploiting plant virus-derived components to achieve in planta expression and for templates for synthetic biology applications. New Phytol 200(1):16–26. https://doi.org/10.1111/nph.12204
9. Steele JF, Peyret H, Saunders K, Castells-Graells R, Marsian J, Meshcheriakova Y, Lomonossoff GP (2017) Synthetic plant virology for nanobiotechnology and nanomedicine. Wiley Interdiscip Rev Nanomed Nanobiotechnol 9(4). https://doi.org/10.1002/wnan.1447
10. Marsian J, Lomonossoff GP (2016) Molecular pharming-VLPs made in plants. Curr Opin Biotechnol 37:201–206
11. Cho C-F, Shukla S, Simpson EJ, Steinmetz NF, Luyt LG, Lewis JD (2014) Molecular targeted viral nanoparticles as tools for imaging cancer. Methods Mol Biol 1108:211–230. https://doi.org/10.1007/978-1-62703-751-8_11. © Springer Science+Business Media, New York
12. Thomas CL, Leh V, Lederer C, Maule AJ (2003) Turnip crinkle virus coat protein mediates suppression of RNA silencing in *Nicotiana benthamiana*. Virology 306:33–41. https://doi.org/10.1016/S0042-6822(02)00018-1
13. Mattanovich D, Rüker F, da Câmara Machado A, Laimer M, Regner F, Stein-kellner H, Himmler G, Katinger H (1989) Efficient transformation of Agrobacterium spp. by electroporation. Nucleic Acids Res 17:6747
14. Peyret H (2015) A protocol for the gentle purification of virus-like particles produced in plants. J Virol Methods 225:59–63

Chapter 2

Isolation and Characterization of Two Distinct Types of Unmodified Spherical Plant Sobemovirus-Like Particles for Diagnostic and Technical Uses

Ina Balke, Gunta Reseviča, and Andris Zeltins

Abstract

Plant virus-like particles (VLPs) structurally resemble their progenitor viruses, but are noninfectious due to absence of viral nucleic acids. Since the 1980s, VLPs have been actively studied with the aim of constructing different nanomaterials, including immunologically active carriers for peptides and whole proteins and proteinaceous shells for the packaging of different ligands.

The technological developments using VLPs require large amounts of purified particles. Here, we describe the laboratory process for isolation and purification of two unmodified plant VLPs, derived from two sobemoviruses, cocksfoot mottle virus (CfMV) and rice yellow mottle virus (RYMV), which is based on cultivation of recombinant *Escherichia coli* cells, VLP precipitation from bacterial extracts and ultracentrifugation. The suggested purification scheme allows the production of 4–45 mg of purified sobemoviral VLPs from a 1 l bacterial culture, depending on the required purity level. Additionally, we provide short protocols for VLP characterization using SDS-PAGE, agarose gel electrophoresis, ultraviolet and mass spectrometry, dynamic light scattering, and electron microscopy.

Key words Virus-like, Sobemovirus, Cocksfoot mottle virus, Rice yellow mottle virus, Ultracentrifugation

1 Introduction

Virus-like particles (VLPs) can be considered as highly organized multimolecular structures of nanometer scale with a very similar overall structure to their corresponding native viruses. These particles are built up from hundreds to thousands of one to few types of viral structural proteins (coat proteins—CPs) and frequently contain host nucleic acids. CPs in VLPs are mostly arranged according to icosahedral or helical symmetry in a highly repetitive manner and can be used as building blocks for the construction of new materials for a variety of different purposes [1].

Christina Wege and George P. Lomonossoff (eds.), *Virus-Derived Nanoparticles for Advanced Technologies: Methods and Protocols*, Methods in Molecular Biology, vol. 1776, https://doi.org/10.1007/978-1-4939-7808-3_2,
© Springer Science+Business Media, LLC, part of Springer Nature 2018

Plant VLPs have been available for more than 30 years, when the formation of artificial viral structures after expression of cDNA from tobacco mosaic virus CP in a bacterial host was demonstrated [2]. Since that time, more than 25 VLPs derived from different plant viruses have been constructed and suggested as immunologically active carriers of peptides and whole proteins, as well as protein containers for the encapsidation of different proteins, nucleic acids, metals, and other substances [3]. Recently, two plant virus-derived VLPs have entered the clinical trial stage as active ingredients in human vaccine formulations (filamentous papaya mosaic virus (PapMV) [4] and icosahedral alfalfa mosaic virus (AlMV) [5]). These examples stress the continuously growing interest in new plant VLP-based technologies for industrial usage.

In laboratory scale experiments, recombinant plant virus-derived VLPs can be obtained from different host systems including bacterial (54% of reported cases), plant (17%), insect (17%), yeast (9%), and mammalian cells (3%) [3]. These data suggest the predominant role of bacterial systems in plant virus VLP production, especially those based on recombinant, commercially available *Escherichia coli* cells. For constructing bacterial expression systems, plant viral structural gene(s) are typically cloned in plasmid vectors under the control of strong promoters (T7, *trp*, *tac*), ensuring a high level of synthesis of recombinant CPs and VLP formation in host cells.

After cell cultivation and disruption at conditions ensuring the structural integrity of the particles, the VLPs have to be identified and purified for further characterization and application [1]. The early identification of newly constructed VLPs is a highly important step in the successful construction of plant virus-based nanoparticles. As a first step, cell extracts containing plant virus CP can be analyzed by ultracentrifugation through sucrose gradients. Alternatively, agarose gel electrophoresis of cell extracts after nuclease treatment can provide the necessary information about the presence of encapsidated, nuclease-protected nucleic acids in the sample, which is characteristic of viruses and VLPs.

After their identification, VLP purification procedures can be developed, depending on the intended downstream application. For the demonstration of VLP formation, acceptable purity levels can be achieved using the above mentioned sucrose gradients or by simple precipitation with polyethylene glycol/NaCl or ammonium sulfate. For such applications as immunological studies and antibody production or targeted packaging of chosen substances inside the VLPs, the required purity level is usually higher; therefore the inclusion of additional purification steps in the isolation protocol is necessary.

The development of new VLP-based nanomaterials requires the use of relatively straightforward procedures to obtain purified virus-derived particles. In this chapter, we describe the methods for early identification of VLPs and provide relatively simple protocols for particle purification from bacterial biomass, using VLP precipitation and ultracentrifugation through "sucrose cushions." As model objects we used two new plant sobemovirus VLPs constructed from cocksfoot mottle virus (CfMV) and rice yellow mottle virus (RYMV) cDNAs. Purified sobemoviral VLPs can be further characterized using SDS-PAGE, agarose gel electrophoresis, ultraviolet (UV) and mass spectrometry (MS), dynamic light scattering (DLS), and electron microscopy (EM). The protocols described here permit the production of the sobemoviral VLPs in amounts sufficient for laboratory scale immunological and nanomaterial studies.

2 Materials

2.1 Instruments and Materials

1. Memmert Incubator IN110.
2. Esco Class II BSC laminar flow hood.
3. TBD-120 thermoblock.
4. HT Ecotron incubator shaker.
5. UP200S ultrasound disintegrator.
6. Boeco S-22 UV/VIS spectrophotometer.
7. NanoDrop ND-1000 spectrophotometer (Thermo Fisher Scientific, USA).
8. Qubit 2.0 fluorometer.
9. Eppendorf 8510R low speed centrifuge.
10. Eppendorf 5418 tabletop centrifuge.
11. Beckman Optima XP-100 ultracentrifuge with swing-out rotor SW-32 and fixed-angle rotor Type 70Ti (Beckman Coulter, Switzerland).
12. JEM-1230 electron microscope (JEOL, Japan).
13. Zetasizer Nano ZS dynamic light scattering (DLS) instrument (Malvern Instruments Ltd., UK); DTS software, version 6.32.
14. SE250 SDS-PAGE chambers 16. Enduro 15.10 agarose gel electrophoresis chambers.
15. AD-1405 laboratory pH-meter.
16. UV illuminator.
17. Amicon-15 ultrafiltration units (100 K, Merck).
18. Mass spectrometry: Autoflex MS (Bruker Daltonik, Germany), MTP Anchor Chip 400/384TF, protein molecular mass calibration standard II (22.3–66.5 kDa; Bruker Daltonik).
19. Qubit Protein Assay Kit (Thermo Fisher Scientific, USA).

2.2 Plasmid DNAs

1. Plasmid vector pET28a(+) containing the T7 polymerase promoter and a kanamycin resistance gene (Novagen, USA, [6]) for constructing the expression plasmids (see below) and for control clones.
2. pACYC-RIL plasmid (Stratagene, USA) encoding the tRNA genes for rare *E. coli* amino acid codons (Arg, Ile, Leu), and containing chloramphenicol resistance gene.
3. CP expression clones for T7 polymerase-based expression in *E. coli*: pET-CCP containing the CfMV *CP* gene (obtained from cDNA of CfMV, GenBank Acc. No. Z48630, kindly provided by Prof. Errki Truve, Tallinn), and pET-RCP containing the RYMV *CP* gene (obtained from cDNA of RYMV, GenBank Acc. No. U23142, kindly provided by Prof. David Baulcombe, Norwich). Detailed descriptions of underlying cloning procedures can be found elsewhere, for example in our recent publication [7].

The corresponding *CP* genes (PCR-amplified by use of cDNA complementary oligonucleotides with NcoI and HindIII restriction site overhangs) are cloned in the NcoI/HindIII sites of pET28a(+), as verified by sequencing. Both expression constructs are available from the authors of this chapter on request.

2.3 E. coli Cells

Escherichia coli BL21(DE3) cells (*fhuA2 [lon] ompT gal (λ DE3) [dcm] ΔhsdS*
λ DE3 = λ sBamHIo ΔEcoRI-B int::(lacI::PlacUV5::T7 gene1) i21 Δnin5; New England Biolabs Inc., USA) as a host for sobemoviral CP synthesis.

2.4 Media

1. LB plates: 1.0% (w/v) tryptone, 0.5% (w/v) yeast extract, 1.0% (w/v) NaCl, 1.5% (w/v) agar supplemented with 25 mg/l kanamycin (for selection of cells transformed with pET-RCP) or with 25 mg/l kanamycin and 25 mg/l chloramphenicol (for selection of cells transformed with pET-CCP/pACYC-RIL) for isolation of single colonies of recombinant *E. coli* BL21(DE3).
2. 2xTY medium: 1.6% (w/v) tryptone, 1% (w/v) yeast extract, 0.5% (w/v) NaCl, containing 25 mg/l kanamycin or kanamycin/chloramphenicol (both 25 mg/l) for cultivation of expression cultures.
3. SOC medium: 2% (w/v) tryptone, 0.5% (w/v) yeast extract, 0.058% (w/v) NaCl, 0.019% (w/v) KCl, 0.203% (w/v) $MgCl_2{\cdot}6H_2O$, 0.12% (w/v) $MgSO_4$, 0.36% (w/v) glucose, sterile filtered (0.2 μm filter) for cell incubation after transformation.
4. For induction: 2.0 M $CaCl_2$, in distilled water, sterile filtered (0.2 μm filter); 2.0 M $MgCl_2$, in distilled water, sterile filtered (0.2 μm filter); 200 mM isopropyl-β-D-1-thiogalactopyranoside (IPTG).

2.5 Buffers and Solutions

1. 0.75 M phosphate buffer (stock solution), pH 5.5: prepare 0.75 M KH_2PO_4 (102 g/l) and 0.75 M K_2HPO_4 (130.5 g/l) separately. Add the K_2HPO_4 solution to the KH_2PO_4 solution until pH 5.5 is reached. Autoclave the resulting stock solution.
2. 15 mM KH_2PO_4 (pH 5.5).
3. 15 mM KH_2PO_4 (pH 5.5), 1% Triton X-100.
4. TBE agarose gel running buffer (10× stock solution): 108 g/l Tris base, 55 g/l boric acid; 7.6 g/l Na_2 EDTA. The pH of resulting solution is approximately 8.5.
5. SDS-PAGE running buffer (10× stock solution): 144 g/l glycine, 30 g/l Tris base, 10 g/l Na dodecyl sulfate (SDS).
6. 2×` Laemmli sample buffer for SDS-PAGE: 100 mM Tris–HCl (pH 7.0), 2% SDS, 50% glycerol, 0.005% Bromophenol Blue, 2% mercaptoethanol.
7. Coomassie G250 protein staining solution: 20% (v/v) ethanol, 2% (w/v) trichloroacetic acid, 0.5 g/l Coomassie G250.
8. Gel destaining solution: 10% (v/v) acetic acid, 10% (v/v) ethanol.
9. Disruption buffer: 15 mM KH_2PO_4, pH 5.5, 5 mM β-mercaptoethanol, 1 mM phenylmethylsulfonyl fluoride (PMSF).
10. Digestion solution: 2 mM $MgCl_2$, 100 mM Tris–HCl, pH 7.0, 5 μl Benzonase (25 U/μl stock solution).
11. Sucrose gradient buffer: 15 mM KH_2PO_4, pH 5.5, 5 mM β-mercaptoethanol, 1 mM PMSF, 1% Triton X-100.
12. Sucrose solutions for sucrose gradient: 20%, 30%, 40%, 50%, and 60% (w/v) in sucrose gradient buffer.
13. Sucrose cushion with Triton X-100: 30% sucrose solution in 15 mM KH_2PO_4 buffer (pH 5.5), 1% Triton X-100.
14. Sucrose cushion without Triton X-100: 30% sucrose solution in 15 mM KH_2PO_4 buffer (pH 5.5).
15. 40% PEG 8000 solution.
16. 4 M NaCl.
17. 2% trifluoroacetic acid.
18. 50 μmol 2,5-dihydroxyacetophenone (2,5-DHAP) matrix solution.
19. 4.2 M ammonium sulfate.

2.6 Gels for Electrophoresis

1. Agarose gels: prepare 0.8% agarose in TBE buffer by melting the agarose in a microwave oven, add Ethidium Bromide to the end concentration of 0.2 μg/ml (*see* **Note 1**).

2. SDS-PAGE gels (12.5%):
   (a) Resolving gel solution (6 ml, sufficient for 2 gels): 2.2 ml of water, acrylamide solution (30% acrylamide–bisacrylamide solution (19:1), 2.5 ml), Tris–HCl buffer (1.875 M, pH 8.8, 1.2 ml), SDS (10%, w/v, 60 μl), ammonium persulfate (APS; 10%, w/v, 60 μl); initiate the acrylamide polymerization with 5 μl of tetramethyl ethylenediamine (TEMED);
   (b) Stacking gel solution (2 ml): water (1.44 ml), acrylamide solution (30%, 0.32 ml), Tris–HCl buffer (1.25 M, pH 6.8, 0.2 ml), SDS (10%, w/v, 20 μl), APS (10%, w/v, 20 μl); initiate the acrylamide polymerization with 8 μl of TEMED.

3 Methods

3.1 Construction of Expression Strains

Perform all manipulations with *E. coli* cells in laminar flow hood.

1. Thaw *E. coli* BL21(DE3) competent cell stocks (100 μl, kept at −70 °C in 1.5 ml tubes) on ice. Add 0.1 μg of corresponding plasmid DNA (pET-CCP or pET-RCP). For pET-CCP sample, add 0.2 μg of pACYC-RIL plasmid (*see* **Note 2**). Incubate on ice at least 15 min.
2. Shock the cells by incubating the tubes at 42 °C for 40 s in the thermoblock. Return the tubes on ice and incubate for an additional 15 min.
3. Add 1 ml of SOC medium to each tube and incubate at 37 °C for 60 min.
4. Plate the cells on LB plates (cells transformed with pET-CCP/pACYC-RIL on kanamycin/chloramphenicol plates, pET-RCP transformed cells on kanamycin plates). Incubate the plates in incubator at 37 °C for 16–20 h.
5. Pick 3–5 individual colonies from each type of cultures and inoculate separately in glass tubes containing 4 ml of 2xTY medium with corresponding antibiotic(s) to obtain starting cultures of the clones. Incubate at 37 °C for 16–20 h without shaking.
6. To determine which clones give the best production level, inoculate 1 ml of starting culture in 20 ml 2xTY with corresponding antibiotic(s) and cultivate as described in Subheading 3.2. Keep the rest of the starting cultures at 4 °C until the experiment is finished.
7. For clones identified (*see* Subheading 3.2) as having the best production level of target protein, mix aliquots of each culture kept at 4 °C (p.6) with glycerol (16% end concentration) and freeze in the stock cultures at −70 °C.

3.2 Synthesis of Sobemoviral Coat Proteins in E. coli

1. Transfer 100 μl of glycerol stock culture (*see* Subheading 3.1, **step** 7) to a 100 ml Erlenmeyer flask with 10 ml 2xTY medium. Add antibiotic(s) to the media (for BL21(DE3)/pET-RCP—kanamycin 25 mg/l, for BL21(DE3)/pET-CCP/pACYC-RIL—kanamycin and chloramphenicol, each 25 mg/l). Cultivate the cells for 16 h in the incubator at 37 °C without shaking.
2. Transfer 10 ml of the starting culture to a 1 l glass Erlenmeyer flask filled with 200 ml 2xTY medium supplemented with appropriate antibiotics and place the flask in a shaking incubator (30 °C, 200 rpm). Monitor the culture growth, using the spectrophotometer at 600 nm, until the density reaches OD (600 nm) = 0.8–1.0.
3. Take a sample before induction for SDS-PAGE analysis: centrifuge 1 ml of cell culture for 1 min at 14,000 rpm (16,873 × *g*), discard the supernatant and resuspend the cell pellet in 70 μl 2× Laemmli buffer. Add 2 mM $CaCl_2$, 5 mM $MgCl_2$, and 0.2 mM IPTG (all final concentrations) to the cultures to induce expression of the gene of interest. Continue the cultivation in incubation shaker for 16 h at 20 °C, 200 rpm (*see* **Note 3**).
4. Measure the optical density of the expression culture after 16 h of cultivation OD(600 nm). Prepare samples for SDS-PAGE: centrifuge 0.15 ml of cell culture for 1 min at 14,000 rpm (16,873 × *g*), discard the supernatant and resuspend the cell pellet in 100 μl 2× Laemmli buffer.
5. Transfer bacterial cultures to 50 ml tubes and collect the cells by low speed centrifugation for 10 min at 6000 rpm (4629 × *g*), 4 °C. Weigh the obtained biomass and keep the tubes in a −70 °C freezer.

3.3 Preparation of Cell Extracts

1. Thaw the preweighted cells from −70 °C on ice. Resuspend the cells in disruption buffer (3 ml buffer/1 g of cells). Transfer cell suspension to an appropriate glass tube. Disrupt the biomass using the ultrasound disintegrator UP200S at period 0.5 and intensity 70% for 16 min (*see* **Note 4**). Keep the tube on ice to prevent the sample from heating.

   Prepare samples for SDS-PAGE: centrifuge 50 μl of disrupted cells in 1.5 ml tube for 10 min at 14,000 rpm (16,873 × *g*), transfer the supernatant to the new 1.5 ml tube and add 150 μl 2× Laemmli buffer. Resuspend the pellet in 200 μl 2× Laemmli buffer by pipetting (store at −20 °C).
2. Clarify the cell lysate by low speed centrifugation at 11,000 rpm (15,557 × *g*) for 10 min at 4 °C in 50 ml tubes with screw-on caps. Transfer the supernatant to new 50 ml tubes for the next VLP purification step. Take an aliquot (100 μl) of the lysate for analysis in agarose gel (store at 4 °C).

3.4 Early Identification of VLPs

3.4.1 Agarose Gel Analysis of Benzonase Treated Samples

1. As the cell lysate contains high amounts of host nucleic acids, pretreatment with Benzonase nuclease is required. Transfer 25 μl of the cell lysate supernatant from aliquot into a 1.5 ml tube and add 25 μl digestion solution. Incubate for 30 min at 37 °C.
2. Load on a 0.8% agarose gel 2.5 μl of untreated supernatant and 5 μl of treated supernatant. Run the gel at 90 V for 45 min. Identify the signals under UV illumination (*see* Fig. 1). After taking a photograph, stain the agarose gel with Coomassie G250 solution by incubating the gel on the orbital shaker at 80 rpm for 30 min at room temperature. Destain the agarose gel by repeated washing with ddH_2O until the gel is transparent and the protein bands become visible (*see* **Note 5**).

3.4.2 Sucrose Gradient Analysis of Samples

1. Parallel to Benzonase treatment, ultracentrifugation of the samples in sucrose gradients can reveal VLPs in cell extracts. Prepare the gradients in tubes (36 ml, for SW-32 rotor, Beckman, USA) by careful filling in 6.2 ml of each of the sucrose solutions in descending order (60–20%), then overlay the sample (4 ml) on the top of the gradient. Avoid mixing of the solutions. Run the ultracentrifuge for 6 h at 25,000 rpm (106,559 × *g*) at 18 °C.
2. After the run, remove approx. 6 ml aliquots from the bottom of the tube with a long needle-syringe. Keep aliquots of gradient samples for SDS-PAGE and agarose gel analysis.
3. Analyze the sucrose gradient fractions on a 0.8% native agarose gel (*see* Subheading 3.4.1, **step 2**) and by SDS-PAGE. For

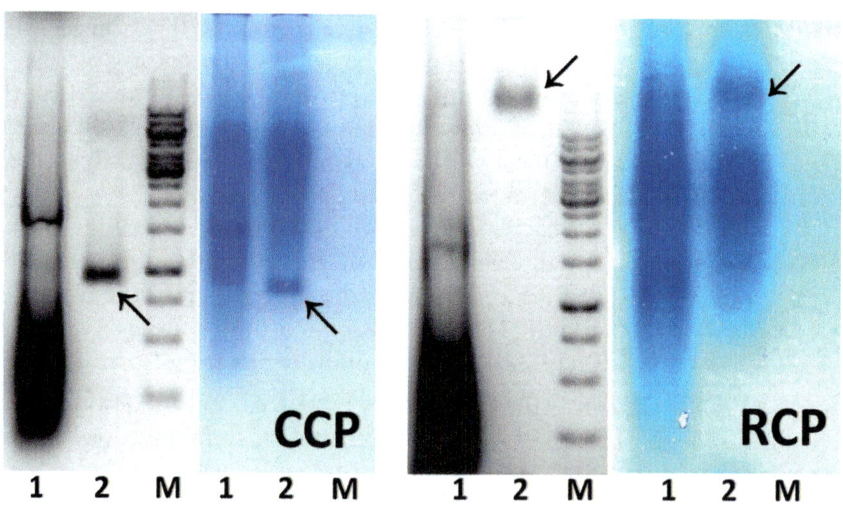

Fig. 1 Agarose gel analysis of crude *E. coli* cell extracts containing sobemoviral CPs. Lane 1—untreated cell extract; lane 2—Benzonase nuclease-treated cell extract, M—1 kb DNA size marker. Panels CCP and RCP denote corresponding extract from cells expressing CfMV or RYMV *CP* genes, respectively. Arrows denote overlapping putative VLP signals, stained with Ethidium Bromide (left) and Coomassie G250 (right)

sample analysis in agarose gel mix 10 μl of sample from each fraction with 2 μl 6×DNA loading dye (Thermo Scientific, USA) and load on gel. Prepare samples for SDS-PAGE by mixing of 25 μl from corresponding sucrose gradient fraction with 25 μl 2× Laemmli buffer. Load 10 μl of prepared and heated samples on gel, including samples from Subheading 3.2 **steps 3** and **4** and from Subheading 3.3, **step 1**. Run gel at 250 V, 20 mA/gel for 50 min in SDS-PAGE running buffer. Fix the gel in destaining buffer, stain with Coomassie G250 and destain the gel with destaining solution (*see* **Note 6** and Fig. 2).

4. If the VLPs are to be used in further experiments, dialyze (12–14 kDa MWCO, 32 mm, Spectra/Por, Spectrum Laboratories, USA) the fractions containing sobemoviral CPs against 100 volumes of 15 mM KH_2PO_4 (pH 5.5) for 16 h at 4 °C. Then, concentrate the VLPs using Amicon-15 ultrafiltration units.

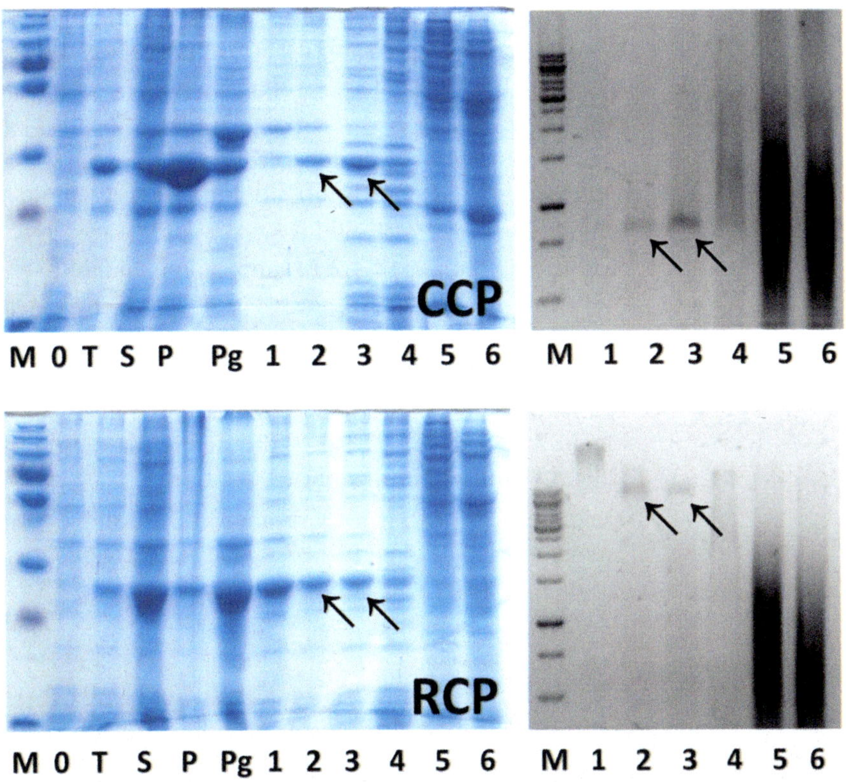

Fig. 2 Analysis of sobemoviral CPs after expression in *E. coli* BL21(DE3) cells and sucrose gradient separation (left—Coomassie-stained polyacrylamide gel; right—Ethidium Bromide-stained agarose gel). M—protein molecular mass marker; 0—total cell proteins before IPTG induction; T—total cell proteins after overnight cultivation; S—soluble proteins in cell lysate after ultrasonic treatment; P—insoluble proteins in cell lysate after ultrasonic treatment; Pg—protein aggregates at the bottom of sucrose gradient tube; 1–6—sucrose gradient fractions (1–60%, 2–50%, 3–40%, 4–30%, 5–20%, 6–0% sucrose). Panels CCP and RCP denote corresponding samples from cells expressing CfMV or RYMV *CP* genes, respectively. Arrows denote the relative localization of sobemoviral protein in the form of putative VLPs in acrylamide and agarose gels, respectively

3.5 VLP Purification

1. As an alternative to sucrose gradients VLPs can be purified by two subsequent precipitations and ultracentrifugation through several "sucrose cushions".

   Precipitate VLPs from extracts (*see* Subheading 3.3) as follows: add to the extract (3 ml) of RCP (RYMV CP), 0.3 ml 40% PEG 8000 solution and 0.225 ml 4 M NaCl (final concentrations: 4% PEG 8000/0.3 M NaCl). Use the same stock solutions to make the 8% PEG 8000 and 1.0 M NaCl, respectively, for CCP (CfMV CP). Incubate the extracts for 4 h on a rotating shaker at 4 °C, 10 rpm (*see* **Note 7**).
2. Take samples for SDS-PAGE analysis: centrifuge 50 μl aliquots in a 1.5 ml tube for 5 min at 14,000 rpm (16,873 × *g*). Transfer the supernatant to a new 1.5 ml tube and add 150 μl 2× Laemmli buffer, then dissolve the pellet in 200 μl 2× Laemmli buffer by vortexing.
3. Centrifuge PEG 8000/NaCl precipitated samples for 10 min at 11,000 rpm (15,557 × *g*), 4 °C. Discard the supernatant and dissolve the pellet in 3 ml of 15 mM phosphate buffer.
4. Take samples for analysis by SDS-PAGE (*see* Subheading 3.5, **step 2**). Centrifuge the extracts for 10 min at 11,000 rpm (15,557 × *g*), 4 °C to remove the insoluble material. Transfer supernatant to the new 50 ml tube. Take sample for 0.8% native agarose gel (*see* Subheading 3.3, **step 2**).
5. Add three volumes of 4.2 M ammonium sulfate to the extract and precipitate for 4 h on a rotating shaker at 4 °C, 10 rpm. Take samples for SDS-PAGE analysis (*see* Subheading 3.5, **step 2**).
6. Centrifuge precipitated samples for 10 min at 11,000 rpm (15,557 × *g*) at 4 °C. Discard the supernatant and dissolve the pellet in 3 ml of 15 mM phosphate buffer.
7. Centrifuge the extracts to remove the insoluble material for 10 min at 11,000 rpm (15,557 × *g*), 4 °C. Take aliquots for analysis in SDS-PAGE (*see* Subheading 3.5, **step 2**), 0.8% native agarose gel (*see* Subheading 3.3, **step 2**) and EM (*see* Fig. 3).
8. Prepare 30% sucrose solution in 15 mM KH2PO4 buffer (pH 5.5), including 1% Triton X-100. Fill 14 ml of 30% sucrose solution in 26 ml ultracentrifuge tubes (for Type 70Ti rotor, Beckman). Carefully overlay 10 ml of VLPs (dissolved and diluted with 15 mM phosphate buffer) on top of the "sucrose cushion." Centrifuge for 4 h, at 50,000 rpm (183,960 × *g*) at 4 °C.
9. Discard the supernatant and dissolve the pellet in the same amount of buffer (10 ml of 15 mM KH_2PO_4 buffer, pH 5.5) overnight on a rotating shaker at 10 rpm, 4 °C.

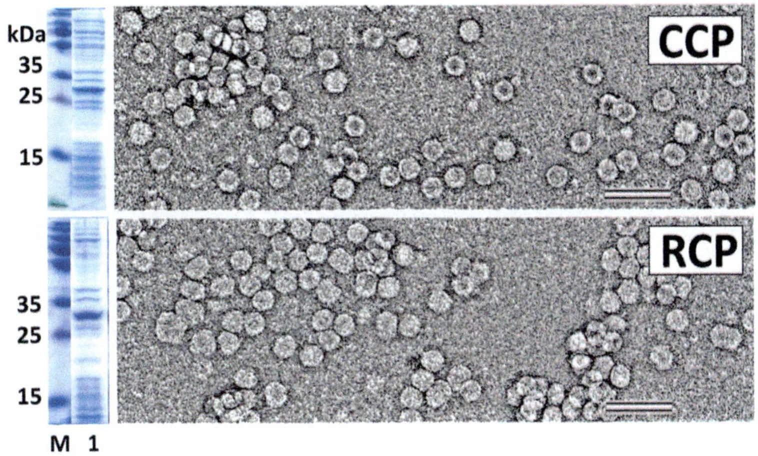

Fig. 3 SDS-PAGE analysis and electron microscopy images of sobemoviral CPs after PEG 8000/NaCl and ammonium sulfate precipitations. Left—SDS-PAGE analysis, right—EM images. M—protein molecular mass marker; 1—protein samples taken for EM analysis; scale bar—100 nm. Panels CCP and RCP denote corresponding samples from cells expressing CfMV or RYMV *CP* genes, respectively

10. Take samples for SDS-PAGE (*see* Subheading 3.5, **step 2**) and 0.8% native agarose gel (*see* Subheading 3.3, **step 2**) analysis. Remove insoluble material by centrifugation for 10 min at 11,000 rpm (15,557 × *g*), 4 °C.
11. Repeat **steps 8–10** with 30% "sucrose cushion" without Triton X-100.
12. Dissolve the VLP-containing pellets in 3 ml of 15 mM KH2PO4 (pH 5.5) overnight on the rotating shaker at 10 rpm, 4 °C.
13. To obtain more purified VLPs, dilute the sample to 10 ml with 15 mM KH2PO4 and repeat the both "sucrose cushion" purifications (**steps 8–11**).
14. Dissolve the VLP-containing pellets in 3 ml of 15 mM KH2PO4 buffer.
15. Run SDS-PAGE (*see* Subheading 3.4.2, **step 3**) and 0.8% native agarose gels (*see* Subheading 3.4.1, **step 2** and Subheading 3.4.2, **step 3**); characterize the VLPs using EM, DLS, and MS (*see* **Note 8** and Fig. 4).

3.6 VLP Characterization

3.6.1 MALDI-TOF Mass Spectrometry

1. Mix 2 μl of purified VLPs (1 mg/ml) with 2 μl of 2% trifluoroacetic acid and then with 2 μl 2,5-DHAP matrix solution and spot 1 μl onto an MTP Anchor Chip 400/384TF. Allow to dry at room temperature.
2. Perform the analysis on Autoflex MS. Use the protein molecular mass calibration standard II for mass calibration. Compare the experimental data with theoretical values of average protein molecular masses using the PeptideMass tool [8]. The results are summarized in Fig. 5 (*see* **Note 9**).

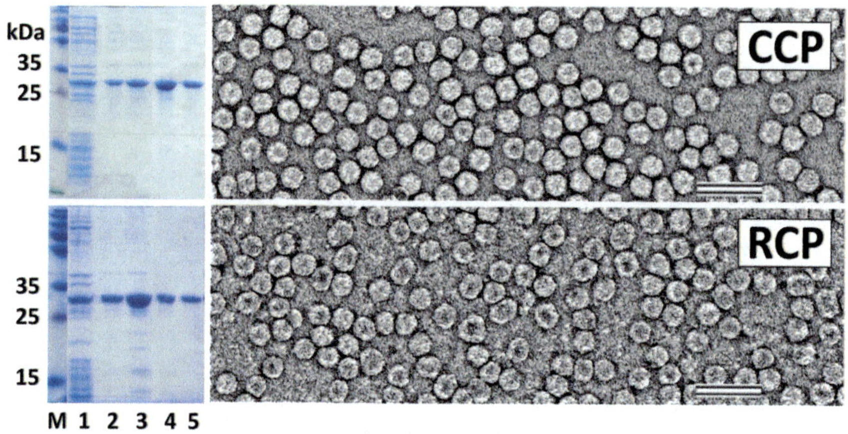

Fig. 4 SDS-PAGE analysis and electron microscopy images of sobemoviral CPs after different stages of purification. Left—SDS-PAGE analysis, right—EM images. M—protein molecular mass marker; 1—samples after PEG 8000/NaCl and ammonium sulfate precipitations; 2—samples after 1st "sucrose cushion"; 3—samples after 2nd "sucrose cushion"; 4—samples after 3rd "sucrose cushion"; 5—samples after 4th "sucrose cushion"; scale bar—100 nm. Panels CCP and RCP denote corresponding samples from cells expressing CfMV or RYMV *CP* genes, respectively

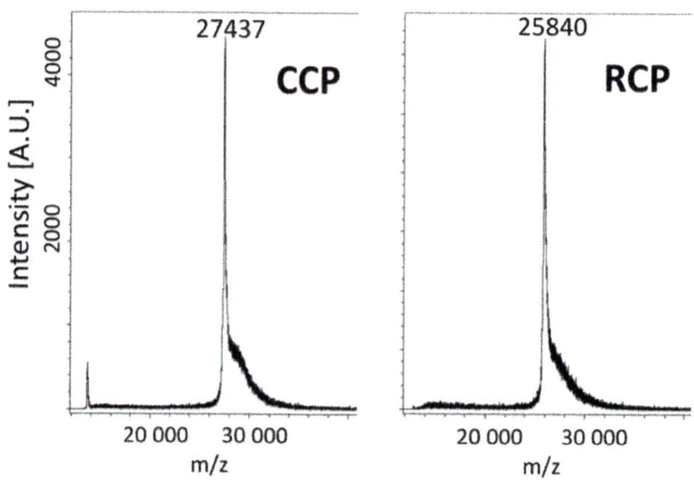

Fig. 5 Mass spectrometric analysis of CfMV-derived VLPs (CCP) and RYMV-derived VLPs (RCP)

3.6.2 UV Spectrophotometry

1. Prepare several dilutions of the purified VLPs in 15 mM phosphate buffer.
2. Pipette 2 μl of the sample solution onto the pedestal of the NanoDrop ND-1000 spectrophotometer. Record the spectrum, using nucleic acid program (*see* **Note 10** and Fig. 6).

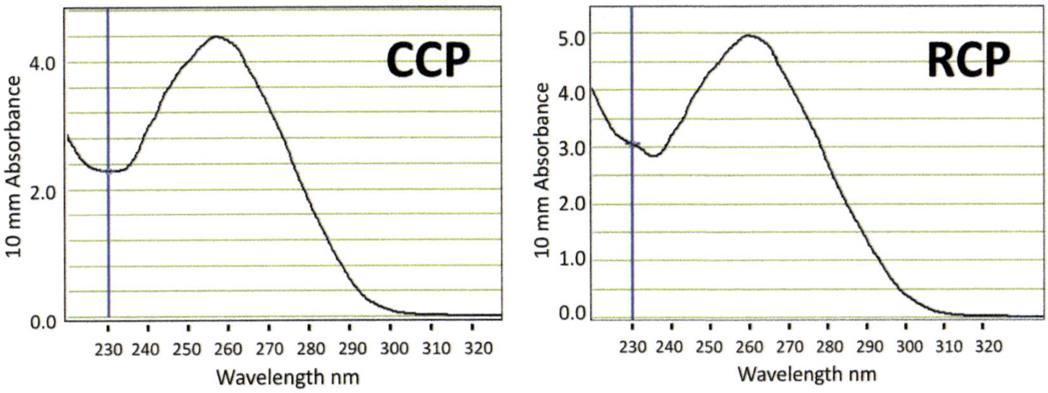

Fig. 6 UV spectra of CfMV-derived VLPs (CCP) and RYMV-derived VLPs (RCP)

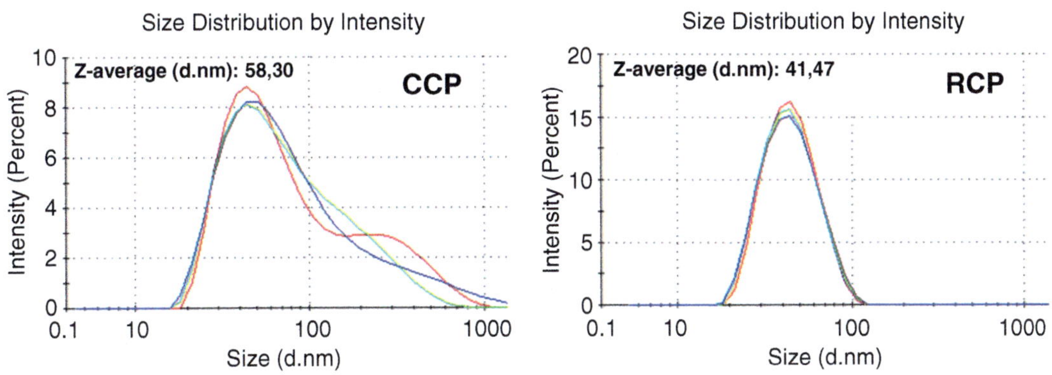

Fig. 7 Dynamic light scattering analysis of CfMV-derived VLPs (CCP) and RYMV-derived VLPs (RCP). Three subsequent measurements for each VLP sample were performed (red, green, and blue lines)

3.6.3 Dynamic Light Scattering Analysis

1. Dilute the VLP samples to approx. 1 mg/ml.
2. Perform the DLS analysis on Zetasizer Nano ZS instrument. Analyze the results by DTS software (*see* Fig. 7 and **Note 11**).

3.6.4 Electron Microscopy

Adsorb purified VLPs (approx. 1 mg/ml) on carbon formvar-coated copper grids and stain negatively with 1% uranyl acetate aqueous solution. Examine the grids using JEM-1230 electron microscope at an accelerating voltage of 100 kV or another type of TEM.

3.6.5 Protein Concentration Determination

Measure the VLP concentrations with Qubit Protein Assay Kit using Qubit 2.0 Fluorometer (*see* **Note 12**). The protocol can be found at the Homepage of the manufacturer [9]. The results of protein concentration determination after different stages of purification are summarized in Table 1.

Table 1
Yield of sobemoviral VLPs. The yields (calculated as mg/l culture) of total proteins were determined in soluble extracts after the corresponding purification step, using Qubit protein concentration measurements

Purification step	CCP [mg/l]	RCP [mg/l]
PEG 8000/NaCl precipitation	331	474
Ammonium sulfate precipitation	187	224
2nd "sucrose cushion" preparation	46.6	35.4
4th "sucrose cushion" preparation	6.4	4.3

4 Notes

1. We typically add the Ethidium Bromide to the agarose gel to stain nucleic acids inside of VLPs. However, other nucleic acid staining reagents can also be used for this purpose, for example Purple Gel loading reagent without SDS (NEB).
2. pACYC-RIL plasmid ensures the synthesis of rare tRNAs in *E. coli* cells. It is necessary for efficient CfMV *CP* gene expression, because the gene contains nine rare Arg codons (AGA and AGG), whereas RYMV *CP* gene contains only two rare Arg codons.
3. It is important to cultivate *E. coli* cells at untypically low temperatures to stimulate the formation of soluble sobemoviral CPs, which is a prerequisite for the VLP formation. At 37 °C, most part of CPs are found in the form of inclusion bodies. VLP formation at low temperature culture conditions have been demonstrated also for other plant CPs [7, 10].
4. Alternatively, contactless ultrasound treatment with S220 device (Covaris, USA) can be used for cell disruption. For unstable VLPs which are sensitive to ultrasound conditions, we suggest the grinding of biomass with Al_2O_3 microcrystals using a mortar and pestle.

   It is important to choose the conditions for cell extract preparation which ensure the VLP structural integrity. As sobemovirus VLPs possibly contain Ca^{2+} ions, similar to the native viruses [11], buffer solutions containing EDTA should be avoided during purification due to potential disassembly of VLPs. Therefore, if new VLPs are under study, the initial conditions for the preparation of cell extracts and particle purification have to be determined in preliminary experiments [1].
5. At the stage of crude cell lysates, the VLPs can be identified as overlapping signals on agarose gels after Ethidium Bromide and subsequent Coomassie staining. As seen in Fig. 1, it is necessary to treat the lysate with Benzonase to

remove host nucleic acids from samples. Note that CCP and RCP VLPs move differently in the electric field, possibly, due to differences in CP amino acid sequences as well as due to different amounts of packaged host nucleic acids.

6. Parallel analysis of samples using agarose and SDS-PAGE gels allows the sucrose gradient fractions which potentially contain VLPs (indicated with arrows in Fig. 2) to be identified. Note that host nucleic acids are localized in the upper sucrose gradient fractions and are separated from the fractions containing the VLPs.
7. The necessary concentrations of PEG8000/NaCl and ammonium sulfate for viral particle precipitations are different for each virus or VLP type (e.g., [12] or [13]) and have to be determined in preliminary experiments.
8. As seen in Fig. 4, VLP precipitation results in only partially purified samples. However, it is possible to obtain homogenous VLP preparations using several subsequent "sucrose cushion" steps.
9. Using MS analysis we obtained experimental molecular masses of 27,437 Da for CCP and 25,840 Da for RCP, respectively. These values are close to theoretical average masses for CCP (27,445 Da) and RCP (25,869 Da), both without N-terminal methionine (Met). Taking into account the amino acid sequence of both proteins the first Met is most probably removed during the protein synthesis in *E. coli* cells [14].
10. VLPs from both viruses contain significant amounts of host nucleic acids. As nucleic acids absorb UV light considerably more strongly than proteins, the UV spectra reflect both the protein and the nucleic acid components of the VLPs shifting peak maximum to the 260 nm range. The UV spectroscopy can be useful for evaluation of viral particle concentrations, as described in [15].
11. DLS can serve as an alternative to EM. It allows the determination of the hydrodynamic sizes of VLPs under study. These are usually greater than the sizes measured by electron micrograph analysis [16], as DLS analysis of VLPs is performed in the solution, whereas the influence of fixation and dehydration upon EM sample preparation on VLP size cannot be excluded. The results of our DLS experiment suggest that RCP gives a nearly homogenous VLP population, whereas CCP VLPs tend to aggregate when analyzed in the tested buffer solution.
12. We typically use Qubit Protein Assay Kit in our laboratory, because it requires low volumes of the sample, is rapid and accurate. However, other protein assay methods, for example BCA Protein Assay Kit (Pierce, USA) can be used for the determination of VLP concentration.

Acknowledgment

We wish to thank Dr. V. Ose for help in the preparation of EM pictures and J. Bogans for DLS experiments. G. Grinberga and V. Zeltina are acknowledged for their technical assistance. The chapter was written with the help of Grant No. SP 672/2014 provided by the Latvian Science Council.

References

1. Zeltins A (2016) Viral nanoparticles: principles of construction and characterization. In: Khudyakov YE, Pumpens P (eds) Viral nanotechnology. CRC Press, Boca Raton, pp 93–119
2. Haynes JR, Cunningham J, von Seefried A et al (1986) Development of a genetically-engineered, candidate polio vaccine employing the self-assembling properties of the tobacco mosaic virus coat protein. Biotechnology 4:637–641
3. Zeltins A (2013) Construction and characterization of virus-like particles: a review. Mol Biotechnol 53:92–107
4. Clinical Trials Identifier: NCT02188810 (2014) Safety and reactogenicity of a PAL combined with seasonal flu vaccine in healthy adults. https://clinicaltrials.gov/ct2/show/NCT02188810. Accessed 3 Apr 2017
5. Clinical Trials Identifier: NCT02013687 (2013) Safety and immunogenicity of plant-derived Pfs25 VLP-FhCMB malaria transmission blocking vaccine in healthy adults. https://clinicaltrials.gov/ct2/show/NCT02013687. Accessed 3 Apr 2017
6. Studier FW, Rosenberg AH, Dunn JJ, Dubendorff JW (1990) Use of T7 RNA polymerase to direct expression of cloned genes. Methods Enzymol 185:60–89
7. Kalnciema I, Balke I, Skrastina D et al (2015) Potato virus M-like nanoparticles: construction and characterization. Mol Biotechnol 57:982–992
8. ExPASy bioinformatics resource portal. Peptide mass. http://web.expasy.org/peptide_mass/. Accessed 3 Apr 2017
9. Qubit® 2.0 Fluorometer https://tools.thermofisher.com/content/sfs/manuals/mp32866.pdf Accessed 3 Apr 2017
10. Kalnciema I, Skrastina D, Ose V et al (2012) Potato virus Y-like particles as a new carrier for the presentation of foreign protein stretches. Mol Biotechnol 52:129–139
11. Tars K, Zeltins A, Liljas L (2003) The three-dimensional structure of cocksfoot mottle virus at 2.7 A resolution. Virology 310:287–297
12. Hesketh EL, Meshcheriakova Y, Dent KC et al (2015) Mechanisms of assembly and genome packaging in an RNA virus revealed by high-resolution cryo-EM. Nat Commun 6:10113
13. Cheng A, Speir JA, Yuan YA et al (2009) Preliminary X-ray data analysis of crystalline hibiscus chlorotic ringspot virus. Acta Crystallogr Sect F Struct Biol Cryst Commun 65:589–593
14. Hirel PH, Schmitter MJ, Dessen P et al (1989) Extent of N-terminal methionine excision from *Escherichia coli* proteins is governed by the side-chain length of the penultimate amino acid. Proc Natl Acad Sci U S A 86:8247–8251
15. Porterfield JZ, Zlotnick A (2010) A simple and general method for determining the protein and nucleic acid content of viruses by UV absorbance. Virology 407:281–288
16. Nikitin N, Trifonova E, Evtushenko E et al (2015) Comparative study of non-enveloped icosahedral viruses size. PLoS One 10(11): e0142415

Chapter 3

RNA-Directed Assembly of Tobacco Mosaic Virus (TMV)-Like Carriers with Tunable Fractions of Differently Addressable Coat Proteins

Sabine Eiben

Abstract

Taking advantage of the ability for in vitro assembly of the plant-infecting virus tobacco mosaic virus (TMV), rod-shaped nanoscale scaffolds presenting different addressable groups can be obtained. We have established procedures resulting in virus-like particles with randomly distributed functional groups, with different groups arranged in striped but randomized structures, and even with distinct groups clustered in adjacent, better-defined domains. The TMV coat protein (CP) variants combined in these approaches can either originate all from TMV mutants propagated *in planta*, or be mixed with CP expressed in *E. coli* (CP_{Ec}). Protocols for expression and purification of a CP_{Ec}.His_6 mutant in *E. coli* as well as the different methods for in vitro assembly and the visualization by decoration of one CP type are explained in detail.

Key words Tobacco mosaic virus (TMV), Multifunctional, In vitro assembly, TMV decoration, Transmission electron microscopy

1 Introduction

One of the reasons that make tobacco mosaic virus (TMV) such an interesting scaffold for nanotechnological applications is its particles' multivalency. More than 2100 identical functional groups, usually one per coat protein (CP), can be displayed and addressed on a single 18 × 300 nm nanorod, with defined distances between each other [1]. The resulting surface properties can be used to mediate mineralization and metallization directed to the outer coat surface [2, 3] or preferentially to the inner channel [4, 5] of TMV, or for chemical coupling of fluorescent dyes [6], polymers [7], and enzymes [8–10]. Another very useful property of TMV is the possibility of assembling TMV virions and virus-like particles (VLP) from their components: the genomic RNA and CPs [11]. Assembly starts at a stem-loop structure in the RNA sequence, the viral origin of assembly (OAs) of about 100 bases length, which nucleates

Christina Wege and George P. Lomonossoff (eds.), *Virus-Derived Nanoparticles for Advanced Technologies: Methods and Protocols*, Methods in Molecular Biology, vol. 1776, https://doi.org/10.1007/978-1-4939-7808-3_3,
© Springer Science+Business Media, LLC, part of Springer Nature 2018

a bidirectional assembly process [12]. In the 5′ direction from the OAs, assembly proceeds fast principally by the addition of two-layered disks, consisting of 34 CPs, in one step. These disks are the prevalent CP oligomer at room temperature, around neutral pH and ionic strength of about 0.1–0.5 M; however, they are always in equilibrium with monomers and lower state oligomers, the so-called A-protein [13]. Assembly at the 3′ end of the RNA occurs more slowly through addition of A-protein oligomers and at least partially after the 5′ end is fully assembled. The equilibrium between A-protein and disks can be shifted to A-protein by decreasing the temperature or increasing the pH in combination with lower ionic strength. The only obligate property of the RNA used for the assembly is the presence of the OAs; this allows a number of manipulations of the VLP composition and structure. Thus VLPs of variable but defined length [14] and, in the presence of more than a single OAs, also kinked structures [15] can be obtained, free in solution or "grown" on surfaces such as gold nanoparticles [16]. In addition, CP mutants with special reactive amino acids such as lysine or cysteine presented on the outer surface, or C-terminally extended sequences such as a His_6-tag can be used to assemble the VLPs [6, 17–19]. A further increase in the number of addressable sites on the viral surface is possible by using CP mutants with more than a single addressable group (*see* Chapter 27 in this book), while by combining different CP variants in one VLP, the amount of differentially addressable sites per rod can be predefined. The mixing of CPs also allows the incorporation of mutant CP variants that cannot efficiently be propagated in the form of virus particles *in planta* and therefore have to be expressed in heterologous hosts such as *E. coli* [18]. Distinct CP types may either be combined by mixing the CPs before initiating assembly of disks and subsequently rods, resulting in an almost even distribution of the CP variants over the whole length of the resulting particle, or by consecutive addition of portions of the different CPs, to generate VLPs with a longitudinal domain structure [6] (*see* Fig. 1). The latter method relies on the observation that assembly of TMV at the chosen CP concentration starts virtually synchronously on all RNAs at the same time and proceeds bidirectionally [12]. Thus adding the first CP variant in a substoichiometric amount results in partially assembled particles of roughly the same length on all RNA strands, which could be confirmed by transmission electron microscopy (TEM)-based length analysis [6]. A second CP type may be used to encapsidate the residual RNA portions protruding at one or both end(s) of the premature particle, yielding complete VLPs with two or three adjacent domains of distinct chemistry, depending on the initial CP amount used. This chapter focuses first on the fundamental methods for preparing CP and RNA educts for all these procedures, including special necessities for the isolation of hexahistidine-terminated TMV CP expressed in

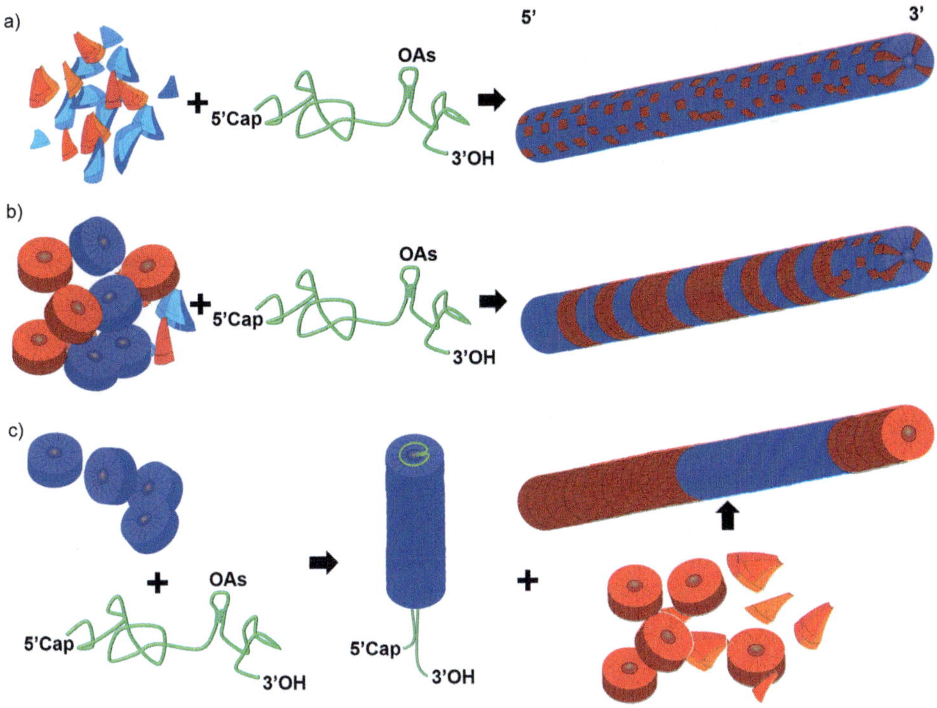

Fig. 1 Scheme showing the different in vitro assembly strategies. (**a**) Starting from mixed CP types (yielding "mixed disks") results in equally mixed rods, (**b**) starting from disks each made up from one CP type only results in narrow stripes, and (**c**) sequential assembly of two CP variants yields domain rods

E. coli. It then describes in detail the assembly of nanorods composed of differently arranged mixtures of two CP types, as developed by Eiben et al. [18], and of nanorods with extended longitudinal domains accessible by the serial use of distinct CP species, initially described by Geiger et al. [6].

The technique can be spatially refined further by combining it with a DNA oligonucleotide-mediated "stop-and-go" approach not included in this chapter. Using "toehold"-equipped DNA "stopper" molecules hybridized to the OAs-containing RNA during the first assembly stage, and their toehold-mediated displacement upon addition of a distinct CP the border between the domains can be accurately defined, see Schneider et al. for details [20].

2 Materials

Prepare all solutions with ultrapure water. Media and buffers should either be autoclaved (20 min, 121 °C) or filter-sterilized (0.2 μm filter). All waste except organic solvents should be autoclaved; virus-contaminated surfaces and labware can be cleaned with highly concentrated alkaline solutions.

2.1 Purification of TMV-CPs

2.1.1 Heterologous Expression and Purification of CP_{Ec}-His_6

1. *E. coli* expression strain: *E. coli* BL21(DE3) (F– ompT hsdSB (rB-mB-) gal dcm (DE3)) containing the plasmid of choice (pET20bCP-His_6 [18] (*see* **Note 1**)).
2. TB medium: 1 g tryptone, 24 g yeast extract, 4 ml glycerol (98%), an 1 L H_2O, autoclave.
3. 100 mg/ml carbenicillin (carb).
4. 50 mM Tris–HCl, pH 7.6.
5. Some means for effective cell disruption of bacterial cells, e.g., a sonifier, cell homogenizer or freeze–thaw cycles (please refer to the respective device manuals for exact conditions)
6. 5 ml His Trap™ FF crude column (GE Healthcare Life Sciences).
7. *Fast protein liquid chromatography (FPLC) system,* e.g., *Äkta system (*GE Healthcare Life Sciences).
8. 0.45 μM cellulose acetate syringe filter membranes, nonsterile.
9. 1 M NaCl.
10. 0.5 M Tris–HCl, pH 8.0.
11. Solvent A: 20 mM Tris–HCl, pH 8.0, 0.5 M NaCl.
12. Solvent B: 1 M imidazole in 20 mM Tris–HCl, pH 8.0, 0.5 M NaCl.
13. Materials for SDS-PAGE analysis.
14. Dialysis tubing with ≤8 kDa MWCO (Spectra/Por 6 dialysis tubing, regenerated cellulose, SpectrumLabs).
15. 75 mM sodium potassium phosphate (SPP) buffer, pH 7.2.
16. 1% (w/v) sodium azide.

2.1.2 Purification of Second TMV-CP from TMV Particles

1. TMV particles from plants (e.g., TMV_{Lys}, TMV_{Cys}, or wt TMV [6, 17]) in 10 mM SPP, pH 7.4. (In this book, chapters 23, 24, and 27 provide protocols for TMV purification.)
2. Glacial acetic acid.
3. Dialysis tubing with MWCO of 8 kDa.
4. 75 mM SPP, pH 7.2.
5. 1% (w/v) sodium azide.

2.2 RNA Isolation from TMV Particles (See Notes 2 and 3)

1. TMV particles (e.g., TMV_{Lys}, TMV_{Cys}, or wt TMV) in 10 mM SPP, pH 7.4.
2. 5× RNA extraction buffer: 0.5 M NaCl, 5 mM EDTA, 5% (w/v) sodium dodecyl sulfate (SDS), 0.1 M Tris–HCl, pH 8.0.
3. CI: chloroform–isoamyl alcohol 24:1.
4. PCI: buffer-saturated phenol: CI 1:1.

5. 3 M Na-acetate, pH 5.2.
6. Ethanol absolute.
7. 70% (v/v) ethanol.
8. Dimethyldicarbonate (DMDC)-treated H_2O: 0.1% (v/v) DMDC.

2.3 Assembly of Different CP Mixtures with TMV RNA

1. CP_{Ec}-His_6 in 75 mM SPP (from Subheading 3.1, **step 13**).
2. CP_{Lys}, CP_{Cys}, or wt CP in 75 mM SPP (from Subheading 3.1.2, **step 8** and **9**).
3. RNA containing the OAs (from Subheading 3.2, **step 11** or in vitro transcribed).

2.4 VLP Analysis

2.4.1 Native Agarose Gels for VLP Separation

1. Mini L 'Revolution' electrophoresis chamber (PeqLab).
2. 5× Particle loading dye (PLD): 50 mM SPP, pH 7.4, 60% (w/v) glycerol, 0.01% (w/v) Bromophenol Blue.
3. Agarose (EEO = 0.05–0.13).
4. 50 mM SPP, pH 7.4.
5. Destaining solution: 10% (v/v) acetic acid, 40% (v/v) ethanol.
6. Staining solution: 0.1% (w/v) Coomassie Brilliant Blue R 250 in destaining solution.

2.4.2 Decoration of VLPs with Proteins via Biotin Coupling

1. EZ-Link™ Maleimide PEG11-Biotin (Thermo Fisher Scientific).
2. 75 mM SPP, pH 7.2.
3. 4× PEG–NaCl: 16% (w/v) PEG 6000, 16% (w/v) NaCl.
4. Strepavidin, avidin, or protein conjugate with either of these.
5. Materials for TEM analysis including stain such as 2% (w/v) uranyl acetate.

3 Methods

3.1 Purification of TMV-CPs

3.1.1 Heterologous Expression and Purification of TMV CP_{Ec}-His_6 in E. coli

Unless otherwise described, standard molecular biology methods according to Green et al. [21] are used. Except for centrifugation of the *E. coli* culture and lysate, all centrifugation steps were performed in tabletop centrifuges for 1.5/2 ml reaction tubes.

1. Inoculate 5 ml TB medium containing 100 µg/ml carb with BL21(DE3) cells containing pET20bCP-His_6 and grow the culture over night at 37 °C in a shaking incubator.
2. Inoculate 50 ml TB medium containing 100 µg/ml carb with 2 ml of the overnight-culture.

3. Incubate in a shaking incubator at 30 °C for 8 h without induction by isopropyl-β-D-thiogalactopyranoside for expression (*see* **Note 4**).
4. Harvest the cells by centrifugation (20 min, 4500 × *g* and 4 °C), and disrupt them for example by sonication in 10 ml 50 mM Tris–HCl, pH 7.6. Separate into "soluble" and "insoluble" fraction by centrifugation (30 min, 10,000 × *g* and 4 °C, SS34 rotor) (*see* **Note 5**). If possible use immediately; however, the cell lysate can be stored for up to 2 days in the fridge before purification of the CP by immobilized metal affinity chromatography (FPLC) (*see* **Note 6**).
5. Adjust the soluble fraction containing the CP_{Ec}-His_6 to a NaCl concentration of 0.5 M and 20 mM imidazole using the stock solutions and filter through a 0.45 μm syringe filter membrane (about 10 ml total volume).
6. Preequilibrate the His Trap™ FF crude column with 4 column volumes solvent A containing 20 mM imidazole by using the automatic buffer mixer of the FPLC and solvent B (2% solvent B, constant flow of 1 ml/min). All solutions are used at room temperature. Use constant A280 measurement.
7. Apply the whole sample using a superloop or manual injection.
8. Remove unbound sample by washing with 2 column volumes equilibration buffer followed by 2 column volumes solvent A with 40 mM imidazole.
9. Elute with a linear gradient (40–1000 mM imidazole in solvent A) over 2 column volumes and harvest 2 ml fractions.
10. Analyze those fractions with highest A280 absorption by SDS polyacrylamide gel electrophoresis.
11. Combine the CP_{Ec}-His_6 fractions with highest yield and purity, as determined by electrophoresis, and dialyze against H_2O at 4–8 °C (MWCO ≤ 8 kDa) changing the water several times until a white precipitate is formed (2–3 times over ~24 h).
12. Centrifuge the solution to harvest the CP_{Ec}-His_6 (20 min, 20,000 × *g*, 4 °C).
13. Resuspend the pellet fraction in 1 ml 75 mM SPP, pH 7.2 supplemented with 0.02% sodium azide (w/v) (*see* **Note 7**).
14. Determine the protein concentration spectrophotometrically: 1 mg/ml has an A280 of about 1.21 (calculated using the formula: ε = (number of tryptophan residues × 5500) + (number of tyrosine residues × 1490)).

3.1.2 Purification of Different CP Variants from Plant Derived TMV Particles

Many TMV mutants can be propagated in planta, e.g., in *Nicotiana tabacum* 'Samsun' nn, and are easily purified with high yields using a protocol adapted from Gooding and Hebert [22]. Variants accessible by this protocol include TMV_{Lys} (T158 K), TMV_{Cys} (S3C), and wt TMV.

The CP preparation is achieved by an acetic acid-based method [23].

1. Mix gently one volume of TMV particles (10 mg/ml in 10 mM SPP, pH 7.4) with two volumes (600 μl TMV + 1200 μl acetic acid) of ice cold glacial acetic acid by turning the tube upside down (2–3 times) and incubate on ice for 15–20 min (precipitating RNA should be visible as schlieren).
2. Precipitate the RNA by centrifugation at 20,000 × *g*, 4 °C for 20 min. A clear pellet should be visible.
3. Take the supernatant and dialyze against ultrapure water at 4–8 °C (MWCO ≤ 8 kDa) with several changes to fresh water until a white precipitate appears (*see* **Note 8**).
4. Take out the solution and spin at 20,000 × *g* for 20 min at 4 °C.
5. Remove the supernatant and add half the volume of your initial TMV solution of 75 mM SPP buffer containing 0.02% Na-azide (w/v). Let the CP dissolve at room temperature for 3–4 h or overnight.
6. Mix the CP by pipetting the solution up and down.
7. Remove insoluble, aggregated CP by spinning at 10,000 × *g* for 20 min at 18 °C.
8. Make dilutions of 1:5 and 1:10 in 75 mM SPP and determine the CP concentration photometrically at 280 nm (1 mg/ml (CPwt, CP_{Lys} and CP_{Cys}) have an OD_{280} of 1.3). The yield should be in the range of 50–60%.
9. For direct assembly with RNA, leave the CP at room temperature to induce disk formation, otherwise it can be stored at 4–8 °C (*see* **Note 9**).

3.2 RNA Isolation from TMV Particles

The easiest way to get RNA suitable for in vitro assembly of TMV-like particles is to isolate the RNA from purified TMV particles. The method used here was already described by Chapman et al. [24].

In brief:

1. Mix 800 μl (10 mg/ml) TMV with 200 μl 5× RNA extraction buffer in a 2 ml reaction tube and extract (mix by vortexing followed by 1 min centrifugation at 10,000 × *g*) three times with 1 ml PCI, always work with the upper aqueous phase.
2. Extract once with pure CI.

3. Take the aqueous phase (~700 μl), divide it into two reaction tubes and add 0.1 volumes 3 M Na-acetate, pH 5.2, mix and add 2.5 volumes ethanol.
4. Precipitate the RNA by incubation at −20 °C for 15 min.
5. Centrifuge at 20,000 × *g*, 15 min at 4 °C.
6. A white pellet can be observed, carefully remove the complete supernatant.
7. Wash pellet once with 500 μl 70% (v/v) ethanol.
8. Let the pellet dry for 5–10 min, then add 70 μl DMDC water to each pellet, do not mix.
9. Allow the RNA to self-dissolve by incubation on ice for 1 h, then resuspend gently.
10. Dilute 2 μl RNA solution with 8 μl DMDC water and determine the concentration at 260 nm photometrically (40 mg/ml RNA has an OD_{260} of 1.0).
11. Adjust the concentration to 1 μg/μl and store in aliquots at −80 °C if possible (otherwise at −20 °C).
12. Check the integrity and length (6.4 kb) of the RNA after denaturation in a formamide containing RNA-sample buffer by standard RNA agarose gel electrophoresis (e.g., 1% denaturing MOPS gel) using an appropriate RNA ladder (e.g., 0.5–9 kB) as reference.

3.3 Assembly of Different CP Mixtures

The assembly state of the CP, i.e., A protein, disks or helices, depends not only on pH and ionic strength of the buffer, but also on the temperature. It was shown by Richards and Williams [25] that at temperatures below 15 °C the A protein is the prevalent form of the CPs. Thus incubating mixtures of different CPs on ice prior to disk formation as described in Subheading 3.3.1 allows an even distribution of the different CP variants (*see* Fig. 1a). Combining CP variants predominantly in the disk state after preincubation overnight at room temperature will result in particles with 5′-domains exhibiting narrow stripes in a randomized order (*see* Fig. 1b). As a helix portion formed from a single disk is roughly 5 nm in length, the majority of such stripes can be expected to exhibit this size. By combining substoichiometric amounts of a first CP type with an RNA scaffold prior to adding a second CP variant will produce rods with two or three longitudinal domains, as described by Geiger et al. [6] (*see* Fig. 1c). In Subheading 3.3.2 a protocol is provided that uses 20% of CP1 and 80% of CP2. Since most of the nanorods will grow quite simultaneously in a bidirectional manner, the majority of products is expected to exhibit a 60–90 nm domain of CP1 and, on both of its sides, two domains consisting of CP2 if genomic TMV RNA is used (for details, refer to [6]). TMV is composed of 5% (w/w) RNA and 95% (w/w) CP, and thus 1 μg RNA gets encapsidated by 20 μg CP. An excess of CP over RNA in the range of 1.2–1.4 (24–28 μg CP per 1 μg RNA) results in highest yields of assembled particles.

3.3.1 In Vitro Mixed Assembly; General Procedure to Yield Mixed or Striped Rods (see Fig. 1a, b)

1. Combine the desired CP variants at concentrations of 1–5 mg/ml in 75 mM SPP (*see* **Note 10**).
2. Incubate on ice for 3–4 h in order to disassemble preformed disks. Shortly vortex the sample, spin the solution down and allow for disk formation of the mixed CP at room temperature at least overnight. This step has to be omitted if striped particles are to be produced.
3. Add the desired RNA, either isolated from TMV or in vitro transcribed, at a concentration of 1 μg RNA per 24 μg of CP. Incubate 6 h at 30 °C or overnight at room temperature (*see* **Note 11**).

3.3.2 In Vitro Domain Assembly (see Fig. 1c)

1. Incubate the desired CP variants at least overnight at room temperature to induce disk formation.
2. To assemble the first CP variant, mix 0.075 μg/μl of RNA (isolated from TMV or in vitro transcribed) with 0.4 μg/μl of the first CP in 75 mM SPP and incubate 30 min at 30 °C to fully consume all of the assembly-competent CP (*see* **Note 12**).
3. Now add the second CP in four times the amount of the first one (final concentration of the second CP~1 μg/μl). Incubate at 30 °C overnight to complete RNA encapsidation into fully assembled nanotubes, typically terminated by two domains of the second CP at both ends.

3.4 Analysis

3.4.1 Native Agarose Gel Electrophoresis of Particles

Native agarose gel electrophoresis of virus particles is an easy method to investigate the combination of CP variants which differ significantly in their isoelectric point (pI), such as wt CP and CP_{Cys}, both pI = 5.09, compared to the T158 K mutant (CP_{Lys}) of pI = 5.41 or the CP_{Ec}-His_6 mutant of pI = 6.03. Due to the additional positive charge on every CP subunit, TMV_{Lys} particles migrate much slower than wt TMV or TMV_{Cys} (*see* **Note 13**). Figure 2 shows the resulting native agarose gels of several mixed assembly approaches using either in vitro transcribed RNA (7 kb and 3 kb), or the native RNA (6.4 kb) in combination with wt CP, CP_{Lys} and CP_{Ec}-His_6. In addition, in Fig. 3a a native agarose gel of domain particles can be seen.

1. Melt 1% NEEO low melting agarose in 50 mM SPP buffer (for the Mini L 'Revolution' electrophoresis chamber use 100 ml) in a microwave oven. Let the solution cool down to approx. 60 °C and cast the gel. Immediately add the desired comb.
2. Mix 15 μl sample (10–20 μg TMV) with 3 μl PLD and load it onto the gel. Suitable reference samples are for example in vitro

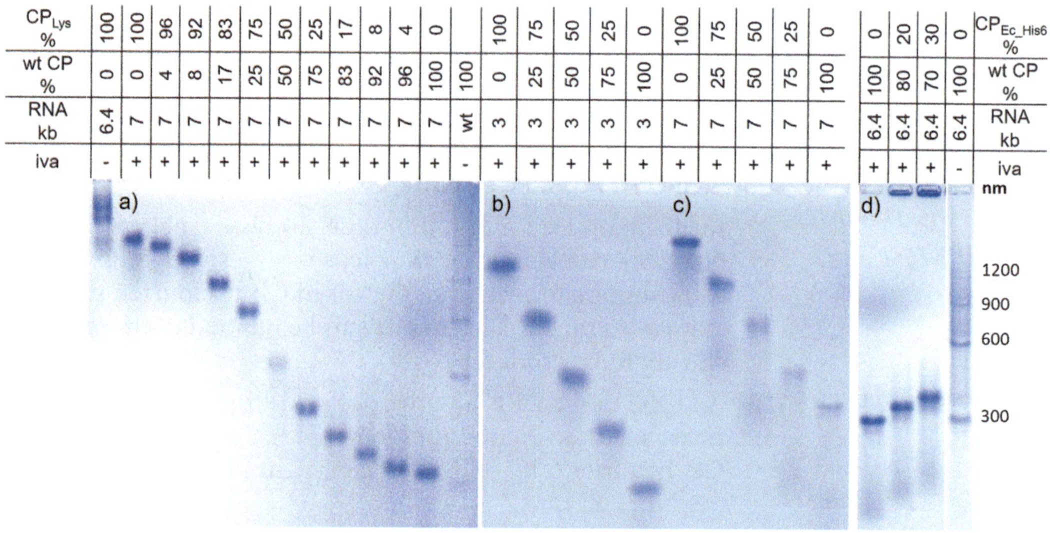

Fig. 2 Native agarose gel electrophoresis of in vitro assembled VLPs. (**a**) Mixed assembly of CP_{Lys} and wt CP with an in vitro transcribed RNA 7000 (*see* ref. 18 for detailed description of the RNA) in comparison to TMV_{Lys} and wt TMV isolated from tobacco. (**b**) Shows VLPs resulting from mixtures of CP_{Lys} and wt CP assembled with a 3000 nt long RNA while in (**c**) preformed disks of CP_{Lys} and wt CP were combined with RNA 7000. (**d**) The native TMV RNA of 6.4 kb was used for a mixed assembly of wt CP and $CP_{Ec\_His_6}$. For the plant-derived wt TMV in the right lane, the corresponding sizes of the TMV multimers are shown. Percentages of CP types and RNA types are indicated above. iva: in vitro assembly

assembled particles with one type of CP or the plant-derived virus particles. Please note that in the latter case, 30–50 μg of TMV should be applied on the gel as the plant-derived TMV particles are prone to head to tail attachment and thus a ladder of up to five different aggregates (300–1500 nm) can be observed (*see* Fig. 2).

3. Perform the electrophoresis for 16 h at 1.6 V/cm and 25 °C (*see* **Note 14**).

3.4.2 Coomassie Staining

All incubation steps are performed at room temperature under gentle agitation (*see* **Note 15**).

1. Incubate the gel 15 min in destaining solution. The gel should always be immersed.
2. Discard the destaining solution and incubate the gel in staining solution for 45–60 min.
3. Pour off the staining solution; it can be reused several times. Apply destaining solution and change it when it gets bluish. The destaining solution can be reactivated by removal of the dye using active carbon.
4. The destaining process of agarose gels takes some time. Although it might be possible to see first results after 5–6 h, the gels look best after approximately 3 days (*see* **Note 16**).

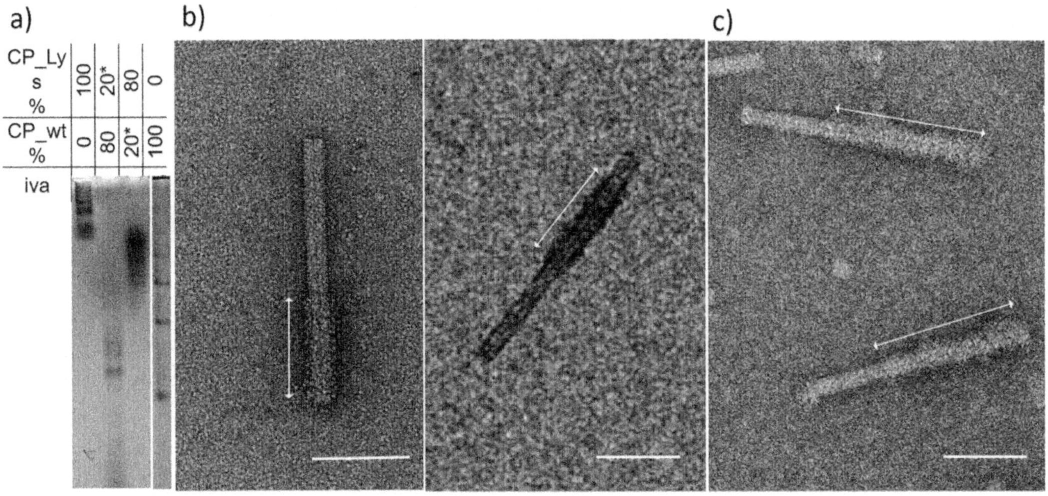

Fig. 3 Native agarose gel and TEM images of sequentially assembled TMV domain particles. (**a**) Sequentially assembled TMV particles with the native TMV RNA (6.3 kb), the first CP variant is designated with an asterisk. (**b** and **c**) TMV particles with domains of CP_{Lys} and CP_{Cys}, the arrows show the cysteine domain, which is decorated with streptavidin-coupled horseradish peroxidase using a bifunctional maleimide-biotin linker. (**b**) CP_{Cys} was first assembled; (**c**) CP_{Lys} was first assembled. The bars represent 100 nm

3.4.3 Decoration of VLPs with Proteins via Biotin Coupling

Some TMV CP mutants such as S3C (TMV_{Cys}) and T158K (TMV_{Lys}) can be specifically addressed by classical chemical coupling of for example peptides or fluorescent dyes. The coupling of these molecules can prove the presence of distinct TMV variants in the mixed particles, add new functionalities to the TMV used as scaffold and help to visualize the domain structure. In the following, the coupling of a biotin linker used to decorate the domain particles with proteins for TEM analysis is explained in detail. In this book, more information on coupling strategies applicable to TMV CP can be found for example in Chapters 26–28, 35, and 37.

1. Incubate 0.9 μg/μl (0.05 nmol) TMV with 2.5 nmol/μl PEG11-Biotin linker (45 × excess linker to CP) in 75 mM SPP, pH 7.2 at room temperature overnight (*see* **Note 17**).
2. Remove the unbound linker by addition of 1/3 volume 4× PEG–NaCl solution and 15 min incubation on ice, followed by 30 min centrifugation at 20,000 × *g* and 4 °C.
3. Resuspend the precipitate in the same volume 75 mM SPP, pH 7.2 as before and add avidin, streptavidin, or (strept)avidin-coupled proteins at a molar ratio of ~5:1 to the CP. Incubate for 2 h at room temperature.
4. Purify the decorated TMV particles by PEG–NaCl precipitation as described before (*see* **step 2**).
5. Investigation of the assembled rods is carried out using TEM (samples negatively stained with 2% (w/v) uranyl acetate) (*see* **Note 18**).

4 Notes

1. Due to the cloning strategy a leucine and a glutamine followed by the plasmid-encoded His_6-tag are added to CP as a translational fusion [18].
2. All solutions for RNA work have to be RNase-free, therefore all buffers without amino-groups should be treated for at least 3 h under rigorous stirring with 0.1% dimethyldicarbonate (DMDC) (diethylpyrocarbonate can also be used in all cases where DMDC is written) followed by autoclaving for 3 h. Buffers such as Tris–HCl containing amino groups have to be prepared using DMDC-pretreated water. Never touch anything that is used for RNA work with bare hands. Remove RNases from pipettes and surfaces using commercial RNase decontamination solutions. Electrophoresis trays can be cleaned by incubation in 3% H_2O_2 for 30 min followed by rinsing with DMDC H_2O.
3. It is possible to perform all mixed assembly strategies with in vitro transcribed RNA, thus particles with a modified, desired length can be obtained using standard molecular biology protocols. In addition, protocols for in vitro transcription of TMV RNA can be found in this book in Chapters 11 and 24.
4. The T7 promoter is quite leaky and the CP expression even in the absence of IPTG is so high that IPTG induction results mainly in a reduction in bacterial growth and thus a lower CP yield [18].
5. The CP_{Ec}-His_6 should be in the soluble fraction; however, those CP variants that accumulate in the precipitate can be solubilized using urea (4–8 M). Urea-treated CPs retain their self-assembly properties after dialysis against 75 mM SPP pH 7.4.
6. Other CP variants expressed in *E. coli* can be purified by anion exchange chromatography or by isoelectric point precipitation through dialysis [26, 27].
7. The CP_{Ec}-His_6 has a high tendency for aggregation, and therefore the concentration should not exceed 2 mg/ml. To remove higher aggregates prior to mixed assembly, these can be sedimented by centrifugation for 1 min at 1000 × *g*.
8. Precipitation of CPs during dialysis after acetic acid treatment relies on the pI of the proteins, thus dialysis time depends on the CP mutant and the pH of the water (due to acidification by dissolved CO_2). After acetic acid precipitation the pH is very low and during dialysis it becomes closer to neutral. For CP_{Lys} it takes longer for the dialysis solution to reach the pH

corresponding to its higher pI as for example with wt CP. If the CP was purified by chromatography in a neutral buffer, the pI of for example wt CP cannot be reached by dialysis against water; instead, a lightly acidic buffer such as 10 mM Na-acetate pH 4.8 can be used.

9. The CP preparation can in most cases be used up to 6 months, although fresh CP assembles a little better.
10. In the range of 1–5 mg/ml CP, VLP assembly works equally well. The employed concentration depends on the initial concentration of the CP variants and the further intended work flow.
11. TMV assembly is quite fast, under optimum conditions using freshly prepared CPs and fully solubilized RNA. wt RNA is fully packaged at 30 °C after roughly 2 h. The times given here are to be on the safe side and for convenience, e.g., start assembly in the morning and perform the native agarose gel overnight.
12. To investigate the assembly process at this stage (by native agarose gel electrophoresis or TEM), the sample can be divided into two aliquots before addition of the second CP. Add the same volume SPP buffer that is used upon addition of CP2 to the "domain rod sample" to the other ("control sample") and store it on ice to stop the assembly process at this partial state. The ratio of CP1 to CP2 in the protocol corresponds to 20% of CP1 to 80% CP2.
13. Partially assembled particles can be investigated as well: They always migrate faster than complete TMV-like rods, due to the highly negative charge of the protruding RNA portion. In addition, the coupling of linker molecules or fluorescent dyes can be monitored. It is important to take into account the charge of the attached molecule, the eventual loss of charge if for example lysine is addressed by the coupling, and the increased size due to the attached molecules—these factors all have an impact on the electrophoretic mobility.
14. The Mini L 'Revolution' electrophoresis chambers have the advantage of built-in buffer circulation; alternatively, the buffer can be circulated by a pump. If there is no such circulation, the buffer gets hot and it is better to run the gels in a coldroom to prevent melting of the agarose.
15. As the agarose gels are much thicker than polyacrylamide gels, a lower Coomassie Blue concentration than for polyacrylamide gels is favorable for staining. It is also advantageous to include an incubation step in destaining solution prior to staining, to remove the SPP buffer in the gels which would otherwise change the pH of the staining solution reducing its reusability.

16. After 24 h in destaining solution it often helps to exchange it for water for further destaining. Due to the size of the TMV particles, the bands stay sharp over a long period of time (at least 2 weeks), hence pictures of different states of destaining can be made.
17. The ratio of linker to CP given here is only a guideline, as each linker is different and the reactivity of the coupling groups declines with time. If possible, use plant-derived TMV particles or in vitro assembled particles of the CP mutant used for coupling and try out several linker-to-CP ratios before starting with for example domain particles. The amount of coupling can be determined by the linker-induced shift in mobility of the CP on a 15% SDS-PAGE. For decoration with proteins, 50% coupling efficiency are enough to visualize the domains.
18. For TEM investigations, the in vitro assembled rods were attached to hydrophilized (either by glow-discharge or short incubation of the grid in 100% ethanol) Formvar©/carbon-coated 400 mesh copper grids, negatively stained with 2% (w/v) uranyl acetate, and analyzed at 120 kV in a magnification range of 10,000–100,000×. In Fig. 3b, c, sequentially assembled and decorated TMV particles are shown for the combination of CP_{Lys}/CP_{Cys} and vice versa.

Acknowledgments

I would like to thank Fania Geiger and Fabian Eber for establishing the protocols for sequential assembly and decoration of TMV, respectively. I am also grateful for the work of Diether Gotthardt, our gardener, and Sigrid Kober for taking care of the plants and virus isolations. Special thanks to Holger Jeske and Christina Wege for their great support and without whom there would be no plant virus nanotechnology in Stuttgart. This work was financed in part by the DFG PAK 410 and SPP1569, the Zeiss foundation, "Projekthaus" NanoBioMater as well as the Baden Württemberg Stiftung in the course of the Network of Competence "Functional Nanostructures."

References

1. Fan XZ, Pomerantseva E, Gnerlich M, Brown A, Gerasopoulos K, McCarthy M, Culver J, Ghodssi R (2013) Tobacco mosaic virus: a biological building block for micro/nano/bio systems. J Vac Sci Technol A 31(5):050815. https://doi.org/10.1116/1.4816584
2. Altintoprak K, Seidenstücker A, Welle A, Eiben S, Atanasova P, Stitz N, Plettl A, Bill J, Gliemann H, Jeske H, Rothenstein D, Geiger F, Wege C (2015) Peptide-equipped tobacco mosaic virus templates for selective and controllable biomineral deposition. Beilstein J Nanotechnol 6:1399–1412. https://doi.org/10.3762/bjnano.6.145
3. Shenton W, Douglas T, Young M, Stubbs G, Mann S (1999) Inorganic-organic nanotube

composites from template mineralization of tobacco mosaic virus. Adv Mater 11(3):253–256. https://doi.org/10.1002/(Sici)1521-4095(199903)11:3<253::Aid-Adma253>3.0.Co;2-7

4. Knez M, Bittner AM, Boes F, Wege C, Jeske H, Maiss E, Kern K (2003) Biotemplate synthesis of 3-nm nickel and cobalt nanowires. Nano Lett 3(8):1079–1082. https://doi.org/10.1021/nl0342545
5. Zhou K, Zhang J, Wang Q (2015) Site-selective nucleation and controlled growth of gold nanostructures in tobacco mosaic virus nanotubulars. Small 11(21):2505–2509. https://doi.org/10.1002/smll.201401512
6. Geiger FC, Eber FJ, Eiben S, Mueller A, Jeske H, Spatz JP, Wege C (2013) TMV nanorods with programmed longitudinal domains of differently addressable coat proteins. Nanoscale 5(9):3808–3816. https://doi.org/10.1039/c3nr33724c
7. Holder PG, Finley DT, Stephanopoulos N, Walton R, Clark DS, Francis MB (2010) Dramatic thermal stability of virus-polymer conjugates in hydrophobic solvents. Langmuir 26(22):17383–17388. https://doi.org/10.1021/la1039305
8. Koch C, Wabbel K, Eber FJ, Krolla-Sidenstein P, Azucena C, Gliemann H, Eiben S, Geiger F, Wege C (2015) Modified TMV particles as beneficial scaffolds to present sensor enzymes. Front Plant Sci 6:1137. https://doi.org/10.3389/fpls.2015.01137
9. Bäcker M, Koch C, Eiben S, Geiger F, Eber F, Gliemann H, Poghossian A, Wege C, Schöning MJ (2017) Tobacco mosaic virus as enzyme nanocarrier for electrochemical biosensors. Sensors Actuators B Chem 238:716–722. https://doi.org/10.1016/j.snb.2016.07.096
10. Koch C, Eber FJ, Azucena C, Forste A, Walheim S, Schimmel T, Bittner AM, Jeske H, Gliemann H, Eiben S, Geiger FC, Wege C (2016) Novel roles for well-known players: from tobacco mosaic virus pests to enzymatically active assemblies. Beilstein J Nanotechnol 7:613–629. https://doi.org/10.3762/bjnano.7.54
11. Fraenkel-Conrat H, Williams RC (1955) Reconstitution of active tobacco mosaic virus from its inactive protein and nucleic acid components. Proc Natl Acad Sci U S A 41(10):690–698. https://doi.org/10.1073/pnas.41.10.690
12. Butler PJ (1999) Self-assembly of tobacco mosaic virus: the role of an intermediate aggregate in generating both specificity and speed. Philos Trans R Soc Lond Ser B Biol Sci 354(1383):537–550. https://doi.org/10.1098/rstb.1999.0405
13. Kegel WK, van der Schoot P (2006) Physical regulation of the self-assembly of tobacco mosaic virus coat protein. Biophys J 91(4):1501–1512. https://doi.org/10.1529/biophysj.105.072603
14. Wu ZY, Mueller A, Degenhard S, Ruff SE, Geiger F, Bittner AM, Wege C, Krill CE (2010) Enhancing the magnetoviscosity of ferrofluids by the addition of biological nanotubes. ACS Nano 4(8):4531–4538. https://doi.org/10.1021/nn100645e
15. Eber FJ, Eiben S, Jeske H, Wege C (2015) RNA-controlled assembly of tobacco mosaic virus-derived complex structures: from nanoboomerangs to tetrapods. Nanoscale 7(1):344–355. https://doi.org/10.1039/c4nr05434b
16. Eber FJ, Eiben S, Jeske H, Wege C (2013) Bottom-up-assembled nanostar colloids of gold cores and tubes derived from tobacco mosaic virus. Angew Chem Int Edit 52(28):7203–7207. https://doi.org/10.1002/anie.201300834
17. Kadri A, Maiss E, Amsharov N, Bittner AM, Balci S, Kern K, Jeske H, Wege C (2011) Engineered tobacco mosaic virus mutants with distinct physical characteristics in planta and enhanced metallization properties. Virus Res 157(1):35–46
18. Eiben S, Stitz N, Eber F, Wagner J, Atanasova P, Bill J, Wege C, Jeske H (2014) Tailoring the surface properties of tobacco mosaic virions by the integration of bacterially expressed mutant coat protein. Virus Res 180:92–96. https://doi.org/10.1016/j.virusres.2013.11.019
19. Gerasopoulos K, McCarthy M, Royston E, Culver JN, Ghodssi R (2008) Nanostructured nickel electrodes using the tobacco mosaic virus for microbattery applications. J Micromech Microeng 18(10):104003. https://doi.org/10.1088/0960-1317/18/10/104003
20. Schneider A, Eber FJ, Wenz NL, Altintoprak K, Jeske H, Eiben S, Wege C (2016) Dynamic DNA-controlled "stop-and-go" assembly of well-defined protein domains on RNA-scaffolded TMV-like nanotubes. Nanoscale 8(47):19853–19866
21. Green MR, Sambrook J (2012) Molecular cloning. A laboratory manual, 4th edn. Cold Spring Harbor Laboratory Press, New York
22. Gooding GV, Hebert TT (1967) A simple technique for purification of tobacco mosaic virus in large quantities. Phytopathology 57:1285

23. Fraenkel-Conrat H (1957) Degradation of tobacco mosaic virus with acetic acid. Virology 4(1):1–4
24. Chapman SN (1998) Tobamovirus isolation and RNA extraction. Methods Mol Biol 81:123–129. https://doi.org/10.1385/0-89603-385-6:123
25. Richards KE, Williams RC (1972) Assembly of tobacco mosaic virus in vitro: effect of state of polymerization of the protein component. Proc Natl Acad Sci U S A 69(5):1121–1124
26. Bruckman MA, Soto CM, McDowell H, Liu JL, Ratna BR, Korpany KV, Zahr OK, Blum AS (2011) Role of hexahistidine in directed nanoassemblies of tobacco mosaic virus coat protein. ACS Nano 5(3):1606–1616
27. Hwang DJ, Roberts IM, Wilson TMA (1994) Expression of tobacco mosaic virus coat protein and assembly of pseudovirus particles in Escherichia coli. Proc Natl Acad Sci U S A 91(19):9067–9071

Chapter 4

Fabrication of Tobacco Mosaic Virus-Like Nanorods for Peptide Display

Emily J. Larkin, Adam D. Brown, and James N. Culver

Abstract

Virus-like particles (VLPs) are genome-free protein shells assembled from virus coat proteins (CPs). The uniform and nanoscale structure of VLPs combined with their noninfectious nature have made them ideal candidates for the display of functional peptides. While the vast majority of VLPs are derived from spherical viruses, tobacco mosaic virus (TMV) produces a rod-shaped particle with a hollow central channel. However, under physiological conditions the TMV CP forms only disk-shaped macromolecules. Here, we describe the design, construction, purification, and processing of rod-shaped TMV-VLPs using a simple bacterial expression system. The robust nature of this system allows for the display of functional peptides and molecules on the outer surface of this novel VLP.

Key words Virus-like particle, Nanoparticle, Peptide display, Self-assembly, Epitope display

1 Introduction

Virus-like particles (VLPs) are derived from the genome-free assembly of virus capsid proteins (CPs). The structure of these particles can provide uniform and repeating molecular patterns that are ideal for the nanoscale display of functional peptides and molecules. As a consequence VLPs have become powerful tools in a range of applications, including vaccines, drug delivery, biomaterials, and sensors [1–3]. However, not every virus CP is capable of self-assembling into a VLP, and there are limitations in the sizes and shapes available for VLP construction. Thus, additional systems that encode for nanostructures of unique shape and function are needed.

Tobacco mosaic virus (TMV) is an extensively studied and well-characterized filamentous plant virus. The TMV virion is a hollow, tube-shaped rod consisting of ~2130 coat protein (CP) subunits encasing a single-stranded, plus-sense RNA genome. The high aspect ratio of the TMV particle (300 nm length × 18 nm

Christina Wege and George P. Lomonossoff (eds.), *Virus-Derived Nanoparticles for Advanced Technologies: Methods and Protocols*, Methods in Molecular Biology, vol. 1776, https://doi.org/10.1007/978-1-4939-7808-3_4,
© Springer Science+Business Media, LLC, part of Springer Nature 2018

diameter) makes it an attractive substrate for the formation of novel nanostructured materials that include electrodes, superhydrophobic surfaces, and antireflective current collectors [4–7]. However, problems associated with virus recombination and genome assembly interfere in the production of TMV mutants designed to display functional peptides. One way to avoid these problems is through the development of a VLP system for this unique rod-shaped virus. While previous studies have expressed TMV CP and a circular CP permutant via bacterial systems at physiological conditions (neutral pH), these purified CPs form only lower order assemblies (small aggregates and disks) and not rod-shaped VLPs, thus limiting their potential use [8, 9]. However, it is possible to modify the viral CP by substituting charge-neutralizing amino acids Q and N at positions E50 and D77 respectively in the virus CP. E50 and D77 form an intersubunit carboxylate pair that functions to control virion assembly [10, 11]. Neutralization of these negatively charged amino acids through the corresponding mutations E50Q and D77N negates the repulsive intersubunit interactions and stabilizes the quaternary structure of the helical rod, even in the absence of viral RNA. This system allows for the assembly of rod-shaped VLPs directly within a bacterial expression system [12]. Here we describe the design, expression, and purification protocols for TMV-based, rod-shaped VLPs.

2 Materials

Prepare all solutions using ultrapure deionized water and analytical-grade reagents. Store all reagents at room temperature unless indicated otherwise. Follow safety guidelines for each reagent as outlined in the corresponding material safety data sheets as well as for autoclave usage and waste disposal.

2.1 Preparing a System Expressing VLP CP

1. The gene sequence for the modified TMV CP must encode the E50Q and D77N substitutions, which are required to drive self-assembly (*see* Fig. 1). Peptide sequences to be displayed on the VLP surface may be added at the C-terminus of the CP (*see* **Notes 1** and **2**). Furthermore, addition of a cysteine codon within the CP N-terminus can direct the vertical assembly of VLPs onto a variety of surfaces [12]. Peptide sequences must also contain N and C-terminal restriction enzyme sites suitable for cloning into the desired expression vector. Additionally, the CP sequence should be codon optimized for the desired expression system such as *Escherichia coli* (*E. coli*).
2. The pET 21a(+) (Clontech-Laboratories Inc.) or comparable expression vector.
3. BL21-CodonPlus cells (*E. coli*) (Agilent Technologies).

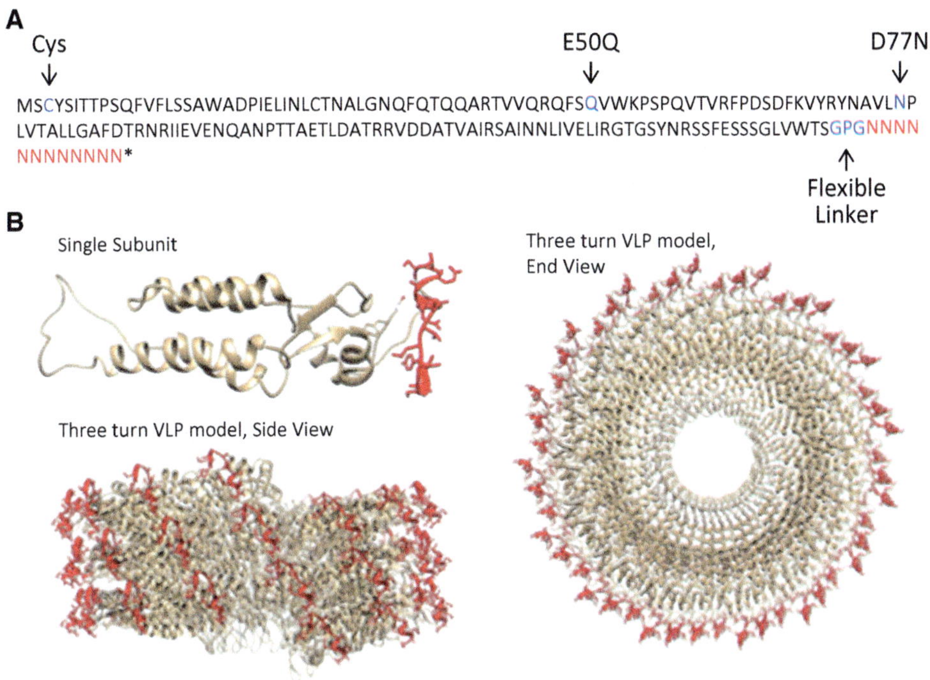

Fig. 1 TMV-VLP design for C-terminus peptide display. (**a**) TMV-VLP CP sequence showing locations of surface binding cysteine residue, helical rod stabilizing E50Q and D77N mutations, and N terminal GPG linker (bold/blue). Peptide sequences of choice can be added to the C-terminus of the VLP CP (bold/red). (**b**) Structural models of a single TMV-VLP CP subunit and helical assemblies showing the location of a 15-amino-acid peptide (showing side chains/red)

4. 100 mg/ml ampicillin (stock solution) in 70% ethanol (for maintenance of the pET expression vector).
5. 20 mg/ml chloramphenicol (stock solution) in 100% ethanol (for maintenance of BL21-CodonPlus cells).
6. Luria–Bertani (LB) agar plates and broth with ampicillin and chloramphenicol: 10 g/l tryptone, 5 g/l yeast extract, and 10 g/l NaCl. For plates add 1.5 g/l agar. Autoclave at 121 °C for 15 min to sterilize. Cool to 45–50 °C before adding ampicillin to a final concentration of 100 μg/ml and chloramphenicol to a final concentration of 20 μg/ml.
7. Wizard Plus DNA purification system (Promega).
8. Dimethyl sulfoxide (DMSO) (to produce −80 °C freezer stocks.).

2.2 Culture and Expression Conditions

1. 100 mg/ml ampicillin (stock solution) in 70% ethanol (*see* Subheading 2.1, **item 4**).
2. 20 mg/ml chloramphenicol (stock solution) in 100% ethanol (*see* Subheading 2.1, **item 5**).

3. Luria–Bertani (LB) agar plates and broth with ampicillin and chloramphenicol (*see* Subheading 2.1, **item 6**).
4. Bacterial inoculum is maintained as −80 °C freezer stocks (*see* **Note 3** and Subheading 2.1, **item 5**).
5. Sterile culture tubes with caps (15 × 125 mm).
6. Sterile 150 ml Erlenmeyer flasks.
7. 1 M isopropyl-β-D-1-thiogalactopyranoside (IPTG) (stock solution). Store at −20 °C. Thaw and add to bacterial culture at a final concentration of 1 mM (*see* **Note 4**).

2.3 VLP Purification

1. 50 ml polypropylene conical centrifuge tubes.
2. Bugbuster HT Protein Extraction Reagent (EMD Millipore). Store at 4 °C (*see* **Note 5**).
3. Lysonase Bioprocessing Reagent (EMD Millipore) Store at −20 °C.
4. 1 M dithiothreitol (DTT). Store at −20 °C (*see* **Note 6**). Only needed if the CP contains N-terminal cysteine residue required for vertical surface assemblies [12].
5. Screw-cap polycarbonate (25 × 89 mm) ultracentrifuge bottles (Beckman Coulter Inc.).
6. 0.1 M phosphate buffer pH 7.0: 57.7 ml 1 M Na_2HPO_4, 42.3 ml 1 M NaH_2PO_4. Bring to 1 l with water. Autoclave at 121 °C for 15 min to sterilize (*see* **Note** 7).
7. 10–40% sucrose gradient: 50 g sucrose in 150 ml of 0.1 M phosphate buffer, pH 7. Mix until dissolved and distribute into open-top (25 × 89 mm) polycarbonate ultracentrifuge tubes (Nalgene). Freeze tubes upright at −20 °C, and thaw before use (*see* **Note 8**).
8. Wide bore cannula and 10 ml disposable syringe for removal of VLP band from sucrose gradient.
9. NanoDrop 1000 Spectrophotometer (Thermo Scientific).
10. Ultracentrifuge.

2.4 Additional Purification Steps for Improved VLP Purity

2.4.1 Chloroform Addition

1. Polypropylene or glass centrifuge tube.
2. Chloroform.

2.4.2 PEG Precipitation

1. Polyethylene glycol (PEG) 8000.
2. KCl.
3. 0.1 M Phosphate buffer pH 7.0 (*see* Subheading 2.3, **item 6**).

3 Methods

Below we describe a simple system for the purification of VLPs from a bacterial expression system. This system works well for the unmodified TMV-VLP as well as many VLPs that display unique peptides. However, some VLP peptide constructs are more difficult to obtain in pure form, presumably due to poor assembly or interactions with cellular components. We also provide two additional steps that we have used to improve the purity of such VLP constructs.

3.1 Preparing a System to Express VLP CP

1. We recommend using a commercial vendor, such as Thermo Scientific to synthesize codon-optimized VLP CP sequences. Synthesized CP sequences can be obtained either cloned into the desired expression vector or with the appropriate N and C terminal restriction sites for vector cloning. For the pET 21a(+) vector we have used N terminal *Nde*I and C terminal *Xho*I restriction sites to insert synthesized CP sequences using standard molecular cloning methods (*see* **Note 9**).
2. Expression vectors containing the CP sequence can be transformed into chemically competent BL21-CodonPlus cells via heat shock. Transformations should be plated onto LB agar plates containing ampicillin and chloramphenicol and incubated at 37 °C overnight.
3. Transformed colonies should be amplified overnight in LB medium under ampicillin and chloramphenicol selection. The CP expression vector is purified using Wizard Plus DNA purification system.
4. BL21-CodonPlus purified CP expression vectors should be sequence-confirmed prior to proceeding with expression.
5. Freezer stocks of BL21-CodonPlus cell containing CP sequence-confirmed expression vector can be prepared by adding 70 μl dimethyl sulfoxide to 1 ml culture and immediately freezing at −80 °C.

3.2 Culture and Expression Conditions

1. Inoculate 5 ml LB medium + ampicillin + chloramphenicol with bacteria from −80 °C freezer stocks in 15 × 125 mm tubes with caps. Incubate overnight at 37 °C on a shaker set for at least 150 rpm.
2. The next morning prewarm a sterile 150 ml Erlenmeyer flask containing 50 ml LB medium + ampicillin + chloramphenicol to 37 °C.
3. Inoculate 200 μl of bacterial culture from the overnight culture into the prewarmed medium.

4. Place the 50 ml culture in a 37 °C incubator and shake at 150 rpm.
5. Grow the culture(s) to an OD_{600} of 0.5 (*see* **Note 10**).
6. Remove the 50 ml culture from the incubator and add 50 μl of 1 M IPTG.
7. Place the flasks on a shaker at 150 rpm and incubate overnight at room temperature (*see* **Note 11**).

3.3 VLP Purification

1. After overnight incubation, transfer the culture to a 50 ml polypropylene conical tube and centrifuge at 16,000 × *g* for 10 min.
2. Pour off supernatant taking care not to disturb the pellet.
3. Following the manufacturer's protocol add 2.5 ml Bugbuster HT Protein Extraction Reagent and incubate at room temperature with gentle shaking (20 rpm), until the pellet is fully dissolved.
4. Add 1 μl of Lysonase Bioprocessing Reagent to the dissolved pellet and incubate for 5 min at room temperature with gentle shaking. Add 2 μl of 1 M DTT if the VLP CP contains the additional N-terminal cysteine residue required for vertical surface attachment.
5. Transfer dissolved pellet to screw-cap polycarbonate (25 × 89 mm) ultracentrifuge tubes and balance using 0.1 M phosphate buffer pH 7.
6. Centrifuge at 91,000 × *g* for 1 h at 4 °C.
7. Pour off the supernatant and add 4 ml 0.1 M phosphate buffer to the screw-cap ultracentrifuge tubes.
8. Place screw-cap ultracentrifuge tube, pellet facing downward, on a shaker at 4 °C, to dissolve the VLP-containing pellet overnight (*see* **Note 12**).
9. The next morning, thaw the sucrose gradients prepared in Subheading 2.3 at room temperature.
10. Gently layer 2 ml of the resuspended pellet onto the top of the fully thawed sucrose gradient (*see* **Note 13**).
11. Centrifuge the gradient at 91,000 × *g* for 1.5 h at 14 °C. The VLPs will form a broad translucent band approximately midway down the gradient (*see* **Note 14** and Fig. 2a).
12. Using a syringe with cannula carefully remove the VLP band and transfer it to a screw-cap ultracentrifuge tube.
13. Centrifuge at 92,000 × *g* at 4 °C for 1.5 h to pellet the VLPs.
14. Carefully discard the supernatant and gently resuspend the pellet in 0.2–1 ml 0.1 M phosphate buffer pH 7.0.

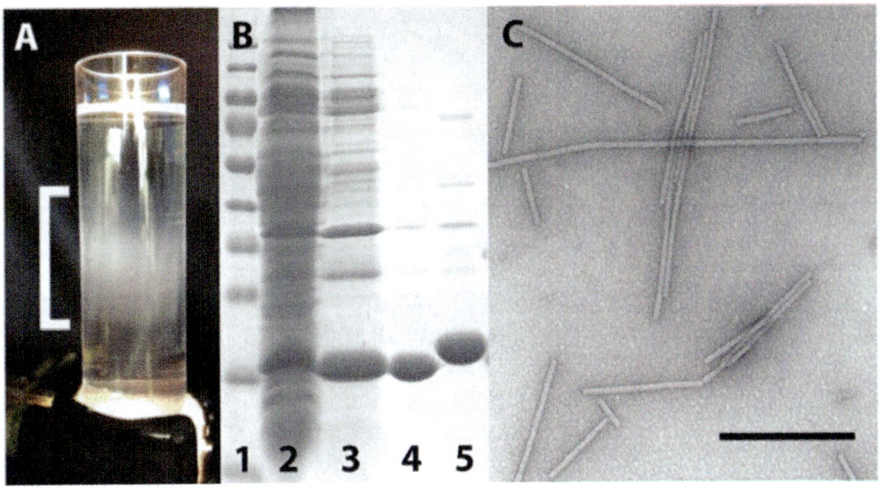

Fig. 2 TMV-VLP purification. (**a**) VLPs form a band (indicated by white bracket) after a 1.5 h centrifugation in a 10–40% (w/v) sucrose gradient. (**b**) SDS-PAGE gel shows increasing purity of samples taken from various steps of the VLP purification procedure (lane 1: Protein ladder, lane 2: bacterial lysate containing VLPs (Subheading 3.3, **step 4**), lane 3: resuspended pellet of first high speed centrifugation (Subheading 3.3, **step 8**), lane 4: purified VLPs (Subheading 3.3, **step 13**), lane 5: purified VLPs with HA peptide tag, showing higher MW. (**c**) Transmission electron micrograph of purified TMV VLPs (bar = 500 nm)

15. Use UV absorbance at 280 nm, corrected for light scattering (*see* **Note 15)** and a calculated extinction coefficient for the specific CP (*see* **Note 16**) to determine VLP concentrations in solution [13, 14]. We typically see between 2 and 5 mg of VLP per 50 ml culture.

3.4 Additional Purification Steps for Improved VLP Purity

3.4.1 Chloroform Addition

Chloroform allows for the removal of organic soluble compounds from VLP preparations [12].

1. At Subheading 3.3, **step 8**, after resuspending the pellet in 4 ml 0.1 M phosphate buffer (pH 7) transfer the solution to a polypropylene or glass centrifuge tube and add 10% chloroform by volume (*see* **Note 17**). Mix gently for 5 min and centrifuge at 17,000 × *g* for 10 min.
2. Carefully collect the upper aqueous phase taking care not to disturb the lower chloroform phase.
3. Resume VLP purification protocol at Subheading 3.3, **step 9**. Alternatively, if additional purification and concentration is needed, proceed to PEG precipitation, Subheading 3.4.2, **step 1**.

3.4.2 PEG Precipitation

Polyethylene glycol 8000 (PEG 8000) treatments can be used to precipitate the VLPs away from other soluble contaminants.

1. Add 6% (w/v) PEG and 1.5% (w/v) KCl to the partially purified VLP solution (from Subheading 3.4.1, **step 3** or Subheading 3.3, **step 8**) and mix thoroughly for 30 min at 4 °C.

2. Centrifuge at 17,000 × *g* for 10 min and discard the supernatant.
3. Resuspend pellet in 4 ml 0.1 M phosphate buffer and proceed to gradient loading (*see* Subheading 3.3, **step 9**).

4 Notes

1. Peptides up to 17 amino acids in length have been displayed from the C terminus of the CP with the inclusion of GPG or GGGGSGGGGS linker peptides [15]. Successfully displayed peptides that have maintained their functionality include those recognized by antibodies and those binding target molecules such as fluorescent imaging agents and 2,4,6-trinitrotoluene (TNT) [12, 16, 17]. For surface attachment the addition of a cysteine codon at position three at the N-terminus will allow for VLP attachment at one end of the rod to gold or other metal surfaces as previously described [17, 18].
2. While many peptides can be displayed and readily assemble into VLPs, we have found others that are more difficult to express or fail to form the rod-shaped VLPs due to steric interference. One solution involves adding an amber stop codon (TAG) between the C-terminus of the CP and the added peptide and using a bacterial expression system such as JM109(DE3) that contains the supE44 amber nonsense mutation suppressor [12, 19]. This will result in the production of two forms of the CP, one formed from termination at the TAG stop codon and another that results from a read-through of the TAG stop codon by addition of a glutamine residue and translation of the added peptide. This may allow the CP to assemble into rods at the expense of reducing the number of CP subunits displaying the added peptide sequence.
3. For a bacterial inoculum we generally scrape a few ice crystals from DMSO −80 °C freezer stocks.
4. For every milliliter of culture to be induced use 1 μl of 1 M IPTG.
5. A precipitate may form in the Bugbuster reagent when stored at 4 °C. If this occurs, leave at room temperature until the precipitate dissolves before using.
6. DTT functions as a reducing agent and is important to keep the cysteine residue at position three, if present, reduced and suitable for surface attachment.
7. Always use a pH meter to confirm the final pH of a buffer.
8. Thawing a 25% (w/v) sucrose solution inside the ultracentrifuge tubes at room temperature will produce a gradient of approximately 10–40% (w/v) that is sufficient to isolate TMV-VLPs. However, linear gradient makers can also be used.

9. The pET 21a(+) expression vector used in this protocol contains ampicillin resistance, and the BL21-CodonPlus bacterial cells used as an optimized protein expression cell line have chloramphenicol resistance. This means that both chloramphenicol and ampicillin need to be present in the LB media when expressing the VLPs using the ampicillin and chloramphenicol concentrations described in Subheading 2.
10. For bacterial growth it usually requires between 3.5 and 4.5 h at 37 °C to reach an optical density of 0.5 at 600 nm.
11. We have found that incubating VLP cultures + IPTG at 20–25 °C produces the highest level of assembled VLPs. Incubations at 37 °C generally result in excess levels of improperly folded or nonassembled CP.
12. If the pellet is not fully dissolved after overnight incubation use a 1 ml micropipette tip to gently break up the pellet. Allow any remaining pellet to settle to the bottom of the tube before loading the supernatant onto the gradient.
13. Use of a wide bore pipette tip will minimize disturbance of the sucrose gradient when applying the VLP solution.
14. Since the VLPs do not contain nor require nucleic acid for assembly there is no size control for their length. However we have found that VLP purification and processing typically results in a size range between 100 nm and 1 μm in length.
15. Light scattering caused by the VLPs should be compensated for in order to accurately determine their concentrations. Rayleigh light scattering is proportional to λ^{-4}. The observed A_{325}, assumed to be solely due to scattering, can be used to correct the A_{280} prior to determining VLP concentration. Essentially, scattering at 280 nm is 1.8 times the scattering observed at 325 nm. The corrected A_{280} is equal to the observed A_{280} minus 1.8 times the observed A_{325}. Thus, VLP concentration in mg/ml equals the corrected A_{280} divided by the calculated mass extinction coefficient for the measured VLP CP in l/(g cm) and adjusting for the path length,

$$C = \frac{\mathrm{OD}_{280} - (1.8 \times \mathrm{OD}_{325})}{\varepsilon_{280} \times l},$$

where ε_{280} is the mass extinction coefficient of the VLP CP at 280 nm and l is the path length in cm.
16. There are several web-based calculators for determining the extinction coefficient of specific peptide sequences. The mass extinction coefficient can be calculated by dividing the molar extinction coefficient by the molecular weight of the CP.
17. Owing to incompatibilities do not add chloroform to polycarbonate tubes.

Acknowledgments

This work was supported in part by the Biochemistry Program of the Army Research Office, award W911NF1410286.

References

1. Fan XZ, Pomerantseva E, Gnerlich M, Brown A, Gerasopoulos K, McCarthy M, Culver J, Ghodssi R (2013) Tobacco mosaic virus: a biological building block for micro/nano/bio systems. J Vac Sci Technol A 31:1–24
2. Culver JN, Brown AD, Zang F, Gnerlich M, Gerasopoulos K, Ghodssi R (2015) Plant virus directed fabrication of nanoscale materials and devices. Virology 479-480:200–212
3. Lee KL, Twyman RM, Fiering S, Steinmetz NF (2016) Virus-based nanoparticles as platform technologies for modern vaccines. Wiley Interdiscip Rev Nanomed Nanobiotechnol 8:554–578
4. Bittner AM, Alonso JM, Gorzny ML, Wege C (2013) Nanoscale science and technology with plant viruses and bacteriophages. Subcell Biochem 68:667–702
5. Ghosh A, Guo JC, Brown AD, Royston E, Wang CS, Kofinas P, Culver JN (2012) Virus-assembled flexible electrode-electrolyte interfaces for enhanced polymer-based battery applications. J Nanomater 2012:1–6
6. McCarthy M, Gerasopoulos K, Enright R, Culver JN, Ghodssi R, Wang EN (2012) Biotemplated hierarchical surfaces and the role of dual length scales on the repellency of impacting droplets. Appl Phys Lett 100:1–5
7. Chiang CY, Epstein J, Brown A, Munday JN, Culver JN, Ehrman S (2012) Biological templates for antireflective current collectors for photoelectrochemical cell applications. Nano Lett 12:6005–6011
8. Dedeo MT, Duderstadt KE, Berger JM, Francis MB (2010) Nanoscale protein assemblies from a circular permutant of the tobacco mosaic virus. Nano Lett 10:181–186
9. Hwang DJ, Roberts IM, Wilson TM (1994) Expression of tobacco mosaic virus coat protein and assembly of pseudovirus particles in Escherichia Coli. Proc Natl Acad Sci U S A 91:9067–9071
10. Lu B, Stubbs G, Culver JN (1996) Carboxylate interactions involved in the disassembly of tobacco mosaic tobamovirus. Virology 225:11–20
11. Namba K, Pattanayek R, Stubbs G (1989) Visualization of protein-nucleic acid interactions in a virus–refined structure of intact tobacco mosaic virus at 2.9 a resolution by X-ray fiber diffraction. J Mol Biol 208:307–325
12. Brown AD, Naves L, Wang X, Ghodssi R, Culver JN (2013) Carboxylate-directed in vivo assembly of virus-like nanorods and tubes for the display of functional peptides and residues. Biomacromolecules 14:3123–3129
13. Freifelder DM (1982) Physical biochemistry, applications to biochemistry and molecular biology vol. W. H. Freeman and Company, San Francisco
14. Porterfield JZ, Zlotnick A (2010) A simple and general method for determining the protein and nucleic acid content of viruses by UV absorbance. Virology 407:281–288
15. Lu P, Feng MG (2008) Bifunctional enhancement of a beta-glucanase-xylanase fusion enzyme by optimization of peptide linkers. Appl Microbiol Biotechnol 79:579–587
16. Zang FH, Gerasopoulos K, Fan XZ, Brown AD, Culver JN, Ghodssi R (2014) An electrochemical sensor for selective TNT sensing based on tobacco mosaic virus-like particle binding agents. Chem Commun (Camb) 50:12977–12980
17. Fan XZ, Naves L, Siwak NP, Brown A, Culver J, Ghodssi R (2015) Integration of genetically modified virus-like-particles with an optical resonator for selective bio-detection. Nanotechnology 26:1–9
18. Royston E, Ghosh A, Kofinas P, Harris MT, Culver JN (2008) Self-assembly of virus-structured high surface area nanomaterials and their application as battery electrodes. Langmuir 24:906–912
19. Inokuchi H, Yamao F, Sakano H, Ozeki H (1979) Identification of transfer RNA suppressors in Escherichia Coli. I. Amber suppressor su+2, an anticodon mutant of tRNA2Gln. J Mol Biol 132:649–662

Chapter 5

In Planta Production of Fluorescent Filamentous Plant Virus-Based Nanoparticles

Sourabh Shukla, Christina Dickmeis, Rainer Fischer, Ulrich Commandeur, and Nicole F. Steinmetz

Abstract

Viral nanoparticles are attractive platforms for biomedical applications and are frequently employed for optical imaging in tissue culture and preclinical animal models as fluorescent probes. Chemical modification with organic dyes remains the most common strategy to develop such fluorescent probes. Here we report a genetic engineering approach to incorporate fluorescent proteins in viral nanoparticles, which can be propagated in their plant host. The fluorescent viral nanoparticles so obtained obviate post-harvest modifications and thereby maximize yields. Our engineering approach transforms filamentous potato virus X (PVX) to display green fluorescent protein (GFP) or mCherry as N-terminal coat protein (CP) fusions at a 1:3 fusion protein to CP ratio through integration of the foot-and-mouth disease 2A sequence. The *in planta* propagation of recombinant GFP-PVX or mCherry-PVX thus produced in *Nicotiana benthamiana* can be easily documented using fluorescence imaging. Molecular farming protocols can be accordingly optimized by monitoring chimera stability over the course of the infection cycle. Moreover, we also demonstrate the utility of recombinant mCherry-PVX in optical imaging of human cancer cells and tumor tissue in preclinical mice model. Together, these features make genetically engineered fluorescent PVX particles ideally suited for molecular imaging applications.

Key words Viral nanoparticles, Potato virus X, Genetic engineering, Ribosome skip, FMDV 2A sequence, mCherry, Tumor homing

1 Introduction

Plant virus-based nanomaterials are fast emerging as multifaceted tools for fields as diverse as materials, chemistry, and medicine [1–4]. The self-assembling three-dimensional architecture of viral nanoparticles (VNPs) offers a robust scaffold for presentation of a precisely controlled ensemble of ligands. For biomedical applications their structural properties and programmability make VNPs efficient carriers for therapeutic payloads enabling tissue-targeted drug delivery, contrast agents for molecular imaging, as well as epitopes or immunostimulatory molecules for applications in vaccines or immuno-

Christina Wege and George P. Lomonossoff (eds.), *Virus-Derived Nanoparticles for Advanced Technologies: Methods and Protocols*, Methods in Molecular Biology, vol. 1776, https://doi.org/10.1007/978-1-4939-7808-3_5,
© Springer Science+Business Media, LLC, part of Springer Nature 2018

therapies [1, 2, 5–8]. A wide array of chemical ligation strategies has been established to conjugate these payloads to viral coat proteins through distinct functional groups presented by amino acid residues, both internally and externally [1, 2]. However, it is the possibility of genetic manipulation of structure and function through a programmable genome that distinguishes VNPs from other synthetic nanomaterials [9–12]. VNPs can be genetically engineered to present alternate functionalities by substituting amino acid side chains or to stably display foreign proteins and polypeptides on coat proteins expressed from the viral genome. Such chimeric VNPs can be propagated through molecular farming to produce engineered particles with high precision and reproducibility [13, 14], and this provides an inherent advantage over the batch-to-batch variations of a typical chemical conjugation process.

Potato virus X (PVX) is a filamentous member of the potexvirus family of plant viruses, measuring 515 × 13 nm. Structurally, it is composed of a capsid made up of 1270 identical copies of a 25 kDa coat protein (CP) wrapped around a single stranded 6.4 kb RNA genome. Each CP subunit offers a ligation handle in form of a reactive lysine residue that has been used to conjugate a high payload of fluorescent probes [15, 16], polymers [17], epitopes [18, 19] or targeting ligands on PVX [20, 21].

In addition to chemically addressable amino acid side chains, the surface-exposed N-terminus of PVX CP enables genetic fusion of peptides and proteins. Using such genetic manipulations, several PVX chimeras have been previously constructed. These include PVX-based vaccines expressing HIV-1 epitopes gp41 [22], *S. aureus* D2 FnBP epitopes [23], 16E7 epitopes derived from human papillomavirus (HPV) [24] and R9 epitopes from hepatitis C virus (HCV) [18]. Similarly, fluorescent PVX particles have also been prepared in plants through expression of fluorescent proteins as N-terminal CP fusion [25]. Efficient assembly of chimeric filamentous PVX particles exposing either GFP or mCherry proteins is achieved through coexpression of free and fusion protein through insertion of an intervening 2A sequence from the foot-and-mouth disease virus (FMDV) (Fig. 1). The 2A sequence induces a ribosomal skip that leads to coexpression of free and fusion proteins, the ratio of which can be defined through sequence selection [25–27].

Propagation of genetically engineered VNPs in plants through molecular farming is advantageous as it reduces the number of downstream, post-harvest, processing, and purification steps and therefore minimizes losses and maximizes yields. Molecular farming of functionalized nanoparticles in leaf tissue could also facilitate production of therapeutics, contrast agents, or vaccines in underdeveloped nations where poor infrastructure is a key challenge in manufacturing and processing of such emerging platform technologies [14, 28, 29].

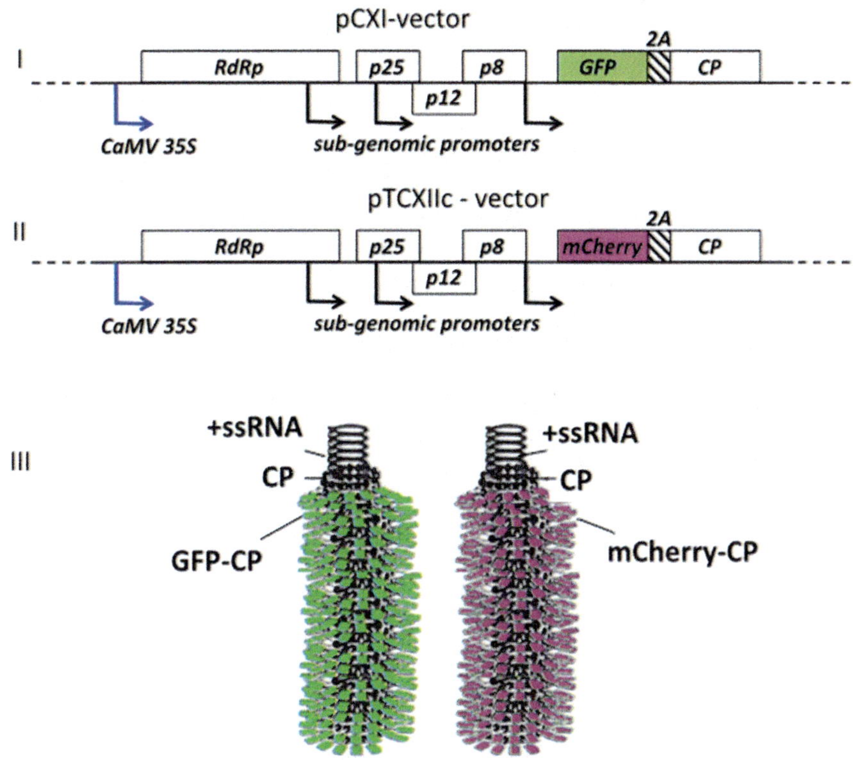

Fig. 1 Schematic representation of the pCXI (I) and pTCXIIc expression vectors (II) containing the full-length genome of PVX under control of the cauliflower mosaic virus (CaMV) 35S promoter: RdRp = RNA-dependent RNA polymerase; p25, p12, and p8 form the PVX triple gene block encoding protein required for movement; 2A sequence from FMDV induces ribosomal skip leading to the production of fusion GFP/mCherry-CP and free CP. Schematic representation of filamentous GFP-PVX and mCherry-PVX depicting assembled CP and fusion proteins encapsulating (+ssRNA) genome (III)

We describe herein the genetic engineering methods used to create GFP-PVX or mCherry-PVX stably expressing the fluorescent probes and the methods employed to monitor the genetic stability of chimeras toward optimization of molecular farming protocols. Also described are the methods used to demonstrate the application of these genetically engineered probes for optical imaging applications in vitro and ex vivo using cancer cells and a preclinical mice model, respectively. With proven enhanced tumor homing and penetration abilities [30], and prolonged circulation through evasion of phagocytic clearance, such molecular probes based on genetically engineered PVX have enormous potential for optical imaging applications. In this chapter, we focus on fluorescent PVX formulations; however, it should be noted that the described techniques could be applied to the study of PVX chimeras with other medical payloads or ligands of interest in biotechnology and energy.

2 Materials

2.1 Enzymes and Cloning Reagents

1. Restriction enzymes *Nhe*I and *Bsp*EI.
2. MSB® Spin PCRapace Kit (Invitek, STRATEC Biomedical AG).
3. *Pfu* DNA polymerase for PCR experiments and GoTaq® DNA polymerase (Promega) for control PCRs.
4. Calf intestinal phosphatase (CIP) for dephosphorylation of vectors after digestion.
5. T4 DNA ligase (Promega).
6. 1× TAE buffer: 40 mM Tris base, 20 mM acetic acid, 1 mM EDTA, pH 7.6.
7. 1.2% (w/v) agarose in 1× TAE buffer.
8. 0.3 mg/l ethidium bromide.
9. GeneRuler™ 100 bp plus and 1-kb ladders (Fermentas, Thermo Fisher Scientific).
10. Wizard® SV Gel and PCR Cleanup System (Promega).

2.2 Plants, Plasmids, and Bacterial Strains

1. *E. coli* DH5α cells.
2. LB medium: 0.5% (w/v) yeast extract, 1% (w/v) NaCl, 0.5% (w/v) tryptone/peptone, pH 7.4.
3. LB plates: 1.5% (w/v) agar in LB medium supplemented with suitable antibiotics, e.g., 100 μg/ml ampicillin.
4. Pure Yield™ Plasmid Miniprep Kit for isolation of small amounts of DNA and the Pure Yield™ Midiprep Kit for larger amounts (Promega).
5. Plasmids: pCXI encoding the PVX genome including the GFP-2A-CP fusion sequence [9, 25], pTRAkc-ERH-mCherry-his6 [31].
6. Target sequence of a fluorescent protein of choice.
7. *Nicotiana benthamiana* plants.

2.3 Expression and Purification of Control Fluorescent Proteins

1. *Agrobacterium tumefaciens* GV3101 pMP90RK (DSMZ, Germany) transformed with pTRAkc-ERH-mCherry-his6 plasmid.
2. YEP medium: 0.5% (w/v) beef extract, 0.1% (w/v) yeast extract, 0.5% (w/v) peptone, 0.5% (w/v) sucrose, 2 mM $MgSO_4$, pH 7.4.
3. YEP plates: 1.5% (w/v) agar in YEP medium supplemented with 50 μg/ml rifampicin, 100 μg/ml carbenicillin, and 50 μg/ml kanamycin.
4. 1 M MES (2-(N-morpholino)-ethanesulfonic acid), pH 5.6, adjust pH with 1 M KOH.

5. 40% (w/v) glucose.
6. 200 mM acetosyringone.
7. 2× infiltration medium: 10% (w/v) sucrose, 0.36% (w/v) glucose, 0.86% (w/v) Murashige and Skoog (MS) salts, pH 5.6.
8. 4–6-week-old *N. benthamiana* plants.
9. 1 ml needleless syringe.
10. Ni-NTA agarose (Qiagen) columns prepared according to the manufacturer's instructions.
11. Phosphate buffered saline (1× PBS): 137 mM NaCl, 2.7 mM KCl, 10 mM Na_2HPO_4, 1.8 mM KH_2PO_4, pH 7.4 (adjust pH with NaOH).
12. 10 mM imidazole in PBS.
13. 150–300 mM imidazole solutions in PBS.
14. 0.01 M phosphate buffer (pH 7.2).
15. Dialysis tubing.
16. Vivaspin 6 columns (Sartorius).

2.4 Plant Virus Purification

1. Celite 545.
2. 0.2 M phosphate buffers: Prepare 0.2 M Na_2HPO_4 and 0.2 M NaH_2PO_4 and prepare various phosphate buffers according to Table 1.
3. 0.05 M phosphate buffer pH 8.0 with 1% (v/v) Triton X-100 (Table 1).
4. 0.01 M phosphate buffer pH 7.2, *see* Table 1.
5. Extraction buffer: 0.1 M phosphate buffer, pH 8.9, 0.2% (v/v) 2-mercaptoethanol, 10% (v/v) ethanol (Table 1).
6. Blender with glass container (Waring™).
7. Miracloth (Calbiochem, CAT 475855).
8. Triton X-100.

Table 1
Phosphate buffer composition (for 100 ml final volume)

pH at 25 °C	x ml 0.2 M Na_2HPO_4	y ml 0.2 M NaH_2PO_4
7.0	30.5	19.5
7.2	36.0	14.0
7.4	40.5	9.5
7.6	43.5	6.5
7.8	45.75	4.25
8.0	47.35	2.65

9. 1 M NaCl/20% (w/v) PEG: solve 58.44 g NaCl, 100 g PEG 6000, and 100 g PEG 8000, in 1 l deionized H_2O while stirring. Autoclave for sterilization.
10. Process buffer: 0.05 M phosphate buffer pH 8.0 (Table 1), 1% (v/v) Triton X-100.
11. Sucrose gradient: Prepare 10% (w/v) and 45% (w/v) sucrose in 0.01 M phosphate buffer pH 7.2 (Table 1), with 0.01 M EDTA. Use 12 ml 10% (w/v) and 12 ml 45% (w/v) sucrose solution with a gradient mixer to generate the gradient.
12. Ultracentrifuge XPN-80 (Beckman Coulter) with swinging bucket rotors SW41Ti and SW 32Ti.
13. Ultra-Clear centrifuge tubes (14 × 89 mm for SW 41Ti and 25 × 89 mm for SW32Ti) (Beckman Coulter).

2.5 SDS-PAGE/Western Blotting

1. Loading dye (5× reducing buffer): 62.5 mM Tris–HCl pH 6.8, 30% (v/v) glycerol, 4% (w/v) sodium dodecyl sulfate (SDS), 10% (v/v) 2-mercaptoethanol, 0.05% (w/v) Bromophenol Blue.
2. 12% polyacrylamide–SDS gels: stacking gel total monomer concentration (T) = 4%, crosslinking degree (C) = 2.7%, pH 6.8; resolving gel T = 12%, C = 2.7%, pH 8.8, 0.4% SDS.
3. 30% (w/v) acrylamide/bisacrylamide solution (37.5:1) (Roth, Germany). Store in the dark at 4 °C.
4. 10% ammonium persulfate. Store stock solution at −20 °C.
5. N,N,N,N′-tetramethyl-ethylenediamine (TEMED) (Roth). Store stock solution in the dark at 4 °C.
6. 5× SDS-PAGE running buffer (pH 8.3): 125 mM Tris base, 950 mM glycine, 0.5% (w/v) SDS. Store at room temperature.
7. ColorPlus Prestained Protein Ladder Broad Range (P7712, New England Biolabs).
8. Coomassie Brilliant Blue staining solution: 0.25% (w/v) Coomassie Brilliant Blue G-250, 50% (v/v) methanol, 10% (v/v) acetic acid.
9. Coomassie destaining solution: 5% (v/v) methanol, 7.5% (v/v) acetic acid.
10. 1x PBS: 137 mM NaCl, 2.7 mM KCl, 10 mM Na_2HPO_4, 1.8 mM KH_2PO_4, pH 7.4 (adjust pH with NaOH).
11. 5% (w/v) skimmed milk in PBS.
12. Amersham™ Protran™ Nitrocellulose membrane 0.45 μm (GE Healthcare).
13. Trans-Blot Turbo (Bio-Rad).
14. Nitroblue tetrazolium chloride–5-bromo-4-chloro-3-indolylphosphate p-toluidine salt (NBT–BCIP) stock solution: 3.33% (w/v) NBT and 1.65% (w/v) BCIP in dimethylformamide.

15. Semidry blotting buffer (pH 9.6): 48 mM Tris base, 39 mM glycine, 20% (v/v) methanol.
16. Alkaline phosphatase (AP) buffer: 100 mM Tris–HCl, pH 9.6, 100 mM NaCl, 5 mM $MgCl_2$.
17. Primary antibodies: α-PVX (DSMZ, Germany), α-mCherry (either a rabbit polyclonal α-DsRed antibody, GeneTex GTX59862 that reacts also against mCherry, or a mouse monoclonal α-mCherry antibody, GeneTex GTX630189 [not used here]), and an α-GFP antibody (GeneTex).
18. Detection antibodies: monoclonal alkaline phosphatase (AP)-conjugated goat anti-rabbit secondary antibody (GAR^{AP}) (Dianova).

2.6 Transmission Electron Microscopy (TEM) and Immunosorbent Transmission Electron Microscopy (ISEM)

1. Pioloform-coated 400-mesh nickel grids (Plano GmbH, Germany).
2. 0.5% (w/v) bovine serum albumin (BSA) in PBS.
3. PBST: 137 mM NaCl, 2.7 mM KCl, 10.1 mM Na_2HPO_4, 1.8 mM KH_2PO_4, 0.05% (v/v) Tween 20, pH 7.4.
4. 1% (w/v) uranyl acetate.
5. Antibodies: α-PVX (DSMZ, Germany), α-mCherry (α-DsRed antibody, GeneTex, USA), and α-GFP antibodies (GeneTex, USA), goat anti-rabbit$^{15nm\ gold}$ (GAR^{15nm}) (BBI Solutions, GB).
6. Zeiss EM 10 TEM.

2.7 Imaging

1. Handheld UV lamp (7000 μW, Novodirect, Germany).
2. Green light (515 nm): KL 2500 LCD lamp for stereomicroscopy (Schott AG) with LEE color film primary red (number 106, Thomann Germany).
3. UV light (260 nm) Blak-Ray® B-100 YP UV Lamp (UVP, USA).
4. Blue light (450 nm): Optimax™ 450 blue LED lamp (Spectronics Corporation, USA) with LEE color film yellow (number 101).
5. Nikon Coolpix 5400 camera (Nikon).
6. Biorevo BZ-9000 fluorescence microscope (Keyence).
7. Leica TCS SP2 confocal laser scanning microscope (Leica Microsystems GmbH).

2.8 RNA Analysis

1. RNeasy Plant Mini Kit (Qiagen).
2. *DNase* I.
3. 0.5 ml Eppendorf reaction tubes.
4. Coating buffer: 15 mM Na_2CO_3, 35 mM $NaHCO_3$, pH 9.6 (not adjusted).
5. PBS-T: 0.05% (v/v) Tween 20 in PBS.

6. M-MLV Reverse Transcriptase *RNase*H Minus, Point Mutant (Promega Corporation).
7. *RNase* H (Roth).
8. RNase-free water.
9. Primers (e.g., oligo-dT).
10. dNTPs.
11. TAE buffer.
12. Primers for cDNA amplification (*see* Table 2).

2.9 Cell Culture

1. HT-29 Cancer cell line.
2. McCoy's5A medium supplemented with 10% (v/v) fetal bovine serum (FBS) and 1% (v/v) penicillin/streptomycin (Life Technologies).
3. Fetal bovine serum (Thermo Scientific).
4. Sterile 1× PBS (Thermo Scientific).
5. 2.5% (w/v) trypsin–EDTA, phenol red (Thermo Scientific).
6. 0.4% (w/v) Trypan Blue (Thermo Scientific).
7. Cooling centrifuge (Beckman Coulter).
8. Serological pipettes and dispenser.
9. Micropipettes.
10. Tissue culture treated flasks (Corning™, Fisher Scientific).
11. 15 ml falcon tubes.
12. Hemocytometer (for cell counting) (Fisher Scientific).
13. Tissue culture hood (Biosafety Type A2).
14. Humidified CO_2 incubator (37 °C and 5% CO_2).

2.10 Cell Imaging

1. 24-well untreated polycarbonate suspension culture plate for imaging study (USA Scientific).
2. Microscope cover glass, 0.15 mm thickness.
3. Fixative solution: 4% (v/v) paraformaldehyde, 0.3% (v/v) glutaraldehyde in DPBS.
4. Sterile 1× DPBS.

Table 2
Suggested primers for cDNA amplification

Primer name	Binding site	Sequence (5′–3′)
CX1	3′ end of CP gene	TTGAAGAAGTCGAATGCAGC
CX4	5′ end of CP gene	CGGGCTGTACTAAAGAAATC
TGB3-fw	3′ end of TGB3 gene	AAGGGCCATTGCCGATCTCAAGC

5. Permeabilization solution: 0.2% (v/v) Tween 20 in PBS.
6. 10% (v/v) goat serum in Dulbecco's Phosphate-Buffered Saline (DPBS).
7. Wheat Germ Agglutinin conjugated with Alexa Fluor 647 (Life Technologies) in 5% (v/v) goat serum (Gibco).
8. Confocal Microscope (Olympus FluoView FV1000 LSCM).
9. Image J 1.44o software (http://imagej.nih.gov/ij).

2.11 Cell Viability Assay

1. Clear flat-bottom 96 well polycarbonate plates (USA Scientific).
2. XTT cell proliferation assay kit (ATCC).
3. TECAN Infinite® 200 PRO multimode plate reader.

2.12 Biodistribution in Healthy Mice

1. 6-week-old C57BL/6 mice (a total of 9 mice; $n = 3$ for 6 h, 24 h and 7 days biodistribution), maintained on alfalfa-free rodent diet (Teklad 2018S, Envigo).
2. 1 ml insulin syringe with 28 gauge insulin needle (BD Micro-Fine™ IV Insulin Syringes, Fisher Scientific).
3. Rodent tail vein restrainer (Braintree Scientific Inc., Fisher Scientific).
4. Dissection tools - scissors, scalpel, dissection forceps (Braintree Scientific Inc., Fisher Scientific).
5. Maestro fluorescence imager (PerkinElmer).

2.13 Imaging in Tumor Mouse Models

1. 6-week-old male NCr nu/nu nude mice maintained on alfalfa-free rodent diet (Teklad 2018S, Envigo).
2. Matrigel for tumor inoculation (BD, Biosciences).
3. Microlitre Hamilton 22-gauge syringe (Hamilton).
4. Stainless steel Vernier Calipers for tumor volume measurement (Fisher Scientific).

2.14 Immuno-fluorescence and Histology

1. OCT cryo-embedding medium (Tissue-Tek, Sakura Finetek, VWR).
2. Leica CM1850 cryostat (Thermo Fisher) for tissue sectioning.
3. Ice-cold 95% (v/v) ethanol for fixation of tissue sections.
4. DPBS (Gibco, Thermo Fisher Scientific).
5. Permeabilization solution: 0.2% (v/v) Triton X-100 in PBS.
6. Blocking solution: 10% (v/v) goat serum (GS) (Life Technologies, Thermo Fisher Scientific,) in PBS.
7. Macrophage staining: primary antibody: Rat anti-mouse F4/80 (Biolegend) 1:250 with 1% (v/v) goat serum in PBS; secondary antibody: Alexa Fluor 488-conjugated goat anti-rat IgG (Life Technologies, Thermo Fisher Scientific) in 1% (v/v) goat serum in PBS.

8. Tumor vasculature staining: FITC-conjugated anti-mouse CD31 antibody (Biolegend).
9. Fluoroshield with DAPI histology mounting medium (Sigma-Aldrich).
10. Hematoxylin (Richard-Allen Scientific Inc., Thermo Fisher Scientific).
11. 70%, 80%, and 100% (v/v) ethanol.
12. Zeiss Axio Z1 motorized FL inverted microscope (Carl Zeiss AG).

3 Methods

All experiments with mice were carried out according to the Case Western Reserve University's Institutional Animal Care and Use Committee (IACUC) approved protocols.

3.1 Genetic Fusions of Fluorescent Proteins to PVX CP

1. The fluorescent proteins mCherry (28.8 kDa) and GFP (27 kDa) are generally too large for a direct fusion to the PVX CP maintaining particle assembly. Therefore fusion via the 2A sequence of the foot-and-mouth disease virus (FDMV) is advisable [25–27]. This leads to the production of the fusion product as well as free CP, and allows an assembly of PVX particles with large fusion proteins (Fig. 1). For all cloning procedures follow standard guidelines provided by Sambrook and Russell [32].
2. Clone the target sequence as an N-terminal CP fusion by replacing the GFP gene in pCXI with the gene of the target fluorescent protein (mCherry from pTRAkc-ERH-mCherry-his6) [9]. The replacement can be achieved by excising the GFP sequence with restriction enzymes NheI and BspEI (*see* **Note 1**).
3. Dephosphorylate vectors with calf intestinal phosphatase after digestion and ligate with T4 ligase overnight at 16 °C.
4. Transform ligation products into *E. coli* DH5α cells and select colonies on LB plates supplemented with ampicillin overnight at 37 °C.
5. Isolate plasmids with Pure Yield™ Plasmid Miniprep or Midiprep Kit from overnight liquid cultures and sequence the modified DNA.

3.2 Expression and Purification of Control Fluorescent Proteins

If noncoupled fluorescent proteins are needed for control experiments, transiently express the mCherry or GFP in *N. benthamiana* following agroinfiltration with *A. tumefaciens* strain GV3101 carrying vector pTRAkc-ERH-mCherry-his6 or pTRAkc-ERH-GFP-his6. In these vectors the target gene is under control of the 35S promoter [10] and is expressed with C-terminal His_6 tag for purification and KDEL tag directing the protein to accumulate in the endoplasmic reticulum for higher yields.

1. Cultivate *A. tumefaciens* bacteria at 26 °C in YEP medium overnight.
2. Supplement cultures with 10 mM MES, 10 mM glucose and 20 μM acetosyringone after 24 h and incubate for a further day.
3. Adjust OD 600 nm to 0.5–1.0 with 2× infiltration medium, supplement with 200 μM acetosyringone and incubate for 30 min at room temperature.
4. Infiltrate 5–6 leaves of 4–6-week-old plants with a needleless syringe and incubate in a phyto-chamber with constant light (25,000–30,000 lux) at 26 °C for 12 h and in the dark at 20 °C for 12 h.
5. Harvest the infiltrated leaves at 5–6 days post inoculation (dpi) (10–20 g).

   Purify the fluorescent protein by immobilized metal ion affinity chromatography (IMAC) (*see* **Note 2**).
6. Grind plant material in two volumes of PBS and centrifuge for 15 min at 4 °C and 15,000 × *g*.
7. Dilute the supernatant 1:10 with PBS and apply to 500 μl Ni-NTA agarose columns (*see* **Note 3**). Collect a sample from the flow through.
8. Wash columns twice with 10 volumes (v/v Ni-NTA agarose) 10 mM imidazole in PBS.
9. Elute with 1 volume (v/v Ni-NTA agarose) 150–300 mM imidazole in PBS (*see* **Note 4**).
10. Dialyze the eluate against 0.01 M phosphate buffer (pH 7.2) to remove the imidazole and concentrate with Vivaspin 6 columns (Sartorius if needed).

3.3 Chimeric PVX Propagation and Purification

1. Prepare 10 μg DNA from Subheading 3.1, **step 4** in 100 μl water for each leaf that should be inoculated. Rub 4-week-old *N. benthamiana* plants dusted by Celite 545 (3–4 mature leaves per plant equally distributed, ca. eight plants for 100 g infected leave material) for inoculation and incubate the plants in a phyto-chamber with constant light (25,000–30,000 lux) at 26 °C for 12 h and in the dark at 20 °C for 12 h.
2. Harvest plant material 14–21 dpi depending on the infection status by symptom distribution and fluorescence (*see* Subheading 3.4.1 and Fig. 2)) and use 100 g of plant material for virus purification with a modified protocol from the International Potato Center, Lima, Peru [9, 18].

Fig. 2 Time course of GFP-PVX infection in *N. benthamiana*. Plants were imaged under daylight (left panel) and UV (right panel) over time (I–VI). Inoculated leaves are indicated by pink arrows at 1 dpi (days post infection); green arrow 8 dpi shows the first signs of GFP-PVX appearance in sink leaves and hence, of systemic infection. In the inoculated leaves, GFP-PVX spreads via cell-to-cell movement 6 dpi (VII), whereas long distance transition occurs when particles reach a vein 10 dpi (VIII) leading to systemic infection 16 dpi (IX). A hydroponic culture at 21 dpi shows fluorescence in roots (X); infected and noninfected plants can be clearly identified by leaves (XI) or roots (XII)

3. Homogenize 100 g leaves with two volumes (w/v) ice-cold extraction buffer in a blender and filter the extract through three layers of Miracloth.
4. Clarify filtrate by centrifugation at 7800 × *g* for 20 min at 4 °C.
5. Process the supernatant by adding 1% (v/v) Triton X-100, stir for 1 h at 4 °C and clarify by centrifugation at 5500 × *g* for 20 min at 4 °C.
6. Process supernatant by adding NaCl/PEG solution (*see* Subheading 2.4, **item 9**) to a final concentration of 0.2 M NaCl and 4% (w/v) PEG (MW 6000–8000) and stir for 1 h at 4 °C and then incubate for 1 h at room temperature (*see* **Note 5**).
7. Precipitate the viral particles by centrifugation at 7800 × *g* for 10 min at 4 °C.
8. Resuspend the pellet carefully in 4 ml 0.05 M phosphate buffer pH 8.0 with 1% (v/v) Triton X-100, rinse the tubes immediately with 2 ml of the same buffer, and combine the samples.
9. Clarify mixture by centrifugation at 7800 × *g* for 10 min at 4 °C.
10. Prepare a 24 ml sucrose gradient (10–45% (w/v)) with a gradient mixer in 25 × 89 mm tubes and carefully load the samples on top.
11. Centrifuge in a swinging bucket rotor at 96,500 × *g* for 75 min at 4 °C (SW32Ti, 23,700 rpm).
12. Collect 1.5 ml gradient fractions from bottom to top (*see* **Note 6**).
13. Analyze all fractions by SDS PAGE (*see* **Note 7**).
14. Select and combine fractions with highest concentration of the desired fusion proteins in 14 × 89 mm tubes and dilute combined fractions at least in the same volume of 0.01 M phosphate buffer pH 7.2.
15. Sediment virus particles by 3–5 h of ultracentrifugation at 102,600 × *g* at 4 °C (SW41Ti, 38,000 rpm).
16. Carefully remove supernatant and resuspend the pellet in 0.2 ml 0.01 M phosphate buffer pH 7.2 and stir overnight at 4 °C.
17. Clarify the solution by centrifugation at 5000 × *g*, 10 min at 4 °C.
18. Determine plant virus concentration in the supernatant by measuring the OD at 260 nm and using the extinction coefficient ε_{PVX} of 2.97 ml mg^{-1} cm^{-1} at 260 nm.

3.4 Chimeric PVX Characterization

3.4.1 Visualization of Fluorescence (Fig. 2)

1. Monitor inoculated plants (*see* Subheading 3.3) daily and analyze for fluorescence using a suitable excitation source (*see* Table 3).
2. Carry out sodium dodecylsulfate polyacrylamide gel electrophoresis (SDS-PAGE). Prepare samples by mixing with 5× loading dye.
3. To visualize fluorescence of samples containing GFP or mCherry in the SDS gel, do not boil the samples prior to loading.
4. Prepare 12% (w/v) polyacrylamide resolving gels and resolve samples by SDS-PAGE [33].
5. Image gel under UV light.

3.4.2 Gel Electrophoresis and Immunoblotting (Fig. 3 V–VI)

1. Resolve protein samples by SDS-PAGE [33] after mixing plant sap (Subheading 3.2, **step 5** or Subheading 3.3, **step 2**) or purified particles (Subheading 3.4, **step 18**) (1–4 μg) with 5× loading dye and boiling for 5–10 min.
2. Stain one gel with Coomassie Brilliant Blue staining solution gently shaking the gel for 20 min.
3. Remove the staining solution for reuse and wash gel with water.
4. Destain background with destaining solution. Change destaining solution when it gets dark blue until the gel is sufficiently destained.
5. For the specific detection of proteins, blot a second unstained polyacrylamide gel onto a nitrocellulose membrane e.g., by semidry blotting according to standard procedures or the transfer cell manufacturer's instructions.
6. Block membranes for 1 h with 5% (w/v) skimmed milk in PBS.
7. Incubate membranes with primary antibodies: α-PVX, α-GFP or α-mCherry antibodies diluted 1:5000 in PBS for at least 1 h.
8. Wash membranes with PBS and incubate with secondary antibody: alkaline phosphatase-labeled anti-rabbit polyclonal secondary antibody diluted 1:5000 in PBS for 2 h.

Table 3
Fluorescent proteins and corresponding excitation and emission properties

Fluorescent protein	Excitation max (nm)	Emission max (nm)	Excitation method
GFP	397/475	508	UV lamp (260 nm)
mCherry	587	610	Green lamp (515 nm)

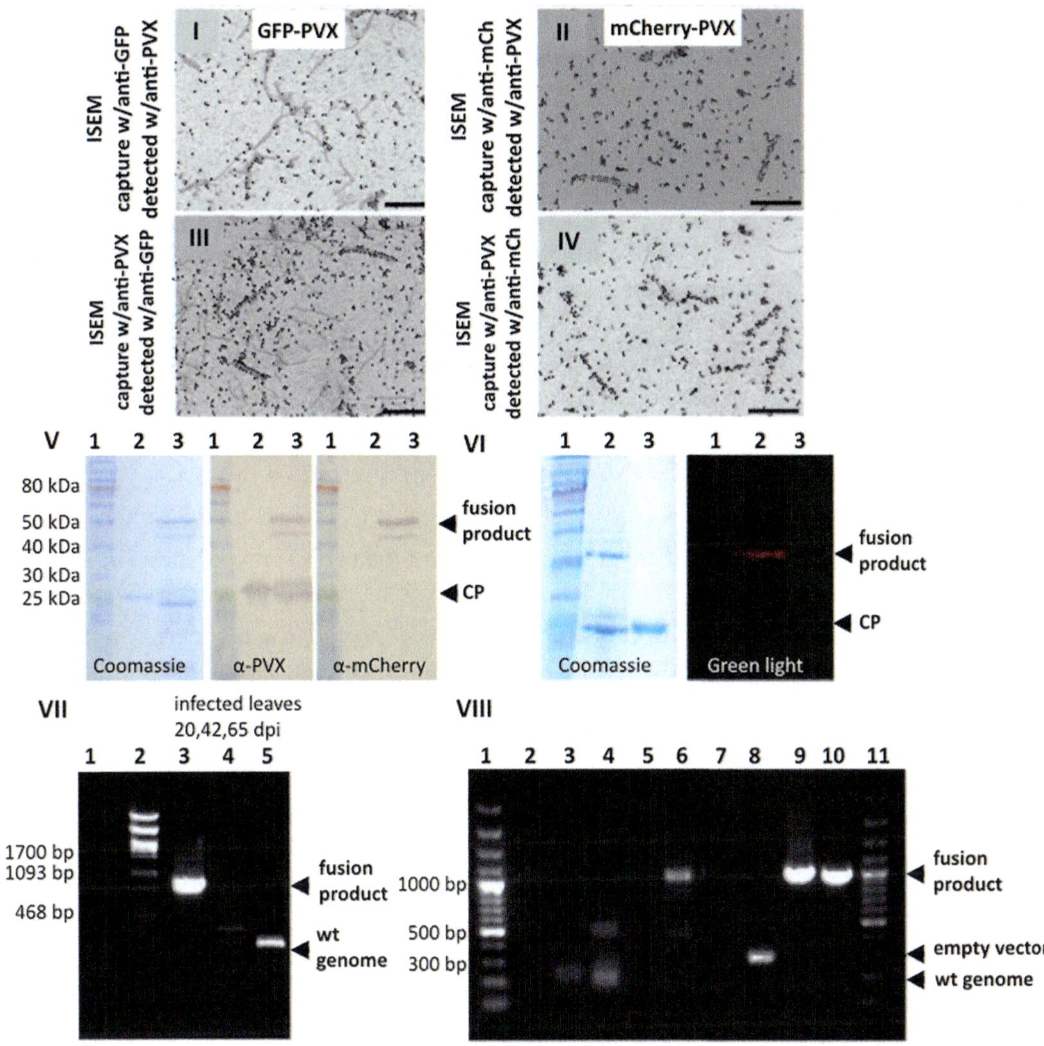

Fig. 3 Analysis of mutant PVX particles from infected plants. Immunosorbent TEM (ISEM) analysis of GFP-PVX and mCherry-PVX was carried out by capturing the particles with α-GFP or α-mCherry antibodies followed by staining with α-PVX antibodies (I, II) or vice-versa (III, IV). Intact PVX filaments were observed in all cases, and expression of fusion proteins at high density was observed (scale bars are 500 nm). SDS-PAGE (left panel) and Western blots of 1 = ColorPlus Prestained Protein Ladder (NEB), 2 = PVX (control), 3 = purified mCherry-PVX, and stained with Coomassie or α-PVX (middle panel) and α-mCherry antibodies (right panel) (V). SDS-PAGE of 1 = ColorPlus Prestained Protein Ladder (NEB), 2 = purified mCherry-PVX, and 3 = PVX (control) visualized by staining with Coomassie Brilliant Blue and direct green light illumination with a red filter (VI). Agarose gel electrophoresis of amplified cDNA after RT-PCR. 1 = *N. benthamiana*, noninfected plant; 2 = marker, lambda DNA digested with PstI, 3–5 = symptomatic, fluorescent leaves from a GFP-PVX infected plant at 20, 42, and 65 dpi (VII). Agarose gel electrophoresis of amplified cDNA after RT-PCR, at 21 dpi, 1 and 11 = GeneRuler™ 100 Plus DNA Ladder (Fermentas), 2 = *N. benthamiana* noninfected, 3 = plant infected with PVX0018 (expected band 216 bp), 4 = plant infected with pTCXI, 5 = empty, 6 = plant infected with pTCXIIc, 7 = PCR negative control (no template), 8 = empty PVX cloning vector (305 bp), 9 = plasmid pTCXI (1011 bp), 10 = plasmid pTCXIIc (1009 bp) (VIII)

9. Wash membranes several times with PBS and incubate in AP buffer for 5 min.
10. Develop the blot with 100 μl NBT/BCIP in 10 ml AP buffer. Stop reaction by removing the NBT/BCIP solution and washing with water.

3.4.3 Transmission Electron Microscopy (TEM) and Immunosorbent Transmission Electron Microscopy (ISEM) (Fig. 3 I–IV).

1. For direct adsorption, prepare drops of around 40 μl with 10 μg of purified particles (from Subheading 3.3, **step 18**) and incubate pioloform-coated nickel grids for 20 min in these drops. Directly proceed with **step 9**.
2. For ISEM, coat the grids with a monoclonal PVX-antibody diluted 1:10 in PBS for 20 min.
3. Wash unbound antibodies off, by carefully incubating the grid in PBST droplets for a few minutes 2–3 times.
4. Block grids with 0.5% (v/v) BSA in PBS for 15 min.
5. Incubate grids with 10 μg of the particle preparation (*see* **step 1**) for 20 min, followed by washing with PBS (*see* **step 3**).
6. Captured particles can be analyzed for proteins exposed on the surface by immunogold staining. Therefore incubate the preparations for 2 h with a polyclonal antibody against the fluorescent protein diluted 1:100 in PBS.
7. Incubate grids overnight at room temperature with goat anti-rabbit secondary antibodies labeled with 15-nm gold particles diluted 1:50 in PBS. Place the samples under a glass cover and include wet towels to avoid the evaporation of the drops.
8. Wash grids thoroughly with PBS and then with distilled water like in **step 3**.
9. Counterstain with five drops of 1% (w/v) uranyl acetate (pH 4.3), remove the liquid carefully on Whatman paper and let the grids dry on air before analysis with a Zeiss EM 10 TEM.

3.4.4 UV-VIS Spectroscopy

1. Measure the absorbance of 2 μl of PVX sample (from Subheading 3.3, **step 18**) using the NanoDrop spectrophotometer.
2. Read absorbance at 260 nm and 280 nm.
3. Compare A260:280 ratio to determine sample purity. A ratio of 1.2 ± 0.1 indicates pure PVX preparations.
4. Determine the VNP concentration using the Beer-Lambert law ($A = \varepsilon\ c\ l$, where A is the absorbance, ε is the extinction coefficient, c is the concentration, and l is the path length). The path length is 0.1 cm for the NanoDrop. The extinction coefficient for PVX is 2.97 ml cm^{-1} mg^{-1} (at 260 nm).

3.4.5 RNA Analysis (RT-PCR, qPCR) (Fig. 3 VII–VIII)

You can either analyze the isolated total RNA from plants (**steps 1–2**) or the encapsulated RNA (**steps 3–6**) by RT-PCR and agarose gel-electrophoresis (**steps 7–11**).

1. Characterize total virus RNA by grinding 300 mg plant material (*see* Subheading 3.3, **step 2**) under liquid nitrogen and isolate total RNA using the RNeasy Plant Mini Kit (Qiagen) according to the manufacturer's guidelines and determine the concentration at 260 nm.
2. Prepare samples containing 3 μg RNA and incubate them with DNaseI to avoid DNA contamination.
3. For characterization of RNA encapsidated in virus particles, virions can be immunocaptured in 0.5-ml Eppendorf tubes. Therefore, coat tubes for at least 4 h at 30 °C with 100 μl α-PVX diluted 1:100 (v/v) in coating buffer.
4. Discard the coating solution and wash the tubes three times with 500 μl PBST before adding 75 μl plant sap (*see* Subheading 3.3, **step 2**) and incubate overnight at 4 °C.
5. Wash the tubes sequentially with 500 μl PBST, 500 μl PBS and 500 μl distilled water.
6. The isolated particles can be used directly for RT-PCR.
7. Reverse-transcribe isolated RNA (from **step 2** or **step 6**) using M-MLV Reverse Transcriptase RNase H Minus, Point Mutant.
8. Mix samples with 27 μl distilled RNase-free water and 0.2 μM primer (e.g., oligo-dT) and denature for 10 min at 80 °C before cooling on ice for primer annealing.
9. Supplement mixture with M-MLV RT 5× Reaction Buffer, 0.5 mM dNTPs and 100 U M-MLV, and top up to 50 μl with distilled water. Perform RT reaction as recommended by manufacturer.
10. Digest RNA with RNase H following cDNA synthesis.
11. Amplify desired region on the viral genome (*see* Table 2) and resolve products by agarose gel electrophoresis in 1× TAE buffer.

3.5 Cell Uptake Studies (Fig. 4 I)

1. Maintain and cultivate HT-29 cells on McCoy's 5A media supplemented with 10% (v/v) Fetal bovine serum (FBS) and 1% (v/v) penicillin-streptomycin at 37 °C, 5% CO_2 (*see* **Note 8**).
2. Culture 25,000 cells/well on glass coverslips placed on the bottom of each well in a 24-well nontreated tissue culture plate for 24 h with 250 μl McCoy's 5A media.
3. Wash the cells with sterile PBS. Add 5 μg mCherry-PVX (from Subheading 3.3, **step 18**) in 100 μl of fresh McCoy's 5A medium to each well and incubate for 3 h. Post incubation, wash the cells with PBS to remove unbound particles (*see* **Note 9**).

4. Fix the cells using fixative solution for 5 min at room temperature. Wash cells three times with tissue culture-grade DPBS.
5. Permeabilize with 0.2% (v/v) Triton X-100 for 2 min and block the cells with 10% (v/v) goat serum for 1 h.
6. Stain cell membrane using Alexa Fluor 647-conjugated wheat germ agglutinin (1:1000 dilution) with 5% (v/v) goat serum for 45 min in the dark (*see* **Note 10**).
7. Following washing with DPBS, mount the coverslips on glass slides with Fluoroshield with DAPI mounting medium (Sigma), for nuclear staining. Seal the coverslips using nail polish, and store the slides protected from light at −20 °C until ready to image.

3.6 Cell Viability Assay

1. Cell viability assay must be done in triplicate for each sample and repeated twice.
2. Seed HT-29 cells at 25,000 cells/well in 100 μl McCoy's 5A media in 96 well plates, and culture the cells for 24 h at 37 °C.
3. Remove medium and incubate the cells with 1 μg, 10 μg or 30 μg mCherry-PVX particles in 100 μl of fresh McCoy's 5A medium for 3 h (*see* **Note 11**).
4. Wash the cells three times with PBS and incubate for additional 24 h with 100 μl fresh medium.
5. Add 50 μl of XTT reagent to each well and incubate for 2–3 h for color development (as per instructions provided by the manufacturer). Record the absorbance on a plate reader at 450 nm and 650 nm and analyze data as per instructions in XTT kit.

3.7 Biodistribution Studies in Mice (Fig. 4 II, III)

1. Maintain C57BL/6 mice on alfalfa-free diet for 2 weeks prior to study to reduce tissue auto-fluorescence.
2. Administer 10 mg/kg of body weight mCherry-PVX particles in 100 μl of sterile PBS through intravenous tail-vein injection using 1-ml 28 gauge insulin syringe. For control group, injected mice with sterile PBS similarly.
3. Sacrifice mice at 6 h, 24 h and 7 days following injections to recover major tissues including brain, heart, lungs, spleen, kidneys and liver (*see* **Note 12**).
4. Image fresh tissues with Maestro fluorescence imager using green excitation and emission filters at 800 ms exposure. Analyze the relative fluorescence intensity using Image J software (*see* **Note 13**).

3.8 Tumor Homing Studies in Mice

1. Maintain NCr *nu/nu* mice on alfalfa free diet for 2 weeks before imaging.

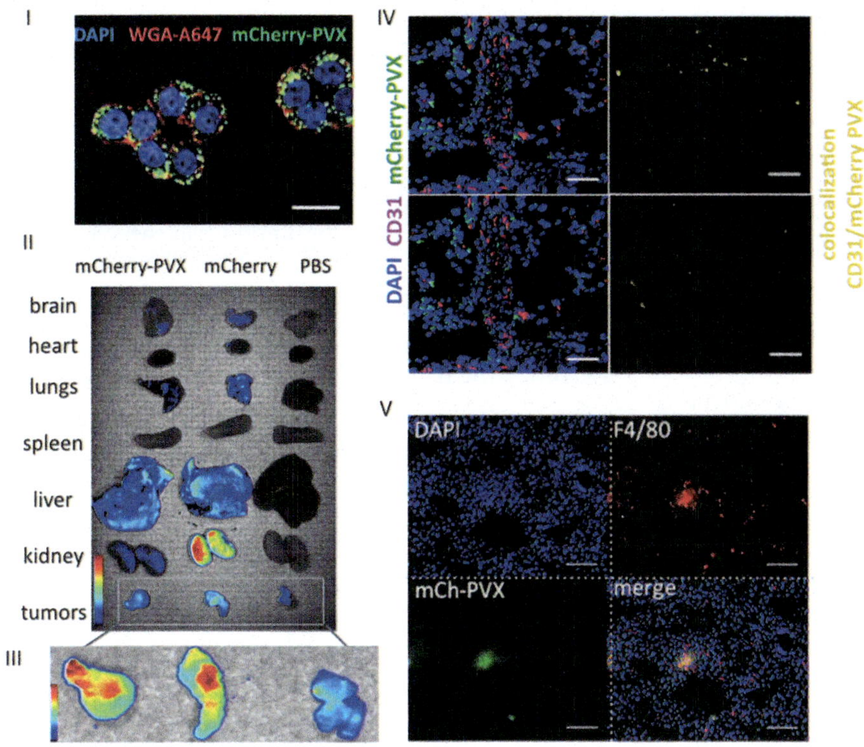

Fig. 4 Cell uptake, biodistribution, and tumor homing study of mCherry-PVX. Cellular uptake of mCherry-PVX showed in HT-29 cancer cells (I). Comparative biodistribution of free mCherry and mCherry-PVX in tumor bearing NCR *nu/nu* mice indicates accumulation in liver, kidneys and tumor tissues (II, III). Intratumoral localization of mCherry-PVX in stained tumor sections suggests penetration into tissues as compared to localization of free mCherry in CD31 labeled vasculature (IV). Immunofluorescence of liver section indicates uptake of mCherry-PVX by F4/80 labeled liver macrophages (V). Scale bars are 20 μm in I, 50 μm in IV, 100 μm in V

2. Inoculate tumor xenografts by injecting 2.5×10^6 HT-29 cells in 50 μl of McCoy's 5A media and equal volume of Matrigel subcutaneously in the right frank of mice using a 22-gauge syringe (*see* **Notes 14** and **15**).
3. Monitor the tumor growth. Commence the imaging study when the tumors have reached 20 mm^3 (typically 12 days after inoculations) (*see* **Note 16**).
4. Administer 10 mg/kg mCherry-PVX and 0.25 mg/kg free mCherry in 100 μl sterile PBS (keeping the mCherry dose constant) and 100 μl sterile PBS as control through intravenous injection via tail vein.
5. Euthanize the mice at 12 h post injections and harvest major organs and tumors (*see* **Note 17**).
6. Image the tissues with Maestro imager and analyze the relative fluorescence intensities using Image J software (*see* **Note 18**).

3.9 Immunofluorescence and Histology of Frozen Tissues (Fig. 4 IV, V)

1. Preserve dissected liver and tumor tissues by embedding in OCT cryomedium and store at −80 °C until sectioning and staining (*see* **Note 19**).
2. Prepare 10 μm thick sections from liver and tumor tissue using a Leica cryostat. Store tissue sections at −20 °C.
3. For immunofluorescence, fix the frozen sections in ice cold 95% (v/v) ethanol for 20 min and wash three times with cold PBS.
4. Permeabilize the sections using 0.2% (v/v) Triton X-100 in PBS for 2 min at room temperature and wash three times with PBS. Block the tissues with 10% (v/v) goat serum for 1 h.
5. For macrophage staining, incubate the liver sections with rat anti-mouse F4/80 antibody at a 1:250 dilution in PBS supplemented with 1% (v/v) goat serum for 1 h at room temperature. Wash the primary antibody off with PBS and incubate tissue section with Alexa Fluor 488-conjugated goat anti-rat IgG (at 1: 500 dilution in PBS) with 1% (v/v) goat serum for 1 h at room temperature (*see* **Note 20**).
6. Wash the secondary antibody off with PBS and mount the slides using Fluoroshield with DAPI mounting medium for nuclear staining. Seal the slides with nail polish and store at −20 °C.
7. Do fixation, permeabilization, and blocking steps as described in **steps 2–4**. Then stain the frozen tumor sections with FITC labeled anti mouse CD-31 antibody (1:250 dilution in PBS) with 1% (v/v) goat serum to highlight the tumor vasculature. Mount the slides as described in **step 6** using Fluoroshield with DAPI, seal with nail polish, and store at −20 °C until imaging.
8. Image the slides using a Zeiss Axio Observer Z1 fluorescence microscope and a Fluoview FV1000 LCSM confocal microscope. Process and analyze the images using Image J software.
9. For histology, stain the fixed liver sections with hematoxylin for 45 s. Following this, sequentially wash the slides with 70%, 80%, and 100% (v/v) ethanol for 3 min each and stain with eosin for 1 min.
10. Mount the tissues with mounting medium and image using Zeiss Axio Observer Z1 fluorescence microscope.

4 Notes

1. *Nhe*I and *Bsp*EI purchased by NEB are not functional in the same buffer, thus a sequential restriction is recommended. The DNA was purified between the two restriction steps with MSB® Spin PCRapace Kit (Invitek).

2. Analyze samples from each step using SDS-PAGE. The purification procedure can be improved by washing with different concentrations of imidazole in the washing steps and by different concentrations during the elution steps. Thus it must be tested which imidazole concentrations are tolerated by the target protein on the column during the washing step. The recommended procedure is to increase the amount of imidazole from 10 to 50 mM during different washing.
3. Undiluted plant sap frequently plugs the column, thus a dilution is recommended.
4. The amount of imidazole needed to elute the target protein from the column can differ from protein to protein. The minimal amount of imidazole needed should be used and must be tested. Thus, it is recommended to prepare different solutions ranging from 150 to 300 mM imidazole.
5. The precipitation can be improved by incubation of the supernatant with NaCl and PEG overnight, if your fluorescent protein is stable under these conditions.
6. Gradient fractions can be collected by puncturing the bottom of the centrifugation tubes with a needle connected to a flexible tube. Collect the drowning liquid in 1.5 ml reaction tubes.
7. For identification of fractions with functional fluorescent protein/CP fusions, avoid boiling the samples for SDS-PAGE and visualize the fluorescence directly in the gel before Coomassie-staining.
8. Cell lines: We have used HT-29 cell line in our studies. However, other cell lines may be used similarly, with culture media, supplements and conditions optimized for that particular cell line.
9. Cell uptake: We studied cell uptake for 3 h, however for detailed insights into uptake kinetics, varying time points can be evaluated.
10. Cell uptake: To decipher cellular uptake mechanisms, other cellular markers such as endosomal LAMP-1 can be stained with fluorescent anti-LAMP-1 antibodies to colocalize fluorescent viral nanoparticles within.
11. Cell viability assay: Controls should include wells with just media and wells with cells only without particles to which XTT reagents must be added. Readouts from these wells must be subtracted from PVX particles incubated wells.
12. Biodistribution studies: While the intravenous route is a preferred administration route for drug delivery, other routes such as subcutaneous, oral and intramuscular can be evaluated. In that case, other tissues such as draining lymph nodes (for subcutaneous injections), intestine and stomach (for oral

administration), muscle tissue (for intramuscular injection) should be harvested and compared for fluorescence studies as well.

13. Maestro imaging: Post dissection, tissues should be placed in PBS and not allowed to dry before imaging under Maestro. Tissues collected at various time points can be stored at 4 °C in PBS in the dark and imaged together.
14. Tumor inoculation: The number of cells used for tumor inoculation has been optimized for HT-29 cells. For other cell lines, these numbers can vary and should be optimized.
15. Tumor inoculation: Cells and Matrigel should be maintained at 4 °C and mixed immediately prior to inoculation.
16. Tumor growth: Tumor size should be monitored every day using calipers. Tumor volumes are measured using the formula $(a \times b^2)/2$, where a and b are two dimensions.
17. Tumor homing was studied at 12 h post administration of nanoparticles. However, different time points can be studied using a larger set of mice and euthanizing them at varying time points.
18. Tumor imaging: High signal intensity from liver can mask signals from all other tissues. Dissected tumor tissues should also be imaged separately from other tissues to highlight the fluorescence signal in tumors.
19. Freshly dissected tissues should be embedded in cyrogel immediately for best preservation of tissue structure.
20. Antibody dilutions used here have been optimized for the given set of antibodies. For antibodies from other manufacturers or other markers, these dilutions must be optimized first.

Acknowledgments

This work was supported by a grant from the National Science Foundation (NSF CMMI NM 1333651) to NFS; a grant from Susan G. Komen Foundation (CCR14298962) to NFS; and a Scholarship RFwN from RWTH Aachen University to CD.

References

1. Pokorski JK, Steinmetz NF (2011) The art of engineering viral nanoparticles. Mol Pharm 8(1): 29–43. https://doi.org/10.1021/mp100225y
2. Steinmetz NF (2010) Viral nanoparticles as platforms for next-generation therapeutics and imaging devices. Nanomedicine 6(5):634–641. https://doi.org/10.1016/j.nano.2010.04.005
3. Yusibov V, Rabindran S, Commandeur U, Twyman RM, Fischer R (2006) The potential of plant virus vectors for vaccine production. Drugs R D 7(4):203–217
4. Steinmetz NF, Manchester M (2011) Viral nanoparticles: tools for materials science and biomedicine. Pan Stanford Publishing Pte. Ltd, Singapore, pp 1–273

5. Plummer EM, Manchester M (2010) Viral nanoparticles and virus-like particles: platforms for contemporary vaccine design. Wiley Interdiscip Rev Nanomed Nanobiotechnol 3(2):174–196. https://doi.org/10.1002/wnan.119
6. Jobsri J, Allen A, Rajagopal D, Shipton M, Kanyuka K, Lomonossoff GP, Ottensmeier C, Diebold SS, Stevenson FK, Savelyeva N (2015) Plant virus particles carrying tumour antigen activate TLR7 and induce high levels of protective antibody. PLoS One 10(2):e0118096. https://doi.org/10.1371/journal.pone.0118096
7. Lizotte PH, Wen AM, Sheen MR, Fields J, Rojanasopondist P, Steinmetz NF, Fiering S (2015) In situ vaccination with cowpea mosaic virus nanoparticles suppresses metastatic cancer. Nat Nanotechnol 11(3):295–303. https://doi.org/10.1038/nnano.2015.292
8. Bruckman MA, Jiang K, Simpson EJ, Randolph LN, Luyt LG, Yu X, Steinmetz NF (2014) Dual-modal magnetic resonance and fluorescence imaging of atherosclerotic plaques in vivo using VCAM-1 targeted tobacco mosaic virus. Nano Lett 14(3):1551–1558. https://doi.org/10.1021/nl404816m
9. Lee KL, Uhde-Holzem K, Fischer R, Commandeur U, Steinmetz NF (2014) Genetic engineering and chemical conjugation of potato virus X. Methods Mol Biol 1108:3–21. https://doi.org/10.1007/978-1-62703-751-8_1
10. Maclean J, Koekemoer M, Olivier AJ, Stewart D, Hitzeroth II, Rademacher T, Fischer R, Williamson AL, Rybicki EP (2007) Optimization of human papillomavirus type 16 (HPV-16) L1 expression in plants: comparison of the suitability of different HPV-16 L1 gene variants and different cell-compartment localization. J Gen Virol 88(Pt 5):1460–1469. https://doi.org/10.1099/vir.0.82718-0
11. McCormick AA, Palmer KE (2008) Genetically engineered tobacco mosaic virus as nanoparticle vaccines. Expert Rev Vaccines 7(1):33–41. https://doi.org/10.1586/14760584.7.1.33
12. Shukla S, Eber FJ, Nagarajan AS, DiFranco NA, Schmidt N, Wen AM, Eiben S, Twyman RM, Wege C, Steinmetz NF (2015) The impact of aspect ratio on the biodistribution and tumor homing of rigid soft-matter nanorods. Adv Healthc Mater 4(6):874–882. https://doi.org/10.1002/adhm.201400641
13. Canizares MC, Lomonossoff GP, Nicholson L (2005) Development of cowpea mosaic virus-based vectors for the production of vaccines in plants. Expert Rev Vaccines 4(5):687–697. https://doi.org/10.1586/14760584.4.5.687
14. Canizares MC, Nicholson L, Lomonossoff GP (2005) Use of viral vectors for vaccine production in plants. Immunol Cell Biol 83(3):263–270. https://doi.org/10.1111/j.1440-1711.2005.01339.x
15. Steinmetz NF, Mertens ME, Taurog RE, Johnson JE, Commandeur U, Fischer R, Manchester M (2010) Potato virus X as a novel platform for potential biomedical applications. Nano Lett 10(1):305–312. https://doi.org/10.1021/nl9035753
16. Shukla S, Wen AM, Ayat NR, Commandeur U, Gopalkrishnan R, Broome AM, Lozada KW, Keri RA, Steinmetz NF (2014) Biodistribution and clearance of a filamentous plant virus in healthy and tumor-bearing mice. Nanomedicine (Lond) 9(2):221–235. https://doi.org/10.2217/nnm.13.75
17. Lee KL, Shukla S, Wu M, Ayat NR, El Sanadi CE, Wen AM, Edelbrock JF, Pokorski JK, Commandeur U, Dubyak GR, Steinmetz NF (2015) Stealth filaments: polymer chain length and conformation affect the in vivo fate of PEGylated potato virus X. Acta Biomater 19:166–179. https://doi.org/10.1016/j.actbio.2015.03.001
18. Uhde-Holzem K, Schlosser V, Viazov S, Fischer R, Commandeur U (2010) Immunogenic properties of chimeric potato virus X particles displaying the hepatitis C virus hypervariable region I peptide R9. J Virol Methods 166(1-2):12–20. https://doi.org/10.1016/j.jviromet.2010.01.017
19. Shukla S, Wen AM, Commandeur U, Steinmetz NF (2014) Presentation of HER2 epitopes using a filamentous plant virus-based vaccination platform. J Mater Chem B 2(37):6249–6258. https://doi.org/10.1039/c4tb00749b
20. Lico C, Benvenuto E, Baschieri S (2015) The two-faced potato virus X: from plant pathogen to smart nanoparticle. Front Plant Sci 6:1009. https://doi.org/10.3389/fpls.2015.01009
21. Shukla S, DiFranco NA, Wen AM, Commandeur U, Steinmetz NF (2015) To target or not to target: active vs. passive tumor homing of filamentous nanoparticles based on potato virus X. Cell Mol Bioeng 8(3):433–444. https://doi.org/10.1007/s12195-015-0388-5
22. Marusic C, Rizza P, Lattanzi L, Mancini C, Spada M, Belardelli F, Benvenuto E, Capone I (2001) Chimeric plant virus particles as immunogens for inducing murine and human immune responses against human immunodeficiency virus type 1. J Virol 75(18):8434–8439
23. Brennan FR, Jones TD, Longstaff M, Chapman S, Bellaby T, Smith H, Xu F, Hamilton WD, Flock JI (1999) Immunogenicity of peptides

derived from a fibronectin-binding protein of S. Aureus expressed on two different plant viruses. Vaccine 17(15-16):1846–1857

24. Massa S, Simeone P, Muller A, Benvenuto E, Venuti A, Franconi R (2008) Antitumor activity of DNA vaccines based on the human papillomavirus-16 E7 protein genetically fused to a plant virus coat protein. Hum Gene Ther 19(4):354–364. https://doi.org/10.1089/hum.2007.122
25. Cruz SS, Chapman S, Roberts AG, Roberts IM, Prior DA, Oparka KJ (1996) Assembly and movement of a plant virus carrying a green fluorescent protein overcoat. Proc Natl Acad Sci U S A 93(13):6286–6290
26. Donnelly ML, Luke G, Mehrotra A, Li X, Hughes LE, Gani D, Ryan MD (2001) Analysis of the aphthovirus 2A/2B polyprotein 'cleavage' mechanism indicates not a proteolytic reaction, but a novel translational effect: a putative ribosomal 'skip. J Gen Virol 82(Pt 5):1013–1025. https://doi.org/10.1099/0022-1317-82-5-1013
27. Donnelly ML, Hughes LE, Luke G, Mendoza H, ten Dam E, Gani D, Ryan MD (2001) The 'cleavage' activities of foot-and-mouth disease virus 2A site-directed mutants and naturally occurring '2A-like' sequences. J Gen Virol 82(Pt 5):1027–1041. https://doi.org/10.1099/0022-1317-82-5-1027
28. Hefferon K (2013) Plant-derived pharmaceuticals for the developing world. Biotechnol J 8(10):1193–1202. https://doi.org/10.1002/biot.201300162
29. Aboul-Ata AA, Vitti A, Nuzzaci M, El-Attar AK, Piazzolla G, Tortorella C, Harandi AM, Olson O, Wright SA, Piazzolla P (2014) Plant-based vaccines: novel and low-cost possible route for Mediterranean innovative vaccination strategies. Adv Virus Res 89:1–37. https://doi.org/10.1016/B978-0-12-800172-1.00001-X
30. Shukla S, Ablack AL, Wen AM, Lee KL, Lewis JD, Steinmetz NF (2013) Increased tumor homing and tissue penetration of the filamentous plant viral nanoparticle potato virus X. Mol Pharm 10(1):33–42. https://doi.org/10.1021/mp300240m
31. Shukla S, Dickmeis C, Nagarajan AS, Fischer R, Commandeur U, Steinmetz NF (2014) Molecular farming of fluorescent virus-based nanoparticles for optical imaging in plants, human cells and mouse models. Biomater Sci 2(5):784–797. https://doi.org/10.1039/C3BM60277J
32. Sambrook J, Russell DW (2001) Molecular cloning: a laboratory manual. Cold Spring Harbor Laboratory Press, Cold Spring Harbor,N.Y. doi:citeulike-article-id:3334416
33. Laemmli UK (1970) Cleavage of structural proteins during the assembly of the head of bacteriophage T4. Nature 227(5259):680–685

Chapter 6

Self-Assembling Plant-Derived Vaccines Against Papillomaviruses

Emanuela Noris

Abstract

Virus-like particles (VLPs) can be used as antiviral vaccines as they mimic the structure of virus particles, with preserved conformation and immunogenicity characteristics. L1, the major capsid protein of papillomaviruses (PV) can self-assemble into VLPs currently used as highly effective vaccines. VLPs can be produced in heterologous systems, including plants. Here, a method for the expression of the L1 protein of human papillomavirus 16 (HPV 16) and the production of highly purified preparations of HPV 16 L1 VLPs is described. The method relies on the transient expression of HPV 16 L1 in *Nicotiana benthamiana* plants using a nonreplicating vector and on the purification of VLPs by different centrifugation steps followed by a cesium sulfate gradient. Such a procedure has also been successfully applied to other HPVs and to bovine papillomavirus 1.

Key words Virus-like particles (VLPs), Transient expression, Recombinant protein, L1 protein, *Nicotiana benthamiana*, Agroinfiltration, Plant-made vaccines, Papillomaviruses

1 Introduction

Traditional vaccines are commonly based on peptides/proteins, naked DNA vectors, and attenuated or inactivated microorganisms. Virus-like particles (VLPs) constitute attractive vaccine platforms in view of their safety and ease of production. In fact, VLPs are not only devoid of an infectious viral genome but, contrary to inactivated or attenuated viruses that must be prepared in mammalian cell lines, they can be produced in heterologous systems, such as bacteria, yeasts, insect cells, or plants. VLPs are highly immunogenic and can induce elevated titers of neutralizing antibodies, even without adjuvants, thanks to the highly ordered structure presenting repetitive epitopes to the immune system cells [1]. Moreover, VLPs offer interesting biotechnological advantages as they can serve as scaffolds for presenting heterologous antigens capable of inducing immune responses against other infectious diseases.

Christina Wege and George P. Lomonossoff (eds.), *Virus-Derived Nanoparticles for Advanced Technologies: Methods and Protocols*, Methods in Molecular Biology, vol. 1776, https://doi.org/10.1007/978-1-4939-7808-3_6,
© Springer Science+Business Media, LLC, part of Springer Nature 2018

Papillomaviruses (PV) are a large group of nonenveloped viruses with an icosahedral capsid of approximately 50–60 nm in diameter. PV virions harbor a circular double-stranded DNA genome of about 7800 base pairs, encoding several early genes necessary for viral transcription and replication and two late genes encoding the major and minor structural proteins (L1 and L2, respectively). Native virions are composed of 72 capsomers, each made of five monomers of the 55-kDa L1 protein. The virion also includes about 12 copies of the 74-kDa L2 protein. Though unnecessary for virion assembly, L2 participates in viral DNA encapsidation and virus entry into cells [2]. VLPs with a T = 7 symmetry are morphologically indistinguishable from native virions. In heterologous systems, L1 also self-assembles into small VLPs of about 30 nm that have a T = 1 symmetry and are composed of 12 capsomers only [3, 4].

Many PVs are tumorigenic; in particular, a group of high-risk human PV (HPVs), including HPV 16 and 18, is responsible for the majority of cervical cancers in women [5]. Prophylactic vaccines against high-risk HPVs consist of VLPs made of L1. Commercial anti-HPV vaccines, produced in yeast or insect cells, induce a strong humoral response in vaccinated individuals and prevent virus entry to cervical epithelial cells [6].

After the initial achievement of HPV VLP assembly in transgenic plants [7] and the demonstration of their immunogenicity [8, 9], a variety of different approaches led to great yield improvements. Nowadays, several transient expression strategies are available, avoiding the time-consuming production of transgenic plants [10]. Up to 0.5 g/kg fresh leaf weight were reported for chloroplast-targeted HPV 16 L1 protein, transiently expressed from a human codon-optimized gene [11], or with an autonomously replicating geminivirus vector [12]. Nonetheless, the highest yields (3 g/kg) were obtained in transplantomic tobacco plants [13].

Due to the strong stimulation of the immune system, VLPs offer the biotechnological advantage of providing immunity against foreign protein epitopes exposed on their surface. Chimeric VLPs can be obtained by chemical linkage or genetic fusion of the foreign protein epitope to the VLP; in the latter case, it is essential that domain rearrangements or substitutions do not affect VLP assembly [14].

Here, a method to express the recombinant L1 protein of HPV 16 in *Nicotiana benthamiana* plants using transient means is outlined. This protocol relies on the use of a nonreplicating vector that contains regulatory sequences derived from cowpea mosaic virus; high yields are further ensured by the coexpression of a gene silencing suppressor protein [15]. In the case described here, a synthetic biology approach with a codon-optimized gene was followed [16]. A protocol for obtaining an enriched/purified preparation of HPV 16 VLPs from leaf tissue expressing the recombinant L1 protein, using cesium-sulfate gradient ultracentrifugation is also described. The VLPs obtained by this procedure have discrete sizes of 55 or 30 nm when observed at the electron microscope (EM) and are compatible with either T = 7 or T = 1 symmetry. Capsomers of about 12 nm are also present in the EM preparations. Moreover, HPV 16 L1-based VLPs obtained from plants using this procedure were recognized by a panel of monoclonal antibodies targeting linear and conformational epitopes of the native virion [16, 17], indicating that they are potentially suitable for immunization purposes.

The method described here was successfully employed to obtain L1 proteins and L1-based VLPs of other PVs, such as HPV 8, a virus with a cutaneous tropism [18] and bovine papillomavirus 1 [19]. This method can, in principle, be used to obtain the L1 protein of other PVs; if other replicative or nonreplicative vectors [20] are used, only slight modifications (e.g., concentration of agrobacteria, addition of constructs encoding silencing suppressors) are required. In principle, it is possible to apply the VLP purification method described here to any tissue expressing a PV L1 protein, including transgenic plants. This protocol was also successfully used to obtain chimeric VLPs expressing heterologous epitopes [16].

2 Materials

All solutions are prepared with ultrapure water (18 M Ω cm, 25 °C). All reagents are stored at room temperature, unless specified otherwise.

2.1 Plant Growth

1. *Nicotiana benthamiana* seeds.
2. Soil–perlite mixture (3:1).
3. Plastic pots (8 × 8 cm).
4. Plant growth chamber with temperature control.

2.2 Agrobacterium Growth

1. Antibiotics: 50 mg/ml kanamycin dissolved in water, 50 mg/ml rifampicin dissolved in ethanol–water (1:1 v/v). Both filter-sterilized (0.2 μm). Sterile solutions are aliquoted and stored at −20 °C. Liquid growth media containing antibiotics are prepared immediately before use. Solid growth media (plates) containing antibiotics can be prepared in advance and stored at 4 °C (*see* **Note 1**).
2. 0.1 M acetosyringone (3′5′dimethoxy-4′-hydroxy-acetophenone). Dissolve 196 mg of acetosyringone in 0.5–1 ml 70% ethanol and bring the volume to 10 ml with water. Filter-sterilize (0.2 μm). Aliquot and store at −20 °C.

3. YEB medium: 10 g/l yeast extract, 10 g/l Bacto peptone, 5 g/l NaCl, pH 7.0, 50 μg/ml kanamycin, 50 μg/ml rifampicin, 40 μM acetosyringone, autoclaved at 121 °C for 15 min. For solid YEB (plates), add 15 g/l agar before autoclaving (*see* **Note 2**).
4. Orbital shaker with controlled temperature.
5. Benchtop centrifuge (e.g., RT 6000, Sorvall, or Eppendorf minifuge).
6. Empty pEAQ-*HT* vector.
7. pEAQ-HT plasmid carrying the HPV 16 L1 gene.
8. *Agrobacterium tumefaciens* LBA4404.
9. *Agrobacterium tumefaciens* LBA4404 transformed with pEAQ-*HT* carrying the HPV16-L1 gene.

2.3 Agroinfiltration

1. Agroinfiltration buffer: 10 mM N-morpholino ethanesulfonic acid (MES) buffer, 10 mM $MgCl_2$, pH 5.6 containing 100 μM acetosyringone. This solution is made freshly from filter-sterilized stocks of 0.5 M MES–KOH buffer, pH 5.6, 0.5 M $MgCl_2$, and 0.1 M acetosyringone.
2. 1 ml needleless syringes.
3. Spectrophotometer.
4. Plastic cuvettes.

2.4 Protein Analysis

1. Liquid nitrogen.
2. Sterile Eppendorf pestles.
3. Hot plate.
4. Mini-PROTEAN® Electrophoresis System (Bio-Rad, Richmond, CA).
5. Mini-PROTEAN® TGX™ Precast Gels, 7.5% acrylamide (Bio-Rad).
6. Mini Trans-Blot® Electrophoretic Transfer Cell (Bio-Rad).
7. Laemmli sample buffer (LSB): 50 mM Tris–HCl, pH 6.8, 2% SDS, 10% glycerol, 0.1% Bromophenol Blue, 100 mM dithiothreitol. Aliquots are stored at −20 °C.
8. Tris–glycine SDS-PAGE buffer: 250 mM glycine, 25 mM Tris, 0.1% SDS, pH 8.3.
9. Prestained Protein Standard Marker (All Blue, Bio-Rad).
10. Bio-Safe™ Coomassie Stain (Bio-Rad).
11. Polyvinylidene difluoride (PVDF) membranes (Millipore, Billerica, MA).
12. 100% ethanol, HPLC grade.
13. Western blot transfer buffer: 192 mM glycine, 25 mM Tris, 20% ethanol.

14. Phosphate buffered saline (PBS): 137 mM NaCl, 2.7 mM KCl, 10 mM Na_2HPO_4, 2 mM KH_2PO_4, pH 7.4. A 10× PBS stock solution autoclaved at 121 °C is used to prepare fresh PBS, following dilution with sterile water.
15. PBS containing 0.05% Tween 20 (PBS-T).
16. Blocking solution: 5% nonfat dry milk in PBST. Made freshly before use.
17. Antibody blocking solution: 1% bovine serum albumin (BSA) in PBST. Made freshly before use.
18. Primary antibody: anti-HPV16 L1 monoclonal antibody (CamVir-1) (Abcam, Cambridge, UK).
19. Secondary antibody: goat anti-mouse IgG, horseradish peroxidase (HRP)-conjugated (Sigma-Aldrich, St. Louis, MO).
20. Plastic film (to wrap/seal membrane).
21. SuperSignal™ West Pico Chemiluminescent Substrate (ThermoFisher Scientific, Waltham, MA).
22. Autoradiography films and development solutions.
23. Glass or plastic container.

2.5 VLP Purification

1. Extraction buffer: 0.5 M Phosphate Buffer (PB), pH 6.0, 10 mM Na_2SO_3, 2.5 mM EDTA, 0.1% β-mercaptoethanol.
2. Triton X-100.
3. Driselase® from *Basidiomycetes* sp., powder formulation (D9515 Sigma-Aldrich).
4. 0.5 M K-phosphate buffer (PB), pH 7.0, 2.5 mM EDTA.
5. 0.1 M PB (pH 7.0).
6. Chloroform, HPLC grade.
7. Cs_2SO_4 gradient: 20–50% Cs_2SO_4 density gradient in 0.5 M PB, pH 7.0, 2.5 mM EDTA. Prepare 50–40–30–20% Cs_2SO_4 stock solutions; to obtain the gradient, carefully pipette 2 ml of each solution in the ultracentrifuge tube (Ultra-Clear™ tubes (14 × 89 mm). Gradients must be prepared immediately before use.
8. Magnetic stirrer.
9. Supercentrifuge (e.g., Sorvall RC 6+ centrifuge, Thermo Scientific) equipped with a Fiberlite™ F14-6 × 250y Fixed Angle Rotor.
10. Ultracentrifuge (e.g., L8 Beckman) equipped with a fixed angle 55.2 Ti rotor and an SW41 rotor (Beckman Coulter, Brea, CA) (*see* **Note 3**).
11. Benchtop ultracentrifuge (TL100 Tabletop, Beckman) equipped with a TLA 100 rotor.

2.6 Electron Microscopy

1. Carbon-coated grids.
2. 0.5% uranyl acetate.
3. CM 10 electron microscope (Philips, The Netherlands).

3 Methods

3.1 Agroinfiltration

1. Grow *N. benthamiana* plants in pots filled with soil–perlite in a growth chamber, under a 16 h light (2500 lx is sufficient) and 8 h dark regimen; irrigate daily. Plants are ready for agroinfiltration when approximately four true expanded leaves are present.
2. Multiply the pEAQ-*HT* plasmid carrying the HPV 16 L1 gene (*see* **Note 4**) in *Agrobacterium tumefaciens* LBA4404 from a fresh YEB agar plate in YEB medium containing kanamycin and rifampicin and acetosyringone. Multiply the empty pEAQ-*HT* vector [15], as control. Grow at 28 °C for 2 days.
3. Centrifuge bacterial cultures using either an Eppendorf centrifuge (10,000 × *g*, 1 min, for small culture volumes) or a supercentrifuge (Sorvall RC 6+, Thermo Scientific; 5000 × *g*, 5 min, for higher volumes).
4. Discard the supernatant and resuspend the bacteria in an appropriate volume of freshly prepared agroinfiltration buffer, to reach an OD_{600} = 0.8 (*see* **Note 5**).
5. Incubate the suspension at room temperature for 2–3 h.
6. Infiltrate the agrobacterium suspension into the leaf parenchyma with a needleless 1 ml syringe. Gently place the syringe on the lower surface of a fully expanded leaf and apply a counterpressure with the finger placed on the upper surface of the leaf opposite the syringe. The suspension will penetrate the intercellular spaces of the leaf parenchyma via the stomata. Repeat the infiltration so that the whole leaf is infiltrated. Infiltrate a few leaves of a control plant with the empty vector pEAQ-*HT* (*see* **Note 6**).
7. Maintain plants at 21–22 °C for 2–9 days after agroinfiltration (*see* **Note 7**).

3.2 Protein Analysis

1. Collect leaf samples. For analytical purposes, a suitable sample consists of a leaf disk collected with the Eppendorf cap. Extract proteins immediately or freeze the sample at −80 °C until required. In either case, grind the leaf disk directly in the Eppendorf tube with the aid of liquid nitrogen and a sterile Eppendorf pestle.
2. To solubilize proteins, add 75 μl LSB per leaf disk, homogenize, mix well and heat to 100 °C for 2 min in a heat block.
3. Centrifuge at 15,300 × *g*, 3 min in an Eppendorf centrifuge and transfer the supernatant into a new tube.

4. Load 5 μl of each sample on a 7.5% mini-PROTEAN TGX™ precast gel in Tris–glycine SDS-PAGE buffer. Load an appropriate aliquot of a protein standard marker for molecular weight determination. Run at 150 V until the dye front reaches the bottom of the gel. Disassemble the apparatus, place the gel in a glass or plastic container and soak it twice in about 50 ml H_2O (15 min each). Discard the liquid and stain the gel with Bio-Safe™ Coomassie Stain for 30 min (*see* **Note 8**).
5. Load a second gel with equal or normalized amounts of sample. Normalization can be achieved using the intensity of the small RuBisCO subunit band (14 kDa) as reference. Load the Prestained Protein Standard Marker and electrophorese at 150 V until the dye front reaches the bottom of the gel. Disassemble the apparatus and electroblot the gel onto a PVDF membrane, previously activated with 100% ethanol and soaked in Protein Transfer Buffer. Carry out electrotransfer in Protein Transfer Buffer at 50 V, 2 h, in a cold room or with the inner reservoir filled with ice. Disassemble the apparatus and proceed with protein immune-detection.
6. Soak the blotted membrane in 10 ml PBS-T for 3 min and then in 15 ml of Blocking solution for 1 h at 4 °C. Wash three times for 5 min, with 15 ml PBS-T. Soak the membrane in 1% BSA in PBS-T, containing the Primary antibody at the appropriate dilution, for 1 h at room temperature. Wash three times for 5 min, with 15 ml PBS-T. Soak the membrane in 1% BSA in PBS-T containing the HRP-conjugated Secondary antibody for 1 h, at room temperature. Wash three times for 5 min, with 20 ml PBS-T. All these steps are carried out at room temperature, on a rotary shaker.
7. Detect the HRP signal using the West Pico Chemiluminescent detection kit. Place the membrane on a plastic film on a flat surface. Add the freshly prepared mixture (1:1) of the Luminol/Enhancer and Stable Peroxide solutions (about 2 ml for a membrane) and incubate for 7 min. Drain the excess liquid and expose the plastic-wrapped membrane to autoradiography films for 1–30 min. Develop with appropriate chemicals (*see* Fig. 1 and **Note 9**).

3.3 VLP Purification

During this step, keep samples on ice and perform all centrifugation steps at 4 °C.

1. Grind leaves expressing the recombinant HPV L1 protein in a mortar, using liquid nitrogen and pestle. Reduce the leaf tissue to a fine powder and homogenize in five volumes of extraction buffer (*see* **Note 10**).
2. Transfer the homogenate to a 250 ml Sorvall RC 6+ centrifugation tube and add Triton X-100 and Driselase powder to 1%

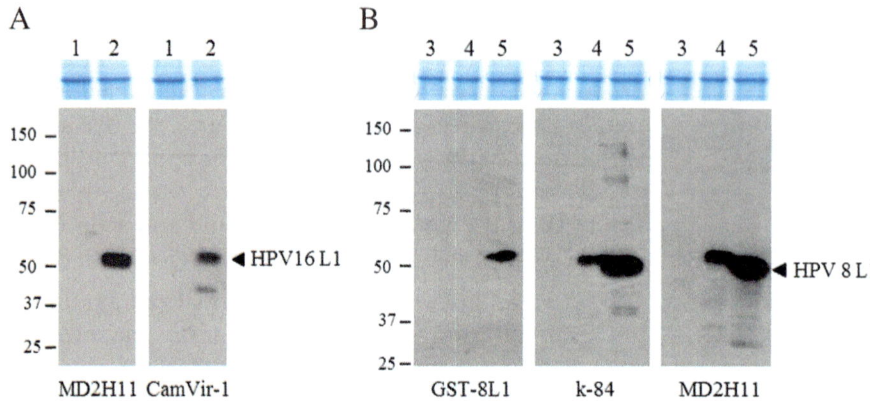

Fig. 1 Immunoblots of extracts from leaves expressing HPV L1 proteins, detected with specific antibodies. Extracts were obtained from leaf tissue infiltrated with agrobacteria carrying different constructs. (**a**) Immunoblots for HPV 16. Lane 1, empty pEAQ-***HT*** vector; lane 2, HPV 16 L1ΔC22 (with a C-terminal deletion of 22 amino acids). (**b**) Immunoblots for HPV 8. Lane 3, empty pEAQ-***HT*** vector; lane 4, HPV8 L1; lane 5, HPV 8 L1ΔC22 (with a C-terminal deletion of 22 amino acids). The primary antibodies used for immuno-detection are indicated below each blot. The monoclonal antibodies MD2H11 and CamVir-1 were diluted 1:5000; both polyclonal sera GST-8L1, prepared at IPSP-CNR using as antigen the HPV8 L1 cS-transferase (GST), produced in ***Escherichia coli*** (Matić et al. [17]) and k-84, a gift from Dr. Pawlita, Heidelberg, Germany, were diluted 1:1000. A portion of the Coomassie-stained gels where the major band is RuBisCo is shown as loading control. Molecular weight markers (in kDa) are indicated on the left of each panel

(v/v) and 0.1% (w/v) final concentration, respectively. Incubate overnight at 4 °C, with stirring.

3. Emulsify the homogenate with 15% (v/v) final concentration chloroform by stirring for 15 min and centrifuge at 8000 × *g* for 15 min.
4. Collect the upper aqueous phase; divide it in two aliquots and centrifuge at 205000 × *g*, 2 h (55.2 Ti rotor).
5. Resuspend both pellets in 0.5 M PB, pH 7.0, 2.5 mM EDTA (2 ml total volume). Centrifuge at 8000 × *g*, 15 min (Fiberlite™ F14-6 × 250y rotor, with rubber adaptors).
6. Collect the supernatant and repeat with another aliquot of 2 ml of 0.5 M PB, pH 7.0, 2.5 mM EDTA.
7. Pool the supernatants. Load onto a 50–20% Cs_2SO_4 density gradient in 0.5 M PB, pH 7.0, 2.5 mM EDTA. Centrifuge at 160,000 × *g*, 5 h (SW41 rotor).
8. Submerge the gradient tube in a glass beaker filled with water and fix it with a hemostat. Illuminate from the bottom of the beaker with an electric portable light, in a dark room, at room temperature, to visualize bands. Collect fractions (max 1.5 ml each) from the top of the gradient tube using a syringe equipped with a hooked needle.
9. Dilute each fraction with 5–10 volumes of 0.1 M PB (pH 7.0). Centrifuge at 390,000 × *g*, 15 min (Beckman TLA 100 rotor).

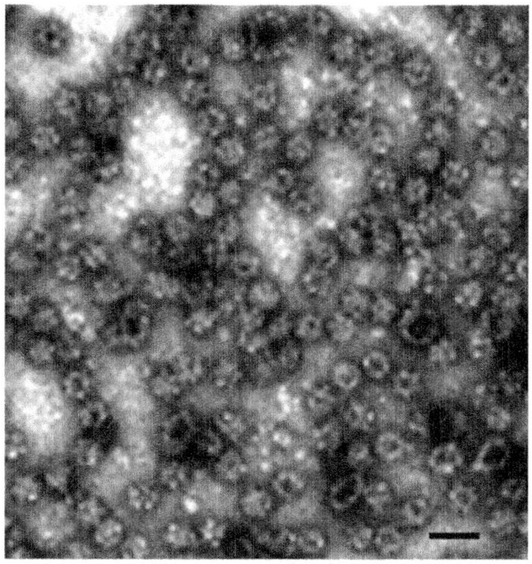

Fig. 2 Photomicrograph of purified HPV-8 L1ΔC22-based VLPs. The leaf tissue was infiltrated with pEAQ-***HT*** carrying the HPV-8 L1ΔC22 encoding gene. VPs (55 nm) and small VLPs (30 nm) are visible. Scale bar = 50 nm

10. Resuspend each pellet in 0.1 M PB (pH 7.0) with a small glass rod, using a total volume of 50 μl. Take small aliquots for EM observation and L1 protein (immune) detection, in order to select the fraction(s) containing the highest amount of L1 and VLPs.

3.4 Electron Microscopy

Adsorb aliquots of the above obtained fractions onto carbon-coated grid for 1–3 min. Remove excess fluid with filter paper and negatively stain the samples with 0.5% uranyl acetate. Observe using a CM 10 electron microscope (*see* Fig. 2).

4 Notes

1. If expression vectors other than pEAQ-*HT* are used, check the antibiotic resistance genes present in the vector. In addition, if agrobacterium strains other than LBA4404 are used (e.g., GV3101, C58C1), other antibiotic combinations might be necessary. However, best results are usually obtained with strains LBA4404 or AGL-1.
2. Agrobacteria can also be grown in lysogeny broth (LB) medium: 10 g/l Bacto tryptone, 10 g/l NaCl, 5 g/l Bacto yeast extract, autoclaved at 121 °C for 15 min.
3. If different centrifugation rotors are available, use the Rotor Calculation program available at https://www.beckmancoulter.com to adjust speed and centrifugation time.

4. L1 genes of other PVs can be potentially expressed by this procedure, either in their native form or following gene modifications. These can include, in principle, any deletion or mutation that does not affect protein expression and subsequent protein folding/self-assembling. Synthetic genes with optimized codon usages and sequences can be used.
5. To reach an $OD_{600} = 0.8$, approximately 2–3 volumes of agro-infiltration buffer with respect to the initial culture volume are required. Estimate the volume of agrobacteria suspension required for each experiment and then calculate the necessary volume of the bacterial culture. Consider approximately 300 μl of agrobacterium suspension for every fully infiltrated leaf.
6. Water plants a few hours before infiltration, since sometimes this is the cause of leaves being recalcitrant to infiltration. Use well-expanded leaves; avoid small young leaves and lower old leaves (they generally do not express detectable amount of heterologous proteins). The infiltrated patch will be visible for about 30 min as a darker area around the injection site. To facilitate further sampling or collection, mark infiltrated areas with a permanent marker on the upper surface; if several leaves are fully infiltrated, cut the apex of each infiltrated leaf with a scalpel.
7. A time-course experiment is useful to follow the level of expression of the heterologous gene that is maximal within 3–10 days after infiltration. For HPV 16 L1, highest yields are obtained at 5–6 days after infiltration.
8. The HPV L1 protein migrates with an apparent MW of 55 kDa, similarly to the large subunit of the RuBisCO. Thus, the L1 band is not detectable after Coomassie staining, but requires an immune-detection technique.
9. As primary antibody, the monoclonal antibody CamVir-1 or MD2H11 can be used, but other antibodies are commercially available. As secondary antibody, any anti-mouse or anti-rabbit antibody suitable for Western blot, conjugated to HRP, can be used. Use of other enzyme conjugates such as alkaline phosphatase, is also possible; in such case, chemiluminescent or colorimetric substrates (e.g., CSPD or NBT/BCIP, respectively) can be used. With colorimetric detection, no dark room and photographic equipment are required.
10. The experiment described is for about 10 g of fresh leaf of material. If greater amounts of leaf material are used, "scaling up" must take into consideration the volumes of the solutions used during the ultracentrifugation steps and, consequently, the availability of suitable equipment (tubes and rotors).

References

1. Wang JW, Roden RB (2013) Virus-like particles for the prevention of human papillomavirus-associated malignancies. Expert Rev Vaccines 12:129–141
2. Buck CB, Cheng N, Thompson CD, Lowy DR, Steven AC, Schiller JT, Trus BL (2008) Arrangement of L2 within the papillomavirus capsid. J Virol 82:5190–5197
3. Baker TS, Newcomb WW, Olson NH, Cowsert LM, Olson C, Brown JC (1991) Structures of bovine and human papillomaviruses. Analysis by cryoelectron microscopy and three-dimensional image reconstruction. Biophys J 60:1445–1456
4. Chen X, Garcea R, Goldberg I, Casini G, Harrison SC (2000) Structure of small virus-like particles assembled from the L1 protein of human papillomavirus 16. Mol Cell 5:557–567
5. Buck CB, Trus BL (2012) The papillomavirus virion: a machine built to hide molecular Achilles' heels. Adv Exp Med Biol 726:403–422
6. McKee SJ, Bergot AS, Leggatt GR (2015) Recent progress in vaccination against human papillomavirus-mediated cervical cancer. Rev Med Virol 1:54–71
7. Sohn U, Nam HG, Park DH, Kim KH (2002) Recombinant human papillomavirus vaccine expressed in transgenic plants. US patent 6,444,805
8. Varsani A, Williamson AL, Rose RC, Jaffer M, Rybicki EP (2003) Expression of human papillomavirus type 16 major capsid protein in transgenic *Nicotiana tabacum* cv. Xanthi. Arch Virol 148:1771–1786
9. Biemelt S, Sonnewald U, Galmbacher P, Willmitzer L, Müller M (2003) Production of human papillomavirus type 16 virus-like particles in transgenic plants. J Virol 77:9211–9220
10. Maclean J, Koekemoer M, Olivier AJ, Stewart D, Hitzeroth II, Rademacher T, Fischer R, Williamson AL, Rybicki EP (2007) Optimization of human papillomavirus type 16 (HPV-16) L1 expression in plants: comparison of the suitability of different HPV-16 L1 gene variants and different cell compartment localization. J Gen Virol 88:1460–1469
11. Regnard GL, Halley-Stott RP, Tanzer FL, Hitzeroth II, Rybicki EP (2010) High level protein expression in plants through the use of a novel autonomously replicating geminivirus shuttle vector. Plant Biotechnol J 8:38–46
12. Fernández-San Millán A, Ortigosa SM, Hervás-Stubbs S, Corral-Martínez P, Seguí-Simarro JM, Gaétan J, Coursaget P, Veramendi J (2008) Human papillomavirus L1 protein expressed in tobacco chloroplasts self-assembles into virus-like particles that are highly immunogenic. Plant Biotechnol J 6:427–441
13. Shirbaghaee Z, Bolhassani A (2015) Different applications of virus-like particles in biology and medicine:vaccination and delivery systems. Biopolymers 105:113–132
14. Sainsbury F, Thuenemann EC, Lomonossoff GP (2009) pEAQ: versatile expression vectors for easy and quick transient expression of heterologous proteins in plants. Plant Biotechnol J 7:682–693
15. Christensen ND, Dillner J, Eklund C, Carter JJ, Wipf GC, Reed CA, Cladel NM, Galloway DA (1996) Surface conformational and linear epitopes on HPV-16 and HPV-18 L1 virus-like particles as defined by monoclonal antibodies. Virology 223:174–184
16. Matić S, Rinaldi R, Masenga V, Noris E (2011) Efficient production of chimeric human papillomavirus 16 L1 protein bearing the M2e influenza epitope in *Nicotiana benthamiana* plants. BMC Biotechnol 11:106
17. Matić S, Masenga V, Poli A, Rinaldi R, Milne RG, Vecchiati M, Noris E (2012) Comparative analysis of recombinant human papillomavirus 8 L1 production in plants by a variety of expression systems and purification methods. Plant Biotechnol J 10:410–421
18. Love AJ, Chapman SN, Matic S, Noris E, Lomonossoff GP, Taliansky M (2012) *In planta* production of a candidate vaccine against bovine papillomavirus type 1. Planta 236:1305–1313
19. Thuenemann EC, Lenzi P, Love AJ, Taliansky M, Bécares M, Zuñiga S, Enjuanes L, Zahmanova GG, Minkov IN, Matić S, Noris E, Meyers A, Hattingh A, Rybicki EP, Kiselev OI, Ravin NV, Eldarov MA, Skryabin KG, Lomonossoff GP (2013) The use of transient expression systems for the rapid production of virus-like particles in plants. Curr Pharm Des 19:5564–5573
20. Peyret H, Lomonossoff GP (2015) When plant virology met agrobacterium: the rise of the deconstructed clones. Plant Biotechnol J 13:1121–1135

Chapter 7

Recombinant Expression of Tandem-HBc Virus-Like Particles (VLPs)

Sam L. Stephen, Lucy Beales, Hadrien Peyret, Amy Roe, Nicola J. Stonehouse, and David J. Rowlands

Abstract

The hepatitis B virus (HBV) core protein (HBc) has formed the building block for virus-like particle (VLP) production for more than 30 years. The ease of production of the protein, the robust ability of the core monomers to dimerize and assemble into intact core particles, and the strong immune responses they elicit when presenting antigenic epitopes all demonstrate its promise for vaccine development (reviewed in Pumpens and Grens (Intervirology 44: 98–114, 2001)). HBc has been modified in a number of ways in attempts to expand its potential as a novel vaccine platform. The HBc protein is predominantly α-helical in structure and folds to form an L-shaped molecule. The structural subunit of the HBc particle is a dimer of monomeric HBc proteins which together form an inverted T-shaped structure. In the assembled HBc particle the four-helix bundle formed at each dimer interface appears at the surface as a prominent "spike." The tips of the "spikes" are the preferred sites for the insertion of foreign sequences for vaccine purposes as they are the most highly exposed regions of the assembled particles. In the tandem-core modification two copies of the HBc protein are covalently linked by a flexible amino acid sequence which allows the fused dimer to fold correctly and assemble into HBc particles. The advantage of the modified structure is that the assembly of the dimeric subunits is defined and not formed by random association. This facilitates the introduction of single, larger sequences at the tip of each surface "spike," thus overcoming the conformational clashes contingent on insertion of large structures into monomeric HBc proteins.

Differences in inserted sequences influence the assembly characteristics of the modified proteins, and it is important to optimize the design of each novel construct to maximize efficiency of assembly into regular VLPs. In addition to optimization of the construct, the expression system used can also influence the ability of recombinant structures to assemble into regular isometric particles. Here, we describe the production of recombinant tandem-core particles in bacterial, yeast and plant expression systems.

Key words Hepatitis B core (HBc), HBc VLPs, *Escherichia coli* expression, *Pichia pastoris* expression, Plant expression

1 Introduction

1.1 Production of Tandem-HBc VLPs in E. coli

The ease and speed of production and the large volumes of culture that can be readily handled highlight the key advantages of producing hepatitis B virus (HBV) core protein (HBc) virus-like

Christina Wege and George P. Lomonossoff (eds.), *Virus-Derived Nanoparticles for Advanced Technologies: Methods and Protocols*, Methods in Molecular Biology, vol. 1776, https://doi.org/10.1007/978-1-4939-7808-3_7,
© Springer Science+Business Media, LLC, part of Springer Nature 2018

particles (HBc VLPs) in *E. coli*. Following transformation with the expression plasmid, the bacteria can be grown to large scale prior to induction of protein expression. After collection by centrifugation, the bacterial cells are ruptured by sonication and the proteins then precipitated from the clarified lysate. The resuspended precipitated material is centrifuged through a sucrose cushion to concentrate and partially purify particulate material. After resuspension, the VLPs are further purified by centrifugation through a sucrose gradient. Gradient fractions are analyzed by SDS-PAGE followed by Coomassie Brilliant Blue staining and/or silver staining and relevant fractions pooled, dialyzed, and the protein content quantified prior to further analysis (*see* Fig. 1 for outline of procedures). The proteins can be produced at 37 °C or at lower temperatures to enhance particle assembly if necessary.

1.2 Production of Tandem-HBc VLPs in Yeast

Recombinant expression of protein in yeast offers a cost-effective approach to producing VLPs in a simple eukaryotic expression system. Genetic manipulation of yeast cells is relatively straightforward, and transformed cells can be grown to very high densities before induction of recombinant protein expression. Thus, large quantities of VLPs can be produced in industrial-scale fermenters. This section describes methods used to express tandem-HBc VLPs in *Pichia pastoris* transformed using a pPICZ C (Invitrogen) vector containing the tandem-HBc sequence. Briefly, transformed cells are grown in a simple medium to a high cell density before induction by addition of methanol. Three to four days after induction, the cells are harvested, pelleted by centrifugation and lysed by vortex mixing with glass beads. The soluble material is clarified by centrifugation and filtration, and VLPs purified by size exclusion chromatography or by centrifugation through sucrose gradients as for VLPs produced in *E. coli* (*see* Fig. 2 for outline of procedures). Yields of up to 40 mg HBc VLP have been achieved from 1 l culture.

1.3 Production of Tandem-HBc VLPs in *Nicotiana benthamiana*

Plants are a useful alternative for the production of tandem-HBc particles, both for the quality of the particles obtained and for the wide-ranging modifications to the basic tandem-HBc particle that are compatible with correct assembly [2]. The expression host used is the solanaceous plant *Nicotiana benthamiana*, and the particles are expressed using transient expression initiated by *Agrobacterium*-mediated gene transfer. The pEAQ vector system [3, 4] is the expression system of choice, and the extraction and purification of tandem core particles is based on the method described in [5]. The leaf material is processed using a blender, and after clarification of the crude extract the particles are concentrated and partially purified by sedimentation through a double-layered sucrose cushion. The sample can then be further purified by Nycodenz-based density gradient centrifugation (as described in Subheading 3.3) (*see* Figs. 3, 4, and 5 for outline of procedures). Yields of 250 mg HBc VLP have been achieved from 1 kg of fresh-weight leaf tissue.

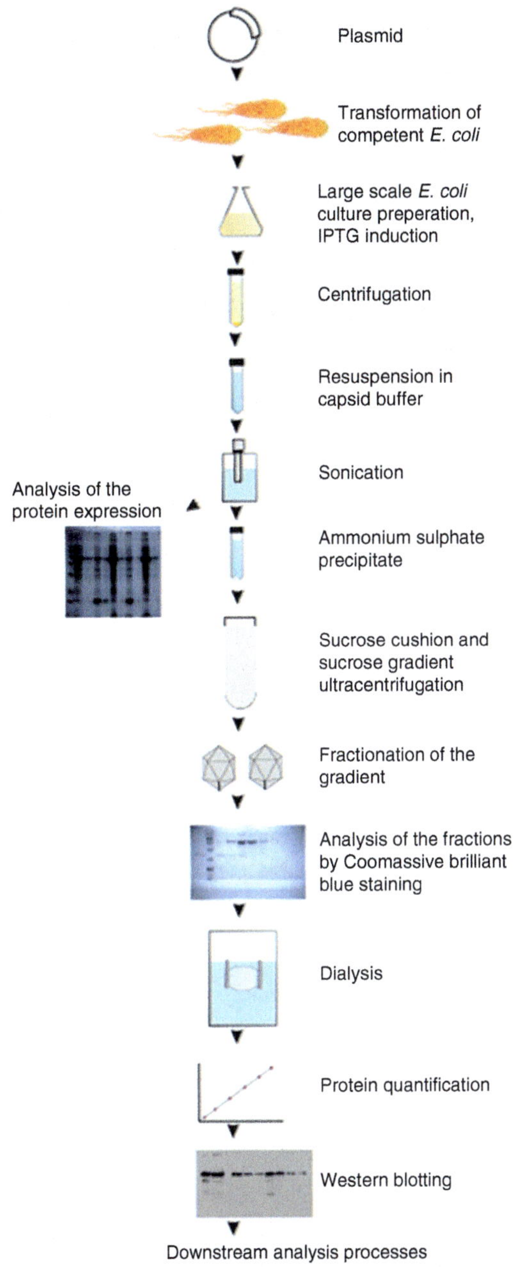

Fig. 1 The *E. coli* HBc expression and analysis pipeline scheme. The tandem-HBc gene is cloned into a bacterial expression vector and propagated in the appropriate strain of *E. coli*. Protein expression in large scale bacterial cultures is induced by IPTG and the culture is centrifuged to recover the cell pellet. This is resuspended in the capsid buffer, sonicated and the proteins precipitated with ammonium sulfate. The tandem-HBc cores are purified by sucrose cushion and sucrose gradient ultracentrifugation. Fractions of the gradient are taken and analyzed for the presence of the VLPs by Coomassie Brilliant Blue staining. The selected fractions are purified by dialysis and the protein is quantified and analyzed by other downstream methods

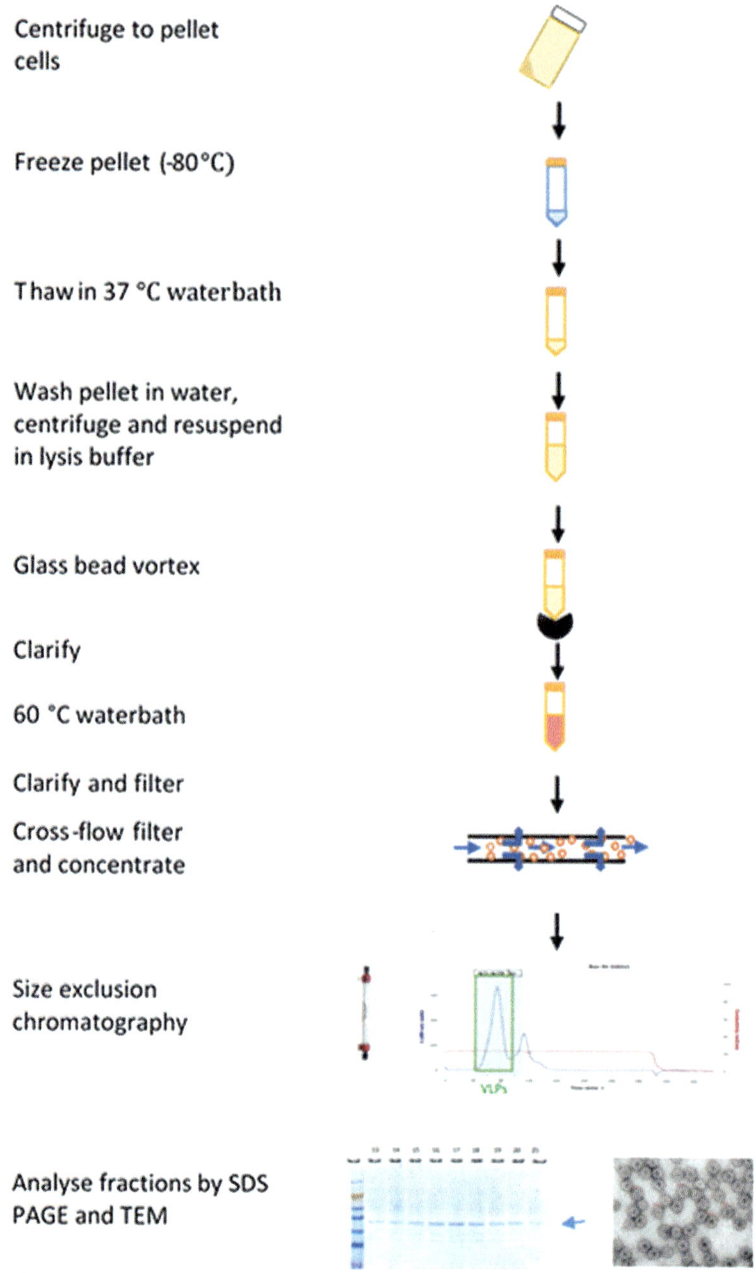

Fig. 2 *P. pastoris* expression, purification, and analysis pipeline. After methanol induced expression, cells are frozen at 80 °C. To release VLPs, the thawed cells are washed and lysed by glass bead vortexing. The clarified lysate is heat treated to precipitate host protein and centrifuged to clarify. Syringe filtration (0.22 μm) is followed by 1 MegaDalton cross-flow filtration to wash and concentrate the VLPs. The retentate containing VLPs is fractionated by size exclusion chromatography and collected fractions are analyzed by SDS-PAGE and negative stain TEM prior to pooling

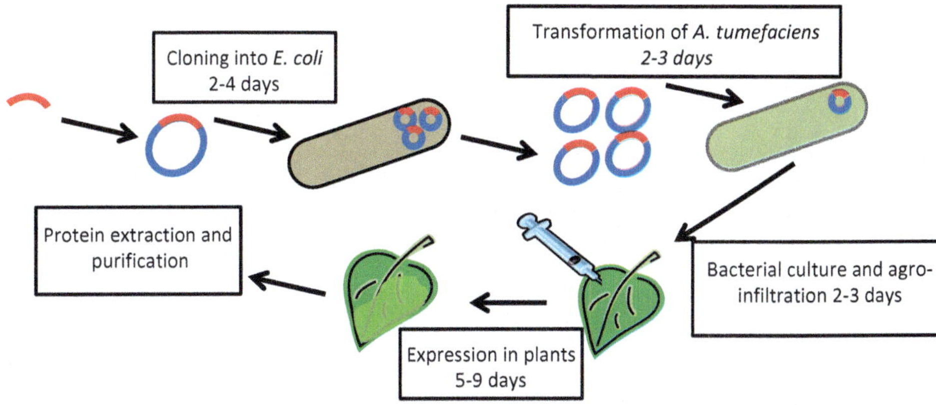

Fig. 3 The plant transient expression pipeline. The tandem-HBc gene is cloned into a plant expression vector and propagated in *E. coli*, then transformed into *Agrobacterium tumefaciens* for agroinfiltration into *N. benthamiana* leaves. VLPs can be extracted from the leaves 5–9 days later. Figure reproduced from Peyret and Lomonossoff [4]

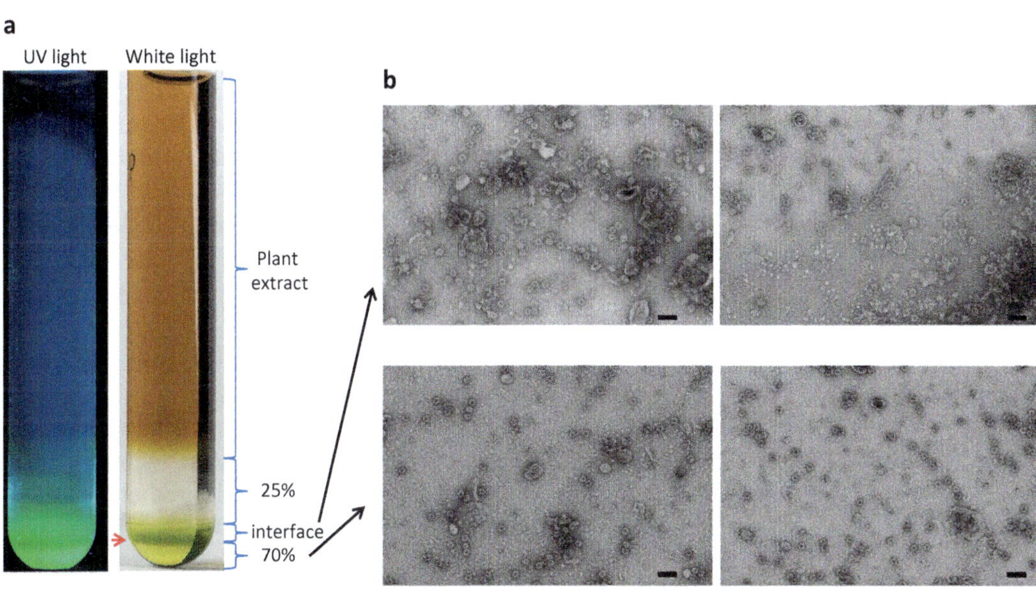

Fig. 4 Using a double sucrose cushion to concentrate and partially purify tandem-HBc VLPs. (**a**) UV light (left) and white light (right) photographs of a 14 × 89 mm ultracentrifuge tube after ultracentrifugation. A visible band of green impurities (red arrow) will sediment at the interface between the 25% and 70% sucrose layers. VLPs will sediment within that interface layer and in the 70% sucrose layer below. For clarity, the VLPs used here are fluorescently labeled and can be visualized under UV light. (**b**) Comparison of the interface (top two micrographs) and 70% sucrose fractions (bottom two micrographs). While VLPs are found in both, the 70% sucrose fraction is noticeably cleaner. All scale bars are 100 nm. Figure reproduced from Peyret [5]

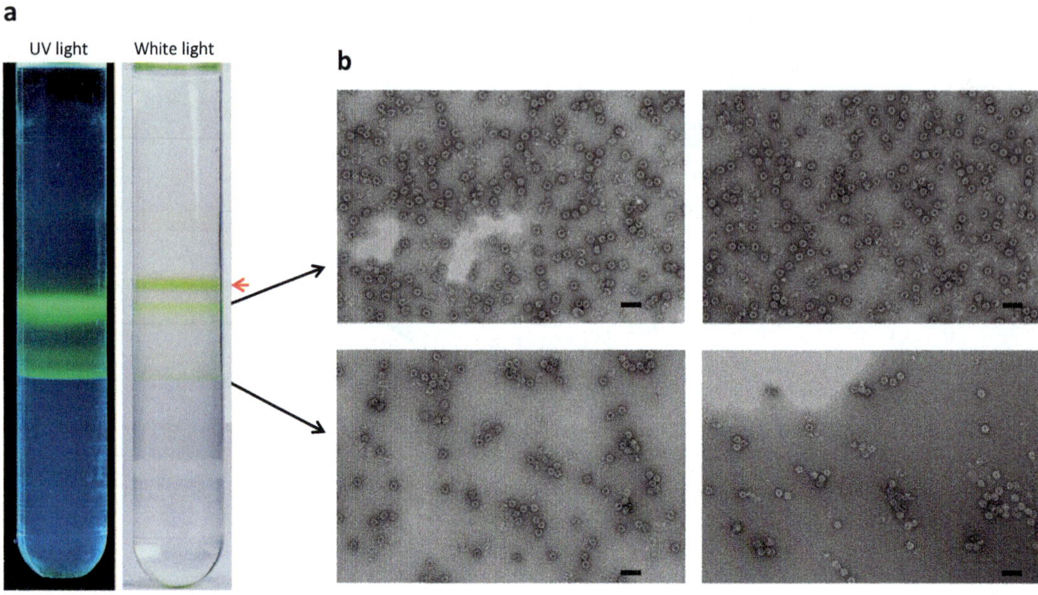

Fig. 5 Using a Nycodenz gradient to purify tandem-HBc VLPs. (**a**) UV light (left) and white light (right) photographs of a 14 × 89 mm ultracentrifuge tube after ultracentrifugation. The impurities will form a green band (red arrow), which will be separate from the VLPs (fluorescent bands), allowing greater purification than the sucrose cushion alone. Because this method of purification is based on density, the Nycodenz gradient can also separate subpopulations of VLPs present in a sample, as seen here with two distinct fluorescent bands. (**b**) TEM analysis of these bands (after dialysis against ammonium bicarbonate) reveals that both contain VLPs. The difference between the two subpopulations is due to nucleic acid content. Note that TEM analysis of the Nycodenz-purified VLPs indicates that they are cleaner than after the double sucrose cushion alone (***see*** Fig. 7). All scale bars are 100 nm. Figure reproduced from Peyret [5]

1.4 Characterization of Tandem-HBc VLPs

The ultimate goal of designing, constructing and expressing VLPs for assessment as potential vaccines is to provide particulate products which are pure, defined, uniform in structure, and which possess desired antigenic features. Hence it is necessary to analyze the expressed VLPs to ensure that they have these desired characteristics. A number of methods are available to analyze VLPs, and the most widely used and suitable are described here.

For downstream characterization of the tandem cores, it is imperative that the sucrose and solutes in the preparation are removed. The procedure to do this is described in Subheading 3.4.1. We follow the procedure laid down in [6].

Transmission electron microscopy (TEM) is used to examine the size and architecture of the electron-dense material, and to check for the presence of $T = 3$ and $T = 4$ HBc particles (*see* Fig. 6 and Subheading 3.4.2.).

Western blot analysis can demonstrate the presence of the HBc protein, but to confirm the presence of assembled tandem-HBc particles, immunogold TEM analysis of the sample is required.

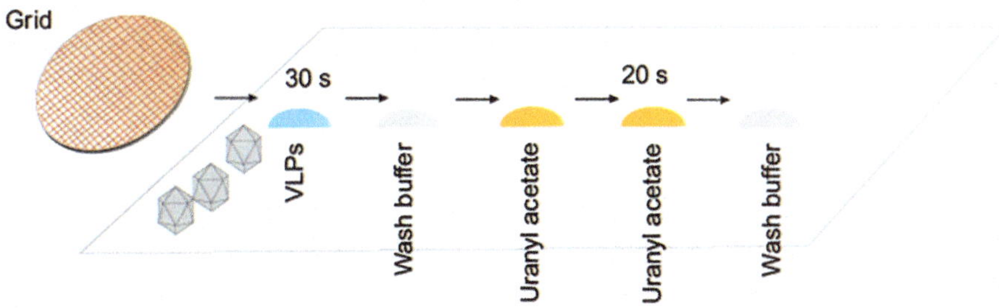

Fig. 6 Scheme of TEM analysis of tandem-HBc VLPs. The purified VLPs are pipetted on to Parafilm and the freshly glow discharged grids are exposed to them. The grids are washed and exposed to two droplets of the stain (2% uranyl acetate) and washed a final time and dried prior to analysis

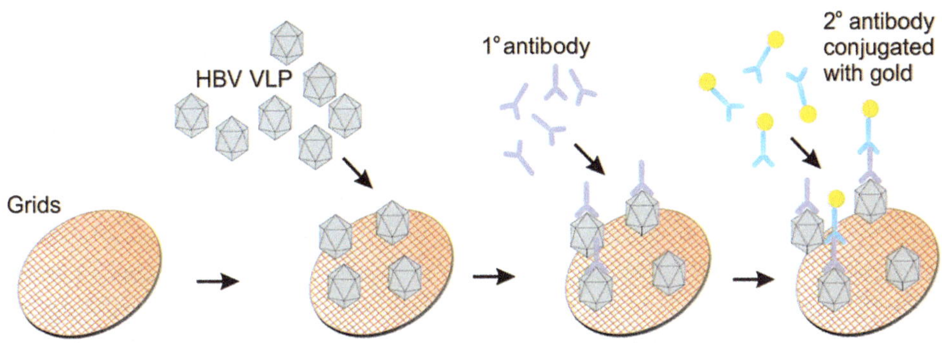

Fig. 7 Scheme of immunogold staining of tandem-HBc VLPs. The sample is placed onto the grid and then the grid is washed prior to blocking. The blocking solution is washed off and the grid is exposed to the primary antibody. After incubation, the grid is washed and exposed to the secondary antibody prior to another wash and exposure to the uranyl acetate stain

The sample is applied to the grid and is then incubated with an HBc-specific primary antibody, followed by incubation with a secondary antibody conjugated with gold (*see* Fig. 7 and Subheading 3.4.3.).

Should the sample contain electron-dense material of unexpected architecture, an immunoadsorbent TEM assay can be used to quickly screen for immunoreactive electron-dense material (adapted from [7]). The grid is incubated with the primary antibody, exposed to the sample, and then washed (*see* Fig. 8 and Subheading 3.4.4.).

Dynamic light scattering (DLS), where the scattered backlight of a beam passing through the solution is analyzed, is useful for both confirming the size of the HBc VLPs and calculating the polydispersity of the preparation, and is currently routinely used for analysing VLP preparations [8] (*see* Subheading 3.4.5.).

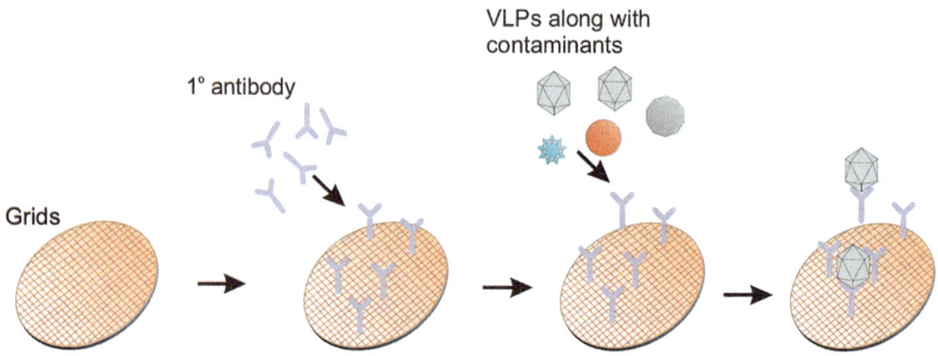

Fig. 8 Scheme of immunoadsorbent assay of tandem-HBc VLPs. The grid is placed in the blocking solution and then washed prior to incubation with the primary antibody. This is then washed off and the grid is exposed to the sample. After washing the grid, it is stained with uranyl acetate

2 Materials

All solutions are prepared using deionized water, if not stated otherwise.

2.1 Production of Tandem-HBc VLPs in E. coli

2.1.1 Equipment

1. High speed centrifuge (Avanti J-26XP, Beckman Coulter or Heraeus Megafuge 16 R, Thermo Scientific) with appropriate bucket and rotor (e.g., SW40 or SW32) (*see* **Note 1**).
2. Sonicator (e.g., Soniprep 150, MSS150, MSE with an MSE-SH100 process timer) (*see* **Note 2**).
3. Equipment for SDS-PAGE and staining of the gels (*see* **Note 3**).
4. Equipment for western blotting (Membrane: nitrocellulose or PVDF membranes, 0.2 μm pore size).
5. Gradient maker (e.g., either a Gradient station *ip*, BIOC MP {Cat. # 153-002} or a dual chamber {SG 30 [product code 80619780] Gradient Maker or SG 15 Gradient Maker [SG15] depending on the rotor}).

2.1.2 Reagents

1. LB medium: 10 g/l tryptone, 5 g/l yeast extract, 10 g/l NaCl; adjust to pH 7.0–7.5 using NaOH. Autoclave; if needed add appropriate antibiotics (30 μg/ml kanamycin) directly before usage.
2. Plasmid: The tandem core constructs in our laboratory are cloned into the protein expressing pET28b backbone, which has a kanamycin resistance selection cassette.
3. Competent bacteria: BL21 DE3, Rosetta II (*see* **Note 4**).
4. 30 mg/μl kanamycin stock (or other appropriate antibiotics) (*see* **Note 5**).

5. LB agar: 10 g/l tryptone, 5 g/l yeast extract, 10 g/l NaCl; adjust to pH 7.0–7.5 using NaOH, add 5 g/l agar. Autoclave; add 30 μg/ml kanamycin directly before pouring plates.
6. Lysis buffer: 20 mM HEPES, 250 mM sodium chloride in deionized water; adjust to pH 7.5 with NaOH and autoclave, then add one protease inhibitor tablet (Roche Complete, Mini, EDTA-free) and 1 μl benzonase nuclease (Novagen, 70746) per 10 ml buffer.
7. Capsid buffer: 20 mM HEPES, 250 mM sodium chloride in deionized water; adjust to pH 7.5 with NaOH and autoclave, then add 2 mM dithiothreitol (DTT) (*see* **Note 6**).
8. 20% and 60% (w/v) sucrose in capsid buffer.
9. 1 M IPTG (S02122, Glycon) stock, frozen (*see* **Note 7**).
10. 4.32 M ammonium sulfate solution: saturated aqueous solution of ammonium sulfate (*see* **Note 8**).
11. 1× TBS-T: 10 mM Tris pH 7.5, 150 mM NaCl, 0.1% Tween 20.
12. Blocking buffer: 5% (w/v) milk powder in 1× TBS-T.
13. Primary antibody: Mouse mAb IgG1 antibody 10E11 (anti-hepatitis B virus core antigen antibody ab8639, Abcam), diluted 1:1000 in blocking buffer.
14. Secondary antibody: Anti mouse IgG peroxidase, produced in goat (product number A 5278, Sigma Aldrich), diluted 1:2000 in blocking buffer.
15. Reagents for Coomassie Brilliant Blue and/or silver staining (ProteoSilver Silver Stain Kit, PROTSIL1-1KT, Sigma) of SDS-PAGE gels (*see* **Note 9**).

2.2 Production of Tandem-HBc VLPs in Yeast

2.2.1 Equipment

1. Sterile McCartney bottles.
2. Autoclave.
3. Water baths.
4. Sterile 250 ml and 2 l baffled flasks.
5. Shaking incubator (25 °C).
6. Sterile 1.5 ml microtubes.
7. Electroporator (e.g., MicroPulser, Bio Rad).
8. 2 mm gap electrocuvettes.
9. Centrifuge tubes.
10. Superspeed centrifuge (e.g., Beckman Avanti J-25 or equivalent) with appropriate bucket and rotor (*see* **Note 1**).
11. Microcentrifuge.
12. Vortex mixer (or multivortexer—clamped vortex mixer for tube racks).

13. Glass beads—Retsch 0.75–1.00 μm (VWR).
14. 0.22 PES filter.
15. MegaDalton cross-flow filtration cartridge: Pelicon XL Ultrafiltration Module, Biomax 1000 kDa.
16. Bio Rad NGC/Akta FPLC system.
17. GE HiPrep 16/60 Sephacryl S-500 HR column.
18. Centrifugal concentration device (optional).

2.2.2 Reagents

1. *Pichia pastoris* KM71H cells.
2. Plasmid: pPICZC-HBcAg.
3. Restriction Enzyme: *Pme I*.
4. HEPES stock solution (pH 4.0).
5. YEPD broth (1 l): 20 g Bacto peptone, 10 g yeast extract; adjust to pH 4.0 with HCl, then add further deionized water to 900 ml. Autoclave, then add 100 ml sterilized 20% (w/v) glucose (200 mg/l).
6. YEPD agar (1 l): 20 g Bacto peptone, 10 g yeast extract; Add deionized water to 800 ml adjust to pH 7.0 with NaOH if necessary, then add 16 g agar and deionized water to 900 ml. Autoclave, then add 100 ml sterilized 20% (w/v) glucose (200 mg/l). Add 100 mg Zeocin™ (100 μg/μl) at 50 °C just prior to pouring plates.
7. YEPDS broth and agar: supplement YEPD broth and agar with 0.5 M sorbitol (91 g/l sorbitol).
8. 1 M DTT in deionized water (*see* **Note 6**).
9. 70% (v/v) ethanol.
10. Isopropanol.
11. 99.9% (v/v) methanol.
12. 1 M sorbitol.
13. BMMY induction medium (1 l): 10 g yeast extract, 20 g peptone, 2.0 g yeast nitrogen base (DIFCO 239210), add deionized water to 890 ml. Autoclave and then add 100 ml sterile 1 M potassium phosphate (pH 6.0), 2.0 ml 0.02% biotin (w/v), 10 ml methanol.
14. BMGY expression medium (1 l): 10 g yeast extract, 20 g peptone, 2.0 g yeast nitrogen base (DIFCO 239210), add deionized water to 800 ml. Autoclave, then add 100 ml sterile 1 M potassium phosphate (pH 6.0), 2.0 ml 0.02% biotin (w/v), 100 ml filter-sterilized 10% glycerol (v/v).
15. Lysis buffer: 20 mM Tris (pH 8.5), 5 mM EDTA, 5 mM DTT, Pierce protease inhibitor tablets (EDTA-free) (*see* **Note 6**).
16. 1× PBS (pH 7.4): Life Technologies PBS tablets (ThermoFisher #003002) prepared according to manufacturer's instructions.

2.3 Production of Tandem-HBc VLPs in *Nicotiana benthamiana*

2.3.1 Equipment

1. Razor blade or scalpel.
2. Waring blender or equivalent.
3. Miracloth (Merck Millipore) or equivalent.
4. Needle.
5. 0.45 μm and 0.2 μm syringe filters.
6. Long needle (e.g., Sigma steel 304 syringe needle, or equivalent).
7. 1 ml needle-less syringe.
8. Ultracentrifuge (Thermo Scientific Sorvall WX floor ultracentrifuge or equivalent) (*see* **Note 10**).
9. Ultracentrifuge swing-out rotor (e.g., TH641 or Surespin 630/36 from Thermo Scientific) (*see* **Note 11**).
10. Ultra-Clear ultracentrifuge tubes (Beckman Coulter).
11. 13 ml ultracentrifuge tubes (Ultra-Clear 14 × 89 mm).
12. Dialysis equipment (such as 100 kDa molecular-weight cut-off Float-a-Lyzer from Spectrum Labs).
13. SpeedVac vacuum concentrator or equivalent.
14. Equipment for SDS-PAGE and staining of the gels (*see* **Note 3**).
15. Equipment for western blotting (Membrane: Amersham nitrocellulose membrane from GE Life Sciences).
16. Optional: Equipment for TEM.

2.3.2 Reagents

1. Competent strain of *E. coli* (such as TOP 10) and suitable growth medium (such as LB broth or agar, *see* Subheading 2.1.2, **item 1** or **5**).
2. Competent *A. tumefaciens* strain LBA4404 and suitable growth medium: LB broth or agar, *see* Subheading 2.1.2, **item 1** or **5 complemented with rifampicin,** or 2 X YT: (31 g of 2 X YT broth from Formedium in 1 l deionized water, pH adjusted to 7.4 with NaOH).
3. pEAQ-*HT* system [3] or other suitable plant expression vector.
4. Kanamycin (use at 50 μg/ml from a 50 mg/ml stock made in water) and rifampicin (use at 50 μg/ml from a 50 mg/ml stock made in DMSO) for selection of transformed bacteria.
5. *N. benthamiana* plants (3–4 weeks after pricking out, typically 1 week before the start of flowering).
6. MMA buffer: 10 mM MES (pH 5.6), 10 mM magnesium chloride, 0.1 mM acetosyringone.
7. 0.1 M sodium phosphate buffer: 3.1 g/l $NaH_2PO_4 \times H_2O$, 10.9 g/l Na_2HPO_4; which will give sodium phosphate at pH 7.2–7.5. Chill to 4 °C prior to usage.

8. Protease inhibitor (such as cOmplete EDTA-free protease inhibitor cocktail tablets from Roche).
9. 25% and 70% (w/v) sucrose in sodium phosphate buffer.
10. 20 mM ammonium bicarbonate solution, pH 8.5.
11. 20%, 30%, 40%, 50% and 60% (w/v) Nycodenz (Axis-Shield PoC AS) solutions in sodium phosphate buffer or PBS (*see* Subheading 2.2.2, **item 17**).
12. 1× PBS (pH 7.4): Life Technologies PBS tablets (ThermoFisher #003002) prepared according to manufacturer's instructions.
13. PBS-T: PBS supplemented with Tween 20 at 0.1% (v/v).
14. Blocking buffer: 5% (w/v) milk powder in 1× TBS-T.
15. Primary antibody: mouse mAb IgG1 antibody 10E11 (anti-hepatitis B virus core antigen antibody ab8639, Abcam), diluted 1:6000 in blocking buffer.
16. Secondary antibody: anti mouse IgG peroxidase, produced in goat (product number A 5278, Sigma Aldrich), diluted 1:10,000 in blocking buffer.
17. Reagents for Coomassie Brilliant Blue and/or silver staining (ProteoSilver Silver Stain Kit, PROTSIL1-1KT, Sigma) of SDS-PAGE gels (*see* **Note 9**).

2.4 Characterization of Tandem-HBc VLPs

2.4.1 Removal of Sucrose and Salts by Dialysis of Tandem-HBc VLPs

1. Spectra/Por1 dialysis membrane standard RC tubing (MWCO: 6–8 kD, part number 132655).
2. Dialysis clips.
3. Foam floaters.
4. Dialysis buffer: 10 mM HEPES, 100 mM sodium chloride 1 mM ethylenediaminetetraacetic acid (EDTA) (*see* **Note 12**). Adjust pH to 7.2 with NaOH and autoclave, then add 1 mM dithiothreitol (DTT) (*see* **Notes 6** and **13**).

2.4.2 Examination of Tandem-HBc VLPs by Transmission Electron Microscopy (TEM)

1. Freshly glow-discharged formvar/carbon-coated 400 mesh grids (copper) (S162-4, Agar Scientific) (*see* **Note 14**).
2. Parafilm.
3. Grid box.
4. Tweezers (either antiparallel or regular with an O-ring, dumoxel or titanium number 7 or equivalent) (*see* **Note 15**).
5. Whatman No. 1 filter paper ((L4164, Agar Scientific), torn into small triangles.
6. Automatic pipettes.
7. Stop watch (or watch with second hand).
8. Transmission electron microscope.

9. Wash buffer: 20 mM Tris(hydroxymethyl)aminomethane (Tris), adjusted to pH 8.0 with HCl.
10. 2% (w/v) aqueous solution of uranyl acetate (*see* **Notes 16** and **17**).

2.4.3 Immunogold TEM Analysis

1. Tweezers (either antiparallel or regular with an O-ring, dumoxel or titanium number 7 or equivalent) (*see* **Note 15**).
2. Freshly glow-discharged carbon/formvar-coated 400 mesh nickel grids (S162-4/S162N-4, Agar Scientific) (*see* **Notes 14** and **18**).
3. Automatic pipettes.
4. Stop watch (or watch with second hand).
5. Grid box.
6. Whatman No. 1 filter paper (L4164, Agar Scientific), torn into small triangles.
7. Fine-tipped Pasteur pipettes.
8. Parafilm.
9. Sample dilution buffer: 20 mM Tris(hydroxymethyl)aminomethane (Tris), adjusted to pH 8.0 with HCl.
10. TBS: 10 mM Tris pH 7.5, 150 mM NaCl (24.2 g Tris, 80 g NaCl adjusted to pH 7.5 with HCl for 1 l of 10× TBS.
11. Immunogold block buffer (IBB): 0.05% (w/v) cold water fish skin gelatin (G7765, Sigma), 0.025% Tween 20 (w/v) in TBS.
12. Immunogold diluent (ID): 0.005% (w/v) cold water fish skin gelatin (G7765, Sigma) in PBS (*see* Subheading 2.2.2, **item 17**), 2% (w/v) aqueous solution of uranyl acetate (*see* **Note 16**).
13. Primary antibody: mouse mAb IgG1 antibody 10E11 (anti-hepatitis B virus core antigen antibody ab8639, Abcam), diluted in 1:25 in ID (*see* **Notes 19** and **20**).
14. Secondary antibody: goat anti-mouse IgG colloidal gold conjugate antibody 10 nM gold (EM.GMHL 10, British Biocell) diluted 1:25 in ID (*see* **Note 19**).
15. Deionized water.
16. Negative stain: 2% (w/v) aqueous solution of uranyl acetate (*see* **Notes 16** and **17**).
17. Transmission electron microscope.

2.4.4 Immunoadsorbent TEM Assay

1. Tweezers (either antiparallel or regular with an O-ring, dumoxel or titanium number 7 or equivalent) (*see* **Note 15**).
2. Freshly glow-discharged carbon/formvarcoated 400 mesh nickel (or copper) grids (S162-4/S162N-4, Agar Scientific) (*see* **Notes 14** and **18**).

3. Automatic pipettes.
4. Grid box.
5. Whatman No. 1 Filter paper (L4164, Agar Scientific), torn into small triangles.
6. Fine-tipped Pasteur pipettes.
7. Parafilm.
8. Sample dilution buffer: 20 mM Tris(hydroxymethyl)aminomethane (Tris) (adjusted to pH 8.0 with HCl).
9. Immunogold block buffer (IBB): 0.05% (w/v) cold water fish skin gelatin (G7765, Sigma), 0.025% (w/v) Tween 20 in TBS (*see* Subheading 3.4.3, **step 10**).
10. Immunogold diluent (ID): 0.005% (w/v) cold water fish skin gelatin (G7765, Sigma) in PBS (*see* Subheading 2.2.2, **item 17**), 2% (w/v) aqueous solution of uranyl acetate (*see* **Note 16**).
11. Primary antibody: mouse Mab IgG1 antibody 10E11 (anti-hepatitis B virus core antigen antibody ab8639, Abcam), diluted in 1:25 in ID (*see* **Notes 19** and **20**).
12. Negative Stain: 2% (w/v) aqueous solution of uranyl acetate (*see* **Notes 16** and **17**).
13. Transmission electron microscope.

2.4.5 Dynamic Light Scattering (DLS)

1. Malvern Zetasizer Nano series (*see* **Note 21**).
2. DTS0012 cuvette.
3. Sample to be measured (minimum total volume 1 ml, minimum concentration 2.5 ng/μl of VLPs, diluted in 1× PBS (*see* Subheading 2.2.2, **item 16**)) (*see* **Note 22**).

3 Methods

Always use appropriate personal protective equipment when handling bacteria and antibiotics.

3.1 Production of Tandem-HBc VLPs in E. coli

1. Transform the plasmid into competent bacteria for protein expression. Unlike the standard process for streaking transformed bacteria, plate the culture as for transformation of a ligation reaction (with an incubation time of at least 1 h in SOC or LB medium). Incubate transformants on LB agar supplemented with the appropriate antibiotic (typically kanamycin) overnight at 37 °C.
2. Pick a colony and grow as a starter culture in 10 ml LB medium with the appropriate antibiotic (30 μg/ml for kanamycin-R plasmid) overnight (37 °C, 180 rpm shaking).

3. Add 2.5 ml of starter culture to 250 ml LB medium containing the antibiotic (30 μg/ml for kanamycin-R plasmid) (*see* **Note 23**). Measure optical density (OD) at 600 nm every hour after inoculation (*see* **Note 24**); use LB medium as blank. When the OD 600 is 0.6–0.8 (*see* **Note 25**), remove 25 ml and culture separately overnight (16 °C) or for 3 h (37 °C). This is the uninduced control culture. To the remainder, add 1 mM IPTG and incubate at 37 °C for 3 h or 16 °C overnight. This is the induced culture.
4. Chill the cultures on ice, then pellet the bacteria by centrifuging at 7000 × *g* for 20 min. Pellets may be stored at −20 °C as cell paste; otherwise resuspend in lysis buffer (4 ml lysis buffer to the pellet obtained from 100 ml bacterial culture). Sonicate the resuspended pellet on ice for 30 s with 30 s intervals at 10 mA, repeating the cycle until the solution clarifies (*see* **Notes 26** and **27**). Centrifuge the sonicated culture at 7000 × *g* for 20 min to separate cell debris from soluble protein and recentrifuge the collected supernatant at 7000 × *g* for 20 min to further clarify. Slowly add saturated ammonium sulfate solution (to a final concentration of 40% saturated) to the supernatant over ice before incubating at 4 °C overnight to allow precipitation. Analyze samples of the uninduced and induced pellets and supernatants by PAGE to check for expression and solubility of the required protein.
5. Centrifuge at 7000 × *g* for 20 min, then discard the supernatant and resuspend the pellet in capsid buffer (1 ml capsid buffer to 100 ml bacterial culture). Clarify the suspension by further centrifugation at 17,000 × *g* for 10 min. Layer the suspension over a 60% sucrose cushion (sucrose in capsid buffer), then centrifuge at 151,000 × *g* for 3 h (*see* **Note 1**). Discard the supernatant and resuspend the pellet overnight in capsid buffer.
6. Layer the suspension over a 20–60% continuous sucrose gradient (sucrose in capsid buffer) and centrifuge at 150,000 × *g* for 3 h using the appropriate bucket and rotor (*see* **Note 1**). Collect 1 ml (SW40) or 2 ml (SW32) fractions. Separate the fractions by SDS-PAGE on a gel of appropriate acrylamide percentage (12.5 or 15% depending on the size of the recombinant tandem core construct being analyzed), then stain with Coomassie Blue. The same gel can be destained and used for silver staining according to the manufacturer's instructions. Analyze appropriate fractions by Western blotting: following transfer onto a membrane, block with 5% milk powder in 1× TBS-T prior to application of the primary and secondary antibody.

3.2 Production of Tandem-HBc VLPs in Yeast

1. For transformation of *P. pastoris* with pPICZC-HBcAg, linearize at least 1 μg of plasmid with restriction enzyme *Pme I* to completion. Precipitate the linearized plasmid by adding 0.7

volumes of isopropanol and place at 4 °C for at least 30 min. Centrifuge at ≥12,000 × *g* for 10 min. Remove the liquid very carefully, taking care not to disturb the very small, translucent pellet that will have formed on the outer wall of the bottom of the microtube. Wash the pellet with 1 ml 70% ethanol (do not aspirate the pellet) and repeat centrifugation as previously. Carefully remove the ethanol and allow the pellet to dry, then resuspend the dried pellet in 10 μl ddH_2O.

2. To prepare *Pichia* cells for transformation, inoculate 5 ml YEPD medium with a single colony of *P. pastoris* KM71H cells and grow overnight at 25 °C. The following afternoon, inoculate 40 ml YEPD medium with a range of volumes of the overnight culture from 0.1 to 0.7 ml (e.g., set up four flasks and inoculate with 0.1, 0.3, 0.5, and 0.7 ml) to ensure that the correct cell density is available for transformation. Again, incubate overnight. For transformation, select a flask from the overnight culture that does not contain flocculated yeast with an OD_{600} of 1.3–1.5. Pellet the cells by centrifugation at 4000 × *g* for 3 min and resuspend in freshly prepared YEPDS + 20 mM HEPES (pH 4). Transfer cells to a 1.5 ml microtube, add 35 μl 1 M DTT, mix gently by inversion and incubate at room temperature for 30 min. Pulse spin (10–15 s) in a microcentrifuge at full speed to pellet cells. Remove supernatant and resuspend in 750 μl 1 M sorbitol by vortex mixing. Repeat the sorbitol wash three times. After the final centrifugation step, resuspend cells in a volume of 1 M sorbitol equal to the cell pellet (80–200 μl). Cells are now ready for electroporation.
3. Mix 40 μl of the *Pichia* cells with 100–600 ng of linearized plasmid DNA and incubate for 15 min at room temperature. Transfer the cells and DNA to a prechilled 2 mm gap electrocuvette (+4 °C) taking care to tap the yeast–DNA mix to the bottom of the cuvette gently and dry the sides of the cuvette of moisture. Electroporate at 2.0 kV and immediately add 1 ml YEPDS medium, transfer to a sterile microtube and incubate at room temperature for 1 h. Plate 200 μl of transformed yeast and 20 μl of control yeast (no DNA) on separate YEPDS agar + Zeocin plates. Incubate at room temperature for 2 h, invert and transfer to 25 °C for a further three-day incubation. Compare colonies on control and transformed cell plates to evaluate the success of transformation.
4. To express the tandem-HBc VLPs, pick a colony of pPICZC-HBcAg-transformed *P. pastoris* KM71H cells and grow as a starter culture in 10 ml YEPD medium (*see* **Note 28**) in McCartney bottles overnight (25 °C, 180 rpm shaking). Use a 0.8 ml aliquot of the starter culture to inoculate 800 ml BMGY expression medium in 2 l baffled conical flasks, incubating on

an orbital shaker (180 rpm) at 25 °C for 3 days (*see* **Notes 29** and **30**). Harvest cells by centrifugation at 2000 ×*g* for 4 min and transfer to 500 ml BMMY induction medium in 2 l baffled conical flasks. Incubate at 25 °C and 180 rpm on an orbital shaker, supplementing the cultures with a further 4 ml methanol after 24, 48, and 72 h. After completion of the 96 h induction period, harvest cells by centrifugation at 2000 × *g* for 4 min (*see* **Note 31**). Remove the supernatant, then resuspend the cell pellet in water (5 ml water to 50 ml culture). Transfer to 50 ml Falcon tubes and store at −80 °C.

5. Purification procedures are outlined in Fig. 5. Thaw frozen cells rapidly in warm (approximately 37 °C) water and centrifuge at 2000 ×*g* for 3 min in 50 ml Falcon tubes. Discard the supernatant, then add 2 g glass beads per tube and resuspend the pellet to 11 ml with ice-cold lysis buffer. Vortex the mix for 15 min on a multivortexer and cool on ice for 5 min to prevent overheating. Repeat this cycle four times, then centrifuge the tubes at 2000 ×*g* for 30 min. Warm the supernatant in a water bath (60 °C) for 30 min (*see* **Note 32**) before clarification by centrifugation at 20,000 ×*g* for 30 min. Remove, pool, and retain the supernatant (lysate), putting aside a 100 μl aliquot for future analysis, then filter the pooled lysate through a 0.22 PES filter. Put aside a 100 μl aliquot of filtered lysate for future analysis.
6. Using a MegaDalton cross-flow filtration cartridge, concentrate the equivalent lysate of 200 ml culture to 5 ml. Dilute this to 40 ml in PBS, then concentrate again to 5 ml. Load the cross-flow retentate onto a GE HiPrep 16/60 Sephacryl S-500 HR column and elute with PBS at a flow rate of 1 ml/minute (*see* **Note 33**). Collect 2.5 ml fractions in order to capture the peak of VLPs (*see* **Notes 34** and **35**). If necessary, concentrate the eluted VLPs using a centrifugal concentration device.

3.3 Production of Tandem-HBc VLPs in *Nicotiana benthamiana*

1. Transform the expression plasmid (pEAQ-*HT*) into a suitable strain of *E. coli* (TOP10) to allow for sequencing, then transform into *A. tumefaciens* (LBA4404). Allow both to grow: *E. coli* at 37 °C overnight on LB agar or broth containing kanamycin (50 μg/ml) for plasmid selection; *A. tumefaciens* at 28 °C for 2–3 days on LB agar, broth, or 2 X YT containing kanamycin (50 μg/ml) for plasmid selection and rifampicin (50 μg/ml) for *A. tumefaciens* selection.
2. Prepare liquid cultures of positive clones of *A. tumefaciens* by growing them to stationary phase at 28 °C in either LB or 2 X YT supplemented with kanamycin and rifampicin (50 μg/ml each) as described in [3] (*see* **Note 36**).
3. Centrifuge the cultures at 1100 ×*g* for 20 min, then resuspend the pellet in MMA buffer to an OD_{600} of 0.4.

4. Carry out syringe agroinfiltration of prepared *N. benthamiana* plants (3–4 weeks after pricking out, typically 1 week before the start of flowering) by lightly scratching the surface of a leaf with a needle, then using a 1 ml needle-less syringe to infiltrate the intercellular space of the leaf with the *A. tumefaciens* suspension. Agroinfiltration of 3–4 leaves from 5 plants is typically sufficient for small-scale experiments. Leave the plants to grow in a glasshouse or growth cabinet at 25 °C for 6–8 days, providing water and 16 h of daylight (artificial if necessary) daily (*see* Fig. 3).
5. Harvest the agroinfiltrated leaves, using a razor blade or scalpel to remove the areas of the leaves that were not agroinfiltrated (*see* **Note 37**). Weigh the remaining leaf material, then use a blender to homogenize the leaf tissue with three volumes of chilled sodium phosphate buffer supplemented with protease inhibitor (e.g., 60 g of leaf tissue would require 180 ml buffer. If using cOmplete EDTA-free protease inhibitor cocktail tablets from Roche, use one tablet for every 50 ml of buffer). Filter the homogenate through a layer of Miracloth, then centrifuge the primary filtrate at 15,000 × *g* for 20 min at 4 °C. Retain the pellet (insoluble fraction) for further analysis if desired, then filter the supernatant through a 0.45 μm syringe filter (*see* **Note 38**).
6. Purification is carried out using a double-layer sucrose cushion. Pour the plant extract into a suitable ultracentrifuge tube, then use a long needle to underlay a volume of 25% sucrose solution approximately equal to 1/6 of the volume of plant extract. Below that, underlay a smaller volume of 70% sucrose solution (1/5–1/10 of the volume of the 25% sucrose solution, *see* **Note 11** for precise volumes). Ultracentrifuge at maximum speed for 2.5–3 h at 4 °C, after which a thick green band is visible at the interface between the sucrose layers. Pierce the bottom of the tube with a needle and allow the sample to drip into a collection tube. The 70% sucrose and the interface fractions combined should include all of the VLPs present in the sample, whereas collection of the 70% fraction only will result in a cleaner sample with a lower yield (*see* Fig. 4). In both cases, dialyze the recovered sample against 20 mM ammonium bicarbonate (pH 8.5) (*see* **Note 39**). Clarify the dialysate by centrifugation at 15,000 ×*g* for 20 min at 4 °C, then filter through a 0.2 μm syringe filter and concentrate to 2 ml in a vacuum concentrator, being careful not to concentrate the sample to dryness (*see* **Notes 40–42**).
7. Layer 2 ml fractions of 20%, 30%, 40%, 50%, and 60% solutions of Nycodenz in a 13 ml ultracentrifuge tube then overlay with the concentrated sample (*see* **Note 43**). Using a TH641 ultracentrifuge rotor, centrifuge at maximum speed

(~274,000 ×*g*) for a minimum of 3 h at 4 °C. VLPs will appear as either a distinct iridescent band or a more diffuse brown-grey band below the sedimented green impurities (*see* Fig. 5) (*see* **Note 44**). Recover the VLP fractions by piercing the bottom of the tube with a needle and collecting 1 ml fractions, or by piercing the side of the tube just below the desired band with a needle and aspirating the band with a syringe. Analyze the fractions by SDS-PAGE or Western blotting (*see* Subheading 4). Fractions containing recombinant protein can be examined for the presence of VLPs by transmission electron microscopy (TEM), or dialyzed against sodium phosphate buffer or PBS (using 100 kDA MWCO Float-a-Lyzer dialysis cassettes from Spectrum Labs) for long term storage in a refrigerator or cold room.

3.4 Characterization of Tandem-HBc VLPs

3.4.1 Removal of Sucrose and Salts by Dialysis of Tandem-HBc VLPs

1. Cut the required length of dialysis membrane and allow it to soak in the dialysis buffer for 5 min before teasing it open and folding one of the ends thrice. Seal the membrane with one dialysis clip at one end, then pipette in the VLP solution through the other (*see* **Note 45**). Press out any bubbles, then seal the distal end by folding it thrice and clipping with a dialysis clip. Connect the floats at either end and dialyze for 1 h at room temperature, using 1.5 l of dialysis buffer per dialysis tube (*see* **Note 46**).
2. Exchange the buffer with new dialysis buffer, then dialyze for 2 h at room temperature. Exchange the buffer with new dialysis buffer, then dialyze for an additional 5–6 h at room temperature (or overnight at 4 °C). Unclip one of the dialysis clips and carefully pipette out the solution, which may then be stored at 4 °C.

3.4.2 Examination of Tandem-HBc VLPs by Transmission Electron Microscopy (TEM)

1. Secure Parafilm on a designated surface, then, in a row, pipette out 10 μl of sample (*see* **Note 47**), 20 μl 20 mM Tris wash buffer, 20 μl uranyl acetate solution, another 20 μl uranyl acetate solution, and 20 μl wash buffer. Using a pair of forceps, incubate a freshly glow-discharged grid on the surface of the sample for 30 s, then dip the grid once in the first droplet of uranyl acetate solution and touch dry on blotting paper (*see* **Note 48**). Dip and incubate the grid for 20 s in the second droplet of the uranyl acetate solution and touch dry on blotting paper, then dip into the second droplet of wash buffer and touch dry on blotting paper. Place the grid on labeled blotting paper to dry, then store the grid in a sample box, recording the number of the alcove where the grid is stored.
2. Examine the grid according to the instructions of the TEM microscope, following the local guidelines.

3.4.3 Immunogold TEM Analysis

1. If necessary, dilute the sample in 10 mM Tris pH 7.5. As controls, incubate the sample with an unrelated primary antibody of the same species or with only ID, followed by incubation with the secondary antibody. Ensure grids are freshly glow-discharged before starting.
2. Place a strip of Parafilm on a smooth surface (*see* **Note 49**), then place an appropriate number (for the number of samples to be labeled) of 20–30 μl droplets of IBB onto the Parafilm. Using the tweezers, carefully take a freshly glow-discharged grid and, using an automatic pipette, load a 5–10 μl droplet of sample onto the grid and incubate for 30–60 s. With a triangle of filter paper in one hand and a minimum of 40 μl of deionized water in a Pasteur pipette in the other, carefully add the water to the grid and wash by touching the filter paper to the side of the droplet, being sure not to touch the grid except at the very edge. As soon as the droplet is removed, place the grid sample side down onto a droplet of IBB and incubate for 45–60 min.
3. Place an appropriate number of 20–30 μl droplets of ID on the Parafilm, then lift the grid and blot as before, by carefully touching the filter paper to the side of the remaining IBB droplet. As soon as the droplet is removed, place the grid sample side down onto a droplet of ID. Place an appropriate number of 20–30 μl droplets of diluted primary antibody on the Parafilm, then lift the grid and blot as before, by carefully touching the filter paper to the side of the remaining ID droplet. As soon as the droplet is removed, place the grid sample side down onto a droplet of primary antibody and incubate for 1.5–2 h.
4. Place an appropriate number of 20–30 μl droplets of ID (5 droplets for each grid) on the Parafilm, then lift the grid and blot as before, by carefully touching the filter paper to the side of the remaining primary antibody droplet. As soon as the droplet is removed, place the grid sample side down onto a droplet of ID. Incubate for 3–5 min, then blot and transfer to the second droplet (*see* **Note 50**). Repeat this process for a total of five washes. Following the final wash, place an appropriate number of 20–30 μl droplets of diluted secondary antibody onto the Parafilm, then lift the grid and blot as before, by carefully touching the filter paper to the side of the remaining ID droplet. As soon as the droplet is removed, place the grid sample side down onto a droplet of secondary antibody and incubate for 45–60 min.
5. Wash the grid as before, in five droplets of ID. Then, with a triangle of filter paper in one hand and a minimum of 40 μl of negative stain in a Pasteur pipette in the other, blot the grid as before, by carefully touching the filter paper to the side of the

remaining ID droplet. As soon as the droplet is removed, place approximately 10 μl of negative stain onto the grid and blot without incubation. Immediately repeat this process, leaving the second negative stain droplet on the grid for 15–20 s, then blot the grid as before, by carefully touching the filter paper to the side of the remaining negative stain droplet. Ensure that any additional stain is blotted from the tweezers before proceeding, then carefully place the grid sample side up on a piece of filter paper and allow to air dry for at least 2 min before transferring the grid to a sample box, recording the number of the alcove where the grid is stored.

6. Examine the grid according to the instructions of the TEM microscope, following the local guidelines.

3.4.4 Immunoadsorbent TEM Assay

1. If necessary, dilute the sample in an 10 mM Tris pH 7.5. As controls, incubate the sample with an unrelated primary antibody of the same species or with only ID. Ensure grids are glow-discharged before beginning.
2. Place a strip of Parafilm on a smooth surface (*see* **Note 49**), then place an appropriate number (for the number of samples to be labeled) of 20–30 μl droplets of IBB onto the Parafilm. Using the tweezers, place the grid sample side down onto one of these droplets and incubate for 45–60 min.
3. Place an appropriate number of 30 μl droplets of ID onto the Parafilm, then carefully blot the grid by touching the filter paper to the side of the IBB droplet, being sure not to touch the grid except at the very edge. As soon as the droplet is removed, briefly place the grid "sample side" down onto a droplet of ID. Place an appropriate number (one per grid) of 25 μl droplets of diluted primary antibody on the Parafilm, then lift the grid and blot as before, by carefully touching the filter paper to the side of the remaining ID droplet. As soon as the droplet is removed, place the grid sample side down onto a droplet of primary antibody and incubate for 30 min.
4. Place an appropriate number of 20–30 μl droplets of ID (5 droplets for each grid) on the Parafilm, then lift the grid and blot as before, by carefully touching the filter paper to the side of the remaining primary antibody droplet. As soon as the droplet is removed, place the grid sample side down onto a droplet of ID. Incubate for 3–5 min, then blot and transfer to the second droplet (*see* **Note 50**). Repeat this process for a total of five washes. Following the final wash, place an appropriate number of 20–30 μl droplets of VLP sample onto the Parafilm, then lift the grid and blot as before, by carefully touching the filter paper to the side of the remaining ID droplet. As soon as the droplet is removed, place the grid sample side down onto a droplet of VLP sample and incubate for 30 min.

5. Wash the grid as before, in five droplets of ID. Then, with a triangle of filter paper in one hand and a minimum of 40 μl of negative stain in a Pasteur pipette in the other, blot the grid as before, by carefully touching the filter paper to the side of the remaining ID droplet. As soon as the droplet is removed, place approximately 10 μl of negative stain onto the grid and blot without incubation. Immediately repeat this process, leaving the second negative stain droplet on the grid for 15–20 s, then blot the grid as before, by carefully touching the filter paper to the side of the remaining negative stain droplet. Ensure that any additional stain is blotted from the tweezers before proceeding, then carefully place the grid sample side up on a piece of filter paper and allow to air dry for at least 2 min before transferring the grid to a sample box, recording the number of the alcove where the grid is stored.
6. Examine the grid according to the instructions of the TEM microscope, following the local guidelines.

3.4.5 Dynamic Light Scattering (DLS)

1. Add a minimum of 1 ml sample into DTS0012 cuvette (*see* Subheading 2.4.5, **items 2** and **3**).
2. Log on to computer and open the Zeta Sizer fbs software (*see* **Note 21**), put in the cuvette, and select the following settings: Measure: "Manual," Temperature: "25 °C" (as appropriate), Sample Name: as appropriate (provide the i.d. used till then in the lab book), Cuvette: "DTS0012" (as appropriate), Measurement Type: "Size," Parameter: "Mark Horowitz," Material: "Protein" (later this can be changed in the software if needed), Dispersant: "PBS."
3. Under Measurement: select "Automatic Measurement," "Duration," set the measurement angle to "173 Back Scatter," the number of measurements to "5" (or "3," as appropriate), and the delay to "0."
4. Under Advanced, select "Default Values."
5. Under Data Processing, select: "General," then "Protein Analysis," then "Repeats."
6. Press the "OK" and the green start button and wait for the machine to collect the data—it will beep upon completion (*see* **Note 51**). (It beeps long and twice if higher temperatures are used.)
7. Ensure that "Number" and "PDI" are enabled under the parameters analyzed, and then save the data (as pdf in H drive).
8. In the "Results" section, export the data as "Number" and copy it into an Excel spreadsheet; transpose for ease of use (*see* **Note 52**). Logout of the computer.

4 Notes

1. Ultracentrifuge buckets ought to be weighed accurately to avoid issues caused by imbalance. Depending on the centrifuge tubes used, it is normally recommended to fill them up almost to the brim, to prevent them collapsing.
2. Always wear appropriate ear protection during sonication.
3. Acrylamide is highly toxic and should be handled with utmost care.
4. The efficacy of the bacteria used is crucial; the Rosetta 2 strain seems to work best.
5. Kanamycin is toxic and must be handled appropriately.
6. DTT is toxic and should be handled with care. DTT solution should be freshly prepared or stored in the dark in a −20 °C freezer. DTT or solutions containing it, should not be autoclaved.
7. The induction needs to be carried out with good quality IPTG; as IPTG is not stable in the fridge, the stock should be frozen.
8. The ammonium sulfate used is a saturated solution; there will be undissolved ammonium sulfate in the stock solution.
9. Reagents used in silver staining and Coomassie Brilliant Blue staining can be harmful to the worker or the environment; appropriate measures need to be taken to protect the operator and to dispose of the waste in accordance to the rules and regulations of the work place.
10. Always use centrifuges and ultracentrifuges appropriately, ensuring rotors are correctly balanced.
11. For the double sucrose cushion, any swing-out (swinging-bucket) ultracentrifuge rotor may be used, depending on the volume of extract to be processed. Two example rotors will be given here:
    (a) With a TH-641 ultracentrifuge spin-out rotor (Thermo Scientific), the ideal tubes are Ultra-Clear 13 ml (14 × 89 mm). The double sucrose cushion is prepared by pouring the plant extract in the tube, then carefully underlaying first with 2 ml of 25% sucrose solution and then 0.25 ml of 70% sucrose solution. The sample is then ultracentrifuged at 40,000 rpm (274,000 × *g*) for 2.5 h at 4 °C. The TH-641 rotor has six buckets that each hold 13 ml tubes, so the maximum volume of leaf extract that can be processed simultaneously is about 66 ml (which corresponds to 22 g of leaf tissue).

(b) The larger Surespin 630/36 spin-out rotor (Thermo Scientific) uses Ultra-Clear 36 ml tubes, 25 × 89 mm. The double sucrose cushion is prepared by pouring the plant extract into the tube, then carefully underlaying first with 5 ml of 25% sucrose, then 1 ml of 70% sucrose and ultracentrifuging at 30,000 rpm (167,000 × *g*) for 3 h at 4 °C. The Surespin 630/36 rotor has six buckets that each hold 36 ml tubes, so the maximum volume of leaf extract that can be processed simultaneously is about 180 ml (which corresponds to 60 g of leaf tissue).

12. Using prepared 500 mM EDTA stock solution is recommended.
13. The dialysis buffer, without DTT, can be prepared, autoclaved and stored at 4 °C.
14. The "sample side," the side to which the sample is applied.
15. Crossover tweezers (Dumont) are highly recommended for the experiment.
16. Uranyl acetate is an alpha emitter and is highly toxic. Local rules would be in place regarding the use and disposal of anything contaminated with uranyl acetate.
17. In the absence of 2% uranyl acetate, 1% uranyl acetate works equally well.
18. Nickel grids are recommended for this procedure.
19. Antibodies should be held on ice and diluted during wash step prior to use.
20. The primary antibody described here binds to the N-terminus of the protein, so any modifications in that region might affect the results.
21. This protocol is written for use with the Malvern Zetasizer Nano series. The protocol will need to be adapted should the user have access only to other DLS machines.
22. Whilst starting the first run, it is recommended to use a serial dilution of the sample to determine the optimal amount needed for accurate measurement. It is also recommended to check the samples first with TEM and silver stain to get a general idea about the purity of the preparation. Should the sample be too impure, then it might be difficult to observe any particles that match the VLP architecture due to the background.
23. This can be scaled up.
24. The initial incubation time of at least an hour seems crucial, given that most of the tandem-HBc plasmids are low copy number plasmids with kanamycin resistance cassettes.

25. Typically, it takes approximately 3 h to reach an OD 600 of 0.6–0.8.
26. Typically, it takes approximately 6–10 cycles until the solution clarifies.
27. Sonication is a crucial step and needs to be carried out on ice. Over-sonication might result in protein aggregation. Depending on the sonicator used, it is advisable to visually check the solution between each round of sonication. The moment the suspension clarifies cell disruption is complete and further sonication is not needed.
28. Antibiotic (Zeocin) is not required after initial selection of transformed *Pichia* clones on agar plates.
29. With regard to schedule: cultures are typically set up on Thursday to provide overnight cultures for priming BMGY expression cultures on Friday. Cells are then ready for induction on Monday.
30. Incubation of cultures at temperatures above 25 °C reduces yields of VLP substantially.
31. Expressions may be carried out for 72–96 h. With a 72 h culture, the cells can be processed and VLPs purified by the end of the week, but yields may be slightly lower than from a 96 h culture.
32. Heating lysates to 60 °C for 30 min prior to purification causes precipitation of host cell proteins, allowing removal by centrifugation. The HBc VLPs form the major component of the clarified, heat-treated lysate.
33. While SEC purification of the HBc VLPs is done routinely, anion exchange chromatography on a DEAE column also yields highly purified VLPs.
34. TEM analysis of size exclusion chromatography fractions may reveal VLPs that are slightly smaller in size (~20 nm diameter) than HBV core. Western blot analysis of these fractions reveals that these assemblies do not contain HBc and are likely endogenous VLPs.
35. Fractions are analyzed routinely by TEM and SDS-PAGE. The fractions containing uniformly assembled VLPs of the correct size as identified by TEM are examined on a stained SDS-PAGE gel. Only those fractions containing HBc protein are pooled and concentrated.
36. Positive clones of *E. coli* and *A. tumefaciens* can be stored as glycerol stocks by supplementing an aliquot of liquid culture with glycerol (to a final concentration of 25% (v/v)) and snap-freezing in liquid nitrogen prior to storage at −80 °C.

37. Take care when cutting leaf tissue with a razor blade or scalpel.
38. Note that large volumes may cause the filters to become clogged with large impurities found in the homogenate, so numerous filters will need to be used for large volumes. Also, syringe filters equipped with prefilters (such as Sartorius Minisart NML plus 0.45 μm syringe filters with glass-fiber prefilter) will allow larger volumes to be processed with a single filter.
39. Due to the high osmotic pressure of the sucrose, the volume of the dialysate may increase twofold to threefold.
40. During vacuum concentration, the ammonium bicarbonate will decompose to volatile compounds as the water evaporates, meaning that the buffer will not be significantly concentrated during vacuum evaporation.
41. The sample will spontaneously remain cold while evaporation is taking place, but will heat up rapidly after the end of concentration due to the ambient temperature in the vacuum concentrator. To avoid heat shock, the sample should be placed on ice immediately after concentration.
42. Because impurities will concentrate along with the sample, short (10 min) centrifugation in a microcentrifuge is recommended halfway through concentration to pellet some of the impurities, particularly if the sample is being concentrated more than fivefold.
43. As an alternative to manually layering fractions to form the Nycodenz gradient, the extract can be mixed with a solution of 40% Nycodenz and ultracentrifuged for 24 h: the gradient will form spontaneously. As Nycodenz forms a density gradient during centrifugation, the components within the sample equilibrate at their isopycnic points and remain there irrespective of duration of centrifugation. As such, the sample cannot be centrifuged for too long. The method described in Subheading 3.3 is intended to be time-saving, allowing the gradient to form after only 3 h. However longer centrifugation (16–24 h) may yield better separation of particles from impurities.
44. The position of the particles in the gradient depends on their buoyant density, which is heavily influenced by nucleic acid content of the particles. Heterogeneity in nucleic acid content will result in a diffuse or multiple bands. A high concentration of tandem-core particles will result in a visible band (or multiple bands) in the Nycodenz gradient, which can easily be recovered by piercing the side of the tube and aspirating the band with a needle and syringe. If concentration is low, bands

may not be visible and the entire gradient may need to be fractionated in order to identify which fraction contains the particles.

45. Care should be taken not to fill the entire dialysis tube lest the tube splits during the process.
46. A magnetic stirrer, stirring at low rpm, is recommended for the dialysis.
47. The VLPs should be dialyzed prior to TEM analysis, lest the sucrose interfere with the staining. Should the VLPs be resuspended in other buffers (especially those containing phosphates), they should be dialyzed into the recommended buffer lest the UA precipitate.
48. The time of the initial incubation is crucial, should the incubation be prolonged, the background would be greater.
49. For example the upturned lid of a sandwich box.
50. It is recommended to work in rows when working with multiple grids, ensuring that each grid has sufficient time in ID.
51. Eye protection must be worn whilst using the apparatus.
52. The measurements themselves are saved as "number" because the "intensity" (default measurement) would be proportional to the size of the particles.

References

1. Pumpens P, Grens E (2001) HBV core particles as a carrier for B cell/T cell epitopes. Intervirology 44:98–114
2. Peyret H, Gehin A, Thuenemann EC, Blond D, El Turabi A, Beales L, Clarke D, Gilbert RJ, Fry EE, Stuart DI, Holmes K, Stonehouse NJ, Whelan M, Rosenberg W, Lomonossoff GP, Rowlands DJ (2015) Tandem fusion of hepatitis B core antigen allows assembly of virus-like particles in bacteria and plants with enhanced capacity to accommodate foreign proteins. PLoS One 10:e0120751
3. Sainsbury F, Thuenemann EC, Lomonossoff GP (2009) pEAQ: versatile expression vectors for easy and quick transient expression of heterologous proteins in plants. Plant Biotechnol J 7:682–693
4. Peyret H, Lomonossoff GP (2013) The pEAQ vector series: the easy and quick way to produce recombinant proteins in plants. Plant Mol Biol 83(1–2):51–58
5. Peyret H (2015) A protocol for the gentle purification of virus-like particles produced in plants. J Virol Methods 225:59–63
6. Stonehouse NJ, Stockley PG (1993) Effects of amino acid substitution on the thermal stability of MS2 capsids lacking genomic RNA. FEBS Lett 334:355–359
7. Liljeroos L, Huiskonen JT, Ora A, Susi P, Butcher SJ (2011) Electron cryotomography of measles virus reveals how matrix protein coats the ribonucleocapsid within intact virions. Proc Natl Acad Sci U S A 108:18085–18090
8. Lua LH, Connors NK, Sainsbury F, Chuan YP, Wibowo N, Middelberg AP (2014) Bioengineering virus-like particles as vaccines. Biotechnol Bioeng 111:425–440

Chapter 8

Production and Application of Insect Virus-Based VLPs

Radhika Gopal and Anette Schneemann

Abstract

Virus-like particles (VLPs) are self-assembling platforms composed of viral structural proteins. They are used for a variety of purposes, ranging from the study of virus assembly to vaccine development. VLPs can be produced in plants, bacteria, yeast, and insect and mammalian cells. The baculovirus expression system is one of the most commonly used systems for production of VLPs in eukaryotic cells. This chapter provides a brief overview of the main strategies used to generate recombinant baculoviruses and the applications of insect virus-derived VLPs in basic and applied research. It then describes detailed protocols for generation of recombinant baculoviruses, screening for their expression of VLPs in insect cells, and VLP purification.

Key words Virus-like particle (VLP), Baculovirus, Insect virus, Insect cells, Assembly, Vaccines

1 Introduction

Virus-like particles (VLPs) are noninfectious and nonreplicating particles that are morphologically similar to infectious viruses. They are devoid of viral genetic material and are formed by structural proteins that have an inherent capacity for self-assembly. VLPs can be produced in a variety of cellular systems including those from bacteria, yeast, insects, plants, and mammals using viral vectors. Since their introduction in the 1980s, baculovirus vectors have become one of the most widely used viral gene delivery systems. This is primarily due to several advantages of this system including high levels of protein expression, safety, scalability and eukaryotic posttranslational modifications. The first VLP to be produced using a baculovirus vector was the Gag VLP of feline immunodeficiency virus (FIV) which was used to analyze molecular determinants of FIV particle assembly [1]. Since then, the baculovirus expression vector system (BEVS) has been used to produce VLPs from different viruses to study viral assembly and genome packaging, to produce VLP-based antigens for immunization and for diagnostic assays.

Christina Wege and George P. Lomonossoff (eds.), *Virus-Derived Nanoparticles for Advanced Technologies: Methods and Protocols*, Methods in Molecular Biology, vol. 1776, https://doi.org/10.1007/978-1-4939-7808-3_8,

© Springer Science+Business Media, LLC, part of Springer Nature 2018

Two main approaches are used to generate recombinant baculoviruses. The first involves homologous recombination between a linearized baculovirus genome and a baculovirus transfer vector containing the gene for the protein of interest. This gene is placed under control of a strong baculovirus promoter, most commonly the polyhedrin promoter. To enhance production of the desired recombinant viral progeny, the linearized baculovirus genome carries a deletion in a gene essential for replication, and the transfer vector contains sequences necessary to repair this gene. Upon cotransfection of transfer vector and linearized baculovirus DNA into insect cells, homologous recombination between the two transfers the gene of interest into the viral genome and reactivates the essential gene. This gives rise to replication-competent, recombinant viruses at an efficiency of >90% [2]. Individual viral clones are subsequently isolated by plaque assay from cell supernatants of the transfected cells.

The second strategy is represented by the Bac-to-Bac system, which involves a donor plasmid, an *E. coli* host strain containing a cloned baculoviral genome, the bacmid, and a helper plasmid [3]. The gene of interest is cloned into the donor plasmid under a baculovirus-specific promoter and the plasmid is transformed into the *E. coli* host strain. A recombinant bacmid is generated by site-specific transposition of the gene of interest from the donor plasmid into the bacmid assisted by proteins supplied by the helper plasmid. *E. coli* colonies containing the recombinant bacmid are identified by screening for antibiotic resistance and loss of the lacZ marker. Recombinant baculovirus DNA is then isolated from positive clones and transfected into insect cells. This method eliminates the need for plaque purification since the recombinant viral DNA is not mixed with parental, nonrecombinant virus.

For commercial or clinical applications, heterologous proteins and VLPs are usually expressed from viruses that have been produced by the homologous recombination approach [4–7]. This is due to relative instability of bacmid-derived vectors in insect cells and the presence of bacterial sequences and antibiotic selection markers in the viral DNA [8].

The BEVS has seen several recent developments that further facilitate construction and isolation of recombinant viruses and which expand their utility. The *flash*BAC method combines homologous recombination with a bacmid-based system which eliminates the need for plaque purification and allows implementation of a high-throughput, automated process [9]. The MultiBAC system allows synthesis of multisubunit protein complexes using a single baculovirus vector and has been used to produce human papilloma virus VLPs with a 40-fold improvement in expression [10]. Multilocus baculovirus vectors that enable insertion of foreign genes at several loci within the baculovirus genome [11] have been used successfully to produce VLPs of influenza virus and bluetongue virus [12].

VLPs of insect viruses have been used to study many aspects of viral structure and assembly. The effect of coat protein maturation and the role of specific coat protein regions on particle morphology were studied using VLPs of the nodavirus Flock House virus (FHV) [13, 14]. The role of coat protein on the structural organization of encapsidated RNA was elucidated using VLPs of FHV and the nodavirus Pariacoto virus [15, 16]. FHV VLPs were also important in the investigation of the specific mechanism by which the bipartite nodaviral RNA genome is selected for encapsidation in native FHV particles [17–19]. Capsid protein folding, virus assembly and pH-dependent kinetics of capsid protein cleavage were studied in Tetraviruses (NωV, TaV, and HaSV) using VLPs expressed in insect and yeast cells [20–22]. VLPs of Densovirus [23], Cypovirus [24], and Triatoma virus [25] were used to identify the viral structural proteins required for capsid assembly. Insect virus-based VLPs have also been used to develop vaccine nanoparticles. Chimeric FHV VLPs displaying peptide fragments of HIV-1 on the capsid protein were used to induce neutralizing antibodies against HIV-1 [26]. In another approach, the entire extracellular domain of the anthrax toxin receptor was inserted into the FHV coat protein and displayed in 180 copies on the VLP surface [27]. The resulting particles have dual function as an anthrax antitoxin and vaccine. Display of fragments from the influenza virus HA stem region on FHV VLPs resulted in the induction of cross-reactive anti-HA antibodies [28].

The two most commonly used insect cell lines for the production of insect virus VLPs are *Spodoptera frugiperda* (*Sf9* or *Sf21*) and *Trichoplusia ni* (*T. ni* or High Five) cells. *T. ni* cells produce higher amounts of recombinant protein and may be more suitable for synthesis of some types of VLPs than *Sf9* or *Sf21* cells. In general, it is important to assess VLP synthesis in different insect cells and compare expression levels before scaling up VLP production.

2 Materials

Materials itemized below are required for generating recombinant baculoviruses by homologous recombination between a transfer vector and linearized viral DNA. Specific suppliers are listed for many items, but alternate sources can be used in most cases.

2.1 Preparation of Baculovirus Transfer Vector

1. cDNA of the target gene encoding the protein of interest.
2. Suitable restriction enzymes and buffers (NEB).
3. Baculovirus transfer vector pBacPAK8 or 9 (Clontech; both encode amp^{r}).

4. 20× sodium borate buffer (stock solution): 38.13 g $Na_2B_4O_7 \cdot 10{\cdot}H_2O$ and 25 g boric acid in 1 L H_2O; pH adjustment not necessary.
5. 1% (w/v) agarose gel.
6. QiaexII gel extraction kit (Qiagen).
7. T4 DNA ligase (400 units/μL) and 10× ligase buffer (NEB).
8. DH5α competent cells (Invitrogen-Thermo Fisher Scientific).
9. LB medium: 10 g/L Bacto tryptone, 5 g/L yeast extract, 10 g/L NaCl. Adjust pH to 7.0.
10. LB agar plates containing appropriate antibiotic: add 15 g/L agar.
11. WizardPlus SV Minipreps DNA Purification System (Promega).

2.2 Generation of Recombinant Baculovirus

1. Bsu36I-linearized BD Baculogold Baculovirus DNA (BD Biosciences).
2. Transfectin (Bio-Rad).
3. Bac1 and Bac2 primers (Clontech) (*see* **Note 1**).
4. Polystyrene tubes 12 × 75 mm.

2.3 Cell Culture Reagents and Consumables (Sterile)

1. *Sf21* cells (*Spodoptera frugiperda*).
2. *T. ni* cells (*Trichoplusia ni*).
3. TC-100 medium for *Sf21* cells (Thermo Fisher).
4. Complete TC100 medium: TC100 medium supplemented with 10% (v/v) heat-inactivated fetal bovine serum, 100 units/mL penicillin, and 100 μg/mL streptomycin.
5. ESF921 medium for *T. ni* cells (Expression Systems) (*see* **Note 2**).
6. Fetal bovine serum.
7. Trypan Blue Solution 0.4% (w/v).
8. 35, 60, and 100 mm tissue culture dishes.
9. T175 tissue culture flasks (175 cm^2).
10. 15 and 50 mL conical tubes.
11. 2.8 L Fernbach flask.
12. 250 mL screw-cap Nalgene wide-mouth bottles.
13. SeaKEM ME Agarose (Lonza).
14. 2× Grace's Insect Medium (Gibco, Thermo Fisher Scientific).
15. 100× penicillin–streptomycin solution (Gibco, Thermo Fisher Scientific).
16. 3 mg/mL Thiazolyl Blue Tetrazolium Bromide (MTT) in H_2O.

2.4 Purification of VLPs

1. A buffer in which VLPs are stable. For example, 50 mM HEPES (4-(2-hydroxyethyl)-1-piperazineethanesulfonic acid), pH 7.0 for nodaviruses, used as an example here.
2. 10% (v/v) Nonidet-P40 (NP-40) in H_2O.
3. Sucrose cushion: 30% (w/w) in 50 mM HEPES pH 7.0 (or use other appropriate buffer).
4. 1 mL syringe with 25 or 27 G needle.
5. Beckman SW41 ultracentrifuge tubes.
6. Beckman 50.2 Ti ultracentrifuge tubes.
7. Beckman SW28 ultracentrifuge tubes.
8. Continuous sucrose gradients: 10–40% (w/w) sucrose in 50 mM HEPES pH 7.0 (or other appropriate buffer).
9. Protein assay kit, e.g., BCA assay, to determine concentration.
10. Denaturing polyacrylamide gel to evaluate protein purity, e.g., 12% NuPAGE Bis-Tris precast mini gel (Thermo Fisher).
11. Coomassie-based dye to stain protein gel, e.g., Simply Blue™ Safe Stain (Thermo Fisher).
12. Negative stain transmission electron microscopy is recommended to confirm presence of VLPs.

3 Methods

3.1 Construction of a Recombinant Baculovirus Transfer Vector

1. Inspect the multiple cloning sites of baculovirus transfer vector pBacPAK8 and/or 9 as well as the sequence of the target gene and select a suitable pair of restriction sites for subcloning your gene of interest (*see* **Notes 3** and **4**).
2. Excise the target gene from its parent plasmid using the corresponding restriction enzymes (according to manufacturer's instructions). Alternatively, if the necessary restriction sites are not suitably available in the parent plasmid, prepare a DNA copy of the target gene by PCR using primers containing these sites.
3. Digest the baculovirus transfer vector pBacPAK8 or 9 with the same pair of restriction enzymes.
4. Incubate both digests overnight at 37 °C to ensure complete digestion.
5. Run the entire reactions through a 1% (w/v) agarose gel in 1× sodium borate buffer and excise the digested transfer vector DNA and the digested target gene DNA.
6. Purify the DNA using the QiaexII Gel Extraction kit following protocols provided by the manufacturer.

7. Set up a ligation reaction (10 μL total volume) using 10× T4 DNA ligase buffer (1 μL), T4 DNA ligase (0.5 μL), digested transfer vector, and target gene.
8. Incubate the ligation reaction for at least 3 h at room temperature if it is a sticky end ligation or overnight if it is a blunt end ligation.
9. Transform DH5α competent cells with 5 μL of the ligation reaction mixture following protocols provided by the manufacturer.
10. Spread the recommended volume (20–100 μL) of transformed cells on LB plates containing 100 μg/mL ampicillin.
11. Incubate plates overnight at 37 °C.
12. Inoculate bacteria from single colonies into several tubes containing 4–5 mL LB medium with 100 μg/mL ampicillin.
13. Grow bacteria overnight by shaking the cultures at 225 rpm at 37 °C.
14. Isolate plasmid DNA from the cultures using a plasmid miniprep kit (follow manufacturer's instructions).
15. Screen several independent plasmid DNA isolates for the presence of the target gene by either PCR amplification with primers Bac1 and Bac2, or by performing a restriction digest with the enzymes from **step 1**.
16. Confirm that the gene is present and free of undesired mutations by having it sequenced with Bac1 and Bac2 primers (*see* **Note 5**).

3.2 Generation and Isolation of a Recombinant Baculovirus Expressing the Target Gene

Note: Unless otherwise noted, growth media should be preheated to 21–27 °C.

1. Add 1 × 10^6 *Sf21* cells in 1.5 mL complete TC100 medium to a 35 mm tissue culture dish.
2. Allow the cells to attach and form a monolayer by incubating them at 27 °C for 1 h.
3. Dilute the pBacPAK transfer vector containing the target gene with sterile water to a concentration of 100 ng/μL.
4. After cells have attached, prepare the transfection mixture (*see* Table 1) in a polystyrene tube (*see* **Note 6**).
5. Mix by gently tapping the tube (*see* **Note 7**).
6. Incubate at room temperature for 15 min.
7. Meanwhile, carefully remove medium from the cells and add 1.5 mL of serum-free, antibiotic-free TC-100 medium.
8. Swirl the medium gently across the monolayer, remove it and replace with 1.5 mL of fresh, serum-free, antibiotic-free TC-100 medium (*see* **Note 8**).
9. Add the transfection mixture dropwise to the cells.

10. Incubate the cells at 27 °C in a humidified incubator for 5 h.
11. Add 1.5 mL of complete TC100 to the cells (the total volume is now 3 mL) and continue incubation at 27 °C in a humidified incubator for 3–5 days.
12. Transfer the cell supernatant to a sterile 15 mL conical tube.
13. Centrifuge at 1500 × *g* for 5 min to pellet any remaining cells or cellular debris.
14. Transfer the supernatant to a fresh 15 mL tube and wrap the tube with aluminum foil to protect the transfection supernatant containing the recombinant viruses from light.
15. Store the tube at 4 °C. This stock is stable for several months with minimal loss of infectivity.

3.3 Plaque Assay to Isolate Individual Recombinant Baculovirus Clones

Viruses released into the supernatant of cotransfected *Sf21* cells represent a mixture of the desired recombinants and progeny that do not express the protein or VLP of interest. Although the fraction of the latter is small, the composition of the mixture will change with repeated passage, resulting in unpredictable expression levels of the heterologous protein or VLP. A clonal stock of the desired recombinant virus should therefore be produced to ensure consistency in protein or VLP production. A clonal isolate of a recombinant virus can be obtained by performing a plaque assay. This assay is designed such that each plaque is the result of a cell being infected by a single virus. Four serial dilutions of the cotransfection supernatant are prepared and each dilution is assayed in triplicate.

1. Set a water bath to 42 °C.
2. For each dilution to be tested, add 1.5×10^6 *Sf21* cells in 5 mL of complete TC-100 medium to three 60 mm tissue culture dishes. For example, if using four dilutions, prepare 12 dishes of *Sf21* cells.
3. Allow cells to attach by incubating the dishes at 27 °C for 1 h.

Table 1
Transfection mixture

	Volume (in μL)
pBacPAK Transfer vector (100 ng/μL)	3
Bsu36I-linearized baculovirus DNA	3
Transfectin	3
Sterile water	63
Total	72

4. Meanwhile, prepare serial tenfold dilutions of the cotransfection supernatant in complete TC-100 medium. Specifically, mix 1.8 mL complete TC100 medium with 0.2 mL cotransfection supernatant and designate this dilution as 10^{-1}. Dilute this thrice serially in the same manner to create 10^{-2}, 10^{-3}, and 10^{-4} dilutions.
5. Inspect culture dishes to ensure that the cells have formed an even monolayer.
6. Remove medium from the dishes by aspiration with a sterile pipette.
7. Gently add 0.5 mL of each dilution dropwise to three dishes.
8. Incubate dishes at room temperature for 1 h on a rocker platform at low setting to allow the viruses to infect the cells.
9. During this incubation, prepare the agarose overlay.
   - Using 5 mL/dish, determine the total volume required plus 20 mL extra (e.g., 80 mL for 12 dishes).
   - In a sterile 100 mL flask, mix 40 mL of 2× Grace's insect medium and 8 mL of FBS and prewarm the mixture to 42 °C in a water bath.
   - In another sterile 100 mL flask, prepare 32 mL of 1.5% (w/v) low melting point agarose in sterile water using a microwave oven. Cool the melted agarose to 42 °C by placing it in the water bath.
   - When the contents of both flasks has equilibrated to 42 °C, transfer Grace's insect medium and FBS to the flask containing the agarose, mix well and return the flask to the water bath. Maintain sterility and work quickly so that the agarose does not solidify. Immediately before the agarose overlay is applied to the cells, add 0.8 mL of room temperature 100× penicillin and streptomycin solution to the flask and mix well.
10. Remove virus inoculum from the cells by tilting the dish and aspirating from the edge.
11. Gently add 5 mL of the agarose overlay to each dish. Let the agarose run down the side of the dish, taking care not to disturb the cell lawn (*see* **Note 9**).
12. Allow agarose to set for at least 10 min at room temperature.
13. Incubate dishes at 27 °C in a humidified incubator for 5 days.

3.4 Staining of Cells to Visualize Plaques

1. Add 0.5 mL of complete TC-100 medium to the center of each dish, followed by 0.5 mL of sterile MTT solution (*see* **Note 10**).
2. Incubate the dishes for several hours or overnight at 27 °C for plaques to become visible. Plaques appear as holes against a deep purple cell lawn and are best viewed on a light box (*see* **Note 11**).

3.5 Isolation of Individual Recombinants

1. Choose dishes containing plaques that are well isolated to ensure harvest of a clonal isolate.
2. With the aid of a light box, mark 5–10 plaques with a black marker pen on the bottom side of the dish.
3. For each plaque to be picked, place 1 mL of complete TC100 medium in a sterile tube with cap.
4. Carefully remove most of the liquid from top of the agarose layer with a pipet by tilting the dish very slightly.
5. Push the tip of a sterile Pasteur pipette through the agarose overlay into the plaque and gently suck a plug of agarose into the tip of the pipette.
6. Transfer the agarose plug to 1 mL of complete TC100 medium.
7. Cover the tubes containing the viral isolates with aluminum foil and store them at 4 °C overnight to allow viruses to diffuse out of the agarose. These samples are referred to as the "plaque-picks." The amount of virus in plaque picks is miniscule and needs to be amplified at least twice in order to obtain a large enough stock that can be used to produce the protein or VLP of interest.

3.6 Generation of Pass 1 Virus Stock

1. For each plaque pick to be amplified seed one 100 mm tissue culture dish with 2.5×10^6 *Sf21* cells in 5–10 mL complete TC-100 medium. Incubate at 27 °C for 1 h to let cells attach (*see* **Note 12**).
2. Remove medium from cells by aspiration with a sterile pipette.
3. To each dish gently add an entire 1 mL plaque-pick. Incubate dishes at room temperature on a rocker at low setting for 1 h.
4. Add 5 mL of complete TC100 medium to each dish and incubate them at 27 °C for 4–5 days until cells show cytopathic effects. Infected cells will appear grainy and many will lift off the dish; changes in cell morphology, such as a rough cell border, may be observed.
5. Collect cells and medium from each dish into a separate 15 mL conical tube.
6. Centrifuge tubes at 1500 × *g* for 5 min to pellet cells and debris.
7. Transfer the supernatant to a fresh 15 mL tube and label this as Pass1 virus stock.
8. Wrap tubes with aluminum foil to protect virus stock from light and store at 4 °C.
9. Freeze cell pellets at −20 °C pending further analysis. The cell pellets can be used to obtain preliminary evidence that the Pass1 recombinant virus expresses the protein of interest in the form of VLPs.

3.7 Confirming Expression of the Target Protein and Testing for the Presence of VLPs

Each amplified recombinant baculovirus isolate must be tested for expression of the target protein at this point. Moreover, it is important to obtain preliminary evidence that the protein is present in the form of VLPs. For VLPs that assemble intracellularly and do not contain lipids, this can be tested in a straightforward manner. Clarified cell lysates are subjected to ultracentrifugation and the resulting pellet is evaluated for the presence of the target protein. For VLPs that are excreted into the medium, the cell supernatant would have to be subjected to ultracentrifugation. The method described below has been used in our laboratory to test for the presence of nodavirus and tetravirus VLPs expressed from recombinant baculovirus vectors and can be adapted to VLPs of viruses that have similar structure and size.

1. Resuspend each cell pellet from Subheading 3.6, **step 9** in 0.9 mL of 50 mM HEPES pH 7.0.
2. Lyse the cells by adding 0.1 mL of 10% (v/v) NP40 and incubate on ice for 10 min.
3. Pellet nuclei and other cell debris at highest speed in a microcentrifuge for 10 min at 4 °C.
4. Transfer the supernatant to 5 mL Beckman SW55 ultracentrifuge tubes and underlay with 0.5 mL of 30% (w/w) sucrose in 50 mM HEPES pH 7.0.
5. Fill the tubes to the top with 50 mM HEPES pH 7.0 and centrifuge at 45,000 rpm (244,418 × *g*) in a Beckman SW55 rotor for 45 min at 11 °C.
6. Drain the tubes and resuspend the very small pellet in 50 μL 50 mM HEPES pH 7.0.
7. Run 5–10 μL of each resuspended pellet through a reducing and denaturing protein gel, e.g., a 12% NuPAGE Bis Tris precast mini gel (Thermo Fisher) and stain it with a Coomassie dye-based stain such as Simply Blue™ Safe Stain (Thermo Fisher); the protein of interest should be present as a prominent band (*see* Fig. 1).
8. If the target protein cannot be clearly identified on a stained gel, perform immunoblot analysis with appropriate antibodies to verify its presence.
9. Select one or two isolates that are positive for expression of the protein of interest/VLP for further amplification.

3.8 Preparation of Pass 2 Virus Stock

1. Transfer 15 × 10^6 *Sf21* cells in 15 mL complete TC100 medium to a 162–175 cm^2 screw-capped tissue culture flask.
2. Let the cells attach for at least 30 min. Carefully remove medium and add 15 mL of fresh, complete TC100.

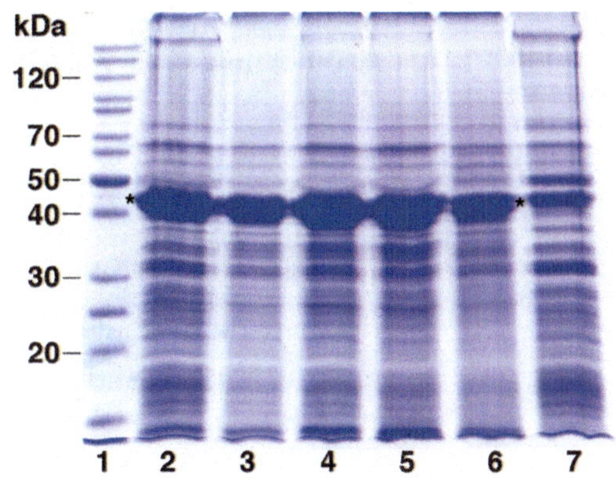

Fig. 1 Protein gel showing presence of FHV coat protein in the pellet after ultracentrifugation of a clarified lysate derived from cells infected with recombinant baculovirus plaque pick. *Sf21* cells infected with six different baculovirus plaque picks were processed as described in Subheading 3.7 and 10 μL of the resuspended pellet was electrophoresed through a reducing and denaturing protein gel followed by staining with Coomassie Brilliant Blue. Lane 1 contains molecular weight markers. Samples in lanes 2–6 were considered positive based on the presence of a prominent band representing FHV coat protein (asterisks). The respective recombinant baculoviruses were further amplified and later shown by transmission electron microscopy to have generated FHV VLPs. The sample shown in lane 7 was not further evaluated

3. Next, add 250 μL of Pass 1 virus stock and place the flask in a 27 °C incubator.
4. After 1 h add another 15 mL of complete TC100 and place back in the incubator.
5. Continue incubation at 27 °C for 4–5 days until the cells show signs of lysis.
6. Transfer the cells and medium (approx. 30 mL) to a 50 mL sterile conical tube.
7. Centrifuge for 5 min at 1500 × *g*.
8. Transfer supernatant to a new sterile 50 mL tube, wrap in aluminum foil, and store at 4 °C. This serves as the inoculum for both small-scale and large-scale production of VLPs (*see* **Notes 13–15**).

3.9 Small-Scale Production of VLPs from Sf21 Cells

The protocol described below is used in our laboratory for purification of nodavirus VLPs, which are $T = 3$ icosahedral VLPs that have a sedimentation coefficient of 135–145 S and a density of approximately 1.3 g/cm^3. For other types of particles, centrifugation conditions may have to be slightly adjusted.

1. Plate 8×10^6 *Sf21* cells in 5–10 mL complete TC100 medium in 8–10 100 mm tissue culture dishes.

2. Infect cells in each dish as described in Subheading 3.6 using as inoculum 0.25 mL of Pass 2 virus in 0.75 mL complete TC100 medium for each dish.
3. VLPs are purified from the culture on day 3–5, when signs of cell lysis are apparent.

   At this point, combine cells and medium from all dishes and add NP-40 to a final concentration of 1% (v/v).
4. Incubate on ice for 10–15 min with occasional agitation.
5. Pellet cell debris and nuclei by low speed centrifugation, e.g., at 10,000 rpm (13,739 × *g*) for 10 min at 4 °C in a Beckman JA17 rotor.
6. Transfer the supernatant to several 11 mL Beckman SW41 ultracentrifuge tubes and underlay with a 1 mL of a 30% (w/w) sucrose cushion in 50 mM HEPES pH 7.0 (or other appropriate buffer).
7. Pellet the VLPs in a Beckman SW41 rotor at 40,000 rpm (273,620 × *g*) for 2.5 h at 11 °C.
8. Drain tubes and resuspend the pellets in 0.5 mL 50 mM HEPES pH 7.0 (or use other appropriate buffer). Pellets usually need to be resuspended by drawing them repeatedly through a 25 or 27 G needle attached to a 1 mL syringe (avoid making bubbles).
9. Remove any insoluble material by centrifugation at highest speed (approx. 16,000 × *g*) in a microcentrifuge for 10 min at 4 °C.
10. Layer the resuspended, clarified pellets on several 10–40% (w/w) continuous sucrose gradients in 50 mM HEPES pH 7.0 (or other appropriate buffer) using Beckman SW41 tubes. Do not load more than 0.6 mL sample per tube.
11. Centrifuge at 40,000 rpm (273,620 × *g*) in a Beckman SW41 rotor for 1.5 h at 11 °C.
12. Following centrifugation, locate the viral band by shining a bright light through the gradient in a darkened room. The band should be white with a bluish hue, located approximately halfway down the tube.
13. Remove the material by puncturing the tube below the band with a 25 G needle attached to a syringe (*see* **Note 16**).
14. Combine fractions containing virus from all gradients. Determine concentration by protein assay and evaluate purity on a protein gel. Image an aliquot of the sample by transmission electron microscopy using negative staining to confirm the presence of VLPs and overall integrity of the particles.

3.10 Large-Scale Production of VLPs from T. ni Cells

Suspension cultures are more practical for large-scale production of VLPs than monolayer cultures. Both *Sf21* cells and *T. ni* cells can be grown in suspension. *T. ni* cells are known to enhance yield dramatically compared to *Sf21* cells [29] and have been used with good success in our laboratory. It is recommended to compare VLP yield and quality from suspension cultures of *Sf21* cells and *T. ni* cells to determine which cell type is best suited for scaling up production of a given type of VLP.

1. Prepare a 1 L suspension culture of *T. ni* cells in ESF921 medium at a density of 2×10^6 cells/mL. Use a 2.8 L Fernbach flask and grow cells on an orbital shaker at approx. 100 rpm at 27 °C.
2. Infect cells with 30 mL of Pass 2 virus stock.
3. Continue incubation of the cells at 27 °C and determine extent of infection daily by Trypan Blue dye exclusion test starting on day 3 post infection. To this end, transfer a small sample (e.g., 50 μL) from the culture to a 1.5 mL microfuge tube and mix with an equal volume of 0.4% (w/v) Trypan Blue solution. View the cells under a light microscope. Infected/dead cells will take up the stain while living cells will not.
4. The culture is typically harvested on day 3 when at least 50% of the cells are dead but not lysed.
5. Distribute the 1 L *T. ni* culture into six 250 mL screw-cap Nalgene wide-mouth bottles and pellet cells at 3840 × *g*, for 10 min at 4 °C (e.g., 5000 rpm in a Beckman JA-14 rotor).
6. Discard the supernatant and resuspend each cell pellet in 40 mL 50 mM HEPES pH 7.0 (or other appropriate buffer).
7. Add NP40 to a final concentration of 1% (v/v).
8. Incubate on ice for 10 min.
9. Pellet cell debris in 250 mL screw-cap Nalgene wide-mouth bottles at 15,300 × *g* for 10 min at 4 °C (e.g., 10,000 rpm in a Beckman JA-14 rotor).
10. Transfer the supernatant containing the VLPs to Beckman 50.2 Ti ultracentrifuge tubes and underlay with 3 mL of 30% (w/w) sucrose in 50 mM HEPES pH 7.0 buffer (or use other appropriate buffer).
11. Centrifuge at 45,000 rpm (244,061 × *g*) in a Beckman 50.2 Ti rotor for 2.5 h at 11 °C.
12. Drain tubes and add 1 mL of 50 mM HEPES pH 7.0 to each pellet.
13. Resuspend the pellets by drawing them repeatedly through a 25 or 27 G needle attached to a 1 mL syringe (avoid making bubbles). Transfer the sample to several 1.5 mL microfuge

tubes and remove insoluble debris by centrifugation for 10 min at the highest speed in a microcentrifuge at 4 °C.

14. Prepare several 10–40% (w/w) continuous sucrose gradients in 50 mM HEPES pH 7.0 (or other appropriate buffer) in Beckman SW28 tubes. Apply no more than 3 mL of the clarified, resuspended VLPs from the previous step.
15. Centrifuge at 28,000 rpm (140,992 × *g*) in a Beckman SW28 rotor at 11 °C for 3 h.
16. After the run, locate the viral band by shining a bright light through the gradient in a darkened room. The band should be white with a bluish hue located approximately halfway down the tube. Remove the material by puncturing the tube below the band with an 18 G needle attached to a 3 mL syringe.
17. Combine virus-containing fractions from all gradients. Determine concentration by protein assay and evaluate purity on a protein gel.

4 Notes

1. It may be more economic to have Bac1 and Bac2 primers custom-synthesized. Also, many companies that provide sequencing services make them available as free universal primers.
2. ESF921 is a serum-free medium that should be supplemented with 100 units/mL penicillin and 100 μg/mL streptomycin.
3. For pBacPAK8 and 9, the target gene must have its own ATG initiation codon and Stop codon. Transcription of the inserted gene is terminated by the polyhedrin polyadenylation signal.
4. For synthesis of VLPs, the target gene must encode a viral structural protein known or expected to assemble in a heterologous expression system.
5. If the gene to be sequenced is very long, an additional primer internal to the target gene may be necessary.
6. The volumes listed in the table are smaller than what is recommended by the manufacturer of Bsu36I-linearized BD Baculogold Baculovirus DNA (BD Biosciences). We have found that these amounts are sufficient for efficient generation of recombinant baculoviruses and conserve this rather expensive reagent.
7. Do not mix by pipetting as this will shear the very long baculovirus DNA (130 kbp).
8. A component in serum inhibits transfection; therefore two washes are optimal to replace complete medium with serum-

free medium before adding the transfection mixture to the cells.

9. Keep the flask containing the agarose overlay in a secondary container with warm water to keep it from solidifying before it is added to all dishes.
10. Most protocols call for staining of the cells with neutral red. We find that plaques are more readily visible against the deep purple color of the cell monolayer after staining with MTT. MTT is toxic and gloves should be worn when handling this reagent.
11. Handle the dishes very gently and keep them level as much as possible as cells in the monolayer may otherwise shift and obscure the plaques.
12. It is recommended to amplify at least five plaque picks, thus a minimum of five dishes of *Sf21* cells should be prepared.
13. Viral stocks are light-sensitive and should be kept in the dark.
14. The virus can be further amplified to a Pass 3 stock using the same procedure. Continuous passage is not recommended, however, as baculovirus mutants and defective interfering particles will accumulate in the stock over time and significantly reduce efficiency of target protein expression.
15. Plaque assays can be performed on the amplified virus stocks to determine their titer. We do not routinely do this since we have found that the titer is usually in the range of 10^7–10^8 pfu/mL.
16. If a band is not visible, fractionate the gradient into 0.5 mL fractions, either manually or with a density gradient fractionation system, and test for the presence of VLPs by running an aliquot of the fractions on a protein gel.

Acknowledgment

This work was supported by NIH grant AI109081.

References

1. Morikawa S, Booth TF, Bishop DH (1991) Analyses of the requirements for the synthesis of virus-like particles by feline immunodeficiency virus gag using baculovirus vectors. Virology 183(1):288–297
2. Kost TA, Condreay JP, Jarvis DL (2005) Baculovirus as versatile vectors for protein expression in insect and mammalian cells. Nat Biotechnol 23(5):567–575. https://doi.org/10.1038/nbt1095
3. Luckow VA, Lee SC, Barry GF, Olins PO (1993) Efficient generation of infectious recombinant baculoviruses by site-specific transposon-mediated insertion of foreign genes into a baculovirus genome propagated in Escherichia coli. J Virol 67(8):4566–4579
4. Felberbaum RS (2015) The baculovirus expression vector system: a commercial manufacturing platform for viral vaccines and gene therapy vectors. Biotechnol J 10(5):702–714

5. Harper DM, Franco EL, Wheeler C, Ferris DG, Jenkins D, Schuind A, Zahaf T, Innis B, Naud P, De Carvalho NS, Roteli-Martins CM, Teixeira J, Blatter MM, Korn AP, Quint W, Dubin G, GlaxoSmithKline HPV Vaccine Study Group (2004) Efficacy of a bivalent L1 virus-like particle vaccine in prevention of infection with human papillomavirus types 16 and 18 in young women: a randomised controlled trial. Lancet 364(9447):1757–1765. https://doi.org/10.1016/S0140-6736(04)17398-4
6. Glenn GM, Smith G, Fries L, Raghunandan R, Lu H, Zhou B, Thomas DN, Hickman SP, Kpamegan E, Boddapati S, Piedra PA (2013) Safety and immunogenicity of a Sf9 insect cell-derived respiratory syncytial virus fusion protein nanoparticle vaccine. Vaccine 31(3):524–532. https://doi.org/10.1016/j.vaccine.2012.11.009
7. Treanor JJ, Atmar RL, Frey SE, Gormley R, Chen WH, Ferreira J, Goodwin R, Borkowski A, Clemens R, Mendelman PM (2014) A novel intramuscular bivalent norovirus virus-like particle vaccine candidate--reactogenicity, safety, and immunogenicity in a phase 1 trial in healthy adults. J Infect Dis 210(11):1763–1771. https://doi.org/10.1093/infdis/jiu337
8. Pijlman GP, van Schijndel JE, Vlak JM (2003) Spontaneous excision of BAC vector sequences from bacmid-derived baculovirus expression vectors upon passage in insect cells. J Gen Virol 84(Pt 10):2669–2678. https://doi.org/10.1099/vir.0.19438-0
9. Hitchman RB, Possee RD, King LA (2012) High-throughput baculovirus expression in insect cells. Methods Mol Biol 824:609–627. https://doi.org/10.1007/978-1-61779-433-9_33
10. Senger T, Schadlich L, Gissmann L, Muller M (2009) Enhanced papillomavirus-like particle production in insect cells. Virology 388(2):344–353. https://doi.org/10.1016/j.virol.2009.04.004
11. Kanai Y, Athmaram TN, Stewart M, Roy P (2013) Multiple large foreign protein expression by a single recombinant baculovirus: a system for production of multivalent vaccines. Protein Expr Purif 91(1):77–84. https://doi.org/10.1016/j.pep.2013.07.005
12. Noad RJ, Stewart M, Boyce M, Celma CC, Willison KR, Roy P (2009) Multigene expression of protein complexes by iterative modification of genomic Bacmid DNA. BMC Mol Biol 10:87. https://doi.org/10.1186/1471-2199-10-87
13. Schneemann A, Dasgupta R, Johnson JE, Rueckert RR (1993) Use of recombinant baculoviruses in synthesis of morphologically distinct viruslike particles of flock house virus, a nodavirus. J Virol 67(5):2756–2763
14. Dong XF, Natarajan P, Tihova M, Johnson JE, Schneemann A (1998) Particle polymorphism caused by deletion of a peptide molecular switch in a quasiequivalent icosahedral virus. J Virol 72(7):6024–6033
15. Tihova M, Dryden KA, Le TV, Harvey SC, Johnson JE, Yeager M, Schneemann A (2004) Nodavirus coat protein imposes dodecahedral RNA structure independent of nucleotide sequence and length. J Virol 78(6):2897–2905
16. Johnson KN, Tang L, Johnson JE, Ball LA (2004) Heterologous RNA encapsidated in Pariacoto virus-like particles forms a dodecahedral cage similar to genomic RNA in wild-type virions. J Virol 78(20):11371–11378. https://doi.org/10.1128/JVI.78.20.11371-11378.2004
17. Krishna NK, Marshall D, Schneemann A (2003) Analysis of RNA packaging in wild-type and mosaic protein capsids of flock house virus using recombinant baculovirus vectors. Virology 305(1):10–24
18. Venter PA, Krishna NK, Schneemann A (2005) Capsid protein synthesis from replicating RNA directs specific packaging of the genome of a multipartite, positive-strand RNA virus. J Virol 79(10):6239–6248. https://doi.org/10.1128/JVI.79.10.6239-6248.2005
19. Venter PA, Schneemann A (2007) Assembly of two independent populations of flock house virus particles with distinct RNA packaging characteristics in the same cell. J Virol 81(2):613–619. https://doi.org/10.1128/JVI.01668-06
20. Taylor DJ, Johnson JE (2005) Folding and particle assembly are disrupted by single-point mutations near the autocatalytic cleavage site of Nudaurelia capensis omega virus capsid protein. Protein Sci 14(2):401–408. https://doi.org/10.1110/ps.041054605
21. Pringle FM, Kalmakoff J, Ward VK (2001) Analysis of the capsid processing strategy of Thosea asigna virus using baculovirus expression of virus-like particles. J Gen Virol 82(Pt 1):259–266. https://doi.org/10.1099/0022-1317-82-1-259
22. Tomasicchio M, Venter PA, Gordon KH, Hanzlik TN, Dorrington RA (2007) Induction of apoptosis in Saccharomyces cerevisiae results in the spontaneous maturation of tetravirus procapsids in vivo. J Gen Virol 88(Pt 5):1576–1582. https://doi.org/10.1099/vir.0.82250-0
23. Croizier L, Jousset FX, Veyrunes JC, Lopez-Ferber M, Bergoin M, Croizier G (2000) Protein requirements for assembly of virus-like particles of Junonia coenia densovirus in insect

cells. J Gen Virol 81(Pt 6):1605–1613. https://doi.org/10.1099/0022-1317-81-6-1605

24. Chakrabarti M, Ghorai S, Mani SK, Ghosh AK (2010) Molecular characterization of genome segments 1 and 3 encoding two capsid proteins of Antheraea mylitta cytoplasmic polyhedrosis virus. Virol J 7:181. https://doi.org/10.1186/1743-422X-7-181
25. Sanchez-Eugenia R, Mendez F, Querido JF, Silva MS, Guerin DM, Rodriguez JF (2015) Triatoma virus structural polyprotein expression, processing and assembly into virus-like particles. J Gen Virol 96(Pt 1):64–73. https://doi.org/10.1099/vir.0.071639-0
26. Scodeller EA, Tisminetzky SG, Porro F, Schiappacassi M, De Rossi A, Chiecco-Bianchi L, Baralle FE (1995) A new epitope presenting system displays a HIV-1 V3 loop sequence and induces neutralizing antibodies. Vaccine 13(13):1233–1239
27. Manayani DJ, Thomas D, Dryden KA, Reddy V, Siladi ME, Marlett JM, Rainey GJ, Pique ME, Scobie HM, Yeager M, Young JA, Manchester M, Schneemann A (2007) A viral nanoparticle with dual function as an anthrax antitoxin and vaccine. PLoS Pathog 3(10):1422–1431. https://doi.org/10.1371/journal.ppat.0030142
28. Schneemann A, Speir JA, Tan GS, Khayat R, Ekiert DC, Matsuoka Y, Wilson IA (2012) A virus-like particle that elicits cross-reactive antibodies to the conserved stem of influenza virus hemagglutinin. J Virol 86(21):11686–11697. https://doi.org/10.1128/JVI.01694-12
29. Krammer F, Schinko T, Palmberger D, Tauer C, Messner P, Grabherr R (2010) Trichoplusia ni cells (High Five) are highly efficient for the production of influenza A virus-like particles: a comparison of two insect cell lines as production platforms for influenza vaccines. Mol Biotechnol 45(3):226–234. https://doi.org/10.1007/s12033-010-9268-3

Chapter 9

Nanomanufacture of Free-Standing, Porous, Janus-Type Films of Polymer–Plant Virus Nanoparticle Arrays

Brylee David B. Tiu, Rigoberto C. Advincula, and Nicole F. Steinmetz

Abstract

We present a facile method for preparing hierarchical assemblies of cowpea mosaic virus (CPMV) nanoparticles adsorbed onto patterned polypyrrole copolymer arrays, which can be released as a freely standing and microporous polymer–protein membrane with a Janus-type structure. The patterning protocol is based on colloidal sphere lithography wherein a sacrificial honeycomb pattern composed of colloidal polystyrene (PS) microspheres is assembled on an electrode. A thin layer of polypyrrole film is electropolymerized within the interstices of the template and monitored using an electrochemical quartz crystal microbalance with dissipation (EC-QCM-D) and microscopy. Dissolving the PS template reveals an inverse opaline pattern capable of electrostatically capturing the CPMV particles. Through an electrochemical trigger, the polypyrrole–CPMV delaminates from the surface producing a self-sustaining polymer–protein membrane that can potentially be used for sensing and nanocargo applications.

Key words Hierarchical assembly, Conducting polymer, Colloidal lithography, Electropolymerization, Cowpea mosaic virus nanoparticles

1 Introduction

Plant-based virus nanoparticles (VNPs) are naturally abundant nanomaterials that exhibit highly symmetric, monodisperse, and precise architectures and are currently breaking new ground in immunotherapeutic [1], sensing [2–4], and photovoltaic applications [5, 6]. In nature, these viruses primarily function as protective cargo vessels for genetic material and, in a sense, as stimuli-responsive sensors since they can specifically recognize their target cells and release their genomes upon entry [7]. Nowadays, wild-type plant virus nanoparticles and their corresponding protein cages are routinely used as nanocarriers for enzymes for nanoreactors [8], drugs [9], agents for magnetic resonance imaging (MRI) [10], light-capturing chromophores [11], and even analyte-selective peptides for sensors [12]. On the other hand, these nanomachines are ideal scaffolds for bioconjugation and symmetric nucleation of inorganic

Christina Wege and George P. Lomonossoff (eds.), *Virus-Derived Nanoparticles for Advanced Technologies: Methods and Protocols*, Methods in Molecular Biology, vol. 1776, https://doi.org/10.1007/978-1-4939-7808-3_9,
© Springer Science+Business Media, LLC, part of Springer Nature 2018

materials including metallic nanoparticles [13], nanotubes and nanoporous networks [14] due to the repeating and programmable protein subunits that comprise their structure.

To further capitalize on these features, protocols for incorporating these virus nanoparticles into devices should be developed in such a way that 2D surface arrangements and higher order assemblies can easily be fabricated with optimum precision and spatial control. Some initial techniques involve simple drop casting and dip coating for initiating the self-assembly of VNPs [15, 16]. More recently, various groups have been employing layer-by-layer deposition (LbL) to create electrostatically favorable surfaces that can form stable interactions with the virus nanoparticles [17–19]. The electrostatic-based LbL technique can easily be combined with various "top-down" fabrication methods such as microcontact printing [20, 21] and capillary force lithography [22] to form distinct patterns of viruses and even induce ordering of the particles [23]. Other techniques include dip pen nanolithography [24, 25] and nucleic acid hybridization protocols [26, 27]. However, particularly for sensors and electrical devices, patterned virus nanoparticles should be directly interfaced with materials that are innately responsive toward electrical signals. In addition, for other practical applications, these arrays should possess sufficient mechanical stability to exist as a freely standing membrane. For these objectives, inherently conducting polymers with highly conjugated structures are ideal candidates. The ability of conducting polymers to electrochemically switch from a conductive and oxidized state to a reduced state thereby changing its optical and mechanical properties is particularly useful for chemical and biological detection and energy storage applications.

In this regard, we fabricated patterned polypyrrole–virus nanoparticle arrays with Janus-type architecture by combining colloidal-templated electrochemical polymerization and the electrostatic adsorption of plant virus nanoparticles [28]. As a model virus nanoparticle, cowpea mosaic virus (CPMV) is a robust icosahedral virus with a diameter of 28 nm and structure that has been defined at 2.8 Å atomic resolution [29]. It can also remain stable in a wide pH range, high temperatures up to 60 °C and persists even in organic solvent–buffer mixtures. As illustrated in Fig. 1, the patterning protocol begins with colloidal lithography wherein sacrificial polystyrene (PS) colloidal microspheres are assembled to form a honeycomb pattern onto a conductive electrode surface [30, 31]. Then, conductive poly(pyrrole-*co*-pyrrole-3-carboxylic acid) is electrochemically polymerized onto the PS-coated substrate. The PS template is insulating while the interstitial spaces are still electrochemically accessible so dissolving the colloidal template reveals a highly ordered inverse opal structure composed of the polypyrrole copolymer. CPMV has an isoelectric point (pI) between 3.4 and 4.5 [17], hence it can gain an overall cationic behavior at pH values less than the recorded

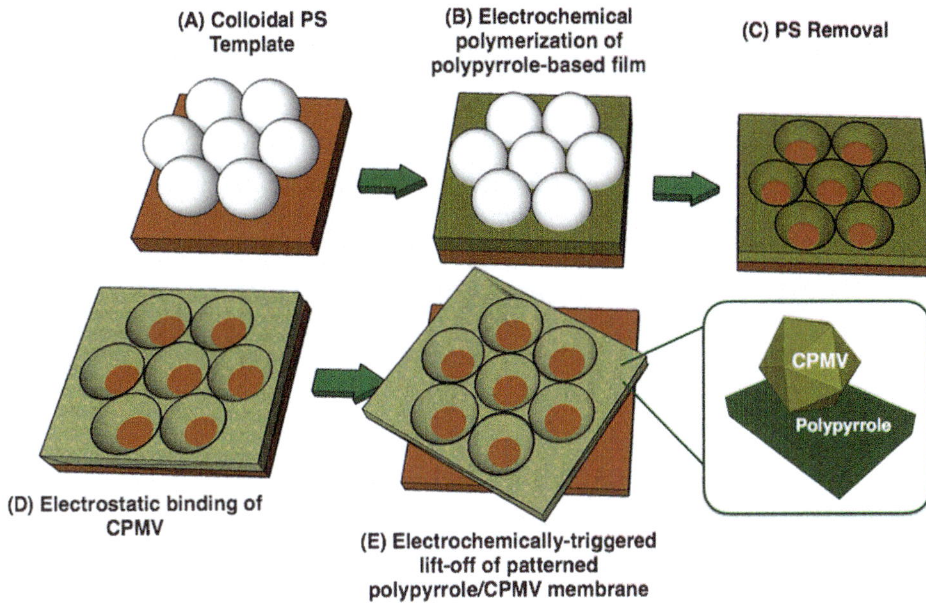

Fig. 1 Schematic illustration of the protocol for fabricating the freely standing, microporous polypyrrole–CPMV arrays which includes (**a**) the assembly of colloidal PS particles on the QCM-D crystal, (**b**) the electrochemical polymerization of the polypyrrole copolymers in the interstitial spaces of the PS pattern, (**c**) the dissolution of the PS monolayer colloidal crystal (MCC) template, (**d**) the electrostatic adsorption of CPMV on the polypyrrole array, and (**e**) the electrochemically stimulated release of the polymer–protein membrane from the surface [28]

pI. Meanwhile, the polypyrrole film has negatively charged carboxylic acid moieties, which can lead to preferential adsorption of CPMV onto the patterned polymer surface. Moreover, polypyrrole can be electrochemically delaminated from the surface due to its responsive behavior due to electrochemomechanical-induced stress leading to the formation of a self-sustaining and microporous polymer–virus membrane [32, 33].

2 Materials

2.1 Reagents

1. *Piranha* solution: Combine sulfuric acid (98% (w/w) assay percent range) and hydrogen peroxide (30% (w/w) in water) at a 3:1 (v/v) H_2SO_4–H_2O_2 ratio. *Caution! Using the Piranha solution requires extreme care due to its highly reactive and corrosive behavior toward organic materials.*
2. Polybead® PS microbeads (Catalog No. 07307-15) (Polysciences, Inc).
3. Sodium *N*-dodecyl sulfate (SDS).
4. Pyrrole (reagent grade, 98%) from Aldrich (St. Louis, MO).
5. Pyrrole-3-carboxylic acid (≥96%) from Aldrich (St. Louis, MO).

6. Lithium perchlorate ($LiClO_4$) from Aldrich (St. Louis, MO).
7. 0.1 M SDS solution (prepared from **item 3**).
8. Potassium phosphate solution: 8.8 mM potassium phosphate monobasic, 1.23 mM phosphoric acid in MilliQ water.
9. 0.5 mg/mL wild-type CPMV nanoparticles (produced as reported elsewhere [34]; *see also* Chapter 23 in this book) in potassium phosphate solution.
10. 0.1 M supporting electrolyte solution: 212.78 mg $LiClO_4$ in 20 mL MilliQ water. Sonicate the solution for at least 15 min.

2.2 Solvents

1. MilliQ Water (>18 MΩ cm).
2. Acetone.
3. Ethanol absolute (200 proof) from Decon Labs, Inc. (Prussia, PA).
4. Tetrahydrofuran.

2.3 Equipment

1. Plasmod GCM 200 oxygen plasma cleaner from March Instruments, Inc. (Concord, CA).
2. FS30 Sonication Bath from Fisher Scientific (Pittsburgh, PA).
3. Dip Coater (*see* Fig. 2a), composed of a T-LSM200A miniature motorized linear stage from Zaber Technologies (Vancouver, BC, Canada) attached to a solid Aluminum optical breadboard from Thorlabs, Inc. (Newton, NJ) (*see* **Note 1**).
4. Vacuum desiccator.

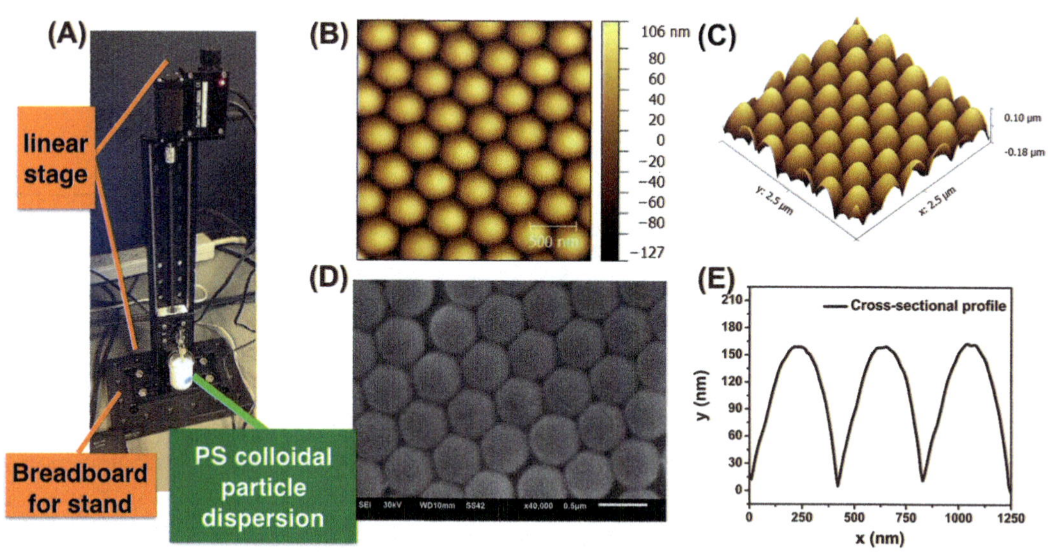

Fig. 2 Colloidal sphere lithography. (**a**) Home-built laboratory dip coater setup for colloidal lithography. (**b**) and (**c**) show 2D and 3D tapping-mode AFM images of the polystyrene monolayer colloidal crystal (MCC) template. (**d**) Scanning electron microscope (SEM) image of the PS MCC. (**e**) Cross-sectional height profile of the PS microspheres in the template

5. 5 MHz Gold-coated Quartz crystal microbalance with dissipation monitoring (QCM-D) crystals from Biolin Scientific (Paramus, NJ).
6. Quartz crystal microbalance with dissipation monitoring E1 or E4 System from Biolin Scientific (Paramus, NJ).
7. Q-Sense QEM-401 electrochemistry module from Biolin Scientific (Paramus, NJ).
8. Q-sense sensor holder from Biolin Scientific (Paramus, NJ).
9. Ismatec IPC-N cassette pump from Ismatec (Wertheim, Germany).
10. PGSTAT 12 Potentiostat from Metrohm (Riverview, FL).
11. PicoScan 2500 atomic force microscope in tapping mode from Agilent Technologies (Sta. Clara, CA).
12. High-resolution NSG30 tapping-mode AFM cantilevers from NT-MDT (Tempe, AZ).
13. JEOL-JSM-6510LV Scanning electron microscope from JEOL USA, Inc. (Peabody, MA).
14. Optical microscope attached to an NX10 Park AFM from Park Systems (Suwon, Korea).

3 Methods

3.1 Colloidal Sphere Lithography

1. Into a small beaker, prepare fresh *Piranha* solution by pouring hydrogen peroxide (30% (w/w)) then sulfuric acid in a 3:1 volume ratio. The mixture is very dangerous and exothermic so handle this with care (*see* **Note 2**).
2. Before any deposition, the Au QCM-D crystal is immersed in the *Piranha* solution for at least 1 min in order to remove any dust particles or organic residues. Orient the QCM chip in a vertical position to protect the gold surface from scratches. If available, arrange the QCM-D crystals in the Q-Sense sensor holder for this cleaning procedure.
3. Afterward, wash the crystal with copious amounts of MilliQ water. Then, rinse it with either acetone or ethanol and dry with nitrogen (*see* **Note 3**).
4. Subject the QCM-D crystals to plasma treatment in an oxygen plasma cleaner for at least 60 s.
5. In a 20 mL plastic vial, add 5.792 mL of the Polybead® PS microbeads colloidal dispersion and 150.1 mg SDS into 9.208 mL MilliQ water. Sonicate the mixture for at least 1 h to ensure optimum dispersion (*see* **Note 4**).
6. The hexagonally close-packed formation of the PS microspheres, which is also known as the monolayer colloidal crystal (MCC), is assembled on the surface of the Au QCM-D crystal

using a dip-coating process. First, while in a vertical orientation, one end of the QCM crystal is attached to a clip of the dip coater (*see* Fig. 2a). Still in a vertical position, the QCM-D crystal is immersed in the freshly sonicated colloidal dispersion of PS microbeads. Using the computer software that controls the dip coater, the QCM crystal is set to move upward at a slow and constant rate of 0.3 mm/min.

7. When most of the QCM surface has emerged from the solution, a white coating of PS colloids on the front and back surfaces of the crystal should be visible. Tilting the PS-coated Au QCM-D crystal should produce rainbow colors on the surface suggesting the successful pattern formation of the PS spheres.
8. Before mounting the QCM-D crystal in the QCM instrument, the colloidal PS particles that are physically adsorbed at the back of the crystal should be removed to ensure electrical contact between the electrodes of the QCM and the crystal. In order to do this, squirt some ethanol or MilliQ water on a piece of tissue paper and carefully wipe the PS particles away from the back surface of the crystal.
9. Store the PS-coated QCM-D crystals in a vacuum desiccator for at least 1 h before use.
10. The honeycomb PS pattern can easily be visualized using either tapping mode atomic force microscopy or scanning electron microscopy (*see* Fig. 2b–e).

3.2 Purification of Pyrrole via Vacuum Distillation

1. Mount an oil bath with a magnetic stir bar on a magnetic hotplate stirrer.
2. Clamp a 100 mL pear-shaped, evaporating flask (which will function as the distilling flask) above the oil bath; the bottom part of the flask should be submerged in the oil bath.
3. Put a magnetic stir bar and approximately 10 mL of the pyrrole monomer into the flask.
4. Attach a three-way adapter on the pear-shaped, evaporating flask and close the top part of the adapter with a glass stopper.
5. Connect a water condenser into the three-way adapter (*see* Fig. 3) and secure it using a clamp and an iron stand.
6. Attach a distilling adapter in the other end of the condenser.
7. Connect the distilling adapter to a 100 mL round bottom flask, which will function as the receiving flask.
8. Clamp the connection between the distilling adapter and the receiving flask and secure it with an iron stand.
9. Put an ice bath under the receiving flask.
10. Use a rubber tubing to connect a water source to one of the openings of the glass condenser that is positioned lower than

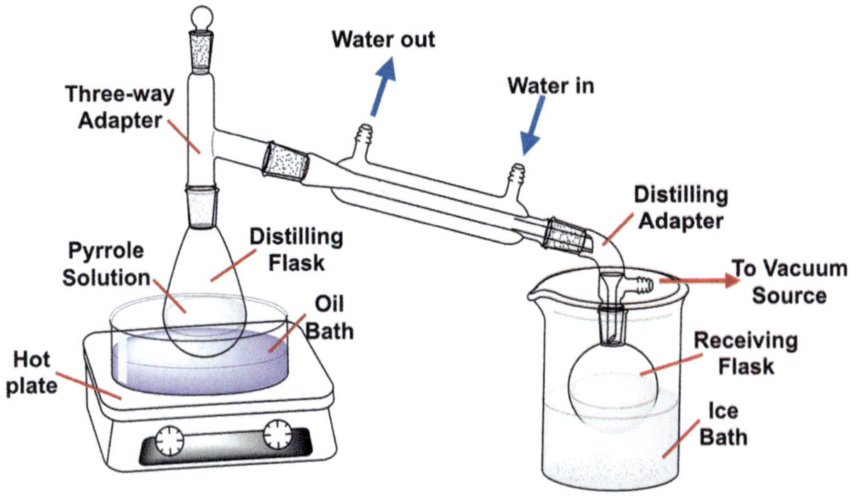

Fig. 3 Schematic of the vacuum distillation setup in purifying the pyrrole monomer

the other (refer to Fig. 3). Attach rubber tubing to the other opening; the end of this tubing will be placed in the sink. Start water flow. Notice that water is flowing from the bottom part of the condenser and out of the upper opening of the condenser in order to make sure that the condenser is completely filled with water and the condensation is more efficient.

11. Use another piece of rubber tubing to connect the opening in the distilling adapter to a vacuum source (i.e., house vacuum or vacuum pump). Turn on vacuum. To make sure that the system does not have leaks, try to remove the glass stopper attached to the three-way adapter and notice the strong pulling force from the vacuum.
12. Turn on the stirring and set the temperature of the hot plate to 120 °C.
13. Keep an eye at the distilling adapter and the receiving round bottom flask. A colorless pyrrole liquid should start forming droplets at a rate of 1 drop per second.
14. Collect the pyrrole monomer in a vial. Seal it and refrigerate the sample if not being used (*see* **Note 5**).

3.3 Fabrication of Polypyrrole Inverse Opals

1. Remove the platinum counter electrode from the Q-sense QCM-D electrochemistry module (QEM 401). Immerse the platinum electrode in freshly prepared *Piranha* solution for at least 15–20 min to remove any organic residues (*see* **Note 2**). Rinse with copious amounts of water and dry with nitrogen.
2. Attach the freshly cleaned platinum counter electrode and the Ag/AgCl reference electrode into the electrochemistry module. Consult the “Handling and Assembly of Electrodes” section in the electrochemistry module manual.

3. After drying the substrate in a vacuum desiccator, mount the PS-coated Au QCM-D crystal onto the electrochemistry module (QEM 401) of the Q-Sense QCM-D instrument (*see* **Note 6**). Then, attach and secure the module base to close the electrochemistry module.
4. Mount the electrochemistry module with the QCM-D chip on the Q-Sense QCM-D chamber. Connect the inlet and outlet tubes on the electrochemistry module.
5. The electrochemical polymerization of polypyrrole is performed using a standard three-electrode electrochemical cell. Connect the SMB cable to the gold port on the module base of the electrochemical cell. The SMB gold connector is connected to the face of the QCM-D crystal, which will act as the working electrode. Attach the other end to the SMB/BNC adaptor, which will be connected to the BNC Banana Binding Poles that will interface with the potentiostat.
6. Afterward, connect the reference electrode alligator clip of the potentiostat to the Ag/AgCl electrode attached to the electrochemistry module.
7. Connect the counter electrode wire of the potentiostat into the jack on right above the inlet and outlet ports.
8. After installing the QCM-D crystal and electrochemistry module, open the QSoft program and set the temperature to stabilize at room temperature (25 °C). Temperature control is very important in achieving stable QCM measurements.
9. Measure and calibrate the QCM-D crystal resonant frequency and dissipation overtones. For the experiment, employing the first (5 MHz), third (15 MHz), fifth (25 MHz), seventh (35 MHz), and ninth (45 MHz) harmonic frequencies are sufficient in characterizing the in situ formation of the polypyrrole film. Afterward, start acquiring data and wait to achieve a stable baseline for at least 5–10 min.
10. While waiting for the baseline, prepare the monomer solution, which will be used for electrochemical polymerization. In 20 mL MilliQ water, add 11.1 μL freshly distilled pyrrole, 66.7 mg pyrrole-3-carboxylic acid, and 212.78 mg lithium perchlorate. Sonicate the mixture for 10 min (*see* **Note 7**).
11. After achieving a stable QCM-D response and preparing the solution, submerge the open end of the inlet tubing into the pyrrole solution and set the peristaltic pump to run at 100 μL/min. Depending on the length of the tubing, monitor the QCM-D response and wait for a sharp decrease in the ΔF signal, which signifies the point when the solution has reached the QCM-D surface. The signal will eventually plateau when the chamber has been completely filled.

12. Stop the pump once the chamber has been completely filled with electropolymerization solution (from **step 10**).
13. Run the potentiostat software to facilitate the electrochemical polymerization. Select and define a cyclic voltammetry (CV) experiment wherein the potential will sweep from 0 to 1.2 V and back at 50 mV/s for 7 cycles.
14. When the CV travels from 0 to 1.2 V, the ΔF signal of the QCM-D should exhibit a sharp decrease. However, when cycling back from 1.2 to 0 V, the signal will slightly increase and then plateau. This behavior persists for each cycle, which eventually leads to the QCM output resembling a staircase. Figure 4a shows representative electrochemical QCM-D data for the electropolymerization. On the other hand, Fig. 4b depicts a typical cyclic voltammogram for the electropolymerization of the polypyrrole film.
15. After the electropolymerization, rinse the system with MilliQ water at 100 μL/min for 10 min. Then, remove the inlet tube from the MilliQ water and allow the pump to run at the same rate without any attached solvent while the inlet tube is exposed to air (to dry the crystal).
16. To clean the instrument, unmount the polypyrrole-coated QCM-D chip and replace with a new QCM-D crystal.
17. Rinse the QCM-D flow cell system by connecting the inlet tube to a vial containing 0.1 M SDS and run the pump at 100 μL/min for at least 10 min. Change the rinsing solvent to MilliQ water and then to ethanol. To dry the flow cell setup, continue running the pump but remove the tubing from the solvent vial and just let air pass through (*see* **Note 8**).
18. Detach the flow cell setup and dismount the QCM crystal. The colloidal PS particles should then be dissolved to finally form the polypyrrole inverse opal films. To do so, place the QCM-D chip in the sensor holder once again and immerse in tetrahydrofuran for 30 min. Afterward, dispose the tetrahydrofuran and replace it with fresh solvent and wait for another 30 min. Dry the substrate with nitrogen and store under vacuum to dry.
19. The inverse honeycomb pattern composed of polypyrrole can be observed using tapping-mode atomic force microscopy (AFM) (*see* Fig. 4c, d).

3.4 Electrostatic Adsorption of CPMV

1. Mount the QCM-D crystal coated with a colloidally patterned polypyrrole inverse opal film onto a precleaned Q-Sense flow cell.
2. Connect the inlet and outlet tubes onto the flow module. Place the cell in the Q-Sense chamber.

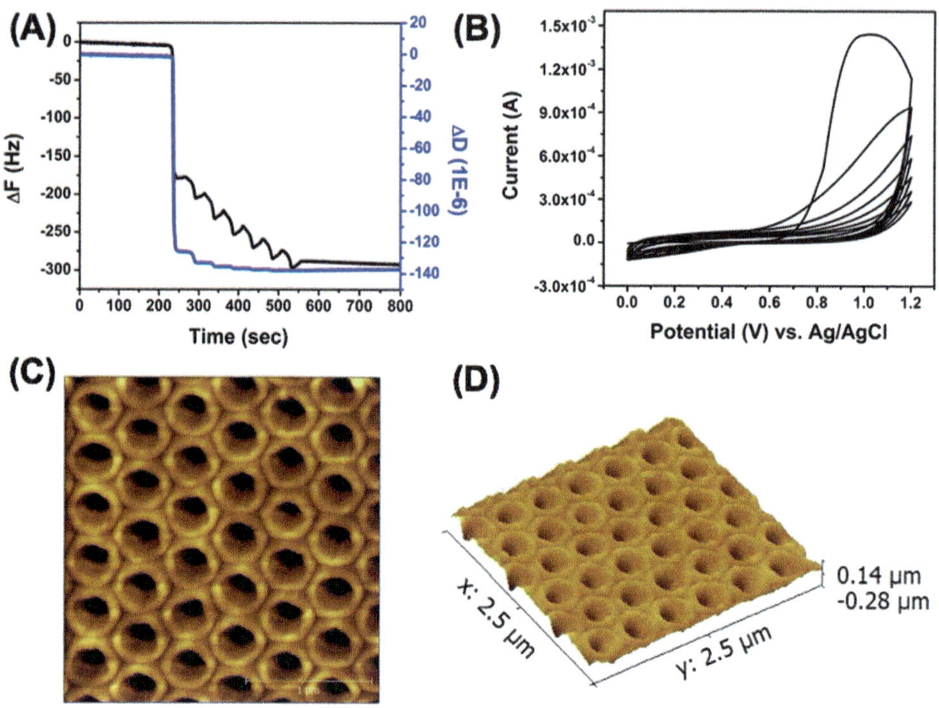

Fig. 4 (**a**) Electrochemical quartz crystal microbalance with dissipation monitoring data gathered from the polypyrrole electropolymerization. (**b**) Cyclic voltammogram corresponding to the electrochemical polymerization of the polypyrrole copolymers through the PS MCC pattern. (**c**, **d**) 2D and 3D topographical AFM images of the polypyrrole inverse honeycomb pattern

3. Open the QSoft software. Turn on the temperature control and find or calibrate the harmonic frequencies of the QCM-D.
4. Immerse the inlet tube into a vial containing a potassium phosphate solution composed of potassium phosphate monobasic (8.8 mM) and phosphoric acid (1.23 mM) in MilliQ water.
5. Set the pump to a flow rate of 100 μL/min and start the flow. Observe the movement of the solvent and make sure that it has completely infiltrated the QCM flow cell.
6. Start data acquisition and wait for 5–10 min to stabilize the frequency and dissipation signals of the QCM-D.
7. Connect the inlet tube to a solution composed of 0.5 mg/mL CPMV in the same potassium phosphate solution (as prepared in Subheading 3.4, **step 4**) and allow the solution to flow through the cell module.
8. Once the CPMV particles reach the QCM-D crystal surface, a sharp decrease in ΔF (increase in ΔD) will be observed.
9. This decrease will eventually start to flatten out suggesting saturation of CPMV adsorption. Stop the cassette pump at this

point to increase the contact time between the virus nanoparticles and the polypyrrole surface.

10. After at least 30 min, allow the buffer to flow through the system again at 100 μL/min to remove loosely bound particles from the surface.
11. Stop the acquisition once the frequency and dissipation shifts stabilize (*see* Fig. 5a). Remove the inlet tube from the vial containing the buffer and allow the pump to run for 10 min without any attached solvent infiltrating the system.
12. Replace the functionalized QCM-D crystal with a new one specifically for cleaning (to avoid exposure of the QCM electrodes to the cleaning solution). Then, allow 0.1 M SDS to fill the system at 100 μL/min for 10 min. Replace with MilliQ water and then with ethanol. Let each solvent run for 10 min.
13. Disconnect the inlet tube from the solvent container. Allow the pump to continuously run in order to dry the flow cell module (*see* **Note 8**).
14. Detach the flow cell module and turn off the instrument.
15. The hierarchical assembly of CPMV onto the patterned polypyrrole surface can be observed using tapping-mode AFM (*see* Fig. 5b, c).

3.5 Electrochemical Release of Free-Standing, Porous Polypyrrole–CPMV Films

1. Similar to **step 1** of Subheading 3.3, clean the platinum counter electrode using *Piranha* solution, rinse with copious amounts of water and dry with nitrogen.
2. Install the platinum counter electrode and Ag/AgCl (3 M potassium chloride (KCl)) reference electrode in the QEM 401 electrochemistry module.
3. Mount the QCM-D crystal coated with polypyrrole micropatterns on the electrochemistry module and screw in the module base to close the electrochemical flow cell.
4. Connect the inlet and outlet tubings for sample introduction and mount the electrochemistry flow module onto the QCM-D chamber. Make sure that the electrical contact between the chamber and the flow cell is secure.
5. Turn on the QCM-D equipment and software and detect the resonant frequency of the QCM-D crystal at various overtones.
6. In the meantime, prepare a 0.1 M supporting electrolyte solution by adding 212.78 mg $LiClO_4$ to a 20 mL MilliQ water. Sonicate the solution for at least 15 min.
7. Connect the inlet tubing of the flow cell to the 0.1 M $LiClO_4$ solution and start the pump's flow at 100 μL/min.

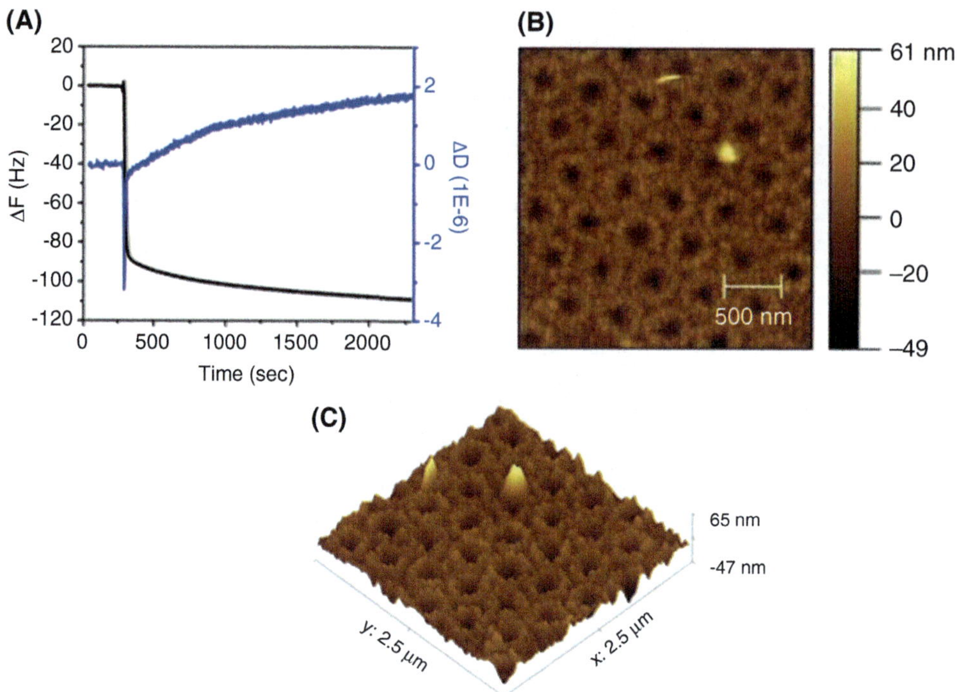

Fig. 5 (**a**) QCM-D adsorption data corresponding to the electrostatic binding of CPMV particles on the negatively charged polypyrrole surface. Topographical 2D (**b**) and 3D (**c**) AFM images of the patterned polypyrrole–CPMV films

8. Once the supporting electrolyte solution has infiltrated the system, start the QCM-D data acquisition; wait for 10–15 min to achieve stable frequency and dissipation readings.
9. While waiting for the QCM-D to stabilize, set up the potentiostatic conditions to electrochemically lift off the colloidally patterned polypyrrole-CPMV membrane.
10. In the potentiostat software, select the chronoamperometric method and set the applied potential at −1.5 V and the application time to 30 min.
11. After reaching stable frequency and dissipation readings in the QCM-D, start the defined chronoamperometric method and wait for 30 min for the experiment to conclude.
12. Stop the data acquisition of the QCM-D and detach the electrochemistry module to unmount the crystal.
13. It is easily observable that most parts of the film are peeled off but the outer areas are still attached. Use a pair of sharp tweezers to carefully detach the film from the crystal and transfer it to a safe container or to a pH 3 solution.
14. The freely standing and microporous polypyrrole–CPMV film can easily be analyzed using an optical microscope or an atomic force microscope (*see* Fig. 6).

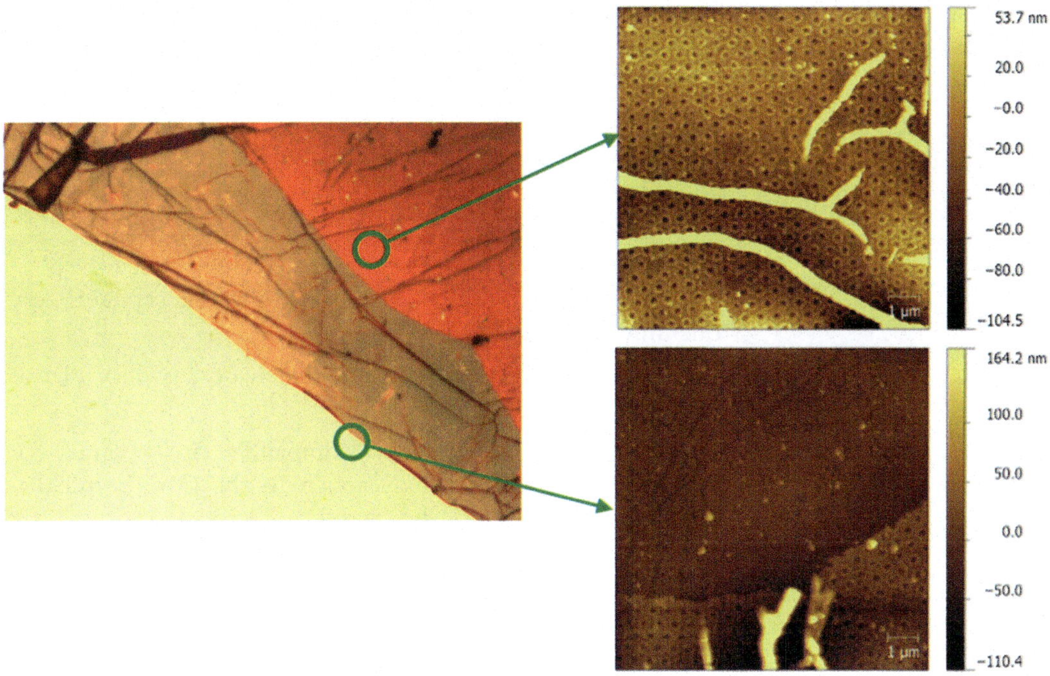

Fig. 6 480 μm × 360 μm optical microscope image of the free-standing polypyrrole–CPMV arrays with the corresponding tapping-mode topographical AFM images on selected regions of the lifted-off membrane

4 Notes

1. The dip coater (Fig. 2) used in the study was not purchased as a ready-to-use instrument. It was assembled using parts that are separately bought. The coater is mainly based on a miniature motorized linear stage with a built-in controller (T-LSM200A) from Zaber Technologies Inc. (Vancouver, British Columbia, Canada). The speed of the 1-dimensional translation of the linear stage can be controlled using the accompanying software Zaber Console. The stage can be attached to a solid aluminum optical breadboard, which can be purchased from Thorlabs, Inc. (Newton, NJ), in order for it to stand in a vertical position.
2. Measuring out the hydrogen peroxide first and then adding sulfuric acid to it is the safer approach. However, slowly adding H_2O_2 into sulfuric acid creates a much more active and effective cleaning agent for the QCM-D crystal.
3. Do not rinse a QCM-D crystal that came directly from *Piranha* solution with an organic solvent. Contact between *Piranha* solution and organic solvents such as ethanol or acetone is highly reactive and extremely dangerous.

4. Add ice cubes into the sonication bath from time to time to avoid the temperature from becoming very hot and cause the colloidal particles to aggregate.
5. The purified pyrrole monomer was used within a week after vacuum distillation and stored in a lab refrigerator if not in use. Unused distilled monomers were discarded after a week or if the monomer starts to turn brownish.
6. Make sure that the sensor O-ring is placed in between the platinum electrode and the QCM chip and that the crystal is only touching the O-ring.
7. Make sure to use the solution immediately because it oxidizes/polymerizes quickly.
8. Do not allow the system to completely dry because the QCM-D crystal will strongly attach to the O-ring and dismounting it might crack the crystal.

Acknowledgments

This work was supported in part by the National Science Foundation (CMMI NM 1333651 to NFS and RCA and STC-0423914 to RCA). Case Farm is acknowledged for scaled-up plant production. The authors also acknowledge technical support from Biolin Scientific Inc. and the Materials for Opto/Electronics Research and Education (MORE) Center at the Case Western Reserve University.

References

1. Lee KL, Shukla S, Wu M et al (2015) Stealth filaments: polymer chain length and conformation affect the in vivo fate of PEGylated potato virus X. Acta Biomater 19:166–179
2. Zang F, Gerasopoulos K, Fan XZ et al (2014) An electrochemical sensor for selective TNT sensing based on Tobacco mosaic virus-like particle binding agents. Chem Commun 50:12977–12980
3. Bruckman MA, Liu J, Koley G et al (2010) Tobacco mosaic virus based thin film sensor for detection of volatile organic compounds. J Mater Chem 20:5715–5719
4. Tiu BDB, Kernan DL, Tiu SB et al (2017) Electrostatic layer-by-layer construction of fibrous TMV biofilms. Nanoscale 9:1580–1590
5. Nam KT, Kim D-W, Yoo PJ et al (2006) Virus-enabled synthesis and assembly of nanowires for lithium ion battery electrodes. Science 312:885–888
6. Dang X, Yi H, Ham M-H et al (2011) Virus-templated self-assembled single-walled carbon nanotubes for highly efficient electron collection in photovoltaic devices. Nat Nanotechnol 6:377–384
7. Culver JN, Brown AD, Zang F et al (2015) Plant virus directed fabrication of nanoscale materials and devices. Virology 479–480:200–212
8. de la Escosura A, Nolte RJM, Cornelissen JJLM (2009) Viruses and protein cages as nanocontainers and nanoreactors. J Mater Chem 19:2274–2278
9. Lu Y, Chan W, Ko BY et al (2015) Assessing sequence plasticity of a virus-like nanoparticle by evolution toward a versatile scaffold for vaccines and drug delivery. Proc Natl Acad Sci 112:12360–12365
10. Bruckman MA, Randolph LN, Gulati NM et al (2015) Silica-coated Gd(DOTA)-loaded protein nanoparticles enable magnetic resonance imaging of macrophages. J Mater Chem B 3:7503–7510
11. Zhou Q, Wu F, Wu M et al (2015) Confined chromophores in tobacco mosaic virus to

mimic green fluorescent protein. Chem Commun 51:15122–15124
12. Uhde-Holzem K, McBurney M, Tiu BDB et al (2016) Production of immunoabsorbent nanoparticles by displaying single-domain protein A on potato virus X. Macromol Biosci 16(2):231–241
13. Aljabali AAA, Lomonossoff GP, Evans DJ (2011) CPMV-polyelectrolyte-templated gold nanoparticles. Biomacromolecules 12:2723–2728
14. Courchesne N-MD, Klug MT, Chen P-Y et al (2014) Assembly of a bacteriophage-based template for the organization of materials into nanoporous networks. Adv Mater 26:3398–3404
15. Evans DJ (2008) The bionanoscience of plant viruses: templates and synthons for new materials. J Mater Chem 18:3746–3754
16. Barick KC, Bahadur D (2010) Self-assembly of colloidal nanoscale particles: fabrication, properties and applications. J Nanosci Nanotechnol 10:668–689
17. Steinmetz NF, Findlay KC, Noel TR et al (2008) Layer-by-layer assembly of viral nanoparticles and polyelectrolytes: the film architecture is different for spheres versus rods. Chembiochem 9:1662–1670
18. Lin Y, Su Z, Niu Z et al (2008) Layer-by-layer assembly of viral capsid for cell adhesion. Acta Biomater 4:838–843
19. Wu Y, Feng S, Zan X et al (2015) Aligned electroactive TMV nanofibers as enabling scaffold for neural tissue engineering. Biomacromolecules 16:3466–3472
20. Balci S, Leinberger DM, Knez M et al (2008) Printing and aligning mesoscale patterns of tobacco mosaic virus on surfaces. Adv Mater 20:2195–2200
21. Yoo SY, Chung W-J, Kim TH et al (2011) Facile patterning of genetically engineered M13 bacteriophage for directional growth of human fibroblast cells. Soft Matter 7:363–368
22. Yoo PJ, Nam KT, Belcher AM et al (2008) Solvent-assisted patterning of polyelectrolyte multilayers and selective deposition of virus assemblies. Nano Lett 8:1081–1089
23. Yoo PJ, Nam KT, Qi J et al (2006) Spontaneous assembly of viruses on multilayered polymer surfaces. Nat Mater 5:234–240
24. Cheung CL, Camarero JA, Woods BW et al (2003) Fabrication of assembled virus nanostructures on templates of chemoselective linkers formed by scanning probe nanolithography. J Am Chem Soc 125:6848–6849
25. Vega RA, Maspoch D, Salaita K et al (2005) Nanoarrays of single virus particles. Angew Chem 117:6167–6169
26. Yi H, Rubloff GW, Culver JN (2007) TMV microarrays: hybridization-based assembly of DNA-programmed viral nanotemplates. Langmuir 23:2663–2667
27. Yi H, Nisar S, Lee S-Y et al (2005) Patterned assembly of genetically modified viral nanotemplates via nucleic acid hybridization. Nano Lett 5:1931–1936
28. Tiu BDB, Tiu SB, Wen AM et al (2016) Free-standing, nanopatterned Janus membranes of conducting polymer–virus nanoparticle arrays. Langmuir 32:6185–6193
29. Lin T, Chen Z, Usha R et al (1999) The refined crystal structure of cowpea mosaic virus at 2.8 Å resolution. Virology 265:20–34
30. Marquez M, Grady BP (2004) The use of surface tension to predict the formation of 2D arrays of latex spheres formed via the langmuir–blodgett-like technique. Langmuir 20:10998–11004
31. Tiu BDB, Pernites RB, Foster EL et al (2015) Conducting polymer-gold co-patterned surfaces via nanosphere lithography. J Colloid Interface Sci 459:86
32. Pyo M, Bohn CC, Smela E et al (2003) Direct strain measurement of polypyrrole actuators controlled by the polymer/gold interface. Chem Mater 15:916–922
33. Kaneto K, Sonoda Y, Takashima W (2000) Direct measurement and mechanism of electro-chemomechanical expansion and contraction in polypyrrole films. Jpn J Appl Phys 39:5918
34. Cho CF, Sourabh S, Simpson EJ et al (2014) Molecular targeted viral nanoparticles as tools for imaging cancer. Methods Mol Biol (Clifton, NJ) 1108:211–230

Chapter 10

Self-Assembly of Rod-Like Bionanoparticles at Interfaces and in Solution

Ye Tian and Zhongwei Niu

Abstract

Rod-like nanoparticles show unique self-assembly behavior benefiting from their anisotropic properties. As a classic example of a one-dimensional (1D) rod-like plant virus, tobacco mosaic virus (TMV) can either assemble in a head-to-tail manner to form 1D long fibers, or align parallel to form crystal-like structures at interfaces or in solution. Here, the self-assembly behaviors of TMV at oil–water or air–liquid interfaces are summarized. In addition, the self-assembly of TMV with polymers in solution is also discussed in this chapter.

Key words Tobacco mosaic virus, Self-assembly, Interfaces, Polyaniline, Polydopamine

1 Introduction

Rod-like nanoparticles, significantly different from spherical nanoparticles, show unique anisotropy in their structural, chemical, electrical, and optical features, and these features are usually aspect ratio-dependent. For instance, an elongated CdSe–CdS core–shell nanorod donor shows a highly enhanced Förster resonance energy transfer (FRET) efficiency compared with its nanodot form [1]. The high aspect ratio and the orientation of metallic nanorods both contribute to a more effective dielectric constant [2]. Moreover, nanorods differ in their endocytosis pathway, uptake efficiency, immunological responses, and biodistribution compared with spherical nanoparticles and these properties are highly influenced by their aspect ratio [3–7]. Based on these properties, self-assembly of rod-like nanoparticles shows great potential in the fields of electronic devices, sensing, imaging, and nanomedicine [8, 9]. Benefitting from their structural and chemical anisotropy, rod-like nanoparticles can self-assemble into more forms than spherically isotropic nanoparticles: they can form long nanofibers through head-to-tail assembly [10, 11]; they can align parallel to one another to generate patterned films [9]; and they can also form 3D bundles

Christina Wege and George P. Lomonossoff (eds.), *Virus-Derived Nanoparticles for Advanced Technologies: Methods and Protocols*, Methods in Molecular Biology, vol. 1776, https://doi.org/10.1007/978-1-4939-7808-3_10,

© Springer Science+Business Media, LLC, part of Springer Nature 2018

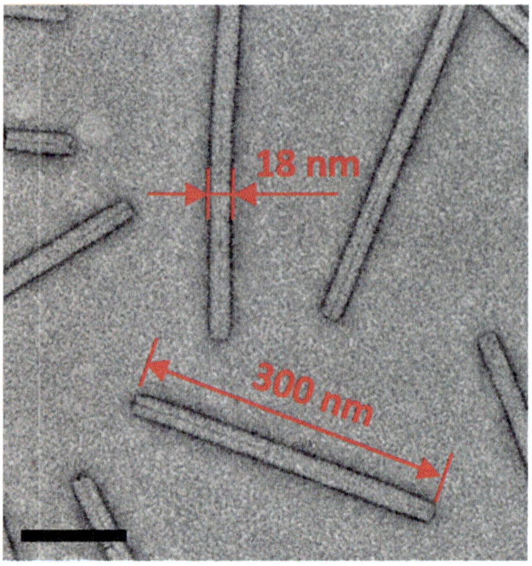

Fig. 1 Transmission electron microscope (TEM) image of TMV following negative staining with UAc. Scale bar: 100 nm

through depletion interaction [12], in which nonadsorbing additives lead to osmotic pressure and push the particles together.

Tobacco mosaic virus (TMV), a 1D rod-like plant virus measuring 300 × 18 nm with a 4 nm central longitudinal cavity, consists of 2130 identical coat proteins helically arranged around a 6.4 kb single-strand of RNA (*see* Fig. 1) [13–15]. TMV has been widely exploited in the fields of light-harvesting systems [16, 17], photoelectrochemical cells [18], electronic devices [19, 20], Li-ion microbatteries [8], and biomedicine [9, 21, 22]. Based on its well-defined structure, size, and shape, TMV provides an ideal template for studying the self-assembly and ordering of rod-like nanoparticles. Acting as a solid stabilizer, TMV can assemble at an oil–water interface to form a Pickering emulsion and shows a concentration-dependent orientation [23]. In a capillary tube, different patterns can be generated because of the interfacial assembly of TMV particles at the air–liquid interface and a pinning-depinning process [9]. In an aqueous solution, via a hierarchical assembly and in situ polymerization process, TMV can assemble with aniline, poly(sulfonated styrene) (PSS), and ammonium persulfate (APS) into conductive composite long fibers [11]. Through a similar method, long hybrid fibers of TMV–polydopamine can be fabricated and used for the in situ reduction of gold ions. In addition, TMV can also assemble with Pluronics F127 into micelles to achieve "PEGylation" of TMV.

In this chapter, we describe protocols for the self-assembly of TMV at oil/water and air–liquid interfaces, TMV self-assembly into long fibers via a hierarchical assembly and in situ polymerization process, and TMV assembly with Pluronics F127 through hydrogen bond or van der Waals forces. Further, we also describe characterization techniques for the assembled particles.

2 Materials

All water used here is ultrapure water (18.2 MΩ•cm). TMV solutions and buffers are stored at 4 °C. Dialysis membranes (biotech cellulose ester, molecular weight cut-off 1000 kDa) are from Spectrum Laboratories, Inc. All TEM samples are negatively stained with 2% (w/v) uranyl acetate (UAc) in aqueous solution.

2.1 Pickering Emulsions at Oil–Water Interfaces

1. 0.01 M potassium phosphate (KH_2PO_4–K_2HPO_4) buffer (pH 7.8): Prepare 0.01 M potassium dihydrogen phosphate (KH_2PO_4) solution and 0.01 M dipotassium hydrogen phosphate (K_2HPO_4) solution in water, then mix them and adjust the volume ratio until the pH value of the buffer reaches 7.8 (*see* **Note 1**).
2. Perfluorodecalin (PFD).
3. TMV solution: (a) 0.2 mg/ml in 0.01 M potassium phosphate buffer (pH 7.8); (b) 0.8 mg/ml in 0.01 M potassium phosphate buffer (pH 7.8) (*see* **Note 2**. In this book, Chapters 23, 24, and 27 provide protocols for TMV purification.).
4. Clean silicon wafers.
5. Carbon-coated copper grids.

2.2 Patterns at Air–Liquid Interfaces

1. Cylindrical open-ended capillary glass tubes (KIMBLE Co.) with an inner diameter of 1.5 mm and length of 22 mm.
2. Piranha solution: 7:3 (v/v) mixture of 98% (w/w) H_2SO_4 and 30% (w/w) H_2O_2.
3. 0.01 M potassium phosphate buffer (pH 7.4): *see* Subheading 2.1, **item 1**. Adjust the volume ratio until the pH value of the buffer reaches 7.4 (*see* **Note 1**).
4. TMV solutions in 0.01 M potassium phosphate buffer (pH 7.4): (a) 0.01 mg/ml; (b) 0.1 mg/ml.

2.3 Water-Soluble Conductive Nanowires: PSS/PANi/TMV Assemblies

1. 0.1 M HCl and 0.1 M NaOH solution in water for pH adjusting.
2. 1.0 mg/ml TMV solution in water.
3. Distilled aniline.
4. 10 mg/ml ammonium persulfate (APS) solution in water.
5. 30% (w/v) poly(sulfonated styrene) (PSS) solution in water (*see* **Note 3**).

2.4 Long Fibers of TMV–Polydopamine–Gold Nanoparticle Composites

1. 0.05 M acetate (NaAc/HAc) buffer (pH 5.5): Prepare 0.05 M sodium acetate (NaAc) solution and 0.075 M acetic acid (HAc) solution in water, then mix them and adjust the volume ratio until the pH value of the buffer reaches 5.5 (*see* **Note 4**).
2. 0.5 mg/ml TMV solution in 0.05 M acetate buffer (pH 5.5).

3. 3,4-dihydroxyphenethylamine hydrochloride (dopamine•HCl).
4. Ammonium persulfate (APS).
5. 0.1 M NaOH aqueous solution for pH value adjustment.
6. 1% (w/v) chloroauric acid ($HAuCl_4$) solution in water.

2.5 "PEGylation" of TMV: TMV–F127 Assemblies

1. 0.01 M potassium phosphate buffer (pH 7.8): *see* Subheading 2.1, **item 1**. Adjust the volume ratio until the pH value of the buffer reaches 7.8 (*see* **Note 1**).
2. 2.5 mg/ml PEG-PPG-PEG (poly(ethylene glycol)-*block*-poly(propylene glycol)-*block*-poly(ethylene glycol)) (Pluronics F127) solution in 0.01 M potassium phosphate buffer (pH 7.8) (*see* **Note 5**).
3. 1 mg/ml TMV solution in 0.01 M potassium phosphate buffer (pH 7.8).
4. Carbon-coated copper grids.

3 Methods

3.1 Pickering Emulsions at Oil–Water Interfaces

1. Shake the TMV solution with 10% (v/v) perfluorodecalin by hand for 2 min at room temperature.
2. Put the emulsions standing at 4 °C for 4–12 h for stabilization (*see* **Note 6**).
3. Remove the aqueous phase and then add fresh potassium phosphate buffer. Repeat this process for three times to remove the excess TMV and the salt from the aqueous phase.
4. Transfer samples of the emulsion onto clean silicon wafers for atomic force microscopy (AFM), or carbon-coated copper grids for transmission electron microscopy (TEM) (*see* **Note 7**).

3.2 Patterns at Air–Liquid Interfaces

1. Clean the glass capillaries using piranha solution at 75 °C for 2 h.
2. Inject 25 μl TMV solution (0.01 mg/ml or 0.1 mg/ml) into the cleaned glass tubes.
3. Place the tubes in a horizontal position on the flat bench at room temperature and 40–60% humidity.
4. Maintain the tubes that way for 3 days.
5. After cracking the tubes, the TMV patterns on the inner surface are analyzed. As visualized by AFM, different patterns are formed depending on the TMV concentration (*see* Fig. 2).

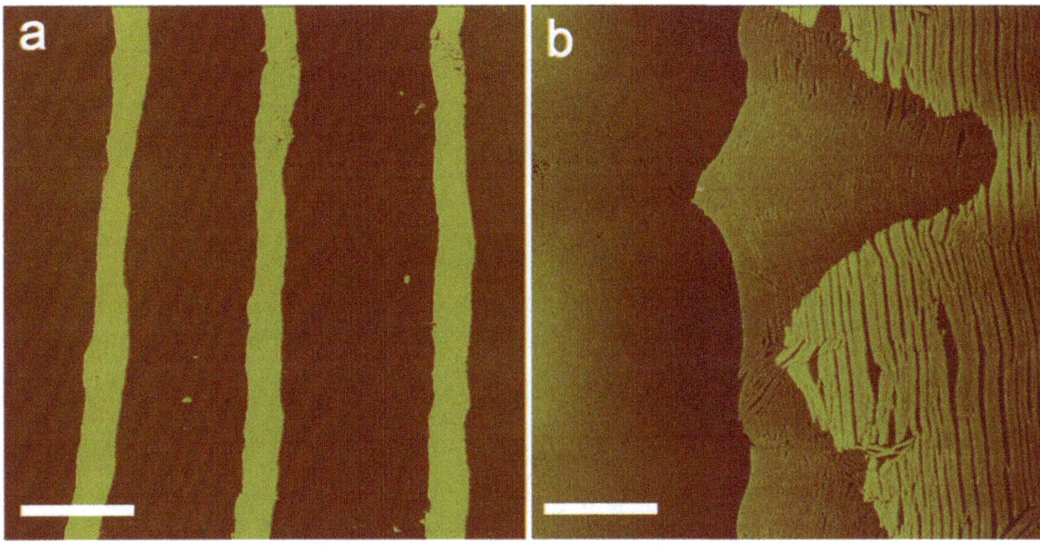

Fig. 2 AFM images of the patterns formed in the glass capillary tubes at different TMV initial concentrations: (**a**) 0.01 mg/ml; (**b**) 0.1 mg/ml. The tubes are placed in a horizontal position on the flat bench at room temperature and 40–60% humidity for three days, and the AFM images are got from the inner glass surface after cracking the tubes. Scale bars: 4 μm [9]

3.3 Water-Soluble Conductive Nanowires: PSS–PANi–TMV Assemblies

3.3.1 Conductive Short Rods

1. Add 10 μl aniline, 1 ml APS solution, and 40 μl PSS solution into 4 ml TMV solution.
2. Adjust the pH of the solution to 4.0 by adding HCl or NaOH solution (*see* **Note 8**).
3. Incubate the reaction mixture at room temperature for 24 h.
4. The solution of PSS–PANi (polyaniline)–TMV composites appears green and transparent (*see* **Note 9**).
5. Dialyze the reaction solution against deionized water to obtain pure PSS–PANi–TMV composite nanorods (TEM images shown in Fig. 3).

3.3.2 Conductive Long Fibers

1. Add 10 μl aniline, 1 ml APS solution into 4 ml TMV solution.
2. Adjust the pH of the solution to 6.5 by adding HCl or NaOH solution.
3. Incubate the reaction mixture at room temperature for 24 h.
4. PANi/TMV long fibers are formed (*see* **Note 10**).
5. Add 40 μl PSS solution to the reaction mixture.
6. Adjust the pH of the solution to 4.0 by adding HCl or NaOH solution.
7. PSS–PANi–TMV long fibers which are water-soluble and conductive are formed.
8. Dialyze the reaction solution against deionized water to obtain pure PSS–PANi–TMV composite long fibers (TEM images shown in Fig. 3) (*see* **Note 11**).

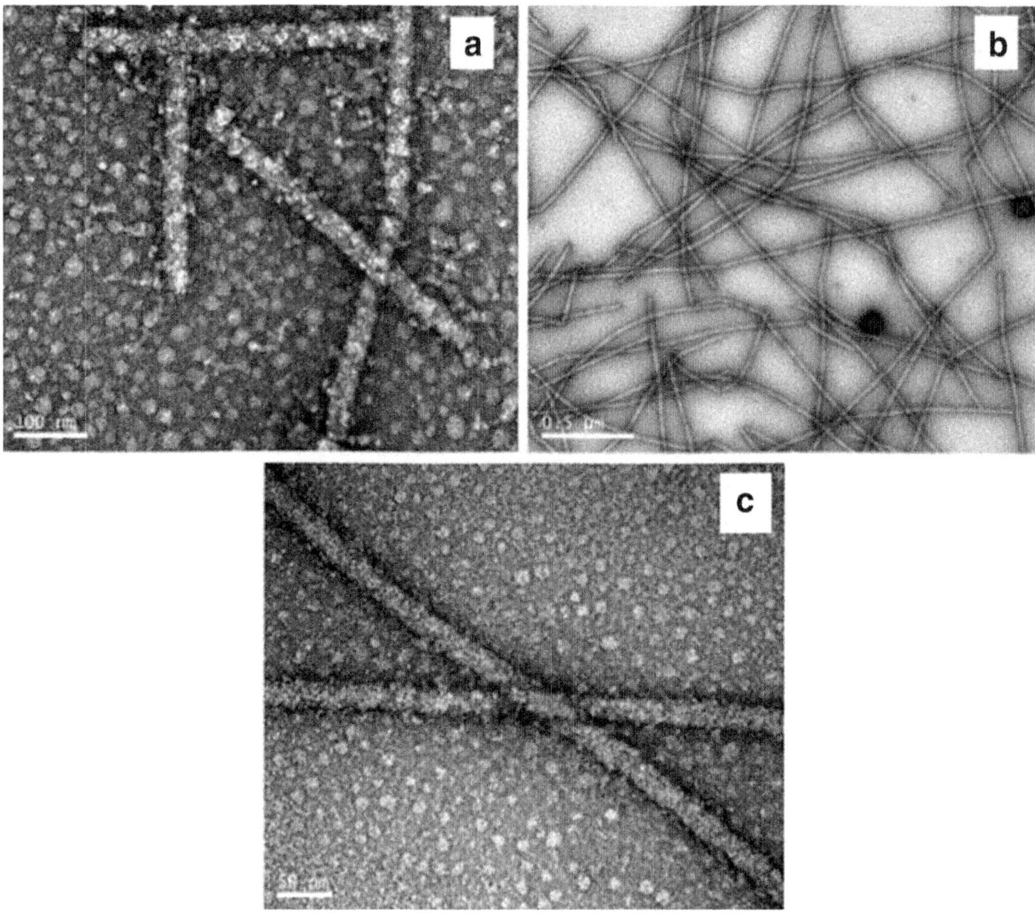

Fig. 3 TEM images of PSS–PANi–TMV conductive short rods (**a**) and PSS–PANi–TMV conductive long fibers (**b**, **c**) following negative staining with UAc. (**c**) is an enlarged image of (**b**). Scale bars: 100 nm in (**a**); 500 nm in (**b**); 50 nm in (**c**) (reproduced from [11] with permission from American Chemical Society)

3.4 Long Fibers of TMV–Polydopamine–Gold Nanoparticle Composites

1. Add 1.25 mg dopamine and 3.02 mg APS into 5 ml TMV solution (*see* **Note 12**).
2. Stir the mixture at room temperature for 4 h (*see* **Note 13**).
3. Adjust the pH value of the reaction solution to 7.8 by adding 0.1 M NaOH.
4. Keep stirring for 2 h.
5. Dialyze the solution against ultrapure water to remove free polydopamine.
6. TMV–polydopamine long fibers are obtained.
7. Mix 2 ml of 0.5 mg/ml TMV–polydopamine solution with 10 μl 1% (w/v) $HAuCl_4$ solution.
8. Shake the mixture by hand for 2 min at room temperature, and then keep it in the dark for 24 h.
9. Long fibers of TMV–polydopamine–gold nanoparticle composites are obtained.

Fig. 4 TEM image of TMV–F127 assemblies following negative staining with UAc. Scale bar: 100 nm [14]

3.5 "PEGylation" of TMV: TMV–F127 Assemblies

1. Add 0.5 ml TMV solution to 4.5 ml F127 solution and mix.
2. Put the mixture to a temperature above 30 °C (*see* **Note 14**).
3. Stir the mixture for 30 min gently.
4. Transfer the mixture onto carbon-coated copper grids for TEM (*see* Fig. 4).

4 Notes

1. To prepare the 0.01 M potassium phosphate buffer, the volume ratio of 0.01 M KH_2PO_4 solution and 0.01 M K_2HPO_4 solution is approximately 9:91 for pH 7.8 buffer, and 20:80 for pH 7.4 buffer.
2. The corresponding TMV number density is 2.88×10^{12} particles/ml (for 0.2 mg TMV/ml) and 1.15×10^{13} particles/ml (for 0.8 mg TMV/ml), respectively.
3. PSS has two functions in this system: (a) It acts as a dopant acid for the PANi to make them water-soluble and conductive. (b) The high electronegativity of PSS provides the composite nanofibers with a charged surface to prohibit their aggregation.
4. To prepare the 0.05 M acetate buffer of pH 5.5, the volume ratio of 0.05 M NaAc solution and 0.075 M HAc solution is approximately 89:11.
5. The critical micelle temperature of F127 at 1 mg/ml is 30 °C.
6. The stabilized emulsion droplets measure 10–100 μm in size.
7. In the Pickering emulsion, the orientation of TMV is depending on the initial TMV concentration. At a TMV concentration of 0.2 mg/ml in the bulk solution, individual

TMV particles are oriented parallel to the perfluorodecalin–water interface with no lateral ordering. At 0.8 mg/ml TMV concentration in the bulk solution, TMV particles at the perfluorodecalin–water interface orient normal to the interface.

8. At a higher pH value (such as pH 5.0), PANi will adopt branched poly or oligo forms, resulting in low conductivity of the solution.
9. If the reaction mixture is adjusted to pH 5.0 at **step 2**, the color of the PANi–TMV composite solution will be yellow, which is attributed to the formation of branched poly or oligo-anilines.
10. At this stage, the PANi–TMV composite long fibers exhibit homogeneous diameter and high aspect ratio, but no conductivity due to the branched structures of PANi.
11. The length of the PSS–PANi–TMV composite long fibers can reach several micrometers.
12. The molar ratio of APS to dopamine is 2:1.
13. The mixture should be exposed to air during the stirring for oxidation.
14. Formation of F127 micelles is a necessary step for the assembly process. Thus below 30 °C, which is the critical micelle temperature of F127, no TMV–F127 assembly structure will appear.

Acknowledgments

This work was supported by the 973 Program of China (Grant No. 2013CB933800) and the National Natural Science Foundation of China (Grant No. 51173198 and 51303191).

References

1. Halivni S, Sitt A, Hadar I et al (2012) Effect of nanoparticle dimensionality on fluorescence resonance energy transfer in nanoparticle–dye conjugated systems. ACS Nano 6: 2758–2765
2. Zheng X, Fontana J, Pevnyi M et al (2012) The effects of nanoparticle shape and orientation on the low frequency dielectric properties of nanocomposites. J Mater Sci 47:4914–4920
3. Yang K, Ma YQ (2010) Computer simulation of the translocation of nanoparticles with different shapes across a lipid bilayer. Nat Nanotechnol 5:579–583
4. Dasgupta S, Auth T, Gompper G (2014) Shape and orientation matter for the cellular uptake of nonspherical particles. Nano Lett 14:687–693
5. Shukla S, Eber FJ, Nagarajan AS et al (2015) The impact of aspect ratio on the biodistribution and tumor homing of rigid soft-matter nanorods. Adv Healthc Mater 4:874–882
6. Agarwal R, Singh V, Jurney P et al (2013) Mammalian cells preferentially internalize hydrogel nanodiscs over nanorods and use shape-specific uptake mechanisms. Proc Natl Acad Sci U S A 110:17247–17252
7. Niikura K, Matsunaga T, Suzuki T et al (2013) Gold nanoparticles as a vaccine platform: influence of size and shape on immunological responses in vitro and in vivo. ACS Nano 7:3926–3938
8. Liu Y, Zhang W, Zhu Y et al (2013) Architecturing hierarchical function layers on self-assembled viral templates as 3D nano-array

electrodes for integrated Li-ion microbatteries. Nano Lett 13:293–300
9. Lin Y, Balizan E, Lee LA et al (2010) Self-assembly of rodlike bio-nanoparticles in capillary tubes. Angew Chem Int Ed 49:868–872
10. Niu Z, Bruckman M, Kotakadi VS et al (2006) Study and characterization of tobacco mosaic virus head-to-tail assembly assisted by aniline polymerization. Chem Commun:3019–3021
11. Niu Z, Liu J, Lee LA et al (2007) Biological templated synthesis of water-soluble conductive polymeric nanowires. Nano Lett 7:3729–3733
12. Li T, Zan X, Winans RE et al (2013) Biomolecular assembly of thermoresponsive superlattices of the tobacco mosaic virus with large tunable interparticle distances. Angew Chem Int Ed Engl 52:6638–6642
13. Liu Z, Qiao J, Niu Z et al (2012) Natural supramolecular building blocks: from virus coat proteins to viral nanoparticles. Chem Soc Rev 41:6178–6194
14. Liu Z, Gu J, Wu M et al (2012) Nonionic block copolymers assemble on the surface of protein bionanoparticle. Langmuir 28:11957–11961
15. Wu M, Shi J, Fan D et al (2013) Biobehavior in normal and tumor-bearing mice of tobacco mosaic virus. Biomacromolecules 14:4032–4037
16. Miller RA, Presley AD, Francis MB (2007) Self-assembling light-harvesting systems from synthetically modified tobacco mosaic virus coat proteins. J Am Chem Soc 129:3104–3109
17. Miller RA, Stephanopoulos N, McFarland JM et al (2010) Impact of assembly state on the defect tolerance of TMV-based light harvesting arrays. J Am Chem Soc 132:6068–6074
18. Chiang CY, Epstein J, Brown A et al (2012) Biological templates for antireflective current collectors for photoelectrochemical cell applications. Nano Lett 12(11):6005
19. Kobayashi M, Seki M, Tabata H et al (2010) Fabrication of aligned magnetic nanoparticles using tobamoviruses. Nano Lett 10:773–776
20. Lim JS, Kim SM, Lee SY et al (2010) Biotemplated aqueous-phase palladium crystallization in the absence of external reducing agents. Nano Lett 10:3863–3867
21. Zan X, Feng S, Balizan E et al (2013) Facile method for large scale alignment of one dimensional nanoparticles and control over myoblast orientation and differentiation. ACS Nano 7:8385–8396
22. Bruckman MA, Jiang K, Simpson EJ et al (2014) Dual-modal magnetic resonance and fluorescence imaging of atherosclerotic plaques in vivo using VCAM-1 targeted tobacco mosaic virus. Nano Lett 14:1551–1558
23. He J, Niu Z, Tangirala R et al (2009) Self-assembly of tobacco mosaic virus at oil/water interfaces. Langmuir 25:4979–4987

Chapter 11

Bottom-Up Assembly of TMV-Based Nucleoprotein Architectures on Solid Supports

Christina Wege and Fabian J. Eber

Abstract

RNA-guided self-assembly of tobacco mosaic virus (TMV)-like nucleoprotein nanotubes is possible using 3′-terminally surface-linked scaffold RNAs containing the viral origin of assembly (OAS). In combination with TMV coat protein (CP) preparations, these scaffold RNAs can direct the growth of selectively addressable multivalent carrier particles directly at sites of interest on demand. Serving as adapter templates for the installation of functional molecules, they may promote an integration of active units into miniaturized technical devices, or enable their presentation on soft-matter nanotube systems at high surface densities advantageous for, for example, biodetection or purification applications. This chapter describes all procedures essential for the bottom-up fabrication of "nanostar" colloids with gold cores and multiple TMV-like arms, immobilized in a programmable manner by way of hybridization of the RNA scaffolds to oligodeoxynucleotides exposed on the gold beads.

Key words Tobacco mosaic virus (TMV), Biotemplate, Self-assembly, Bottom-up, Coat protein (CP), RNA, Origin of assembly (OAS), Hybridization, Programmable, Nanostar

1 Introduction

The multivalent ribonucleoprotein particles of tobacco mosaic virus (TMV) offer a number of unique opportunities for the design and fabrication of nanotubular components adapted to specific applications, which range from technical uses in composites with inorganic compounds to biomedical, purification, and detection purposes [1–4]. Major contributors to the versatile processing capacities of TMV are its simple and robust rod-shaped capsid structure and a self-assembly mechanism that enables an externally controlled growth of virus-derived architectures also in vitro.

Natural virions of TMV consist of about 2130 identical coat protein (CP) subunits incorporating a single genomic RNA strand of 6395 nucleotides length, sandwiched and protected between a right-handed CP helix. The 300 nm × 18 nm tubes include a 4 nm wide longitudinal channel lined with the internal loops of the CP

Christina Wege and George P. Lomonossoff (eds.), *Virus-Derived Nanoparticles for Advanced Technologies: Methods and Protocols*, Methods in Molecular Biology, vol. 1776, https://doi.org/10.1007/978-1-4939-7808-3_11,

© Springer Science+Business Media, LLC, part of Springer Nature 2018

molecules, whereas both their N- and C-termini are exposed on the outer nanotube surface [5–7]. In addition to chemical reactions allowing functionalization of the wildtype (wt) TMV CP subunits, several mutant virus variants have been generated that promote the facile coupling of linker and effector molecules to the nanotube surfaces (in this book, *see* e.g., Chapters 3, 24–28, 35, and 37). This enables the immobilization of, for example, signal-generating, substrate-converting or target-anchoring units at high surface densities and with precise spacing on the 700 CPs accessible per 100 nm biotemplate length. As TMV may be produced in large amounts from plants, and as the physicochemical properties of the viral surface seem to be advantageous for a stable performance of certain protein functions [8], applications tested with TMV-based carrier templates focus on the presentation of bioactive molecules both in vitro and in vivo to an increasing extent [1].

The RNA-directed self-organization of TMV and its building blocks can take place also in vitro [9], giving these plant viral templates additional value, especially for uses in nanostructured bioinorganic hybrid layouts, e.g., for array-based procedures. The TMV capsid exhibits an “open” shape as it may incorporate not only cognate viral RNA, but also heterologous or shortened sequences ([10], and references therein, [11, 12]), which will eventually delimit the final length of the resulting particle. Hence, appropriate RNA molecules may guide an efficient production of virus-resembling tubes on demand and, optionally, of tailored aspect ratio. Furthermore, TMV in vitro assembly proceeds in a bidirectional and nonsymmetric mode, which opens intriguing possibilities for its redesign. Not only may nonlinear kinked and branched products be established [13], the mechanism will also work on 3′-terminally immobilized RNA strands to grow nucleoprotein helices bottom-up directly at sites of interest on solid supports [14–16]. If covalently coupled RNA is employed [15, 16], such procedures give rise to biotemplate sticks stably interconnected with the support. Upon use of hybridization-mediated RNA linkage, distinct assembly sites may be programmed by way of the respective target sequences, to be equipped with, for example, TMV-like arms of different length [14].

This chapter describes the protocols for the bottom-up fabrication of “nanostar” colloids with gold cores and adjustable numbers of soft-matter arms accessible to subsequent functionalization, in two different length classes (*see* Fig. 1). Hybridization-mediated RNA immobilization makes use of sequences approved in this context [14]. The methods may be easily transferred to related layouts, be it flat gold-covered substrates, RNA scaffolds ligated covalently at their 3′-terminus to DNA anchors [15], other OAS-containing RNA species, and/or different TMV CP variants from plants or heterologous expression hosts, optionally employed in blends to

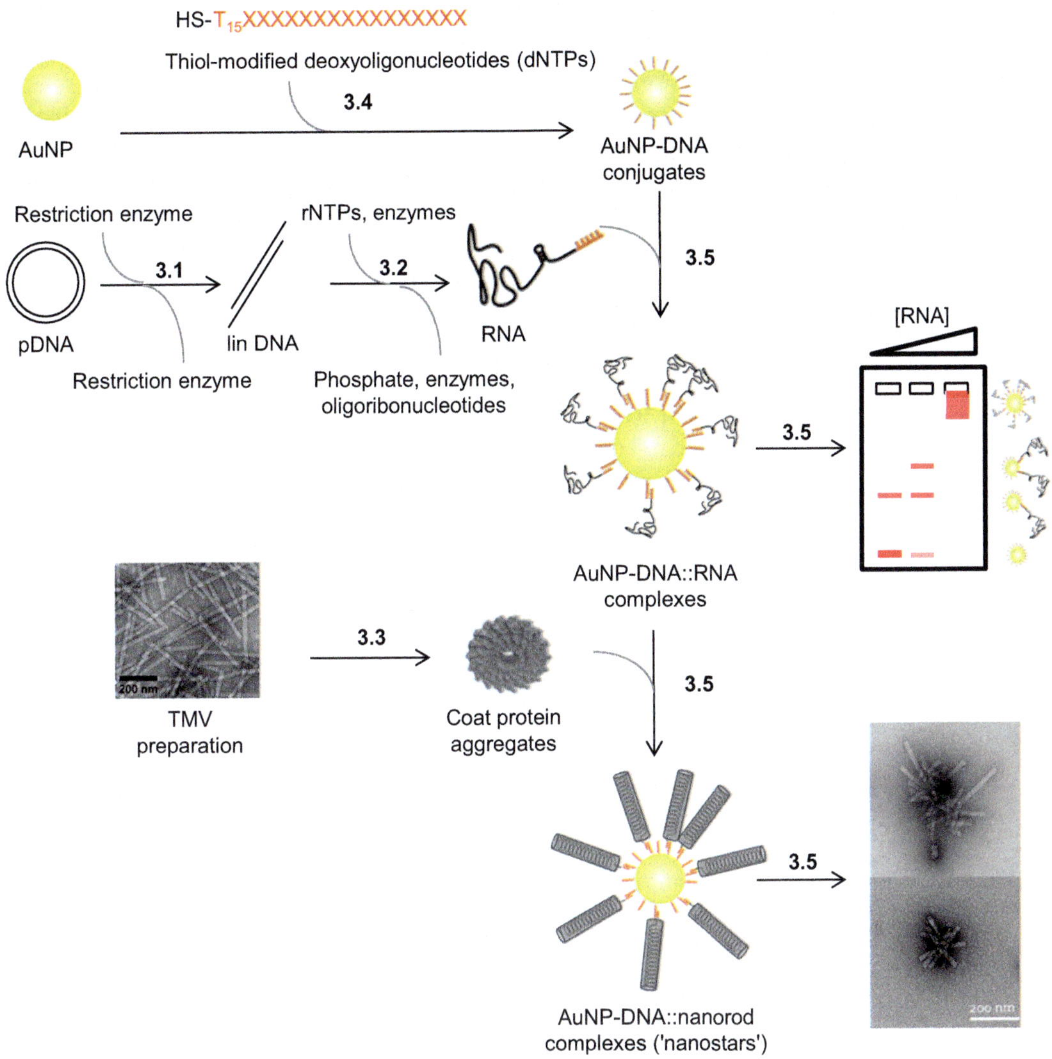

Fig. 1 Workflow for the production of colloidal AuNP-DNA::nanorod complexes ("nanostars" composed of gold cores and TMV-like soft-matter arms). Individual process steps are marked with a reference to the respective section of this chapter

integrate distinct functionalities [10–12, 17–19] in individual surface-linked nanotubes.

The following paragraph describes a few mechanistic details resolved for the hierarchical in vitro assembly stages of TMV-like nucleoprotein particles from CP and RNA components, which are essential for enabling their bottom-up fabrication on solid supports. TMV-like tubes may be grown from suitable CP preparations in the presence of RNA strands provided that the respective constructs contain the viral origin of assembly (OAS) sequence [20–22]. A specific OAS RNA stem-loop was shown to nucleate TMV particle formation most efficiently in combination with CP preincubated under conditions advancing its oligomerization into ring-shaped two-layered 34-meric aggregates (named "disks"), which coexist with various further types of CP oligomers in the respective preparations [9]. A selective interaction of assembly-competent CP intermediates, most likely in the disk state, with the OAS hairpin-loop initiates a nonsymmetric bidirectional growth of a TMV-like CP-RNA helix. According to the prevalent model, a conformational transition of the disk into a short protohelix or "lockwasher" leads to an integration of the RNA strand between the CP helix, which precedes a fast tube elongation toward the 5′-end of the RNA and a concomitant slow growth in 3′-direction [23]. The fast process results from a serially repeated, probably cooperative addition of further disks, along with a stepwise incorporation of adjacent RNA portions 5′ of the OAS start site. The exposed RNA at the "head" of the nascent capsid folds into a "traveling loop," while the rest of the nonencapsidated RNA 5′-segment resides within the central channel of the growing tube and is pulled up until its very 5′ end is packaged. In contrast to this high-speed process, the 3′-portion of the RNA undergoes a severalfold slower 5′ to 3′ packaging, which uses small CP oligomers to progress to its 3′-terminus. For this reason, RNA scaffolds bound to appropriate anchor sites covalently or by means of nucleic acid hybridization, may direct a bottom-up assembly of TMV-like tubes efficiently, employing the methods detailed below.

2 Materials

2.1 Plasmids

1. Depending on the required length of the TMV-like tubes on the gold surface and whether you want to include controls for selective hybridization, select one or more plasmids from the following list:
2. pGEM/TMV4407-6191 containing part of the cDNA sequence of TMV (nts4407-6191) including the OAS, yielding RNA_s-SeqA: a 2253 nts RNA with the sequence SeqA at its 3′-end if linearized with *Pci*I prior to in vitro transcription with T7 RNA polymerase [14].
3. pGEM/TMV4407-6395 containing part of the cDNA sequence of TMV (nts4407-6395) including the OAS, yielding RNA_s-SeqB: a 1989 nts RNA with the sequence SeqB at its 3′-end if linearized with *Bsi*WI prior to in vitro transcription with T7 RNA polymerase [14].
4. p843pe35TMVr.1 containing the cDNA of the complete TMV genome [24] including the OAS, yielding RNA_l-SeqB: a 7025 nts RNA with the sequence SeqB at its 3′-end if linearized with *Bsi*WI prior to in vitro transcription with T3 RNA polymerase [14].
5. For a fast overview of RNA preparation/application treatments, *see* Table 1.

Table 1
Materials and enzymes to be used for linearization of plasmid DNA templates for in vitro transcription of three TMV-derived assembly-competent RNAs with different 3′ end sequences [14]

Plasmid	Restriction enzyme	RNA polymerase	RNA	RNA 3′ end sequence name	RNA end sequence
pGEM/ TMV4407-6191	*Pci*I	T7	RNA_s-SeqA	SeqA	5′-... CCACAGAAUCAGGGGAUAACGCAGGAAAGAACAUG-3′
pGEM/ TMV4407-6395	*Bsi*WI	T7	RNA_s-SeqB	SeqB	5′-...GAUUGUGUCCGUAAUCACACGUGGUGCGUAC-3′
p843pe35TMVr.1	*Bsi*WI	T3	RNA_l-SeqB	SeqB	5′-...GAUUGUGUCCGUAAUCACACGUGGUGCGUAC-3′
pGEM/ TMV4407-6395	*Bsi*WI	T7	RNA_s-SeqB	SeqB	5′-...GAUUGUGUCCGUAAUCACACGUGGUGCGUAC-3′

2.2 Thiol-Modified Oligonucleotides

1. HS-T_{15}-Seq$\bar{A}$: 5′-HS-T_{15}-CATGTTCTTTCCTGCGTTATCC CCTGATTCTGTGG-3′ (with its 3′-terminal sequence portion Seq$\bar{A}$ reverse-complementary to the 3′-end of the 2253 nts RNA).
2. HS-T_{15}-Seq$\bar{B}$: 5′-HS-T_{15}-GTACGCACCACGTGTGATTAC GGACACAATC-3′ (with its 3′-terminal sequence portion Seq$\bar{B}$ reverse-complementary to the 3′-end of the 1989 nts and the 7025 nts RNA).

2.3 Citrate-Coated Gold Nanoparticles (AuNPs)

10 nm and 50 nm citrate-coated AuNPs [e.g., from Sigma-Aldrich (Munich, Germany) or BBInternational (Cardiff, UK)].

2.4 TMV Particles from Plants

10 mg/ml TMV (please refer for example to [25] and/or to Chapters 23, 24, and 27 of this book for details about the preparation of TMV particles from plants, e.g., from *Nicotiana tabacum* 'Samsun' nn leaf tissue infected with TMV).

2.5 Enzymes, Commercial Kits, and Molecular Weight Standards

1. Plasmid DNA preparation kit: QIAprep Spin Miniprep kit (Qiagen, Hilden, Germany).
2. T7 RNA polymerase-based in vitro transcription kit: MEGAscript® T7 High Yield Transcription Kit (Ambion, Austin, USA/ThermoFisher Scientific). The kit contains the following components required for RNA preparation with this protocol: T7 Enzyme Mix, 10× Reaction Buffer, ATP Solution, CTP Solution, GTP Solution, UTP Solution, TURBO DNase, Lithium Chloride Precipitation Solution (7.5 M LiCl, 50 mM EDTA), and nuclease-free water.
3. T3 RNA polymerase-based in vitro transcription kit: MEGAscript® T3 High Yield Transcription Kit (Ambion, Austin, USA/ThermoFisher Scientific). The kit contains the same components as listed previously, except for T3 (instead of T7) Enzyme Mix.
4. *Bsi*WI and *Pci*I restriction enzymes with the buffer recommended by the manufacturer.
5. DNA molecular weight standard (e.g., Lambda DNA, *Hin*dIII or *Pst*I-digested).
6. ssRNA molecular weight standard (e.g., ssRNA ladder #N0362S, New England Biolabs).

2.6 Buffers, Solutions, and Instrumentation

All buffers and solutions are prepared with molecular biology grade reagents and ultrapure water (18.2 MΩ•cm at 25 °C) unless otherwise stated. Except for amine-containing compounds, all solutions should be treated with DMDC to inactivate RNases (*see* **Note 1** for exceptions). For DMDC-treatment, add 1 ml of DMDC to 1 l of ultrapure water or buffer and stir vigorously overnight. Autoclave at 121 °C for 20 min to remove residual DMDC. Prepare stock solutions according to standard molecular biology procedures [26]. Solutions contained in the commercial kits (*see* **items** in Subheading 2.5) are not listed separately.

1. Ultrapure water (18.2 MΩ·cm at 25 °C).
2. DMDC-treated ultrapure water.
3. 0.5 M ethylenediaminetetraacetic acid (EDTA) pH 8.0; pH adjusted with NaOH.
4. 3 M sodium acetate, pH 4.8; pH adjusted with glacial acetic acid.
5. 3 M sodium acetate, pH 4.8; pH adjusted with glacial acetic acid, treated with DMDC.
6. Ethanol (Rotisol®, Carl Roth).
7. 70% (v/v) ethanol.
8. 70% (v/v) ethanol in DMDC-treated ultrapure water.
9. 10% (w/v) sodium dodecyl sulfate (SDS).
10. 5 mg/ml Proteinase K.
11. Dimethyl dicarbonate (DMDC).
12. Phenol, saturated with 10 mM Tris–HCl buffer pH 8.0 and 1 mM EDTA (e.g., Sigma Phenol BioUltra, for molecular biology, TE-saturated, ~73%; for hazards and safety precautions, *see* **Note 2**).
13. Chloroform.
14. Isoamyl alcohol.
15. PCI: phenol–chloroform–isoamyl alcohol mixture, 25:24:1 (v/v/v): prepare by mixing 1 volume of buffer-saturated phenol with 1 volume of chloroform–isoamyl alcohol (24:1 (v/v)); use lower phase separated after mixing.
16. Glacial acetic acid (ice cold).
17. Dialysis tubing (10 kDa molecular weight cut-off), wetted/pretreated according to the manufacturer's instruction: wash with ultrapure water before use.
18. 1% (w/v) sodium azide (NaN_3).
19. 10× MOPS buffer: 20 mM EDTA, 200 mM 3-(N-morpholino)-propane-sulfonic acid (MOPS), 50 mM sodium acetate, adjust pH to 7.0 using 10 M NaOH. Dilute to 1× MOPS with DMDC-treated ultrapure water.
20. 37% (v/v) formaldehyde stock solution.
21. 10 mg/ml ethidium bromide.
22. RNA denaturation solution: 25 μl 10× MOPS buffer, 125 μl formamide (deionized), 44 μl formaldehyde (of 37% (v/v)

stock solution), 0.75 µl ethidium bromide (of 10 mg/ml stock solution), 8 µl DMDC-treated ultrapure water.

23. Gel loading solution: 30% (v/v) glycerol (f.c.), 50 mM EDTA, pH 8 (f.c.; added from 0.5 M EDTA pH 8.0 stock solution), 0.001% (w/v) bromophenol blue (f.c.).
24. 5× TBE buffer: 54 g/l of Tris base, 27.5 g/l of boric acid, 20 ml/l of 0.5 M EDTA (pH 8.0). Dilute to 1× TBE and 0.5× TBE with water.
25. 10× AuNP loading buffer: 25% (w/v) Ficoll (f.c.) in 5× TBE. Dilute to 1× AuNP loading buffer with DMDC-treated ultrapure water.
26. Agarose (NEEO ultra-quality, e.g., Carl Roth).
27. 75 mM SPP buffer pH 7.2.
28. 0.1 M SPP buffer pH 7.0.
29. 1 M NaCl.
30. 2× Sodium salt citrate buffer (SSC): 0.3 M NaCl, 0.03 M sodium citrate (pH 7.0 with HCl), treat with DMDC.
31. Formvar®-carbon-coated copper grids (400 mesh, Ted Pella Inc., Redding, USA).
32. Staining solution: 1% (w/v) uranyl acetate, 250 µg/ml bacitracin, prepare freshly before use from 2× concentrated separate stock solutions by mixing equal volumes of each stock component.
33. PCR cycler.
34. Parafilm®, filter paper.

3 Methods

3.1 Template Preparation for In Vitro Transcription

In vitro transcription of assembly-competent RNA may be performed from linearized plasmid DNA (pDNA) or PCR products. For linearization of pDNA (isolated from bacterial cultures by means of a standard plasmid preparation kit), restriction enzymes are used. To remove potential RNase contaminations brought into the reaction by the restriction enzyme preparation, the linearized pDNA is purified by phenol-chloroform extraction and ethanol precipitation.

Depending on the RNA to be transcribed and the promoter sequence encoded in the plasmid, different restriction enzymes and RNA polymerases have to be used (*see* Table 1).

3.1.1 Plasmid Linearization

1. Use 20 µg of each of the three pDNA types for linearization (*see* Table 1). Every plasmid is treated in a total volume of 500 µl with 50 U of restriction enzyme (*Bsi*WI or *Pci*I according to Table 1) in the buffer recommended by the manufacturer. Incubate for 3 or 16 h, respectively.

2. Prepare a 1% (w/v) agarose gel in 1× TBE supplemented with 0.5 μg/ml ethidium bromide.
3. Add 2 μl of gel loading solution and 6 μl ultrapure water to 2 μl of every digestion mixture, incubate for 5 min at 60 °C and load onto the 1% agarose gel.
4. Prepare and load an appropriate DNA molecular weight standard in parallel.
5. Run the gel for 30–60 min at 5 V/cm and visualize under UV light. If only one single band is visible per reaction mixture, the reaction is complete.
6. Add 25 μl 0.5 M EDTA and 53 μl 3 M sodium acetate pH 4.8 to every reaction mixture containing linearized plasmid DNA, mix, add 1 ml EtOH and mix again.
7. Incubate at −20 °C for 30 min up to 1 h.
8. Centrifuge at 20,000 × *g* and 4 °C for 15 min.
9. Discard the supernatants and wash the precipitates by adding 1 ml 70% EtOH.
10. Centrifuge (as in **step 8**), remove supernatants and shortly centrifuge again for 1 min. Remove all residual liquid with a 10 μl pipet and dry every pellet for a few minutes until its appearance has changed from white to transparent.
11. Dissolve pellet in 100 μl ultrapure water.

3.1.2 Phenol/Chloroform Extraction

1. Add 5 μl SDS (10%) and 3 μl Proteinase K to 100 μl linearized plasmid (from Subheading 3.1.1, **step 9**).
2. Incubate at 50 °C for 30 min.
3. Add 75 μl DMDC-treated ultrapure water and 175 μl PCI. Mix by shaking and inverting the tube several times and centrifuge at 20,000 × *g* and 20 °C for 1 min.
4. Carefully transfer the upper phase into a fresh tube and add 175 μl PCI, mix and centrifuge (as in **step 3**).
5. Carefully transfer the upper phase into a fresh tube and add 175 μl chloroform, mix and centrifuge (as in **step 3**).
6. Repeat **step 5** once, transfer the upper phase into a fresh tube and add 15 μl sodium acetate (treated with DMDC) and, after mixing, 330 μl ethanol.
7. Precipitate the DNA at 20 °C for 30 min to 1 h.
8. Centrifuge at 20,000 × *g* and 20 °C for 15 min.
9. Discard the supernatant and wash the precipitate with 500 μl 70% ethanol (prepared with DMDC-treated ultrapure water).
10. Centrifuge (as in **step 8**), remove supernatant and shortly centrifuge again for 1 min. Remove all residual liquid with a 10 μl

pipet and dry the pellet until its appearance has changed from white to transparent.

11. Dissolve the pellet in 15 μl DMDC-treated ultrapure water and determine the concentration and quality of template DNA by absorption measurement in a microvolume spectrophotometer (*see* **Note 3**).

3.2 RNA In Vitro Production and Analysis

Bacteriophage RNA polymerases are used to transcribe RNA from appropriate DNA templates in vitro, with ribonucleoside triphosphates as educts. Here, bacteriophage T7 RNA polymerase is applied for the generation of two TMV-like particle-scaffolding RNAs of about 2000 nts length (the two "short" RNA_s-SeqA and RNA_s-SeqB) with distinct 3′-terminal ends, as these TMV OAS-containing cDNA sequences are cloned downstream the T7 promoter in the plasmid constructs. Bacteriophage T3 polymerase is used for transcription of an ≈7000 nts RNA scaffold with OAS (the "long" RNA_l-SeqB, *see* **Note 4**). The presence of pyrophosphatase in the MEGAscript® High Yield Transcription Kits increases the RNA yield because it hydrolyzes the RNA polymerase-inhibiting pyrophosphate generated upon RNA elongation. After transcription, RNAs longer than about 200 nts are precipitated in the presence of lithium chloride.

3.2.1 In Vitro Transcription

1. Thaw all components of the MEGAscript® T7 or T3 (*see* Table 1 for correct choice) High Yield Transcription Kit and centrifuge tubes before use. The buffer must be kept at room temperature and mixed before use to avoid the formation of precipitates.
2. The reaction is assembled at room temperature. A typical 20 μl reaction is composed of 2 μl ATP, 2 μl CTP, 2 μl GTP, 2 μl UTP (concentration of the stock solutions: 75 mM each), 2 μl of buffer (10× concentrate), 1 μg of linear DNA template (from Subheading 3.1.2, **step 11**) and 2 μl of the enzyme mixture, according to the kit manufacturer's instructions.
3. After flipping the tube to mix the components and a short centrifugation, transcription is allowed to proceed at 37 °C for about 3 h.
4. Stop the reaction by adding 1 μl TURBO DNase solution (from the MEGAscript® kit) and incubation at 37 °C for another 15 min.
5. Add 30 μl Lithium Chloride Precipitation Solution (from the MEGAscript® kit) to the reaction and precipitate the RNA at −20 °C for 16 h.
6. After centrifugation (4 °C, 20,000 × *g*, 15 min) remove the supernatant and wash the pellet with 500 μl of 70% ethanol (prepared with DMDC-treated water).

7. After centrifugation (4 °C, 20,000 × *g*, 15 min), remove the ethanol, centrifuge the tube again, and remove residual ethanol with a 10 μl pipet.
8. Dry the RNA pellet for a few minutes until its appearance has changed from white to transparent.
9. Dissolve the RNA in 50 μl of DMDC-treated ultrapure water.
10. Dilute 1 μl of RNA solution with 4 μl DMDC-treated ultrapure water and determine the concentration of the RNA (*see* **Note 5**). Apply 1.5 μl of the diluted RNA to a denaturing RNA gel (*see* Subheading 3.2.2).
11. Use the concentration of the diluted RNA to calculate the concentration of the undiluted RNA solution (the RNA stock solution). Adjust the RNA stock solution to 1 μg/μl with DMDC-treated ultrapure water and store in aliquots of 10 μl at −20 °C.

3.2.2 Denaturing Agarose Gel Electrophoresis of RNA

1. For 50 ml of a 0.75% (w/v) agarose gel, suspend 0.4 g agarose in 36 ml DMDC-treated ultrapure water, wait for 2 min to hydrate, and heat at suitable intervals in a microwave oven until the agarose is completely dissolved.
2. Cool the agarose solution to about 50 °C and add 5 ml 10× MOPS buffer, keep at 50 °C in a water bath.
3. Add 9 ml formaldehyde solution (37%) and mix with a stirrer until the agarose preparation is homogeneous.
4. Pour the gel into an electrophoresis tray, add a comb and wait for gelation. Chilling the gel at 4 °C helps to complete gelation.
5. Remove the comb, place gel into electrophoresis chamber filled with 1× MOPS buffer and perform a "prerun" for 5 min (5 V/cm).
6. Supplement 1–5 μl per sample with 15 μl RNA denaturation solution and incubate at 65 °C for 10 min. After heat denaturation of RNA secondary structures, the solution is shock-cooled on ice.
7. Treat an appropriate RNA molecular weight standard according to the manufacturer's instructions in parallel.
8. Add 2 μl of gel loading solution to every denatured RNA sample, mix and load the gel after the "prerun."
9. RNAs are separated for about 3 h at 3–4 V/cm. Buffer has to be transferred at regular intervals from the anode to the cathode reservoir and vice versa (every 30 min, e.g., with a Pasteur pipet; *see* **Note 6**).
10. Ethidium bromide-stained nucleic acid bands are visualized and documented under UV light. For in vitro transcripts, only a single band should be detectable (*see* Subheading 2.1 for expected molecule lengths).

3.3 Preparation of Assembly-Competent TMV CP from Purified Virus Particles

TMV CP for nanotube assembly on RNA scaffolds is prepared from virus particles by an acetic acid-based method [27]. The TMV particles may be isolated from infected *Nicotiana tabacum* "Samsun" nn leaves with high yields following an adapted protocol according to Gooding and Hebert [25] (in this book, *see* Chapters 23 and 27 for descriptions of the isolation method).

1. To a solution of one volume TMV particles (10 mg/ml, typically 600 μl) add two volumes of ice-cold glacial acetic acid (typically 1200 μl) and invert the tube gently until there are no streaks anymore. Incubate on ice for about 20 min.
2. Sediment the RNA of the virus at 20,000 × *g*, 4 °C for 20 min. The CP fraction remains in the supernatant.
3. Fill dialysis tubing with the supernatant, close with two clips and dialyze for about 6–16 h against ultrapure water. Change the water and dialyze again for 6–16 h. Repeat the latter step once more. Check the solution in the dialysis tubing for the occurrence of a white precipitate.
4. When a white precipitate (protein) is visible, remove the solution from the dialysis tubing and centrifuge at 20,000 × *g*, 4 °C for 20 min.
5. Remove the supernatant and add to the pellet 0.5 volumes (compared to the initial TMV solution; typically 300 μl) of 75 mM SPP buffer containing 0.02% sodium azide. Let the CP dissolve at room temperature overnight.
6. Mix the CP by pipetting the solution up and down.
7. Precipitate aggregated CP by centrifuging at 10,000 × *g* for 20 min at 18 °C and use the supernatant for further experiments.
8. Determine the CP concentration of a fivefold and a tenfold dilution in 75 mM SPP photometrically at 280 nm (*see* **Note 7**). Calculate the protein concentration with the extinction factor of

$$\varepsilon = 1.3\ \mu l/(\mu g\ cm).$$

9. For direct assembly leave the CP at room temperature; otherwise it may be stored at 4–8 °C (*see* **Note 8**).

3.4 Coupling Thiol-Modified Oligonucleotides to Gold Nanoparticles (AuNPs)

Commercial preparations of gold nanoparticles (AuNPs in the following) are usually stabilized by negatively charged citrate shells adsorbed on the gold surface. Thiol groups react with elementary gold to form a stable bond between the two compounds [28], a reaction that can be used to couple thiol-terminated deoxyoligonucleotides to AuNPs [29]. If this is applied to citrate-stabilized AuNPs, the citrate adlayer is partly replaced by a nucleic acid shell. To promote formation of gold–thiol bonds between the two partner components, the salt concentration is gradually increased [30].

1. Bring the AuNPs (10 or 50 nm diameter, depending on the intended product type) to a concentration of about 15 nM (typically by centrifugation at 4000 × *g* for 15 min and dissolving the AuNPs in one fifth of the original volume in DMDC-treated ultrapure water) (*see* **Note 9**).
2. Dissolve the thiol-modified oligonucleotide of choice (to hybridize with the intended scaffold RNA, or not to hybridize in the case of control reactions) to a concentration of 100 μM in DMDC-treated ultrapure water.
3. Mix 13.5 nM AuNPs and 5 μM oligonucleotides (both final concentration) and incubate in a rotational mixer for 8–16 h.
4. Adjust the mixture of AuNPs and oligonucleotides to 10 mM SPP pH 7.0 (f.c.) and 58 mM NaCl (f.c.) using 0.1 M SPP pH 7.0 and 1 M NaCl solutions and mix again in a rotational mixer for 8–16 h.
5. Adjust the mixture of AuNPs and oligonucleotides to 10 mM SPP pH 7.0 and 100 mM NaCl as described above and mix again in a rotational mixer for 8–16 h.
6. Adjust the mixture of AuNPs and oligonucleotides to 10 mM SPP pH 7.0 and 300 mM NaCl as described above and 0.01% sodium azide (with 1% sodium azide solution) and mix again in a rotational mixer for 8–16 h. This solution may be stored at room temperature for several days.

3.5 Stepwise Assembly of RNA Scaffolds on AuNP, and TMV CP on the Immobilized RNA

To produce nanostar-shaped AuNP-DNA::nanorod complexes, first an OAs-containing scaffold RNA is hybridized to the oligonucleotide-modified AuNPs [31, 32]. The RNA will hybridize if its 3′ end is complementary to the DNA oligomer on the AuNP (e.g., RNA_s-SeqA will hybridize to AuNPs modified with HS-T_{15}-Seq$\bar{A}$). Use RNAs with a 3′ end noncomplementary to the DNA oligomer on the AuNPs (e.g., RNA_s-SeqB with AuNPs modified with HS-T_{15}-Seq$\bar{A}$) as a negative control to detect unspecific RNA binding. The resulting AuNP-DNA::RNA complexes are analyzed by native agarose gel electrophoresis to determine the amount of RNA molecules bound to the AuNP-DNA conjugates. Afterward, TMV CP in its assembly-competent state is added to the AuNP-DNA::RNA complexes to promote the bottom-up formation of nanorods through RNA-directed assembly. TMV-like nanrod (i.e., soft-matter arm) length will depend on that of the RNA scaffolds, with shorter RNAs yielding more homogeneous arm length distributions than longer ones [14]. To demonstrate that there is no unspecific binding of nanotubes that were assembled on free RNA (which may be present at a low concentration in the AuNP-DNA::RNA preparation) to DNA-modified AuNPs, a further control reaction is set up with "preassembled" nanotubes and DNA-modified AuNPs. To visualize the nonmetallic biological structures in the hybrid objects, the soft-matter components of the AuNP-DNA::nanorod complexes are visualized via electron microscopy (uranyl acetate is used as a negative-contrasting agent) [17].

3.5.1 Hybridization of RNA to Oligonucleotide-Modified AuNPs

1. Centrifuge oligonucleotide-modified AuNPs (from Subheading 3.4, **step 6**) at 4000 × *g* for 15 min. Resuspend the pellet in 2× SSC (use the original volume of the oligonucleotide-modified AuNPs). Repeat this step two times and determine the concentration of AuNPs (*see* **Note 11**).
2. Mix 15–20 nM AuNP (f.c.) and 200 nM RNA (f.c.) (from Subheading 3.2.1, **step 11**) in 2× SSC (*see* **Note 10**, f.c. refers to the final concentration of AuNP and RNA in the mixture). You may also prepare a negative control sample without RNA.
3. The hybridization procedure is performed in a PCR cycler. Heat the mixture to 65 °C for 5 min and slowly cool down to 30 °C with a rate of 0.01 °C/s. Then keep the mixture at 30 °C for 5 h.
4. After hybridization, wash the AuNP-DNA::RNA complexes three times with 75 mM SPP buffer (pH 7.2) by help of centrifugation, as described in **step 1**.
5. Dissolve the AuNP-DNA::RNA complexes in 75 mM SPP pH 7.2 to yield a concentration of about 20 nM in respect to the AuNPs (*see* **Note 11**).

3.5.2 Native Agarose Gel Electrophoresis of AuNP-DNA::RNA Complexes

1. Prepare a 1% agarose gel supplemented with 0.6 mg/ml ethidium bromide in 0.5× TBE buffer.
2. Apply 0.2 pmol AuNP-DNA::RNA complexes in 1× AuNP loading buffer to the 1% agarose gel.
3. Run the gel at 5 V/cm for about 2 h. The gel is inspected under white light (AuNPs bands appear red) and under UV light (to detect ethidium bromide-stained bands with nucleic acid). The negative control sample containing oligonucleotide-modified AuNPs only is expected to show under both UV-light and white light a single band. In comparison, AuNP-DNA::RNA complexes migrate with reduced velocity during electrophoresis. The higher the number of RNAs bound to the oligonucleotide-modified AuNPs, the lower the electrophoretic mobility. Thus, if hybridization was successful, retarded bands in reference to the oligonucleotide-modified AuNPs will indicate samples containing AuNP-DNA::RNA complexes. At very high surface densities of RNA strands on the gold beads, only a single band or smear of negligible electrophoretic mobility occurs (Fig. 1, step 3.5).

3.5.3 Bottom-Up Assembly of AuNP::Nanorod Complexes ("Nanostars")

1. To initiate the assembly of TMV-like nanorod arms, mix 0.2 pmol AuNP-DNA::RNA with about 10 μg assembly-competent TMV CPs dissolved in 75 mM SPP pH 7.2 (from Subheading 3.3, **step 9**) in a total volume of 10–15 μl in 75 mM SPP buffer pH 7.2.
2. Incubate the mixture for at least 5 h at room temperature.

3.5.4 Preparation of a Negative Control

This control experiment is carried out to detect potential unspecific interactions of "preassembled" nanotubes with AuNP-DNA.

1. Mix assembly-competent CPs with RNA to a final concentration of 2.6 mg/ml CP and 0.1 mg/ml RNA (corresponding to about 160 nM of a 2000 nts long RNA) in 75 mM SPP buffer pH 7.2 (typical reaction volume: 10 μl).
2. Incubate the mixture for at least 5 h at room temperature.
3. Add 1 volume (v/v) of AuNP-DNA (from Subheading 3.4, **step 6**) with a concentration of 30–40 nM. This yields a mixture with 15–20 nM AuNP-DNA and preassembled nanotubes with a concentration of about 80 nM RNA.
4. Incubate this (control) mixture for at least 5 h at room temperature.

3.5.5 Sample Preparation for Electron Microscopy

1. Dilute 2 μl of the sample from Subheading 3.5, **step 3** (or from Subheading 3.5, **step 4** for the negative control) with 8 μl water and adsorb for 2 min to a Formvar®-carbon-coated copper grid: pipet 10 μl of the respective sample on a sheet of Parafilm®, dip the coated copper grid in absolute ethanol (to hydrophilize the carbon surface), remove ethanol from the grid with a small piece of filter paper and place the grid "overhead" onto the sample droplet.
2. Remove residual sample liquid from the grid with a filter paper.
3. Place the grid on a drop (10 μl) of ultrapure water for a few seconds, remove water and repeat this washing procedure twice.
4. Place the grid on a drop (10 μl) of staining solution for a few seconds, remove staining solution with a filter paper and repeat this "prestaining" procedure twice.
5. Place the grid on a drop (10 μl) of staining solution for 5 min. Remove the staining solution with filter paper.
6. Visualize stained AuNP-DNA::nanorod complexes in an electron microscope.

4 Notes

1. DMDC reacts with amines. Thus, amine-containing buffer solutions such as Tris-based buffers cannot be treated with DMDC. In this case, to prepare RNase-free buffers, use DMDC-treated ultrapure water, certified RNase-free buffer substances, spatula baked for 3 h at 200 °C and stirring bars and glass bottles from the preparation of DMDC-treated ultrapure water (after autoclaving).
2. Phenol is a hazardous corrosive and toxic compound. Before use, read the supplier's documentation carefully and follow all safety instructions recommended. Ensure to work under a hood, to wear protective equipment including goggles and gloves, and to have in close vicinity an appropriate polyethylene glycol (PEG-300 or –400) stock for fast first-aid treatment in case of dermal exposure. After PCI has been mixed from the individual compounds, an upper phase will separate from the major mixed PCI fraction within a few minutes. For nucleic acid purification, use the lower PCI phase.
3. Nucleic acids absorb at 260 nm, phenol additionally absorbs at 230 nm. Thus, a ratio of A260 to A230 lower than 2 is an indicator for residual phenol in the preparation, which may lead to aberrant in vitro transcription.
4. The RNA polymerase to be used for transcription depends on the RNA polymerase promoter present on the DNA template. Here, two different RNA polymerases are used, because the starting plasmid p843pe35TMVr.1 carries a T3 promoter and because the vector used for cloning (pGEM-T) carries a T7 promoter.
5. Nucleic acid concentrations are determined by UV absorption measurement at 260 nm. A 40 μg/ml RNA (or a 50 μg/ml double stranded DNA) solution absorbs 1 A260/cm.
6. Alternatively, use an electrophoresis chamber with buffer circulation.
7. The typical yield of the preparation of TMV CP is 50–60%.
8. The CP may be used up to 6 months although fresh CP assembles a little bit better. When stored at 4 °C (which we recommend for periods of more than a few days), bring the CP preparation to, and keep it at room temperature for at least 16 h before performing the assembly, to let the CP adopt an assembly-competent state. Centrifuge at 10,000 × *g* at 18 °C for 20 min and check the concentration before adding the CP solution to the assembly reaction.
9. Gold nanoparticles <100 nm have a characteristic red color due to surface plasmon resonance. When the stabilizing shell

is disrupted, e.g., due to an increase in the salt concentration during the coupling reaction, the nanoparticles may aggregate. This aggregation is easily detected by loss of the red color of the solution. Aggregated gold nanoparticles cannot be used for further production and must be discarded. To avoid aggregation during the coupling reaction (which requires an increase in the salt concentration), the salt concentration is gradually increased.

10. The RNA concentration in the hybridization reaction may be varied (e.g., from 50 to 500 nM) to yield variable surface densities of RNA on the AuNPs (after hybridization) and thus variable surface densities of nanorods (after assembly), resulting in "nanostars" with different numbers of soft-matter arms.
11. The concentration of gold nanoparticles is determined by absorption measurement at 520 nm. A 10^{-8} M or 5×10^{-11} M solution of 10 nm or 50 nm AuNPs, respectively, shows an absorption of 1 A520 per 1 cm pathlength.

References

1. Koch C, Eber FJ, Azucena C, Förste A, Walheim S, Schimmel T, Bittner A, Jeske H, Gliemann H, Eiben S, Geiger F, Wege C (2016) Novel roles for well-known players: from tobacco mosaic virus pests to enzymatically active assemblies. Beilstein J Nanotechnol 7:613–629. https://doi.org/10.3762/bjnano.7.54
2. Culver JN, Brown AD, Zang F, Gnerlich M, Gerasopoulos K, Ghodssi R (2015) Plant virus directed fabrication of nanoscale materials and devices. Virology 479–480:200–212. https://doi.org/10.1016/j.virol.2015.03.008
3. Bittner AM, Alonso JM, Górzny ML, Wege C (2013) Nanoscale science and technology with plant viruses and bacteriophages. In: Mateu MG (ed) Structure and physics of viruses: an integrated textbook, Subcellular biochemistry, vol 68. Springer Science+Business Media, Dordrecht, pp 667–702. https://doi.org/10.1007/978-94-007-6552-8_22. (ISSN: 0306-0225)
4. Wen AM, Steinmetz NF (2016) Design of virus-based nanomaterials for medicine, biotechnology, and energy. Chem Soc Rev 45:4074–4126. https://doi.org/10.1039/C5CS00287G
5. Caspar DLD (1963) Assembly and stability of the tobacco mosaic virus particle. Adv Protein Chem 18:37–121. https://doi.org/10.1016/S0065-3233(08)60268-5
6. Namba K, Pattanayek R, Stubbs G (1989) Visualization of protein-nucleic acid interactions in a virus. Refined structure of intact tobacco mosaic virus at 2.9 a resolution by X-ray fiber diffraction. J Mol Biol 208(2):307–325. https://doi.org/10.1016/0022-2836(89)90391-4
7. Fromm SA, Bharat TAM, Jakobi AJ, Hagen WJH, Sachse C (2015) Seeing tobacco mosaic virus through direct electron detectors. J Struct Biol 189(2):87–97. https://doi.org/10.1016/j.jsb.2014.12.002
8. Koch C, Wabbel K, Eber FJ, Krolla-Sidenstein P, Azucena C, Gliemann H, Eiben S, Geiger F, Wege C (2015) Modified TMV particles as beneficial scaffolds to present sensor enzymes. Front Plant Sci 6:1137. https://doi.org/10.3389/fpls.2015.01137
9. Butler PJG (1999) Self-assembly of tobacco mosaic virus: the role of an intermediate aggregate in generating both specificity and speed. Philos Trans R Soc Lond B Biol Sci 354(1383):537–550. https://doi.org/10.1098/rstb.1999.0405
10. Schneider A, Eber FJ, Wenz N, Altintoprak K, Jeske H, Eiben S, Wege C (2016) Dynamic DNA-controlled "stop-and-go" assembly of well-defined protein domains on RNA-scaffolded TMV-like nanotubes. Nanoscale 8:19853–19866. https://doi.org/10.1039/C6NR03897B
11. Wu Z, Mueller A, Degenhard S, Ruff SE, Geiger F, Bittner AM, Wege C, Krill CE 3rd (2010) Enhancing the magnetoviscosity of ferrofluids by the addition of biological nanotubes. ACS Nano 4(8):4531–4538. https://doi.org/10.1021/nn100645e
12. Altintoprak K, Seidenstücker A, Krolla-Sidenstein P, Plettl A, Jeske H, Gliemann H,

Wege C (2017) RNA-stabilized protein nanorings: high-precision adapters for biohybrid design. Bioinspir Biomim Nan 6:208–223. https://doi.org/10.1680/jbibn.16.00047
13. Eber FJ, Eiben S, Jeske H, Wege C (2015) RNA-controlled assembly of tobacco mosaic virus-derived complex structures: from nanoboomerangs to tetrapods. Nanoscale 7(1):344–355. https://doi.org/10.1039/c4nr05434b
14. Eber FJ, Eiben S, Jeske H, Wege C (2013) Bottom-up-assembled nanostar colloids of gold cores and tubes derived from tobacco mosaic virus. Angew Chem Int Ed Engl 52(28):7203–7207. https://doi.org/10.1002/anie.201300834
15. Mueller A, Eber FJ, Azucena C, Petershans A, Bittner AM, Gliemann H, Jeske H, Wege C (2011) Inducible site-selective bottom-up assembly of virus-derived nanotube arrays on RNA-equipped wafers. ACS Nano 5(6):4512–4520. https://doi.org/10.1021/nn103557s
16. Azucena C, Eber FJ, Trouillet V, Hirtz M, Heissler S, Franzreb M, Fuchs H, Wege C, Gliemann H (2012) New approaches for bottom-up assembly of tobacco mosaic virus-derived nucleoprotein tubes on defined patterns on silica- and polymer-based substrates. Langmuir 28(42):14867–14877. https://doi.org/10.1021/la302774h
17. Mueller A, Kadri A, Jeske H, Wege C (2010) In vitro assembly of tobacco mosaic virus coat protein variants derived from fission yeast expression clones or plants. J Virol Methods 166(1-2):77–85. https://doi.org/10.1016/j.jviromet.2010.02.026
18. Geiger FC, Eber FJ, Eiben S, Mueller A, Jeske H, Spatz JP, Wege C (2013) TMV nanorods with programmed longitudinal domains of differently addressable coat proteins. Nanoscale 5(9):3808–3816. https://doi.org/10.1039/c3nr33724c
19. Eiben S, Stitz N, Eber F, Wagner J, Atanasova P, Bill J, Wege C, Jeske H (2014) Tailoring the surface properties of tobacco mosaic virions by the integration of bacterially expressed mutant coat protein. Virus Res 180:92–96. https://doi.org/10.1016/j.virusres.2013.11.019
20. Zimmern D (1983) An extended secondary structure model for the TMV assembly origin, and its correlation with protection studies and an assembly defective mutant. EMBO J 2(11):1901–1907. https://doi.org/10.1002/j.1460-2075.1983.tb01677.x
21. Zimmern D (1977) The nucleotide sequence at the origin for assembly on tobacco mosaic virus RNA. Cell 11(3):463–482. https://doi.org/10.1016/0092-8674(77)90065-4
22. Zimmern D, Butler PJG (1977) The isolation of tobacco mosaic virus RNA fragments containing the origin for viral assembly. Cell 11(3):455–462. https://doi.org/10.1016/0092-8674(77)90064-2
23. Butler PJ, Lomonossoff GP (1980) RNA-protein interactions in the assembly of tobacco mosaic virus. Biophys J 32(1):295–312. https://doi.org/10.1016/S0006-3495(80)84958-7
24. Kadri A, Maiss E, Amsharov N, Bittner AM, Balci S, Kern K, Jeske H, Wege C (2011) Engineered tobacco mosaic virus mutants with distinct physical characteristics in planta and enhanced metallization properties. Virus Res 157(1):35–46. https://doi.org/10.1016/j.virusres.2011.01.014
25. Gooding GV, Hebert TT (1967) A simple technique for purification of tobacco mosaic virus in large quantities. Phytopathology 57:1285
26. Green MR, Sambrook J (2012) Molecular cloning : a laboratory manual, 4th edn. Cold Spring Harbor Laboratory Press, New York
27. Fraenkel-Conrat H (1957) Degradation of tobacco mosaic virus with acetic acid. Virology 4(1):1–4. https://doi.org/10.1016/0042-6822(57)90038-7
28. Hakkinen H (2012) The gold-sulfur interface at the nanoscale. Nat Chem 4(6):443–455. https://doi.org/10.1038/nchem.1352
29. Demers LM, Mirkin CA, Mucic RC, Reynolds RA 3rd, Letsinger RL, Elghanian R, Viswanadham G (2000) A fluorescence-based method for determining the surface coverage and hybridization efficiency of thiol-capped oligonucleotides bound to gold thin films and nanoparticles. Anal Chem 72(22):5535–5541. https://doi.org/10.1021/ac0006627
30. Taton TA (2002) Preparation of gold nanoparticle-DNA conjugates. Curr Protoc Nucleic Acid Chem 12.12.11–12.12.12. https://doi.org/10.1002/0471142700.nc1202s09
31. Giljohann DA, Seferos DS, Daniel WL, Massich MD, Patel PC, Mirkin CA (2010) Gold nanoparticles for biology and medicine. Angew Chem Int Ed 49(19):3280–3294. https://doi.org/10.1002/anie.200904359
32. Giljohann DA, Seferos DS, Daniel WL, Massich MD, Patel PC, Mirkin CA (2010) Goldnanopartikel in Biologie und Medizin. Angew Chem 122(19):3252–3266. https://doi.org/10.1002/ange.200904359

Part II

Functional Compounds in Virus-Based Containers: From Preparation to Applications of Hybrid Nanoobjects

Chapter 12

Internal Deposition of Cobalt Metal and Iron Oxide Within CPMV eVLPs

Alaa A. A. Aljabali and David J. Evans

Abstract

Empty (containing no genomic material) CPMV virus-like particles are loaded within the virus capsid with metal or metal oxide. Metal ions are allowed to diffuse through pores in the capsid surface and are reduced or hydrolyzed to metallic nanoparticles. The external surface of the virus-like particles remains amenable to further chemical modification.

Key words Protein cages, Iron oxide, Cobalt, Cowpea mosaic virus-like particles, Templated mineralization

1 Introduction

The cowpea mosaic virus (CPMV) capsid is composed of 60 asymmetric units, each made up of two different types of coat proteins: the small (S) subunit with one domain and the large (L) with two domains. The asymmetric units assemble into a $T = 3$ pseudoicosahedral symmetry capsid of approximately 30 nm diameter [1–3]. Empty CPMV virus-like particles (eVLPs) are produced by the cleavage of the coat protein precursor (VP60) by the action of the virus-encoded 24 K proteinase. This method utilizes the highly efficient plant transient expression system pEAQ-*HT* [4–7]. The pEAQ-*HT* system is used to simultaneously express the VP60 coat protein precursor and the 24 K proteinase in plants via an agro-infiltration process. Efficient processing of VP60 to the L and S proteins occurs, leading to the formation of CPMV eVLPs [8, 9]. The generated VLPs are identical in their capsid structure to those of the wild-type (*see* Fig. 1).

Most icosahedral viruses contain pores in their capsids through which small molecules can be loaded by diffusion (*see* Fig. 2). CPMV eVLPs have been used as nanocontainers for the encapsulation of various molecules [10], and for the formation of cobalt and

Christina Wege and George P. Lomonossoff (eds.), *Virus-Derived Nanoparticles for Advanced Technologies: Methods and Protocols*, Methods in Molecular Biology, vol. 1776, https://doi.org/10.1007/978-1-4939-7808-3_12,
© Springer Science+Business Media, LLC, part of Springer Nature 2018

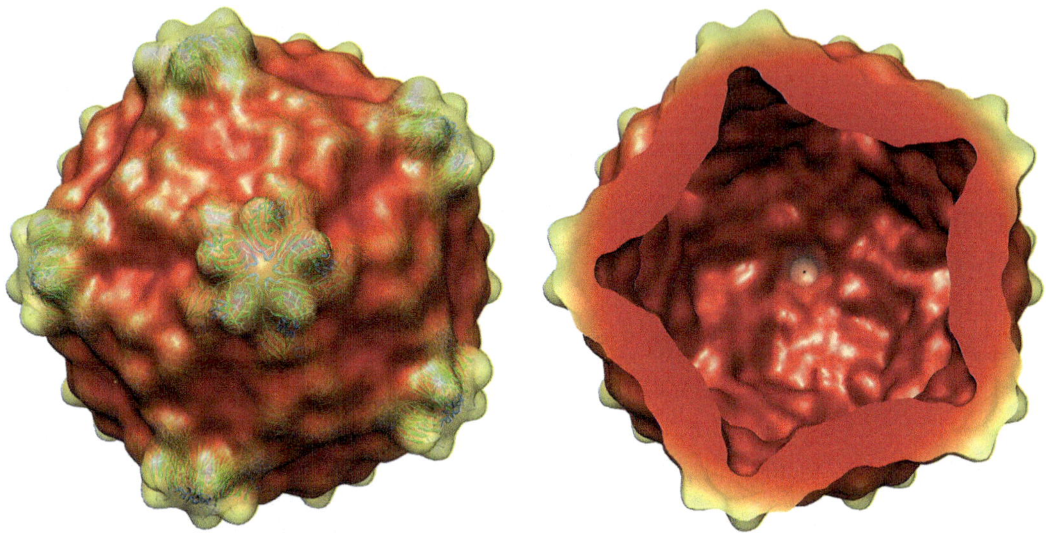

Fig. 1 Cryo-electron microscope 3D image reconstruction of rapidly frozen, at 100% relative humidity in liquid-nitrogen-cooled liquid ethane, CPMV eVLPs, imaged at 84,900× magnification and 300 kV on a Leica-EM-GP, showing the exterior and the empty interior of the virus capsid. Image reconstructions of the reordered images were accomplished by iterative-projection-matching using a spherical matching model and the published crystal structure of CPMV [1]. 3 μL of suspended particles in 10 mM sodium phosphate buffer supplemented with 50 mM $CaCl_2$ were applied to Quantifoil R2/1200-mesh Cu grids that had been glow discharged in air. CPMV capsids have ***T*** = 3 pseudoicosahedral symmetry and an external diameter of approximately 28 nm and an inner cavity of approximately 25–26 nm in diameter. The yellow color depicts the small protein subunit and the orange is for the large protein subunit. Copyright (2015) from 'Viral Nanotechnology' by Khudyakov Y, Pumpens P, eds. Reproduced by permission of Taylor and Francis Group, LLC, a division of Informa plc

iron oxide nanoparticles within the virus-like shells [11]. Cobalt and iron salts can diffuse into the cavity and, once inside, they can interact with the negatively charged interior. Reduction or alkaline hydrolysis of the metal cations led to the formation of metallic cores within the virus capsid. The external surface of the CPMV eVLPs can be further modified with desired moieties. Internal loading and external modification enables CPMV eVLPs to perform as multifunctional nanoparticles with potential applications in, for example, diseased cell targeting, imaging, and therapy.

Previously, we have investigated the role of the carboxyl (C) terminus of the S coat protein in controlling access to the interior of CPMV eVLPs by determining the efficiency of internal mineralization. The presence of the C-terminal 24-amino acid peptide of the S protein was found to inhibit internal mineralization, an effect that could be eliminated by enzymatic removal of this region [12].

To load with cobalt, CPMV eVLPs were simply incubated with cobalt chloride aqueous solution followed by reduction with sodium borohydride. To load the particles with iron oxide (magnetite, Fe_3O_4), they were incubated with a 1:2 mixture of iron (II) chloride and iron (III) chloride, and the pH was raised to approximately 10.2 for 1 h at room temperature while gently

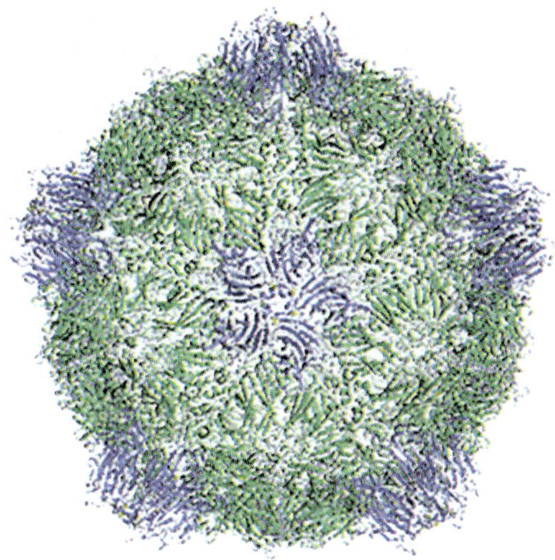

Fig. 2 Schematic presentation of one pentamer of the CPMV capsid showing the fivefold symmetry axes with one of the 12 pores (center). The small subunit is shown in cyan and the large subunit in green. The sequence was obtained from the Protein Data Bank, ID 1NY7, the image was constructed in PyMol software 1.7.4.5 Edu

stirring. The alkaline hydrolysis resulted in the synthesis of iron oxide within the virus capsid. The metal cores can be observed by transmission electron microscopy (TEM) of unstained samples.

Observing the loaded nanoparticles by TEM provides information about the presence of metal inside the virus but not about its structure. However, atomic force microscopy analysis of cobalt-loaded VLPs provided more information about the 3D topographies. We concluded that cobalt inside CPMV eVLPs forms a discontinuous structure that does not completely fill the virus cavity and reaches only about 10% of its volume [13], rather than forming solid clusters.

2 Materials

All reagents were used without further purification. Reactions were performed using Milli-Q water (18.2 MΩ cm).

1. CPMV eVLPs: The propagation and purification of CPMV eVLPs were performed by standard procedures ([14]; *see* also Chapter 23). Purified virions were stored at 4 °C in 10 mM sodium phosphate buffer pH 7.0. CPMV concentration was determined spectrophotometrically using a Lambda 25 UV-Vis spectrophotometer connected to UV WinLab software (PerkinElmer). The Beer–Lambert law was used to calculate particle concentration from the absorbance measurements at λ_{max} = 280 nm, using a molar extinction coefficient of 1.28 mL mg^{-1} cm^{-1}.

2. 10 mM sodium phosphate buffer pH 7.0: Mix 0.2 M Na_2HPO_4 (prepare 27.0 g in 1 L) and 0.2 M NaH_2PO_4 (prepare 28.39 g in 1 L) to give the stock solutions, autoclave before mixing. For the preparation of 0.1 M buffer solution use 305 mL of Na_2HPO_4 solution and 195 mL of NaH_2PO_4 solution, add 500 mL of Milli-Q water to give a solution of pH of 7.0–7.4.
3. 10 mM cobalt chloride hexahydrate solution, prepare freshly prior to use. The solution appeared purple in color.
4. Disposable PD-10 desalting columns with Sephadex G-25 resin of 2.5 mL sample volume (GE Healthcare).
5. Amicon Ultra-15 centrifugal filter unit with Ultracel-100 regenerated cellulose membrane, 100 kDa molecular weight cutoff (Merck Millipore).
6. 5 mM sodium borohydride solution (Sigma-Aldrich ≥98%), prepare freshly prior to use. Solutions that were kept for more than 8 h at room temperature were not effective.
7. Sucrose gradient: 50%, 40%, 30%, 20%, and 10% (w/v) sucrose solutions in 10 mM sodium phosphate buffer pH 7.4, store at 4 °C and use within 2 weeks.
8. 100 and 300 kDa molecular weight cutoff dialysis Biotech grade cellulose ester (CE) membranes (Spectrum Labs).
9. 5 mM Iron (II) chloride aqueous solution.
10. 10 mM iron (III) chloride aqueous solution.
11. 1 mM sodium hydroxide solution.
12. Blocking solution (PBSTM): 5% (w/v) skimmed-milk powder and 0.025% (v/v) Tween 20 dissolved in 1× phosphate buffer saline (10× PBS: 1.37 M NaCl, 27 mM KCl, 81 mM Na_2HPO_4, 15 mM KH_2PO_4, pH 7.4, adjust pH with HCl).
13. Rabbit polyclonal antibody specific to CPMV (G49, John Innes Centre, UK).
14. Horseradish peroxidase-conjugated anti-rabbit IgG antibody (Sigma-Aldrich).
15. Chemiluminescence: SuperSignal West Dura substrate kit (Thermo Scientific).
16. HyperfilmTM ECL (Amersham Biosciences).
17. Curix 60 film processor (Agfa Gevaert).
18. Nitrocellulose membrane (Amersham).
19. Cobalt stain: 20 mM 1-nitroso-2-naphthol in 1:1 (v/v) methanol–MilliQ water. Prepare stain freshly as needed.
20. Destaining: 10 mM NaOH as reported previously [11].
21. Iron stain (Prussian Blue): Equal volumes of 20% (v/v) hydrochloric acid in Milli-Q water and 10% (w/v) potassium ferrocyanide in Milli-Q water; mix just prior to use.

22. Biotin-*N*-hydroxysuccinimide ester (Aldrich) dissolved in dimethylformamide (Aldrich).
23. Streptavidin-coated Series S Sensor Chip SA (GE Healthcare Life Sciences).
24. For examination of results: UV–visible spectrophotometry (yield), SDS-PAGE (particle integrity), TEM (FEI Tecnai2, particle integrity), dynamic light scattering (DynaPro Titan, Wyatt Technology Corporation, particle size), zeta potential measurements (Malvern Instruments Zetasizer–Nano ZS, surface charge), energy-dispersive X-ray spectroscopy (Oxford Instruments INCA Energy 200Premium) and dot-blot analysis (virus integrity, metal content), and surface plasmon resonance (Biacore T200 instrument, binding of biotinylated VLPs).

3 Methods

3.1 Generation of Cobalt Within CPMV eVLPs

1. Incubate 1 mL of CPMV empty virus-like particles (1 mg/mL, suspended in 10 mM sodium phosphate buffer pH 7.0) with freshly prepared aqueous cobalt (II) chloride hexahydrate solution (to a final concentration of 10 mM by the addition of extra volumes of 10 mM sodium phosphate buffer pH 7.0 if needed). Prepare the cobalt chloride solution freshly just prior to use.
2. Allow the reaction to proceed at room temperature for 30–40 min while gently stirring at 500 rpm on mixing rollers.
3. Purify the modified particles on Sephadex G-25 PD-10 desalting columns preequilibrated with 10 mM sodium phosphate buffer pH 7.0 (*see* **Note 1**).
4. Wash the eluted sample twice with Milli-Q water (3 mL each time) on Ultra-15 centrifugal filters. This step is essential to remove all of the cobalt cations electrostatically bound to the external surface of the eVLP (*see* **Note 2**).
5. The particles are further concentrated on Ultra-15 centrifugal filter units by centrifugal ultrafiltration 5000 × *g* at 4 °C for 5–10 min (*see* **Note 2**).
6. Incubate the sample with freshly prepared sodium borohydride aqueous solution at a final concentration of 5 mM for 30 min at room temperature while stirring at 1000 rpm (*see* **Notes 3** and **4**). The addition of sodium borohydride was dropwise to maintain the reaction pH.
7. Purify the cobalt-loaded particles on a 10–50% sucrose step gradient as an extra purification step to ensure that no unmodified particles remain (*see* **Note 5**).

8. Dialyze the cobalt-loaded particles overnight (10–14 h) using 100 kDa molecular weight cutoff dialysis membranes against 10 mM sodium phosphate buffer pH 7.0 at 4 °C (cold room). Exchange buffer after 2 and 4 h.
9. The yield of cobalt-loaded VLPs after further concentration on Ultra-15 centrifugal filter units can be determined by UV–visible spectrophotometry and usually is about 70% based on the initial virus concentration (*see* **Note 6**).
10. A control experiment, under identical conditions but without eVLPs, gives visible bulk precipitation of metal.
11. The particle integrity can be established by SDS-PAGE gel electrophoresis (*see* Fig. 3) and TEM (*see* Fig. 4). Dynamic light scattering and zeta potential measurements can confirm that particle size and surface charge has not changed as a result of loading. This also confirms that the particle surface remains intact and accessible for further modification. The metal content can be confirmed by energy-dispersive X-ray spectroscopy and dot-blot analysis (*see* Subheading 3.4).

3.2 Generation of Iron Oxide Within CPMV eVLPs

1. Suspend CPMV eVLPs (1–1.2 mg/mL, 1 mL) in 10 mM sodium phosphate buffer pH 7.0. Add freshly prepared aqueous solutions of iron (II) chloride and iron (III) chloride to give a final concentration of 5 mM and 10 mM, respectively (this ratio favors the formation of Fe_3O_4). The pH after addition is between 3.5 and 3.8. Different quantities of iron salts result in the formation of large iron oxide nanoparticles outside the virus capsid (*see* **Note 7**).
2. Incubate the virus particles with the metal salt solutions overnight (~15 h) while gently stirring at 4 °C (cold room).
3. Wash the loaded virus particles twice (3 mL each) with Milli-Q water at room temperature using 100 kDa molecular weight cutoff centrifugal membrane filters.
4. Recover the sample from the previous step in 10 mM sodium phosphate buffer pH 7.0 and transfer to a clean sterilized vial, raise the pH slowly to 10.2 by dropwise addition of 1 mM aqueous sodium hydroxide solution.
5. Leave the reaction to proceed for another 60–70 min at room temperature while gently shaking at 500 rpm on a roller shaker.
6. Purify the loaded virus particles on a 10–50% sucrose step gradient to ensure that no unmodified particles remain (*see* **Note 5**)
7. The iron oxide loaded VLPs that were purified on sucrose gradient (1 mL) are transferred into a dialysis Float-A-Lyzer G2 (100 kDa molecular weight cutoff membrane ready-to-use dialysis device). The dialysis device is placed in a 1 L beaker

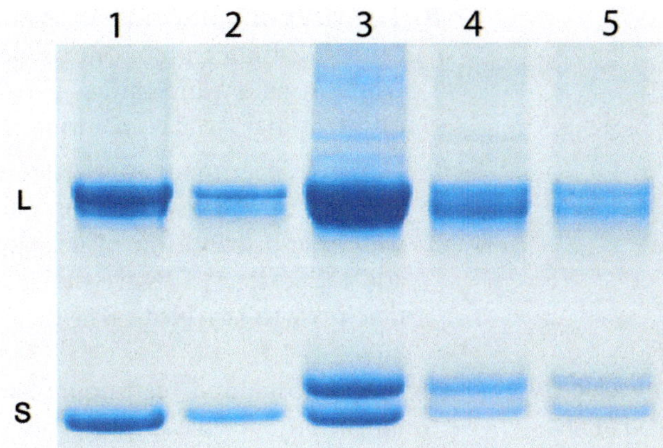

Fig. 3 SDS-PAGE gel stained with Coomassie Blue using 12% (w/v) Bis-Tris NuPAGE® gels (Invitrogen). Lane 1: wild-type CPMV; lane 2: wild-type CPMV "top component" (devoid of RNA); lane 3: CPMV eVLPs; lane 4: cobalt-loaded VLPs; lane 5: iron oxide-loaded VLPs. Bands for the large (L) and small (S) coat proteins have been detected in all samples. Differences in the bands in lanes 3, 4, and 5 are due to differences in the processing of the C-terminus of the S coat protein. Reproduced from Small, 2010, with permission from John Wiley & Sons

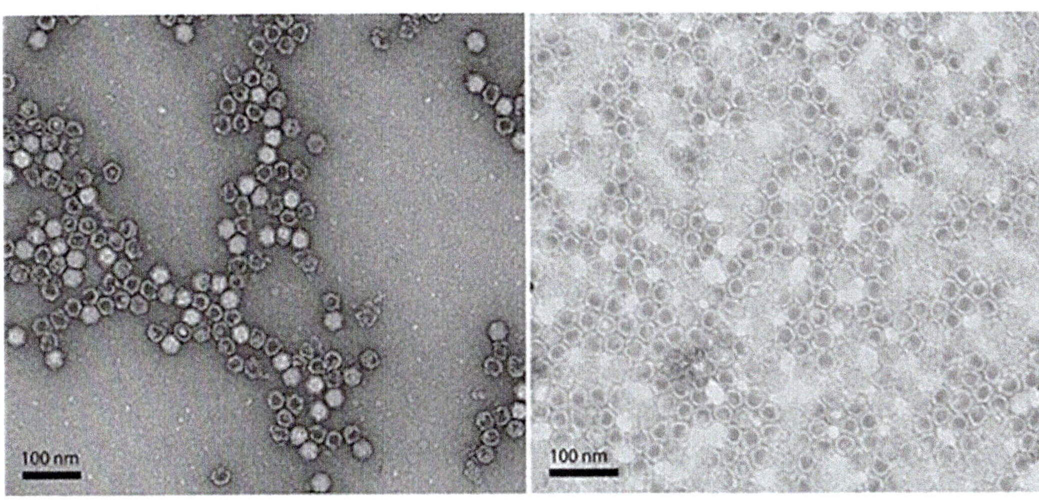

Fig. 4 TEM images of uranyl acetate stained CPMV eVLPs (left panel) and unstained cobalt-loaded VLPs showing the protein shell as bright ring around the darker metal core (right panel). Reprinted with permission from Jaafar M, Aljabali AA, Berlanga I, et al. Structural insights into magnetic clusters grown inside virus capsids. ACS Applied Materials & Interfaces 6(23):20,936–20,942. Copyright (2014) American Chemical Society

filled with 10 mM sodium phosphate buffer pH 7.0 and left for 15 h in the cold room. To ensure all small impurities are removed from the sample, buffer was replaced at 2 and 4 h to disrupt the equilibrium of contaminants from inside and outside the dialysis device. The concentration of small contaminants and sucrose within the dialysis device can be decreased to acceptable or negligible levels.

8. The yield of iron oxide-VLPs after concentration on Ultra-15 centrifugal filter units can be determined by UV–visible spectrophotometry and ranges usually between 40 and 45% based on the initial virus concentration (*see* **Note 6**).
9. Control experiments under identical conditions except for the absence of eVLPs give nonspecific bulk precipitation with a wide size distribution of nanoparticles as observable by TEM and DLS; the eVLPs are thus essential for controlled nanoparticle growth (*see* **Note 8**).

3.3 Immunological Detection of CPMV Coat Proteins

The immunostaining technique uses antibodies to detect specific protein in any given sample. This relies on the ability of CPMV antibodies to recognize specifically and bind to the virus coat protein. The test confirms the preservation of the coat protein and that there is no significant metal or metal oxide coating of the external virus surface.

1. Spot 4 μL of each particle suspension (eVLP, cobalt-loaded VLP, and iron oxide-loaded VLP, in 10 mM sodium phosphate buffer pH 7.0) onto a nitrocellulose membrane (Amersham).
2. Allow membranes to air-dry at room temperature for 30 min.
3. Use blocking solution (1× PBSTM) to block nonspecific binding sites of the CPMV antibodies. Leave the blocking reaction to proceed at room temperature for 2 h or overnight at 4 °C.
4. Probe the membrane with a rabbit polyclonal antibody specific to CPMV (anti-CPMV antibody/1:1000 in 1× PBSTM (20 μL of G49 in 20 mL freshly prepared buffer)). Follow with an anti-rabbit antibody (IgG–HRP/1:2000 in 1× PBSTM (2 μL)) as a secondary antibody. Leave the reaction to proceed for 2 h at room temperature or overnight at 4 °C.
5. Use SuperSignal West Dura substrate kit (Thermo Scientific) according to the manufacturer's instructions to generate chemiluminescence and capture this on a radiographic film (HyperfilmTM ECL, Amersham Biosciences) developed using a Curix 60 film processor (Agfa Gevaert) (*see* Fig. 5).

3.4 Staining for Cobalt and Iron

1. Spot 4 μL aliquots of each particle suspension (eVLP, cobalt-loaded VLP, and iron oxide-loaded VLP, suspended in 10 mM sodium phosphate buffer pH 7.0) onto a nitrocellulose membrane (Amersham).
2. Allow membranes to air-dry at room temperature for 30 min.
3. Expose the membranes to either cobalt stain followed by treatment with destaining solution or iron stain (*see* Fig. 5).

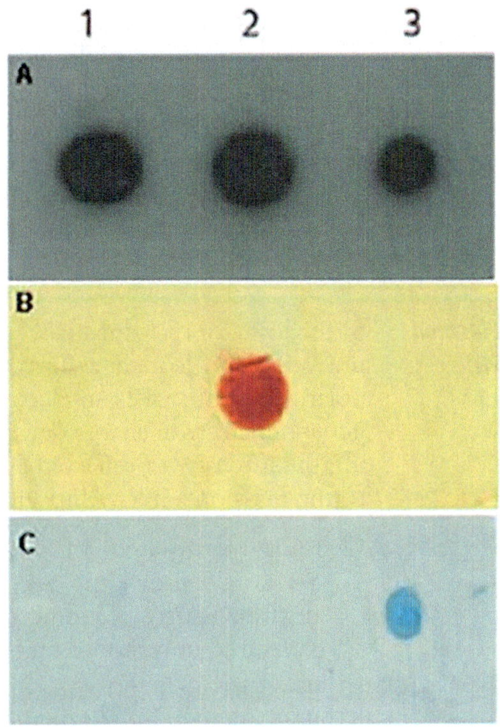

Fig. 5 Images of intact VLPs spotted on a nitrocellulose membrane probed with: (**a**) polyclonal antibodies raised against CPMV particles—dark spots show immunoreactive CPMV coat protein; (**b**) 1-nitroso-2-naphthol stain for cobalt; (**c**) Prussian Blue stain for iron. Column 1, eVLPs; column 2, cobalt-loaded VLPs; column 3, iron oxide-VLPs. Reproduced from Small, 2010, with permission from John Wiley & Sons

3.5 Functionalization of the Exterior Capsid of CPMV eVLPs and Cobalt-Loaded VLPs

CPMV eVLPs and cobalt-loaded VLPs are biotinylated using an adaptation of our standard procedure [15].

1. To the virus particles suspended in 10 mM sodium phosphate buffer pH 7.0 add a 2000 molar excess of biotin-*N*-hydroxysuccinimide ester solution in DMF; the final concentration of DMF to water is adjusted to 20% (v/v). The number of moles of the virus were determined using the Beer–Lambert law based on absorption maximum at λ = 280 nm with molar extinction coefficient of ε = 1.28 mL mg^{-1} cm^{-1} (*see* **Note 6**). The number of molecules was determined by the multiplication of moles by Avogadro's number. For each mole, there are 6.022×10^{23} particles.
2. Incubate the reaction overnight with continuous magnetic stirring (500 rpm) at 4 °C, then dialyze the CPMV eVLP biotinylated particles using 100 kDa molecular weight cutoff dialysis membranes against 10 mM sodium phosphate buffer. Typically 1 L of buffer is used and exchanged at 2 and 4 h with

continuous stirring at 500–1000 rpm at room temperature, followed by overnight incubation at 4 °C (15–16 h in total).

3. Concentrate the sample on Ultra-15 centrifugal filters, and determine the final concentration of VLPs by measuring the absorption at 280 nm. The binding of each of biotinylated-eVLPs and biotinylated-cobalt-loaded VLPs to a streptavidin modified chip can be monitored by surface plasmon resonance (*see* Subheading 3.6).

3.6 Surface Plasmon Resonance (SPR) Determined Using a Biacore T200

SPR allows for real-time, label-free detection of virus particles both unloaded and loaded with metal. SPR phenomenon occurs after polarized light strikes an electrically conducting chip surface. This generates electron charge density waves known as plasmons, reducing the intensity of reflected light at a specific angle in proportion to the mass on a microchip surface.

1. Spot biotinylated VLP samples onto a streptavidin-coated Series S Sensor Chip SA. The chip has a carboxymethylated dextran matrix preimmobilized with streptavidin. Binding is measured in response units (RU).
2. The Biacore T200 instrument has a four-flow cell system (each flow cell volume is 0.06 μL) equipped with a temperature control unit and an optical system to measure the molecular interaction occurring on the sensor chip in real-time.
3. Use the SPR instrument in manual run mode with a flow rate of 10 μL/min, at room temperature, with PBS pH 7.4 as running buffer.
4. Load the cells as follows: Flow cell 1: negative control with biotin-*N*-hydroxysuccinimide ester solution, Flow cell 2: positive control with multiinjection steps using biotinylated-eVLPs with a contact time of 1 min until each run reaches saturation (the overall time for chip saturation is 360 s), Flow cell 3: biotinylated-eVLPs, Flow cell 4: biotinylated-cobalt-loaded VLPs (the particle concentration for each of the samples is the same as determined spectrophotometrically). The external surface of cobalt-loaded VLPs remains amenable to chemical modification) and the cobalt-loaded VLPs show an enhanced response, due to their higher mass, on the surface plasmon resonance sensorgram compared to biotinylated-eVLP (*see* Fig. 6).

4 Notes

1. Equilibrate PD-10 columns with 10 mM sodium phosphate buffer pH 7.0 just prior to use. First wash the columns with Milli-Q water to remove the ethanol that comes within the columns, continue with 3–4 washes (each 5 mL) using 10 mM

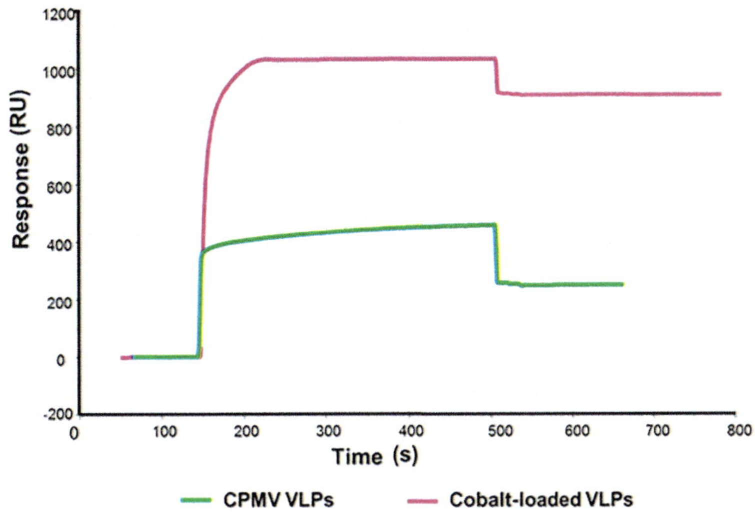

Fig. 6 Sensorgram showing the evolution of resonance response units (RU) versus time during association and dissociation measurements of biotinylated-eVLPs (green) and biotinylated cobalt-loaded-VLPs (pink) on a streptavidin-modified chip. Both samples have the same concentration as confirmed by UV–visible spectrophotometry. A delay of 100 s was applied and the association was allowed to occur during 360 s before the flow is switched to buffer for the dissociation step. Reproduced from Small, 2010, with permission from John Wiley & Sons

sodium phosphate buffer pH 7.0, and then load the sample onto the column. The column void volume is 3.5 mL, collect 500 μL fractions, and determine the CPMV eVLP content spectrophotometrically (*see* **Note 6**).

2. It is crucial not to allow the buffer to pass totally through the filter leaving the particles to dry on top of it. This results in very low particle recovery and difficulties in retrieving the sample. The length of the spin should be adjusted accordingly and the centrifuge stopped to check that the sample is not dry. This allows the optimum length and speed of the centrifugation to be determined.

3. Freshly prepared aqueous solutions of sodium borohydride are necessary because $NaBH_4$ reacts quickly with water at room temperature and the exposure of sodium borohydride to air will lead to the loss of reducing activity.

4. This is a crucial step and can affect the percentage of eVLP recovery after mineralization. To ensure the sodium borohydride is active check for evolution of gas, which leads to the lid of the reaction vessel popping open 2–3 min after mixing. If it takes longer than this, discard the solution and start again.

5. Sucrose step gradients are prepared by underlying sucrose solutions (50%, 40%, 30%, 20%, and 10% (w/v)) of increasing

density (175 or 500 μL) in a centrifuge tube of 2.1 mL or 5 mL, respectively, and carefully overlaying the sample to fill the tube (normally 300–500 μL). Centrifuge the gradients in an ultracentrifuge in a swing-out rotor (Thermo Scientific, AH-650) and centrifuged at 137,000 × *g* for 1.5–2.5 h at 4 °C. Collect 175–300 μL fractions containing the modified particles (in a typical experiment they are fractions 5–8 from the top as determined by the absorbance at 280 nm) and exchange the buffer with 10 mM sodium phosphate buffer pH 7.4 (*see* Subheading 3.1, **step 8**).

6. UV–visible spectroscopy of CPMV eVLPs: particles have an absorption maximum at a wavelength of λ = 280 nm (derived mainly from the amino acids tryptophan, tyrosine, and cysteines) with a molar extinction coefficient of $\varepsilon = 1.28$ mL mg^{-1} cm^{-1} as determined previously [16]. The Beer–Lambert law can be used to calculate the concentration:

$$A = \varepsilon \cdot c \cdot d$$

where A is the absorbance, c is the concentration of the particles in mg/mL, d is the length of the light path in cm, and ε is the molar extinction coefficient in mL mg^{-1} cm^{-1}.

7. Larger quantities (double or triple) of both iron salts result in the formation of a wide distribution of iron oxide nanoparticle sizes outside the virus capsid. These particles are difficult to remove from the loaded-virus samples.

8. As a control experiment, unmodified wild-type CPMV particles are incubated with the same concentration of iron salts for 1 h, prior to raising the pH to 10.2 and mixing for another hour, followed by purification by gel filtration columns, dialysis, and sucrose gradients. No templated mineralization should be observed.

Acknowledgment

This work was supported by the Biotechnology and Biological Sciences Research Council, UK (Core Strategic Grant to the John Innes Centre, D.J.E., and JIC DTG, A.A.A.A.).

References

1. Lin T, Chen Z, Usha R et al (1999) The refined crystal structure of cowpea mosaic virus at 2.8 Å resolution. Virology 265(1):20–34
2. Lin T, Johnson JE (2003) Structures of picorna-like plant viruses: implications and applications. Adv Virus Res 62:167–239
3. Lomonossoff GP, Johnson JE (1991) The synthesis and structure of comovirus capsids. Prog Biophys Mol Biol 55(2):107–137
4. Saunders K, Sainsbury F, Lomonossoff GP (2009) Efficient generation of cowpea mosaic virus empty virus-like particles by the proteo-

lytic processing of precursors in insect cells and plants. Virology 393(2):329–337

5. Sainsbury F, Thuenemann EC, Lomonossoff GP (2009) pEAQ: versatile expression vectors for easy and quick transient expression of heterologous proteins in plants. Plant Biotechnol J 7(7):682–693
6. Sainsbury F, Canizares MC, Lomonossoff GP (2010) Cowpea mosaic virus: the plant virus-based biotechnology workhorse. Annu Rev Phytopathol 48:437–455
7. Sainsbury F, Saxena P, Aljabali AA, Saunders K, Evans DJ, Lomonossoff GP (2014) Genetic engineering and characterization of cowpea mosaic virus empty virus-like particles. Methods Mol Biol 1108:139–153
8. Sainsbury F, Lomonossoff GP (2008) Extremely high-level and rapid transient protein production in plants without the use of viral replication. Plant Physiol 148(3):1212–1218
9. Hesketh EL, Meshcheriakova Y, Dent KC, Saxena P, Thompson RF, Cockburn JJ, Lomonossoff GP, Ranson NA (2015) Mechanisms of assembly and genome packaging in an RNA virus revealed by high-resolution cryo-EM. Nat Commun 6:10113–10119
10. Wen AM, Shukla S, Saxena P et al (2012) Interior engineering of a viral nanoparticle and its tumor homing properties. Biomacromolecules 13(12):3990–4001
11. Aljabali AAA, Sainsbury F, Lomonossoff G et al (2010) Cowpea mosaic virus unmodified empty viruslike particles loaded with metal and metal oxide. Small 6(7):818–821
12. Sainsbury F, Saunders K, Aljabali A et al (2011) Peptide-controlled access to the interior surface of empty virus nanoparticles. ChemBioChem 12(16):2435–2440
13. Jaafar M, Aljabali AA, Berlanga I et al (2014) Structural insights into magnetic clusters grown inside virus capsids. ACS Appl Mater Interfaces 6(23):20936–20942
14. Wellink J (1998) Comovirus isolation and RNA extraction. Methods Mol Biol 81:205–209
15. Steinmetz N, Calder G, Lomonossoff G et al (2006) Plant viral capsids as nanobuilding blocks: construction of arrays on solid supports. Langmuir 22(24):10032–10037
16. van Kammen A (1967) Purification and properties of the components of cowpea mosaic virus. Virology 31(4):633–642

Chapter 13

Plant Virus-Based Nanoparticles for the Delivery of Agronomic Compounds as a Suspension Concentrate

Richard H. Guenther, Steven A. Lommel, Charles H. Opperman, and Tim L. Sit

Abstract

Nanoparticle formulations of agrichemicals may enhance their performance while simultaneously mitigating any adverse environmental effects. Red clover necrotic mosaic virus (RCNMV) is a soil-transmitted plant virus with many inherent attributes that allow it to function as a plant virus-based nanoparticle (PVN) when loaded with biologically active ingredients. Here we describe how to formulate a PVN loaded with the nematicide abamectin (Abm) beginning with the propagation of the virus through the formulation, deactivation, and characterization of the finished product.

Key words Plant pathogenic nematode, Nematicide, Agrichemical, Plant viral nanoparticle, Nanoparticle formulation

1 Introduction

Due to emerging resistance among many pathogens, the high cost and long lead times to develop new compounds along with increased environmental concerns, alternative formulations of existing agrichemicals are being sought. One approach being evaluated is the use of plant viral capsids as nanovessels. For agrichemical applications, a plant pathogen like Red clover necrotic mosaic virus (RCNMV) that is transmitted naturally through the soil may possess attributes that would make it an ideal plant virus-based nanoparticle (PVN) candidate to formulate with agrichemicals. RCNMV is not a significant commercial pathogen and can be chemically inactivated making it suitable for use in agricultural applications. RCNMV is a positive strand RNA virus with a 37 nm spherical capsid. This particle size is small enough to remain mobile and suspended in the microlayer of water that coats soil particles. Yet it is large enough to reduce interactions with and binding to the clay and organic components of soil, thus increasing the time it

Christina Wege and George P. Lomonossoff (eds.), *Virus-Derived Nanoparticles for Advanced Technologies: Methods and Protocols*, Methods in Molecular Biology, vol. 1776, https://doi.org/10.1007/978-1-4939-7808-3_13,
© Springer Science+Business Media, LLC, part of Springer Nature 2018

remains bioavailable in soils. The RCNMV capsid is very robust (even when formulated as a PVN) and is capable of withstanding long incubation periods in soil without degradation. This feature can be utilized to reduce the unwanted binding of an agrichemical to organic matter in soil as well as control the rate of release of the agrichemical. This combination of attributes may allow for repurposing of agrichemicals or lower the amount of agrichemical used. RCNMV capsids form a stable suspension at concentrations greater than 30 mg/ml in aqueous solutions. Thus, for some low solubility agrichemicals, formulation as a PVN will allow them to be applied at a higher effective rate.

It was discovered that RCNMV capsids could be loaded with small molecules by following a defined series of steps that alter buffer conditions and manipulate the capsid structure in a controlled manner [1]. At low and neutral pH values, the capsid is rigid and impervious while exposure to divalent cation chelators and/or elevated pH values causes the capsid structure to relax and the interior becomes solvent accessible [2]. Under these accessible conditions, small molecules will infuse into the capsids. The capsids (now referred to as PVNs after loading) are then returned to a low pH value and the chelators are removed. The PVNs retain most of the initial inherent properties of the capsids including their morphology, stability, and electrostatic ones. After formulation, we have found that the PVNs slowly release their cargo at a rate that varies between different cargos and the conditions under which the PVNs are stored.

One initial application we have been exploring for RCNMV PVNs is their use as a platform for the control of plant parasitic nematodes (PPN). PPN are a worldwide problem, causing losses in excess of $125 billion annually [3]. Compounding the PPN problem is the restriction of traditional agrichemical approaches due to environmental concerns with their usage. In an initial survey of four effective nematicides [abamectin (Abm), levamisole, mebendazole, and piperazine], all formulations were found to be bioactive as PVN formulations. Additional testing determined that only Abm retained bioactivity for periods longer than 1 month. A further advantage of Abm is its effectiveness against a broad range of nematodes and its existing approval for agricultural applications (Avicta® and Avid®). When the RCNMV PVN formulated with Abm was tested in soil, it was found to provide crop protection for tomatoes against root-knot nematodes (RKN; [4]). While PVNs are most well suited for soil rather than foliar application, they are not limited to delivery of only nematicides. It is envisioned that PVNs could also be used as a delivery platform for micronutrients, pesticides and fungicides, and possibly formulated to carry multiple active ingredients. Besides agrichemicals, PVNs can also be exploited for other diverse nanotechnology applications [5–7].

2 Materials

Prepare all solutions using deionized water (DI dH_2O) and analytical grade reagents. Prepare and store all reagents at room temperature (unless indicated otherwise). Follow all waste disposal regulations when disposing of waste materials.

2.1 Plant Propagation

1. Host plants/seeds: *Nicotiana clevelandii.*
2. Germination promoting solution: 30 mM gibberellic acid (Sigma-Aldrich Cat # G-7645) in 95% ethanol (v/v) (*see* **Note 1**).
3. Soil: Farfard® 2 Growing Mix (Sun Gro Horticulture).
4. Pots: 4″ and 6″ sterile clay pots.
5. Nitrogen rich fertilizer: Miracle-Gro Lawn Food (Scotts Miracle-Gro).

2.2 Viral Propagation

1. 600 mesh carborundum (Kramer Industries).
2. Infectious RNA transcripts of RCNMV genomic RNAs (*see* **Note 2**).
3. 50 ml mortar and pestle.
4. Inoculation buffer: 10 mM sodium phosphate buffer pH 7.2 (*see* **Note 3**).
5. 3 × 3 × 0.5 cm squares of foam rubber (Jo-Ann Fabric and Craft).

2.3 Viral Extraction and Purification

1. Balance.
2. Blender: Waring Xtreme MX1000XTX Commercial Blender.
3. Grade 560 cheesecloth (Fisher Scientific Cat # AS240).
4. MiraCloth (EMD Millipore).
5. Extraction buffer: 0.2 M sodium acetate, pH 5.3, 0.1% β-mercaptoethanol (v/v) (*see* **Note 4**).
6. Virus precipitation solution: 40% PEG 8000 (w/v), 1 M NaCl (*see* **Note 5**).
7. Virus resuspension buffer: 40 mM sodium acetate pH 5.3 prepared by 1:50 dilution of 2 M sodium acetate solution (*see* **Note 4**).
8. Ultracentrifugation pad solution: 30% (w/v) sucrose solution.
9. 50 ml Oakridge centrifuge tubes and 250 ml polypropylene bottles (Nalgene).
10. 60 ml syringe.
11. 9” Pasteur pipet.
12. 3 ml polyethylene transfer pipet.

13. Ultracentrifuge tubes: 13 × 51 mm Ultra-Clear centrifuge tubes (Cat # 344057), 25 × 89 mm Quick-Seal centrifuge tubes (Cat # 342414) (Beckman Coulter).
14. Graduated cylinder.
15. 2 l beaker.
16. Pierce Protein Assay Kit (Cat # 1856210) (Thermo Fisher Scientific).

2.4 PVN Formulation

1. Formulation loading buffer: 0.2 M phosphate buffer pH 9.76 (*see* **Note 6**).
2. EDTA loading solution: 5 mM EDTA solution (*see* **Note 7**).
3. Formulation finishing buffer: 0.2 M sodium phosphate pH 3.7 (*see* **Note 8**).
4. Cargo: 2 mg/ml technical Abm (abamectin) in acetonitrile (*see* **Note 9**).
5. Deactivation reagent: β-propiolactone (BPL) (Sigma-Aldrich, Cat # P5648) (*see* **Note 10**).
6. Purification columns: Illustra NAP-10 (GE Healthcare).

2.5 PVN Characterization

1. HPLC grade acetonitrile.
2. HPLC system: Agilent 1260.
3. Agilent Poroshell 120 EC-C18, 2.7 μm, 3 × 100 mm column.
4. Mobile phase: 20:80 H_2O (ammonia)–acetonitrile.
5. *Caenorhabditis elegans,* strain N2 (Caenorhabditis Genetics Center, University of Minnesota, Minneapolis, MN).
6. 96-well microtiter plate: Flat bottom 96-well plate.
7. M9 culture buffer (*see* **Note 11**).
8. Rainin L200 LTS multichannel pipet.

2.6 Recommended Equipment

1. Ultracentrifuge: Beckman L8-M with Ti 70 and SW55 rotors (Beckman Coulter).
2. Floor model centrifuge: Sorvall RC-5B with GSA and SS-34 rotors (Thermo Fisher Scientific).
3. NanoDrop 1000 Spectrophotometer (Thermo Fisher Scientific).
4. 50× dissecting scope: Zeiss Model 47 50 52 9901 Stereo microscope (Zeiss).
5. pH meter: Mettler Seven Compact (Mettler Toledo).
6. Stir plate: Barnstead Model 131325 (Thermo Fisher Scientific).
7. Microcentrifuge: Eppendorf 5424 Centrifuge (Eppendorf North America).

8. Rocking platform shaker: VWR Rocking Platform 200 (VWR).
9. Grow lights: Sun System 3 growlights (Sunlight Supply).
10. Analytical Balance: Denver M-220 (Denver Instrument/Sartorius).

3 Methods

3.1 Plant Propagation

1. The host plant used for RCNMV propagation is *Nicotiana clevelandii*. Pre-treat the seeds by soaking overnight in a giberillic acid solution (with agitation) to promote germination. For germination, grow the seeds in bulk in a 6″ clay pot containing Farfard® 2 Growing Mix.
2. Grow the plants in a heated greenhouse with a photoperiod of 12 h with light supplemented by growth lights.
3. Transplant the seedlings when they reach a diameter between 1 and 2 cm (about 2 weeks) into individual 4″ clay pots containing the same soil mix. The seedlings are grown for an additional 2–3 weeks (10–15 cm diameter). The plants are fertilized weekly with nitrogen-rich Scotts Miracle-Gro Lawn Food.

3.2 Viral Propagation

For optimum virus yield the host plants should be inoculated prior to senescence/budding (4–6 weeks old). Three to four plants are initially inoculated by mechanical transmission of infectious RNA transcripts (*see* **Note 2**).

1. Dust 3–4 leaves per plant with 600 mesh carborundum.
2. Prepare a transcript mixture (1 μl of each genomic RNA transcript with 100 μl inoculation buffer) for each plant and evenly spot onto the leaves. Gently rub the RNA solution with gloved fingers into the leaves and rinse off with water.
3. After symptoms appear (typically 3–5 days) these plants serve as the source material for scale up by sap transmission to more plants. Select infected leaves from three different plants showing the characteristic ringspots. Macerate the leaves with a mortar and pestle in 5 ml of inoculation buffer.
4. Dust leaves of plants to be infected with carborundum. Soak a 3 cm square of foam rubber with the solution and then gently rub each leaf 3–5 times in one direction. After inoculation, rinse the excess carborundum/inoculum from the plant leaves with water and culture the plants in a temperature-controlled glass house (18–26 °C). In winter, the photoperiod is maintained at 12 h by using growth lights.
5. The virus is allowed to replicate in the plants for 7–14 days post inoculation (dpi). Subsequent infections are propagated by use of infected tissue. The symptoms of viral infection are generally visible 3–5 dpi. Figure 1 shows healthy and infected plants 10 dpi.

Fig. 1 *Nicotiana clevelandii* is a species of wild tobacco used as the host for virus propagation. Panel A shows a healthy 6-week-old plant in a 4″ clay pot. Panel B shows an RCNMV infected plant 10 days post inoculation with the characteristic ringspot necrosis visible over the entire plant.

3.3 Viral Extraction and Purification

The method for virus purification is based on a previously published one [8].

1. Weigh the harvested tissue and then homogenize with a blender using a 1:4 (w/v) ratio of tissue to extraction buffer.
2. Decant the homogenized tissue into a 2 l beaker which is placed on ice and slowly stir for 30 min before passing through three layers of fine cheesecloth.
3. Place the filtered extract in 250 ml centrifuge bottles. Insoluble solids are removed by centrifugation for 15 min at 6455 × *g*, 5 °C in a GSA rotor. Pass the supernatant through one layer of Miracloth and determine the volume in a graduated cylinder.
4. Pour the extracted solution into a 2 l beaker. Precipitate the virus by the addition of the virus precipitation solution in a 4:1 (v/v) ratio of plant extract to virus precipitation solution and stir the mixture on ice for 2 h. Pellet the virus by decanting the mixture into 250 ml centrifuge bottles and centrifuge for 15 min at 6455 × *g*, 5 °C in a GSA rotor.
5. Discard the supernatant. Resuspend the pellet in virus resuspension buffer by repeated flushing with a 3 ml transfer pipet. The ratio of resuspension buffer to extract solution is 1:8 (30 ml for each 250 ml bottle).
6. Place the resuspended pellet solution in several 50 ml Oakridge tubes and centrifuge for 20 min at 11,000 × *g*, 5 °C in a SS-34 rotor.

7. Collect the supernatant containing the extracted virus in a 250 ml bottle and allow to rest overnight at 4 °C.
8. Concentrate the extracted virus by ultracentrifugation using 25 × 89 mm Quick-Seal centrifuge tubes with each tube containing ~35 ml extract and a 3 ml 30% (w/v) sucrose pad. The extract is added to the tube with a 60 ml syringe and the 30% (w/v) sucrose pad is added to the bottom through a 9" Pasteur pipet placed at the bottom of the tube. Once filled, the tubes are balanced and sealed. Place the tubes in a Ti 70 rotor and centrifuge for 2 h at 237,000 × *g*, 5 °C.
9. Discard the supernatant immediately after the spin and resuspend the pellet in 500 μl of virus resuspension buffer. Incubate overnight at 4 °C.
10. Complete the resuspension with 2 × 250 μl rinses of the tube with virus resuspension buffer and place into a 1.5 ml Eppendorf tube. Any insoluble contaminants are removed by centrifugation for 3 min at 9391 × *g*, 20 °C.
11. To increase purity, perform a second ultracentrifugation. Pour the virus solution from the first ultracentrifugation into a 13 × 51 mm Ultraclear centrifuge tube, place in a SW 55 rotor and centrifuge for 2 h at 279,000 × *g*, 5 °C.
12. Discard the supernatant immediately after the spin and resuspend the pellet in the same manner as the first ultracentrifugation step (*see* **step 9**).
13. Determine the concentration of the purified virions by UV measurement in a NanoDrop spectrophotometer. A 1 mg/ml virus suspension has an OD_{260} of 6.46 [8]. The yield by this method is typically 70–100 μg virions per gram of infected tissue.

3.4 PVN Formulation

For optimal loading, the formulation protocol of a PVN must be varied depending on the chemical properties of the cargo to be loaded. This formulation protocol has been optimized for the loading of Abm to the PVN. While the following protocol describes a 1 ml formulation volume, the proportions used are scalable from 100 μl to 20 ml volumes. While a virus concentration of 20 mg/ml is the greatest optimal starting concentration, lower concentrations will perform with the same efficiency.

1. Treat 600 μl of virus solution (from Subheading 3.3, **step 12**) with 100 μl of EDTA loading solution and 100 μl of formulation loading buffer for 15 min at room temperature (*see* **Note 12**).
2. Add 100 μl of Abm in acetonitrile (2 mg/ml) and allow the mixture to gently rock for 12–16 h at room temperature.

3. After the loading period, lower the pH of the solution back to 5.3 by addition of 100 μl formulation finishing buffer. At this point treat the PVN with deactivation reagent (4 μl per ml) for 15 min to deactivate the virus (*see* **Note 13**).
4. After incubation, the nonloaded cargo and unreacted deactivating reagent is removed by size exclusion chromatography. For a 1 ml formulation volume, commercially prepacked GE Illustra NAP columns work quite well (*see* **Note 14**). Columns are used according to manufacturer's recommended protocols.

3.5 PVN Characterization

After the PVN is fully formulated, it is characterized to determine its concentration, cargo loading and bioactivity of the cargo.

3.5.1 Determination of PVN Concentration

1. Determine the PVN concentration by measuring the sample's absorbance at 260 nm and using a conversion of 6.78 OD = 1.0 mg PVN/ml (*see* **Note 15**).
2. If the cargo has strong UV absorbance, an alternative method of quantification may be needed to determine PVN concentration.

3.5.2 Determination of PVN Cargo Loading

The amount of Abm in the formulation is determined by HPLC analysis.

1. Place a 10 μl aliquot of PVN (from Subheading 3.4, **step 4**) in 190 μl of HPLC grade acetonitrile and incubate for at least 8 h before analysis (*see* **Note 16**).
2. The HPLC system used is an Agilent 1260 monitored at a wavelength of 245 nm. Separation is performed using an Agilent Poroshell 120 EC-C18, 2.7 μm, 3 × 100 mm column at a temperature of 40 °C. The separation is isocratic using a 20:80% v/v H_2O (ammonia)–acetonitrile mobile phase. The time between injections is 6 min with a retention time of ~1.9 min.
3. Calculate the number of molecules loaded per capsid by dividing the sample concentration of Abm by the virus concentration. The formula weights for the calculations are 887 g/mole for Abm and 1×10^7 g/mole for the PVN.

3.5.3 Determination of PVN Bioactivity

The bioactivity of PVN Abm is determined in a 96 well plate acute toxicity assay using *C. elegans* as the test species (*see* **Notes 17** and **18**). This method can be used to quantify the bioactive concentration of Abm in a PVN formulation. This assay is also useful for evaluating the delivery performance of the PVN in test samples gathered in soil mobility and cargo release studies (*see* **Note 19**). The method described is an adaptation to a 96-well format of a previously reported method [9].

1. Array the samples on the plate depending on the bioactive concentrations to be determined or by grouping of fractions from soil affinity and mobility tests to survey for the presence of Abm. In both cases, each test plate has eight wells of untreated *C. elegans* to serve as negative controls and eight wells containing a threefold serial dilution series of an Abm analytical standard to serve as a positive/performance control.
2. When the concentration of Abm in a sample is determined, 150 μl aliquots of test samples are arrayed in the A row of the plate. 100 μl M9 buffer is placed in the B-H rows of the plate and a 50 μl threefold serial dilution of the samples is made, preferably with a multichannel pipet, down the columns of the plate. The 50 μl excess from the last well in the dilution series is discarded.
3. When the assay is used to survey a collection of test samples, 100 μl of each sample is arrayed onto the plate followed by addition of 20 μl M9 buffer suspension containing 100 ± 25 *C. elegans* into each well. At 2, 24, and 48 h post mixing, nematode motility is visually assessed using a Zeiss stereo microscope. At each time point, the motility of the nematodes is scored as follows: "motile" (movement of the nematodes in the test solution is comparable to those in the control wells), "impaired" (most nematodes are immotile but detectible movement is observed for at least 10 individuals in a test well) or "immotile" (movement of fewer than 5 nematodes in a well is observed). Scoring of the motility is aided by the phenotypic response of nematodes exposed to the paralytic Abm: normal, healthy nematodes exhibit a curved or S shape, while a nematode exposed to a toxic concentration of Abm will present a rod- or pencil-like appearance.
4. The concentration of bioactive Abm in a test sample that has been titrated is assigned in the following manner: wells in which the transition from motile to impaired occurs are identified and assigned a concentration of 0.2 μg/ml Abm [10]. Corrections are then made for sample dilution to derive the bioactive concentration of Abm (*see* **Note 20**).

4 Notes

1. The germination promoting solution is prepared in 10 ml aliquots by adding 100 mg of gibberellic acid to 10 ml of 95% ethanol (v/v) in a 50 ml polypropylene tube. The solution is stored at −20 °C. Fresh solutions should be prepared every 6 months.

2. Viral RNA transcripts are prepared from linearized plasmid DNA templates of the infectious cDNA constructs for RCNMV RNA-1 and RNA-2 using the MEGAscript T7 Transcription Kit (Ambion, Carlsbad, CA). RCNMV genomic RNAs do not possess 5′ cap structures [11] so cap analogs are not required during the transcription process. 1 μg of linearized plasmid DNA is transcribed in a 20 μl total reaction volume. RNA transcript quality was assessed by agarose gel electrophoresis prior to usage for plant inoculation without any further purification steps.
3. Inoculation buffer is prepared by dilution of a 100× 1 M sodium phosphate pH 7.2 stock solution. The 100× stock is prepared by dissolving 112.8 g NaH_2PO_4 and 44.3 g Na_2HPO_4 in a final 1 l volume using DI dH_2O.
4. Extraction buffer is prepared fresh for each extraction. Once the mass of tissue to be extracted is determined, the extraction buffer is prepared at a 1:4 (w/v) ratio of tissue–buffer. For example, if 250 g of tissue is to be extracted, 1 l of extraction buffer will be prepared by adding 100 ml of a 10× 2 M sodium acetate pH 5.3 stock and 1 ml β-mercaptoethanol to 899 ml DI dH_2O. The 10× sodium acetate stock is prepared at the 1 l scale. In a 2 l beaker 164.06 g of sodium acetate (anhydrous) is dissolved in 900 ml of DI dH_2O. The pH of this solution is then lowered to 5.3 by the addition of glacial acetic acid (about 30 ml). Once the pH has been adjusted the final volume is brought up to 1 l with additional DI dH_2O. This solution is placed in a 1 l bottle and stored at 4 °C for use as needed.
5. Virus precipitation solution is best prepared in 1 l aliquots. 600 ml of DI dH_2O is poured into a 2 l beaker containing a large stir bar. Then 58.44 g of NaCl is added and allowed to dissolve. A 200 g portion of PEG 8000 is added and allowed to dissolve. This generally takes about 10 min and then a second 200 g portion of PEG 8000 is added and the solution is allowed to stir until fully dissolved, generally about 2 h. Once fully dissolved the solution is placed in a 1 l glass bottle and stored at 4 °C.
6. The loading buffer is a 200 mM solution of sodium phosphate dibasic. It is prepared in 100 ml aliquots by adding 5.36 g $Na_2HPO_4*7H_2O$ to 100 ml of DI dH_2O. The pH of this solution is 9.76.
7. EDTA is prepared using EDTA dihydrate at 0.25 M as a saturated solution with a pH of ~5. Preparation in this manner keeps the pH and ionic strength low.
8. The finishing buffer is a 200 mM solution of sodium phosphate monobasic. It is prepared by adding 2.76 g of $NaH_2PO_4*H_2O$ to 100 ml DI dH_2O. The pH of this solution is 3.70.

9. The Abm used in formulation is prepared by weighing 2 mg technical Abm (Alfa Aesar, Cat # AAJ60039–06) into a 1.5 ml microfuge tube and dissolving by the addition of 1 ml HPLC grade acetonitrile.
10. β-propiolactone (BPL) should be treated as a carcinogen. It should be used in an exhaust ventilation hood with waste disposed of as hazardous waste. In aqueous conditions, BPL hydrolyzes within 15 min, reducing its inhalation hazard.
11. The M9 buffer used in the nematode acute toxicity assay is prepared by adding 3 g KH_2PO_4, 6 g Na_2HPO_4, 0.5 g NaCl, and 1 g NH_4Cl to DI dH_2O and bringing the volume up to 1 l.
12. It has been determined that these conditions relax the capsid structure and allow solvent accessibility to the entire particle [2].
13. This treatment has been found to reduce RCNMV infectivity (both virions and extracted viral RNA) > 99.99%.
14. For larger scale formulations, Sephadex G-25 columns can be prepared and used. When optimizing the loading protocol, smaller scale formulations can be purified using GE Illustra Microspin columns.
15. Deactivation with BPL results in a ~5% increase in 260 nm absorbance in comparison to the native virions. The accuracy of quantification by UV measurement was found to be comparable to Coomassie-based protein assays and PAGE gel analysis while being more precise than either.
16. This allows the loaded cargo to release from the particle. Acetonitrile is used for extraction because (1) abamectin is highly soluble in acetonitrile resulting in its full release from PVN and (2) acetonitrile does not interfere in the subsequent HPLC analysis.
17. The published LD50 of Abm for *C. elegans* is 0.1 μg/ml [12]. For this assay, it is more accurate to rate motility using a 90% immotility score rather than a 50% threshold.
18. This assay was also performed using J2 stage RKN (*Meloidogyne hapla*) as the test species, and it was observed that the toxic response to Abm was the same for both species.
19. The plate-to-plate variation of observed toxicity has been quite reproducible with a concentration of technical Abm greater than 0.7 μg/ml at 2 h and 0.2 μg/ml at 24 h capable of immobilizing or impairing greater than 90% of the individuals. These results are similar to those reported by others [9].
20. While the average load concentration determined by HPLC was found to be 177 ± 9 Abm/PVN, the bioactive concentrations of Abm are lower but increased over time as a function of Abm release from the PVN: 20 ± 12 at 2 h, 92 ± 20 at 24 h, and 139 ± 23 at 48 h Abm/PVN.

Acknowledgments

This work was supported with funding from the USDA NIFA Agricultural System and Technology, Nanotechnology for Agricultural and Food System program and the Bill & Melinda Gates Foundation. We would also like to thank the College of Agriculture and Life Sciences Agricultural Research Service at NC State University for providing their facilities and support in this work.

References

1. Loo L, Guenther RH, Lommel SA, Franzen S (2008) Infusion of dye molecules into Red clover necrotic mosaic virus. Chem Commun (Camb) 7:88–90
2. Martin SL, He L, Meilleur F, Guenther RH, Sit TL, Lommel SA, Heller WT (2013) New insight into the structure of RNA in red clover necrotic mosaic virus and the role of divalent cations revealed by small-angle neutron scattering. Arch Virol 158:1661–1669
3. Singh S, Singh B, Singh AP (2015) Nematodes: A threat to sustainability of agriculture. Procedia Environ Sci 29:215–216
4. Cao J, Guenther RH, Sit TL, Lommel SA, Opperman CH, Willoughby JA (2015) Development of abamectin loaded plant virus nanoparticles for efficacious plant parasitic nematode control. ACS Appl Mater Interfaces 7:9546–9553
5. Lockney DM, Guenther RH, Loo L, Overton W, Antonelli R, Clark J, Hu M, Luft C, Lommel SA, Franzen S (2011) The Red clover necrotic mosaic virus capsid as a multifunctional cell targeting plant viral nanoparticle. Bioconjug Chem 22:67–73
6. Honarbakhsh S, Guenther RH, Willoughby JA, Lommel SA, Pourdeyhimi B (2013) Polymeric systems incorporating plant viral nanoparticles for tailored release of therapeutics. Adv Healthc Mater (7):1001–1007
7. Cao J, Guenther RH, Sit TL, Opperman CH, Lommel SA, Willoughby JA (2014) Loading and release mechanism of Red clover necrotic mosaic virus derived plant viral nanoparticles for drug delivery of doxorubicin. Small 10:5126–5136
8. Lommel, SA. (1983). Genome and replicative organization of the plant dianthoviruses and Arkansas bee virus. Thesis (Ph.D. in Plant Pathology), University of California, Berkeley
9. Opperman CH, Chang S (1991) Effects of Aldicarb and Fenamiphos on acetycholinesterase and motility of Caenorhabditis elegans. J Nematol 23:20–27
10. Qiao K, Liu X, Wang H, Xia X, Ji X, Wang K (2012) Effect of abamectin on root-knot nematodes and tomato yield. Pest Manag Sci 68:853–857
11. Mizumoto H, Tatsuta M, Kaido M, Mise K, Okuno T (2003) Cap-independent translational enhancement by the 3′ untranslated region of Red clover necrotic mosaic virus RNA1. J Virol 77:12113–12121
12. Holden-Dye L, Walker RJ. Anthelmintic drugs and nematicides: studies in Caenorhabditis elegans. (http://www.wormbook.org (2014). Accessed 27 Mar 2017)

Chapter 14

Nanowires and Nanoparticle Chains Inside Tubular Viral Templates

Kun Zhou and Qiangbin Wang

Abstract

One-dimensional (1D) inorganic nanomaterials, especially with magnetic and optical properties, are key components in material synthesis for applications in nanoelectronics, catalysis, and sensing. To achieve these objectives, tubular viral templates are emerging as natural anisotropic bioreactors for the control of the synthesis of inorganic materials with spatial confinement. In particular, tobacco mosaic virus (TMV) with a longitudinal cylinder shape provides a defined narrow cavity to direct the controllable synthesis of 1D inorganic nanomaterial. Based on the understanding of biological characteristics of viral capsids, we can introduce genetic modifications to tailor the arrangement of functional motifs for specific electroless deposition. Here we present an overview of methods for the utilization of the TMV-derived interior surface to realize spatially selective chemisorption, nucleation, and growth of nanocrystals into nanowires and nanoparticle chains.

Key words Tobacco mosaic virus, Electroless deposition, Mineralization, Template synthesis, Spatial selectivity, One-dimensional nanomaterial

1 Introduction

In recent years biomacromolecules have received enormous attention in the field of nanotechnology as templates for the heterogeneous synthesis of inorganic material in vitro [1, 2]. Biotemplates such as proteins and nucleic acids have evolved to self-assemble into sophisticated structures with multiple functionalized surfaces, making them good candidates for casting complicated materials [3–6]. Their physical structures can impose characteristics such as specific size, shape, and composition on the products with molecular precision, thereby providing controllability for the synthesis of hybrid organic–inorganic nanomaterials. Moreover, their amenability to practical manipulations such as chemical modifications and genetic engineering enables further optimization through the attachment of reactive ligands or motifs at specific sites for programmable mineralization. In particular, viruses and viral capsids represent a powerful toolkit for

Christina Wege and George P. Lomonossoff (eds.), *Virus-Derived Nanoparticles for Advanced Technologies: Methods and Protocols*, Methods in Molecular Biology, vol. 1776, https://doi.org/10.1007/978-1-4939-7808-3_14,

© Springer Science+Business Media, LLC, part of Springer Nature 2018

template-assisted synthesis due to sufficient understanding of their biochemistry, simple structures, good monodispersity and ready availability. Besides, compared to artificial templates with a size distribution, viral structures of similar type are normally uniform in shape, composition, and their physicochemical properties. This makes them well-defined bioreactors for arranging nucleation and growth of inorganic phases [7–10].

Tobacco mosaic virus (TMV), which is composed of a single strand of RNA and ~2130 identical coat proteins (cp), is one of the most widely studied viruses in the emerging repertoire of nanotechnological applications. Native TMV has a tubular structure with a modal length of 300 nm and with outer and inner diameters of 18 nm and 4 nm respectively, thereby providing a longitudinal cylindrical template for fabrication of one-dimensional (1D) nanomaterials by electroless deposition on its surfaces [11–15]. The exterior surface of TMV has been functionalized by coating with magnetic metals [16], noble metals [17], or semiconducting materials for chemical catalysis [18, 19]. It has also been used as/in/for battery electrodes [20], and field-effect transistors [21] (*see* Chapter 27). On the other hand, it is fascinating to achieve mineralization of 1D inorganic material on the sub-10 nm scale by confinement using the narrow channel of TMV or TMV virus-like particle. By tailoring reagents and buffer conditions during the electroless deposition process, Knez et al. successfully demonstrated the preferential mineralization of 3 nm Ni and Co nanowires in the native TMV channel [22]. However, exclusive metal deposition within the hollow TMV channel is more difficult to achieve than the utilization of its outer surface because of steric hindrance and lack of spatial selectivity on the inner surface [14, 15]. To circumvent these drawbacks and enhance the yield of 1D metal nanomaterials, genetic modifications have been employed to tailor the properties of the inner surface. For example, mutated tomato mosaic virus, a very close relative to TMV, was constructed to enhance coatings of Co/Pt alloy in the inner channel by increasing the number of positively charged sites for adsorption of precursor cations [23]. Despite such developments, some desirable mutations may not be possible due to their negative effects on virus propagation in plants, and more effective ways of utilizing TMV-derived nanochannels for template synthesis of various functional materials need to be explored.

In this chapter, we describe methodology to realize spatially selective mineralization of functional compounds into TMV-derived protein nanochannels. A substitution mutant of cp (T103C) was constructed by substituting a cysteine at position 103, a position lying in the low-radius region of TMVcp [24]. The construct was expressed in an *Escherichia coli* (*E. coli*) system, resulting in stable protein nanotubes with improved self-assembly properties (*see* Fig. 1). The introduced thiol is arrayed on the inner

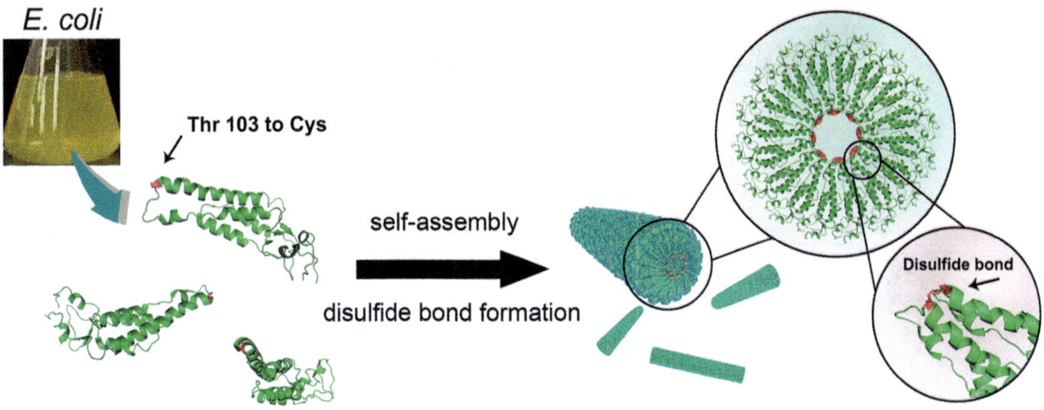

Fig. 1 Self-assembly of TMV-T103Ccp into protein nanotubes with thiol ligands on the inner surface of the nanochannel. The mutant is constructed by the site-directed mutagenesis of Thr 103 to Cys; this amino acid occurs at the low-radius region of the subunit (reproduced from [24] with permission from American Chemical Society)

surface of the nanotubes enabling selective chemisorption and nucleation of noble metal ions [25]; thereby we can achieve controlled growth of noble or magnetic metals confined by the T103Ccp protein nanochannels into 1D conformation to create nanoparticle chains and nanowires (*see* Fig. 2).

2 Materials

In this section, we describe the components necessary to prepare TMV-T103Ccp and achieve mineralization of inorganic materials into the T103C protein nanochannels. This includes devices, buffers, solutions, and corresponding measurement setups (*see* **Note 1**). The buffers and solutions used throughout this protocol are made from ultrapure water obtained from a Millipore Milli-Q Academic water purification system (18.2 MΩ).

2.1 Preparation of TMV-T103Ccp Nanotubes

1. WT-TMVcp gene cDNA fragment with codon usage optimization for expression in *E. coli* (Sangon Biotech, China).
2. Primers:

   P1: 5’-AGATATACATATGAGCTATAGCAT-3’

   P2: 5’-GTTTCCGCGGTACACGGATTGGCCT-3’

   P3: 5’-AGGCCAATCCGTGTACCGCGGAAAC-3’

   P4: 5’-GGAATTCTCAGGTCGCCGGGCC-3’
3. *Nde*I and *Eco*RI restriction enzymes.
4. T4 DNA ligase.

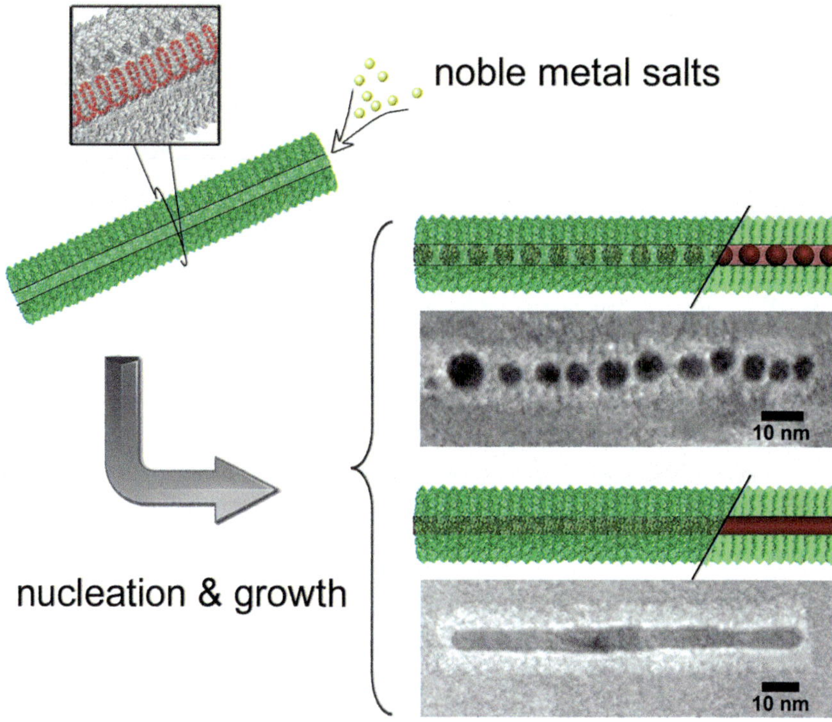

Fig. 2 Schematic illustration of the spatially selective biomineralization inside a TMV-T103Ccp nanochannel. This was achieved by the specific chemisorption and nucleation of noble metal ions, followed by the further growth of metal deposits into nanoparticle chains or nanowires (reproduced from [25] with permission from John Wiley & Sons)

5. Vector plasmid pET32a(+) (Novagen).
6. Chemically competent *E. coli* BL21 (DE3) cells.
7. Glycerol.
8. Luria–Bertani (LB) medium: 10 g/l Bacto tryptone, 5 g/l yeast extract, 10 g/l NaCl.
9. 100 mg/ml ampicillin (stock solution), stored at −20 °C.
10. 1 M isopropyl β-D-1-thiogalactopyranoside (IPTG) (stock solution), stored at −20 °C.
11. 50 mM Tris–HCl, pH 7.0.
12. 0.8 M dithiothreitol (DTT) (stock solution), stored at −20 °C.
13. Buffer A: 20 mM Tris–HCl, pH 7.2, 20 mM NaCl, 20 mM EDTA.
14. Glass beaker.
15. Ammonium sulfate.
16. Carbonate–bicarbonate buffer: 50 mM Na_2CO_3–$NaHCO_3$, pH 9.5. Weigh 1.59 g Na_2CO_3 and 2.94 g $NaHCO_3$, dissolve into 1 l water. Store at 4 °C.
17. Material for SDS-PAGE to evaluate the purity of the expressed protein.

18. Phosphate buffer (PB) (stock solution): 400 mM Na_2HPO_4–NaH_2PO_4, pH 7.0. Add about 700 ml water to a 1 l glass beaker. Weigh 18.24 g NaH_2PO_4 and 88.82 g $Na_2HPO_4{\cdot}12H_2O$, and transfer to the glass beaker and dissolve. Make up to 1 l with water. Store at 4 °C. Dilute with water for different concentrations as needed.
19. Dialysis tubing (molecular weight cutoff (MWCO) of 8000–14,000 Da).
20. 10% (w/w) sucrose solution in 150 mM PB.
21. 50% (w/w) sucrose solution in 150 mM PB.
22. Ultrafiltration device (Amicon Ultra-15, 100 kDa MWCO, Millipore).
23. Gradient Station (BioComp, Canada).
24. Centrifuge (e.g., Thermo scientific, Heraeus Multifuge X3).
25. Ultracentrifuge (Beckmann Optima XPN-90, SW40 Ti rotor).
26. Material for transmission electron microscopy (TEM) analysis.

2.2 Electroless Deposition Inside TMV-T103Ccp Nanotube Channel

2.2.1 Mineralization of Gold Nanoparticle Chains

1. 5 mM and 50 mM tetrachloroauric(III) acid hydrate ($HAuCl_4{\cdot}4H_2O$).
2. 10 mM sodium borohydride ($NaBH_4$); freshly prepared and iced.
3. 2 mM and 20 mM ascorbic acid.
4. Sample vial (Glass, 1.8 ml).
5. Magnetic stirrer.
6. Microcentrifuge tubes: 1.5 ml.
7. Spectrafuge Mini (C1301, Labnet).

2.2.2 Mineralization of Gold Nanowires

1. 50 mM tetrachloroauric(III) acid hydrate ($HAuCl_4{\cdot}4H_2O$).
2. 100 mM sodium borohydride ($NaBH_4$); freshly prepared and iced.
3. 5 mM ascorbic acid.
4. Phosphate buffer (PB): 100 mM Na_2HPO_4-NaH_2PO_4, pH 7.0.
5. Microcentrifuge tubes: 1.5 ml.
6. Eppendorf thermomixer.
7. Ultrafiltration device with 30 kDa MWCO (Amicon Ultra-4, Millipore).
8. Sample vial (Glass, 1.8 ml).
9. Magnetic stirrer.

2.2.3 Mineralization of Platinum Nanoparticles

1. 50 mM tetrachloroauric(III) acid hydrate ($HAuCl_4{\cdot}4H_2O$).
2. 20 mM and 100 mM sodium borohydride ($NaBH_4$); freshly prepared and iced.

3. 10 mM chloroplatinic acid hexahydrate ($H_2PtCl_6 \cdot 6H_2O$).
4. Phosphate buffer (PB): 100 mM Na_2HPO_4-NaH_2PO_4, pH 7.0.
5. Microcentrifuge tubes: 1.5 ml.
6. Eppendorf thermomixer.
7. Ultrasonic Instrument (KQ-300DA, Kunshan Shumei, China).

2.2.4 Mineralization of Nickel Nanowires

1. 5 M NaCl.
2. 50 mM sodium tetrachloropalladate(II) (Na_2PdCl_4).
3. 450 mM nickel(II) acetate tetrahydrate ($NiC_4H_6O_4 \cdot 4H_2O$).
4. 1 M L-(+)-lactic acid.
5. 200 mM dimethylamine borane (DMAB).
6. 1 M NaOH.
7. Nickel reaction buffer: 80 μl $NiC_4H_6O_4 \cdot 4H_2O$ (450 mM), 46 μl lactic acid (1 M), 67 μl DMAB (200 mM), 45 μl of NaOH (1 M) (*see* **Note 2**).
8. Microcentrifuge tubes: 1.5 ml.
9. Eppendorf thermomixer.
10. Dialysis tubing (MWCO of 8000–14,000 Da).
11. Ultrafiltration device with 30 kDa MWCO (Amicon Ultra-4, Millipore).

2.2.5 Additional Experiments

1. Material for TEM analysis.

3 Methods

The methods described below outline the construction of plasmids, the expression and purification of TMV-T103Ccp, the self-assembly and purification of TMV-T103Ccp protein nanotubes (*see* Subheading 3.1) [24], as well as the spatially selective mineralization of gold nanoparticle chains, gold nanowires [25], platinum nanoparticles, and nickel nanowires inside the nanochannels of TMV-T103Ccp nanotubes (*see* Subheading 3.2).

3.1 Preparation of TMV-T103Ccp Protein Nanotubes

3.1.1 Construction of Plasmid pET32a(+)/TMV-T103C

1. The pET32a(+)/TMV-T103C plasmid is constructed according to standard cloning procedures. In short, the mutagenesis of T103C-TMVcp is conducted by overlapping PCR using the two pairs of primers (*see* Subheading 2.1, **Step 2**) and the WT-TMVcp gene sequence. Then, the PCR product is digested with *Nde*I/*Eco*RI and ligated into the similarly digested pET32a(+) vector using T4 DNA ligase.

2. The pET32a(+)/TMV-T103C plasmid is verified by sequencing and transformed into competent *E. coli* BL21 (DE3). Grow the cells in LB medium and prepare glycerol stocks.

3.1.2 Expression and Purification of TMV-T103Ccp

1. Inoculate 5 ml of LB medium containing 100 μg/ml ampicillin with the *E. coli* strain from a glycerol stock and incubate in a shaker at 37 °C and 200 rpm overnight.
2. Inoculate 500 ml of LB medium containing 100 μg/ml ampicillin in a 1 l shaker flask with 5 ml of culture (*see* Subheading 3.1.2, **step 1**) and incubate at 37 °C and 200 rpm for 2–3 h.
3. Add isopropyl β-D-1-thiogalactopyranoside (IPTG) into the culture to give a final concentration of 1 mM and incubate the culture for 8–12 h at 30 °C and 180 rpm.
4. Harvest cells by centrifugation at 10,000 × *g* for 10 min. Wash cells with 50 mM Tris–HCl (pH 7.0) and then harvest cells again by centrifugation. Store the cell pellet at −20 °C.
5. Resuspend the cell pellet in 40 ml of buffer A containing 15 mM DTT (*see* **Note 3**) and lyse cells on ice by ultrasonication (power: 400 W, ultrasonic time: 4 s, interval time: 4 s, total time: 45–60 min).
6. Centrifuge the lysate for 30 min at 13,000 × *g* and 4 °C to remove cell debris and membrane fractions and transfer the supernatant to a glass beaker.
7. Purify T103Ccp by ammonium sulfate fractionation (*see* **Note 4**): While stirring at 4 °C, slowly add solid ammonium sulfate into lysate solution. The protein precipitating between 0–0.15, 0.15–0.25, and 0.25–0.35 ammonium sulfate saturation is removed and collected by centrifugation at 13,000 × *g* at 4 °C for 12 min. Resuspend the pellets in carbonate-bicarbonate buffer (50 mM, pH 9.5) containing 20 mM DTT at 4 °C with gentle shaking overnight. Purity is confirmed by sodium dodecyl sulfate polyacrylamide gel electrophoresis (SDS-PAGE).
8. Store protein at −80 °C after adding glycerol to a final concentration of 10% (v/v).

3.1.3 Self-Assembly and Purification of TMV-T103Ccp Protein Nanotubes

1. Thaw the T103Ccp protein stocks (*see* **Note 5**) and pipette them into dialysis tubing with MWCO of 8000–14,000 Da. Dialyze against 400 mM PB (pH 7.0) for at least 16 h, replacing the buffer three times.
2. Incubate protein samples at room temperature for 5 days to allow self-assembly of T103Ccp into protein nanotubes (*see* **Note 6**).
3. Prepare a sucrose gradient consisting of a 10–50% (w/w) continuous gradient in 150 mM PB (pH 7.0) in polyallomer centrifuge tubes using the Gradient Station according to the manufacturer's instructions.

4. Carefully transfer protein samples onto the sucrose gradient in polyallomer centrifuge tubes (*see* **Note 7**).
5. After ultracentrifugation (SW 40 Ti) at 175,000 × *g* and 15 °C for 3.5 h, collect fractions from the top of the polyallomer centrifuge tubes (*see* **Note 8**).
6. Verify the formation of protein nanotubes by TEM using negative staining, and dialyze against 100 mM PB (pH 7.0) using a dialysis tubing with MWCO of 8000–14,000 Da for at least 24 h, replacing the buffer three times (*see* **Note 9**). Concentrate the protein solution using an ultracentrifugal filter device with 100 kDa MWCO.
7. Incubate T103Ccp protein nanotubes with 20 mM DTT at a final concentration of 0.8 mg/ml for 3 h at room temperature, then dialyze against 100 mM PB (pH 7.0) using a dialysis tubing with MWCO of 8000–14,000 Da, and be ready to undertake the mineralization procedure.

3.2 Electroless Deposition Inside TMV-T103Ccp Nanotube Channel

3.2.1 Mineralization of Gold Nanoparticle Chains

1. Add 2.5 μl of 50 mM $HAuCl_4$ to 150 μl of prepared T103Ccp nanotubes in a glass vial with a small magnetic stir bar (*see* **Note 10**) while gently stirring on a magnetic stirrer, and incubate for 3 h.
2. Quickly inject 20 μl of 10 mM $NaBH_4$ (freshly prepared on ice) into the mixture while stirring at 800 rpm on the magnetic stirrer and incubate for 5 min. The color of the solution turns red immediately.
3. Incubate at room temperature overnight without stirring to water quench the residual $NaBH_4$.
4. Transfer the mixture to a clean 1.5 ml microcentrifuge tube and spin at 2000 × *g* using a Spectrafuge Mini (or similar) for 10 min.
5. Transfer the supernatant to a new 1.5 ml microcentrifuge tube.
6. Add 2 μl of 50 mM $HAuCl_4$ while shaking gently for 5–10 min and then add 10 μl of 20 mM ascorbic acid while shaking gently. Incubate at room temperature for at least 2 h.
7. Centrifuge in a microcentrifuge tube at 10,000 × *g* for 10 min (*see* **Note 11**). Collect the supernatant carefully and transfer it into a new glass vial with a new magnetic stir bar (*see also* **Note 10**).
8. Keep stirring at 200 rpm on the magnetic stirrer, add 2 μl of 5 mM $HAuCl_4$ and incubate for 10 min.
9. Keep stirring at 200 rpm on the magnetic stirrer, add 5 μl of 2 mM ascorbic acid and incubate for 15–20 min.
10. Repeat **steps 8** and **9** for the continuous growth of gold nanoparticle chains inside the protein nanochannels (*see* Fig. 3).

3.2.2 Mineralization of Gold Nanowires

1. Add 5 μl of 50 mM $HAuCl_4$ into 150 μl of prepared T103Ccp nanotubes in a microcentrifuge tube while shaking at 800 rpm on Eppendorf thermomixer for 30 min at room temperature.
2. Ultrafiltrate T103Ccp nanotubes at least three times using a centrifugal filter device (30 kDa MWCO) with supplement of 100 mM PB (pH 7.0) to remove excess $HAuCl_4$.
3. Add 5 μl of $NaBH_4$ (100 mM) into the mixture while shaking at 800 rpm on Eppendorf thermomixer for 5 min, resulting in a clear colorless solution (*see* **Note 12**).
4. Repeat ultrafiltration at least three times using a centrifugal filter device (30 kDa MWCO) to remove excess $NaBH_4$ with supplement of 100 mM PB (pH 7.0).
5. Transfer the nanotubes into a new glass vial with a magnetic stir bar (*see* also **Note 10**) and add 2 μl of $HAuCl_4$ (50 mM) into the mixture and incubate for 30 min while stirring at 200 rpm on a magnetic stirrer.
6. While stirring at 200 rpm on the magnetic stirrer, add ten times 2 μl of ascorbic acid (5 mM) at 15 min intervals (*see* **Note 13**) (*see* Fig. 4).

3.2.3 Mineralization of Platinum Nanoparticles

1. Prepare 200 μl of Au-actived protein nanotubes in a new 1.5 ml microcentrifuge tube following the procedures as previously described (*see* **steps 1–4** in Subheading 3.2.2).
2. Add 10 μl of 10 mM H_2PtCl_6 into the mixture followed by ultrasonic bath for 30 s.
3. Add 5 μl of $NaBH_4$ (20 mM) into the mixture followed by ultrasonic bath for 30 s.
4. Repeat **steps 2** and **3** three times (*see* Fig. 5 a).

3.2.4 Mineralization of Nickel Nanowires

1. Mix 200 μl of T103Ccp nanotube solution with 60 μl of water, 40 μl of 5 M NaCl, and 2 μl of 50 mM Na_2PdCl_4 in a 1.5 ml microcentrifuge tube and incubate at room temperature for 15 min at 600 rpm on an Eppendorf thermomixer.
2. Dialyze against water in a dialysis tubing with MWCO of 8000–14,000 Da for at least 16 h and then concentrate the Pd actived-protein solution to a final volume of 300–400 μl by ultrafiltration using a 30 kDa MWCO filtration device (Millipore).
3. Mix 50 μl of Pd actived-protein sample with an equal volume of nickel reaction buffer in a microcentrifuge tube (*see* **Note 14**). The color of the mixture will turn dark after several minutes accompanied by the formation of gas bubbles.
4. Stop the reaction by adding 1 ml of water, centrifuge for 1 min at 12,000 × *g*, and disperse the pellet in 100 μl of water for TEM characterization (*see* Fig. 5 b).

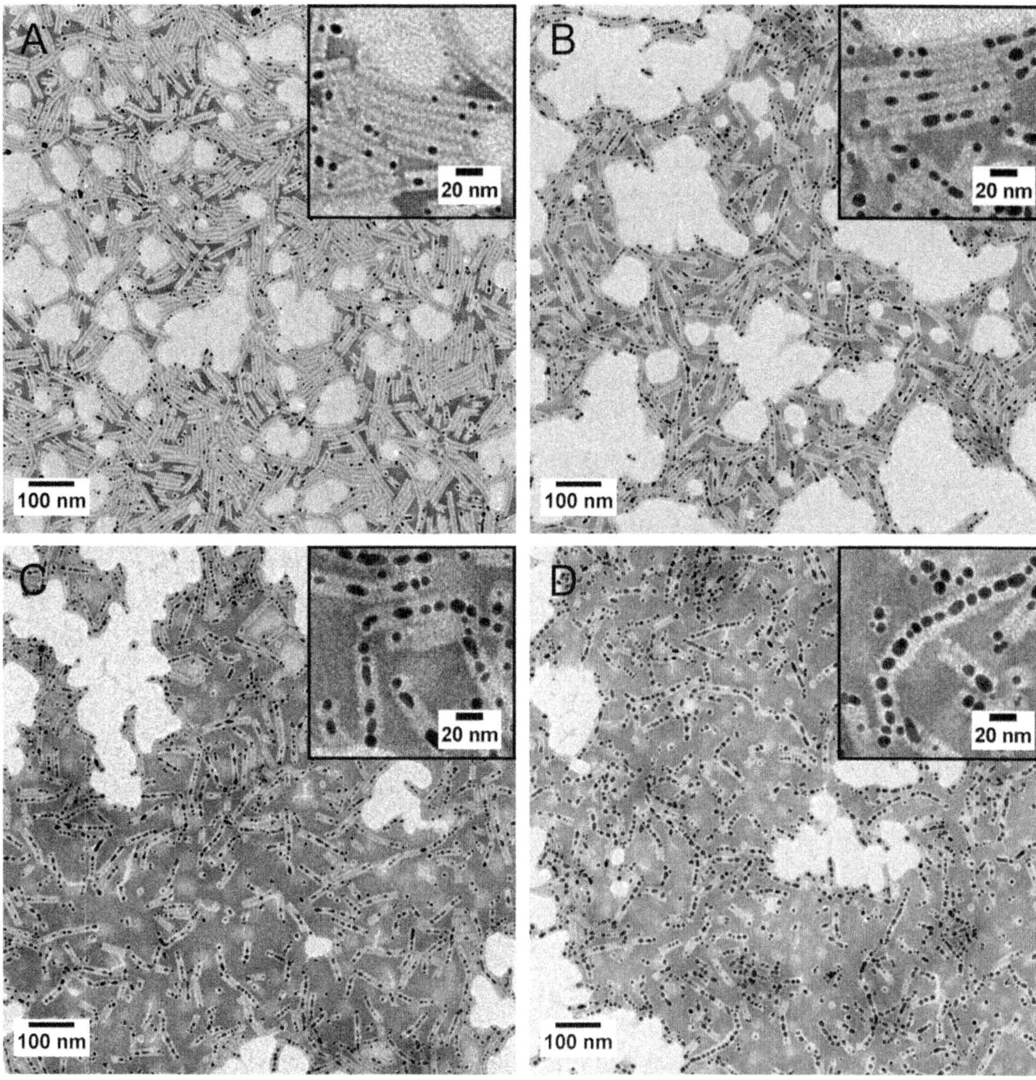

Fig. 3 TEM images of gold nanoparticle chains inside TMV-T103Ccp nanotubes. They were produced with continuous cyclic growth for 0 (**a**), 7 (**b**), 14 (**c**), and 24 (**d**) times (reproduced from [25] with permission from John Wiley & Sons)

4 Notes

1. During the mineralization procedure of protein nanotubes, all experimental devices should be cleaned properly because traces of metals may cause nonspecific nucleation and aggregation of metal nanoparticles. Noble metal salt solutions are light sensitive and should be stored in the dark.
2. The mixture containing 80 μl of 450 mM $NiC_4H_6O_4{\cdot}4H_2O$, 46 μl of 1 M lactic acid, and 67 μl of 200 mM DMAB will produce a white precipitate, which can be completely dissolved

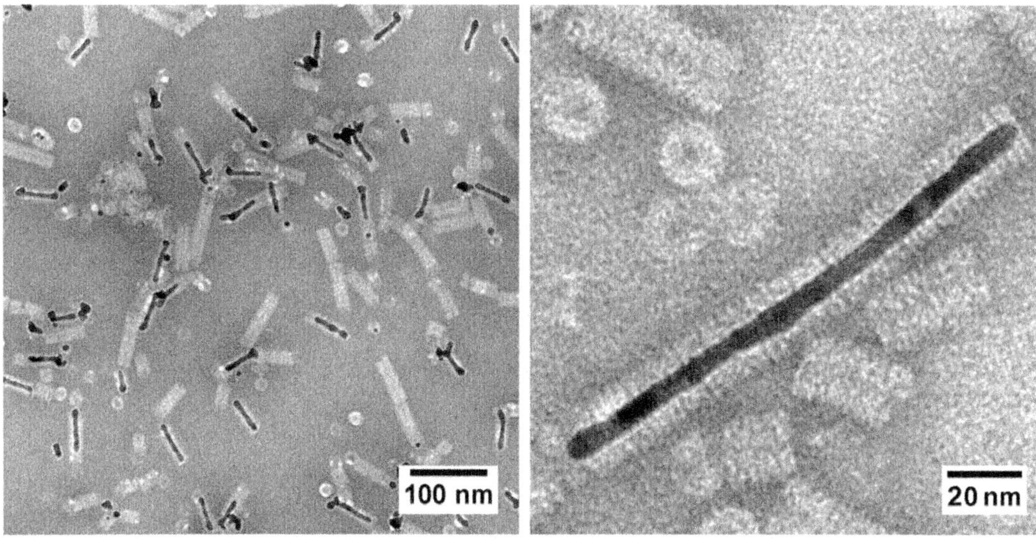

Fig. 4 TEM images of gold nanowires inside TMV-T103Ccp nanotubes (reproduced from [25] with permission from John Wiley & Sons)

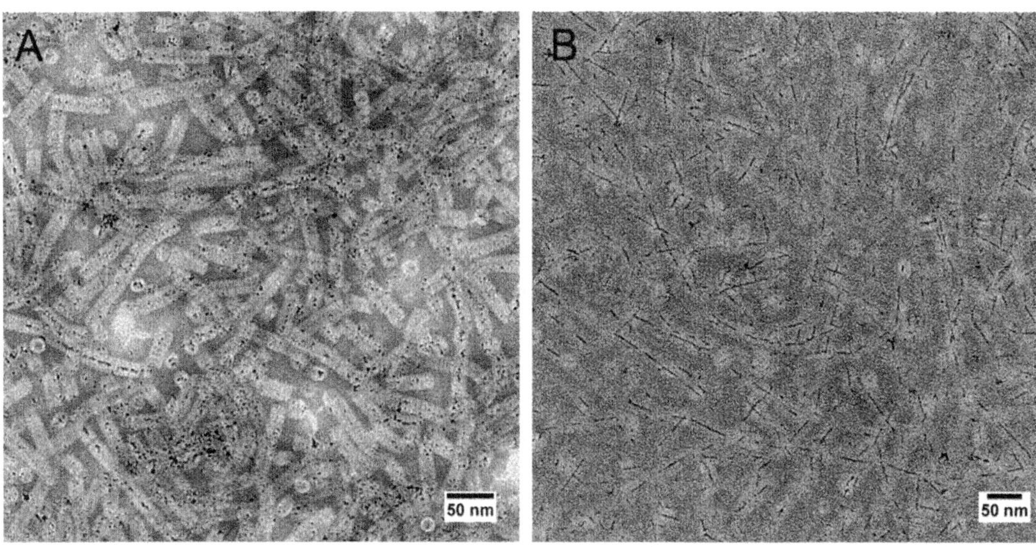

Fig. 5 TEM images of platinum nanoparticles (**a**) and nickel nanowires (**b**) inside TMV-T103Ccp nanotubes

by adding 45 μl of NaOH (1 M). This nickel reaction buffer can be stored at 4 °C for weeks.

3. Take the cell pellet out of the freezer and resuspend it in buffer immediately. After thawing the cell pellet will become extremely viscous and hard to disperse.
4. Normally, ammonium sulfate fractionation precipitation is sufficient to obtain TMV-T103Ccp with high purity of up to 95%. However, the purification result depends on the quality of the cells and the ultrasonication procedure. It is recommended to

adjust the concentration of ammonium sulfate fractionation gradually. As an alternative, ion exchange chromatography is also recommended for further purification, as follows: dialyze protein solution against Bis-Tris buffer (20 mM, pH 6.6), apply to a strong anion exchange column (HiTraP Q HP, GE Healthcare) on an AKTA prime plus FPLC system (GE Healthcare), elute with a 0–800 mM NaCl gradient.

5. It is recommended that the protein stock is thawed using a water bath at room temperature. Thawing the T103Ccp sample at 4 °C may cause irreversible precipitation.
6. Verify the self-assembly of protein nanotubes by TEM. If the yield is not good, increase the incubation time or use a buffer with a lower pH, such as pH 6.2 PB (400 mM), as this can facilitate the formation of protein nanotubes.
7. Do not disturb the sucrose gradient when adding the protein sample.
8. Mark the polyallomer centrifuge tube for division into nine equal sections from top to bottom. The protein nanotubes will be concentrated at fifth, sixth and seventh fractions counted starting at the top after the sucrose density gradient centrifugation step.
9. The presence of sucrose in the sample can result in swelling of dialysis tubing due to the osmotic pressure. Use longer dialysis tubing to prevent the membrane from bursting.
10. Glassware and magnetic stir bars should be cleaned extensively with aqua regia (three parts 12 M HCl, one part 70% HNO_3) for 30 min, rinsing thoroughly with ultrapure water, and drying. Preparation of aqua regia should be carried out in well-ventilated areas under an aspirated fume hood, and gloves and safety glasses should be worn.
11. The selective nucleation process is the crux of achieving spatially specific gold mineralization inside the TMV-T103Ccp nanotubes. However, some factors such as the concentration and the quality of protein nanotubes may give rise to variations in the mineralization yield. It is recommended to adjust centrifugation force and time to remove gold aggregates outside the protein nanotubes.
12. If the solution is not colorless, it means that excess $HAuCl_4$ exists in the solution. Increase ultrafiltration times at **step 2** in Subheading 3.2.2.
13. Centrifugation or density gradient centrifugation is helpful to yield pure gold nanowires inside protein nanotubes.
14. Do not mix the solution by violent pipetting or vortexing.

References

1. Nudelman F, Sommerdijk NAJM (2012) Biomineralization as an inspiration for materials chemistry. Angew Chem Int Ed 51(27):6582–6596
2. Sotiropoulou S, Sierra-Sastre Y, Mark SS, Batt CA (2008) Biotemplated nanostructured materials. Chem Mater 20(3):821–834
3. Sun W, Boulais E, Hakobyan Y, Wang WL, Guan A, Bathe M, Yin P (2014) Casting inorganic structures with DNA molds. Science 346(6210):1258361
4. Uchida M, Klem MT, Allen M, Suci P, Flenniken M, Gillitzer E, Varpness Z, Liepold LO, Young M, Douglas T (2007) Biological containers: protein cages as multifunctional nanoplatforms. Adv Mater 19(8):1025–1042
5. Dickerson MB, Sandhage KH, Naik RR (2008) Protein- and peptide-directed syntheses of inorganic materials. Chem Rev 108(11):4935–4978
6. Boi Hoa S, Sungsu K, Hyun MS, Hyunjoo L, Duk-Young J, Kyeong Kyu K (2011) Platinum nanoparticles encapsulated by aminopeptidase: a multifunctional bioinorganic nanohybrid catalyst. Angew Chem Int Ed 50(50):11924–11929
7. Chuanbin M, Solis DJ, Reiss BD, Kottmann ST, Sweeney RY, Andrew H, George G, Brent I, Belcher AM (2004) Virus-based toolkit for the directed synthesis of magnetic and semiconducting nanowires. Science 303(5655):213–217
8. Young M, Willits D, Uchida M, Douglas T (2008) Plant viruses as biotemplates for materials and their use in nanotechnology. Annu Rev Phytopathol 46:361–384
9. Flynn CE, Lee SW, Peelle BR, Belcher AM (2003) Viruses as vehicles for growth, organization and assembly of materials. Acta Mater 51(19):5867–5880
10. Sun J, DuFort C, Daniel MC, Murali A, Chen C, Gopinath K, Stein B, De M, Rotello VM, Holzenburg A, Kao CC, Dragnea B (2007) Core-controlled polymorphism in virus-like particles. Proc Natl Acad Sci U S A 104(4):1354–1359
11. Tseng RJ, Tsai C, Ma L, Ouyang J, Ozkan CS, Yang Y (2006) Digital memory device based on tobacco mosaic virus conjugated with nanoparticles. Nat Nanotechnol 1(1):72–77
12. Niu Z, Liu J, Lee LA, Bruckman MA, Zhao D, Koley G, Wang Q (2007) Biological templated synthesis of water-soluble conductive polymeric nanowires. Nano Lett 7(12):3729–3733
13. Wayne S, Trevor D, Mark Y, Gerald S, Stephen M (1999) Inorganic–organic nanotube composites from template mineralization of tobacco mosaic virus. Adv Mater 11(3):253–256
14. Knez M, Sumser M, Bittner AM, Wege C, Jeske H, Martin TP, Kern K (2004) Spatially selective nucleation of metal clusters on the tobacco mosaic virus. Adv Funct Mater 14(2):116–124
15. Dujardin E, Peet C, Stubbs G, Culver JN, Mann S (2003) Organization of metallic nanoparticles using tobacco mosaic virus templates. Nano Lett 3(3):413–417
16. Wu Z, Mueller A, Degenhard S, Ruff SE, Geiger F, Bittner AM, Wege C, Krill CE (2010) Enhancing the magnetoviscosity of ferrofluids by the addition of biological nanotubes. ACS Nano 4(8):4531–4538
17. Lim JS, Kim SM, Lee SY, Stach EA, Culver JN, Harris MT (2010) Biotemplated aqueous-phase palladium crystallization in the absence of external reducing agents. Nano Lett 10(10):3863–3867
18. Górzny MŁ, Walton AS, Evans SD (2010) Synthesis of high-surface-area platinum nanotubes using a viral template. Adv Funct Mater 20(8):1295–1300
19. Yang C, Choi CH, Lee CS, Yi H (2013) A facile synthesis-fabrication strategy for integration of catalytically active viral-palladium nanostructures into polymeric hydrogel microparticles via replica molding. ACS Nano 7(6):5032–5044
20. Chen X, Gerasopoulos K, Guo J, Brown A, Wang C, Ghodssi R, Culver JN (2010) Virus-enabled silicon anode for lithium-ion batteries. ACS Nano 4(9):5366–5372
21. Atanasova P, Rothenstein D, Schneider JJ, Hoffmann RC, Dilfer S, Eiben S, Wege C, Jeske H, Bill J (2011) Virus-templated synthesis of ZnO nanostructures and formation of field-effect transistors. Adv Mater 23(42):4918–4922
22. Knez M, Bittner AM, Boes F, Wege C, Jeske H, Maiβ E, Kern K (2003) Biotemplate synthesis of 3-nm nickel and cobalt nanowires. Nano Lett 3(8):1079–1082
23. Kobayashi M, Seki M, Tabata H, Watanabe Y, Yamashita I (2010) Fabrication of aligned magnetic nanoparticles using tobamoviruses. Nano Lett 10(3):773–776
24. Zhou K, Li F, Dai G, Meng C, Wang QB (2013) Disulfide bond: dramatically enhanced assembly capability and structural stability of tobacco mosaic virus nanorods. Biomacromolecules 14(8):2593–2600
25. Zhou K, Zhang JT, Wang QB (2015) Site-selective nucleation and controlled growth of gold nanostructures in tobacco mosaic virus nanotubulars. Small 11(21):2505–2509

Chapter 15

In Vitro-Reassembled Plant Virus-Like Particles of Hibiscus Chlorotic Ringspot Virus (HCRSV) as Nano-Protein Cages for Drugs

Sek-Man Wong and Yupeng Ren

Abstract

Spherical shaped plant viruses require a precise quantity, size, and shape of their coat protein subunits to assemble into virions of identical dimensions. The capsid of spherical plant virus particles typically consists of a precisely shaped protein cage, which in many cases is assembled from identical coat protein subunits. In addition to packaging the viral genome, such protein cages may have the capacity to load foreign compounds, either large molecules (e.g., polymers) or small molecules (e.g., anticancer chemotherapy drugs). Therefore, reassembled protein cages of suitable viruses can serve as carriers for cargo loading, which is what makes them an attractive platform for drug delivery. Here we describe methods to reassemble plant virus-like particles of hibiscus chlorotic ringspot virus (HCRSV) as nano-protein cages including the techniques to purify coat protein, prepare virus-like particles, and load them with foreign compounds.

Key words Coat protein, Protein cage, Reassembly, Hibiscus chlorotic ringspot virus (HCRSV), Virus-like particles, Drug delivery

1 Introduction

Protein cages of plant viruses assembled from identical coat proteins are, in many cases, uniform in size and precise in structure [1–8]. Under specified chemical and/or environmental conditions, the viruses may undergo protein cage disassembly and viral nucleic acids release. In several cases, conditions have also been found which lead to the reassembly of virus-like particles (VLPs) in vitro in the absence of the viral genome [5, 9]. Such reassembled VLPs can serve as carriers for cargo, as foreign compounds could be loaded inside the empty protein cages. The resulting novel properties of the VLPs provide an attractive platform for drug delivery [10].

Here we describe the methods for purifying viral coat proteins, in vitro reassembly of VLPs of hibiscus chlorotic ringspot virus (HCRSV), and loading foreign compounds inside the nucleic

Christina Wege and George P. Lomonossoff (eds.), *Virus-Derived Nanoparticles for Advanced Technologies: Methods and Protocols*, Methods in Molecular Biology, vol. 1776, https://doi.org/10.1007/978-1-4939-7808-3_15,
© Springer Science+Business Media, LLC, part of Springer Nature 2018

acid-free VLPs. To purify the viral coat protein, plant virus particles are dialyzed against a high pH buffer (e.g., pH 8.0 buffer) in the absence of calcium ions to disassemble the virus particles and release viral RNA [9, 11, 12]. This method is gentle enough for maintaining the coat proteins' stability and their conformational structure. Viral RNA can be removed completely and coat proteins of high purity (based on O.D. ratio of 260 nm and 280 nm) are obtained, devoid of any detectable host proteins. The purified coat protein is functional and able to reassemble into protein cages. TEM observation reveals that the empty VLPs formed are identical in both shape and size of the original plant virus particles [2, 9, 12].

For the HCRSV system, negative charge and a high molecular weight were two prerequisites for the functional compound to be loaded successfully inside the VLPs. Polymers with negatively charged acid groups, such as polystyrenesulfonic acid (PSA) and polyacrylic acid (PAA), could be loaded inside the protein cages, while neutral compounds, such as dextrans, could not be loaded, although subjected to the same methodology. However, as small compounds may leak out through openings in the protein cages, only molecules with a molecular radius larger than the gaps in the protein cage can be successfully loaded and retained inside the VLPs, while smaller molecules could escape through the gaps of the protein cage. For example, polyacids with a molecular weight higher than 13 kDa were successfully loaded inside the HCRSV-like particles, while PSA with a molecular weight of 1.4 and 4.3 kDa, respectively, could not be entrapped [12].

To load small molecules inside the VLPs, a method named "polyacid association" was established. This method was established because most anticancer drugs are positively charged molecules with low molecular weight and therefore do not meet the two loading prerequisites. For this method, large molecular weight polyacids are used to form a semistable complex with the anticancer drugs. This semistable complex can mimic negatively charged macromolecules and initiate the formation of cargo-loaded VLPs. The "polyacid association" method not only possesses high efficacy in drug loading but also makes a quick release of the anticancer drug possible. Therefore, the corresponding VLPs show promise as platforms for small molecular weight chemotherapy drug delivery [13].

2 Materials

Prepare all solutions using ultrapure water with a conductivity of 18 MΩ cm. Store the solutions at 4 °C, unless indicated otherwise.

2.1 Buffers and Solutions

1. Purified HCRSV in resuspension buffer.
2. 1 mg/ml "target molecules" in resuspension buffer.

3. 0.6 mg/ml "negatively charged polymer" in resuspension buffer.
4. 0.2 mg/ml "small molecule anticancer drug" in resuspension buffer.
5. Resuspension buffer: 4.1 g/l NaOAc, 2.9 g/l NaCl, 1.5 g/l EDTA, 2.2 g/l $CaCl_2$. Adjust pH to 5.4 using 0.1 M NaOH solution or 0.1 M HCl solution (*see* **Note 1**).
6. Buffer A: 4.1 g/l NaOAc, 2.9 g/l NaCl, 0.58 g/l EDTA, 2.2 g/l $CaCl_2$. Adjust pH to 5.0 using 0.1 M NaOH solution or 0.1 M HCl solution (*see* **Note 2**).
7. Buffer B: 6.05 g/l Tris, 0.31 g/l dithiothreitol (DTT), 35 mg/l phenylmethylsulfonyl fluoride (PMSF), 1.5 g/l EDTA. Adjust pH to 8.0 using 0.1 M NaOH solution or 0.1 M HCl solution (*see* **Note 3**).
8. Buffer C: 6.05 g/l Tris, 0.31 g/l DTT, 35 mg/l PMSF, 1.5 g/l EDTA, and 58 g/l NaCl. Adjust pH to 8.0 using 0.1 M NaOH solution and 0.1 M HCl solution (*see* **Note 4**).
9. Buffer D: 6.05 g/l Tris, 0.31 g/l DTT, 35 mg/l PMSF, 1.5 g/l EDTA, and 58 g/l NaCl. Adjust pH to 7.0 using 0.1 M NaOH solution and 0.1 M HCl solution (*see* **Note 5**).
10. Precipitation buffer: 5.5 M $CaCl_2$ (*see* **Note 6**).
11. 1 M $CaCl_2$.
12. Concentrated NaOAc buffer: 16.4 g/l NaOAc, 2.9 g/l NaCl, 5.8 g/l EDTA, 2.2 g/l $CaCl_2$. Adjust pH to 5.0 using 0.1 M NaOH solution and 0.1 M HCl solution.
13. Sucrose gradient: 10%, 20%, 30%, and 40% (w/v) sucrose in resuspension buffer. Inject the four sucrose solutions successively from low to high concentrations into the bottom of a 30 ml ultracentrifugation tube (*see* Fig. 1). Place the sucrose gradient at 4 °C overnight before use.
14. Sucrose cushion: 15% (w/v) sucrose in resuspension buffer.

2.2 Devices and Consumable Items

1. Dialysis tubing (MWCO 12 kDa).
2. NanoDrop or Beckman UV spectrophotometer (to measure the $OD_{260/280\ nm}$).
3. Amicon® Ultrafiltration tube (MWCO 10 kDa) 50 ml.
4. Beckman-Coulter Optima™ L-ICO X ultracentrifuge.
5. Beckman Model J2-21 high speed centrifuge.
6. Beckman rotors JA-20 (fixed angle) and SW41 (swing out).
7. Beckman polypropylene centrifuge tubes, round bottom 50 ml.
8. Beckman thinwall, Ultra-Clear™ centrifuge tube, round bottom 13.2 ml.

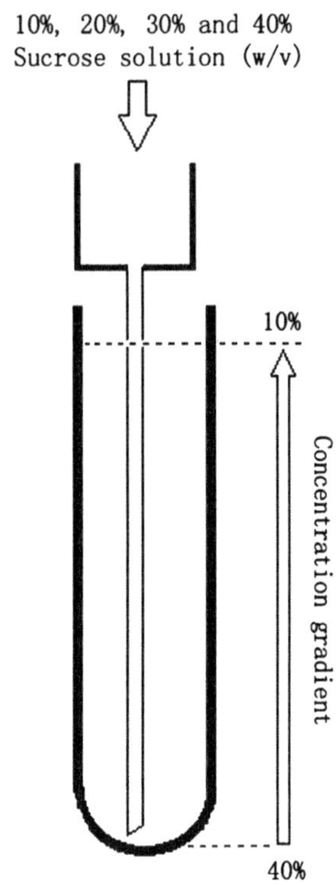

Fig. 1 Inject sucrose solution at concentrations of 10%, 20%, 30%, and 40% into the centrifugation tube to form a sucrose gradient

9. Eppendorf tabletop high-speed centrifuge 5417C.
10. Eppendorf microfuge tubes 1.5 ml.
11. Glass beakers 300 ml, and magnetic stir bar 5 cm.

3 Methods

All procedures are carried out at 4 °C, unless otherwise stated.

There are three alternative protocols for the assembly of VLPs provided; these will result in either empty particles (Subheading 3.2), particles loaded with large molecules (Subheading 3.3), or particles loaded with small molecules (Subheading 3.4).

3.1 Purification of HCRSV Coat Protein

1. Dilute the purified virus solution to a concentration of 10 mg/ml with resuspension buffer (*see* **Note 7**).
2. Transfer about 0.1 ml of the solution into a dialysis tube. Dialyze for 12 h against buffer B (*see* **Note 8**).
3. Transfer the solution from the dialysis tube into a microcentrifuge tube. Mix the solution with 10% (v/v) 5.5 M $CaCl_2$ to achieve a Ca^{2+} concentration of ~0.5 M (*see* **Note 9**).

4. Incubate at 4 °C for 30 min (*see* **Note 10**) and centrifuge at 25,000 × *g* at 4 °C for 30 min. Collect the supernatant.
5. Measure $OD_{280\ nm}$ to determine concentration of the coat protein (*see* **Note 11**).

3.2 In Vitro Reassembly of Empty VLPs

To generate empty VLPs, it is necessary to lower the pH of the solution.

1. Put about 0.5 ml of the purified coat protein solution into a dialysis tube (*see* **Note 12**). Dialyze against 100 ml buffer C for 4 h (*see* **Note 13**).
2. Remove the dialysis tube from the buffer C. Rinse surface of the dialysis tube briefly with water.
3. Put the dialysis tube into a bottle containing 1000 ml buffer A. Dialyze for 15 h (*see* **Note 14**).
4. Transfer the solution onto a 5 ml 15% sucrose cushion in a SW-41 thin wall, Ultra-Clear ultracentrifuge tube. Centrifuge at 100,000 × *g* for 1 h using SW-41 rotor. Remove the supernatant. Resuspend the pellet in 200 μl resuspension buffer (*see* **Note 15**).

3.3 Alternative Method for Preparation of VLPs to be Loaded with Target Molecules

1. Prepare coat protein solution (from Subheading 3.1) with a concentration of about 1 mg/ml.
2. Prepare solution of target molecules (*see* **Note 16**) in resuspension buffer with a concentration of 1 mg/ml.
3. Transfer about 1 ml coat protein solution and 330 μl target molecule solution into a dialysis tube (*see* **Note 17**).
4. Dialyze against buffer A for 15 h.
5. Transfer the solution onto a 10–40% sucrose gradient in SW-41 thinwall Ultra-Clear centrifuge tube. Centrifuge at 100,000 × *g* for 4 h using SW-41 rotor. Collect 1 ml sucrose gradient fractions containing VLPs (*see* **Note 18**).
6. Transfer the solutions containing VLPs into an ultrafiltration tube. Centrifuge at 12,000 × *g* in an Eppendorf tabletop centrifuge and wash with resuspension buffer to remove excess amount of sucrose in the solution (*see* **Note 19**).

3.4 Alternative Method for Preparation of VLPs to be Loaded with Small Molecules

1. Dissolve negatively charged polymer (*see* **Note 20**) and small molecule anticancer drug (*see* **Note 21**) in resuspension buffer. The final concentration of the polymer and small molecules should be 0.6 mg/ml and 0.2 mg/ml, respectively.
2. Transfer 0.5 ml of the polymer–drug mixture into a test tube. Add 1 ml of purified coat protein (1 mg/ml) suspended in buffer D (*see* **Note 22**).

3. Gently mix the solution (*see* **Note 23**).
4. Adjust the pH to 5 using concentrated NaOAc buffer. Add about 10 μl $CaCl_2$ solution (1 M) to a final concentration of 5 mM.
5. Incubate at 4 °C for 15 h (*see* **Note 24**).
6. Place the solution on 15% sucrose cushion in SW-41 thinwall, Ultra-Clear centrifuge tube and centrifuge at 100,000 × *g* for 1 h at 4 °C in a SW-41 rotor. Remove the supernatant and resuspend the pellet in resuspension buffer (*see* **Note 25**).

4 Notes

1. The solution contains 50 mM NaOAc, 50 mM NaCl, 5 mM EDTA, 20 mM $CaCl_2$, with pH 5.4.
2. The solution contains 50 mM NaOAc, 50 mM NaCl, 2 mM EDTA and 20 mM $CaCl_2$, with pH 5.0.
3. The Tris buffer is made up of 50 mM Tris, 2 mM dithiothreitol (DTT), 0.2 mM PMSF, 5 mM EDTA, with pH 8.0. It does not contain Ca^{2+} because EDTA can remove Ca^{2+} from the virus. The buffer will facilitate swelling and disassembly of the virus particles.
4. The Tris buffer with higher concentration of NaCl provides better stability of coat proteins.
5. The Tris buffer with pH 7.0 rather than buffer C is used when loading small molecules into VLPs.
6. The 5.5 M $CaCl_2$ solution is used to precipitate viral RNA to purify the coat protein.
7. The concentration can be measured by detecting the OD_{260nm}.

   Viral concentration (mg/ml) = ($OD_{260\ nm}$ * dilution factor)/extinction coefficient.

   It is estimated that the extinction coefficient is approximately five for most spherical plant viruses.
8. The plant virus particles are not stable under high pH and in the absence of Ca^{2+}. Buffer B provides a high pH environment while the EDTA can remove Ca^{2+} from the virus structure. Therefore, after dialysis for several hours, the virus particles undergo swelling and disassembly to release the genomic and subgenomic RNAs. Our practice is to dialyze the solution overnight.
9. Mix gently and avoid vigorous stirring to keep the protein integrity.
10. After 30 min incubation, Ca^{2+}, genomic and subgenomic RNAs which are negatively charged form a stable precipitate.

The protein does not interact with the Ca^{2+} and remains stable in the solution.

11. Determine the concentration of coat protein by means of the OD at 280 nm. The extinction coefficient (e) of the coat protein at 280 nm can be derived from its amino acid sequence [14].
12. Dialysis tubing with an appropriate molecular weight cutoff should be used. The molecular weight threshold allows free exchange of solvent inside and outside the tube, while keeping the coat protein inside the tube.
13. Cool the buffer on ice before use. The purpose of this step is to remove a high concentration of Ca^{2+} (0.5 M). An intermediate dialysis step with buffer of the same ionic strength (1 M NaCl) is found very essential to stabilize the coat protein.
14. Cool the buffer on ice before use. Reassemble the virus-like particles by dialysis. The low pH and presence of Ca^{2+} are essential for the reassembly.
15. The virus-like particles can be examined by optical density, TEM, zeta-sizer, etc.
16. The compound should be negatively charged with appropriate large molecular weight. For example, accurately weight 10 mg polystyrene sulfonate acid and dissolve in 10 ml resuspension buffer.
17. The virus-like-particles could load approximately 30–40% (w/w) of foreign compound. Therefore, the mass ratio of coat protein and introduced molecules should be about 3:1 (w/w).
18. Measure the OD at 280 nm of the sucrose gradient fractions. Collect the fractions with high OD 280 nm from the middle region of the sucrose gradient.
19. Use an ultrafiltration tube with appropriate molecular weight cutoff. Excessive sucrose is removed by ultrafiltration and the solution is concentrated.
20. Negatively charged polymers, such as PSA, serve as scaffolds to bond with the drug molecules.
21. The drug should be positively charged molecules, such as doxorubicin which can form a complex with the negatively charged polymers.
22. The w/w ratio of the introduced compound and coat protein depends on the loading efficiency of the virus-like particles. For some viruses, such as HCRSV, a ratio of not less than 30% could be achieved.
23. Avoid vigorous agitation to keep the coat protein intact.
24. Dialysis is not appropriate because the small drug molecules can run off. If the polymer is replaced by other negatively

charged molecules (such as tripolyphosphate acid) with low molecular weight, loading of the drug molecules is not successful because the low molecular weight compound could leak through gaps on the surface of the virus-like particles.

25. Loaded small molecules will be released into the resuspension buffer slowly over a period of a few hours.

References

1. Bancroft JB, Hiebert E (1967) Formation of an infectious nucleoprotein from protein and nucleic acid isolated from a small spherical virus. Virology 32:354–356
2. Bancroft JB, Wagner GW, Bracker CE (1968) The self-assembly of a nucleic-acid free pseudo-top component for a small spherical virus. Virology 36:146–149
3. Hiebert E, Bancroft JB, Bracker CE (1968) The assembly in vitro of some small spherical viruses, hybrid viruses, and other nucleoproteins. Virology 34:492–508
4. Hiebert E, Bancroft JB (1969) Factors affecting the assembly of some spherical viruses. Virology 39:296–311
5. Douglas T, Young M (1998) Host-guest encapsulation of materials by assembled virus protein cages. Nature 393:152–155
6. Doan DN, Lee KC, Laurinmaki P et al (2003) Three-dimensional reconstruction of hibiscus chlorotic ringspot virus. J Struct Biol 144:253–261
7. Speir JA, Munshi S, Wang G et al (1995) Structures of the native and swollen forms of cowpea chlorotic mottle virus determined by X-ray crystallography and cryo-electron microscopy. Structure 3:63–78
8. Chatterji A, Burns LL, Taylor SS et al (2002) Cowpea mosaic virus: from the presentation of antigenic peptides to the display of active biomaterials. Intervirology 45:362–370
9. Fox JM, Wang G, Speir JA (1998) Comparison of the native CCMV virion with in vitro assembled CCMV virions by cryoelectron microscopy and image reconstruction. Virology 244:212–218
10. Douglas T, Young M (2006) Viruses: making friends with old foes. Science 312:873–875
11. Zhao X, Fox JM, Olson NH et al (1995) In vitro assembly of cowpea chlorotic mottle virus from coat protein expressed in Escherichia coli and in vitro-transcribed viral cDNA. Virology 207:486–494
12. Ren Y, Wong SM, Lim LY (2006) In vitro-reassembled plant virus-like particles for loading of polyacids. J Gen Virol 87:2749–2754
13. Ren Y, Wong SM, Lim LY (2007) Folic acid-conjugated protein cages of a plant virus: a novel delivery platform for doxoribucin. Bioconjug Chem 18:836–843
14. Mach H, Middaugh CR, Lewis RV (1992) Statistical determination of the average values of the extinction coefficients of tryptophan and tyrosine in native proteins. Anal Biochem 200:74–80

Chapter 16

CCMV-Based Enzymatic Nanoreactors

Mark V. de Ruiter, Rindia M. Putri, and Jeroen J. L. M. Cornelissen

Abstract

Protein-based nanoreactors are generated by encapsulating an enzyme inside the capsid of the cowpea chlorotic mottle virus (CCMV). Here, three different noncovalent methods are described to efficiently incorporate enzymes inside the capsid of these viral protein cages. The methods are based on pH, leucine zippers, and electrostatic interactions respectively, as a driving force for encapsulation. The methods are exclusively described for the enzymes horseradish peroxidase, glucose oxidase, and *Pseudozyma antarctica* lipase B, but they are also applicable for other enzymes.

Key words Cowpea chlorotic mottle virus (CCMV), Functional cargo, Enzyme encapsulation, Virus-like particles, Nanoreactors, Leucine zippers, pH-responsive assembly

1 Introduction

Nature provides excellent examples of effective and controlled chemical pathways that are catalyzed by various enzymes. The improvement of enzyme efficiency inside the cells and organelles, in comparison to the bulk systems, is suggested to be the result of cellular compartmentalization and molecular confinement [1, 2]. Consequently, designing bio-based artificial compartments as enzymatic nanoreactors in order to mimic the natural organelles of cells is of increasing interest [3–7]. Among a plethora of artificial compartments reported, the viral protein cages, specifically the capsid of the plant virus cowpea chlorotic mottle virus (CCMV) [8], are particularly attractive owing to their biocompatibility and ease of modifications (both chemically and genetically).

The CCMV capsid was the first icosahedral viral protein shell to be reassembled in vitro to encapsulate RNA [9, 10]. The 28 nm-sized capsid is naturally stable at around pH 5 and consists of 180 identical subunits that form icosahedral particles with

---

Mark V. de Ruiter and Rindia M. Putri contributed equally to this work.

Christina Wege and George P. Lomonossoff (eds.), *Virus-Derived Nanoparticles for Advanced Technologies: Methods and Protocols*, Methods in Molecular Biology, vol. 1776, https://doi.org/10.1007/978-1-4939-7808-3_16,
© Springer Science+Business Media, LLC, part of Springer Nature 2018

$T = 3$ quasi-symmetry [11]. At a higher pH (>7) and ionic strength (>0.4 M), the capsid disassembles into protein dimers, releasing its RNA cargo, which can be removed by precipitation in the presence of calcium ions [12]. The resultant protein dimers can self-assemble back into their native cage-like structures when the pH is adjusted back to pH 5 (*see* Fig. 1; for additional information *see also* Chapters 17 and 18). Furthermore, structurally, CCMV bears a native N-terminal arginine rich motif (ARM) that is inherently designed to encapsulate negatively charged RNA cargo. Devoid of its native RNA cargo, the protein subunits of CCMV can be reassembled at neutral pH in the presence of a negatively charged template into monodisperse virus-like structures to form $T = 1$/ pseudo-$T = 2$ particles with size ranges between 16 and 22 nm [13–16].

Here we describe efficient noncovalent strategies to encapsulate an enzyme inside a CCMV capsid in order to form artificial nanoreactors. The first strategy solely harnesses the pH-responsiveness of CCMV structures to incorporate an enzyme inside the capsid (*see* Fig. 1a) [12]. The second strategy, which uses leucine zippers (i.e., the E-coil and K-coil), displays a combination of genetic modification and the use of dimerization interactions to selectively encapsulate an enzyme (*see* Fig. 1b) [17]. The third strategy, without any genetic modification needed, emphasizes on a versatile technique of using a nucleic-acid tag to encapsulate the enzyme of interest, based on electrostatic interactions between the negatively charged tag and the capsid (*see* Fig. 1c) [18].

2 Materials

Prepare all solutions using ultrapure water (18 MΩ cm at 25 °C). Keep and store all proteins at 4 °C. When working with bacteria, work according to the appropriate biosafety regulations. Diligently follow all waste disposal regulations when disposing waste materials. Carry out the purification of wild type CCMV and the removal of RNA according to procedures in the literature [12]. Buffers denoted with (A) contain 2 M NaCl, are for the capsid protein production, and denoted with (B) contain 0.3 M NaCl and are for the *Pseudozyma antarctica* lipase B (PalB) enzyme production.

2.1 Components for Encapsulation Induced by pH

1. Horseradish peroxidase (enzyme to be encapsulated, Sigma-Aldrich): 225 μM (10 mg/ml) in Neutral buffer pH 7.5 (*see* **Note 1**).
2. Neutral buffer pH 7.5, 0.05 M Tris, 0.5 M NaCl, 0.001 M DTT. Add 6.06 g Tris, 29.22 g NaCl, and 0.1543 g DTT to a 1-l graduated cylinder. Add water to a volume of 900 ml.

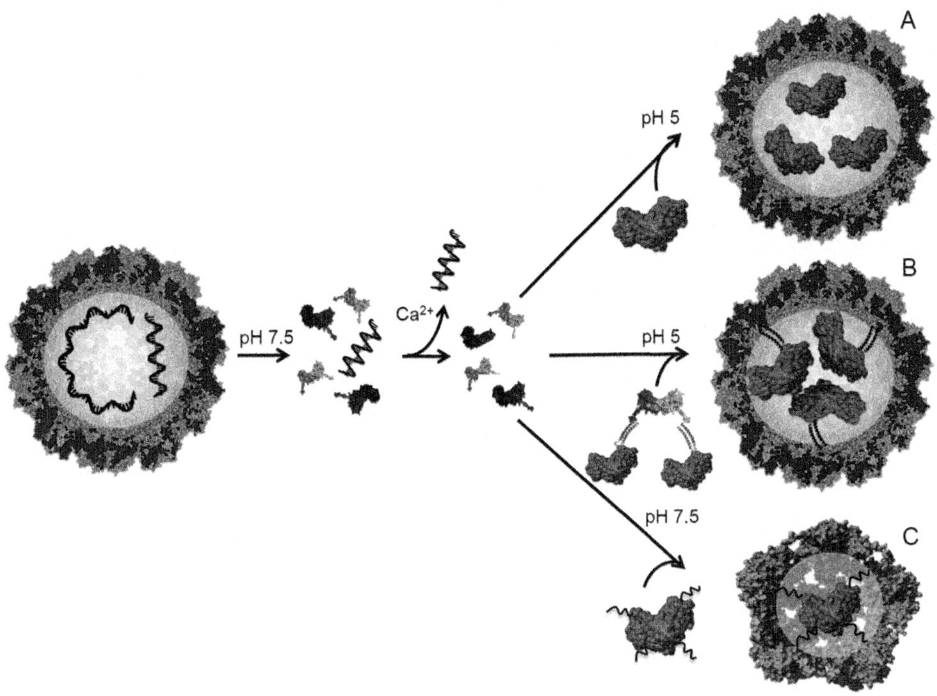

Fig. 1 Overview of the different enzyme encapsulation pathways in the capsid of CCMV. (**a**) shows the pH responsiveness of CCMV structures to incorporate an enzyme into a $T = 3$ sized capsid. (**b**) shows the leucine-zipper approach to encapsulate an enzyme into a $T = 3$ sized capsid. (**c**) shows the use of nucleic-acid tags to encapsulate the enzyme in a $T = 1$ and/or pseudo $T = 2$ sized capsid

Mix and adjust pH with HCl. Make up to 1 l with water. Store at 4 °C.

3. Capsid storage buffer pH 5.0, 0.05 M NaOAc, 1 M NaCl,0.001 M Na-azide. Add 4.10 g NaOAc, 58.44 g NaCl, and 0.065 g Na-azide to a 1 l graduated cylinder. Add water to a volume of 900 ml. Mix and adjust pH with acetic acid. Make up to 1 l with water.
4. CCMV coat protein: around 15 mg/ml in 0.05 M capsid storage buffer pH 5.0 (*see* **Note 1**). Produce the coat protein from around 25 mg/ml of wild type CCMV in the capsid storage buffer by dialyzing overnight at 4 °C against high calcium buffer (refer to [12] for details). Centrifuge at 15,000 × *g* at 4 °C overnight to remove precipitated RNA.
5. Centrifugal filters (100 kDa MWCO, Centricon YM-100, Millipore).
6. Size-exclusion chromatography (SEC), Superose 6 10/100 GL column (GE Healthcare) with 24 ml bed volume and 500 μl injection volume.
7. Dialysis tubing, 12–14 kDa MWCO regenerated cellulose, width: 25 mm (Spectra/Por).

2.2 Components for Encapsulation Induced by Leucine Zippers

1. pET-15b vector containing the DNA sequence encoding the capsid protein with K-coil and His-tag sequences. The following is the amino acid sequence, with K-coil and hexahistidine tag inserted into the N-terminal region (*see* **Notes 2** and **3**).

   Capsid protein with His-tag and K-coil (His-CK) amino acid sequence

   MGSSHHHHHHSGLVPRGSHS**KIAALKEKIAALKEKIAALKE**GMMSTVGTGKLTRAQRRAAARKNKRNTRVVQPVIVEPIASGQGKAIKAWTGYSVSKWTASCAAAEAKVTSAITISLPNELSSERNKQLKVGRVLLWLGLLPSVSGTVKSCVTETQTTAAASFQVALAVADNSKDVVAAMYPEAFKGITLEQLAADLTIYLYSSAALAEGDVIVHLEVEHVRPTFDDSFTPVY

2. pET-22b vector encoding the *Pseudozyma* (formerly *Candida*) *antarctica* lipase B (PalB) with E-coil. The following is the amino acid sequence, with the E-coil and a polyhistidine tag inserted into the C-terminal region (*see* **Notes 2** and **3**).

   PalB with His-tag and E-coil amino acid sequence

   MKYLLPTAAAGLLLLAAQPAMAMGLPSGSDPAFSQPKSVLDAGLTCQGASFSSVSKPILLVPGTGTTGPQSFDSNWIPLSAQLGYTPCWISPPPFMLNDTQVNTEYMVNAITTLYAGSGNNKLPVLPWSQGGLVAQWGLTFFPSIRSKVDRLMAFAPDYKGTVLAGPLDALAVSAPSVWQQTTGSALTTALRNAGGLTQIVPTTNLYSATDEIVQPQVSNSPLDSSYLFNGKNVQAQAVCGPLFVIDHAGSLTSQFSYVVGRSALRSTTGQAFSADYGITDCNPLPANDLTPEQKVAAAALIAPAAAAIVAGPKQNCEPDLMPYARPFAVGKRTCSGIVTPLD**EIAALEKEIAALEKEIAAIEK**LVPRGSVEHHHHHHHH

3. BL21(DE3)pLysS *E. coli* cells (Novagen) containing the appropriate plasmids (*see* **Note 4**).
4. 1 M IPTG (isopropyl-β-D-1-thiogalactopyranoside) stock: dissolve 2.383 g in 10 ml water. Filter-sterilize over a 0.2 μm pore filter and freeze stock at −20 °C in 1 ml aliquots.
5. Fast protein liquid chromatography/Size-exclusion chromatography system, Amersham Ettan LC system, equipped with Superose 6 PC 3.2/30 analytical column and a Superdex 200 PC 3.2/30 analytical column from GE Life Sciences (*see* **Note 5**).
6. Ni-NTA agarose beads.
7. LB medium: 10 g tryptone, 5 g yeast extract, and 10 g NaCl. Make up to 1 l. Autoclave (liquid cycle). When cooled, add antibiotics (Sigma):

   LB medium Amp^+ Cam^+: 0.05 g/l ampicillin and 0.025 g/l chloramphenicol.

   LB medium Amp^+ Tet^+: 0.05 g/l ampicillin and 0.0125 g/l tetracycline (*see* **Note 6**).

8. Lysis buffer pH 8 A and B: 50 mM NaH_2PO_4, 10 mM imidazole, 2 M NaCl (buffer A); or 0.3 M NaCl (buffer B). Add 14.2 g $Na_2HPO_4*2H_2O$, 2.16 g NaH_2PO_4, 0.66 g imidazole, and 116.8 g (A) or 17.5 g (B) NaCl, respectively, to a 1 l graduated cylinder. Add water to a volume of 900 ml. Mix and adjust pH with HCl. Add 10 mg of lysozyme (Fluka). Make up to 1 l with water. Store at 4 °C (*see* **Note 7**).
9. Branson Ultrasonic Sonifier 250 with microtip.
10. RNase I at least 0.1 mg (New England Biolabs) and DNase I at least 50 μg (New England Biolabs).
11. Wash buffer pH 8 (A and B): 50 mM NaH_2PO_4, 25 mM imidazole, and 2 M NaCl (buffer A) or 0.3 M NaCl (buffer B). Prepare as in **step 8,** only add 1.65 g imidazole.
12. Elution buffer pH 8 (A and B): 50 mM NaH_2PO_4, 250 mM imidazole, and 2 M NaCl (A) or 0.3 M NaCl (B). Prepare as in **step 8,** only add 16.5 g imidazole.
13. Dialysis tubing, 12–14 kDa MWCO regenerated cellulose, width: 25 mm (Spectra/Por).
14. Neutral buffer 2 pH 7.5: 0.5 M NaCl, 50 mM Tris–HCl, 10 mM $MgCl_2$, 1 mM EDTA. Add 6.06 g Tris, 29.2 g NaCl, 2 g $MgCl_2$, 0.37 g EDTA and transfer to the cylinder. Add water to a volume of 900 ml. Mix and adjust pH with HCl. Make up to 1 l with water. Store at 4 °C.
15. CCMV coat protein in capsid storage buffer pH 5 (*see* Subheading 2.1, **item 4**).
16. Encapsulation buffer pH 5: 0.5 M NaCl, 0.05 M Na-Acetate, 0.01 M $MgCl_2$, 0.001 M EDTA. Add 2.05 g Na-Acetate, 29.2 g NaCl, 2 g $MgCl_2$ and 0.37 g EDTA to a 1 l graduated cylinder. Add water to a volume of 900 ml. Mix and adjust pH with acetic acid. Make up to 1 l with water. Store at 4 °C.

2.3 Components for Encapsulation Induced by DNA Tags

1. 10 mM phosphate buffered saline (PBS) pH 7.4.
2. 5′-thiolated single-stranded DNA (synthesized by Eurofins, sequence: 5′-HS-$(CH_2)_6$GGGTAGGGCGGG-TTGGGTTT T-3′): 100 μM in 10 mM phosphate buffered saline (PBS) pH 7.4.
3. 2 mM dithiothreitol: in 10 mM phosphate buffered saline (PBS) pH 7.4.
4. Amicon Ultra centrifugal filters (10 kDa, 30 kDa MWCO, Millipore).
5. 100 μM Nickel (II) sulfate in water.
6. 2 μM glucose oxidase (GOx, enzyme to be encapsulated) in 10 mM phosphate buffered saline (PBS) pH 7.4.

7. 100 μM cross-linker N-[ε-maleimidocaproyloxy]sulfosuccinimide ester] (Pierce) in 10 mM phosphate buffered saline (PBS) pH 7.4, prepared freshly before use.
8. UV spectrophotometer (λ = 260 nm).
9. 15–16 mg/ml CCMV coat protein in 0.05 M sodium acetate buffer pH 5.0.
10. The assembly buffer pH 7.5: 250 mM Tris–HCl, 500 mM NaCl, 50 mM $MgCl_2$, 1 mM DTT. Add 30.3 g Tris, 29.2 g NaCl, 10 g $MgCl_2$ to a 1 l graduated cylinder. Add water to a volume of 900 ml. Mix and adjust pH with HCl. Make up to 1 l with water and add 0.1543 g DTT. Store at 4 °C.
11. Dialysis tubing, 12–14 kDa MWCO regenerated cellulose, width: 25 mm (Spectra/Por).

3 Methods

3.1 Encapsulation Induced by pH

1. Prepare the coat proteins of CCMV (500 μl, 15 mg/ml) in a capsid storage buffer with sodium acetate (0.05 M, pH 5.0) (*see* Subheading 2.1, **item 4**).
2. Dialyze 500 μl of the CCMV coat protein in 0.05 M sodium acetate capsid storage buffer pH 5 against 500 ml of the neutral buffer (0.05 M Tris–HCL, pH 7.5) (*see* Subheading 2.1, **item 2),** using the regenerated cellulose dialysis tubing. Under continuous stirring in a beaker at 4 °C, change the buffer every 3 h and repeat this for three times.
3. Add 600 μl of horseradish peroxidase solution (*see* **Note 8**) in the same neutral buffer (0.05 M Tris–HCl, pH 7.5) in a 66-times excess to the capsid solution in an Eppendorf tube (one capsid corresponds to 90 coat protein dimers), incubate the mixture for 1.5 h at 4 °C on a roller bank or a shaker for gentle continuous mixing.
4. Dialyze the incubated solution against the initial capsid storage buffer pH 5 (500 ml, sodium acetate 0.05 M, pH 5.0), change the buffer every 3 h and repeat this for three times.
5. Purify the samples further with size-exclusion chromatography (SEC) over a Superose 6 10/100 GL column with 24 ml bed volume and 500 μl injection volume at room temperature at 0.5 ml/min flow rate of capsid storage buffer pH 5. Monitor at $\lambda = 280$ nm for the protein content and $\lambda = 405$ nm for the HRP (or $\lambda = 530$ if coupled to dyes). The dimers of capsid protein and the HRP both elute around 18 ml, whereas the filled assembled particles elute around 10 ml.
6. Concentrate the encapsulated horseradish peroxidase in CCMV fractions using centrifugal 100 K MWCO filters at 4000 × g in a swing-out bucket. Store samples at 4 °C.

3.2 Encapsulation Induced by Leucine Zippers

1. Scrape with a pipette tip BL21 *E. coli* from glycerol stock from −80 °C and put the tip into 100 ml of LB medium. Use LB Amp^+ Cam^+ for capsid protein with K-coil and LB medium Amp^+ Tet^+ for PalB with E-coil. Grow overnight at 30 °C on a shaking platform.
2. Add the overnight cultures each to 0.9 l of the corresponding LB medium and grow the cultures at 30 °C on a shaking platform to an optical density of OD_{600nm} of 0.6–0.8 A.U.
3. Induce protein expression by addition of 1 ml of IPTG solution. Grow at 30 °C for 5 h for capsid protein with K-coil, and at 25 °C for 20 h for PalB with E-coil. Harvest cells after expression by centrifuging the grown cultures (4000 × *g*, 15 min at 4 °C). Discard the supernatants (*see* **Note 9**).
4. Resuspend the K-coil cells in 10 ml lysis buffer A and the PalB cells in lysis buffer B. Incubate at 4 °C for 30 min. Sonicate the solution five times for 10 s with a duty cycle 40 and output control 6.
5. Add 10 μg/ml RNase and 5 μg/ml DNase and incubate for 15 min at 4 °C (*see* **Note** 7). Centrifuge at 15,000 × *g* for 20–30 min.
6. Incubate each supernatant with 0.5 ml of Ni-NTA agarose beads in a column for 1 h at 4 °C. Wash column with 20 ml of wash buffer A for capsid protein and wash buffer B for the PalB (*see* **Note 10**). Elute column using approximately 10 ml of elution buffer A for capsid protein and elution buffer B for the PalB (*see* **Note 11**).
7. Dialyze overnight against neutral buffer (*see* **Note 12**).
8. Add PalB with E-coil in a two times excess over His-CK, stir for 16 h at 4 °C to form the PalB-capsid protein complex (*see* **Note 10**).
9. Purify using size-exclusion chromatography with a Superdex 200 column with the neutral buffer. Pure PalB with E-coil elutes around $V = 1.6$ ml (monitor $\lambda = 280$ nm absorption), and the PalB-capsid protein complex elutes between $V = 1.3$–1.8 ml.
10. Dialyze the coat protein against 500 ml of the neutral buffer 2. Change the buffer every 3 h and repeat this three times.
11. Add in the desired ratio the PalB-Capsid protein complex to wild type capsid protein (*see* **Note 13**) both in the neutral buffer 2. Mix for 5 min and dialyze overnight against the encapsulation buffer.
12. Use size-exclusion chromatography with a Superose 6 column to purify the encapsulated PalB using the encapsulation buffer pH 5. Monitor at $\lambda = 280$ nm. The product is expected to elute between 8 and 10 ml at the flow rate of 0.5 ml/min. Store samples at 4 °C.

3.3 Encapsulation Induced by DNA Tags

1. Dissolve the thiolated DNA in PBS containing 2 mM DTT. Add a certain amount of PBS (refer to the product sheet) to dissolve the DNA. Incubate the solution for 3 h at room temperature to reduce any disulfide groups.
2. Remove the excess DTT by exchanging the solution six times against PBS buffer using Amicon Ultra centrifugal filters (10 kDa MWCO). Check if all excess DTT was successfully removed by adding 20 μl of nickel (II)-sulfate to 10 μl filtrate; a transparent color indicates a DTT free solution, whereas a brown-yellow color indicates DTT contamination. If the contamination is present, repeat the exchanging against PBS.
3. Prepare a cross-linker modified glucose oxidase (GOx) by incubating a solution of enzyme (9 mg) and sulfo-EMCS (2.5 mg) in 25 ml PBS for 1 h at room temperature.
4. Remove the excess sulfo-EMCS by exchanging the solution five times against PBS buffer using Amicon Ultra centrifugal filters (30 kDa MWCO).
5. Mix the modified GOx with thiolated DNA in a 2:1 (v/v) ratio and incubate for 1 h at room temperature.
6. Remove noncoupled (excess) DNA by exchanging the solution six times against PBS buffer using Amicon Ultra centrifugal filters (30 kDa MWCO). Test the filtrates for DNA contamination by measuring the specific UV-Vis absorbance for DNA at $\lambda = 260$ nm. No signal at $\lambda = 260$ nm indicates that the removal of uncoupled DNA is complete.
7. Dialyze 250 μl of dimeric CCMV coat proteins using the dialysis tubing (12–14 kDa MWCO) against assembly buffer for encapsulation.
8. Exchange the PBS system of GOx-DNA against MilliQ water using Amicon Ultra centrifugal filters (30 kDa MWCO) (*see* **Note 14**).
9. Mix the GOx-DNA (in MilliQ water) and dimeric capsid proteins (in assembly buffer) in a 4:1 (v/v) ratio and incubate for 2 h at 4 °C (store at 4 °C overnight, if necessary).
10. Purify the encapsulated enzyme-DNA using size-exclusion chromatography over a Superose 6 10/100 GL column with 24 ml bed volume and 500 μl injection volume at room temperature (*see* **Note 15**) at a 0.5 ml/min flow of 5× diluted assembly buffer. Monitor at $\lambda = 260$ nm and $\lambda = 280$ nm. The assembled particles elute around 10–12 ml. Store at 4 °C.

4 Notes

1. To both the neutral buffer and the capsid storage buffer, it is possible to add 1 mM EDTA, 10 mM $CaCl_2$, and 0.2 mM phenylmethylsulfonyl fluoride for increased stability of the

proteins. The intact capsid of CCMV, which consists of 90 dimers of 40 kDa each, obtained upon the removal of its RNA material using calcium precipitation, can be stored at 4 °C for a maximum of 1 month in the sodium acetate buffer. At elevated pH, the capsid proteins gradually lose their arginine-rich motifs (ARM) sequence. Use a concentration of the capsid that is as high as possible.

2. The genes can be ordered and made by various companies. For the capsid protein (a 5′-NdeI and a 3′-NdeI) and for PalB (a 5′-XhoI and a 3′-XhoI) restriction sites were added to the different termini with an additional filler sequence, for easy insertion in the respective plasmids using these restriction enzymes, agarose-gel separation and ligation. It is advisable to first insert these "new" plasmids in an *E. coli* strain that can retain and multiply these plasmids, like NovaBlue cells.
3. A different enzyme can be used, however it is advised to use an enzyme that is active around pH 5.0. Additionally the substrate used for reactions in the virus-like particle should not be too large, so that it fits through the pores; that is, the substrate ideally is up to ~1 nm in size, otherwise it could result in a significant decrease in catalytic activity.
4. Plasmids can be inserted into competent BL21 cells according to supplier's protocol. As a check, the plasmids can be extracted and sequenced (from the T7 RNA polymerase promotor or T7 terminator region).
5. Other size exclusion techniques can also be used, but the appropriate elution volumes need to be checked.
6. Do not store the media longer than 2 weeks, due to potential loss of activity of the antibiotics.
7. Add the enzymes from a −20 °C glycerol stock just before use.
8. The HRP used in the experiments was labeled with an Alexa Fluor 532–NHS ester kit (Thermo Fisher Scientific). The unbound dye was removed using size exclusion chromatography (FPLC).
9. The pelleted cells can optionally be stored at −20 °C overnight to continue the following day. Defrost first before the next step.
10. To generate a leucine-zipper complex with a protein that does not have a His-tag, follow the following procedure: after the washing step of the capsid protein on the column, the cell lysate of the enzyme can directly be added to the column and incubated for 1 h at 4 °C, followed by a washing and elution step. This yields a pure fusion product. Subheading 3.2, **step 8** can now be skipped [19].
11. A flow rate of approximately 10 volumes per h is optimal for efficient purification. If flow rate is too fast, more impurities will

contaminate the eluted fractions. Resin can be cleaned with 2 volumes of 6 M guanidine-HCl, 0.2 M acetic acid, washed with water and stored in 25% ethanol. One can also regenerate the Coulomb resin (if the blue color disappears) with nickel sulfate or nickel chloride. Do not use chelating compounds in your buffer, this will remove nickel from the resin.

12. To store the enzyme, it is best to dialyze against PBS. To determine the actual number of encapsulated enzymes it is recommended to label the enzyme with a bright fluorescent dye for instance using the NHS ester of Alexa Fluor® 568 in PBS.
13. The ratio used depends on the desired number of enzymes per CCMV capsid. *See* Fig. 2 as an indication of the ratio and the number of proteins loaded per capsid.
14. The presence of a high amount of salts in the enzyme–DNA system significantly lowers the encapsulation efficiency, and therefore exchange the buffered enzyme–DNA system to water prior to mixing with CCMV coat proteins.
15. Use a 5× diluted assembly buffer as the elution buffer during chromatography for example by FPLC (50 mM Tris–HCl, 100 mM NaCl, 10 mM $MgCl_2$, 0.5 mM DTT, pH 7.5). The best practice is to vacuum-filter all buffers used in SEC

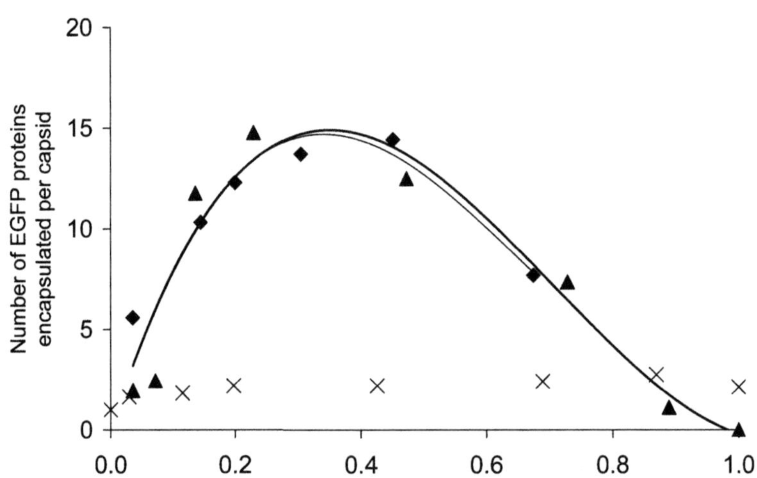

Fig. 2 Number of encapsulated enhanced green fluorescent proteins (EGFPs) per capsid as a function of the capsid protein-E-coil-K-coil complex to total protein ratio. Diamonds and triangles represent data points of duplicate experiments. Crosses represent negative control experiments with noncomplexed capsid protein with K-coil. The thick line represents the polynomial trend line through the data points depicted with a triangular shape; the thin line represents the polynomial trend line through the diamond-shaped data points. Reprinted with permission from [17]. Copyright © 2009, American Chemical Society

Acknowledgment

We acknowledge financial support from the ERC Consolidator Grant (Protcage) and the Indonesia Endowment Fund for Education (LPDP).

References

1. Minton AP, Wilf J (1981) Effect of macromolecular crowding upon the structure and function of an enzyme: glyceraldehyde-3-phosphate dehydrogenase. Biochemistry 20:4821–4826
2. Minton AP (2006) How can biochemical reactions within cells differ from those in test tubes? J Cell Sci 119:2863–2869
3. Peters RJ, Marguet M, Marais S et al (2014) Cascade reactions in multicompartmentalized polymersomes. Angew Chem Int Ed 53:146–150
4. Marguet M, Bonduelle C, Lecommandoux S (2013) Multicompartmentalized polymeric systems: towards biomimetic cellular structure and function. Chem Soc Rev 42:512–529
5. Patterson DP, Schwarz B, Waters RS et al (2014) Encapsulation of an enzyme cascade within the bacteriophage P22 virus-like particle. ACS Chem Biol 9:359–365
6. Patterson DP, Prevelige PE, Douglas T (2012) Nanoreactors by programmed enzyme encapsulation inside the capsid of the bacteriophage P22. ACS Nano 6:5000–5009
7. Minten IJ, Claessen VI, Blank K et al (2011) Catalytic capsids: the art of confinement. Chem Sci 2:358–362
8. Douglas T, Young M (1998) Host-guest encapsulation of materials by assembled virus protein cages. Nature 393:152–155
9. Bancroft JB, Hiebert E (1967) Formation of an infectious nucleoprotein from protein and nucleic acid isolated from a small spherical virus. Virology 32:354–356
10. Bancroft JB, Hills GJ, Markham R (1967) A study of the self-assembly process in a small spherical virus. Formation of organized structures from protein subunits in vitro. Virology 31:354–379
11. Speir JA, Munshi S, Wang G et al (1995) Structures of the native and swollen forms of cowpea chlorotic mottle virus determined by X-ray crystallography and cryo-electron microscopy. Structure 3:63–78
12. Comellas-Aragones M, Engelkamp H, Claessen VI et al (2007) A virus-based single-enzyme nanoreactor. Nat Nanotechnol 2:635–639
13. Kwak M, Minten IJ, Anaya DM et al (2010) Virus-like particles templated by DNA micelles: a general method for loading virus nanocarriers. J Am Chem Soc 132:7834–7835
14. Brasch M, de la Escosura A, Ma Y et al (2011) Encapsulation of phthalocyanine supramolecular stacks into virus-like particles. J Am Chem Soc 133:6878–6881
15. Minten IJ, Ma Y, Hempenius MA et al (2009) CCMV capsid formation induced by a functional negatively charged polymer. Org Biomol Chem 7:4685–4688
16. Comellas-Aragones M, de la Escosura A, Dirks AT et al (2009) Controlled integration of polymers into viral capsids. Biomacromolecules 10:3141–3147
17. Minten IJ, Hendriks LJ, Nolte RJ et al (2009) Controlled encapsulation of multiple proteins in virus capsids. J Am Chem Soc 131:17771–17773
18. Brasch M, Putri RM, de Ruiter MV et al (2017) Assembling enzymatic cascade pathways inside virus-based nanocages using dual-tasking nucleic acid tags. J Am Chem Soc 139:1512–1519
19. Rurup WF, Verbij F, Koay MST et al (2014) Predicting the loading of virus-like particles with fluorescent proteins. Biomacromolecules 15:558–563

Chapter 17

Protocol for Efficient Cell-Free Synthesis of Cowpea Chlorotic Mottle Virus-Like Particles Containing Heterologous RNAs

Rees F. Garmann, Charles M. Knobler, and William M. Gelbart

Abstract

We report a protocol for efficient cell-free synthesis of cowpea chlorotic mottle virus (CCMV)-like particles containing a broad range of lengths and sequences of RNA. Our protocol starts with a purified stock of wild-type CCMV (protocols for harvesting and purifying the virus are detailed elsewhere) and features three basic steps: disassembly of the CCMV and purification of the capsid protein (CP) from the viral RNA; coassembly of the purified CP and an RNA of choice; and characterization of the assembly products. We highlight several key factors that increase the yield of the assembly reaction: the CP should be uncleaved and sufficiently free of viral RNA; the length of the RNA should be between about 100 and 4000 nucleotides; and the stoichiometry of CP and RNA should be 6–1 by mass. Additionally, we point out that separating the assembly reaction into multiple steps—by successively lowering the ionic strength and then the pH of the assembly buffers—results in the highest yields of well-formed, nuclease-resistant, CCMV-like particles. Finally, we describe methods for characterizing the assembly products using native agarose gel electrophoresis and negative-stain transmission electron microscopy.

Key words CCMV, Virus-like particle, Cell-free synthesis, Self-assembly, Heterologous RNA packaging

1 Introduction

The protein capsid of cowpea chlorotic mottle virus (CCMV) is a single-molecule-thick shell that measures about 28 nm in outer diameter and about 20 nm in inner diameter [1]. It is constructed from exactly 180 copies of a single protein—the capsid protein (CP)—that are arranged with $T = 3$ [2] icosahedral symmetry (Fig. 1). In nature, the CCMV capsid is the only barrier between the fragile RNA genome and the outside world and it has evolved to be both a protective barrier against degradation of the genome by nucleases and a conduit for delivery of the genome to the replication machinery of the host cell.

Christina Wege and George P. Lomonossoff (eds.), *Virus-Derived Nanoparticles for Advanced Technologies: Methods and Protocols*, Methods in Molecular Biology, vol. 1776, https://doi.org/10.1007/978-1-4939-7808-3_17,
© Springer Science+Business Media, LLC, part of Springer Nature 2018

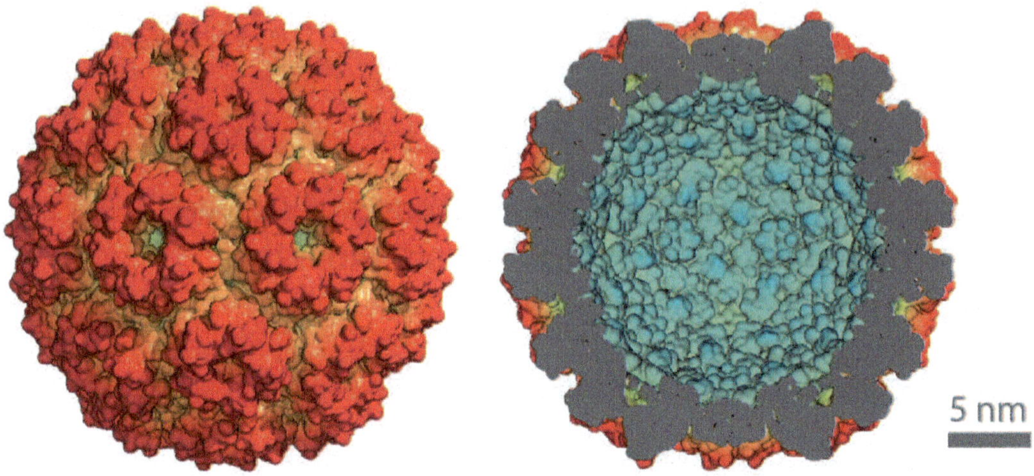

Fig. 1 Physical dimensions of the CCMV capsid. The outer surface is shown in red on the left and a cutaway of the capsid that reveals the inner surface is shown in blue on the right. Models were rendered from crystallographic data obtained from VIPERdb [3] (http://viperdb.scripps.edu)

In the laboratory, CCMV capsids can be synthesized in a range of sizes and shapes [4–7]. CCMV and similar viral capsids have attracted attention from colloid and materials scientists for use as building blocks in the construction of supramolecular architectures with novel properties [8–10] [*see* Chapters 16 and 19], and from biomedical scientists for use as vectors for the delivery of specialized cargos to a variety of host and nonhost cells [11–13] [*see* Chapter 28] and as components of vaccines [14–16]. Such applications are often facilitated by labeling the exterior surface of the capsids with various functional moieties [17–19] [*see* Chapter 40]. In addition to external modifications it is often desirable to control the RNA content within the capsids—for such situations we describe a cell-free protocol for encapsidating heterologous RNA of any origin, as long as the length is between about 100 and 4000 nucleotides (nt) [20]. We note that RNAs with lengths between about 4000 and 2000 nt package into $T = 3$ capsids that are indistinguishable from the wild-type virus, and that RNAs shorter than about 2000 nt are packaged into capsids with "$T = 2$" symmetry [21], and that RNAs shorter than 1000 nt are packaged with multiple copies in the same "$T = 2$" capsid.

Our protocol exploits the long-established [22] fact that CCMV capsids can be disassembled outside of the host cell and then reconstituted around heterologous RNAs by careful control of the buffer solutions. We use the original buffer recipes introduced in the 1960s and 1970s [23, 24] and we report on several key factors [25] that lead to highly efficient encapsidation: in particular, we specify the purity (with respect to RNA contamination and N-terminal cleavage) required of the CP, the lengths of heterologous RNA (about 100–4000 nt) that can be fully encapsidated [20], and the stoichiometry (6:1 by mass) between CP and RNA that leads to essentially complete encapsidation yields [20, 26]. Moreover, we point out that separating the assembly reaction into multiple steps (by successively lowering the ionic strength and then the pH of the assembly buffers) leads to the highest yields of well-formed, nuclease-resistant, capsids [27].

Our protocol is a direct extension of the pioneering works of Bancroft and Hiebert [22, 23], and Adolph and Butler [24, 28], and Zlotnick and coworkers [29, 30], and numerous other investigators [6, 31–33] who have contributed to our current—and still imperfect—understanding of the reconstitution of CCMV. Along these lines, we look forward to upcoming improvements that others will make to the work presented here.

2 Materials

Some general comments on working with RNA and CCMV CP and on preparing sterile buffer solutions are found in **Notes 1–3**. Diligently research and follow all safety guidelines when working with any potentially hazardous materials. And follow all waste disposal regulations when disposing of waste materials. Chemicals used in this protocol that require special attention are concentrated hydrochloric acid, glacial acetic acid, phenylmethane sulfonyl fluoride (PMSF), ethidium bromide, and uranyl acetate.

2.1 Components for Protein Purification

1. Purified CCMV stock solution: 1 mg/ml solution of purified CCMV. For protocols on harvesting and purifying CCMV *see* Rao et al. [34] (also *see* **Note 4**).
2. Disassembly Buffer: 50 mM Tris–HCl (pH 7.5), 500 mM $CaCl_2$, 1 mM ethylenediaminetetraacetic acid (EDTA), 1 mM dithiothreitol (DTT), 0.5 mM PMSF (*see* **Note 5**). Add DTT and PMSF directly before use (*see* **Notes 6** and 7).
3. Assembly Buffer I: 50 mM Tris–HCl (pH 7.2), 1 M NaCl, 1 mM EDTA, 1 mM DTT, 1 mM PMSF (*see* **Note 5**). Add DTT and PMSF directly before use (*see* **Notes 6** and 7).
4. Sterile, 6–8 kDa molecular-weight cutoff (MWCO) dialysis tubing (Spectrum Laboratories) with 10 mm flat width prepared in ultrapure water (*see* **Note 8**).
5. Sterile dialysis clips (both buoyant clips and weighted clips; Spectrum Laboratories).
6. Sterile magnetic stir bars, magnetic stirrer.
7. Sterile 1 l beakers.
8. Sterile 1.5 ml tubes.

9. Ultracentrifuge (e.g., Optima TLX, or equivalent; Beckman Coulter) and compatible fixed-angle rotor capable of spinning 1–2 ml of sample at 100,000 × *g* (e.g., TLA-110, or equivalent; Beckman Coulter), and compatible, thickwall polycarbonate ultracentrifuge tubes (e.g., 13 × 56 mm, or equivalent; Beckman Coulter). Chill both the rotor and tubes to 4 °C before use (*see* **Note 9**).
10. UV-Vis spectrophotometry (required), SDS-PAGE (required), MALDI-TOF mass spectrometry (optional).

2.2 Components for VLP Synthesis

1. Purified single-stranded RNA for encapsidation (*see* **Note 10**).
2. Assembly Buffer I: 50 mM Tris–HCl (pH 7.2), 1 M NaCl, 1 mM EDTA, 1 mM DTT, 1 mM PMSF (*see* **Note 5**). Add DTT and PMSF directly before use (*see* **Notes 6** and **7**).
3. Assembly Buffer II: 50 mM Tris–HCl pH 7.2, 50 mM NaCl, 10 mM KCl, 5 mM $MgCl_2$, 1 mM DTT (*see* **Note 5**). Add DTT directly before use (*see* **Note 6**).
4. Assembly Buffer III: 50 mM sodium acetate, 8 mM magnesium acetate (pH 4.8) (*see* **Note 11**).
5. Sterile 6–8 kDa MWCO dialysis tubing with 10-mm flat width prepared in ultrapure water (*see* **Note 8**).
6. Sterile dialysis clips (both buoyant clips and weighted clips; Spectrum Laboratories,).
7. Sterile magnetic stir bars, stir plate.
8. Sterile 1 l beakers.
9. Sterile 1.5 ml tubes.

2.3 Components for VLP Characterization

2.3.1 Components for Gel-Shift Assay

1. Electrophoresis Buffer: 100 mM sodium acetate–acetic acid (pH 5.5), 1 mM EDTA (*see* **Note 12**).
2. High-purity agarose (e.g., UltraPure agarose; Fisher Scientific).
3. Molecular biology grade glycerol (e.g., ≥99% glycerol; Sigma-Aldrich).
4. Nucleic acid stain (e.g., ethidium bromide) (*see* **Note 13**), and protein stain (e.g., Coomassie Instant Blue; Expedeon).
5. Small horizontal gel electrophoresis apparatus (e.g., Fisher Minigel; Thermo Fisher Scientific): gel casting tray, 6-well (or greater) comb, tank, and a power supply capable of producing a constant 50–100 V field across the apparatus (e.g., FisherBiotech Compact Model FB300Q; Thermo Fisher Scientific).
6. RNase A (Thermo Fisher Scientific).

2.3.2 Components for Negative-Stain Transmission Electron Microscopy

1. Access to an electron imaging facility.
2. Carbon-coated copper 400 mesh TEM grids (Ted Pella).
3. 2% (w/v) aqueous uranyl acetate (*see* **Note 14**).
4. Whatman No. 1 filter paper (GE Healthcare Bio-Sciences).
5. High precision tweezers.

3 Methods

Always wear gloves to avoid contamination of the sample and buffer solutions. Avoid dust.

3.1 CP Purification

The CP of CCMV is obtained from the virus by disrupting the quaternary structure of the capsid and then selectively precipitating the viral RNA. In the first step, CCMV is dialyzed against Disassembly Buffer where the neutral pH and high ionic strength result in disassembly of the capsid into noncovalent dimers of CP, and the high concentration of $CaCl_2$ decreases the solubility of the viral RNA. In the second step, the disassembled CCMV is subjected to ultracentrifugation which causes the viral RNA to precipitate from solution while the majority of the CP dimers remain in the supernatant. In the final step, the purified CP is dialyzed against Assembly Buffer I in preparation for reassembly with heterologous RNA.

1. Prepare a sealed dialysis bag containing 1 ml of 1 mg/ml purified CCMV (*see* **Note 15**).
2. Place the dialysis bag in a beaker containing 1 l of Disassembly Buffer at 4 °C. Add a magnetic stir bar to the beaker and stir gently on a stir plate at 4 °C overnight.
3. Remove the contents of the dialysis bag (i.e., disassembled CCMV in Disassembly Buffer) with a pipette and place in a sterile ultracentrifuge tube on ice (*see* **Note 16**). Add additional Disassembly Buffer to the sample until its total volume is 1.5 ml.
4. Prepare a balance tube by pipetting the appropriate weight (*see* **Note 17**) of Disassembly Buffer into a separate, labeled ultracentrifuge tube. Seal each tube with Parafilm to prevent evaporation and spilling prior to centrifugation.
5. Prepare the centrifuge for a 100-min run at 100,000 × *g* at 4 °C.
6. Remove the Parafilm from each tube, load the tubes and rotor into the centrifuge, and immediately begin centrifugation.
7. Remove the sample tube immediately after centrifugation and quickly fractionate (*see* **Note 18**) the sample solution from top to bottom into separate 300 μl volumes in labeled, sterile 1.5 ml tubes (*see* **Note 19**).

8. Prepare five labeled, sealed, dialysis bags each containing one of the collected fractions (*see* **Note 15**). Place each of the dialysis bags in a single beaker containing 1 l of Asssembly Buffer I at 4 °C. Add a magnetic stir bar to the beaker and stir gently on a stir plate at 4 °C overnight.
9. Collect 1 ml of the Assembly Buffer I dialysate for use in acquiring a blank UV-Vis spectrum in **step 11**.
10. Remove the contents of each dialysis bag (i.e., purified CP in Assembly Buffer I) with a pipette and place each sample in a separate labeled, sterile, 1.5 ml tube on ice.
11. Take UV-Vis spectra of each fraction using the aliquot of Assembly Buffer I dialysate from **step 8** as a blank (*see* **Note 20**).
12. Compare the ratio of the absorbance values at 280 nm (A_{280}) and 260 nm (A_{260}) to determine the purity of the CP in each fraction with respect to viral RNA contamination (*see* Fig. 2a). A_{280}/A_{260} ratios above 1.5 correspond to less than 5% RNA contamination and are suitable for use in reassembly reactions.
13. Calculate the concentration of CP in suitably pure fractions using the formula $(\mathrm{CP})_{\mathrm{mg/ml}} = (\mathrm{Abs280\ nm} * 1197.7 - 24.032)/1000$.

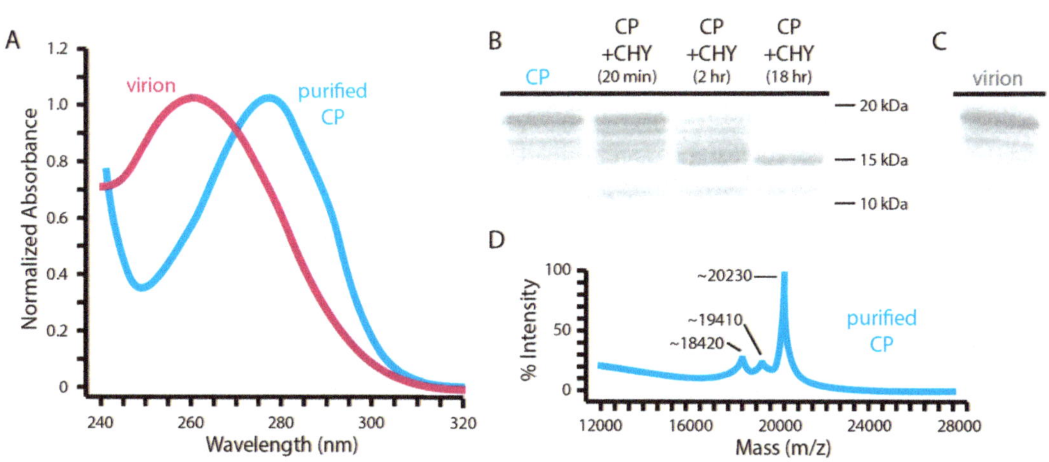

Fig. 2 Determining the purity and integrity of disassembled capsid protein. (**a**) Comparison of the normalized UV-Vis spectrum of CCMV virion (pink) and purified CP (blue) shows a considerable decrease in the absorbance at 260 nm for pure CP, corresponding to removal of the viral RNA. (**b**) SDS-PAGE analysis of purified CP (left-most lane) shows that the majority of signal corresponds to the full-length 20 kDa monomer. However, we also consistently observe a small amount of shorter-than-full-length fragments. The products of limited proteolysis reactions (100 CP: 1 chymotrypsin (CHY), by mass) are also shown (three rightmost lanes) to illustrate severe degradation. (**c**) SDS-PAGE analysis of the original sample of CCMV virion shows identical bands as in the purified CP, indicating that the fragments are not caused by the CP purification process. (**d**) MALDI-TOF mass spectrometry of the purified CP further confirms the presence of small fragments

14. Determine the integrity of the purified CP by 15% (w/v)-SDS-PAGE (*see* **Note 21**). Full-length CP is observed as a single predominant band corresponding to 20 kDa. A number of weaker bands corresponding to shorter fragments may also be present (*see* Fig. 2b). Densitometry traces allow comparison of the intensity of the full-length band relative to the fragment bands. Purified CP fractions with full-length bands contributing greater than 85% of their total intensity are suitable for use in reassembly reactions.
15. Further assessment of CP integrity can be made using MALDI-TOF mass spectrometry. This step is optional.

3.2 VLP Synthesis

VLPs are synthesized by dialyzing an aqueous mixture of CP and RNA against a series of buffer solutions that collectively promote the coassembly of RNA-filled capsids (*see* Fig. 3). In the first step, a specific stoichiometry—6:1 by mass—of CP and RNA is mixed in Assembly buffer I where the neutral pH and high ionic strength disfavor intermolecular interaction. It has been shown [20, 26] that this particular stoichiometry leads to essentially complete encapsidation of any length of RNA by providing electrostatic charge-matching between the cationic N-termini of the CP and the negatively charged phosphate backbone of the RNA. In the second step, the mixture is dialyzed against Assembly Buffer II where the ionic strength is an order of magnitude lower and CP–RNA attraction is strong. These conditions lead to the formation of amorphous structures consisting of a single molecule of RNA decorated by a disordered arrangement of bound CP [27]. This intermediate stage is crucial for avoiding kinetically trapped aggregate structures. In the final step, the mixture is dialyzed against Assembly Buffer III where the acidic (4.8) pH strengthens the CP–CP attraction. Under these conditions well-formed capsids form with sizes that depend on the length of the RNA [20].

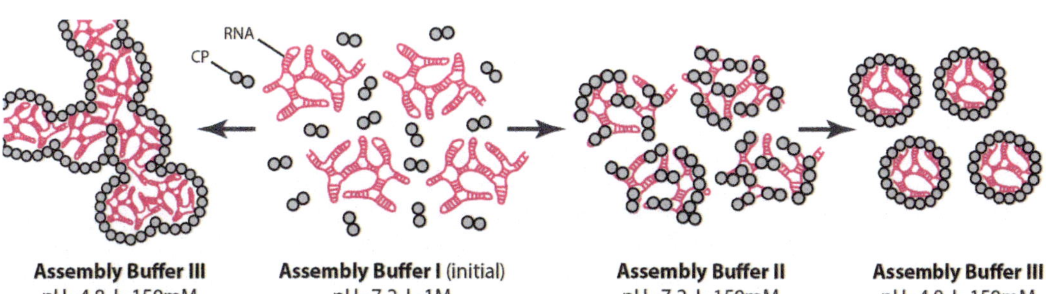

Fig. 3 The two-step assembly process. We start with noninteracting CP and RNA in Assembly Buffer I (second from the left). We then turn on CP–RNA attractions by dialyzing against Assembly Buffer II, yielding amorphous CP–RNA complexes (second from the right). Finally, we turn on CP–CP attractions by dialyzing against Assembly Buffer III, yielding well-formed capsids (rightmost). However, if we instead dialyze the initial mixture of noninteracting CP and RNA directly against Assembly Buffer III, disordered aggregate structures form (leftmost)

1. Mix 54 μg of purified CP with 9 μg of RNA in a total volume of 300 μl of Assembly Buffer I and place in a sealed dialysis bag (*see* **Note 15**). The total volume and the absolute concentrations specified above can be scaled up by a factor of 2 or 3 (higher values have not been systematically tested), or scaled down by more than an order of magnitude, as long as the relative concentration of CP and RNA remains constant at 6 to 1.
2. Place the dialysis bag in a beaker containing 1 l of Assembly Buffer II at 4 °C. Add a magnetic stir bar to the beaker and stir gently on a stir plate at 4 °C for at least 6 h.
3. Remove the dialysis bag from Assembly Buffer II and place it in a beaker containing 1 l of Assembly Buffer III at 4 °C. Add a magnetic stir bar to the beaker and stir gently on a stir plate at 4 °C overnight.

3.3 VLP Characterization

3.3.1 Gel-Shift Assay

A variety of methods exist for characterizing the products of VLP synthesis reactions. Among the most robust and illuminating is the native agarose gel-shift assay [29, 35] which can be used to approximate the yield of RNA encapsidation and the regularity of the resulting VLP structures (*see* Figs. 4 and 5a). The method distinguishes between macromolecular complexes according to their electrophoretic mobility through an agarose gel matrix. In general, naked RNA molecules migrate faster than encapsidated RNA and thus the relative yield of encapsidation can be approximated. Furthermore, because the mobility of CCMV capsids depends only on the surface charge density—and not the charge of the packaged RNA—the structural homogeneity of an ensemble of synthesized VLPs can be determined by comparing the position and sharpness of the VLP band to that of wild-type CCMV.

1. Add 0.01 g agarose per ml of Electrophoresis Buffer to a total volume that corresponds to the size of your particular gel-casting tray. This volume is usually between 35 and 50 ml for a 6–10-well gel tray.
2. Heat the agarose mixture to a boil in a microwave oven that has been designated for lab use (*see* **Notes 22** and **23**).
3. Allow the agarose mixture to cool to about 55 °C (cool enough to be held in bare hands) and pour into the gel-casting tray. Immediately insert the comb to form the loading wells.
4. When the gel has solidified, remove the comb and place the tray in the gel tank housed in a 4 °C chromatography refrigerator or cold room. Fill the gel tank with Electrophoresis Buffer and allow the system to equilibrate at 4 °C.
5. Prepare samples for electrophoresis: make (1) two identical 15 μl (*see* **Note 24**) aliquots containing 0.5 μg of pure RNA in Assembly Buffer III, (2) two 15 μl aliquots containing 0.5-μg-worth-of-RNA in the form of VLP synthesis products,

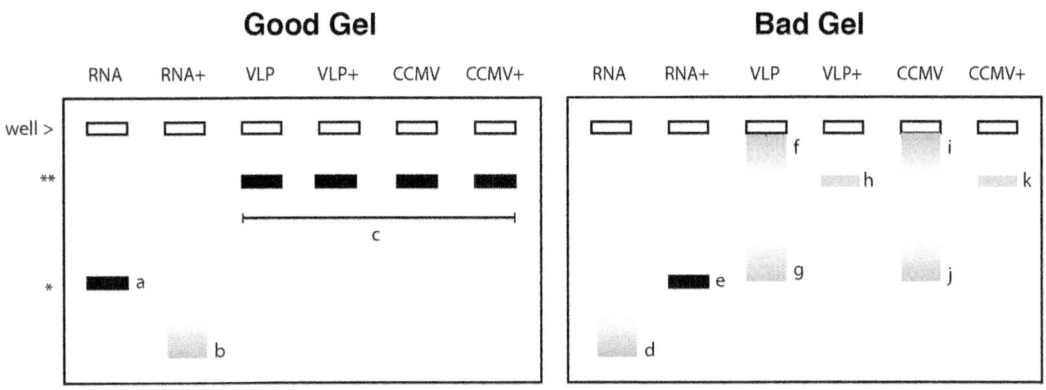

Fig. 4 Schematic examples of desirable ("good") and undesirable ("bad") results of the gel-shift assay. Lanes containing RNase A are denoted "+"; RNA is visualized by ethidium bromide. Good gel: (**a**) pure RNA migrates as a single sharp band of stain intensity with mobility (*) greater than that of CCMV (**); (**b**) digestion by RNase A of the pure RNA sample increases the mobility of the band and decreases—sometimes entirely—the stain intensity; (**c**) VLP synthesis products appear as a single sharp band with approximately the same mobility as CCMV, and no sensitivity to RNase A is observed for either VLPs or CCMV. Note that we sometimes see VLP synthesis products migrating slightly slower than CCMV because of the binding of excess protein to the surface of the VLP. Bad gel: (**d**) pure RNA migrates as a smear of intensity, indicating that the RNA is degraded and should be replaced; (**e**) RNase A has no effect on the pure RNA band, indicating that the RNase is either inactive and should be replaced or is too dilute; (**f**) VLP synthesis products appear as a smear of intensity close to the well, indicating either that the CP:RNA ratio used was too high and should be rechecked or that the pH of Assembly Buffer II was too low and should be rechecked; (**g**) VLP synthesis products appear as a smear of intensity with higher mobility than CCMV, indicating that the CP:RNA ratio used was too low and should be rechecked; (**h**) VLP synthesis products show decreased stain intensity as a result of RNase digestion, indicating that the reconstituted capsids are not nuclease resistant and the pH of Assembly Buffer III should be rechecked; (**i**) and (**j**) CCMV stock does not migrate as a single sharp band of intensity, indicating either that the concentration of the original stock is too high and should be rechecked or that the original stock is no longer viable, or that the pH of the Electrophoresis Buffer is too high or the voltage across the gel was too high; (**k**) CCMV shows decreased stain intensity as a result of RNase digestion, indicating that either the concentration of RNase A used in the digestion was too high and should be rechecked, or that the original stock is no longer viable

(3) two 15 μl aliquots containing 0.5-μg-worth-of-RNA in the form of wild-type CCMV (*see* **Note 25**). Add 1 μl of 10 ng/μl RNase A to one of each pair of 15 μl aliquots (1)–(3) and equilibrate at room temperature for 20 min.

6. Add 5 μl of glycerol to each aliquot (1–3) (*see* **Note 26**).
7. Slowly pipette the sample–glycerol mixtures into separate wells. Care must be taken to ensure that the entire volume ends up in the well.
8. Perform the electrophoresis by applying a constant field of 5 V/cm for 1 or 2 h (*see* **Note 27**).
9. Remove the gel from the gel tank and submerge it for at least 30 min with agitation in roughly 100 ml of ultrapure water containing 0.5 μg/ml ethidium bromide. Ethidium bromide is a fluorescent, intercalating nucleic acid stain used for the detection and quantification of RNA (*see* **Note 13**).

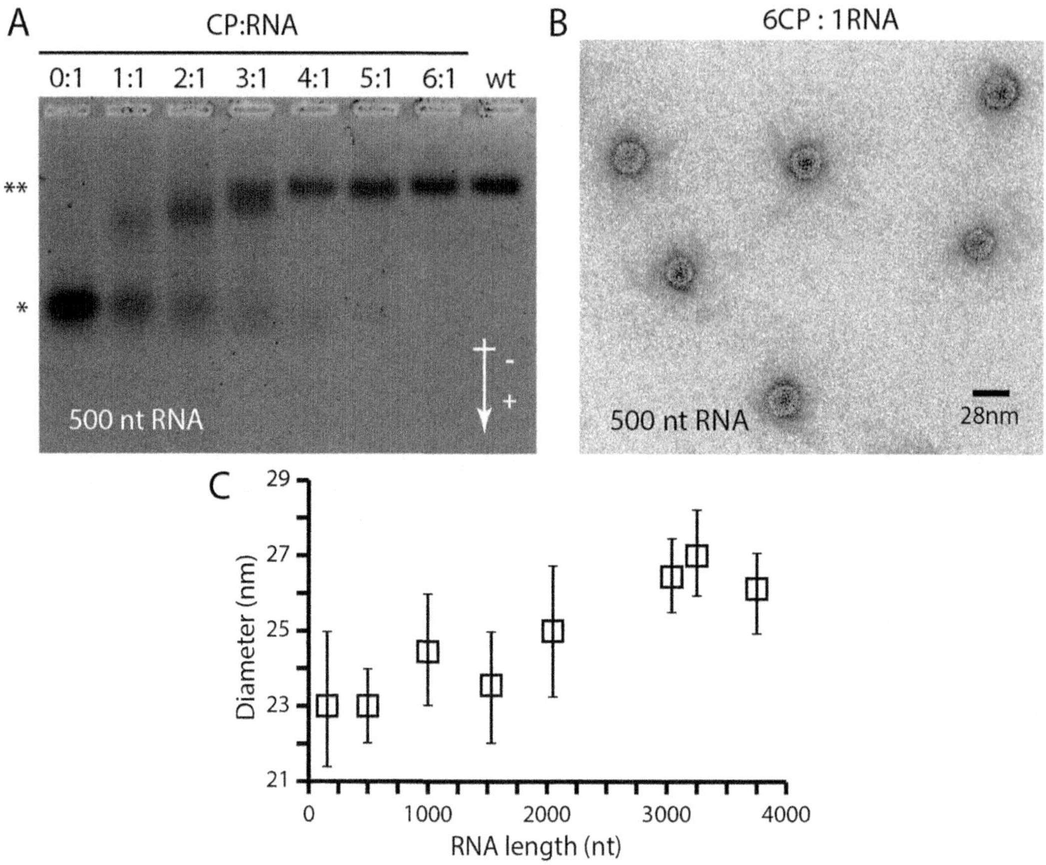

Fig. 5 Determining the yield and size distribution of VLPs. (**a**) Native 1% agarose gel-shift assay stained with ethidium bromide shows the electrophoretic mobility of assembly reactions carried out with increasing CP–RNA ratios. Unencapsidated 500-nt RNA (*) migrates farther than reconstituted virus-like particles (**) which migrate the same distance as wild-type CCMV (rightmost lane). Complete encapsidation is defined to occur at the CP–RNA mass ratio at which all of the ethidium bromide stain intensity migrates with the same mobility as CCMV, that is, 6CP:1RNA. (**b**) Negative-stained transmission electron microscopy (TEM) shows that the assembly products are well-formed spherical particles. (**c**) Plotting the diameter of the assembly products formed from separate assembly reactions containing different lengths of RNA but the same 6:1 mass ratio of CP and RNA shows that the shorter RNAs are packaged into smaller diameter capsids. Error bars show the full width at half maximum of the TEM-measured size distribution histograms

10. Wash the gel with a few hundred milliliters of ultrapure water and then leave submerged in ultrapure water for an additional 30 min with agitation.
11. Visualize the gel by excitation with broadband UV. *See* Fig. 4 for notes on interpreting gel-shift assays.
12. Approximate the yield of RNA encapsidation in the VLP sample by integrating the stain intensity with the same mobility as CCMV and comparing it to the amount of stain intensity throughout the remainder of the lane. Complete encapsidation is expected for VLPs synthesized using a CP:RNA mass ratio of 6–1.

13. Remove the gel from the imaging apparatus and submerge it for at least 2 h with agitation in Coomassie Instant Blue protein stain.
14. Remove background stain intensity by washing the gel with a few hundred milliliters of ultrapure water and then leave submerged in a few hundred milliliters of ultrapure water overnight with mild agitation (*see* **Note 28**).
15. Visualize the gel using white light. Confirm that the protein stain intensity for the VLP synthesis products comigrates with the previously measured RNA stain intensity, and has the same mobility as CCMV.

3.3.2 Transmission Electron Microscopy

Transmission electron microscopy (TEM) provides a quantitative method of determining the distribution of sizes and shapes of synthesized VLPs (*see* Fig. 5b, c). Standard negative-stain transmission electron microscopy is the easiest to perform and provides the highest-resolution structures of individual VLPs. Staining and imaging VLPs by TEM is relatively straightforward, and only a brief outline is given below.

1. Briefly plasma-etch a carbon-coated copper TEM grid, carbon-side up.
2. Hold the grid carbon-side up by its outermost rim using a single pair of tweezers placed securely on a working surface dedicated to uranyl acetate work.
3. Dilute a solution of VLP synthesis products with Assembly Buffer III to a final RNA concentration of 0.01 mg/ml and add a 6 μl drop of this solution to the carbon side of the plasma-etched TEM grid. Leave the drop for 1 min.
4. Blot the drop of solution with Whatman No. 1 filter paper and immediately replace it with a 6 μl drop of 2% (w/v) aqueous uranyl acetate (*see* **Note 14**). Leave the drop for 1 min.
5. Blot the drop of solution with Whatman No. 1 filter paper and allow the grid to dry completely under ambient conditions.
6. Image by TEM (*see* **Note 29**).

4 Notes

1. A general note on RNA: Special—but not excessive—care is required to avoid degradation of long single-stranded RNAs. For a useful introduction to working with RNA *see* Nielsen [36]. Solutions of purified RNA should be stored at −80 °C and used immediately upon thawing. All portions of the protocol involving RNA should be carried out on ice or at 4 °C. It is not necessary to employ diethylpyrocarbonate (DEPC) to destroy nucleases.

2. A general note on the CP: We obtain CP from wild-type CCMV that is harvested from infected California cowpea plants (*Vigna ungiculata* cv. Black Eye) according to the protocol described in Rao et al. [34]. Others have shown that CP can be expressed in *E. coli* [32] and yeast [37]. Whatever the source, CCMV CP is particularly prone to cleavage when it is not assembled as capsids: solutions of CCMV or CCMV VLPs can be stored for years at −80 °C (or −20 °C) prior to use, but purified CP should not be frozen and should be stored at 4 °C and used within 2–3 weeks. All portions of the protocol involving purified CP should be carried out on ice or at 4 °C. In certain steps of the protocol we add PMSF to inhibit proteases.
3. A general note on preparing sterile buffer solutions: Prepare all buffers using glassware that is free of RNases. Use ultrapure water (18.2 MΩ cm) and molecular-biology grade reagents that are certified free of RNases and proteases. Further sterilize all buffer solutions made from dry reagents by filtering through a 0.22 μm porous membrane and autoclaving at 121 °C for at least 15 min. Store buffers at 4 °C (unless indicated otherwise) to minimize growth of microorganisms. Adjust the pH of all buffers at the temperature they will be used (4 °C) and double-check the pH before use.
4. We and others have found that CCMV purification protocols that involve precipitation by polyethylene glycol lead to coprecipitation of proteases and result in high levels of CP cleavage. For this reason we strongly recommend protocols that feature density gradient centrifugation and density-cushion centrifugation instead of polyethylene glycol precipitation.
5. To prepare Disassembly Buffer and Assembly Buffers I and II, combine sterile concentrated aqueous stock solutions of the individual components and then dilute with ultrapure autoclaved water. Some useful stock solutions for this purpose are: 1 M Tris–HCl pH 7.5 (500 ml), 1 M Tris–HCl pH 7.2 (500 ml), 2.5 M $CaCl_2$ (1 l), 2.5 M NaCl (2 l), 1 M $MgCl_2$ (250 ml), 1 M KCl (500 ml), 0.45 M EDTA pH 8.0 (250 ml). To prepare these solutions dissolve the dry reagents in ultrapure water, filter (0.22 μm), and autoclave (121 °C for at least 15 min). The pH of Tris–HCl solutions should be adjusted with concentrated HCl at 4 °C before filtering. Double-check the final pH of any Tris–HCl containing buffer before use. Note that EDTA will not dissolve unless the pH is brought to 8 by the addition of NaOH.
6. To prepare individual single-use aliquots of DTT, make 50 ml of 0.67 M DTT in ultrapure autoclaved water and store as 1.5 ml aliquots at −20 °C. Use single-use aliquots of DTT immediately upon thawing.

7. PMSF is toxic and corrosive and requires special handling and disposal procedures. To prepare individual single-use aliquots of PMSF, make 100 ml of 0.2 M PMSF in isopropanol and store in 1.25 ml single-use aliquots at −20 °C. Use single-use aliquots of PMSF immediately upon thawing.
8. To prepare wet dialysis tubing, cut a roughly half-meter portion of the dry roll into 10 cm lengths and submerge in 1 l of 2% (w/v) sodium bicarbonate and 1 mM EDTA. Heat this mixture on a hot plate and allow it to boil for 10 min. Cool the mixture and rinse the tubing thoroughly with ultrapure water. Add the tubing to a 1 l media bottle filled with ultrapure water. Autoclave at 121 °C for 15 min and store at 4 °C. Dialysis tubing prepared in this way can be used for many weeks. The tubing should be removed from the bottle with a sterile spatula and care should be taken to avoid contamination.
9. To sterilize thickwall polycarbonate ultracentrifuge tubes, scrub with a 2% SDS detergent solution and rinse thoroughly with ultrapure water.
10. The RNA to be encapsidated can be from any source as long as it is between 100 and 4000 nt in length and sufficiently pure. RNAs much shorter than 100 nt do not result in stable capsids (Comas-Garcia, personal communication), and RNAs much longer than 4000 nt are encapsidated by multiple capsids that are "strung" together by exposed RNA [20]. Common impurities associated with in vitro transcribed RNAs are left-over template DNA and free nucleotides—both must be completely removed. Free nucleotides are especially problematic because they are difficult to detect and they interfere with the UV-absorbance measurements used to quantify the concentration of RNA. To remove free nucleotides, dilute the RNA sample fivefold and reconcentrate with a 100 kDa. Centricon filtration unit (EMD Millipore, Billerica, MA, USA)—repeat the dilution-concentration cycle five times to ensure complete removal of free nucleotides.
11. Simple way to prepare Assembly Buffer III: dilute a sterile, 10× concentrated stock solution with ultrapure, autoclaved water. Note that the final pH of the 10× concentrated stock should be adjusted to 4.8 at 4 °C with glacial acetic acid before filtering. Double-check the final pH of Assembly Buffer III before use.
12. To prepare Electrophoresis Buffer, dilute a sterile, 10× concentrated, stock solution with ultrapure autoclaved water. Note that the final pH of the 10× concentrated stock should be adjusted to 5.5 at 4 °C with glacial acetic acid before filtering. Double-check the final pH of Electrophoresis Buffer before use.

13. Ethidium bromide is a potential mutagen that requires special handling and disposal procedures.
14. Uranyl acetate is toxic and radioactive and requires special handling and disposal procedures.
15. To prepare a sealed dialysis bag, fold over a 2 cm portion of one end of precut, sterile, dialysis tubing and fasten the clasp of a weighted dialysis clip over the fold. Orient the clipped end of the tube facing downward and load the sample solution into the top of the tube so that the solution fills the bottom (clipped) end. Fold over a 2 cm portion of the top of the tubing. Leave a slight gap of air between the fold and the top surface of the sample solution. Fasten the clasp of a buoyant dialysis clip over the fold. Make sure that the air trapped in the tubing has not made the dialysis membrane taut. If the membrane is taut, unclip the buoyant clip and release a portion of the air.
16. The contents of the dialysis bag may appear cloudy due to RNA precipitation.
17. Add to the balance tube a volume of Disassembly Buffer that makes the total weight of the filled balance tube equal to that of the filled sample tube to within 0.01 g.
18. Fractionate the sample from top to bottom as quickly as possible without disturbing the amber-colored pellet of RNA at the bottom of the tube. Quick fractionation is crucial as the pellet will continuously release contaminant RNA into the bottommost fractions that contain the most CP.
19. You may choose to discard the topmost fraction as it contains very little CP.
20. Because Assembly Buffer I contains a fraction of oxidized DTT—which absorbs at 280 nm—it is important to prepare a reliable solvent blank. Record a solvent blank with a portion of the aliquot of Assembly Buffer I and then measure the absorbance spectrum from a separate portion of the same aliquot. The measured absorbance should be zero for all wavelengths.
21. For a helpful introduction to performing SDS-PAGE analysis *see* Sambrook and Russell [38].
22. A laboratory microwave should never be used to prepare food.
23. Use heat-resistant gloves when handling boiling agarose mixtures. Handle with caution as superheating of the agarose solution can occur.
24. This volume depends on the maximum capacity of the wells in the gel.
25. About 20% of the mass of CCMV is RNA. Thus 2.5 μg of CCMV corresponds to about 0.5 μg of RNA.

26. The addition of 20% (v/v) glycerol serves to increase the density of the samples so that they remain at the bottom of the well.
27. Electrophoresis poses potential electrical hazards and requires special safety procedures to prevent electrical shock.
28. Staining with Coomassie Instant Blue protein stain makes agarose gels more fragile. Handle with care.
29. For a helpful introduction to TEM imaging *see* De Carlo and Harris [39].

Some useful numbers related to CCMV

Outer diameter of capsid	≈28 nm
Inner diameter of capsid	≈20 nm
Number of CP in capsid	180
Mass of CP	20.3 kDa
Total mass of the CCMV virion	≈4.6 MDa
Fraction of CCMV that is RNA	≈20%
Fraction of CCMV that is CP	≈80%
Absorbance at 260 nm of 1 mg/ml of CCMV	5.88 OD ($A_{260}/A_{280} = 1.6$)
Absorbance at 260 nm of 1 mg/ml of ssRNA	25 OD ($A_{260}/A_{280} > 2$ in TE buffer)
Absorbance at 280 nm of 1 mg/ml of CP	0.81 OD ($A_{260}/A_{280} \le 0.67$)

Acknowledgments

Much of the original work that serves as the basis for this protocol was performed in close collaboration with Mauricio Comas-Garcia and Ruben Cadena-Nava, and was supported by the National Science Foundation grants CHE 0714411 and 1051507.

References

1. Speir JA, Munshi S, Wang G et al (1995) Structures of the native and swollen forms of cowpea chlorotic mottle virus determined by X-ray crystallography and cryo-electron microscopy. Structure 3:63–78. https://doi.org/10.1016/S0969-2126(01)00135-6
2. Caspar DLD, Klug A (1962) Physical Principles in the Construction of Regular Viruses. Cold Spring Harb Symp Quant Biol 27:1–24. https://doi.org/10.1101/SQB.1962.027.001.005
3. Carrillo-Tripp M, Shepherd CM, Borelli IA et al (2009) VIPERdb2: an enhanced and web API enabled relational database for structural virology. Nucleic Acids Res 37:D436–D442
4. Zlotnick A, Aldrich R, Johnson JM et al (2000) Mechanism of capsid assembly for an icosa-

hedral plant virus. Virology 277:450–456. https://doi.org/10.1006/viro.2000.0619

5. Mukherjee S, Pfeifer CM, Johnson JM et al (2006) Redirecting the coat protein of a spherical virus to assemble into tubular nanostructures. J Am Chem Soc 128:2538–2539. https://doi.org/10.1021/ja056656f
6. Lavelle L, Gingery M, Phillips M et al (2009) Phase diagram of self-assembled viral capsid protein polymorphs. J Phys Chem B 113:3813–3819. https://doi.org/10.1021/jp8079765
7. Garmann RF, Sportsman R, Beren C et al (2015) A simple RNA-DNA scaffold templates the assembly of monofunctional virus-like particles. J Am Chem Soc 137:7584–7587. https://doi.org/10.1021/jacs.5b03770
8. Douglas T, Young M (2006) Viruses: making friends with old foes. Science 312:873–875. https://doi.org/10.1126/science.1123223
9. Kostiainen MA, Kasyutich O, Cornelissen JJ, Nolte RJ (2010) Self-assembly and optically triggered disassembly of hierarchical dendron–virus complexes. Nat Chem 2:394–399. https://doi.org/10.1038/nchem.592
10. Tsvetkova IB, Dragnea BG (2015) Encapsulation of nanoparticles in virus protein shells. In: Protein Cages. Springer, New York, pp 1–15. https://doi.org/10.1007/978-1-4939-2131-7_1
11. Destito G, Yeh R, Rae CS et al (2007) Folic acid-mediated targeting of cowpea mosaic virus particles to tumor cells. Chem Biol 14:1152–1162. https://doi.org/10.1016/j.chembiol.2007.08.015
12. Aljabali AA, Shukla S, Lomonossoff GP et al (2012) CPMV-DOX Delivers. Mol Pharm 10:3–10. https://doi.org/10.1021/mp3002057
13. Azizgolshani O, Garmann RF, Cadena-Nava R et al (2013) Reconstituted plant viral capsids can release genes to mammalian cells. Virology 441:12–17. https://doi.org/10.1016/j.virol.2013.03.001
14. Dalsgaard K, Uttenthal A, Jones TD et al (1997) Plant-derived vaccine protects target animals against a viral disease: Abstract: Nature Biotechnology. Nat Biotech 15:248–252. https://doi.org/10.1038/nbt0397-248
15. Steinmetz NF, Lin T, Lomonossoff GP, Johnson JE (2009) Structure-based engineering of an icosahedral virus for nanomedicine and nanotechnology. In: Viruses Nanotechnol. Springer, New York, pp 23–58. https://doi.org/10.1007/978-3-540-69379-6_2
16. Plummer EM, Manchester M (2011) Viral nanoparticles and virus-like particles: platforms for contemporary vaccine design. Wiley Interdiscip Rev Nanomed Nanobiotechnol 3:174–196. https://doi.org/10.1002/wnan.119
17. Wang Q, Kaltgrad E, Lin T et al (2002) Natural supramolecular building blocks: wild-type cowpea mosaic virus. Chem Biol 9: 805–811. https://doi.org/10.1016/S1074-5521(02)00165-5
18. Strable E, Johnson JE, Finn MG (2004) Natural nanochemical building blocks: icosahedral virus particles organized by attached oligonucleotides. Nano Lett 4:1385–1389. https://doi.org/10.1021/nl0493850
19. Steinmetz NF, Evans DJ, Lomonossoff GP (2007) Chemical introduction of reactive thiols into a viral nanoscaffold: a method that avoids virus aggregation. Chembiochem 8:1131–1136. https://doi.org/10.1002/cbic.200700126
20. Cadena-Nava RD, Comas-Garcia M, Garmann RF et al (2012) Self-assembly of viral capsid protein and RNA molecules of different sizes: requirement for a specific high protein/RNA mass ratio. J Virol 86:3318–3326. https://doi.org/10.1128/JVI.06566-11
21. Krol MA, Olson NH, Tate J et al (1999) RNA-controlled polymorphism in the in vivo assembly of 180-subunit and 120-subunit virions from a single capsid protein. Proc Natl Acad Sci 96:13650–13655. https://doi.org/10.1073/pnas.96.24.13650
22. Bancroft JB, Hiebert E (1967) Formation of an infectious nucleoprotein from protein and nucleic acid isolated from a small spherical virus. Virology 32:354–356. https://doi.org/10.1016/0042-6822(67)90284-X
23. Hiebert E, Bancroft JB (1969) Factors affecting the assembly of some spherical viruses. Virology 39:296–311. https://doi.org/10.1016/0042-6822(69)90050-6
24. Adolph KW, Butler PJG (1975) Reassembly of a spherical virus in mild conditions. Nature 255:737–738. https://doi.org/10.1038/255737a0
25. Garmann RF, Comas-Garcia M, Knobler CM, Gelbart WM (2016) Physical principles in the self-assembly of a simple spherical virus. Acc Chem Res 49:48–55. https://doi.org/10.1021/acs.accounts.5b00350
26. Garmann RF, Comas-Garcia M, Koay MS et al (2014) Role of electrostatics in the assembly pathway of a single-stranded RNA virus. J Virol 88:10472–10479. https://doi.org/10.1128/JVI.01044-14
27. Garmann RF, Comas-Garcia M, Gopal A et al (2014) The assembly pathway of an icosahedral single-stranded RNA virus depends on the strength of inter-subunit attractions. J Mol Biol

426:1050–1060. https://doi.org/10.1016/j.jmb.2013.10.017

28. Adolph KW, Butler PJG (1974) Studies on the assembly of a spherical plant virus: I States of aggregation of the isolated protein. J Mol Biol 88:327–341. https://doi.org/10.1016/0022-2836(74)90485-9
29. Johnson JM, Willits DA, Young MJ, Zlotnick A (2004) Interaction with capsid protein alters RNA structure and the pathway for in vitro assembly of cowpea chlorotic mottle virus. J Mol Biol 335:455–464. https://doi.org/10.1016/j.jmb.2003.10.059
30. Johnson JM, Tang J, Nyame Y et al (2005) Regulating self-assembly of spherical oligomers. Nano Lett 5:765–770. https://doi.org/10.1021/nl050274q
31. Verduin BJM, Bancroft JB (1969) The infectivity of tobacco mosaic virus RNA in coat proteins from spherical viruses. Virology 37:501–506. https://doi.org/10.1016/0042-6822(69)90240-2
32. Zhao X, Fox JM, Olson NH et al (1995) In vitro assembly of cowpea chlorotic mottle virus from coat protein expressed in Escherichia coli and in vitro-transcribed viral cDNA. Virology 207:486–494. https://doi.org/10.1006/viro.1995.1108
33. Ni P, Wang Z, Ma X et al (2012) An examination of the electrostatic interactions between the N-Terminal tail of the brome mosaic virus coat protein and encapsidated RNAs. J Mol Biol 419:284–300. https://doi.org/10.1016/j.jmb.2012.03.023
34. Rao A, Duggal R, Lahser F, Hall T (1994) Analysis of RNA replication in plant viruses. In: Adolph KW (ed) Methods Mol. Genet. Mol. Virol. Tech. Academic Press, Orlando, Florida, pp 216–236
35. Zlotnick A, Porterfield JZ, Wang JC-Y (2013) To build a virus on a nucleic acid substrate. Biophys J 104:1595–1604. https://doi.org/10.1016/j.bpj.2013.02.005
36. Nielsen H (2011) Working with RNA. In: Nielsen H (ed) RNA. Humana Press, pp 15–28. https://doi.org/10.1007/978-1-59745-248-9_2
37. Brumfield S, Willits D, Tang L et al (2004) Heterologous expression of the modified coat protein of Cowpea chlorotic mottle bromovirus results in the assembly of protein cages with altered architectures and function. J Gen Virol 85:1049–1053. https://doi.org/10.1099/vir.0.19688-0
38. Sambrook J, Russell DW (2006) SDS-polyacrylamide gel electrophoresis of proteins. CSH Protoc. https://doi.org/10.1101/pdb.prot4540
39. De Carlo S, Harris JR (2011) Negative staining and cryo-negative staining of macromolecules and viruses for TEM. Micron 42:117–131. https://doi.org/10.1016/j.micron.2010.06.003

Chapter 18

Packaging DNA Origami into Viral Protein Cages

Veikko Linko, Joona Mikkilä, and Mauri A. Kostiainen

Abstract

The DNA origami technique is a widely used method to create customized, complex, spatially well-defined two-dimensional (2D) and three-dimensional (3D) DNA nanostructures. These structures have huge potential to serve as smart drug-delivery vehicles and molecular devices in various nanomedical and biotechnological applications. However, so far only little is known about the behavior of these novel structures in living organisms or in cell culture/tissue models. Moreover, enhancing pharmacokinetic bioavailability and transfection properties of such structures still remains a challenge. One intriguing approach to overcome these issues is to coat DNA origami nanostructures with proteins or lipid membranes. Here, we show how cowpea chlorotic mottle virus (CCMV) capsid proteins (CPs) can be used for coating DNA origami nanostructures. We present a method for disassembling native CCMV particles and isolating the pure CP dimers, which can further bind and encapsulate a rectangular DNA origami shape. Owing to the highly programmable nature of DNA origami, packaging of DNA nanostructures into viral protein cages could find imminent uses in enhanced targeting and cellular delivery of various active nano-objects, such as enzymes and drug molecules.

Key words Nucleic acids, DNA nanotechnology, DNA origami, Self-assembly, Virus capsid protein, Electrostatic assembly, CCMV

1 Introduction

The specific binding between nucleobases (Watson–Crick base pairing) can be exploited in creating DNA-based nanomaterials and devices for numerous biomedical applications. The first programmed DNA nanostructures were proposed and demonstrated by Nadrian "Ned" Seeman in the 1980s [1]. Since then, various design strategies for DNA nanoarchitectures have been presented [2, 3]. One of them is DNA origami [4], which has proven to be a remarkably robust and straightforward method. With the DNA origami technique, it is possible to create customized and complex two- and three-dimensional nano-objects with nanometer-scale precision [4–6]. Most commonly, origami structures are formed by folding a long single-stranded DNA scaffold strand into a desired shape with a set of designed oligonucleotides, i.e., staple strands.

Christina Wege and George P. Lomonossoff (eds.), *Virus-Derived Nanoparticles for Advanced Technologies: Methods and Protocols*, Methods in Molecular Biology, vol. 1776, https://doi.org/10.1007/978-1-4939-7808-3_18,
© Springer Science+Business Media, LLC, part of Springer Nature 2018

These staples are unique in sequence and they bind to specific locations at the scaffold strand, thus driving the exact assembly of the desired origami shape. Importantly, DNA origami can be used to arrange different materials such as enzymes [7], metal nanoparticles [8], and carbon nanotubes [9]. The authors have recently demonstrated how DNA origami shapes can be utilized in creating enzymatic nanoreactors [10] and metallic nanostructures [11], and how the structures can be efficiently positioned and anchored on different substrates [12, 13].

Due to their biocompatibility and modularity, DNA origami structures are seen as promising candidates for developing biomedical delivery vehicles carrying drug molecules, functional binding sites and cell-targeting ligands [14, 15]. Although recent studies indicate that, for example, DNA nanotubes [16], CpG-sequence-coated DNA origami [17], and various other shapes containing drug molecules or functional enzymes [18–20] can enter cells to some extent, the highly polar nature of the DNA structures seems to prevent efficient transfection [21]. Besides widely used transfection reagents, designed intercalators [22], cationic polymers [23], and lipid-membrane coatings [24] could be utilized to enhance the stability, transfection efficiency or pharmacokinetic bioavailability of the DNA nanostructures. However, other types of further improved and sophisticated delivery systems are still urgently needed.

One attractive possibility for achieving enhanced transfection and protection of DNA nanostructures is to take advantage of virus particles [25, 26]. Viruses are natural transfection vectors and previous studies have already shown the possibility of using long double-stranded DNA or drug-loaded DNA micelles as templates to produce virus capsid protein-coated structures [27–29]. Here, we show in detail how DNA origami nanostructures can be coated and encapsulated using virus capsid proteins. For this, we utilize cowpea chlorotic mottle virus (CCMV), which has been successfully used for the encapsulation of various macromolecules [30, 31] and as a component in protein-based crystals and superlattices [32–34]. In brief, we combine ligated rectangular DNA origami templates (92 nm × 72 nm × 2 nm) with purified CCMV capsid protein (CP) dimers [35] in conditions where the self-assembly of proteins alone into higher-order structures is prevented [36]. We show that CPs bind effectively to the surface of negatively charged DNA origami due to their positively charged N-termini, and adopt configurations remarkably different from the naturally occurring icosahedral capsid structure.

2 Materials

2.1 DNA Origami Preparation

2.1.1 DNA Origami Materials

1. 7249-nucleotide single-stranded phage DNA (ssDNA), M13mp18 (New England Biolabs or Tilibit Nanosystems), at 100 nM concentration.
2. Set of short staple strands (Integrated DNA Technologies), at 100 μM concentration. For a rectangular DNA origami, sequences of the 128 core staple strands can be found in ref. 4.
3. 10× TAE buffer with magnesium: 400 mM tris(hydroxymethyl) aminomethane (Tris), 190 mM acetic acid, 10 mM ethylenediaminetetraacetic acid (EDTA), and 125 mM magnesium acetate (MgAc) or magnesium chloride ($MgCl_2$), pH ~8.3.

2.1.2 Optional DNA Origami Ligation

1. 10× kinase reaction buffer.
2. T4 polynucleotide kinase.
3. 10× ligase reaction buffer.
4. T4 DNA ligase.

2.1.3 Purification of DNA Origami

1. Millipore Amicon Ultra Centrifugal Filters, 0.5 ml with 100 kDa molecular weight cutoff (MWCO).
2. Origami buffer (HEPES–NaOH buffer): 6.5 mM HEPES, 2.0 mM NaOH, pH 6.8.

2.2 CCMV CP Isolation

1. Slide-A-Lyzer® Mini or similar dialysis units (3.5 kDa MWCO).
2. 20–40 μl of native CCMV solution: 10 mg/ml CCMV in acetate buffer (100 mM acetic acid, 1 mM EDTA, 1 mM NaN_3, pH 5.0 adjusted with NaOH).
3. CCMV capsid disassembly buffer: 50 mM Tris–HCl, 500 mM $CaCl_2$, 1 mM dithiothreitol (DTT), pH 7.5. Make fresh before use.
4. CPs clean buffer: 50 mM Tris–HCl, 150 mM NaCl, 1 mM DTT, pH 7.5. Make fresh before use.

Note that the ready CP solution should be used within 1 week (in order to avoid degradation).

2.3 Concentration and Purity

1. Cuvette: Quartz cuvette with 0.3 cm light path.
2. UV/Vis spectrometer (e.g., PerkinElmer LAMBDA 950).

2.4 Encapsulation Procedure

1. 50 μl of DNA origami solution ~8 nM.
2. 20 μl of CP solution ~50 μM.
3. 1.5 M NaCl solution.

2.5 Characterization

2.5.1 Transmission Electron Microscopy (TEM)

1. TEM grids: Carbon film grids on 300 hex mesh copper (CF300H-Cu, Electron Microscopy Sciences).
2. Negative stain: (0.5% (w/v) uranyl acetate in Milli-Q water).
3. Filter paper.
4. Milli-Q water.

2.5.2 Gel Electrophoretic Mobility Shift Assay (EMSA)

1. Agarose gel: prepare 1% agarose gels by dissolving 1 g of agarose in 100 ml of 1× TAE buffer (40 mM Tris, 19 mM acetic acid, 1 mM EDTA, and 11 mM MgAc or $MgCl_2$). Mix and boil the solution for 30 s, allow to cool and add 30 μl of ethidium bromide (EthBr) stain solution (0.625 mg/ml).
2. Running buffer: 1× TAE (*see* above) with 11 mM MgAc or $MgCl_2$.
3. The DNA origami–CP samples: Prepare the samples 1 h before running the agarose gel.
4. 6× loading dye.
5. 1.5 μl of M13mp18 ssDNA (~100 nM).

3 Methods

3.1 DNA Origami Preparation

The fabricated shape is a 72 nm × 92 nm × 2 nm rectangular origami (*see* **Note 1**). The structure consists of a 7249-base-long M13mp18 plasmid scaffold and 128 short oligonucleotides that fold the scaffold into a desired shape in a thermal annealing process. The staple design, core staple sequences and thermal ramp for folding have been adapted from ref. 4 (*see* **Note 2**). After DNA origami formation by thermal annealing, the origami structures can be optionally ligated using T4 DNA ligase (*see* **Note 3**). The excess staple strands are removed after structure formation by a nondestructive spin-filtering procedure.

3.1.1 DNA Origami Annealing

This is a recipe for a ligated version of origami (if ligation is omitted, kinase reaction buffer and kinase (**step 2**) can be replaced by in total 8 μl of distilled water). The structures are prepared as 50 μl quantities.

1. Prepare staple strand mix by pipetting equal amounts of each 128 staple strands (10 μM) and mixing them together. This results in 0.781 μM concentration for each staple strand in the staple strand mix.
2. Mix 5 μl 10× TAE Mg^{2+} buffer, 25.5 μl distilled water, 5 μl 10× kinase reaction buffer, 6.1 μl staple strand mix (each staple at 0.781 μM concentration), 3 μl T4 polynucleotide kinase (*see* **Note 3**), and 5.4 μl of 93 nM M13mp18 ssDNA. The concentration of the scaffold strand in the total reaction

volume of 50 μl is 10 nM. The staple strands are used in 10 times excess compared to the scaffold strand.

3. Keep the solution for 1 h at 37 °C (optimum operational temperature for kinase). This step can be skipped if ligation is omitted.
4. Immediately afterwards initiate folding of the origamis in an annealing process using a thermal cycler with the following thermal ramp: Allow annealing from 90 °C to 20 °C at a rate of 1 °C/min using 0.1 °C increments (i.e., 0.1 °C decrease every 6 s). After folding, the final concentration of DNA origami is 10 nM (assuming 100% yield in folding).

3.1.2 Optional DNA Origami Ligation Procedure

Ligation is carried out using T4 DNA ligase. The ligation procedure is prepared using 100 μl quantities.

1. Mix in the given order: 9 μl 10× TAE Mg^{2+} buffer, 67 μl distilled water, 9 μl 10× ligase reaction buffer, 10 μl annealed origami solution, and 5 μl T4 DNA ligase.
2. After mixing, protect the solution from light and incubate overnight at room temperature.

3.1.3 Purification of DNA Origami

Most of the excess staples can be removed in a nondestructive spin-filtering process. Simultaneously, the buffer is exchanged (from TAE-based buffer to HEPES/NaOH buffer).

1. Inject 500 μl of ligated DNA origami solution into the filter. If ligation is not used, dilute 50 μl of DNA origami solution to 500 μl using HEPES–NaOH buffer.
2. Centrifuge the solution for 3 min using 14,000 × *g* and discard the flow-through.
3. Add 450 μl of HEPES/NaOH buffer to the filter and centrifuge again for 3 min using 14,000 × *g*.
4. Repeat **steps 2** and **3** three more times (in total four spins).
5. Gently pipette the remaining solution out from the filter unit after the last spinning. Collect the rest of the solution by placing the filter upside down in a fresh container and spinning for 2 min at 1,000 × *g*.

Typically, after filtration the volume of the solution is brought from 500 μl down to 17–20 μl. Dilute the solution back to 50 μl by adding HEPES–NaOH origami buffer. This results in pure DNA origami at approximately 10 nM concentration (assuming 100% yield in spin-filtering) (*see* **Note 4**).

3.2 CCMV CP Isolation

1. Disassemble CCMV by dialyzing 20–40 μl of the CCMV stock solution (10 mg/ml) against disassembly buffer for at least 8 h using Slide-A-Lyzer® Mini or similar dialysis units with 3.5 kDa MWCO.

2. Collect the solution inside the dialysis unit to a 1.5 ml reaction tube and centrifuge it overnight (17,000 × *g*, 4 °C).
3. Collect the resulting supernatant carefully without touching the RNA pellet (note that sometimes the RNA pellet can be relatively faint). Dialyze the supernatant against clean buffer using dialysis units (3.5 kDa MWCO).
4. Storing CP solution: Collect the solution inside the dialysis unit and dilute it with clean buffer if needed (*see* **Note 5**).

3.3 Concentration and Purity

Carry out measurements at room temperature with a UV–Vis spectrometer. Use a quartz cuvette with a 0.3 cm light path. Subtract a reference spectrum (blank buffer) from the spectra of actual samples.

3.3.1 CPs

Determine the CP concentration and purity by using the Beer–Lambert relationship, $A_{280} = \varepsilon_{280} c_{CP} l$, where A_{280} is absorbance at 280 nm wavelength, ε_{280} is the calculated extinction coefficient for CCMV protein monomer (23,590 1/(M cm)) and l is the length of the light path in centimeters (0.3 cm).

CP concentration in different batches typically varies between 70 and 160 μM. Purity can be determined from A_{280}/A_{260} ratio, which can vary between 1.3 and 1.5. The higher the ratio, the cleaner the CP solution.

3.3.2 DNA Origami

DNA origami concentration (c_{DNA}) was estimated using Lambert–Beer relationship, $A_{260} = \varepsilon_{260} c_{DNA} l$, where A_{260} is absorbance at 260 nm wavelength, ε_{260} is the calculated extinction coefficient ($0.93 * 10^8$ 1/(M cm)) [37] and l is the length of the light path in centimeters (0.3 cm).

3.4 Encapsulation Procedure

1. Mix 10 μl DNA origami solution (~8 nM) and 0–10 μl of CP solution (~50 μM or diluted to 5 μM). Prepare several samples with different CP/DNA base pair ratios (marked as x) (*see* **Note 6**). Adjust the final NaCl concentration for all samples to 150 mM by adding an appropriate amount of 1.5 M NaCl (for the amounts of all materials required for the encapsulation, *see* Table 1).
2. Incubate at 4 °C for 1 h.
3. Analyze the samples as described below.

3.5 Characterization

3.5.1 Transmission Electron Microscopy (TEM)

1. Prepare samples on carbon-coated grids by placing a 3 μl drop of the sample solution on the grid. Leave the sample drop on the grid for 1 min after which the excess solution can be blotted away with a piece of filter paper. Allow the samples to dry.
2. Wash the sample with a 3 μl drop of Milli-Q water.

Table 1
Amounts of different materials required for the encapsulation.

Sample number	Ref.	1	2	3	4
x = n(CP)/n(DNA bp)		0.05	0.1	0.5	1
V (HEPES/NaOH, µl)	53	53	53	53	53
V (NaCl 1.5 M, µl)	7	7	7	7	7
V (DNA origami, µl)	10	10	10	10	10
V (clean buffer, µl)	10	5	0	5	0
V (CP diluted 10×, µl)		5	10		
V (CP, µl)				5	10
V (total, µl)	80	80	80	80	80

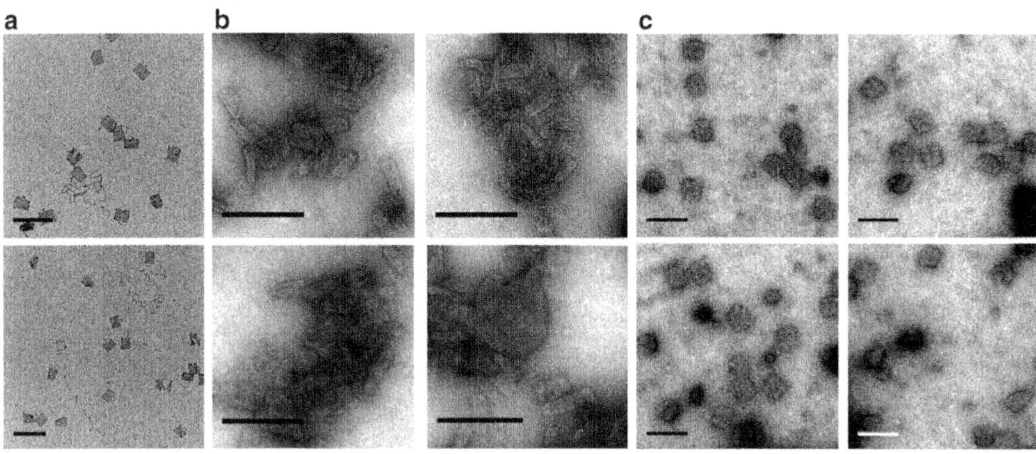

Fig. 1 TEM micrographs from samples x = 0 (**a**), x = 0.08 (**b**), and x = 0.64 (**c**) revealing the interaction between DNA origami and CPs. The scale bar in each figure is 200 nm. Reprinted with permission from Nano Letters 2014, 14, 2196–2200. Copyright 2014 American Chemical Society

3. Stain the samples negatively by applying 3 µl of stain (0.5% (w/v) uranyl acetate in Milli-Q water) onto the grid and remove immediately the excess stain with a piece of filter paper. Repeat the staining procedure once and dry the sample for at least 5 min before imaging (Fig. 1).

3.5.2 Gel Electrophoretic Mobility Shift Assay (EMSA)

1. Sample preparation: Mix 14 µl of the DNA origami-CP complex solution and 1 µl of 6× loading dye. Dilute 1.5 µl of scaffold strand to 14 µl and mix similarly with the loading dye.
2. Load the wells with 15 µl of sample solution.
3. Run the gels with a constant voltage of 90 V for 45 min and image under UV light (Fig. 2).

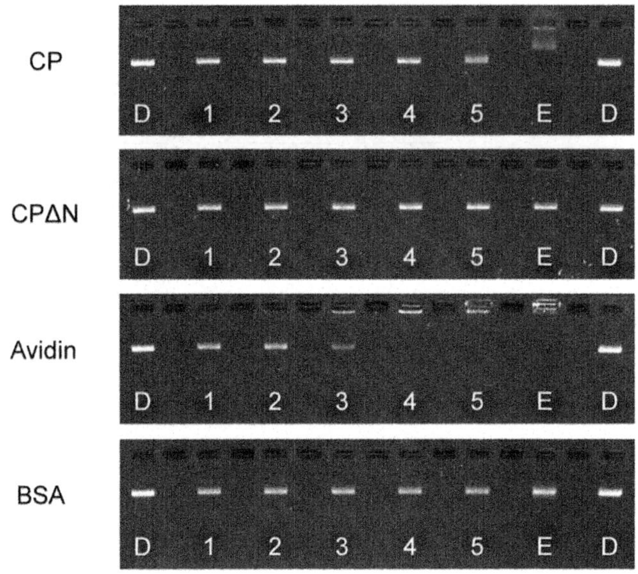

Fig. 2 EMSA gels of four different protein samples mixed with constant amounts of spin-filtered DNA origami structures in 150 mM NaCl solution. Protein concentration gradually increases from lane 1 to lane E ($\boldsymbol{x} = 0$, $\boldsymbol{x} = 0.016$, $\boldsymbol{x} = 0.032$, $\boldsymbol{x} = 0.08$, $\boldsymbol{x} = 0.32$, $\boldsymbol{x} = 3.2$). D denotes the scaffold strand that is used as a reference. Figure shows that CP binds on DNA origami and hinders the electrophoretic mobility of complexes when the ratio is high ($\boldsymbol{x} = 0.32$–3.2) also in 150 mM NaCl solution unlike CP without N-terminus (CPΔN). In the case of avidin, which has a high positive net surface charge, electrophoretic mobility of DNA origami is decreased already at low protein concentrations ($\boldsymbol{x} = 0.032 - 0.08$). In contrast, similarly varied concentrations of bovine serum albumin (BSA) showed no effect on the electrophoretic mobility of origami. Reprinted with permission from Nano Letters 2014, 14, 2196–2200. Copyright 2014 American Chemical Society

4 Notes

1. In this protocol, the formation of the DNA origami–CP complexes is demonstrated using a 2D rectangular DNA origami. However, the coating method can be generalized and equally used for 3D origami shapes. 3D structures might require a longer thermal ramp (e.g., lowering the temperature from 65 °C to 40 °C at a rate of 1 °C per 3 h). In addition, 3D structures usually need more magnesium (~20 mM) than 2D structures (12.5 mM) for the successful folding.
2. The side-strands of the original DNA origami design [4] (strands at the edges of the design) are omitted in order to avoid undesired aggregation of DNA origami, i.e., to prevent their stacking to each other via blunt ends. It is highly recommended taking this stacking interaction into account if different DNA origami designs are used.

3. T4 polynucleotide kinase is used for adding phosphate to the 5′ end of staple strands for the following optional ligation procedure. Ligation might improve the thermal stability of DNA structures [38].
4. To ensure the quality of the DNA origami folding and purification, gel electrophoresis can be used (similarly to as described in Subheadings 2.5.2 and 3.5.2). M13mp18 scaffold strand can be used as a reference for the running speed of the DNA origami sample.
5. It is useful to start the procedure by fabricating the DNA origami first (decent long-term storing properties) and prepare CP solution only after that. This is fairly important, since the CPs start to degrade within 1 week after the preparation (the N-terminus (first ~30 amino acids) of CP is known to degrade within weeks, even if the solution is stored at 4 °C) [39].
6. At a ratio of $x = 0.08$, the CPs start to bend the 2D DNA origami into tube-like conformations. When the ratio is increased to $x = 0.64$, round complexes are observed instead of tube-shaped structures. The tube formation and folding of the rectangles could be attributed to a high flexibility and significantly twisted natural shape of a used DNA origami (twisting occurs when the origami is designed using a square lattice packing [40, 41]). Therefore, if more rigid 3D origami structures are used in the complex formation, twisting and bending might not appear as prominent. The observed phenomena are likely caused by an electrostatic effect: the (positive) charge residues of CPs could effectively reduce repulsion between the DNA helices and further facilitate the complete folding of rectangles (similar folding can be observed when rectangular origamis are anchored onto substrates under localized electric fields [12, 42, 43]).

Acknowledgment

Financial support from the Academy of Finland (grants 263504, 267497, 273645, and 286845), Jane and Aatos Erkko Foundation, Emil Aaltonen Foundation, and Biocentrum Helsinki is gratefully acknowledged. This work was carried out under the Academy of Finland Centers of Excellence Programme (2014–2019) and made use of the Aalto Nanomicroscopy Centre (Aalto NMC) premises.

References

1. Seeman NC (1982) Nucleic acid junctions and lattices. J Theor Biol 99:237–247
2. Seeman NC (2010) Nanomaterials based on DNA. Annu Rev Biochem 79:65–87
3. Linko V, Dietz H (2013) The enabled state of DNA nanotechology. Curr Opin Biotechnol 24:555–561
4. Rothemund PWK (2006) Folding DNA to create nanoscale shapes and patterns. Nature 440:297–302
5. Douglas SM, Dietz H, Liedl T, Högberg B, Graf F, Shih WM (2009) Self-assembly of DNA into nanoscale three-dimensional shapes. Nature 459:414–418
6. Linko V, Kostiainen MA (2016) Automated design of DNA origami. Nat Biotechnol 34:826–827
7. Fu J, Liu M, Liu Y, Woodbury NW, Yan H (2012) Interenzyme substrate diffusion for an enzyme cascade organized on spatially addressable DNA nanostructures. J Am Chem Soc 134:5516–5519
8. Kuzyk A, Schreiber R, Fan Z, Pardatscher G, Roller E-M, Högele A, Simmel FC, Govorov AO, Liedl T (2012) DNA-based self-assembly of chiral plasmonic nanostructures with tailored optical response. Nature 483:311–314
9. Maune HT, Han S-p, Barish RD, Bockrath M, Goddard WA III, Rothemund PWK, Winfree E (2010) Self-assembly of carbon nanotubes into two-dimensional geometries using DNA origami templates. Nature Nanotech 5:61–66
10. Linko V, Eerikäinen M, Kostiainen MA (2015) A modular DNA origami-based enzyme cascade nanoreactor. Chem Commun 51:5351–5354
11. Shen B, Linko V, Tapio K, Kostiainen MA, Toppari JJ (2015) Custom-shaped metal nanostructures based on DNA origami silhouettes. Nanoscale 7:11267–11272
12. Shen B, Linko V, Dietz H, Toppari JJ (2015) Dielectrophoretic trapping of multilayer DNA origami nanostructures and DNA origami-induced local destruction of silicon dioxide. Electrophoresis 36:255–262
13. Linko V, Shen B, Tapio K, Toppari JJ, Kostiainen MA, Tuukkanen S (2015) One-step large-scale deposition of salt-free DNA origami nanostructures. Sci Rep 5:15634
14. Chen Y-J, Groves B, Muscat RA, Seelig G (2015) DNA nanotechnology from the test tube to the cell. Nature Nanotech 10:748–760
15. Linko V, Ora A, Kostiainen MA (2015) DNA nanostructures as smart drug-delivery vehicles and molecular devices. Trends Biotechnol 33:586–594
16. Ko S-H, Liu H, Chen Y, Mao C (2008) DNA nanotubes as combinatorial vehicles for cellular delivery. Biomacromolecules 9:3039–3043
17. Schüller VJ, Heidegger S, Sandholzer N, Nickels PC, Suhartha NA, Endres S, Bourquin C, Liedl T (2011) Cellular immunostimulation by CpG-sequence-coated DNA origami structures. ACS Nano 5:9696–9702
18. Zhao Y-X, Shaw A, Zeng X, Benson E, Nyström AM, Högberg B (2012) DNA origami delivery system for cancer therapy with tunable release properties. ACS Nano 6:8684–8691
19. Zhang Q, Jiang Q, Li N, Dai L, Liu Q, Song L, Wang J, Li Y, Tian J, Ding B, Du Y (2014) DNA origami as an in vivo drug delivery vehicle for cancer therapy. ACS Nano 8:6633–6643
20. Ora A, Järvihaavisto E, Zhang H, Auvinen H, Santos HA, Kostiainen MA, Linko V (2016) Cellular delivery of enzyme-loaded DNA origami. Chem Commun 52:14161–14164
21. Okholm AH, Nielsen JS, Vinther M, Sørensen RS, Schaffert D, Kjems J (2014) Quantification of cellular uptake of DNA nanostructures by qPCR. Methods 67:193–197
22. Brglez J, Nikolov P, Angelin A, Niemeyer CM (2015) Designed intercalators for modification of DNA origami surface properties. Chem Eur J 21:9440–9446
23. Kiviaho JK, Linko V, Ora A, Tianen T, Järvihaavisto E, Mikkilä J, Tenhu H, Nonappa, Kostiainen MA (2016) Cationic polymers for DNA origami coating – examining their binding efficiency and tuning the enzymatic reaction rates. Nanoscale 8:11674–11680
24. Perrault SD, Shih WM (2014) Virus-inspired membrane encapsulation of DNA nanostructures to achieve in vivo stability. ACS Nano 8:5132–5140
25. Ma Y, Nolte RJM, Cornelissen JJLM (2012) Virus-based nanocarriers for drug delivery. Adv Drug Deliv Rev 64:811–825
26. Yildiz I, Shukla S, Steinmetz NF (2011) Applications of viral nanoparticles in medicine. Curr Opin Biotechnol 22:901–908
27. Mukherjee S, Pfeifer CM, Johnson JM, Liu J, Zlotnick A (2006) Redirecting the coat protein of a spherical virus to assemble into tubular nanostructures. J Am Chem Soc 128:2538–2539
28. Kwak M, Minten IJ, Anaya D-M, Musser AJ, Brasch M, Nolte RJM, Müllen K, Cornelissen JJLM, Herrmann A (2010) Virus-like particles

templated by DNA micelles: a general method for loading virus nanocarriers. J Am Chem Soc 132:7834–7835

29. Geiger FC, Eber FJ, Eiben S, Mueller A, Jeske H, Spatz JP, Wege C (2013) TMV nanorods with programmed longitudinal domains of differently addressable coat proteins. Nanoscale 5:3808–3816
30. Douglas T, Young M (1998) Host-guest encapsulation of materials by assembled virus protein cages. Nature 393:152–155
31. Young M, Willits D, Uchida M, Douglas T (2008) Plant viruses as biotemplates for materials and their use in nanotechnology. Annu Rev Phytopathol 46:361–384
32. Kostiainen MA, Hiekkataipale P, Laiho A, Lemieux V, Seitsonen J, Ruokolainen J, Ceci P (2013) Electrostatic assembly of binary nanoparticle superlattices using protein cages. Nature Nanotech 8:52–56
33. Mikkilä J, Rosilo H, Nummelin S, Seitsonen J, Ruokolainen J, Kostiainen MA (2013) Janus-dendrimer-mediated formation of crystalline virus assemblies. ACS Macro Lett 2:720–724
34. Liljeström V, Mikkilä J, Kostiainen MA (2014) Self-assembly and modular functionalization of three-dimensional crystals from oppositely charged proteins. Nat Commun 5:4445
35. Mikkilä J, Eskelinen A-P, Niemelä EH, Linko V, Frilander MJ, Törmä P, Kostiainen MA (2014) Virus-encapsulated DNA origami nanostructures for cellular delivery. Nano Lett 14:2196–2200
36. Bruinsma RF, Gelbart WM, Reguera D, Rudnick J, Zandi R (2003) Viral self-assembly as a thermodynamic process. Phys Rev Lett 90:248101
37. Hung AM, Micheel CM, Bozano LD, Osterbur LW, Wallraff GM, Cha JN (2010) Large-area spatially ordered arrays of gold nanoparticles directed by lithographically confined DNA origami. Nature Nanotech 5:121–126
38. O'Neill P, Rothemund PWK, Kumar A, Fygenson DK (2006) Sturdier nanotubes via ligation. Nano Lett 6:1379–1383
39. Tang J, Johnson JM, Dryden KA, Young MJ, Zlotnick A, Johnson JE (2006) The role of subunit hinges and molecular "switches" in the control of viral capsid polymorphism. J Struct Biol 154:59–67
40. Ke Y, Douglas SM, Liu M, Sharma J, Cheng A, Leung A, Liu Y, Shih WM, Yan H (2009) Multilayer DNA origami packed on a square lattice. J Am Chem Soc 131:15903–15908
41. Castro CE, Kilchherr F, Kim D-N, Shiao EL, Wauer T, Wortmann P, Bathe M, Dietz H (2011) A primer to scaffolded DNA origami. Nat Methods 8:221–229
42. Kuzyk A, Yurke B, Toppari JJ, Linko V, Törmä P (2008) Dielectrophoretic trapping of DNA origami. Small 4:447–450
43. Linko V, Paasonen S-T, Kuzyk A, Törmä P, Toppari JJ (2009) Characterization of the conductance mechanisms of DNA origami by AC impedance spectroscopy. Small 5:2382–2386

Chapter 19

In Vitro Assembly of Virus-Derived Designer Shells Around Inorganic Nanoparticles

Stella E. Vieweger, Irina B. Tsvetkova, and Bogdan G. Dragnea

Abstract

Nanoparticle-templated assembly of virus shells provides a promising approach to the production of hybrid nanomaterials and a potential avenue toward new mechanistic insights in virus phenomena originating in many-body effects, which cannot be understood from examining the properties of molecular subunits alone. This approach complements the successful molecular biology perspective traditionally used in virology, and promises a deeper understanding of viruses and virus-like particles through an expanded methodological toolbox. Here we present protocols for forming a virus coat protein shell around functionalized inorganic nanoparticles.

Key words Virus self-assembly, Coat protein, Hybrid nanoparticles, Virus-like particle, Inorganic cargo

1 Introduction

Viral nanoparticles (VNPs) are derived from virus capsids, the protein shells that normally encase the virus genetic material. VNPs are usually generated by in vitro encapsulation of a foreign cargo. The ability to direct the assembly of virus-based capsids around nonnative cargo has immense potential for applications in medicine [1–3], materials [4–6], and antiviral drug development, specifically for targeting infectious virus particle formation [7–9]. Other possible applications include energy generation [10, 11], catalysis [12], immunotherapy, targeted therapeutics [13–18], and imaging [19–23].

Virus capsids also make unique biomaterial building blocks [24, 25]. Features such as homogenous size distribution, precise subunit copy stoichiometry, structural symmetry, the ability to quickly and efficiently self-assemble, and the relative ease with which both the inner and outer surfaces of capsids can be modified, have fueled their increased use in biomaterial development.

Christina Wege and George P. Lomonossoff (eds.), *Virus-Derived Nanoparticles for Advanced Technologies: Methods and Protocols*, Methods in Molecular Biology, vol. 1776, https://doi.org/10.1007/978-1-4939-7808-3_19,
© Springer Science+Business Media, LLC, part of Springer Nature 2018

The availability of high resolution structural information, access to recombinant and plant-based expression systems for ease of scalability, and the ability to produce virus-like particles (VLPs) lacking the infectious genetic material, make VLPs viable, safe, and well-characterized systems with significant technological potential.

Inorganic nanoparticles (INPs) are synthesized from metal, metal oxide, semiconductor, block copolymer, or some combination of the components mentioned above, to yield novel materials and structures with unique optical, fluorescent, thermal, and conductive characteristics [26–28]. As a result, INPs make highly sensitive optical and spectroscopic probes. Gold nanoparticles (GNPs) exhibit a large optical absorption cross section that renders them capable of photothermal-based sub-10 nm optical detection [29, 30], as well as localized laser ablation of tumors [31]. Quantum dots (QDs) have found a niche as fluorescent imaging tags [32], while magnetic nanoparticles (MNPs) are used as cell sorters and magnetic resonance imaging (MRI) contrast enhancement agents [33].

As nanomaterial platforms advance toward more efficient, multifunctional modalities, combinations of materials are being developed with novel, unique properties. For instance, Tee et al. have reported a self-healing electronic skin produced by mixing nickel nanoparticles with a polymer material [26]. A transistor engineered from a gold nanoparticle–organic molecule blend imitates how neural synapses work [27]. The size and targeting selectivity, immunogenicity, and cellular membrane transport capabilities of VLP/virus capsids combined with highly tunable INP probes yield hybrid platforms with immunomodulatory [34] as well as optical, magnetic, or fluorescence capabilities [20, 35–39].

While it is possible to grow INPs within the lumen of viral protein cages through reactions reminiscent of biomineralization [40], it is often the case that synthetic control of important INP characteristics such as crystallinity, size polydispersity, and morphology are achieved under conditions that are incompatible with biological matter. An alternative is to coassemble INPs and capsid proteins into VNPs. Methods commonly used to encapsulate inorganic materials within plant virus capsids rely either on diffusion across pores and subsequent reactions inside the intact capsid [24, 41, 42], or on self-assembly of the capsids from dissociated capsid subunits around INPs synthesized prior to assembly [36–38, 43]. The former method takes advantage of a favorable electrostatic potential across capsid walls and works well for molecules about the size of capsid pores such as fluorescent labels and small molecule drugs. The latter method, which is our focus, allows for the encapsulation of larger moieties while maintaining capsid integrity.

The work described in this chapter uses the capsid of brome mosaic virus (BMV), an extensively studied plant virus that was one of the first to be self-assembled into native capsids in vitro [44, 45]. BMV [46, 47] is a 28 nm, $T = 3$ icosahedral virus containing

180 identical copies of capsid protein and four single RNA strands, packed into three infectious particles (two of the four are copackaged). The capsid protein is a 189 amino acid polypeptide folded into a β-barrel that drives protein–protein interactions to facilitate capsid assembly and stability. Virus assembly is believed to be initiated by nonspecific electrostatic interactions between the positively charged, flexible N-terminal domain of the coat protein and the negatively charged nucleic acid core. However, it was recently found that at least in the final state after equilibration, RNA protein interactions do exhibit some specificity [48].

Due to the dominance of nonspecific interactions between cargo and protein, VNPs encapsulating INPs can be generated with high efficiency (Fig. 1) [37]. In the following, we provide protocols for a few selected cargo particles which have been encapsulated in BMV coat protein cages. The broad features of coassembly are valid for a variety of icosahedral viruses, especially those that share the canonical coat protein motif [49], but have been found to be operational for other virus proteins, too [50]. Knowledge of the factors that trigger capsid assembly in other viruses should guide specific template modification and reaction optimizations toward encapsulating foreign cargo.

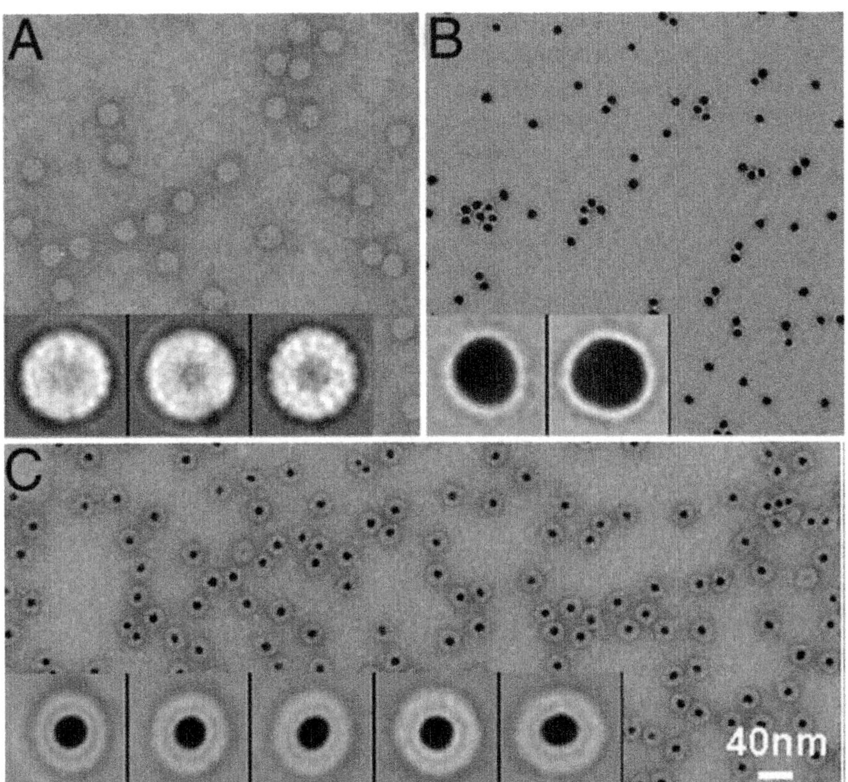

Fig. 1 Typical TEM images of R3BMV (**a**), PEG-Au (**b**), and VLP with gold nanoparticle core (**c**) as used for single-particle image reconstruction (inset: particle classes). Adapted from ref. 37 with permission

2 Materials

For preparation of all solutions use analytical grade reagents (unless specified) and water with 18 MΩ.cm resistivity at room temperature (25 °C). Filter all buffer solutions through 0.2 μm filter and store at 4 °C. Reagents can be stored at room temperature, unless otherwise indicated.

2.1 Gold Nanoparticle Synthesis Reagents

1. Tetrachloroauric acid ($HAuCl_4$).
2. Trisodium citrate ($Na_3C_6H_5O_7$).
3. Tannic acid ($C_{76}H_{52}O_{46}$).
4. 0.0025 M potassium carbonate (K_2CO_3), freshly prepared.
5. TEG ligand: (11-mercaptoundecyl)tetra(ethylene glycol)ethoxyacetic acid ($HS\text{-}(CH_2)_{11}\text{-}(OCH_2CH_2)_4\text{-}OCH_2\text{-}COOH$) (ProChimia Surfaces, Poland).

2.2 CdSe/ZnS Quantum Dots Synthesis Reagents

1. Tri-*n*-octylphosphine oxide (TOPO).
2. Argon supply (~1 atm).
3. Vacuum oven.
4. Cd/Se/TOP stock solution (Selenium shot (Se), tri-*n*-octylphosphine (TOP), dimethylcadmium ($(CH_3)_2Cd$)) (see protocol for details).
5. Zn/S/TOP stock solution (Bis(trimethylsilyl)sulfide ($((CH_3)_3Si)_2S$), tri-*n*-octylphosphine (TOP), 1 M dimethylzinc in heptane ($(CH_3)_2Zn$)).
6. Anhydrous methanol (CH_3OH).
7. Anhydrous chloroform ($CHCl_3$).
8. Thiol poly(ethylene glycol) acid (HS-PEG-COOH, MW 2000, Nanocs Inc., USA).
9. TBE buffer: 0.4 M Tris base, 0.45 M boric acid, 0.01 M EDTA.
10. Equipment for agarose gel electrophoresis.

2.3 Iron Oxide Magnetic Nanoparticles Synthesis Reagents

1. Iron (III) chloride hexahydrate ($FeCl_3.6H_2O$).
2. Sodium oleate ($C_{18}H_{33}NaO_2$).
3. Ethanol (C_2H_5OH).
4. Hexane (C_6H_{14}).
5. Argon supply.
6. Separation funnel.
7. Rotary evaporator.

8. Acetone (C_3H_6O).
9. Vacuum oven.
10. Oleic acid sodium salt ($C_{18}H_{34}O_2Na$).
11. Docosane ($C_{22}H_{46}$).
12. Chloroform ($CHCl_3$).
13. 1,2-distearoyl-*sn*-glycero-3-phosphoethanolamine-*N*-[carboxy-(polyethylene glycol) ammonium salt (HOOC-PEG-PL; 2000 Da, Avanti Lipids, USA).
14. Sonication bath (Branson 1510 or similar).
15. 0.2 μm syringe filter.

2.4 BMV Capsid Dissociation, Protein Purification, and Reassembly Buffers

1. Leaves from *Nicotiana benthamiana* systemically infected with BMV via agroinoculation using protocol previously described [59].
2. Cheesecloth.
3. 10% (w/v) sucrose.
4. 38.5% (w/v) cesium chloride (CsCl) solution.
5. For dialysis: 12–14 kDa MWCO RC membrane.
6. Centrifugal filter devices 10 kDa MWCO.
7. Disassembly buffer: 0.5 M $CaCl_2$, 0.005 M Tris–HCl, pH 7.4.
8. Protein storage buffer: 1 M KCl, 0.01 M Tris–HCl, 0.005 M $MgCl_2$, pH 7.4.
9. Reassembly buffer: 0.05 M Tris–HCl, 0.05 M NaCl, 0.01 M KCl, 0.005 M $MgCl_2$, pH 7.4.
10. Virus buffer: 0.05 M sodium acetate ($C_2H_3O_2Na$), 0.008 M magnesium acetate.

3 Methods

3.1 Synthesis, Functionalization, and Purification of Gold Nanoparticles

The methods used to synthesize citrate-stabilized gold nanoparticles depend on the desired particle size. The Slot and Geuze method [51] builds on Mühlpfordt's tannic acid reduction method [52] to yield sizes in the 5–15 nm range. In this method, the tannic acid reduces the gold and the citrate serves primarily as a stabilizing or capping agent. Frens's modification [53] of Turkevich's protocol [54, 55] yields particle sizes larger than 15 nm using citrate as both a reducing and stabilizing agent. This protocol describes the Slot and Geuze method for 12 nm GNPs. This size is suitable as core for VLPs with native $T = 3$ BMV capsids.

Table 1
Volumes of 1% (w/v) tannic acid and the resulting corresponding sizes of gold nanoparticles

Tannic acid volume (mL)	Gold particle size (nm)
0.022	13.1 ± 0.9
0.470	10.4 ± 0.8
0.900	6.1 ± 0.5

3.1.1 Slot and Geuze Method of Synthesis

1. Ensure that all glassware are cleaned (*see* **Note 1**) and rinsed in copious amounts of water.
2. In first beaker, add 1 mL of freshly prepared 1% w/v tetrachloroauric acid ($HAuCl_4$) to 79 mL of water and bring to 60 °C in a water bath. Label this solution A.
3. While A is heating, in second beaker, mix 4 mL of freshly prepared 1% w/v trisodium citrate with equal volumes of 1% w/v tannic acid and freshly prepared 0.0025 M K_2CO_3 (45 μL each for 12 nm particles; *see* Table 1), and bring the volume to 20 mL with water. Label this solution B and heat it to 60 °C.
4. Once both solutions (A + B) have equilibrated at 60 °C, add a clean stir bar to A and begin vigorous stirring (about 800 rpm).
5. Quickly add contents of B to A and bring the total volume to boil while stirring. Changes in solution color from light grey to orange-red corresponding to particle growth will be observed.
6. Allow the solution to boil for an additional 10 min after the color stabilizes.
7. Allow the solution to cool to room temperature.
8. Store at room temperature or at 4 °C. Particles should be stable for several months.

3.1.2 Functionalization and Characterization of GNPs

Particle functionalization occurs via the displacement method—covalent thiol bonds readily displace adsorbed citrate molecules on the particle surfaces resulting in more stable colloids.

1. To 100 mL of synthesized nanoparticles, add 10 μL of TEG ligand. Allow the mixture to stir at 400–600 rpm for 6–12 h at room temperature to ensure optimum surface functionalization (*see* **Note 2**).
2. Remove excess ligand by three repeated centrifugation and resuspension of the pellet in water. 12 nm particles are centrifuged at 100,000 × *g* for 45 min. Centrifugation parameters should be optimized for the particle size.
3. Prepare a diluted solution (very light pink in color) of the functionalized particles and determine the size and polydisper-

sity using dynamic light scattering (DLS, use Malvern Zetasizer Nano ZS or similar).

4. The charge of the particles can be determined by zeta potential measurements (Malvern Zetasizer Nano ZS or similar). Typical zeta potential value for 12 nm GNP coated with TEG ligand is about −50 mV.
5. Determine the nanoparticle concentration by measuring absorbance at 400 nm and using an extinction coefficient of 12.7 cm^2/mg for the functionalized particles [56].
6. Store functionalized particles in water at room temperature or at 4 °C.

3.2 CdSe/ZnS QD Synthesis, Functionalization, and Purification

Quantum dot particles are synthesized using a method previously reported by Hines and Guyot-Sionnest [57]. Structurally, the resulting nanoparticles comprise a TOPO-capped ZnS shell encasing a CdSe core. A higher bandgap shell fused to a lower bandgap core has been demonstrated to improve luminescence and stability of the nanoprobe.

3.2.1 CdSe/ZnS Synthesis

1. *See* **Note 3** before starting protocol.
2. Use glovebox/inert atmosphere for this step. To prepare Cd/Se stock solution, dissolve 0.2 g (0.0025 mol) Se in 4.5 mL TOP, add 0.25 mL (0.0035 mol) $(CH_3)_2Cd$, and dilute with an additional 19.5 mL TOP.
3. Use glovebox/inert atmosphere for this step. The Zn and S stock solution are prepared with 0.52 mL of Bis-trimethylsilylsulfide ($(TMS)_2S$) in 4.5 mL of TOP followed by addition of dimethyl zinc (Me_2Zn) solution.
4. Heat 12.5 g TOPO under vacuum to 200 °C in a three-neck round bottom flask; dry and degas for 20 min.
5. Purge flask with argon and bring the temperature to 350 °C under argon (~1 atm).
6. Add 0.7 mL of the Cd/Se/TOP stock to the degassed and dried TOPO and remove from heat.
7. Let reaction temperature cool down to 310 °C at which point an aliquot of CdSe nanoparticles may be removed for characterization purposes.
8. At 300 °C, add the Zn/S/TOP stock solution in five 0.55 mL aliquots at 20 s intervals. The final molar ratio should be 1:4 Cd/Se–Zn/S.
9. Cool to 100 °C and stir at this temperature for 1 h.
10. Purify the capped quantum dots by precipitating with anhydrous methanol, centrifugation at 3000 rpm for 30 min and three times wash with anhydrous methanol to remove any excess TOPO.

11. Redisperse and store purified particles in 10 mL anhydrous chloroform. This method produces particles approximately 4 nm in diameter.

3.2.2 CdSe–ZnS QD Functionalization and Characterization

Of the various ligands tested for encapsulation, QDs surface-modified with HS-PEG-COOH yielded the most stable and uniform QD-encapsulated virus-like particles.

1. To 1 mg TOPO-capped QDs in 1 mL chloroform, add 10 mg HS-PEG-COOH and allow to mix overnight between 60 and 70 °C (water bath).
2. Remove any remaining solvent in a vacuum oven at room temperature.
3. Resuspend the residue in 1 mL of water. The reddish residue will turn clear orange in aqueous solution.
4. Prepare a 10 times dilution of this solution and use it to determine the hydrodynamic particle size by DLS (*see* **Note 4**).
5. Confirm ligand attachment by gel electrophoresis on a 2% (w/v) agarose gel in TBE buffer, run at 7.5 V/cm. Good ligand replacement should yield a well migrating band toward the positive electrode.
6. Characterize the fluorescence of the particles by scanning the emission spectrum from 500–700 nm after excitation at 470 nm.

3.3 Iron Oxide MNP Synthesis, Functionalization, and Characterization

3.3.1 Synthesis of Iron Oleate Complex [58]

1. Dissolve 3.24 g (12 mmol) $FeCl_3.6H_2O$ in 12 mL water.
2. Filter the solution with syringe filter and add 10.95 g (40 mmol) sodium oleate, 24 mL ethanol, 6 mL water, and 32 mL hexane.
3. Heat solution to 70 °C. Stir at this temperature for 4 h under argon.
4. After 4 h, separate top layer (reddish-brown) from reaction mixture. This layer contains the iron oleate complex in hexane.
5. Wash three times with 9 mL water in a separation funnel.
6. Evaporate the hexane solvent in the mixture with a rotary evaporator.
7. To remove free oleic acid (formed by hydrolysis of sodium oleate) wash this product twice with ethanol and repeat the wash twice with acetone. Use centrifugation to separate precipitate between washes.
8. Dry the iron oleate complex in a vacuum oven at 70 °C for 24 h. The final purified product will be a waxy solid which can be stored at room temperature for a few weeks.

3.3.2 Synthesis of Iron Oxide MNPs

The method used relies on the thermal degradation of iron oleate to generate 20 nm particles.

1. Pretreat the iron oleate complex by placing in a vacuum oven at 70 °C for 24 h (if not used fresh from previous step).
2. Mix 2.78 g (3 mmol) iron oleate complex, 0.96 mL (3 mmol) oleic acid, and 10 mL docosane (solid at room temp) in a three-neck round bottom flask.
3. Heat mixture to 60 °C to liquefy the hydrocarbon solvent and stir vigorously to dissolve the reagents.
4. While still stirring, increase the temperature steadily to 370 °C at a rate of 3.3 °C/min under reflux. The solution will change color from reddish-brown to brownish-black.
5. Maintain the temperature at 370 °C for 3 min, then remove heater and allow the reaction mixture to cool down to 50 °C.
6. Add 10 mL hexane and 40 mL acetone to the flask. The nanoparticles will precipitate.
7. Purify the nanoparticles by centrifugation (90,000 × *g*, 2 h) followed by three washes with hexane–acetone solvent mixture. Repeat this step one more time.
8. Store the iron oxide particles in chloroform. They should be stable over several months.

3.3.3 Functionalization and Characterization of Iron Oxide MNPs

1. Prepare solution of 20 nm oleic acid-capped particles at concentration 1.7 mg in 1 mL chloroform.
2. To this solution, add 1.4 mg HOOC-PEG-PL and sonicate in bath until completely dissolved (~5 min).
3. Evaporate the chloroform solvent by warming the solution in a water bath thermally equilibrated at 80 °C.
4. Add 1 mL of water to the residue and vigorously stir to resuspend. A uniform and clear dark brown solution should form.
5. Centrifuge three times at 90,000 × *g* for 2 h to remove excess, unreacted HOOC-PEG-PL, resolubilizing in water each time.
6. Characterize functionalized iron oxide particle sizes by preparing a dilution series, sonicating for 10–20 min, filtering with 0.2 μm syringe filter, and measuring DLS intensity distributions.
7. Determine particle concentration by measuring the absorption at 300 nm using an extinction coefficient of 14.25 l/(g cm) [20].

3.4 Purification of Capsid Proteins from Native BMV

BMV is inoculated into *Nicotiana benthamiana* plants using an *Agrobacterium*-mediated protocol [59–61]. Agrobacteria cells are first transformed with DNA plasmid constructs encoding the viral RNA (vRNA), and then cultured in the presence of kanamycin,

MES buffer, and acetosyringone [62]. The culture is then diluted to the OD_{600} = 0.5 and inoculated into the plant leaves by means of syringe agro-infiltration. Systemically infected plant leaves are harvested a week after inoculation and stored at −80 °C.

3.4.1 Purification of Native BMV

All materials and buffers must be kept cold at 4 °C or on ice unless otherwise indicated.

1. Blend a known mass of leaves (50 g in typical preparation) with three volumes (v/w) of virus buffer.
2. Filter the homogenate through 6–8 layers of cheesecloth into a beaker.
3. Centrifuge the filtrate for 25 min at 5000 × *g* and 4 °C to remove any remaining leaf debris. Collect the supernatant.
4. Add 25 mL of the previously collected supernatant to 30 mL thin wall centrifuge tubes.
5. Carefully add 5 mL of 10% sucrose solution to the bottom of each tube and centrifuge at 90,000 × *g* and 4 °C for 3 h to pellet the virus.
6. Resuspend the pellets in 300–400 μL of 38.5% (w/v) CsCl overnight by gentle shaking in a fridge or cold room at 4 °C.
7. Centrifuge this solution at 7500 × *g* and 4 °C for 10 min and place the supernatant in a thermal sealing centrifuge tube.
8. Centrifuge at 145,000 × *g* and 4 °C for 24 h.
9. Extract the visible band (about the bottom third of the tube) of concentrated BMV, dialyze against several changes of virus buffer (*see* **Note 5** for characterization).
10. Aliquot and store at −80 °C.

3.4.2 Purification of Capsid Protein Dimers

All materials and buffers must be kept cold or on ice, unless otherwise indicated. Dialysis must be carried out in a fridge or cold room at 4 °C for a minimum of 6 h.

1. Thaw an aliquot of BMV and dialyze ~0.5 mg of virus against disassembly buffer overnight.
2. Centrifuge at 17,000 × *g* for 30 min to remove vRNA (pellet). The supernatant contains dissociated capsid proteins.
3. Dialyze the supernatant against protein storage buffer overnight and determine the concentration from absorbance measurement at 280 nm and extinction coefficient of 0.82 cm^2/mg (Thermo Scientific NanoDrop or similar). An absorbance ratio (A_{260}/A_{280}) of ~0.6 indicates pure protein.
4. Confirm capsid protein integrity via MALDI-TOF mass spectrometry (~20,300 Da) or if not available, SDS-PAGE.

3.5 In Vitro Reassembly of BMV-Like Shells Around Nanoparticle Cores

Functionalized nanoparticles are typically mixed with purified capsid proteins in the protein storage buffer, followed by a two-step modification of a previously published reassembly protocol [63]. First, the protein–nanoparticle mixture in the neutral pH and high ionic strength protein storage buffer is dialyzed against a low ionic strength buffer also at neutral pH (reassembly buffer). This step promotes association of the positively charged N-terminal protein domains with the negatively charged nanoparticles.

The second step involves dialyzing the resulting nanoparticle–protein complex against a low ionic strength and low pH buffer (virus buffer). The complex condenses into structured capsids around the nanoparticle core and is further stabilized by divalent magnesium ions present in the virus buffer. This route of incorporation results in the highest yield of VLPs.

3.5.1 VLP Assembly Around Gold Nanoparticle Cores

1. Dialyze functionalized particles (prepared in Subheading 3.1) against protein storage buffer at 4 °C overnight.
2. Mix particles and proteins in a ratio of 0.75 equivalent of GNP to 180 equivalents protein monomers ($T = 3$ capsid).
3. Optimize assembly conditions by adjusting the final protein concentration to 0.35–0.5 mg/mL. Typical reaction volume is 100 μL.
4. Dialyze this GNP–BMV protein mixture overnight against reassembly buffer.
5. Follow this up by dialysis against virus buffer overnight.
6. The assembly yields 90–95% of fully assembled VLPs (*see* Fig. 1). If a purer sample is crucial, centrifuge the sample to remove unassembled proteins and nanoparticles over 1 mL 10% (w/v) sucrose cushion in virus buffer in a 1.5 mL conical centrifuge tube (20,000 × *g*, 30 min, 4 °C).
7. Dialyze against virus buffer and store at 4 °C.

3.5.2 VLP Assembly Around CdSe/ZnS QD Cores

1. Mix one equivalent of HS-PEG-COOH-capped QDs (prepared in Subheading 3.2) with 180 equivalents BMV protein monomers in protein storage buffer.
2. Adjust the final protein concentration to 0.5 mg/mL (the typical reaction volume is 100–200 μL) and dialyze against reassembly buffer overnight.
3. Dialyze against virus buffer overnight.
4. Purify VLPs by centrifuging 100 μL of the mixture on a 1 mL bed of 10% (w/v) sucrose solution in virus buffer at 50,000 × *g* and 4 °C for 30 min.
5. Resuspend the VLPs (pellet) in 100 μL virus buffer (*see* Fig. 2). Store sample at 4 °C

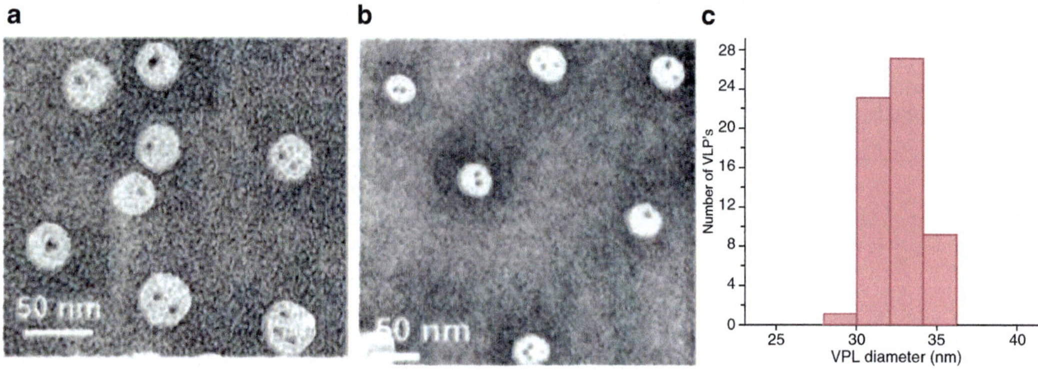

Fig. 2 TEM images of VLPs containing HS-PEG-COOH-coated QDs. The equivalent ratios of QD to CP are 2:1 (**a**) and 3:1 (**b**), respectively. (**c**) Size histogram of VLPs corresponding to (**b**). Adapted from ref. 38 with permission

3.5.3 VLP Assembly Around Iron Oxide Cores

The assembly protocol is modified to exclude dialysis against reassembly buffer because the HOOC-PEG-PL-MNPs are less stable in high ionic strength buffers.

1. Mix one equivalent of functionalized MNPs with 540 equivalents of BMV protein monomers in a total volume of 100 μL and final protein concentration of 0.5 mg/mL.
2. Dialyze against virus buffer at 4 °C for 24 h.
3. To a 1.5 mL centrifuge vial, add 100 μL of assembly mixture to 200 μL of 10% (w/v) sucrose cushion in virus buffer. Centrifuge at 9000 × *g* and 4 °C for 30 min to purify VLPs.
4. Resuspend the pellet in 100 μL virus buffer and store in virus buffer at 4 °C (*see* Fig. 3).

3.6 Characterization of Virus-Like Particles

VLPs are predominantly characterized by negative-stain transmission electron microscopy (TEM). Particle core and capsid uniformity, and the efficiency of encapsulation can readily be visualized and analyzed using digital imaging software. To ensure true representation of particle distributions, images are taken from different sections of a grid and no less than 200 particles are typically analyzed. Particle size measurements made this way are typically complemented with dynamic light scattering (DLS) measurements.

To visualize particle morphologies, EMAN software is used to generate 3D reconstructed models from negatively stained TEM images [64]. Images of these models are then rendered in Chimera software [65]. Structural accuracy is improved the greater the number of particles that are averaged. Typically, several thousand particles are sorted and averaged (*see* Fig. 4).

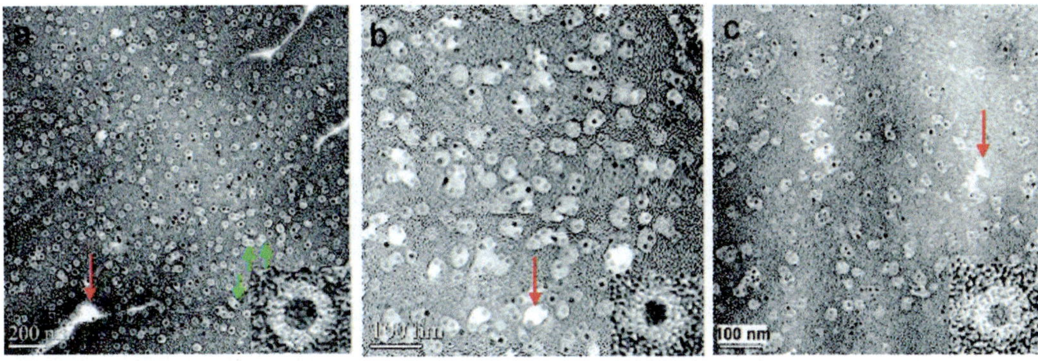

Fig. 3 TEM images of VLPs formed by self-assembling of BMV proteins around 20.1 (**a**), 10.6 (**b**), and 8.5 (**c**) nm spherical MNPs coated with HOOC-PEG-PL. In all images dark circular spots are MNPs. Light colored areas around MNPs are BMV shells. The HOOCPEG-PL shells are not visible. Irregular white spots indicated by red arrows are defects in staining. The green arrows in (**a**) indicate merged (peanut shaped) VLPs. Lower insets show individual VLPs at a higher magnification. Adapted from ref. 20 with permission

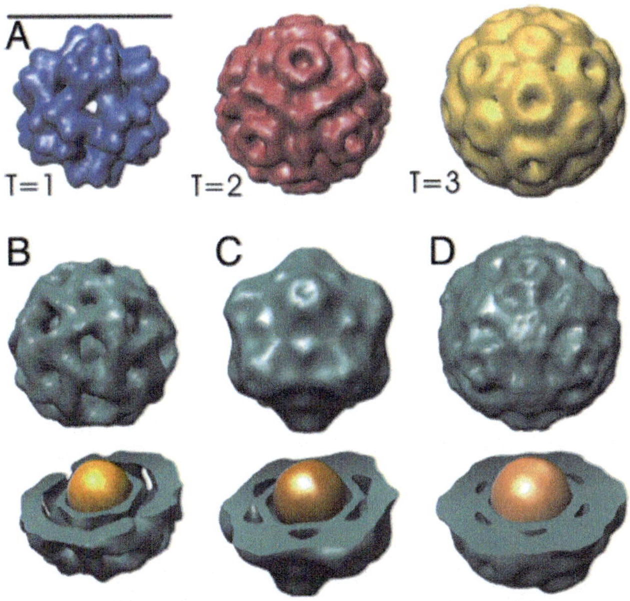

Fig. 4 Three-dimensional single particle reconstructions of VLP from negative stain EM data. (**a**) T = 1, pseudo-T = 2, and T = 3 models of BMV capsids. (**b**) VLP6 with 6 nm core GNPs is characterized by the absence of electron density at the threefold symmetry axes. Its structure and diameter suggest a T = 1 capsid. (**c**) The VLP9 structure, with a 9 nm diameter GNP core, is reminiscent of a pseudo T = 2. The presence of electron density at the threefold axes distinguishes it from the VLP6 structure. (**d**) The VLP12 shape resembles to the spherical shape of R3BMV. Concentric layering is a characteristic of all VLPs. Adapted from ref. 37 with permission

4 Notes

1. To prevent any contaminants from inhibiting or seeding particle growth, ensure that all materials for nanoparticle synthesis, including the stir bars, are clean. Aqua regia (three parts nitric acid to one part hydrochloric acid) is particularly effective for removing stains and deposits from previous syntheses. It is also a highly corrosive and oxidizing solution and appropriate care should be taken prior to handling. Alternatively, particles of discrete sizes with narrow size distributions can also be purchased commercially from companies such as Ted Pella, Inc. (Redding, CA), Nanopartz (Loveland, CO), nanoComposix (San Diego, CA), and Sigma-Aldrich (St. Louis, MO).
2. The amount of ligand added is based on the literature value of 0.152 nm^2 space occupied by a single bound thiol molecule on a spherical surface [66].
3. The reagents used for synthesizing these quantum dots are highly reactive and explosive in air. Conduct experiment in a glove box. Exercise caution; read the Safety Data Sheets and follow all safety recommendations.
4. In general, samples analyzed with DLS should be prepared as a dilution series in buffers of varying ionic strength to optimize particle stability. Aggregates can be redispersed by bath sonicating for 5–15 min. Persistent large aggregates can be removed by filtering the sample using 0.1–0.2 μm membranes that are compatible with sample solvents.
5. The concentration of BMV is calculated using the absorbance at 260 nm and extinction coefficient E (0.1%, 260 nm) of 5.2 cm^2/mg on a NanoDrop instrument. An absorbance ratio (A_{260}/A_{280}) greater than 1.5 is taken as an indication of a pure virus sample. DLS values average 28–30 nm diameter (Malvern Zetasizer).

References

1. Rao VR, Upadhyay AK, Kompella UB (2013) pH shift assembly of adenoviral serotype 5 capsid protein nanosystems for enhanced delivery of nanoparticles, proteins and nucleic acids. J Control Release 172(1):341–350
2. van Kan-Davelaar HE, van Hest JC, Cornelissen JJ, Koay MS (2014) Using viruses as nanomedicines. Br J Pharmacol 171(17):4001–4009
3. Yildiz I, Shukla S, Steinmetz NF (2011) Applications of viral nanoparticles in medicine. Curr Opin Biotechnol 22:901–908
4. Korkmaz N, Kim YJ, Nam CH (2013) Bacteriophages as templates for manufacturing supramolecular structures. Macromol Biosci 13(3):376–387
5. Culver JN et al (2015) Plant virus directed fabrication of nanoscale materials and devices. Virology 479–480:200–212
6. Flenniken ML et al (2009) A library of protein cage architectures as nanomaterials. Curr Top Microbiol Immunol 327:71–93
7. Prevelige P (1998) Inhibiting virus-capsid assembly by altering the polymerisation pathway. Trends Biotechnol 16(2):61–65
8. Katen SP et al (2010) Trapping of hepatitis B virus capsid assembly intermediates by phen-

ylpropenamide assembly accelerators. ACS Chem Biol 5:1125–1136

9. Zlotnick A, Mukhopadhyay S (2011) Virus assembly, allostery and antivirals. Trends Microbiol 19:14–23
10. Oh D, Qi J, Lu YC, Zhang Y, Shao-Horn Y, Belcher AM (2013) Biologically enhanced cathode design for improved capacity and cycle life for lithium-oxygen batteries. Nat Commun 4:2756
11. Neumann O, Urban AS, Day J, Lal S, Nordlander P, Halas NJ (2013) Solar vapor generation enabled by nanoparticles. ACS Nano 7(1):42–49
12. Glasgow J, Tullman-Ercek D (2014) Production and applications of engineered viral capsids. Appl Microbiol Biotechnol 98(13):5847–5858
13. Wu M, Wu M, Sherwin T, Brown WL, Stockley PG (2005) Delivery of antisense oligonucleotides to leukemia cells by RNA bacteriophage capsids. Nanomedicine 1(1):67–76
14. Steinmetz NF, Lin T, Lomonossoff GP, Johnson JE (2009) Structure-based engineering of an icosahedral virus for nanomedicine and nanotechnology. Curr Top Microbiol Immunol 327:23–58
15. Singh R, Kostarelos K (2009) Designer adenoviruses for nanomedicine and nanodiagnostics. Trends Biotechnol 27(4):220–229
16. Franzen S, Lommel SA (2009) Targeting cancer with 'smart bombs': equipping plant virus nanoparticles for a 'seek and destroy' mission. Nanomedicine (Lond) 4(5):575–588
17. Plummer EM, Manchester M (2011) Viral nanoparticles and virus-like particles: platforms for contemporary vaccine design. Wiley Interdiscip Rev Nanomed Nanobiotechnol 3(2):174–196
18. Pokorski JK, Hovlid ML, Finn MG (2011) Cell targeting with hybrid Qβ virus-like particles displaying epidermal growth factor. Chembiochem 12(16):2441–2447
19. Jung B, Anvari B (2013) Virus-mimicking optical nanomaterials: near infrared absorption and fluorescence characteristics and physical stability in biological environments. ACS Appl Mater Interfaces 5(15):7492–7500
20. Huang X, Stein BD, Cheng H, Malyutin A, Tsvetkova IB, Baxter DV, Remmes NB, Verchot J, Kao C, Bronstein LM, Dragnea B (2011) Magnetic virus-like nanoparticles in N. benthamiana plants: a new paradigm for environmental and agronomic biotechnological research. ACS Nano 5(5):4037–4045
21. Li F et al (2010) Viral coat proteins as flexible nano-building-blocks for nanoparticle encapsulation. Small 6(20):2301–2308
22. Li C, Li F, Zhang Y, Zhang W, Zhang XE, Wang Q (2015) Real-time monitoring surface chemistry-dependent in vivo behaviors of protein nanocages via encapsulating an NIR-II Ag2S quantum dot. ACS Nano 9(12):12255–12263
23. Destito G, Yeh R, Rae CS, Finn MG, Manchester M (2007) Folic acid-mediated targeting of cowpea mosaic virus particles to tumor cells. Chem Biol 14(10):1152–1162
24. Liepold LO et al (2005) Structural transitions in Cowpea chlorotic mottle virus (CCMV). Phys Biol 2(4):S166–S172
25. DuFort CC, Dragnea B (2010) Bio-enabled synthesis of metamaterials. Annu Rev Phys Chem 61:323–344
26. Tee BC, Wang C, Allen R, Bao Z (2012) An electrically and mechanically self-healing composite with pressure- and flexion-sensitive properties for electronic skin applications. Nat Nanotechnol 7:825–832
27. Bichler O et al (2012) Pavlov's dog associative learning demonstrated on synaptic-like organic transistors. Neural Comput 25(2):549–566
28. Kelly KL et al (2003) The optical properties of metal nanoparticles: the influence of size, shape, and dielectric environment. J Phys Chem B 107:668–677
29. Boyer D et al (2002) Photothermal imaging of nanometer-sized metal particles among scatterers. Science 297(5584):1160–1163
30. Vieweger M et al (2011) Photothermal imaging and measurement of protein shell stoichiometry of single HIV-1 Gag virus-like nanoparticles. ACS Nano 5(9):7324–7333
31. Ayala-Orozco C et al (2014) Au nanomatryoshkas as efficient near-infrared photothermal transducers for cancer treatment: benchmarking against nanoshells. ACS Nano 8(6):6372–6381
32. Wegner KD, Hildebrandt N (2015) Quantum dots: bright and versatile in vitro and in vivo fluorescence imaging biosensors. Chem Soc Rev 44(14):4792–4834
33. Akbarzadeh A, Samiei M, Davaran S (2012) Magnetic nanoparticles: preparation, physical properties, and applications in biomedicine. Nanoscale Res Lett 7(1):144
34. Luckanagul JA et al (2015) Plant virus incorporated hydrogels as scaffolds for tissue engineering possess low immunogenicity in vivo. J Biomed Mater Res A 103(3):887–895
35. Goicochea NL et al (2011) Structure and stoichiometry of template-directed recombinant HIV-1 Gag particles. J Mol Biol 410(4):667–680
36. Huang X et al (2007) Self-assembled virus-like particles with magnetic cores. Nano Lett 7(8):2407–2416

37. Sun J et al (2007) Core-controlled polymorphism in virus-like particles. Proc Natl Acad Sci U S A 104(4):1354–1359
38. Dixit SK et al (2006) Quantum dot encapsulation in viral capsids. Nano Lett 6(9):1993–1999
39. Dragnea B et al (2003) Gold nanoparticles as spectroscopic enhancers for in vitro studies on single viruses. J Am Chem Soc 125(21):6374–6375
40. Douglas T, Young M (1998) Host-guest encapsulation of materials by assembled virus protein cages. Nature 393:152–155
41. Cardinale D, Carette N, Michon T (2012) Virus scaffolds as enzyme nano-carriers. Trends Biotechnol 30(7):369–376
42. Loo, L., et al., Infusion of dye molecules into Red clover necrotic mosaic virus. (1359–7345 (Print)).
43. Loo LN, Guenther RH, Basnayake VR, Lommel SA, Franzen S (2006) Controlled encapsulation of gold nanoparticles by a viral protein shell. J Am Chem Soc 128:4502–4503
44. Hiebert E, Bancroft JB, Bracker CE (1968) The assembly in vitro of some small spherical viruses, hybrid viruses, and other nucleoproteins. Virology 34:492–508
45. Cuillel M, Berthet-Colominas C, Timmins PA, Zulauf M (1987) Reassembly of brome mosaic virus from dissociated virus. Eur Biophys J 15(3):169–176
46. Lucas RR (2002) The crystallographic structure of brome mosaic virus. J Mol Biol 317(1):95–108
47. Lane L (1974) The bromoviruses. Adv Virus Res 19:151–220
48. Vaughan R et al (2014) The tripartite virions of the brome mosaic virus have distinct physical properties that affect the timing of the infection process. J Virol 88:6483–6491
49. Casjens S (1985) Virus structure and assembly
50. He L et al (2013) Hepatitis virus capsid polymorph stability depends on encapsulated cargo size. ACS Nano 7:8447–8454
51. Slot JW, Geuze HJ (1985) A new method of preparing gold probes for multiple-labeling cytochemistry. Eur J Cell Biol 38(1):87–93
52. Muhlpfordt H (1982) The preparation of colloidal gold particles using tannic acid as an additional reducing agent. Exp Dermatol 38:1127–1128
53. Frens G (1973) Controlled nucleation for the regulation of the particle size in monodisperse gold suspensions. Nature (London) Phys Sci 241(105):20–22
54. Turkevich J, Stevenson PC, Hillier J (1953) The formation of colloidal gold. J Phys Chem 57:670–673
55. Turkevich J, Stevenson PC, Hillier J (1951) The nucleation and growth processes in the synthesis of colloidal gold. Discuss Faraday Soc 11:55–75
56. Tsvetkova IB, Dragnea BG (2015) Encapsulation of nanoparticles in virus protein shells. Methods Mol Biol 1252:1–15
57. Hines MA, Guyot-Sionnest P (1996) Synthesis and characterization of strongly luminescing ZnS-capped CdSe nanocrystals. J Phys Chem 100(2):468–471
58. Bronstein LM, Huang X, Retrum J, Schmucker A, Pink M, Stein BD, Dragnea B (2007) Influence of iron oleate complex structure on iron oxide nanoparticle formation. Chem Mater 19:3624–3632
59. Gopinath K, Dragnea B, Kao C (2005) Interaction between brome mosaic virus proteins and RNAs: effects on RNA replication, protein expression, and RNA stability. J Virol 79(22):14222–14234
60. Marillonnet S, Thoeringer C, Kandzia R, Klimyuk V, Gleba Y (2005) Systemic Agrobacterium tumefaciens-mediated transfection of viral replicons for efficient transient expression in plants. Nat Biotechnol 23(6):718–723
61. Dillen W, De Clercq J, Kapila J, Zambre M, Van Montagu M, Angenon G (1997) The effect of temperature on Agrobacterium tumefaciens-mediated gene transfer to plants. Plant J 12(6):1459–1463
62. Sheikholeslam SN, Weeks DP (1987) Acetosyringone promotes high efficiency transformation of Arabidopsis thaliana explants by Agrobacterium tumefaciens. Plant Mol Biol 8(4):291–298
63. Cuillel M, Berthet-Colominas C, Timmins PA, Zulauf M (1987) Reassembly of brome mosaic virus from dissociated virus. Eur Biophys J 15(3):169–176
64. Ludtke SJ, Baldwin PR, Chiu W (1999) EMAN: semiautomated software for high-resolution single-particle reconstructions. J Struct Biol 128(1):82–97
65. Pettersen EF, Goddard TD, Huang CC, Couch GS, Greenblatt DM, Meng EC, Ferrin TE (2004) UCSF Chimera - a visualization system for exploratory research and analysis. J Comput Chem 25(13):1605–1612
66. Chen S, Kimura K (1999) Synthesis and characterization of carboxylate-modified gold nanoparticle powders dispersible in water. Langmuir 15:1075–1082

Chapter 20

In Vivo Packaging of Protein Cargo Inside of Virus-Like Particle P22

Kimberly McCoy and Trevor Douglas

Abstract

Protein cages are ubiquitous in nature and have been manipulated to encapsulate a range of nonnative cargos including organic, inorganic, and small molecules. Many protein cages are derived from virus capsids that have been rendered noninfectious through the preferential production and use of proteins that are solely involved in capsid assembly, but which do not encapsulate genetic material and therefore do not contribute to infectivity. Here, we describe the production of protein cargo(s) encapsulated inside of P22 virus-like particles (VLPs), derived from bacteriophage P22. This is achieved via genetic fusion of the cargo to a scaffolding protein, which becomes encapsulated in the P22 VLP during templated assembly of the protein cage.

Key words P22, Virus-like particle, VLP, Encapsulate, Protein cargo, Enzyme encapsulation, Heterologous expression

1 Introduction

Virus-like particles (VLPs) provide a unique platform for encapsulation [1] of protein cargo, which have uses in catalysis, vaccine development, and crowding studies [2]. VLPs are an ideal choice of nanocontainer because they are easily produced from minimal components, from which they self-assemble into highly homogenous populations. Production of nanometer sized VLPs often results in high yields, and they are easier to sustainably produce than their synthetic counterparts due to their largely proteinaceous makeup. Importantly, these nanocontainers protect cargo from the environment while providing multiple interfaces for further functionalization [3].

The VLP P22, or P22 capsid, is derived from the P22 bacteriophage and is assembled from two proteins: the coat protein (CP) and the scaffolding protein (SP) [4]. The latter can be severely truncated and still direct the assembly of 420 CP to form $T = 7$ icosahedrons [5]. The SP is also tolerant of genetic fusions, as it

Christina Wege and George P. Lomonossoff (eds.), *Virus-Derived Nanoparticles for Advanced Technologies: Methods and Protocols*, Methods in Molecular Biology, vol. 1776, https://doi.org/10.1007/978-1-4939-7808-3_20,

© Springer Science+Business Media, LLC, part of Springer Nature 2018

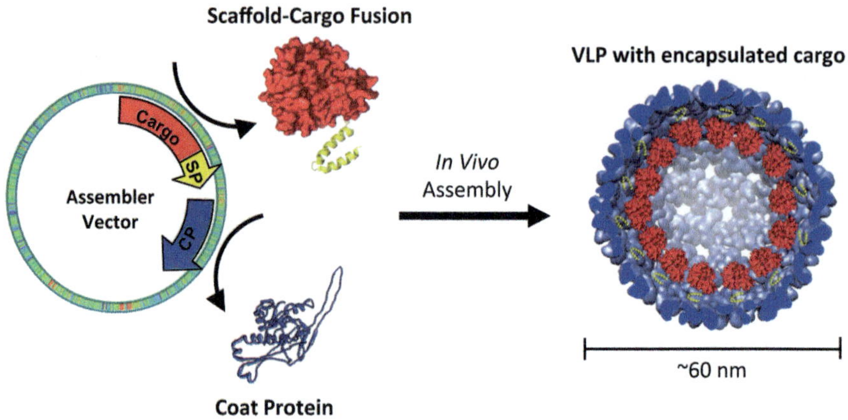

Fig. 1 Expression of both the P22 coat protein (CP, blue) and scaffolding protein (SP, yellow)-cargo (red) fusion results in the self-assembly of T = 7 virus-like particles (VLPs) that are indistinguishable from wild-type VLPs

does not disrupt the assembly process when gene products of interest such as enzymes or fluorescent proteins are genetically fused to either the N- or C-terminus [6, 7]. This fusion directs cargo encapsulation into the P22 capsid once the CP is sequentially expressed or coexpressed heterologously in an *E. coli* host (*see* Fig. 1) [8].

We have created a library of protein-encapsulated P22 VLPs in order to study cargo protection, enzyme immobilization, proximity effects, enzyme intermediate diffusion, and antigen presentation [9–12]. To date, cargo sizes ranging from 20 kDa to 180 kDa have been effectively encapsulated. In this chapter we describe the heterologous expression and purification of protein cargo encapsulated inside P22 VLPs.

2 Materials

2.1 Cloning and Transformation

1. Sequences of P22 coat protein (NCBI database Gene ID: 1262831) and cargo fused to the scaffolding protein (Gene ID: 2944242) cloned on the same or separate plasmids (plasmids have to be IPTG/L-arabinose inducible) (*see* Subheading 3.1, **step 1**).
2. BL21 DE3 chemically competent *E. coli* cells.
3. Super Optimal broth with Catabolite repression (SOC) medium: 20 g tryptone, 5 g yeast extract, 0.584 g NaCl, 0.186 g KCl, and 2.4 g $MgSO_4$; dissolve in 1.0 l deionized water. Cover with foil and autoclaved at 121 °C, 20 min. Once cooled, add 20 ml of filter-sterilized 20% glucose solution.
4. Stock solution of appropriate antibiotic(s) (depending on the plasmids used).

5. LB-agar plates: 10 g tryptone, 5 g yeast extract, 10 g sodium chloride, 15 g agar; dissolve in 1.0 l deionized water. Cover with foil and autoclaved at 121 °C, 20 min. Once cooled to approximately 55 °C, add appropriate antibiotic(s) (which is dependent on the antibiotic resistance provided by the plasmid) on (*see* **Note 1**). Slowly pour agar into plastic petri dishes so that the bottom is fully covered. Place lids 70% over bottoms and let cool. Once solidified, store inverted at 4 °C for up to 1 month.
6. LB medium: 10 g NaCl, 10 g tryptone, 5 g yeast extract; dissolve in 1.0 l deionized water. Cover with foil and autoclaved at 121 °C, 20 min.
7. Liquid nitrogen.
8. 50% (v/v) glycerol.

2.2 Heterologous Expression

1. Competent cells (agar plate or glycerol stock) transformed with plasmid(s) containing genes that encode the P22 coat protein, and the scaffolding protein fused with desired protein gene (from Subheading 3.1).
2. LB medium: 10 g NaCl, 10 g tryptone, 5 g yeast extract; dissolve in 1.0 l deionized water. Cover with foil and autoclaved at 121 °C, 20 min.
3. MaxQ 4000 Thermo Scientific shaker (or comparable).
4. 1000× (0.3 M) isopropyl β-D-1-thiogalactopyranoside (IPTG): Dissolve 71.49 mg IPTG in 1.0 ml deionized water. Use 1 ml stock solution per each liter of culture. 1.0 ml aliquots can be stored at −20 °C.
5. 100× (1.33 M) L-arabinose: Dissolve 2 g L-arabinose in 10 ml deionized water. Use 10 ml stock solution per each liter of culture. Make this solution fresh.

2.3 Purification and Transformation

1. Phosphate buffer: 50 mM sodium phosphate, 100 mM sodium chloride, pH 7.0. Weigh 2.92 g $NaH_2PO_4*H_2O$ sodium phosphate monobasic monohydrate, 7.73 g $Na_2HPO_4*7H_2O$ sodium phosphate dibasic heptahydrate, and 5.84 g NaCl sodium chloride and dissolve in 900 ml deionized water. Adjust pH to 7.0 using 2 M NaOH. Add water to a final volume of 1.0 l. Filter buffer through a 0.22 μm membrane filter.
2. Lysing agents: Weigh 20 mg DNase, 30 mg RNase, and 15 mg lysozyme and dissolve in 1.0 ml deionized water. Freeze 100 μl aliquots until required.
3. Ice water bath.
4. Sonifier (with a macro tip).
5. 0.45 μm and 0.22 μm syringe filter.

6. Luer-lock syringe with needle extension.
7. 35% (w/v) sucrose.
8. Ultracentrifuge, rotor, and tubes.
9. HiPrep 16/60 Sephacryl S-500 HR size exclusion column (or any column with a separation range between 4×10^4 and 2×10^7 Da).
10. BioLogic DuoFlow Medium Pressure Chromatography Systems (or comparable).

2.4 Characterization

1. Materials for casting and running SDS-PAGE.
2. For TEM negative stain: uranyl acetate solution 2%.
3. Carbon film coated copper mesh grid.
4. Transmission electron microscope.
5. For multiangle light scattering: S-200 column; Elution buffer: 50 mM $NaPO_4$, 100 mM NaCl, 200 ppm azide, pH 7.2.
6. Multiangle static light scattering (MALS) detector.
7. Differential refractive index (dRI) detector.
8. High pressure liquid chromatography system.

3 Methods

3.1 Cloning and Transformation

1. Sequence information for the P22 Coat Protein and Scaffolding Protein can be obtained from the NCBI database. Fusion of cargo to the Scaffolding Protein (*see* **Note 2**) can be via the N- or C-terminus. Incorporation of each gene can be accomplished using the well-established Gibson cloning protocol [13] into multiple cloning sites on the same or separate plasmid(s) of interest (*see* **Note 3**).
2. Add 1 μl of each plasmid (between 10 ng/μl and 100 ng/μl) to 80 μl chemically competent cells (*see* **Note 4**) (in a 0.5–2 μl Eppendorf tube) and incubate on ice for 5 min. Heat cells at 42 °C for 30 s. Incubate on ice for 5 min, then add 250 μl SOC medium and incubate at 37 °C with shaking for 1 h. Plate 5–200 μl (*see* **Note 5**) on an agar plate containing appropriate antibiotic, invert, and incubate at 37 °C for 18 h.
3. Inoculate a pipette tip with cells from a single colony from the agar plate and dispense tip into 1.0 ml LB containing appropriate antibiotic. Shake culture at 37 °C. At OD_{600} = 0.6–0.8, or after overnight growth, mix 300 μl culture with 700 μl 50% glycerol. Flash-freeze by immersing in liquid nitrogen. Store at −80 °C. Store plates at 4 °C for up to 1 month.

3.2 Heterologous Expression

1. Inoculate 3–5 ml of LB medium containing the appropriate antibiotic (*see* **Notes 1** and **6**) with *E. coli* cells, either from the agar plate or glycerol stock, containing the P22 scaffolding protein (SP)-protein fusion and coat protein (CP) genes on the same or separate plasmids overnight (18 h) at 37 °C with continuous shaking at 250 rpm.
2. Use one overnight culture to inoculate 1.0 l of 37 °C LB medium containing the appropriate antibiotics and grow at 37 °C with continuous shaking at 250 rpm until OD_{600} reaches 0.6–0.8 (4–8 separate l cultures are recommended).
3. If both CP and SP genes are on a single plasmid, induce with EITHER (*see* **Note 7**) 1.0 ml 0.3 M IPTG to a final concentration of 0.3 mM per liter of culture OR with 10 ml 1.33 M L-arabinose to a final concentration of 13.3 mM per liter of culture.

   If genes are on separate plasmids, induce cargo-SP gene first (*see* **Note 7**) with 10 ml 1.33 M L-arabinose (usually) to a final concentration of 13.3 mM per liter of culture. Express for an optimal amount of time (*see* **Note 8**), then induce the CP with IPTG to a final concentration of 0.3 mM (usually) per liter of culture. Allow CP to express 2 h.
4. Pellet cells by centrifugation at 3700 × *g* for 20 min or 4500 × *g* for 10 min. Transfer cells to 50 ml conical tubes (5–10 g per tube). Freeze at −80 °C with a small volume of phosphate buffer (enough to resuspend cells in the tube) until ready for purification.

3.3 Purification

1. Allow cell pellets to thaw (*see* **Note 9**), then resuspend in 5–10 ml phosphate buffer per gram of cell pellet (*see* **Notes 10** and **11**). Add lysing agents at approximately one 100 μl aliquot per liter of original culture. Rock resuspension at room temperature for 0.5–1 h.
2. Place the conical tubes in an ice water bath. Sonicate (*see* **Note 12**) ~40 ml aliquots with a macro-tip at 50% amplitude for 2.5 min pulsing 0.3 s on, 0.7 s off. Let solution cool to 4 °C.
3. Pellet cellular debris by centrifuging at 12,000 × *g* for 45 min. Slowly remove supernatant via electric pipet being careful not to pull in debris. Pass supernatant through a 0.45 μm syringe filter.
4. Fill ultracentrifuge tubes with approximately 20 ml of supernatant. Using a luer-lock syringe with needle extension, slowly inject 35% (w/v) sucrose beneath the supernatant until the liquid fills the tubes completely. Balance the tubes to within 5 mg using phosphate buffer or water and spin in ultracentrifuge for 50 min at 45,000 × *g*. Remove the tubes immediately after the spin has been completed and pour off the superna-

tant, being sure to remove all liquid. Resuspend remaining pellet in phosphate buffer (*see* **Note 13**) with approximately 1 ml per 1 cm of pellet diameter (or 1 ml per 20 mg protein). Rock at 4 °C to resuspend for 1–2 h, taking care that the rocking liquid completely covers the pellet. Spin resuspended pellet(s) in a tabletop centrifuge at 17,000 × *g* for 3 min to remove aggregated material. Slowly pipet supernatant out of the tubes, then filter through a 0.22 μm syringe filter.

5. Equilibrate FPLC-connected Sephacryl S-500 column with phosphate buffer (flow rate = 1 ml/min). Load 1 ml aliquots of P22 sample onto the column via sample loading loop and elute with isocratic flow for 120 min. Collect appropriate fractions in 4 ml aliquots. P22 VLPs elute around 60 min (*see* **Notes 14** and **15**).

3.4 Characterization

Particle size distribution, sample purity, and molecular weight are assessed using size exclusion chromatogram (SEC), SDS-PAGE (*see* **Note 16**), negative stain TEM, and multiangle light scattering (MALS). For MALS characterization, 100 μl of 1 mg/ml purified and filtered P22 is loaded onto a S-200 column and eluted with 50 mM $NaPO_4$, 100 mM NaCl, 200 ppm azide pH 7.2 at a flow rate of 0.7 ml/min.

4 Notes

1. Common antibiotics used and their appropriate final concentrations:

   Ampicillin: 50 μg/ml; Kanamycin: 30 μg/ml; Chloramphenicol: 34 μg/ml.

2. Full length SP can be used. Alternatively, a truncated version, which only includes amino acid residues 142–303, can be used to increase the VLP internal volume.
3. The principle reason genes are placed on separate plasmids is for control over induction time. If the cargo protein(s) needs time for maturation before encapsulation, the genes should be ligated into separate plasmids under the control of different promoters.
4. Competent cells used are generally Novagen kits but other chemically or electro-competent cells may be used.
5. Volume of cells plated is determined experimentally based on transformation efficiency.
6. Check the plasmid map to determine the proper antibiotic resistance.
7. Check the plasmid map to determine the proper inducer.

8. Optimal expression times vary for different constructs but can be determined experimentally by
   (a) Expression levels: Grow inoculated cells on a small scale (1–5 ml) and remove 200 μl culture aliquots after expression for 2 h, 3 h, 4 h, 5 h, 6 h, and overnight. Spin aliquots down on a tabletop centrifuge at 17,000 × *g*, remove supernatant, and freeze pellets at −20 °C or go on directly to resuspending. Resuspend thawed pellets in 20 μl deionized water and 20 μl 4× SDS-PAGE loading dye. Boil samples for 5 min (no longer or the sample will be too viscous to load) and briefly centrifuge (5 s to collect liquid to the tube bottom). Load 15 μl of each sample and 5 μl appropriate molecular weight standard ladder on a 12% (w/v) denaturing SDS-PAGE gel for 30–60 min (depending on expected protein molecular weight) to determine optimal expression amount as a function of expression time.
   (b) Enzyme activity levels: Grow 4–8 l of culture and express as normal, but spin down 1.0 l each of culture after 2 h, 3 h, 4 h, 6 h, and O/N of expression (NOT including CP expression times). Purify as directed above, then use assay(s) specific to encapsulated enzyme(s) to determine optimal activity as a function of expression time.
9. To speed up thawing the cell pellet, place conical tube in a lukewarm water bath with stir bar. When pellet is partially thawed, the addition of buffer at room temperature can also help resuspension.
10. Addition of cofactors such as metals or small molecules may be necessary for enzyme activity or protein stability. Refer to literature on protein of interest. Cofactors can be added into lysis buffer or after purification but may or may not be incorporated after folding, depending on the protein.
11. Protease inhibitor cocktail tablets or PMSF to 2 mM final concentration can be added at this step if working with a protease-labile construct. However, this step is not commonly performed for standard P22 encapsulated cargos.
12. Immerse sonicator tip in as much solution as possible without contacting the bottom of the tube.
13. P22 is stable in a range of buffers but is often stored and assayed in PBS. Buffer choice should largely be based on stability and activity of encapsulated cargos.
14. Standard fractions are collected from 30–90 min, but it is recommended that all fractions be collected during initial purification. Depending on the protein loaded, the peak width can be large.

15. Often, a peak that is not fully separated from the P22 peak elutes first. This often represents an aberrantly assembled P22 population and incorporation of this peak in the final sample should be avoided.
16. Often, a bacterial vesicle protein, Omp, copurifies with the P22 capsid as visualized by SDS-PAGE at ~33 kDa.

References

1. Douglas T, Young M (1998) Host-guest encapsulation of materials by assembled virus protein cages. Nature 393(6681):152–155. https://doi.org/10.1038/30211
2. Khudyakov Y, Pumpens P (eds) (2016) Viral Nanotechnology. CRC Press, Boca Raton, FL
3. Douglas T, Young M (2006) Viruses: making friends with old foes. Science 312(5775):873–875. https://doi.org/10.1126/science.1123223
4. Prevelige PE, Thomas D, King J (1988) Scaffolding protein regulates the polymerization of P22 coat subunits into icosahedral shells in vitro. J Mol Biol 202(4):743–757. https://doi.org/10.1016/0022-2836(88)90555-4
5. Parker MH, Casjens S, Prevelige PE (1998) Functional domains of bacteriophage P22 scaffolding protein. J Mol Biol 281(1):69–79. https://doi.org/10.1006/jmbi.1998.1917
6. O'Neil A, Reichhardt C, Johnson B, Prevelige PE, Douglas T (2011) Genetically programmed in vivo packaging of protein cargo and its controlled release from bacteriophage P22. Angewandte Chemie-International Edition 50(32):7425–7428. https://doi.org/10.1002/anie.201102036
7. Qazi S, Miettinen HM, Wilkinson RA, McCoy K, Douglas T, Wiedenheft B (2016) Programmed self-assembly of an active P22-Cas9 nanocarrier system. Mol Pharm 13(3):1191–1196. https://doi.org/10.1021/acs.molpharmaceut.5b00822
8. Jordan PC, Patterson DP, Saboda KN, Edwards EJ, Miettinen HM, Basu G, Thielges MC, Douglas T (2016) Self-assembling biomolecular catalysts for hydrogen production. Nat Chem 8(2):179–185. https://doi.org/10.1038/nchem.2416
9. O'Neil A, Prevelige PE, Douglas T (2013) Stabilizing viral nano-reactors for nerve-agent degradation. Biomater Sci 1(8):881–886. https://doi.org/10.1039/c3bm60063g
10. Patterson DP, McCoy K, Fijen C, Douglas T (2014) Constructing catalytic antimicrobial nanoparticles by encapsulation of hydrogen peroxide producing enzyme inside the P22 VLP. J Mater Chem B 2(36):5948–5951. https://doi.org/10.1039/c4tb00983e
11. Patterson DP, Schwarz B, Waters RS, Gedeon T, Douglas T (2014) Encapsulation of an enzyme cascade within the bacteriophage P22 virus-like particle. ACS Chem Biol 9(2):359–365. https://doi.org/10.1021/cb4006529
12. Schwarz B, Madden P, Avera J, Gordon B, Larson K, Miettinen HM, Uchida M, LaFrance B, Basu G, Rynda-Apple A, Douglas T (2015) Symmetry controlled, genetic presentation of bioactive proteins on the P22 virus-like particle using an external decoration protein. ACS Nano 9(9):9134–9147. https://doi.org/10.1021/acsnano.5b03360
13. Gibson DG, Young L, Chuang R-Y, Venter JC, Hutchison CA III, Smith HO (2009) Enzymatic assembly of DNA molecules up to several hundred kilobases. Nat Methods 6(5):343–U341. https://doi.org/10.1038/nmeth.1318

Chapter 21

Encapsulation of Negatively Charged Cargo in MS2 Viral Capsids

Ioana L. Aanei, Jeff E. Glasgow, Stacy L. Capehart, and Matthew B. Francis

Abstract

Encapsulation into virus-like particles is an efficient way of loading cargo of interest for delivery applications. Here, we describe the encapsulation of proteins with tags comprising anionic amino acids or DNA and gold nanoparticles with negative surface charges inside MS2 bacteriophage capsids to obtain homogeneous nanoparticles with a diameter of 27 nm.

Key words MS2 bacteriophage, Encapsulation, Virus-like particles, Delivery vehicles, Nanoparticles, Self-assembly

1 Introduction

Nanoparticles have received much attention in recent decades due to their potential for efficient drug delivery to tumor sites based on the enhanced permeability and retention (EPR) effect [1]. Relative to carriers of other compositions, virus-like particles (VLPs) have the advantages of homogeneous particle sizes, precise composition, and biodegradability. The self-assembly of viral coat proteins provides unique opportunities to develop novel materials, often through the covalent attachment of synthetic groups to the outer [2] and inner [3] surfaces of the resulting supramolecular structures. This has proven useful for the installation of magnetic resonance imaging (MRI) contrast enhancement agents [4, 5], radionuclides for positron emission tomography (PET) [6], drug molecules [7], and singlet oxygen-generating porphyrins [8]. As a complementary strategy, several groups have introduced larger cargo into the spherical structures during the self-assembly step [9]. This process generally takes advantage of specific interactions of encapsulation "tags" with the interior-facing surfaces of the coat protein monomers, thus initiating the assembly of the capsids.

Christina Wege and George P. Lomonossoff (eds.), *Virus-Derived Nanoparticles for Advanced Technologies: Methods and Protocols*, Methods in Molecular Biology, vol. 1776, https://doi.org/10.1007/978-1-4939-7808-3_21,
© Springer Science+Business Media, LLC, part of Springer Nature 2018

As one area of particular interest, engineered particles that can house enzymes have attracted a lot of attention recently, due to their advantages over free catalysts (enzymes or small molecules). Viral capsids can protect the enzyme from environmental factors and increase the local substrate concentration. For example, researchers showed that the enzyme, *Candida antarctica* lipase B had increased activity upon encapsulation within the shell of cowpea chlorotic mottle virus (CCMV) [10]. Other examples of protein encapsulation have been demonstrated using the Qβ and P22 capsids [11–14].

Previous efforts to encapsulate cargo inside the bacteriophage MS2 capsid have taken advantage of the specific interaction of a short RNA sequence with the interior surface of the capsid coat protein [15, 16]. This strategy consists of covalently attaching the RNA to the molecule of interest to initiate assembly of the protein coat dimers around cargo such as the ricin A chain, quantum dots, doxorubicin, and siRNA [16]. Although the technique is fairly versatile and has proven to be successful, the instability of the RNA used and the cost of production limit the practicality of encapsulation on larger synthetic scales. To address some of these issues, our group has developed several alternative methods for the encapsulation of proteins inside bacteriophage MS2 viral capsids. The first approach is to attach a negatively charged molecule (DNA, RNA, or polymers) to the protein of interest through a covalent linkage and then incubate it with the disassembled MS2 capsid. This strategy capitalizes on the positively charged interior surfaces that many viral capsids have evolved to interact favorably with the negatively charged genetic material (DNA or RNA). The second approach is to use genetic methods to append a negatively charged peptide tag to the protein to be encapsulated in order to induce capsid reassembly. This method is more cost-effective, scalable, and versatile, since a variety of proteins could be genetically modified and prepared for encapsulation. In both cases, the addition of a protein-stabilizing osmolyte, trimethylamine *N*-oxide (TMAO), significantly increased the yields of reassembly (*see* **Note 1**). It is important to note that some naturally occurring negatively charged proteins could potentially be encapsulated into the MS2 capsid without further modification, although we have not thoroughly investigated this avenue.

Inorganic nanoparticles are also of great interest for encapsulation, due to their utility as imaging agents for fluorescence microscopy, electron microscopy, and magnetic resonance imaging. Several groups have reported the encapsulation of metal nanoparticles, such as quantum dots [16, 17], gold [18–23], and iron oxide nanoparticles inside of protein cages (also *see* Chapters 12, 19 and references 17, 24). In this vein, our lab has employed a strategy similar to that used for protein encapsulation in MS2 to encapsulate gold nanoparticles (AuNP) with anionic surface groups. We noted that, in the case of the AuNPs, the presence of the osmolyte was not required, possibly due to the very high amount of negative charge on the metal surfaces.

We present herein an overview of the different strategies developed in our lab [25–29] to encapsulate large cargo inside bacteriophage MS2 viral capsids. These nanoparticles could find numerous applications in the fields of biocatalysis, protein stabilization, vaccine development, and drug delivery.

2 Materials

Prepare all solutions using ultrapure water (18 MΩ·cm at 25 °C) and analytical grade reagents. Diligently follow all waste disposal regulations when disposing of waste materials.

2.1 Buffers

1. Lysis buffer: 20 mM taurine (adjust the pH to 9 with NaOH), 6.5 mM DTT, 6 mM $MgCl_2$, and 10 μg/mL each of DNase I and RNase A from bovine source, make fresh before use.
2. Anion exchange buffer (AXB): 20 mM taurine (adjust the pH to 9 with NaOH), 6.5 mM DTT, make fresh before use.
3. High-salt anion exchange buffer: 20 mM taurine (adjust the pH to 9 with NaOH), 1.7 M NaCl.
4. Ammonium sulfate saturated solution: 500 g $(NH_4)_2SO_4$ dissolved in 700 mL water, store at 4 °C.
5. Phosphate buffers: potassium phosphate monobasic at different ionic strengths and pH values: 20 mM (pH 6.5), 50 mM (pH 6.5), 500 mM (pH 6.5), 50 mM (pH 7), 10 mM (pH 7.2), 100 mM (pH 7.2), 50 mM (pH 7.5), 20 mM (pH 8), 100 mM (pH 8).
6. Chilled glacial acetic acid.
7. 1 mM acetic acid (chilled).
8. 0.05% (w/v) Bromophenol Blue solution.
9. Elution buffer for His Gravitrap column (see below): 50 mM phosphate buffer, pH 7.5, 300 mM NaCl, 300 mM imidazole.
10. 50 mM bis-Tris buffer, pH 6.
11. 1.8 M TMAO solution in 50 mM bis-Tris, pH 6.0. Make fresh before use.
12. 100 mM isatoic anhydride solution (stock solution) in anhydrous DMSO. Make fresh before use.
13. Phenylene-diamine NHS ester [30] 50 mM stock in a 1:1 dimethyl formamide (DMF)–water mixture.
14. 50:1 (v/v) chloroform–acetic acid.

15. 50 mM sodium periodate solution ($NaIO_4$) (stock solution) in water. Make fresh before use.
16. 20 mM bis-Tris solution at pH 6.
17. 1 M NaCl in 20 mM bis-Tris solution at pH 6.
18. 250 mM 3-(4-aminophenyl) propionic acid in DMF.
19. 313 mM 1-ethyl-3-(3-dimethylaminopropyl) carbodiimide (EDC·HCl) in 0.5 M phosphate buffer pH 6.5.
20. 1.2 M *N*-hydroxysulfosuccinimide sodium salt (sulfo-NHS).
21. β-mercaptoethanol.
22. 20 mM sodium bicarbonate–carbonate buffer pH 9.
23. Aminophenol-poly(isobutylene-alt-maleic anhydride) (PIBMA).
24. Bis(p-sulfonatophenyl)phenylphosphine dehydrate dipotassium salt (BSPP).
25. 100 mM phosphate buffer, pH 7.2, 0.5 M NaCl.
26. Methanol.
27. Polyethylene glycol with a molecular weight of 6000 (PEG-6000).
28. Sodium dithionite: make fresh before use, 50 mM stock solution in 500 mM phosphate buffer pH 6.5 (*see* **Note 2**)

2.2 Chromatography Columns

1. DEAE-Sephadex anion exchange column (GE Healthcare).
2. Sephacryl S1000 size exclusion chromatography column (GE Healthcare).
3. NAP-5 desalting columns (GE Healthcare).
4. His Gravitrap column for purification of proteins with His tag (GE Healthcare).
5. Polysep GFC-P-5000 gel filtration column (Phenomenex) or Biosep SEC-S-4000 gel filtration column (Phenomenex).
6. HiTrap Q XL strong anion exchange column (GE Healthcare).
7. HiPrep DEAE anion exchange columns (GE Healthcare).

2.3 Other Materials

1. *E. coli* cell pellets (3–8 g wet weight) obtained after induction and protein expression of either T19pAF MS2 or T19pAF/N87C thawed on ice for 2 h (*see* **3.1**).
2. Sonicator (Fisher Scientific, Model 120, 1/8 inch probe).
3. SDS-PAGE equipment: 10% bis-Tris NuPage gels and Mini Gel Tank apparatus (Thermo).
4. Amicon Ultra centrifugal filters with 3, 10, 30, or 100 kDa molecular weight cutoff (Millipore).
5. Spectrophotometer.

6. Plasmid containing a construct of the protein of interest with a negatively charged tag and a tag for purification (e.g., His_6-GFP-Neg).
7. *E. coli* cells optimized for protein production (such as BL21 Codon+RIL, Agilent) and appropriate medium for selection of transformed cells.
8. ssDNA (20 base pairs) containing a 5′ amino group (arbitrary sequence, Integrated DNA Technologies), 1 mM stock solution in water.
9. ssDNA with a 5′ thiol C6 S-S modification (5′-$(CH_2)_6$-S-S-$(CH_2)_6$-phosphodiester bond-oligonucleotide-3′), arbitrary sequence, 20–30 base pairs, (Integrated DNA Technologies).
10. Unconjugated gold colloid solutions (AuNPs): 5 nm diameter (5×10^{13} particles/ mL), 10 nm diameter (5.7×10^{12} particles/mL) (Ted Pella, Inc.).
11. Transmission electron microscope (TEM) and dynamic light scattering system (DLS) for characterization of the particles.
12. Native agarose gel electrophoresis (2.5% w/v agarose gel, TBE running buffer).

3 Methods

Carry out all procedures on ice or at 4 °C, unless otherwise specified.

3.1 Expression and Purification of Genome-Free MS2 Viral Capsids

The plasmid production of bacteriophage MS2 T19pAF has been previously reported [31]. The tRNA- and tRNA-synthetase-encoding plasmids necessary for *p*-aminophenylalanine (pAF) incorporation [32] were provided by the Schultz lab (Scripps Research Institute, La Jolla, CA). To obtain the T19pAF N87C MS2 mutant, the amino acid at position 87 was mutated into a cysteine using the following forward and reverse primers. This mutation allows for efficient modification through a cysteine alkylation reaction with maleimides.

Forward: 5′ – AGCCGCATGGCGTTCGTACTTATGTATG GAACTAACCATTC – 3’.

Reverse: 5′ – GAATGGTTAGTTCCATACATAAGTACGAA CGCCATGCGGCT – 3’.

The expression of the T19pAF MS2 and the T19pAF/N87C double mutant can be carried out in autoinducing media following the published protocol [31]. For all other mutants, lysogeny broth (LB medium) can be used during the protein expression protocol.

The purification of the genome-free MS2 capsids can be accomplished as follows:

1. Thaw the *E. coli* cell pellets obtained after induction and protein expression on ice for 2 h.
2. Resuspend the cells in 20 mL lysis buffer.
3. Sonicate the cells for 10 min on ice (*see* **Note 3**).
4. Spin down the cell lysate for 45 min at 8000 × *g* at 4 °C.
5. Load the supernatant on a weak anion exchange column (e.g., DEAE-Sephadex) following the manufacturer's instructions. Use AXB as the eluent for the column. Analyze the fractions by sodium dodecyl sulfate–polyacrylamide gel electrophoresis (SDS-PAGE) and collect the fractions that contain MS2 (MW ~13,800 Da). Clean the column by washing with 2–3 column volumes of the high-salt anion exchange buffer (or follow instructions for the particular column you are using).
6. Precipitate the proteins with large molecular weights from the anion exchange column fractions by using an equal volume of saturated ammonium sulfate and incubating at 4 °C for 2–12 h.
7. Pellet the precipitated proteins by centrifugation at 11,000 × *g* for 10 min at 4 °C.
8. Resuspend the pellet in ~10 mL of 10 mM phosphate buffer (pH 7.2) and load onto a size exclusion chromatography (e.g., Sephacryl S1000) column. Perform chromatography following the manufacturer's instructions. Use 10 mM pH 7.2 phosphate buffer as the eluent for the column. Analyze the fractions by SDS-PAGE and combine the fractions that contain MS2.
9. Concentrate the fractions of interest using 100 kDa molecular weight cutoff (MWCO) centrifugal concentrators. Yields of ~10 mg/L culture can be obtained for the T19pAF N87C MS2 double mutant after one anion exchange and two size exclusion columns. Higher yields can be obtained for T19pAF and other mutants.

3.2 Disassembly of MS2 Viral Capsids

1. Concentrate the MS2 protein solution to 10 mg/ mL (~720 μM MS2 monomer, *see* **Note 4**) by using 100 kDa MWCO spin concentrators.
2. Add 2 volumes of chilled glacial acetic acid to 1 volume protein solution for a final concentration of acetic acid of 66% (v/v). The solution will turn cloudy, as the nucleic acids precipitate. Keep the solution on ice for 30 min.
3. Spin the disassembly mixture at 14,000 × *g* for 10 min in a refrigerated centrifuge at 4 °C. A white pellet should form at the bottom of the tube.
4. Carefully pipet the supernatant onto a desalting column pre-equilibrated with chilled 1 mM acetic acid to exchange the buffer to a higher pH. Perform desalting following the manufacturer's instructions. Elute with 1 mM acetic acid solu-

tion. Collect 100 μL fractions as the sample is coming off the column. Check the pH of the fractions by using 2 μL of the fraction and 1 μL of the Bromophenol Blue solution. Collect only the fractions that are blue, indicating a less acidic solution (*see* **Note 5**).

5. Combine the fractions of interest and centrifuge again at 14,000 × *g* to remove aggregates. Measure the absorbance at 280 nm of the pooled fractions (*see* **Note 6**) and adjust the concentration as needed. Use the MS2 dimer solution as soon as possible. If needed, the solution can be stored on ice for short period of times (less than 2 h).

3.3 Encapsulation of Proteins Inside MS2 Viral Capsids

The model protein presented herein is the green fluorescent protein (GFP). The protocol can also be used for other proteins of interest.

3.3.1 Encapsulation of Negatively Charged GFP

Express and purify your protein of interest with a negatively charged tag. Herein, we present a monomeric Enhanced Green Fluorescent Protein (mEGFP; *see* **Note 7**) with an increased negative charge due to the addition of a C-terminal peptide tag, EEEEDDDEDDDDEEDD (Neg tag, 16 additional charges). An N-terminal 6 × His tag was also added for purification purposes.

1. Transform *E. coli* cells optimized for protein production (such as BL21 Codon+RIL) with the plasmid containing the protein construct with a negative tag and a tag for purification purposes, such as 6 × His (optional). For our design of a His6-GFP-Neg construct, please see [25]. The purification includes the use of a His Gravitrap column and elution in His Gravitrap column elution buffer. The bright green fractions are pooled and concentrated and stored at 4 °C for encapsulation assays.
2. Mix a solution of MS2 coat protein dimers (15 μM, from Subheading 3.2, **step 5**) with the His6-GFP-Neg (10 μM) and 1.8 M TMAO in 50 mM bis-Tris (pH 6). Incubate at 4 °C for 48 h.
3. To purify the reassembled capsids, precipitate the protein with an equal volume of saturated ammonium sulfate solution. Resuspend the precipitated protein in 10 mM phosphate buffer (pH 7.2) and centrifuge at 14,000 × *g* for 10 min to remove any aggregated protein. Alternatively, a 0.22 μm centrifugal filter can be used to remove aggregates.
4. Assay for assembly by HPLC SEC using a Polysep GFC-P-5000 column or a Biosep SEC-S-4000 column with 1 mL /min 10 mM phosphate buffer (pH 7.2) as eluent (*see* **Note 8**). Reassembly efficiency under these conditions is typically greater than 75%, as indicated by the area of the SEC peaks. The intact capsid elutes significantly earlier than coat protein dimers.

5. Characterize the reassembled capsids by TEM and DLS. Typical diameters measured are around 27 nm. Some small particles and large aggregates can be observed.

3.3.2 Encapsulation of GFP–DNA

1. To synthesize a GFP–aniline conjugate, incubate isatoic anhydride (100 mM stock solution in DMSO, final concentration 1 mM) with a 1 mg/ mL solution of GFP in 20 mM phosphate buffer (pH 8) at room temperature for 1 h. Remove excess isatoic anhydride by using a NAP desalting column followed by five rounds of centrifugal concentration through a 10 kDa MWCO centrifugal filter.
2. To synthesize the phenylene diamine–DNA, incubate DNA (500 μM final concentration, *see* **Note 9**) containing a 5′ amino group with 20-fold molar excess phenylene diamine NHS ester [32] in 1:1 v/v dimethyl formamide (DMF) and water for 2 h. A typical reaction volume is 100 μL. Extract the DMF and the excess linker into 3 volumes (300 μL) 50:1 chloroform–acetic acid 3 times. Add the solvent mixture, vortex, and centrifuge briefly at maximum speed to separate the layers. The labeled DNA remains in the aqueous layer. The remaining phenylene diamine–DNA can be further purified using five rounds of spin concentration against a 3 kDa MWCO centrifugal filter.
3. Mix aniline–GFP and phenylene diamine–DNA in 20 mM phosphate buffer (pH 6.5) in a 1:15 molar ratio. Add the sodium periodate solution as a freshly prepared solution in water (final concentration of 5 mM) and react for 1 h at room temperature.
4. Load the reaction mixture onto a 1 mL HiTrap Q XL strong anion exchange column. Perform anion exchange following the manufacturer's instructions. Elute using a linear gradient of 0–1 M NaCl in 20 mM bis-Tris solution at pH 6. Pool fractions containing the GFP–DNA conjugate, concentrate using a 30 kDa centrifugal filter, and store at 4 °C.
5. Characterize the conjugate by SDS-PAGE and quantifying the band corresponding to the modified protein, following the protocol from the manufacturer. The protein electrophoresis samples were heated for 10 min at 95 °C in the presence of β-mercaptoethanol to ensure reduction of any disulfide bonds. Gels were run for 35–60 min at 150–200 V in 2-(N-morpholino) ethanesulfonic acid (MES)–SDS buffer to allow good separation of the bands. Commercially available markers (Bio-Rad) were applied to at least one lane of each gel for assignment of apparent molecular masses. Visualization of protein bands was accomplished by staining with Coomassie Brilliant Blue R-250 (Bio-Rad). Quantification of the degree of modification was obtained by imaging on a Gel Doc™ EZ imager (Bio-Rad) and subsequent optical densitometry using the ImageJ (National

Institutes of Health, Bethesda, MD) or Image Lab (Bio-Rad) software.

6. Usual levels of modification are around 35%.
7. Encapsulate the DNAGFP conjugate into MS2 by following **steps 2–5** in Subheading 3.2, **step 1** (*see* **Note 10**).

3.3.3 Encapsulation of GFP–Polymer

1. To install aniline moieties onto the amines of the GFP, mix solutions of 250 mM 3-(4-aminophenyl)propionic acid, 313 mM EDC·HCl, and 1.2 M sulfo-NHS, in a 5:4:1 ratio. Allow to react for 15 min at room temperature. Quench the unreacted EDC by adding β-mercaptoethanol to a final concentration of 200 mM. Add 10 μL of the reaction mixture to 90 μL of GFP solution (1 mg/ mL in 20 mM bicarbonate buffer pH 9). Incubate at room temperature overnight with gentle agitation (vortex on lowest setting). Purify by using NAP-5 desalting columns and five rounds of spin concentration against a 10 kDa MWCO centrifugal filter (*see* **Note 11**).
2. Synthesize 3-nitrotyramine by mixing a 3-hydroxyphenylpropionic acid (tyramine, 1 equiv.) solution in 50% acetic acid with fuming nitric acid (5 equiv.). The solution will quickly turn bright orange. After 15 min, add the reaction mixture to ice water, filter, and dry to yield a yellow powder (*see* **Note 12**).
3. Charge a flame-dried round bottom flask with PIBMA (average molecular weight 6000, 1 equiv.) and *N,N*-diisopropylethylamine (DIPEA, 0.3 equiv. to acidic groups) in dry tetrahydrofuran (THF). Stir the solution for 1 h at 60 °C. Reduce the temperature and add 3-nitrotyramine (0.05 equiv.) along with additional DIPEA (0.087 equiv.) dissolved in THF (*see* **Note 13**). Stir overnight at room temperature.
4. Remove the solvent and excess DIPEA under reduced pressure and dissolve the resulting residue in aqueous bicarbonate (adjust the pH to 8). Dialyze the solution against several changes of deionized water over multiple days. Lyophilize to get nitrophenol–PIBMA.
5. Reduce the nitrophenol–PIBMA by adding an equal volume of freshly prepared sodium dithionite (50 mM in 500 mM phosphate buffer pH 6.5) to a solution of 1 mM polymer in water. Immediate change of color from pale yellow (nitrophenol) to colorless (aminophenol) should be observed. Allow the reaction to proceed for 15 min at room temperature. Desalt the polymer against a 3 kDa MWCO centrifugal filter and wash 5 times with water. Quantify the aminophenol concentration by measuring the absorbance at 280 nm ($\varepsilon = 2.76\ mM^{-1}\cdot cm^{-1}$).

6. Mix one equivalent of the aniline-modified GFP (20 μM in 50 mM phosphate buffer pH 6.5) with five equivalents of the aminophenol–PIBMA. Add the oxidant (freshly prepared 50 mM sodium periodate in H_2O) to a final concentration of 5 mM, and incubate the reaction at room temperature for 10 min.
7. Encapsulate the polymer–GFP conjugate into MS2 by following **steps 2–6** in Subheading 3.2, **step 1.**

3.4 Encapsulation of Gold Nanoparticles Inside MS2 Viral Capsids

To improve the stability of the gold nanoparticles during the encapsulation process, AuNPs should be modified with single stranded DNA (ssDNA). Modification of AuNPs with ssDNA is a two-step process. First, the surface ligands on the AuNPs are exchanged to bis(*p*-sulfonatophenyl)phenylphosphine dehydrate dipotassium salt (BSPP) ligands [33]. The BSPP-stabilized AuNPs are then modified with thiol-terminated ssDNA. We found that 5 nm and 10 nm diameter gold nanoparticles encapsulate more efficiently than larger diameter AuNPs (15 or 20 nm particles). Therefore, the protocol presented herein will focus on the encapsulation of 5 nm and 10 nm nanoparticles.

1. Mix BSPP (5 mg) with the AuNPs (10 mL at different concentrations, *see* **Note 14**) and stir at room temperature overnight. Add NaCl gradually until the color of the solution changes from deep red to purple and the AuNPs crash out. Centrifuge the solution at 2500 × *g* for 10 min and redissolve the pellet in an aqueous solution of BSPP (25 mg in 100 mL dd-H_2O). Precipitate again with methanol, until the color changes from deep red to purple. Centrifuge for 10 min at 2500 × *g* and remove the supernatant. Dissolve the pellet in an aqueous solution of BSPP (25 mg in 100 mL dd-H_2O), and store at room temperature until use. The AuNP concentration can be determined by measuring the absorbance at 520 nm (*see* **Note 15**).
2. Modification of BSPP-stabilized AuNPs with thiol-terminated ssDNA can then be achieved following a previously reported procedure [34]. Briefly, lyophilized oligonucleotides containing a 5′ thiol protecting group are resuspended in dd-H_2O at a final concentration of 100 μM. The disulfide functionality is cleaved by incubation with DTT (0.1 M) at RT in 100 mM phosphate buffer pH 8.0 for 1 h. Excess DTT is removed by passing the solution through a NAP-5 gel filtration column preequilibrated with dd-H_2O and following the manufacturer's instructions. The solution containing the ssDNA is immediately added to BSPP-stabilized AuNPs (9 nM) to yield a final oligonucleotide concentration of 10 μM. The resulting solution is then incubated at RT for 20 min. The concentration of NaCl is increased by increments of 0.05 M to a final concentration of 0.1 M through successive additions of 2 M NaCl. Following each addition of 2 M NaCl, the alkanethiol oligonucleotide-modified

AuNP solution is sonicated for 10 s, followed by agitation at RT for 20 min. After reaching a final concentration of 0.1 M NaCl, the AuNP solution is incubated at RT overnight with no agitation. Excess oligonucleotides can then be removed through multiple rounds (up to five) of centrifugal filtration with 100 kDa MWCO filter (Millipore) using 50 mM phosphate buffer, pH 7.0. This procedure will lead to a high coverage of ssDNA on the surface of the AuNPs (*see* **Note 16**).

3. Mix ssDNA-modified AuNPs in 100 mM phosphate buffer, pH 7.2 containing 100 mM NaCl with MS2 coat protein dimer solution in a final molar ratio of 1:1 AuNPs:MS2 capsids. Allow the reassembly to proceed for 48 h at 4 °C.
4. At the end of the incubation period, precipitate the samples using a final concentration of 0.5 M NaCl, 10 (w/v)% PEG_{6k} for 1 h at 4 °C. Centrifuge at 2500 × *g* for 20 min, then resuspend in 50 mM phosphate buffer, pH 7.0.
5. Confirm reassembly through DLS, TEM, and native agarose gel electrophoresis (2.5% w/v agarose gel, TBE running buffer, 2 h at 40 V, on ice).

4 Notes

1. To promote the reassembly process, other osmolytes (glycine, arginine, proline, urea, guanidinium chloride) were also tested. TMAO gave the most promising results; however, at high concentrations of TMAO, the yield of the reassembly decreases, probably due to a "salting out" effect [25, 35, 36].
2. It is very important to store the dithionite tightly closed in a dry and well-ventilated place. Handle and store under inert gas and never allow the dithionite to get in contact with water during storage since the compound is air-, heat-, and moisture-sensitive. Please note that, if stored at 4 °C, dithionite should be allowed to equilibrate to room temperature before opening the container.
3. We recommend sonicating with a pulse sequence that allows for cooling off of the cell lysate in between energy pulses (e.g., 2 s on, 4 s off).
4. Since the MS2 expressed recombinantly in *E. coli* contains a small amount of adventitious nucleic acids, the extinction coefficient at 280 nm predicted for the protein is not accurate for calculating the concentration of the solution. We have found empirically that a value of $\varepsilon_{280} = 62{,}500\ M^{-1}\ cm^{-1}$ gives fairly accurate results. A shortcut can be multiplying the A_{280} by 16 and getting the concentration of the solution in μM (if a path length of 1 cm is used). Other methods of calculating the concentration of the solution (bicinchoninic

acid or Bradford assays, optical densitometry, etc.) can also be used.

5. Bromophenol Blue is a pH indicator with a working range of pH 3–4.6. When mixing a small amount of the fractions with the Bromophenol Blue solution, the observed color can be blue (pH above 4.6) or yellow (pH below 3). Do not use the acidic fractions since they will lower the overall pH of the reassembly reaction mixture, leading to lower reassembly yields.
6. After disassembly, most of the nucleic acids will be removed, and thus the extinction coefficient predicted based on the amino acid sequence (ε_{280} = 18,500 M^{-1} cm^{-1}) can be used to calculate the concentration of the protein solution. Bicinchoninic acid or Bradford assays can also be used to determine the protein concentration.
7. The monomeric form of enhanced GFP (mEGFP) [37] has a charge of approximately −7 at pH 7.2 and incubation with MS2 coat protein dimers leads to reassembly yields of ~10%. A control experiment meant to show the necessity of the negative charge for efficient reassembly included the use of lysozyme, since the protein does not have an overall negative charge. No significant levels of reassembly were observed [25].
8. The reassembly yield can be quantified using the Trp fluorescence peak (excitation at 280 nm, emission at 330 nm) at 7.9–8.3 min (Polysep column) or 5.9–6.3 min (Biosep column), which corresponds to the assembled MS2 capsids.
9. Although the RNA translational repressor sequence has been shown to trigger capsid reassembly specifically [38], work in our lab has found that in the presence of TMAO, even an arbitrary DNA sequence can promote reassembly [25]. At high DNA concentrations, the reassembly yields are decreased, presumably due to a kinetic trap [25, 39]. Replacing the RNA with DNA for reassembly purposes greatly reduces cost while increasing nucleic acid stability, making large scale encapsulation more feasible.
10. GFP–DNA was able to initiate ~35% of the MS2 dimers to assemble. Based on the fluorescence signal coeluting with the capsid, there was an average of 6.5 ± 1.9 GFP copies per capsid, corresponding to a concentration of ~2 mM GFP within the capsid volume [25].
11. The number of aniline groups installed under these conditions was found to be 1–7 per protein by electrospray ionization-time of flight mass spectrometry (ESI-TOF MS) [25].
12. The scale of the reaction commonly used is 900 mg 3-hydroxyphenylpropionic acid (tyramine, 6.3 mmol) reacted

with 50% acetic acid and 1.5 mL fuming nitric acid (33 mmol). After 15 min of reaction time, the product can be isolated by adding the reaction mixture to ice water, filtering, and drying the resulting yellow powder. Usual yields are ~30%.

13. A typical reaction setup would be as follows: 200 mg of PIBMA (0.033 mmol) in 100 mL dry THF were mixed with 68 μL DIPEA (0.39 mmol) and stirred for 1 h at 60 °C. After reducing the temperature, 12 mg of 3-nitrotyramine (0.065 mmol) and 20 μL of DIPEA (0.114 mmol) dissolved in THF were added dropwise. The solution was allowed to stir overnight at room temperature.
14. The stock concentrations for the AuNP used are as follows [40–42]: 5 nm AuNP 5×10^{13} particles/mL; 10 nm AuNP 5.7×10^{12} particles/mL.
15. The extinction coefficients for the AuNP used are as follows [40–42]: 5 nm AuNP $\varepsilon_{520\ \text{nm}} = 9.3 * 10^6\ \text{M}^{-1}\ \text{cm}^{-1}$; 10 nm $\varepsilon_{520\ \text{nm}} = 8.1 * 10^7\ \text{M}^{-1}\ \text{cm}^{-1}$.
16. The exact sequence of the ssDNA does not seem to influence the reassembly efficiency. The ssDNA sequences we typically use are between 20 and 30 bp.

Acknowledgments

This work was generously supported by the Office of Science, Materials Sciences and Engineering Division, of the US Department of Energy under Contract No. DEAC02-05CH11231. SLC was supported by an NSF graduate research fellowship (NSF GRFP). JEG was supported by the Hellman Family Faculty Fund and the UC Berkeley Chemical Biology Graduate Program (NIH Training Grant 1 T32 GMO66698).

References

1. Maeda H, Wu J, Sawa T, Matsumura Y, Hori K (2000) Vascular permeability and the EPR effect in macromolecular therapeutics: a review. J Control Release 65:271–284
2. Wang Q, Lin T, Tang L, Johnson JE, Finn MG (2002) Icosahedral virus particles as addressable nanoscale building blocks. Angew Chem Int Ed 41:459–462
3. Hooker JM, Kovacs EW, Francis MB (2004) Interior surface modification of bacteriophage MS2. J Am Chem Soc 126:3718–3719
4. Datta A, Hooker JM, Botta M, Francis MB, Aime S, Raymond KN (2008) High relaxivity gadolinium hydroxypyridonate-viral capsid conjugates: nanosized MRI contrast agents. J Am Chem Soc 130:2546–2552
5. Garimella PD, Datta A, Romanini DW, Raymond KN, Francis MB (2011) Multivalent, high-relaxivity MRI contrast agents using rigid cysteine-reactive gadolinium complexes. J Am Chem Soc 133:14704–14709
6. Farkas ME, Aanei IL, Behrens CR, Tong GJ, Murphy ST, O'Neil JP, Francis MB (2013) PET imaging and biodistribution of chemically modified bacteriophage MS2. Mol Pharm 10:69–76

7. Wu W, Hsiao SC, Carrico ZM, Francis MB (2009) Genome-free viral capsids as multivalent carriers for taxol delivery. Angew Chem Int Ed 48:9493–9497
8. Stephanopoulos N, Tong GJ, Hsiao SC, Francis MB (2010) Dual-surface modified virus capsids for targeted delivery of photodynamic agents to cancer cells. ACS Nano 4:6014–6020
9. Li F, Wang Q (2014) Fabrication of nanoarchitectures templated by virus-based nanoparticles: strategies and applications. Small 10:230–245
10. Minten IJ, Claessen VI, Blank K, Rowan AE, Nolte RJM, Cornelissen JJLM (2011) Catalytic capsids: the art of confinement. Chem Sci 2:358–362
11. Fiedler JD, Brown SD, Lau JL, Finn MG (2010) RNA-directed packaging of enzymes within virus-like particles. Angew Chem Int Ed 49:9648–9651
12. O'Neil A, Reichhardt C, Johnson B, Prevelige PE, Douglas T (2011) Genetically programmed in vivo packaging of protein cargo and its controlled release from bacteriophage P22. Angew Chem Int Ed 50:7425–7428
13. Patterson DP, Prevelige PE, Douglas T (2012) Nanoreactors by programmed enzyme encapsulation inside the capsid of the bacteriophage P22. ACS Nano 6:5000–5009
14. Patterson DP, Schwarz B, El-Boubbou K, van der Oost J, Prevelige PE, Douglas T (2012) Virus-like particle nanoreactors: programmed encapsulation of the thermostable CelB glycosidase inside the P22 capsid. Soft Matter 8:10158–10166
15. Wu M, Brown W, Stockley P (1995) Cell-specific delivery of bacteriophage-encapsidated Ricin A chain. Bioconjug Chem 6:587–595
16. Ashley CE, Carnes EC, Phillips GK, Durfee PN, Buley MD, Lino CA, Padilla DP, Phillips B, Carter MB, Willman CL, Brinker CJ, Caldeira Jdo C, Chackerian B, Wharton W, Peabody DS (2011) Cell-specific delivery of diverse cargos by bacteriophage MS2 virus-like particles. ACS Nano 5:5729–5745
17. Loo L, Guenther RH, Lommel SA, Franzen S (2007) Encapsidation of nanoparticles by red clover necrotic mosaic virus. J Am Chem Soc 129:11111–11117
18. Dragnea B, Chen C, Kwak ES, Stein B, Kao CC (2003) Gold nanoparticles as spectroscopic enhancers for in vitro studies on single viruses. J Am Chem Soc 125:6374–6377
19. Chen C, Kwak ES, Stein B, Kao CC, Dragnea B (2005) Packaging of gold particles in viral capsids. J Nanosci Nanotechnol 5:2029–2033
20. Loo L, Guenther RH, Basnayake VR, Lommel SA, Franzen S (2006) Controlled encapsidation of gold nanoparticles by a viral protein shell. J Am Chem Soc 128:4502–4503
21. Chen C, Daniel MC, Quinkert ZT, De M, Stein B, Bowman VD, Chipman PR, Rotello VM, Kao CC, Dragnea B (2006) Nanoparticle-templated assembly of viral protein cages. Nano Lett 6:611–615
22. Aniagyei SE, Kennedy CJ, Stein B, Willits DA, Douglas T, Young MJ, De M, Rotello VM, Srisathiyanarayanan D, Kao CC, Dragnea B (2009) Synergistic effects of mutations and nanoparticle templating in the self-assembly of cowpea chlorotic mottle virus capsids. Nano Lett 9:393–398
23. Daniel MC, Tsvetkova IB, Quinkert ZT, Murali A, De M, Rotello VM, Kao CC, Dragnea B (2010) Role of surface charge density in nanoparticle-templated assembly of bromovirus protein cages. ACS Nano 4:3853–3860
24. Young M, Debbie W, Uchida M, Douglas T (2008) Plant viruses as biotemplates for materials and their use in nanotechnology. Annu Rev Phytopathol 46:361–384
25. Glasgow JE (2014) *Encapsulation of biomolecules in bacteriophage MS2 viral capsids.* Dissertation, University of California, Berkeley
26. Capehart S L (2014) *New synthetic methods for integrating metal nanoparticles with biomolecules.* Dissertation, University of California, Berkeley
27. Glasgow JE, Capehart SL, Francis MB, Tullman-Ercek D (2012) Osmolyte-mediated encapsulation of proteins inside MS2 viral capsid. ACS Nano 6:8658–8664
28. Glasgow JE, Asensio MA, Jakobson CM, Francis MB, Tullman-Ercek D (2015) The influence of electrostatics on small molecule flux through a protein nanoreactor. ACS Synth Biol 4:1011–1019
29. Capehart SL, Coyle MP, Glasgow JE, Francis MB (2013) Controlled integration of gold nanoparticles and organic fluorophores using synthetically modified MS2 viral capsids. J Am Chem Soc 135:3011–3016
30. Hooker JM, Esser-Kahn AP, Francis MB (2006) Modification of aniline containing proteins using an oxidative coupling strategy. J Am Chem Soc 128:15558–15559
31. Carrico ZM, Romanini DW, Mehl RA, Francis MB (2008) Oxidative coupling of peptides to a virus capsid containing unnatural amino acids. Chem Commun 10:1205–1207. https://doi.org/10.1039/b717826c

32. Mehl RA, Anderson JC, Santoro SW, Wan L, Martin AB, King DS, Horn DM, Schultz PG (2003) Generation of a bacterium with a 21 amino acid genetic code. J Am Chem Soc 125:935–939
33. Loweth CJ, Caldwell WB, Peng X, Alivisatos AP, Schultz PG (1999) DNA-based assembly of gold nanocrystals. Angew Chem Int Ed 38:1808–1812
34. Hurst SJ, Lytton-Jean AKR, Mirkin CA (2006) Maximizing DNA loading on a range of gold nanoparticle sizes. Anal Chem 78:8313–8318
35. Street TO, Bolen DW, Rose GD (2006) A molecular mechanism for osmolyte-induced protein stability. Proc Nat Acad Sci USA 103:13997–14002
36. Zhang Y, Cremer PS (2010) Chemistry of Hofmeister anions and osmolytes. Ann Rev Phys Chem 61:63–83
37. Zacharias DA, Violin JD, Newton AC, Tsien RY (2002) Partitioning of lipid-modified monomeric GFPs into membrane microdomains of live cells. Science 296:913–916
38. ElSawy K, Caves L, Twarock R (2010) The impact of viral RNA on the association rates of capsid protein assembly: bacteriophage MS2 as a case study. J Mol Biol 400:935–947
39. Rolfsson O, Toropova K, Ranson NA, Stockley PG (2010) Mutually-induced conformational switching of RNA and coat protein underpins efficient assembly of a viral capsid. J Mol Biol 401:309–322
40. https://www.tedpella.com/gold_html/gold-sols.htm, Accessed 28 Apr 2017
41. Claridge SA, Goh SL, Fréchet JMJ, Williams SC, Micheel CM, Alivisatos AP (2005) Directed assembly of discrete gold nanoparticle groupings using branched DNA scaffolds. Chem Mater 17:1628–1635
42. Claridge SA, Liang HW, Basu SR, Fréchet JMJ, Alivisatos AP (2008) Isolation of discrete nanoparticle-DNA conjugates for plasmonic applications. Nano Lett 8:1202–1206

Chapter 22

Delivering Cargo: Plant-Based Production of Bluetongue Virus Core-Like and Virus-Like Particles Containing Fluorescent Proteins

Eva C. Thuenemann and George P. Lomonossoff

Abstract

This chapter provides a practical guide to the *in planta* transient production of bluetongue virus-like particles containing a fluorescent cargo protein. Bluetongue virus (BTV) particles are icosahedral, multishelled entities of a relatively large size. Heterologous expression of the four main structural proteins of BTV results in the assembly of empty virus-like particles which resemble the native virus externally, but are devoid of nucleic acid. The space within the particles is sufficient to allow incorporation of relatively large cargo proteins, such as green fluorescent protein (GFP), by genetic fusion to the structural protein VP3. The method described utilizes the pEAQ vectors for high-level transient expression of such particles in *Nicotiana benthamiana*.

Key words Bluetongue virus, Chimeric, Fluorescent, pEAQ vectors, *HyperTrans*, Transient, Heterologous, Plant

1 Introduction

In recent years, virus-like particles (VLPs) have been the focus of much research and attention [1]. These particles are composed of the structural components of a virus, but lack the viral genome thereby making them noninfectious and unable to replicate. In mimicking the virus in structure and antigenicity, VLPs have been shown to be useful as safe and effective vaccines, with the first VLP vaccines approved for human use in 2007 (Cervarix, GlaxoSmithKline).

Beyond their use as vaccines, VLPs have also proved to be a useful tool in nanotechnology. Many rod-shaped and icosahedral viruses form highly regular structures of a defined size. When expressed as a VLP, these structures can be used as scaffolds for presentation of peptides or small molecules on the outside or inside surface of the particle [2]. Modification of particle surfaces

Christina Wege and George P. Lomonossoff (eds.), *Virus-Derived Nanoparticles for Advanced Technologies: Methods and Protocols*, Methods in Molecular Biology, vol. 1776, https://doi.org/10.1007/978-1-4939-7808-3_22,
© Springer Science+Business Media, LLC, part of Springer Nature 2018

has also enabled researchers to perform metallization, thereby producing nanowires or metal nanoparticles of a defined size [3]. Such modified VLPs have the potential to be used in a wide array of fields from medicine to live-cell imaging and electronics.

Modification of the inside surface of VLPs can be a particularly powerful tool because it leaves the outside surface largely unmodified and therefore produces particles which still resemble the native virus [4]. However, not all VLPs lend themselves to such internal modification, especially in the case of genetic fusion of a cargo protein to the particle structure. The VLP needs to be able to assemble in the absence of nucleic acid, thereby leaving a cavity large enough to accommodate the cargo protein [5]. Also, the number of copies of the viral protein needed to form the virus capsid must be low enough so that fusion of another protein does not cause steric hindrance and abolish particle formation. Both in terms of size of cavity and copy number of the innermost structural protein, bluetongue virus (BTV) lends itself particularly well to such modifications and has been shown to accommodate green fluorescent protein (GFP) by fusion to VP3 [6].

Bluetongue virus is a nonenveloped, icosahedral virus of the *Reoviridae* family. Expression of the structural proteins of BTV in heterologous systems such as insect cells and plants has been shown to produce virus-like particles and core-like particles (CLPs) which are both devoid of nucleic acid [7–9]. BTV has four main structural proteins which assemble in three concentric shells to form the full VLP, or two concentric shells to form a CLP (*see* Fig. 1). The innermost layer is composed of 120 copies of VP3 (103 kDa) which assemble into a thin icosahedral shell with an internal diameter of 46 nm. Upon this VP3 scaffold, a layer of 780 copies of VP7 (37 kDa) is assembled in the form of 260 trimers which give the CLP a characteristic spikey appearance when visualized using transmission electron microscopy (TEM). CLPs consisting of VP3 and VP7 are stable structures whose assembly is independent of the presence of further structural proteins. In order to form a full VLP which resembles the native virus, a further layer consisting of 180 copies of VP2 (111 kDa) and 360 copies of VP5 (59 kDa) must be assembled. Full VLPs are larger than CLPs and have a visibly thicker protein coat when imaged using TEM [7].

Particularly well assembled and stable BTV VLPs and CLPs have been made in plants using the pEAQ vectors [7, 10], and these methods have been extended to the production of particles incorporating GFP and other proteins of interest [11]. The pEAQ vector series allows for rapid, high-level, transient expression of heterologous proteins in leaf tissue using *HyperTrans* (*HT*) technology [12]. Transient expression is a method which

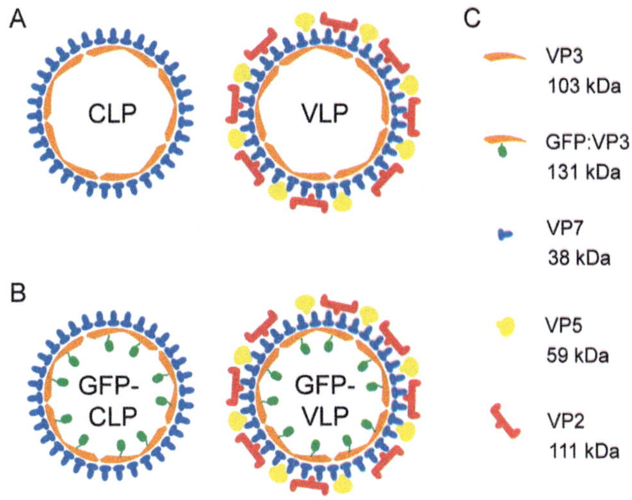

Fig. 1 Schematic representation of bluetongue virus-like particle structure. (**a**) Multishelled structure of core-like particle (CLP) and virus-like particle (VLP). (**b**) Structure of particles encapsulating green fluorescent protein (GFP-CLP and GFP-VLP). (**c**) Names and molecular weights of proteins contained in particles

allows expression of a protein in as little as 4–7 days by means of infiltration of leaf tissue with a solution of *Agrobacterium tumefaciens* containing the gene of interest. Background and other applications of *HT* and the pEAQ vectors have been described in recent reviews [1, 13].

The following method details all the steps necessary to express, extract, purify, and analyze fluorescent BTV VLPs and CLPs from plants. Detailed instructions are also provided to allow replacement of the sequence encoding GFP with any other gene of interest.

2 Materials

2.1 Cloning and Transformation

2.1.1 Enzymes, Kits, Reagents, and Equipment

1. Plant expression plasmids containing the genes for BTV VP3, VP7, VP5, and VP2, as well as GFP linked to VP3. Here, we use pEAQ-GFP:VP3*HT* (*see* **Note 5**), pEAQ-VP7*HT*, and pEAQex-VP5*HT*-VP2*HT* [7].These all contain genes of BTV serotype 8 which have been codon-optimized for plant expression.
2. Phusion® High-Fidelity DNA polymerase (New England Biolabs).
3. Go Taq® Green Master Mix (Promega).
4. Restriction enzymes: *Nru*I-HF, *Age*I-HF, *Xho*I, *Bsp*EI (New England Biolabs) (*see* **Note 1**).
5. PCR Purification Kit (Qiagen).
6. Gel Extraction Kit (Qiagen).

7. Plasmid MiniPrep Kit (Qiagen).
8. T4 DNA ligase.
9. Antibiotics: rifampicin (10 mg/ml stock in methanol), kanamycin (50 mg/ml stock in deionized water).
10. Equipment for PCR and agarose gel electrophoresis.
11. Spectrophotometer.
12. 50% (v/v) glycerol.
13. Electroporation equipment.

2.1.2 Bacterial Strains and Plants

1. OneShot®TOP10 Chemically Competent *E. coli* (Invitrogen).
2. Electrocompetent *Agrobacterium tumefaciens* strain LBA4404 (*see* **Note 2**).
3. *Nicotiana benthamiana* plants grown in a glasshouse with supplemental lighting (16 h per day) and a constant temperature of 24 °C. Plants should be used when they have three fully expanded mature leaves but before flowering commences (4–7 weeks old).

2.1.3 Media and Buffers

1. Luria-Bertani (LB) medium: 10 g/l tryptone, 10 g/l NaCl, 5 g/l yeast extract, pH 7.0 with NaOH. Made with deionized water, autoclaved.
2. LB agar: LB medium with 10 g/l agar, containing appropriate antibiotics.
3. SOC medium: 20 g/l tryptone, 5 g/l yeast extract, 0.58 g/l NaCl, 0.19 g/l KCl, 2.03 g/l $MgCl_2$, 2.46 g/l magnesium sulfate 7-hydrate, 3.6 g/l glucose. Made with deionized water, autoclaved.
4. 2× YT medium: 16 g/l tryptone, 10 g/l yeast extract, 5 g/l NaCl, pH 7.4 with 5 M NaOH. Made with deionized water, autoclaved.

2.2 Agroinfiltration

1. Luria–Bertani (LB) medium (*see* Subheading 2.1.3, **item 1**).
2. MMA medium: 10 mM MES (2-[N-morpholino]ethanesulfonic acid), pH 5.6 with NaOH, 10 mM $MgCl_2$, 100 μM acetosyringone (4′-hydroxy-3′,5′-dimethoxyacetophenone).
3. For syringe infiltration: use 1 ml disposable syringes and syringe needles.
4. For vacuum infiltration: use a vacuum desiccator, beaker, vacuum pump, and vacuum gauge.

2.3 Extraction and Gradient Centrifugation

1. CLP extraction buffer: 50 mM Bicine, pH 8.4 with NaOH, 140 mM NaCl, 0.1% (w/v) NLS (N-lauroylsarcosine) sodium salt, 1 mM DTT (dithiothreitol), Complete Protease Inhibitor Cocktail (Roche) (*see* **Note 3**).

2. CLP purification buffer: 20 mM Tris–HCl, pH 8.4, 140 mM NaCl.
3. CLP 60% sucrose: 20 mM Tris–HCl, pH 8.4, 140 mM NaCl, 60% (w/v) sucrose.
4. VLP extraction buffer: 50 mM Bicine, pH 8.4 with NaOH, 20 mM NaCl, 0.1% (w/v) NLS (N-lauroylsarcosine) sodium salt, 1 mM DTT (dithiothreitol), Complete Protease Inhibitor Cocktail (Roche) (*see* **Note 3**).
5. VLP purification buffer: 20 mM Tris–HCl, pH 8.4, 20 mM NaCl.
6. VLP 60% sucrose: 20 mM Tris–HCl, pH 8.4, 20 mM NaCl (for CLPs, use 120 mM NaCl), 60% (w/v) sucrose. Use this solution to prepare 50%, 40%, and 30% sucrose solutions by diluting in the same buffer.
7. The ultracentrifugation method can be adapted and scaled to any available equipment (*see* **Note 4**). In this protocol, Thinwall Ultra-Clear™ 38.5 ml tubes (Beckman Coulter) are used in a Surespin 630 rotor (Thermo Scientific).
8. Blender and funnel lined with two layers of Miracloth.

2.4 Analysis

2.4.1 Polyacrylamide Gel Electrophoresis (SDS-PAGE)

Any available SDS-PAGE gel making and running equipment may be used. In this protocol, the following are used:

1. NuPAGE™ No.5vex™ 4–12% (w/v) Bis-Tris Protein Gels (Invitrogen).
2. NuPAGE™ MOPS Running Buffer (Invitrogen).
3. Reducing Sample buffer: 750 μl NuPage® LDS Sample Buffer 4× (Invitrogen), 250 μl β-mercaptoethanol. Used as a 3× stock.
4. InstantBlue™ Protein stain (Expedeon).

2.4.2 Transmission Electron Microscopy (TEM)

1. Pyroxylin and carbon-coated 400-mesh copper TEM grids.
2. 2% (w/v) uranyl acetate for negative staining.

3 Methods

3.1 Construct Design (See Fig. 2)

To allow for efficient expression of a foreign protein attached to the N-terminus of BTV VP3, a linker was designed to provide a flexible glycine-glycine-serine amino acid segment between the protein of interest and VP3 (*see* Fig. 2c). This linker also contains the unique restriction site *Bsp*EI. From the plasmid pEAQ-GFP:VP3*HT* (*see* Note 5), the sequence encoding GFP can be excised using *Age*I/*Nru*I and *Bsp*EI (*see* Note 6), and replaced with another gene sequence flanked by *Age*I/*Nru*I and *Bsp*EI sites.

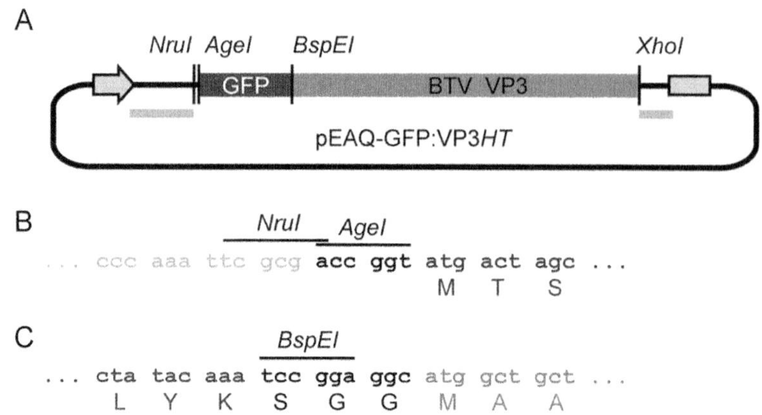

Fig. 2 Schematic representation pEAQ-GFP:VP3HT and cloning site details. (**a**) Structure of the HyperTrans expression cassette within pEAQ-GFP:VP3HT with key restriction sites indicated. Arrow: 35S promoter; grey lines: HyperTrans 5′- and 3′ untranslated regions; grey box: Nos terminator. (**b–c**) DNA sequence of the 5′- and 3′ cloning sites for the replacement of the GFP-encoding sequence with a gene of interest. Amino acids of the coding sequences are indicated beneath each codon. Key restriction sites are indicated above. Shades of grey refer to those used in (**a**)

3.2 Cloning

The following method can be used to link a protein of interest to the N-terminus of BTV VP3.

3.2.1 Insert Preparation

The sequence for the gene of interest is usually derived from another plasmid by PCR.

1. Amplify the gene of interest using high fidelity polymerase (such as Phusion®) and end-tailoring primers to encode *Age*I/*Nru*I and a start codon at the 5′ end, and *Bsp*EI at the 3′ end (thereby eliminating the stop codon, if necessary).
2. Check for successful amplification by resolving a small amount (5 μl) of product on an agarose gel.
3. Purify the PCR product using a PCR purification kit, according to the manufacturer's instructions (*see* **Note 7**).
4. Digest the purified PCR product using *Age*I/*Nru*I and *Bsp*EI (*see* **Note 6**). (10 min at room temperature is sufficient.)
5. Purify the digested PCR product by running it on an agarose gel and using a gel extraction kit, according to the manufacturer's instructions.
6. The purified insert can be quantified spectrophotometrically. Expected concentrations should be >5 ng/μl.

3.2.2 Vector Preparation

1. Digest 1–5 μg of plasmid pEAQ-GFP:VP3*HT* using *Age*I/*Nru*I and *Bsp*EI (*see* **Note 6**).
2. Purify the digested vector by separating the digestion products on an agarose gel and using a gel extraction kit according to the manufacturer's instructions.
3. The purified vector can be quantified spectrophotometrically. Expected concentrations should be >10 ng/μl.

3.2.3 Ligation and Transformation

The *Age*I/*Nru*I and *Bsp*EI digestions have produced fragments and vectors with at least one cohesive end, which allows for efficient ligation of the vector and insert.

1. Set up a ligation reaction using T4 DNA ligase according to the manufacturer's instructions. Vector and insert should be present at a molar ratio of 1:3, with a total DNA content of 50–100 ng in a 10 μl reaction.
2. After 10–30 min of ligation time at room temperature, add 1 μl of the ligation reaction to an aliquot of chemically competent *E.coli* on ice. Leave on ice for 30 min.
3. Heat shock cells at 42 °C for 30 s, followed by 2 min on ice. Immediately add 250 μl of sterile SOC medium and transfer to a shaker at 37 °C for 1 h.
4. Plate 50 μl and 250 μl of the reaction on two LB agar plates containing 50 μg/ml kanamycin. Incubate at 37 °C overnight.

3.2.4 Confirmation of Successful Cloning

1. Perform a colony PCR on a sufficient number of colonies to test for positive insertion of the gene of interest. If *Age*I was used for cloning, it will also be necessary to check for correct orientation of the insert (*see* **Notes 6** and **8**).
2. Grow positive clones at 37 °C overnight in 5 ml 2× YT medium containing 50 μg/ml kanamycin (*see* **Note 9**).
3. Prepare plasmid from the overnight culture using a QIAgen MiniPrep Kit (or equivalent), according to the manufacturer's instructions. Submit this plasmid for sequencing using appropriate gene-specific or backbone-specific primers.
4. This liquid culture can be used to prepare a glycerol stock by mixing 400 μl of 50% glycerol with 700 μl liquid culture, flash-freezing and storing at −80 °C.

3.3 Agrobacterium Transformation

When the construct has been confirmed by DNA sequencing, it can be transformed into electrocompetent *Agrobacterium tumefaciens* (*see* **Note 2**). The pEAQ vectors give good expression when used in *A. tumefaciens* strain LBA4404.

1. Add 0.5 μl of plasmid DNA to an aliquot (50 μl) of electrocompetent LBA4404 on ice. Transfer cells and DNA to a

precooled electroporation cuvette. Tap the cuvette gently on a hard surface to dislodge any air bubbles.

2. Electroporate at 2.5 kV. Immediately add 500 μl of SOC medium to the electroporated cells, transfer to a sterile 1.5 ml microcentrifuge tube and place in a shaking incubator at 28 °C for 1 h.
3. Spread 50 μl and 250 μl onto two LB-agar plates each containing 50 μg/ml kanamycin and 50 μg/ml rifampicin. Incubate at 28 °C for 2–3 days (*see* **Note 10**).
4. Colonies can be screened using colony PCR as described under Subheading 3.2.3. Once confirmed by PCR, use a positive colony to inoculate a 5 ml liquid culture (LB containing 50 μg/ml kanamycin and 50 μg/ml rifampicin) and grow overnight at 28 °C in a shaking incubator (200 rpm).
5. Use the liquid culture to prepare a glycerol stock by mixing 400 μl of 50% glycerol with 700 μl liquid culture and storing at −80 °C.
6. The liquid culture can also be used to inoculate a larger liquid culture for use in inoculum preparation (*see* Subheading 3.4).

3.4 Inoculum Preparation

1. Prepare a 5 ml liquid culture (LB with 50 μg/ml kanamycin and 50 μg/ml rifampicin) by inoculating from a colony, as above, or straight from glycerol stock. In addition to pEAQ-GFP:VP3*HT*, a separate *Agrobacterium* culture of construct pEAQ-VP7*HT* will be needed to produce CLPs, and pEAQ-VP7*HT* as well as pEAQex-VP5*HT*-VP2*HT* (*see* **Note 11**) will be needed to produce VLPs [7].
2. Use the overnight 5 ml liquid cultures to inoculate larger cultures for inoculum preparation. The volume of culture needed will depend on the number of plants that will be infiltrated. As a rule of thumb (for syringe infiltration), 5 ml of inoculum will be needed per plant, which can be prepared from 1 ml of grown liquid culture. This is a conservative estimate, ensuring that enough culture and inoculum will be available for the experiment (*see* **Note 12**).
3. Measure the optical density (OD_{600}) of each culture, then use the following formula to calculate the volume of culture needed to prepare the desired volume of inoculum at the desired OD_{600}:

$$V_{\text{culture}} = \frac{V_{\text{inoculum}} \times \text{OD}_{\text{inoculum}}}{\text{OD}_{\text{culture}}}$$

where V_{culture} is the volume of culture needed, V_{inoculum} is the volume of inoculum to be prepared, $\text{OD}_{\text{inoculum}}$ is the desired

OD_{600} of the inoculum, $OD_{culture}$ is the OD_{600} of the grown culture (*see* **Notes 13** and **14**).

4. Transfer the required volume of culture into a 50 ml centrifuge tube. When performing a coinfiltration, mix the required volumes of each constituent at this stage (*see* **Note 15**). Centrifuge at 4000 × *g* for 10 min to pellet the bacteria, then resuspend the pellets in MMA medium (*see* **Note 16**).
5. Leave the inoculum to rest at room temperature for 0.5–3 h before infiltration.

3.5 Agroinfiltration

Agroinfiltration can be performed in two ways: syringe infiltration or vacuum infiltration. The syringe infiltration method is useful for small-scale experiments (1–20 plants) as well as experiments that compare several different constructs in parallel. Vacuum infiltration can be useful for scale-up of a single construct (or combination of constructs) as multiple whole plants may be infiltrated simultaneously, depending on the equipment used. A description of the processes is provided below and useful videos of the methods have been published [14].

3.5.1 Syringe Infiltration

Prepare at least 5 ml of inoculum per plant to be infiltrated.

1. Make a small wound in the top layer of leaf tissue using a needle; try to avoid making a hole right through the leaf.
2. Fill a 1 ml syringe (without needle) with inoculum, making sure to remove bubbles.
3. Place a finger onto the underside of the leaf, just beneath the wound. Cover the wound with the end of the syringe, applying gentle pressure to ensure a seal between the syringe and the leaf surface, while not pressing too hard as this would compress the leaf tissue.
4. Gently inject the leaf with inoculum by depressing the plunger. Leaf tissue that has been infiltrated is distinctly darker than tissue that has not been infiltrated.
5. When the infiltrated area no longer increases in size, stop and make a new wound in an area of the leaf which has not yet been infiltrated.
6. Repeat with two more leaves (*see* **Notes 17** and **18**).

3.5.2 Vacuum Infiltration

This procedure requires a basic vacuum infiltration apparatus, which can consist of a vacuum chamber, beaker, vacuum pump and vacuum gauge.

1. Prepare enough inoculum to fill the beaker, plus extra to top up the level between plants. Place the beaker inside the vacuum chamber.

2. Invert a plant and submerge the leaves and stem below the inoculum. Note that some method of holding the pot and soil will need to be devised. Infiltration is most efficient if the leaves are completely submerged (*see* **Note 19**).
3. Close the vacuum chamber and evacuate using a vacuum pump until approx. 100–200 mbar (i.e., ca. 80–90% vacuum) is achieved. Maintain the vacuum for 2 min.
4. Gently release the vacuum. While the vacuum is released, any intercellular spaces will be filled with inoculum.
5. At the end of the procedure, return the plant to an upright position, rinse the plant with water and continue to water and grow as normal.

3.6 Harvesting and Extraction (See Fig. 3)

To purify BTV particles, leaf tissue is harvested at 8 days post-infiltration. When expressing a GFP-containing particle, fluorescence can be detected using a UV lamp (*see* Fig. 3b).

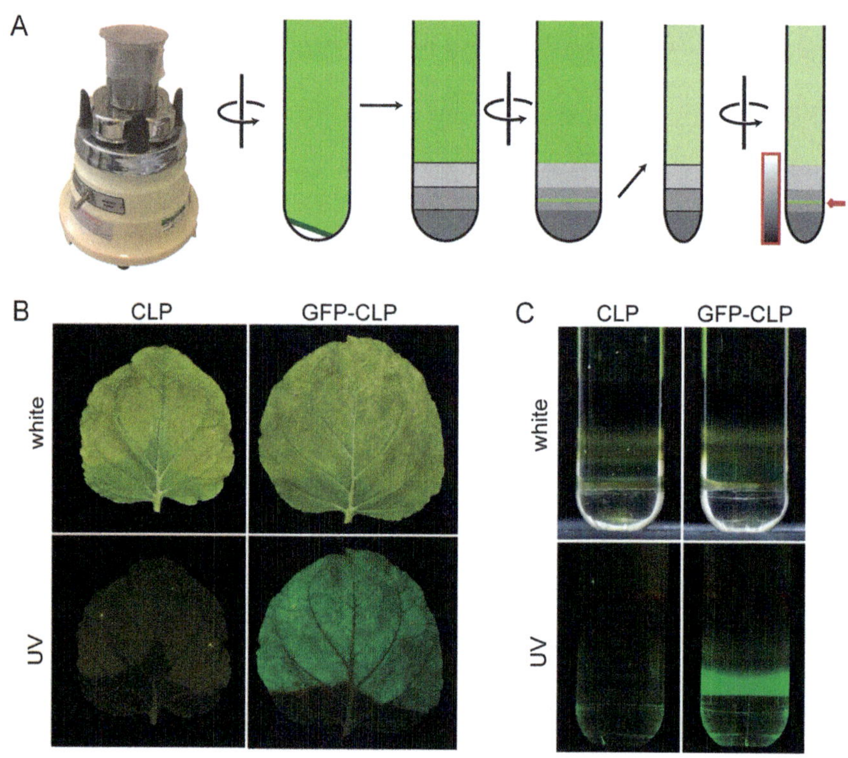

Fig. 3 Method for purification of BTV particles. (**a**) Schematic of the main steps of the purification protocol. Leaves are blended, then subjected to clarification followed by gradient ultracentrifugation. The fractions containing the banded particles are diluted and subjected to a second round of gradient ultracentrifugation. (**b**) Photographs of leaves expressing CLP (VP3 + VP7) and fluorescent GFP-CLP (GFP:VP3 + VP7). Leaves are imaged under white light (top) and UV light (bottom). (**c**) Photographs of gradient ultracentrifugation tubes from the first round of centrifugation, showing sedimentation of CLP and GFP:CLP. Tubes are imaged under white light (top) and UV light (bottom)

1. Collect any infiltrated leaves and excise uninfiltrated areas, if necessary.
2. Extract leaf tissue in a blender with three volumes of BTV extraction buffer (CLP or VLP, depending on infiltrated constructs). Blend until homogeneous (30–60 s at 22,000 rpm; *see* Fig. 3a).
3. Pour the homogenate through a funnel lined with two layers of Miracloth. Carefully squeeze the Miracloth to wring out remaining extract. Discard the plant debris-containing Miracloth.
4. Clarify the crude extract by centrifugation at 4200 × *g*, 10 °C for 5 min. The supernatant constitutes the clarified extract.

3.7 Gradient Ultracentrifugation (See Fig. 3)

The BTV particles are partially purified on sucrose step gradients, containing 60%, 50%, 40%, and 30% (w/v) sucrose.

1. Prepare the step gradient by layering 3 ml of each of the sucrose solutions in a 36 ml ultracentrifuge tube. This is easily done by starting with the 30% sucrose solution, then using a syringe and long canula or needle to underlay the 40% solution, then 50% and finally 60%.
2. Finally, carefully overlay ~26 ml of clarified extract until the tube is full (*see* **Note 20**).
3. Centrifuge gradients at 86,000 × *g*, 10 °C for 3 h using a SureSpin 360 swing-out rotor with 36 ml buckets.
4. Fractionate the gradients by piercing the bottom of the tube and collecting ten fractions of 1 ml each. The remaining supernatant can be discarded (*see* **Note 4**).

The most intact BTV particles are normally contained within the first four fractions. For particles containing GFP, these can be easily located using a handheld UV lamp to examine the gradients and/or fractions (*see* Fig. 3c).

To obtain a CLP or VLP sample of higher purity, a second round of ultracentrifugation can be performed. For this, the fractions containing the majority of the particles from the first ultracentrifugation are pooled, diluted to <30% sucrose and applied to another sucrose gradient. This second gradient can be prepared in a smaller tube (*see* Fig. 3a).

3.8 Analysis

3.8.1 Polyacrylamide Gel Electrophoresis (See Fig. 4a, b)

SDS polyacrylamide gel electrophoresis (SDS-PAGE) is a useful tool for assessing the presence and ratio of BTV proteins in the various gradient fractions. Fully formed BTV particles characteristically band at a higher density than most plant contaminants, therefore fractions tend to be quite clean and BTV proteins are easy to identify.

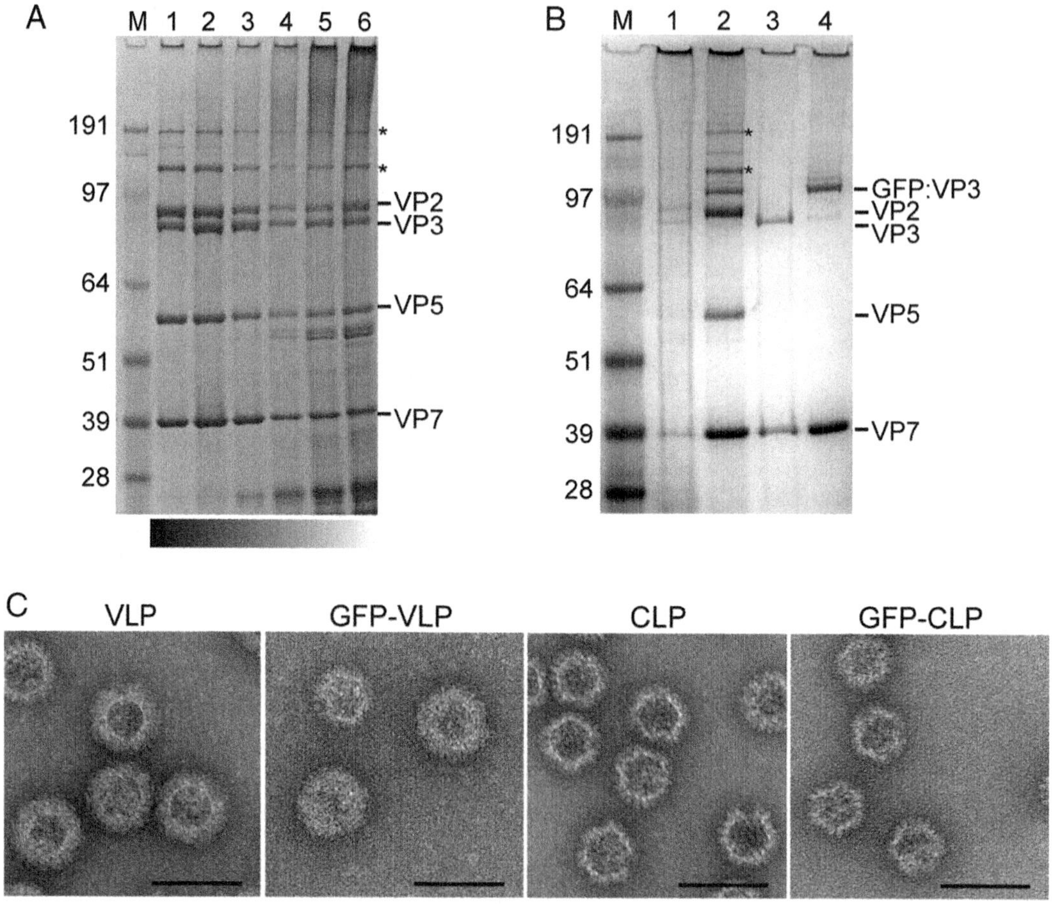

Fig. 4 Examples of expected results from analyses of particles. (**a**–**b**) Polyacrylamide gel electrophoresis and Coomassie Blue staining gradient purified particles (**a**) Sucrose gradient fractions (1–6, 50%–30% sucrose) of a VLP sample. (**b**) Purified samples of VLP (1), GFP-VLP (2), CLP (3) and GFP-CLP (4). Structural proteins are indicated on the right. M: Molecular weight marker; *: these bands have been identified by mass spectrometry as multimers of VP5. (**c**) Transmission electron micrographs of different particles, as indicated above. Scale bar = 100 nm. pEAQ-GFP:VP3 was used in these experiments (*see* **Note 5**)

1. Prepare samples by boiling in reducing sample buffer for 5 min. Load 10 μl per lane on a polyacrylamide gel.
2. After separation, stain the gel with Coomassie Blue-based stain, such as Instant Blue.
3. Subsequent densitometric analysis may provide more information on the ratio of the proteins present in each sample.

3.8.2 Transmission Electron Microscopy (See Fig. 4c)

Transmission electron microscopy (TEM) is a crucial tool for assessing the integrity of particles in each sample. Morphological differences can be observed between assembled CLPs and VLPs, and particle size is also characteristic between the two types of particle.

1. Dilute the sample twofold in CLP or VLP purification buffer. Spot 20 μl of sample onto a plastic and carbon-coated TEM grid (400 mesh).
2. After 20 s, rinse the grid to remove surplus sample and sucrose. This can be done by gently flushing the grid or by floating it on a sequence of 3–5 water droplets, and blotting in between. Take care not to dry the grid out completely.
3. Stain by adding a drop of uranyl acetate for 20 s, then blot to dry.

A good magnification for analysis is in the range of 20,000–50,000×.

3.8.3 Other Analysis Tools

A series of other analysis techniques can be useful for further characterizing the particles.

1. Western blotting and immunodetection using an antibody specific to the protein of interest fused to VP3. This will show that the protein of interest is present and whether any of it has been cleaved off.
2. Agarose gel electrophoresis can be used to determine particle migration in agarose. Gels can be stained with Coomassie Blue and/or Ethidium Bromide and can reveal whether or not nucleic acid is contained within the particles.
3. Dynamic light scattering (DLS). This will reveal whether a suspension of particles is monodisperse or includes clumps of particles. It can also be used to give an indication of particle size.
4. Absorbance at 280 nm (A280). This can be used to determine the concentration of a pure preparation.

4 Notes

1. These restriction enzymes all work in the same conditions (buffer, temperature) and can be used in a double digest.
2. Electrocompetent *Agrobacterium tumefaciens* strain LBA4404 can be prepared using any conventional protocol for preparation of electrocompetent cells. Grow cells at 28 °C and use rifampicin (50 μg/ml) to avoid contamination and maintain its functionality. *Agrobacterium* is slow-growing compared to other bacteria, with a doubling time of 2–4 h depending on growth conditions. Alternatively, electrocompetent LBA4404 can be purchased from a range of suppliers.

3. This buffer can be made in advance; however, DTT and Protease Inhibitor cocktail should be added fresh before use.
4. The ultracentrifugation steps can be scaled to any available ultracentrifuge rotors and tubes. For instance, for tubes that have half of the volume of that described here, reduce gradient step volumes by half. At any given speed, the centrifugal force will be dependent on the diameter of the rotor, therefore the speed of centrifugation may need to be adapted. When trying this protocol in a rotor assembly for the first time, it is best to fractionate and analyze the entire gradient to determine where the particles have banded.
5. pEAQ-GFP:VP3*HT* is based on pEAQ-*HT* (GenBank accession GQ497234) and contains the *HyperTrans* mutation in the 5′ untranslated region which enhances translation efficiency thereby producing high protein yields [11]. It has been found that such high overexpression can lead to the formation of large numbers of subcore-like particles (SCLPs) consisting of only VP3 (or GFP:VP3 in this case) and thereby producing a mixture of SCLPs, CLPs, and VLPs. This can be suppressed by using a version of this plasmid without the *HyperTrans* mutation, termed pEAQ-GFP:VP3 [7].
6. *Age*I and *Bsp*EI produce compatible cohesive ends. If both of these enzymes are used, it will be necessary to screen resulting colonies both for presence of the insert as well as its orientation. An alternative is to use *Nru*I and *Bsp*EI to produce a fragment with one blunt end (*Nru*I) and one cohesive end (*Bsp*EI). This will ensure cloning in the desired orientation (*see* Fig. 2).
7. Depending on how the PCR product was resolved on the gel, it may be necessary to perform a gel purification at this step.
8. To check for correct orientation of the insert during colony PCR, it is possible to perform the PCR with a pair of primers where one binds on the vector backbone close to the insertion site and the other binds within the insert.
9. The pEAQ vectors generally give relatively low yield in plasmid preps. However, it has been found that 2×YT media gives higher plasmid yields (200–300 ng/μl) than LB media.
10. It is most time-efficient to perform this step on a Friday, allowing colonies to form over the weekend ready for colony PCR on Monday morning.
11. pEAQex-VP5*HT*-VP2*HT* is a plasmid based on pEAQexpress which enables two *HyperTrans* expression cassettes to be cloned onto the same binary vector [10]. Using this plas-

mid, bluetongue virus VP5 and VP2 can be coexpressed from the same *Agrobacterium*.

12. *Agrobacterium* has a doubling time of approx. 2 h, therefore inoculation of a culture at 6 pm with an OD_{600} of 0.01 will yield a grown culture of approx. OD_{600} 2.5 at 10 am on the following morning.

13. To ensure efficient transfer of T-DNA into each plant cell, a minimum optical density (OD_{600}) of 0.1 should be used. Ideally, the optical density of each *Agrobacterium* strain should be in the range of 0.2–0.4.

14. For example, to prepare 100 ml of inoculum ($V_{inoculum}$) at OD_{600} = 0.2 ($OD_{inoculum}$) from a culture of OD_{600} = 2.5 ($OD_{culture}$), use 8 ml [(100 ml × 0.2)/2.5] of grown culture.

15. When coinfiltrating, the optical density of each constituent *Agrobacterium* strain is kept at a level that would ensure efficient transfer into each plant cell (e.g., OD_{600} of 0.2). Therefore, the total OD of the inoculum will be a multiple of this OD. For example, coinfiltration of three constructs at OD_{600} = 0.2 would result in an overall OD of 0.6.

16. It is easiest to resuspend the pellet in a small volume (1–5 ml) first using a pipette and gently washing over the pellet repeatedly. When the pellet is fully resuspended without clumps, the remaining volume of MMA required for the desired inoculum volume can be added.

17. The efficiency and ease of infiltration can vary depending on growth conditions (lighting, temperature, humidity). If plants are grown in a greenhouse, for example, they may be more difficult to infiltrate on a particularly warm and sunny day. To ease infiltration, it can be helpful to water the plants and to place them in the shade before infiltration. This will increase the local humidity around the plants and allow stomata to open.

18. Typically, older leaves (near the bottom of the plant) are easier to infiltrate but give lower expression levels than younger leaves (top of the plant).

19. It has been found that when even a small part of a leaf is in contact with the inoculum surface, the whole leaf may not be infiltrated efficiently. It may therefore be necessary to devise a mechanism of keeping the leaves fully submerged.

20. Layering of gradients can be done in a number of different ways, and the presented method is only one example. For instance, one could also start with the densest solution and then overlay progressively less dense solutions by careful pipetting. Alternatively, a gradient maker could be used.

Acknowledgments

This work was supported by the UK Biotechnological and Biological Sciences Research Council (BBSRC) Institute Strategic Programme Grant "Understanding and Exploiting Plant and Microbial Secondary Metabolism" (BB/J004596/1), the BBSRC Synthetic Biology Research Centre 'OpenPlant' award (BB/L014130/1), and the John Innes Foundation. We thank Andrew Davis for photography.

References

1. Marsian J, Lomonossoff GP (2016) Molecular pharming—VLPs made in plants. Curr Opin Biotech 37:201–206. https://doi.org/10.1016/j.copbio.2015.12.007
2. Peyret H, Gehin A, Thuenemann EC, Blond D, El Turabi A, Beales L, Clarke D, Gilbert RJ, Fry EE, Stuart DI, Holmes K, Stonehouse NJ, Whelan M, Rosenberg W, Lomonossoff GP, Rowlands DJ (2015) Tandem fusion of hepatitis B core antigen allows assembly of virus-like particles in bacteria and plants with enhanced capacity to accommodate foreign proteins. PLoS One 10(4):e0120751. https://doi.org/10.1371/journal.pone.0120751
3. Aljabali AAA, Barclay JE, Lomonossoff GP, Evans DJ (2010) Virus templated metallic nanoparticles. Nanoscale 2(12):2596–2600. https://doi.org/10.1039/c0nr00525h
4. Wen AM, Shukla S, Saxena P, Aljabali AAA, Yildiz I, Dey S, Mealy JE, Yang AC, Evans DJ, Lomonossoff GP, Steinmetz NF (2012) Interior engineering of a viral nanoparticle and its tumor homing properties. Biomacromolecules 13(12):3990–4001. https://doi.org/10.1021/bm301278f
5. Rurup WF, Verbij F, Koay MST, Blum C, Subramaniam V, Cornelissen JJLM (2014) Predicting the loading of virus-like particles with fluorescent proteins. Biomacromolecules 15(2):558–563. https://doi.org/10.1021/bm4015792
6. Kar AK, Iwatani N, Roy P (2005) Assembly and intracellular localization of the bluetongue virus core protein VP3. J Virol 79(17):11487–11495. https://doi.org/10.1128/Jvi.79.17.11487-11495.2005
7. Thuenemann EC, Meyers AE, Verwey J, Rybicki EP, Lomonossoff GP (2013) A method for rapid production of heteromultimeric protein complexes in plants: assembly of protective bluetongue virus-like particles. Plant Biotechnol J 11(7):839–846. https://doi.org/10.1111/pbi.12076
8. Hewat EA, Booth TF, Loudon PT, Roy P (1992) 3-Dimensional reconstruction of baculovirus expressed bluetongue virus core-like particles by cryoelectron microscopy. Virology 189(1):10–20
9. Hewat EA, Booth TF, Roy P (1994) Structure of correctly self-assembled bluetongue virus-like particles. J Struct Biol 112(3):183–191
10. Sainsbury F, Thuenemann EC, Lomonossoff GP (2009) pEAQ: versatile expression vectors for easy and quick transient expression of heterologous proteins in plants. Plant Biotechnol J 7(7):682–693. https://doi.org/10.1111/j.1467-7652.2009.00434.x
11. Sainsbury F, Lomonossoff GP (2008) Extremely high-level and rapid transient protein production in plants without the use of viral replication. Plant Physiol 148(3):1212–1218. https://doi.org/10.1104/pp.108.126284
12. Peyret H, Lomonossoff GP (2013) The pEAQ vector series: the easy and quick way to produce recombinant proteins in plants. Plant Mol Biol 83(1–2):51–58. https://doi.org/10.1007/s11103-013-0036-1
13. Leuzinger K, Dent M, Hurtado J, Stahnke J, Lai HF, Zhou XH, Chen Q (2013) Efficient agroinfiltration of plants for high-level transient expression of recombinant proteins. J Vis Exp (77). https://doi.org/10.3791/50521
14. Brillault L, Jutras PV, Dashti N, Thuenemann EC, Morgan G, Lomonossoff GP, Landsberg MJ, Sainsbury F (2017) Engineering Recombinant Virus-like Nanoparticles from Plants for Cellular Delivery. ACS Nano 11 (4):3476–3484

Part III

Functions Exposed on Virus Backbones and Combined Approaches: From Fabrication to Technical Uses of Hybrid Particles, Films, and Materials

Chapter 23

Bioinspired Silica Mineralization on Viral Templates

Christina Dickmeis, Klara Altintoprak, Patrick van Rijn, Christina Wege, and Ulrich Commandeur

Abstract

Plant virus capsids are attractive entities for nanotechnological applications because of their variation in shape and natural assembly ability. This chapter describes the production and modification of three differently shaped plant virus capsids for silica mineralization purposes. The chosen plant viruses exhibit either an icosahedral (cowpea mosaic virus, CPMV), or a flexuous rod-like structure (potato virus X, PVX), or a rigid rod-like shape (tobacco mosaic virus, TMV), and are well-known and frequently used plant viruses for biotechnological applications. We describe the production (including genetic or chemical modification) and purification of the plant viruses or of empty virus-like particles in the case of CPMV, as well as the characterization of these harvested templates. The mineralization procedures and differences in the protocols specific to the distinct viruses are described, and the analyses of the mineralization results are explained.

Key words Plant virus, Coat protein, CPMV, PVX, TMV, Virus-like particles (VLPs), Nanoparticles, Functionalization, Mineralization, Silica

1 Introduction

Capsids of plant viruses are excellent tools for nanotechnological applications. They consist of multiple copies of identical protein subunits, which assemble with or without their genomic information (DNA or RNA) into virus or virus-like particles (VLPs), respectively. The coat protein (CP) subunits can be modified by genetic engineering, chemical conjugation, or both, thus providing different possibilities for a selective attachment and presentation of organic and inorganic molecules. Such functional molecules may include metals or semiconductors, carbohydrates, polypeptides, and proteins [1–4]. The virus particles can be produced in large amounts using plants as bioreactors with inexpensive culture

Christina Dickmeis and Klara Altintoprak contributed equally to this chapter.

Christina Wege and George P. Lomonossoff (eds.), *Virus-Derived Nanoparticles for Advanced Technologies: Methods and Protocols*, Methods in Molecular Biology, vol. 1776, https://doi.org/10.1007/978-1-4939-7808-3_23,

© Springer Science+Business Media, LLC, part of Springer Nature 2018

conditions [5–9]. Several different plant viruses have been developed as peptide-presentation systems to date. The most widely used so far are particles of cowpea mosaic virus (CPMV) [10] and tobacco mosaic virus (TMV) [2, 11].

CPMV is a plus-strand RNA virus with a bipartite genome comprising RNA-1 (5.8 kb) and RNA-2 (3.3 kb). CPMV particles consist of two different CPs, the large (L) and small (S) subunit, which are encoded by RNA-2 and processed from the same precursor molecule by the RNA-1-encoded viral proteinase. Sixty copies of each CP are arranged into an icosahedral particle with pseudo $T = 3$ ($P = 3$) symmetry and around 28 nm diameter [12]. Three different kinds of particles can be isolated from infected plants by density gradient centrifugation: Top (T), Middle (M), and Bottom (B) components. These represent empty virus-like particles (eVLPs) (T), particles containing RNA-2 (M), and particles containing RNA-1 (B) [13, 14]. eVLPs of CMPV can be easily produced recombinantly by the coexpression of the proteinase and the L and S precursor, VP60 [15]. They can be used for the functionalization and presentation of peptides on the particle surface, as well as for loading of the particle interior [15, 16].

TMV particles comprise 2130 CP subunits assembled with a single viral RNA to form the virion structure [17]. The virions are 300 × 18 nm rods with a central channel 4 nm in diameter [18–20]. TMV particles have been used to present foreign peptides mainly as C-terminal CP fusions. A leaky stop codon can be used to ensure the coexpression of native CPs simultaneously with the engineered CP species, resulting in TMV nanotubes equipped with a subset of peptide-fashioned CP subunits [21]. Peptide presentations can also be achieved by inserting the foreign sequence into an N-terminal loop region of the CP [22, 4]. TMV CPs are also in the focus of research as building blocks for nanomaterials because of their ability to form virus-like particles (VLPs) in the absence of viral RNA [23–26]. These structures are suitable for metallization [27, 28] and functionalization with organic and inorganic molecules [29]. This is promoted by the possibility of genetically introducing selectively addressable amino acids into the outer TMV CP surface, allowing simple chemical coupling reactions. The strategy also enables efficient linker-mediated immobilization of peptides to moderately engineered TMV variants such as TMV_{Lys} [30, 31], which is the fabrication route for the TMV-based templates employed in the following protocol.

Rod-shaped as well as filamentous virus particles are in several cases advantageous over icosahedral viruses because they have no intrinsic size limitations or RNA packaging constraints, expanding the possibilities for generating novel CP fusion proteins [32] and particles of altered length or even shape [33, 34]. In this regard, another promising plant virus for peptide and protein presentation is potato virus X (PVX) [35]. Approximately 1270 CP subunits assemble with the genomic plus-strand RNA to form virus particles as 515 × 13.5 nm flexuous rods [36].

In the last two decades, several protocols for a site-selective silica (silicon oxide, i.e., SiO_2) mineralization on the outer surface of TMV particles have been developed [37–40, 31]. Inspired by naturally existing protein-based biomineralization systems such as silaffins [41, 42], TMV and TMV derivatives have served as templates for SiO_2 deposition in a number of laboratories [37–40, 43, 31, 44]. The choice of the most suitable method depends on the desired outcome, e.g., low by-product amounts, silica shell thickness, surface quality of the mineral (e.g., rough or smooth), monodispersity or structural integrity of the templates.

The SiO_2 mineralization of TMV is mainly accomplished by the use of a sol–gel process in the presence of silica precursors. In a first step, alkoxysilane educts are hydrolyzed under acidic conditions resulting in silanols and silicic acid. In the following reaction under neutral or basic conditions, silicic acid molecules condensate into an amorphous silica gel (*see* Fig. 1). In contrast to the Stöber process [45], where the condensation reaction of silicic acid is catalyzed by the addition of ammonium hydroxide, the template is the catalytically active and structuring element. Commonly, alkoxysilanes like tetraethyl orthosilane (TEOS), tetramethyl orthosilane (TMOS), or (3-aminopropyl)triethoxysilane (APTES) are used as precursors for silica deposition on the templates. In the presence of an appropriate template, silicic acid molecules polymerize at low concentration solely at the template's surface, while unspecific condensation of the silica precursors in solution is minimal. Consequently, lower amounts of by-products, such as spherical particles, are formed. However, the applied template should be stable in ethanol or methanol for a certain time as is the case for TMV [39, 46]. Alternative approaches without the addition of alcohol have been established to mineralize TMV particles or TMV-derived disk-shaped coat protein rings by the use of a TEOS–APTES mixture, or a silicic acid precursor generated from TMOS [38, 44].

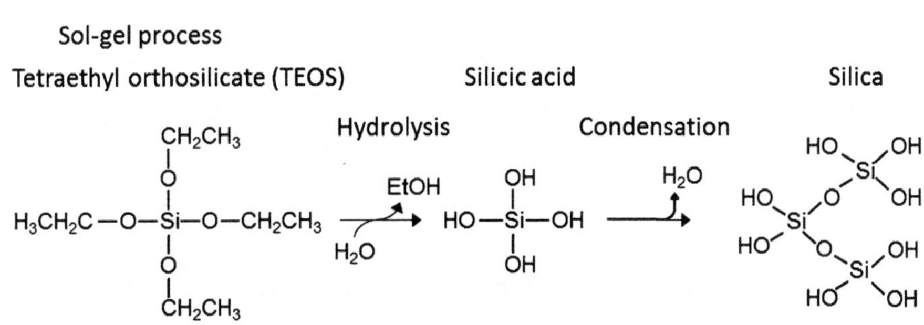

Fig. 1 Reaction mechanism of silica formation with the silicic acid precursor TEOS

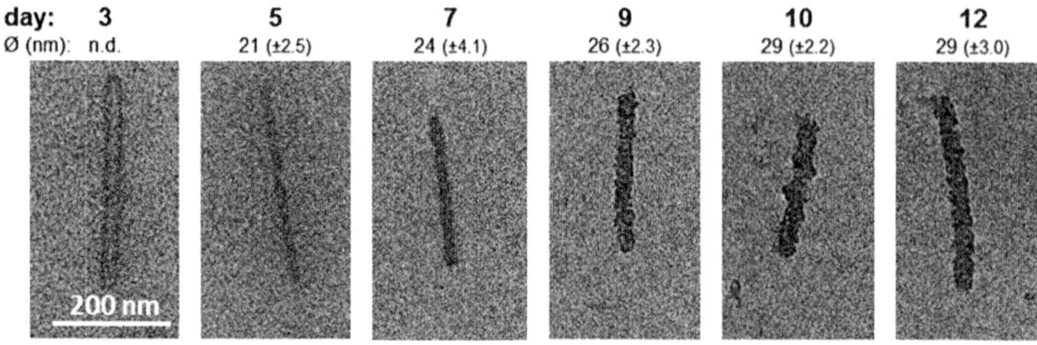

Fig. 2 TEM sample series of TMV particles functionalized with $(KD)_{10}C$ undergoing continuously proceeding mineralization, collected from the reaction mixture at different time points. The silica shell increases time-dependently with a deposition rate of ≈0.6 nm per day

To our experience, the silica shell thickness can be controlled most tightly in a time dependent manner if the template surface is appropriately tailored with suitable peptides, and mineralized in a water–ethanol–TEOS mixture with a 4:4:1 ratio. The deposition of a homogeneous mineral layer is a balancing act between thickness and roughness of the silica shell. In general, the roughness increases with growing silica deposition (*see* Fig. 2).

As described previously, different peptides with sequences either adopted from earlier studies or newly designed on the basis of published data [31] were used for chemical modification of moderately bioengineered TMV particles (TMV_{Lys}) [30]. The peptides were conjugated via thiol groups of terminal cysteines at high surface densities to linker-fashioned TMV_{Lys} [31]. For several applications, homogeneously distributed silicified particles are desired. The monodispersity mainly depends on both the surface properties [47] and the concentration of the templates. For example, peptides containing the amino acid residue histidine tend to induce accumulation of TMV particles into bundles.

For the icosahedral CPMV and flexuous PVX particles, fewer studies on their use as mineralization templates have been published. CPMV was modified for mineralization by introduction of the amino acid sequence YSDQPTQSSQRP into the surface-exposed βB-βC loop of the S subunit of the capsid [48]. Silicification was achieved by a sol–gel process with TEOS and APTES resulting in evenly coated spheres. The average coating of silica on each particle measured approximately 2 nm (*see* Fig. 3). Unmodified CPMV was also used to create hollow silica nanoparticles by mineralization of TEOS and APTES and removal of the CPMV capsids prior to drug loading [49].

PVX particles prefer reactions under mild conditions (pH-neutral aqueous solution at room temperature). Genetically modified PVX particles with the amino acid sequence YSDQPTQSSQRP N-terminally fused to the CP for presentation on the surface showed a mineralization process at room temperature with TEOS [50].

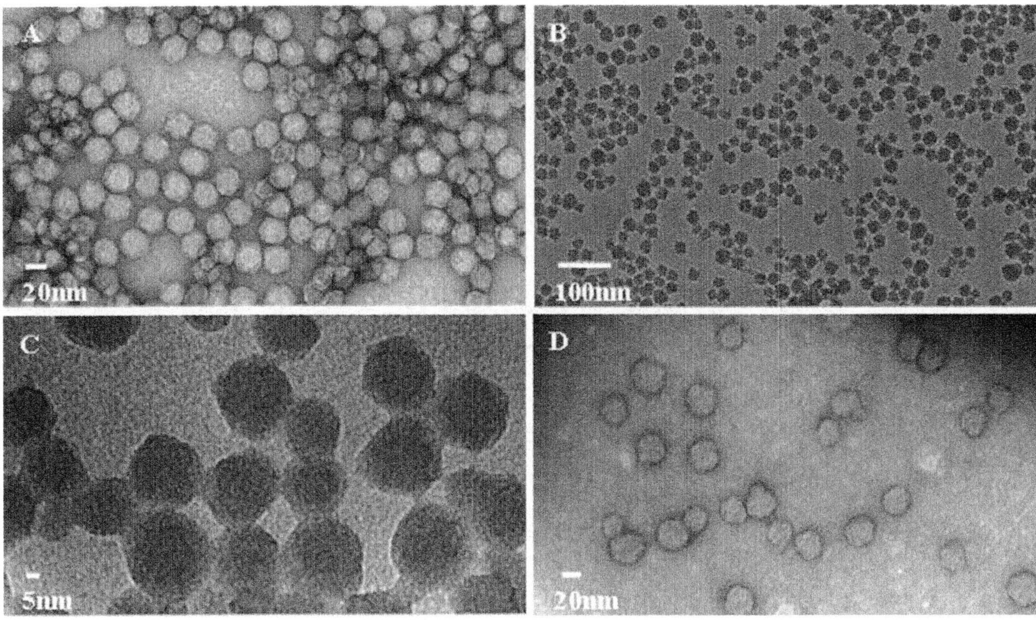

Fig. 3 TEM images: (**a**) CPMVsilica-chimaera particles before mineralization, stained with uranyl acetate. (**b, c**) Unstained silicified-CPMVsilica showing densely mineralized particles. (**d**) Uranyl acetate-stained silicified-CPMVsilica. Reproduced from Steinmetz et al., 2009 with permission from Wiley-VCH [48]

To grant access to a set of plant virus-templated silica nanoparticles of distinct shapes, this chapter describes protocols for derivatives of CPMV, PVX, and TMV. They make use of either CPMV eVLP or PVX mutants displaying appropriate peptides genetically fused to their CP subunits, or a suitably engineered TMV_{Lys} variant that can be equipped with commercially synthesized mineralization-directing peptides by linker-mediated chemical coupling. Thus, spherical, elongated curved or tubular silica nanostructures can be generated through the methods detailed in the following.

2 Materials

Prepare all solutions with deionized water (ddH_2O; 18.3 MΩ cm; e.g., purified by a membraPure system (Aquintus) or other comparable water quality).

2.1 Genetic Engineering

These enzymes and cloning materials are used for genetic engineering of mineralization-directing peptide sequences into viral CPs (in CPMV eVLP agroinfiltration clones and PVX infectious cDNA clones).

1. All restriction enzymes for digestion of plasmids can be purchased from New England Biolabs (NEB).
2. If combinations of the restriction enzymes cannot be used in the same buffer, digest plasmid DNA sequentially and purify

after each reaction using for example an MSB® Spin PCRapace Kit (#1020220300, Invitek by STRATEC Biomedical AG).

3. *Pfu* DNA polymerase (#M7745 Promega Corporation) is used for PCR experiments, and control PCRs are carried out using GoTaq® DNA polymerase (#M7845, Promega).
4. Calf intestinal phosphatase (CIP, #M0290S, NEB), T4 DNA ligase (#M1801, Promega).
5. 1.2% (w/v) agarose in 1× TAE buffer: 40 mM Tris base, 20 mM acetic acid, 1 mM ethylenediamine tetraacetic acid disodium salt dihydrate (EDTA).
6. 0.3 mg/l ethidium bromide.
7. GeneRuler™ 100 bp plus (#SM0321) and 1-kb (#SM0311) ladders (Fermentas by Thermo Fisher Scientific).
8. Wizard® SV Gel and PCR Clean-Up System (#A9282, Promega).

2.1.1 Genetic Modification of CPMV eVLPs

1. Plasmid, e.g., pEAQexpress-VP60–24 K [51] for CPMV eVLPs.
2. Primer containing the sequence of mineralization-directing peptide (*see* also Subheading 3.1.1.) including suitable restriction enzyme recognition sites.
3. *E. coli* DH5α cells (genotype *fhuA2 lac(del)U169 phoA glnV44 Φ80' lacZ(del)M15 gyrA96 recA1 relA1 endA1 thi-1 hsdR17*).
4. LB medium: 0.5% (w/v) yeast extract, 1% (w/v) NaCl, 0.5% (w/v) tryptone/peptone.
5. LB plates: LB medium, 1.5% (w/v) agar supplemented with suitable antibiotics (depending on plasmid), e.g., 50 μg/ml kanamycin or 100 μg/ml ampicillin.
6. Pure Yield™ Plasmid Miniprep Kit (#A1222, Promega) for isolation of small amounts of DNA and the Pure Yield™ Midiprep Kit (#A2495, Promega) for larger amounts.

2.1.2 Genetic Modification of PVX

E. coli cells, materials for bacteria cultivation and plasmid purification as listed for modification of CPMV eVLPs.

1. Plasmid, e.g., pCXI [52] for PVX inoculation.
2. Primer containing the sequence of mineralization-directing peptide (*see* also Subheading 3.1.2.) including suitable restriction enzyme recognition sites.

2.2 Propagation of eVLPs and Virus Particles

2.2.1 Peptide-Exposing CPMV eVLPs

1. pEAQexpress-VP60–24 K derivative, yielding CPMV CP-peptide fusion protein after agroinfiltration.
2. Four-to-six-week-old *Nicotiana benthamiana* plants.
3. *Agrobacterium tumefaciens* GV2260 [53] or LBA4404 [54].

4. YEP plates: YEP medium (see next item), 1.5% (w/v) agar supplemented with 50 μg/ml rifampicin, 100 μg/ml carbenicillin (for GV2260) or 50 μg/ml streptomycin (for LBA4404), and 50 μg/ml kanamycin (for selection of pEAQ-based plasmids).
5. YEP medium: 0.5% (w/v) beef extract, 0.1% (w/v) yeast extract, 0.5% (w/v) peptone, 0.5% (w/v) sucrose, 2 mM $MgSO_4$.
6. 1 M 2-(N-morpholino)ethanesulfonic acid (MES), pH 5.6, adjust with KOH.
7. 40% (w/v) glucose.
8. 200 mM acetosyringone.
9. 2× infiltration medium: 10% (w/v) sucrose, 0.36% (w/v) glucose, 0.86% (w/v) Murashige and Skoog (MS) salts, pH 5.6, adjust with KOH.

2.2.2 Peptide-Exposing PVX

1. pCXI [52] derivative, i.e., plant-infectious PVX cDNA clone yielding virions exposing peptide of choice.
2. Four-week-old *N. benthamiana* plants.
3. Celite 545 (#0011.2, Carl Roth).

2.2.3 Chemically Addressable TMV_{Lys}

1. Plasmid: e.g., p843TMV_{Lys} [30], i.e., plant-infectious TMV_{Lys} cDNA clone yielding chemically addressable virions.
2. *N. tabacum* 'Samsun' nn plants in the two-to-three leaf stage (or four-week-old *N. benthamiana* plants).
3. Celite 545 (#0011.2, Carl Roth).

2.3 eVLP or Plant Virus Purification

2.3.1 Peptide-Exposing CPMV eVLPs

1. 20–50 g of Agro-inoculated leaves.
2. Laboratory blender (e.g., #8010 EB, Waring™).
3. 0.1 M sodium phosphate buffer pH 7.0.
4. Polyvinylpolypyrrolidone (#107302, Merck Millipore).
5. 1 M NaCl/20% (w/v) PEG: solve 58.44 g NaCl, 100 g PEG 6000 (#0158, Carl Roth), and 100 g PEG 8000 (#0263, Carl Roth), in 1 l deionized H_2O while stirring.
6. 10 mM sodium phosphate buffer pH 7.0.
7. Miracloth (#475855, Merck Millipore).
8. Ultracentrifuge XPN-80 (Beckman Coulter) with swinging bucket rotors SW41Ti and SW32Ti.
9. Ultra-Clear centrifuge tubes (14 × 89 mm for SW41Ti and 25 × 89 mm for SW32Ti) (Beckman Coulter).
10. Benchtop centrifuge.
11. Spectrophotometer.

2.3.2 Peptide-Exposing PVX

1. 50–100 g infected plant material (freshly harvested or stored at −80 °C).
2. Laboratory blender (e.g., #8010 EB, Waring™).
3. Extraction buffer: 0.1 M sodium phosphate buffer, pH 8.0, 0.2% (v/v) 2-mercaptoethanol (#8.05740, Merck Millipore), 10% (v/v) ethanol (cooled).
4. Triton X-100 (#3051, Carl Roth).
5. 1 M NaCl/20% (w/v) PEG: solve 58.44 g NaCl, 100 g PEG 6000 (#0158, Carl Roth) and 100 g PEG 8000 (#0263, Carl Roth), in 1 l deionized H_2O while stirring.
6. Sucrose gradient: Prepare 10% (w/v) and 45% (w/v) sucrose in 0.01 M sodium phosphate buffer pH 7.2, with 0.01 M EDTA. Use 12 ml 10% (w/v) and 12 ml 45% (w/v) sucrose solution with a gradient mixer to generate the gradient.
7. 0.05 M sodium phosphate buffer pH 8.0 with 1% (v/v) Triton X-100.
8. 0.01 M sodium phosphate buffer pH 7.2.
9. Miracloth (#475855, Merck Millipore).
10. Ultracentrifuge XPN-80 (Beckman Coulter) with swinging bucket rotors SW41Ti and SW 32Ti.
11. Ultra-Clear centrifuge tubes (14 × 89 mm for SW 41Ti and 25 × 89 mm for SW32Ti) (Beckman Coulter).
12. Benchtop centrifuge.
13. Spectrophotometer.

2.3.3 Chemically Addressable TMV_{Lys}

1. 20 g *N. tabacum* 'Samsun' nn (or *N. benthamiana*) leaves systemically infected with TMV_{Lys} or a suitable coupling-competent TMV variant; stored at −20 °C or lower temperature after harvesting.
2. Laboratory blender (e.g., #8010 EB, Waring™).
3. 0.5 M sodium potassium phosphate (SPP) buffer, pH 7.2 containing 1% (v/v) β-mercaptoethanol (#8.05740, Merck Millipore).
4. 99.5% 1-butanol ROTIPURAN® (#7171, Carl Roth).
5. 10 mM SPP buffer, pH 7.2.
6. PEG 6000 (#8.17007, Merck Millipore) and NaCl.
7. Miracloth (#475855, Merck Millipore) and filter paper (e.g., #WH10311651, Schleicher & Schuell, Munich, Germany).
8. Ultracentrifuge (e.g., Optima L-90 K ultracentrifuge, Beckman Coulter).
9. Spectrophotometer.

2.4 Chemical Functionalization of TMV Particles with Mineralization-Inducing Peptides

1. Ultracentrifugation tubes (#357448, Microfuge® Tube, Polypropylene, Beckman Coulter).
2. ThermoMixer C™ with a ThermoTop® (Eppendorf).
3. TMV particle solution: 5 mg/ml TMV_{Lys} [30] in 10 mM SPP buffer at pH 7.2.
4. Heterobifunctional cross-linker stock solution: 1 M succinimidyl-(N-maleimidopropionamido)-tetraethyleneglycol ester) ($SM(PEG)_4$, #22104, Thermo Scientific) dissolved in dimethyl sulfoxide (DMSO, purity 99.5%). Store at −20 °C.
5. Peptide stock solution: 3.3 mg/ml $(KD)_{10}C$ (purity 95%, GeneCust) dissolved in dimethyl formamide (DMF, purity 99.8%). Store at −20 °C.
6. Reaction tubes: LLG microcentrifugation tubes 1.5 ml clear (#9.409.024, Faust).
7. MilliQ water for the resuspension of virus particles after ultracentrifugation.

2.5 Virus(-like) Particle Analysis

1. Standard equipment for SDS-PAGE and western blotting.
2. All antibodies (*see* Table 1) were purchased from DSMZ (Braunschweig, Germany), Bioreba (Reinach, Switzerland), Dianova (Hamburg, Germany) or BBI Solutions (Cardiff, GB).

2.6 Native Agarose Gel Electrophoresis

1. Agarose NEEO ultra-quality (#2267.4, Roth).
2. 5× TBE: 490 mM Tris–HCl pH 8.0, 445 mM boric acid, 10 mM EDTA.
3. 5× Native sample buffer: 50 mM SPP pH 7.2, 0.5% Bromophenol Blue, 50 (v/v) % glycerol stored at room temperature.
4. Staining solution: 0.1% (w/v) Coomassie® Brilliant Blue R250, 40% (v/v) ethanol, 10% (v/v) acetic acid.
5. Destaining solution: 40% (v/v) ethanol, 10% (v/v) acetic acid.

Table 1
Antibodies used for western blot and TEM analyses

Name	Type	Label	Source
α-PVX	Polyclonal, rabbit	–	#AS-0126, DSMZ
α-CPMV	Polyclonal, rabbit	–	#AS-0012, DSMZ
α-TMV	Polyclonal, rabbit	–	#190412, Bioreba
Goat anti-rabbit^{AP} (GAR^{AP})	Polyclonal, goat	Alkaline phosphatase	#111–055-008, Dianova
Goat anti-rabbit^{15nm} (GAR^{15nm})	Polyclonal, goat	15 nm gold particles	#EM.GAR15, BBI solutions

2.7 Peptide-Directed Particle Mineralization

1. 98% tetraethyl orthosilicate (TEOS, Sigma-Aldrich), generally used as precursor for SiO_2 coating on virus particles. Alternatively, (3-aminopropyl)triethoxysilane (APTES, Sigma-Aldrich) provides access to meso-superstructures of PVX-SIL-SiO_2.
2. MilliQ water rather than buffers is used to dilute virus particles to 1 mg/ml concentration for the mineralization experiments.
3. Vortexer IKA Genius 3 (2500 rpm) for vortex mixing (IKA-Werke).
4. 100 kDa MW cutoff centrifugal columns, e.g., Roti®-Spin MINI 100 (#CL15.1, Roth).
5. Reaction tubes: LLG microcentrifugation tubes 1.5 ml clear (#9.409.024, Faust).
6. 99.8% Ethanol ROTIPURAN® (Roth).
7. 50% (v/v) ethanol.
8. Ultrasonic bath (35 kHz): Sonorex Digiplus DL 255 H (Bandelin).

2.8 TEM and Cryo-TEM

2.8.1 TEM, SEM

1. Clean surface (e.g., Parafilm® or grid holder pad).
2. Pioloform-coated 400-mesh nickel grids (#SP162N4, Plano) or 400-mesh copper grids (#G2400C, Plano) covered with formvar 15/95E (#F6164, Sigma-Aldrich) and a carbon film.
3. n-Si wafer (5 × 10 mm; CrysTec, Berlin, Germany) for SEM analysis as supporting substrate.
4. 99.8% (v/v) ethanol.
5. Filter paper (fiber free).
6. 1% (w/v) uranyl acetate (pH 4.3; optionally supplemented with 250 μg/ml bacitracin for improved even staining and background).
7. Zeiss EM 10 TEM or Tecnai G2 Sphera electron microscope (FEI, Hillsboro, Oregon, USA) with a Tietz F214 camera (TVIPS, Munich, Germany).

2.8.2 Immun-osorbent TEM

1. Clean surface (e.g., Parafilm® or grid holder pad).
2. Pioloform-coated 400-mesh nickel grids (#SP162N4, Plano).
3. Filter paper (fiber free).
4. 0.5% (v/v) bovine serum albumin (BSA) in PBS.
5. PBST: 137 mM NaCl, 2.7 mM KCl, 10.1 mM Na_2HPO_4, 1.8 mM KH_2PO_4, 0.05% (v/v) Tween 20, pH 7.4.
6. 1% (w/v) uranyl acetate (pH 4.3, optionally supplemented with 250 μg/ml bacitracin as above).

7. Use primary antibodies as listed in Table 1 and goat anti-rabbit secondary antibodies labeled with 15-nm gold (BBI Solution) for detection.
8. Zeiss EM 10 TEM or Tecnai G2 Sphera electron microscope (FEI, Hillsboro, Oregon, USA) with a Tietz F214 camera (TVIPS, Munich, Germany).

2.8.3 Cryo-TEM

1. Holey carbon coated grids (Quantifoil 3.5/1, Quantifoil Micro Tools, Großlöbichau, Germany).
2. Standard Vitribot filter paper.
3. FEI Vitribot (Ted Pella, Inc.).
4. Philips Tecnai 20 cryo-electron microscope (Philips) equipped with a Gatan model 626 cryo-stage (Gatan), operating at 200 kV (slow recording under low dose conditions).
5. Instrumentation for elemental analysis by energy dispersive X-ray spectroscopy (EDX): Oxford X-Max 80 Silicon drift EDX detector with a regular TEM. Data analysis was performed using INCA software.

3 Methods

3.1 Genetic Engineering

This section describes the generation of CPMV eVLP agroinfiltration clones and PVX infectious cDNA clones, both with genetically engineered CPs, displaying mineralization-directing peptides. These are used for mineralization after purification from leaf tissues.

The peptide sequence YSDQPTQSSQRP was successfully used for the silicification of CPMV eVLPs [48] and PVX virions [50]. Other suitable peptide sequences tested for mineralization can be found in [55, 31] (out of which silica deposition by means of a distinct peptide has been optimized on TMV, as described in Subheading 3.4 below). For details of all cloning procedures, follow standard guidelines as provided by Sambrook and Russell [56].

3.1.1 Genetic Modification of CPMV eVLPs

1. For engineering CPMV eVLPs transiently produced after agroinfiltration, choose peptide sequence of interest (*see* **Note 1**). Fusions should meet the following requirements [57]: Peptide length < 40 amino acids; isoelectric point of inserted peptide <9. Clone peptide cDNA sequence into S or L protein-encoding sequence as desired, e.g., via suitable restriction enzyme sites as previously described [15] and outlined in the following.
2. Dephosphorylate vector pEAQexpress-VP60–24 K with calf intestinal phosphatase after digestion with appropriate restriction enzymes (e.g., with PacI and AscI [15]) and ligate

overnight at 16 °C with the peptide cDNA-containing CPMV target cDNA (see the previous step) according to standard protocols.

3. Transform ligation products into *E. coli* DH5α cells and select colonies on LB plates supplemented with kanamycin or ampicillin (depending on plasmid) overnight at 37 °C.
4. Isolate plasmids from bacterial overnight cultures and sequence the modified DNA using Pure Yield™ Plasmid Miniprep Kit (#A1222, Promega) for isolation of small amounts of DNA and the Pure Yield™ Midiprep Kit (#A2495, Promega) for larger amounts.
5. Transform the verified plasmid into *A. tumefaciens* for agroinfiltration and transient production of CPMV empty VLPs.

3.1.2 Genetic Modification of PVX

1. For engineering PVX, choose peptide of interest (*see* **Note 2**). Direct fusions of the peptide-encoding cDNA sequence to that of the CP should meet the following requirements: Peptide length < 15 amino acids [58]; isoelectric point (pI) of the resulting particle between 5 and 9 [59]. Avoid high content of tryptophan and positively charged amino acids [59, 58]. If the peptide of interest does not meet the requirements try a fusion via the 2A sequence of the foot-and-mouth-disease virus (FMDV) [60] (*see* **Note 3**). Clone the target sequence as N-terminal fusion to the CP, e.g., by SOE-PCR as previously described [52].

3.2 Propagation of eVLPs and Virus Particles

3.2.1 Peptide-Exposing CPMV eVLPs

1. Prepare *A. tumefaciens* infiltration cultures (*see* **Note 4**) by cultivating *A. tumefaciens* bacteria containing an appropriate plasmid at 26 °C in YEP medium overnight. Supplement cultures with 10 mM MES (pH 5.6), 10 mM glucose and 20 μM acetosyringone after 24 h and incubate for a further day. Adjust $OD_{600\ nm}$ to 0.5–1.0 with 2× infiltration medium, supplement with 200 μM acetosyringone and incubate for 30 min at room temperature.
2. Infiltrate leaves of 4–6-week-old *N. benthamiana* plants with a needleless syringe and incubate in a phytochamber with constant light (25,000–30,000 lux) at 26 °C for 12 h and in the dark at 20 °C for 12 h (*see* Chapter 4, Subheading 3.2).
3. Harvest the infiltrated leaves at 5–6 dpi and use freshly or store at −80 °C until further use (*see* **Note 5**).

3.2.2 Peptide-Exposing PVX

1. Rub four-week-old *N. benthamiana* plants with 10 μg DNA and Celite 545 for inoculation. Chose 3–4 equally distributed leaves per plant and spare the youngest ones. Rinse leaves with water for around 10 min after rubbing and incubate the plants in a phytochamber with constant light (25,000–30,000 lux) at 26 °C for 12 h and in the dark at 20 °C for 12 h.

2. Harvest plant material at 14–21 dpi depending on the infection status, and use 50 g–100 g of plant material for virus purification freshly or store the material at −80 °C until further use (*see* **Note 5**).

3.2.3 Chemically Addressable TMV$_{Lys}$

1. Dust 1–2 leaves of 1 *N. tabacum* or *N. benthamiana* plants with Celite 545 and mechanically inoculate with ≈7 μg plasmid DNA per leaf; rinse with water. Cultivate the plants in a day–night regime with light (150–200 μE/(m^{2}s)) at 22 °C for 16 h and in the dark at 18 °C for 8 h.
2. Harvest systemically infected symptomatic leaves at 28 dpi, and use either freshly or store at −20 °C or at lower temperatures for up to a few months. 20 g of plant material are typically used for virus purification (*see* **Note 5**).

3.3 eVLP or Plant Virus Purification

Purify CPMV eVLPs as follows. The method has previously been described in detail [15].

3.3.1 Peptide-Exposing CPMV eVLPs

1. Homogenize leaf tissue with three volumes of 0.1 M sodium phosphate buffer pH 7.0.
2. Add 2% (w/v) PVPP to improve purity of the final eVLPs.
3. Filter homogenate through three layers of Miracloth and spin at 13,000 × g for 20 min at 4 °C to remove cell debris.
4. Add NaCl to a final concentration of 0.2 M and PEG 6000 to 4 (w/v) % and stir overnight.
5. Collect the precipitate by centrifugation at 13,000 × *g* for 10 min at 4 °C and resuspend pellet in 10 mM sodium phosphate buffer pH 7 (0.5 ml/g leaf material).
6. Spin at 27,000 × *g* for 20 min at 4 °C, transfer supernatant to ultracentrifugation tubes and spin at 118,700 × *g* for 2.5 h at 4 °C.
7. Resuspend pellet in approximately 500 μl of 10 mM sodium phosphate buffer and spin again at 10,000 × *g* to remove last particle aggregates. The CPMV eVLPs are found in the supernatant.
8. Determine CPMV eVLP concentration with UV spectroscopy at a wavelength of 260 nm. Use the extinction coefficient for CPMV eVLPs particles of 1.28 ml/(mg cm) [13].

3.3.2 Peptide-Exposing PVX

Purification of PVX particles is achieved by a modified protocol from International Potato Center (Lima, Peru), as described in detail in Chapter 5 and previously published [52, 58] (*see* **Note 6**).

1. Determine plant virus concentration by measuring the OD_{260nm} and divide by the extinction coefficient of 2.97 ml/(mg cm) of PVX.

3.3.3 Chemically Addressable TMV_{Lys}

This protocol for TMV purification essentially follows the method published by Gooding & Hebert [61]. All steps are performed at room temperature unless otherwise indicated.

1. Homogenize 20 g infected *N. tabacum* 'Samsun' nn leaves (*see* **Note 7**) in 50 ml 0.5 M SPP (pH 7.2) containing 1% β-mercaptoethanol in a blender for 5 min (*see* **Note 8**).
2. Filtrate the leaf homogenate through a Miracloth sheet (15 × 15 cm) and subsequently through a filter paper (*see* **Note 9**).
3. Repeat filtration through a filter paper.
4. Add 4 ml butanol per 50 ml homogenized leaves drop by drop to the filtrate while stirring with a magnetic stirrer. Wait until chloroplasts are coagulated. Thereafter, stir for further 15 min.
5. Transfer the solution to a centrifugation tube and centrifuge for 30 min at 10,000 × *g* and room temperature (*see* **Note 10**).
6. Filter the lower liquid phase through Miracloth and a filter paper into a fresh vessel. Avoid the transfer of the upper liquid phase or pellet.
7. Supplement the filtrate with 2 g PEG 6000 per 50 ml while stirring. Continue stirring for 10–15 min until the PEG 6000 platelets are completely dissolved (*see* **Note 11**).
8. Transfer the solution to a centrifugation tube and centrifuge for 15 min at 10,000 × *g* and room temperature. Discard the supernatant.
9. Resuspend the white pellet in 20 ml 10 mM SPP (pH 7.2).
10. Supplement the resuspended pellet with 2 g PEG 6000 and 2 g NaCl per 50 ml while stirring. Continue stirring for 10–15 min.
11. Transfer the solution to a centrifugation tube and centrifuge for 15 min at 10,000 × *g* and room temperature. Discard the supernatant.
12. Resuspend the pellet in 10 ml 10 mM SPP (pH 7.2).
13. Transfer the solution to a centrifugation tube and centrifuge for 5 min at 10,000 × *g* and 4 °C.
14. Transfer the supernatant into an ultracentrifugation tube. Centrifuge for 1.5 h at 120,000 × *g* and 4 °C. Discard the supernatant.
15. Resuspend the pellet in 2 ml 10 mM SPP (pH 7.4) at RT.
16. Determine the TMV concentration by UV spectroscopy at a wavelength of 260 nm. Use the extinction coefficient for TMV particles (3 ml/(mg cm)) [62].

3.4 Chemical Functionalization of TMV Particles with Mineralization-Inducing Peptides

The amino group-exposing TMV variant TMV_{Lys} is equipped with a different type of peptide than the other viral templates and by a distinct production route: The repetitively arranged charge-relay-exerting sequence $(KD)_{10}C$ is mounted on the outer nanotube surface by chemical conjugation [31]. The latter is achieved by means of a heterobifunctional PEG_4 chain cross-linker [$SM(PEG)_4$] addressing the accessible amino groups of TMV_{Lys} via its succinimidyl ester moiety [S], and the thiol group of a C-terminal cysteine present in the silicification-guiding peptide via its opposite maleimide terminus [M]. Thereby, more than a thousand peptides can be immobilized on every TMV particle's surface, given that 50% of its CP subunits will be fashioned with linker molecules retaining reactivity for coupling to the peptides' thiol groups (which is typically the minimum yield obtained by use of this protocol).

1. Add 1.5 μl 1 M $SM(PEG)_4$ bifunctional linker to 200 μl 5 mg/ml TMV_{Lys} in 10 mM SPP (pH 7.2) and mix thoroughly with a vortexer. Incubate the TMV_{Lys}–$SM(PEG)_4$ reaction mixture at 37 °C for 2 h under horizontal agitation (500 rpm).
2. Transfer the TMV_{Lys}-$SM(PEG)_4$-containing solution into an ultracentrifugation tube and sediment the linker-functionalized particles (named TMV_{Lys}-PEG in the following) at 90,500 × *g* for 1.5 h at 4 °C.
3. Discard supernatant and wash the pellet (without resuspending it) with 1 ml 10 mM SPP (pH 7.2).
4. Completely resuspend the pellet in 100 μl 10 mM SPP (pH 7.2). This TMV_{Lys}-PEG solution is used for the conjugation of the peptides via the exposed maleimide-terminus of the PEG linker.
5. Add 800 μl 10 mM SPP (pH 7.2) and 40 μl 3.3 mg/ml peptide $(KD)_{10}C$ stock solution to 100 μl of the TMV_{Lys}-PEG solution and mix thoroughly with a vortexer. Incubate at 30 °C for 2 h under horizontal agitation.
6. Centrifuge as above (*see* **step 2**).
7. After centrifugation, wash the pellet with 1 ml deionized water without resuspending the pellet.
8. Resuspend the pellet in 100 μl deionized water.
9. Before starting the mineralization reaction, allow standing at least for 10 h at room temperature for complete dispersion of the peptide-equipped TMV_{Lys} particles (*see* **Note 12**).

3.5 Virus (–like) Particle Analysis

These and the following methods for characterizing plant virus-based particles can be applied to all template species employed here: CPMV eVLPs, PVX, and TMV_{Lys} (both without and with linkers/peptides coupled), to determine the integrity and purity of the particle preparations. Among the techniques, SDS-PAGE is most important in order to verify the success and efficiency of peptide–CP fusion protein integration into CPMV and PVX particles, as obvious from these proteins' molecular weights increased by about 1.4 kDa for an 12-amino-acid peptide in comparison to native CPs (native CPMV CP L subunit: 41.2 kDa, S subunit: 23.7 kDa [63], PVX CP: 25.0 kDa [64]). For TMV_{Lys} (native TMV CP: ≈17.5 kDa [65]), SDS-PAGE will not resolve the presence of the additional amino group compared to wild-type TMV, but can reveal coupling of the bifunctional linker alone as well as the efficiency of peptide conjugation thereafter (*see* Subheading 3.4 above). In addition to electrophoretic and western blot analyses, electron microscopy (EM) is recommended to visualize the viral particle structures. As this method is also essential to analyze the mineralization products, EM techniques are described in Subheading 3.8. at the end of the chapter.

3.5.1 SDS-PAGE

Resolve protein samples by SDS-PAGE [66] on two gels, after mixing plant sap or purified particles (1–4 μg) with 5× reducing buffer and boiling for 5–10 min (*see* **Note 13**). Alternatively, dilute the solution of functionalized TMV particles 1:10 and mix 2 μl with 2 μl 5× sample buffer. Heat to 95 °C for 5 min, centrifuge at 10,000 × *g* for 1 min and load the samples and the protein molecular weight calibration ladder into the wells of the stacking gel.

3.5.2 Western Blot Analyses

For the specific detection of proteins, blot the polyacrylamide gels onto a nitrocellulose membrane, e.g., by semidry blotting using the appropriate buffer. It is also possible to visualize the proteins transferred to the membrane by means of Ponçeau S-staining (with subsequent destaining again), as described for example in Chapter 24. All subsequent steps are carried out at room temperature under gentle agitation.

3.6 Native Gel Electrophoresis

Native gel electrophoresis is a powerful technique shedding light on virus and virus-like particle integrity, size, and surface charge, within its limits of resolution. It can thus be applied to discriminate between peptide-fashioned and plain particles according to their mobility under native conditions, especially, if a charge shift due to the peptides is pronounced. The following protocol has been optimized for TMV and its derivatives (with native TMV having an isoelectric point [IEP] of about 3.5), but is suitable for the other viral particles after minor adaptations (CPMV: IEP between 3.7 and 4.5 [63]; PVX: IEP of 6.7 [67, 68]).

1. Prepare 0.9% (w/v) agarose in 1× TBE as follows: Suspend 0.36 g agarose in 40 ml 1× TBE. Boil the solution preferably in a microwave oven until the agarose powder is completely dissolved. Cast the solution into a flat-bed mold of for example

7 × 8 cm and place for example a 12-well comb in the solution. Wait at least for 45 min before loading the samples. Place the gel tray into an electrophoresis chamber and fill the chamber with 1× TBE to completely submerge the gel. Remove the comb carefully to prevent gel damage or air bubble formation in the wells.

2. For sample preparation, supplement 1.2–3.0 μl of the peptide-equipped virus or virus-like particles with 1× native sample buffer and load the complete volume into the well (*see* **Note 14**).
3. Run the gel at 5 V/cm for 4 h.
4. Fix the proteins with destaining solution in the gel matrix for 15 min.
5. Stain the gel with staining solution for 1–8 h.
6. Destain the gel with destaining solution changes every 1 h (*see* **Note 15**).

3.7 Peptide-Directed Particle Mineralization

3.7.1 CPMV eVLPs

1. Mix 15 mg/ml peptide-exposing CPMV eVLPs with 1 ml TEOS and APTES (90:10 mol).
2. Leave the reaction mixture for 2–4 days at room temperature.
3. Heat the mixture for 2 days at 45 °C (*see* **Note 16**).
4. Recover particles by centrifugation (15,000 × *g*, 10 min).
5. Pass supernatant containing the particles through 100 kDa MW cut-off columns and wash three times with MilliQ water.
6. Use purified particles for further TEM analyses (*see* Fig. 3).

3.7.2 PVX

1. Mix 1 mg/ml purified recombinant peptide-exposing PVX particles with 2 μl/ml TEOS or 20 μl/ml TEOS (*see* **Note 17** and **Note 18**). Alternatively, mix 1 mg/ml genetically engineered PVX with 2 μl/ml TEOS–APTES (50:50 mol) (*see* **Note 19**).
2. Sonicate the mixture three times for 10 s interspersed with vortexing.
3. Leave the reaction mixture undisturbed for 3 days at room temperature.
4. Directly use the mixture for further TEM and cryo-TEM analyses (*see* Fig. 4) or recover particles by spinning at 15,000 × *g* for 20 min at 4 °C for later analysis (*see* **Note 20**).

3.7.3 TMV

1. Mix 10 μl TEOS with 40 μl 99.8% (v/v) ethanol thoroughly (*see* **Note 21**). Add 40 μl 10 mg/ml (*see* **Note 22**) peptide fashioned TMV_{Lys} particles (TMV_{Lys}-PEG-peptide suspended in deionized water) and mix thoroughly. Incubate the mineralization reaction for 1–10 days under agitation (horizontal shaking at 500 rpm) at 25 °C (*see* **Note 23**).

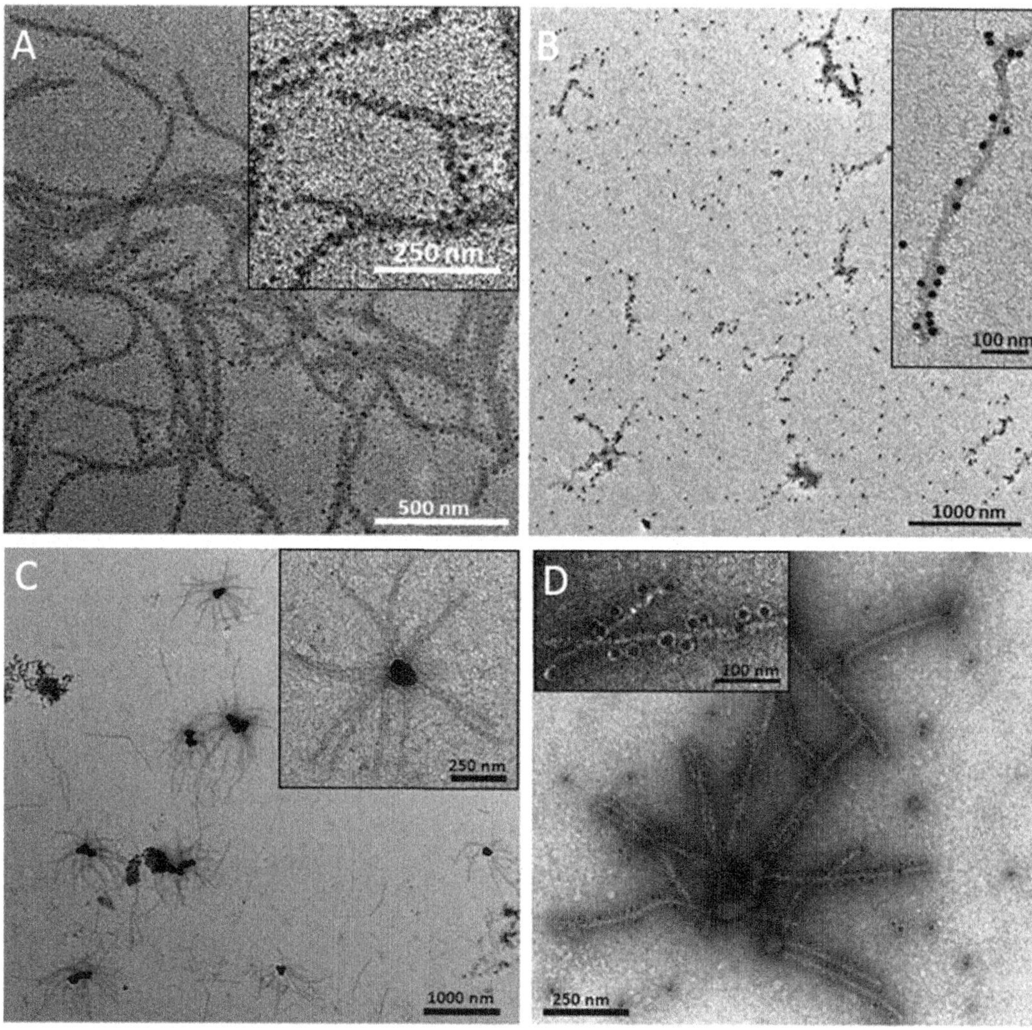

Fig. 4 TEM analysis of double-labeled PVX-SIL virus particles (**a** and **b**) and PVX-SiO_2 superstructures (**c** and **d**). PVX-SIL (1 mg/ml in MilliQ-water) was mixed with 2 μl/ml TEOS, sonicated briefly and left for 3 days (**a**). The coated particles were then immunolabeled with gold nanoparticles to generate a triple-hybrid virus–silica–gold structure (**b**).The defined SiO_2 particles cannot be distinguished separately due to the high contrast of the gold nanoparticles. PVX-SIL particles (1 mg/ml in MilliQ-water) were mixed with 2 μl/ml of a 1:1 mixture of TEOS and APTES, sonicated briefly three times and left for 3 days at room temperature. TEM analysis revealed large structures with a dense core and radiating PVX-SIL filaments (**c**) that were still accessible for immunogold labeling (**d**). Reproduced from van Rijn et al., 2015 with permission from Wiley-VCH [50]

2. Sediment the mineralized TMV_{Lys}-PEG-peptide particles by centrifugation for 15 min at 20,000 × *g* at 18–20 °C. Discard supernatant.
3. Wash the pellet with 200 μl 50% (v/v) ethanol without resuspending.
4. Resuspend the pellet in 50 μl deionized water and treat them in an ultrasonic bath at 100% intensity for 15 min (*see* **Note 24**).

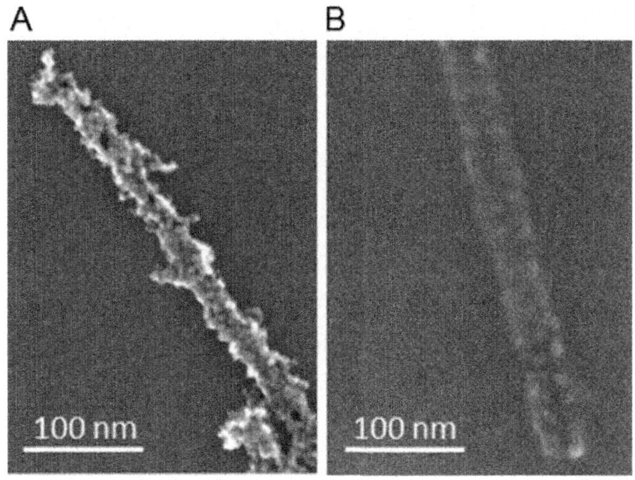

Fig. 5 Influence of solvent on the surface appearance of air-dried silicified TMV particles shown by SEM analysis. Particles diluted in ddH_2O (**a**) or EtOH (**b**)

3.8 TEM and Cryo-TEM

Electron microscopy (EM) is commonly applied to analyze the quality and quantity of deposited inorganic material on the surface of the viral particles. The electron density of the silicon atoms is sufficient to detect low amounts of deposited silica with layer thicknesses of 2–4 nm by TEM analysis without additional staining (e.g. uranyl acetate) as shown in Fig. 2. SEM provides interesting insights into the texture (e.g., roughness, smoothness) of the silica shell (*see* Fig. 5). For SEM analysis the silica shell thickness should be in a range of 6 nm to give an adequate electron contrast.

3.8.1 TEM and SEM

1. For direct adsorption, prepare drops of around 40 μl PBS with 10 μg of purified particles onto a clean surface and incubate Pioloform-coated 400-mesh nickel grids for 20 min in these drops.
2. Counterstain the loaded grids (from **step 1**) with 1% (w/v) uranyl acetate (pH 4.3, optionally supplemented with bacitracin) before analysis with for example a Zeiss EM 10 TEM.
3. Alternatively, immerse a carbon formvar-covered copper grid into 99.8% (v/v) ethanol. Remove excess ethanol with a filter paper.
4. Place the grids onto a clean surface with the carbon formvar film pointing upward and pipette 3 μl of the mineralized TMV_{Lys}-PEG-peptide particles in deionized water on the grid. Incubate for 3–5 min and remove excess solution with a filter paper (*see* **Note 25**).
5. Wash the grids with five droplets deionized water and let them dry at the air or in a desiccator.
6. Analyze the unstained TMV_{Lys}-PEG-peptide particles on the grid with a Tecnai G2 Sphera electron microscope (FEI,

Hillsboro, Oregon, USA) equipped with a Tietz F214 camera (TVIPS, Munich, Germany) at 120 kV. To prevent destruction of the carbon film it is recommended to use a spot size of 5–6.

7. For SEM analysis use n-Si wafer (5 × 10 mm; CrysTec, Berlin, Germany) as substrate.
8. Drop 20 μl of a 1:250 dilution of the mineralized TMV_{Lys}-PEG-peptide particles solution with deionized water on an n-Si wafer and let the droplet air-dry (*see* **Note 26**).
9. To achieve reasonable resolution, use a high-resolution field emission SEM (e.g., FE-SEM, S-5200, Hitachi Ltd., Tokyo, Japan) at 30 kV.

3.8.2 Immunosorbent Transmission Electron Microscopy (ISEM)

1. For ISEM, coat Pioloform-coated 400-mesh nickel grids with an anti-virus antibody or anti-serum diluted 1:10 in PBS for 20 min.
2. Wash unbound antibodies off with PBST by dropping approximately 3 ml solution carefully on the grid which is held by a forceps.
3. Block grids with 0.5% (v/v) BSA in PBS for 15 min.
4. Incubate grids with 10 μg of the particle preparation in 40 μl PBS for 20 min, followed by washing with PBS.
5. Captured particles can be analyzed for engineered peptides exposed on the surface by immunogold staining. Therefore incubate the preparations for 2 h with a virus-specific antibody from a different origin as the capture antibody or specific anti-peptide antibody diluted 1:100 in PBS.
6. Incubate grids overnight with secondary antibodies labeled with 15-nm gold particles diluted 1:50 in PBS.
7. Wash grids thoroughly with PBS and then with distilled water.
8. Contrast with 1% (w/v) uranyl acetate (pH 4.3; optionally supplemented with bacitracin for improved staining) before analysis with for example a Zeiss EM 10 TEM.

3.8.3 Cryo-TEM

1. Deposit a few microliters of virus on holey carbon-coated grids.
2. Blot the excess liquid and vitrify the grids in liquid ethane in a Vitrobot (FEI).
3. Analyze grids in for example a Philips Tecnai 20 cryo-electron microscope equipped with a Gatan model 626 cryo-stage, operating at 200 kV.
4. Record images under low-dose conditions with a slow-scan CCD camera.

4 Notes

1. The βB-βC loop is the most exposed loop on the particle surface and has been used in many cases for peptide presentation [57]. Other possibilities for peptide introduction are the βC'-βC" loop of the S protein or the βE-αB loop of the L protein.
2. N-terminal fusions are recommended because only in this configuration, the foreign peptide is presented on the outer surface of the particles [69].
3. PVX needs its genomic RNA for particle assembly [70], and thus a transient expression of the CP alone, as shown for CPMV and TMV, will not lead to the production of VLPs.
4. A coexpression of the silencing suppressor p19 of tomato bushy stunt virus can improve the product yields in *N. benthamiana* [71].
5. Plant material can be stored after harvesting by immediately freezing it to −80 °C.
6. For some recombinant particles the yields can be improved by modifying the precipitation steps. PEG precipitation can be carried out as described in the original protocol, and the centrifugation step on a sucrose cushion can be skipped because too many particles get lost during this step. If you decide to skip the sucrose cushion, the centrifugation of the pooled fractions of the sucrose gradient should be done for at least 3 h. Additionally, an adaptation of the buffer pH to the isoelectric point of the modified particles can further improve the particle yields.
7. Use leaves 28 days post infection of *N. tabacum* plants, which were inoculated at the two-to-three leaf stage.
8. Mix 50 ml 0.5 M SPP (pH 7.2) with 0.5 ml β-mercaptoethanol in a measuring cylinder. Place the frozen leaves into the blender's container and subsequently pour the homogenization solution into the blender. While homogenizing leaves, stop every 1 min and wipe back the shredded leaves attached to the container wall with a rubber spatula to the bottom.
9. To save time, place the filter paper into a funnel and on top the Miracloth sheet. Pour the homogenized leaves into the Miracloth-filter-equipped funnel. For higher TMV yield, wring out the residual liquid from the Miracloth carefully.
10. After centrifugation two liquid phases and one pellet are visible: The upper liquid phase is the green, chloroplasts-containing butanol fraction. The lower liquid phase is the TMV-containing buffer. The white pellet contains cell debris.

11. After the addition of PEG 6000, the color of the solution turns brownish.
12. As some sort of peptides influence the absorbance at 260 nm, the concentration of the TMV_{Lys}-PEG-peptide cannot be determined by UV-spectroscopy exactly. Hence, compare the chemically modified coat proteins to a known amount of TMV coat protein by SDS-PAGE.
13. Use wild-type/unmodified virus particles as controls.
14. Load at least 12 μg of TMV particles per lane.
15. Destain until the staining background is one half less than at the beginning of the destaining. Now destain with 0.5× destaining solution to remove residual background dye. Low dye background in the gel can be removed with deionized water with several water changes. The complete destaining of an agarose gel takes at least 1 week.
16. Check before if your particles are stable at 45 °C for 2 days. CPMV is in general stable at 45 °C [72], but for example PVX particles are not [50].
17. Always use unmodified particles as control.
18. Higher concentrations of TEOS generate a hydrogel-like material (unpublished).
19. For PVX particles an additional formation of superstructures was observed when a TEOS–APTES mixture was used, which was not observed with TEOS alone [50]. Mixtures of TEOS with an equal volume of APTES give rise to viral-silicon dioxide superstructures. The mineralization process is not confined to the surface of the virus particles. Instead, independent nucleation sites can arise spontaneously, to which the ends of the PVX-SIL particles become attached and form a mesoporous silicon dioxide core with radiating PVX tentacles.
20. A complex 3D superstructure of 1–2 μm in diameter will form, comprising several PVX rods radiating from a central mesoporous silicon dioxide core. The PVX template is not uniformly coated with SiO_2; instead, isolated SiO_2 nanoparticles are formed, leaving parts of the surface of the virus particle still accessible. This allows for other targeting methods to be applied, such as immunolabeling with monoclonal gold-conjugated antibodies to target the PVX surface.
21. The use of a TEOS/APTES-based method, however, does not result in well-dispersed nanorods, but in striking mesoscale arrays of TMV-templated silica, arranged in parallel or even star-shaped orientations depending on the precursor concentrations and reaction conditions [38]. In this approach, thick

silica shells of around 30 nm are achieved, but without full control over the final product characteristics. In contrast, the silicic acid-based method, applied to TMV-derived coat protein disks [73], comes out to reproducibly silicify the outer, peptide-equipped disk rim, with, however, no detailed analysis of the silica deposition kinetics so far [44].

22. The concentration of TMV has a major influence on the appearance of mesostructures versus colloidal well-dispersed particles. Concentrations above 10 mg/ml TMV were shown to arrange into hexagonally ordered mesoporous silica-TMV hybrid materials as demonstrated for a 10 mg/ml TMV solution supplemented with 90 mol-% TEOS and 10 mol-% APTES [38]. Under these conditions, TMV particles aligned in parallel into well-ordered lattice fringes. For concentrations below 1 mg/ml, single separate TMV particles could be obtained [37, 39].
23. The TEOS–ethanol mixture should be prepared before combining it with the viral particles to prevent their disassembly. The silica shell thickness is time-dependently deposited. Hence, to obtain thicker silica shells deposited around the viral templates, expand reaction times. However, reaction times longer than 10 days result in unspecific silica particle formation. To obtain well-dispersed mineralized TMV particles, the TMV_{Lys}-PEG-peptide concentration can be reduced. However, the TEOS content has to be adapted to maintain the same ratio of TMV_{Lys}-PEG-peptide to TEOS. For example, mix 5 μl TEOS with 40 μl 99.8% (v/v) ethanol and 5 μl deionized water. Add 40 μl 5 mg/ml TMV_{Lys}-PEG-peptide in deionized water to the TEOS–ethanol–water mixture.
24. After ultrasonication of the mineralized viral particles grid preparation should follow without long storage time. Storage of longer than 3 days results in the formation of spherical silica particles.
25. If the concentration of mineralized TMV_{Lys}-PEG-peptide is too high, a concentration series can be prepared. Furthermore, the sample preparation for TEM analysis can be improved by placing 15 μl droplets of the mineralized TMV_{Lys}-PEG-peptide solution onto a piece of Parafilm® and placing the ethanol-treated grid on top of the droplets. Incubate for 5 min and remove excess solution with a filter paper.
26. If the concentration is too high or too low, test different dilutions. Alternatively, the dilution can also be carried out with 99.8% (v/v) ethanol. However, the contrast of these particles is lower than that of mineralized TMV_{Lys}-PEG-peptide particles diluted in deionized water (*see* Fig. 5).

References

1. Steinmetz NF, Lomonossoff GP, Evans DJ (2006) Cowpea mosaic virus for material fabrication: addressable carboxylate groups on a programmable nanoscaffold. Langmuir 22(8):3488–3490
2. Culver JN, Brown AD, Zang F, Gnerlich M, Gerasopoulos K, Ghodssi R (2015) Plant virus directed fabrication of nanoscale materials and devices. Virology 479-480:200–212
3. Wen AM, Steinmetz NF (2016) Design of virus-based nanomaterials for medicine, biotechnology, and energy. Chem Soc Rev 45:4074–4126
4. Bittner AM, Alonso JM, Gorzny ML, Wege C (2013) Nanoscale science and technology with plant viruses and bacteriophages. Subcell Biochem 68:667–702
5. Ma JK, Barros E, Bock R, Christou P, Dale PJ, Dix PJ, Fischer R, Irwin J, Mahoney R, Pezzotti M, Schillberg S, Sparrow P, Stoger E, Twyman RM, European Union Framework 6 Pharma-Planta Consortium (2005) Molecular farming for new drugs and vaccines. Current perspectives on the production of pharmaceuticals in transgenic plants. EMBO Rep 6(7):593–599
6. Marsian J, Lomonossoff GP (2016) Molecular pharming—Vlps made in plants. Curr Opin Biotechnol 37:201–206
7. Rosales-Mendoza S, Angulo C, Meza B (2016) Food-grade organisms as vaccine biofactories and oral delivery vehicles. Trends Biotechnol 34(2):124–136
8. Sack M, Hofbauer A, Fischer R, Stoger E (2015) The increasing value of plant-made proteins. Curr Opin Biotechnol 32:163–170
9. Paul M (2015) International society for plant molecular farming: 1st conference, Berlin-Dahlem, Germany, 17th-19th June 2014. Transgenic Res 24(2):375–380
10. Sainsbury F, Canizares MC, Lomonossoff GP (2010) Cowpea mosaic virus: the plant virus-based biotechnology workhorse. Annu Rev Phytopathol 48:437–455
11. Koch C, Eber FJ, Azucena C, Forste A, Walheim S, Schimmel T, Bittner AM, Jeske H, Gliemann H, Eiben S, Geiger FC, Wege C (2016) Novel roles for well-known players: from tobacco mosaic virus pests to enzymatically active assemblies. Beilstein J Nanotechnol 7:613–629
12. Lomonossoff GP, Johnson JE (1991) The synthesis and structure of comovirus capsids. Prog Biophys Mol Biol 55(2):107–137
13. van Kammen A (1967) Purification and properties of the components of cowpea mosaic virus. Virology 31(4):633–642
14. van Kammen A (1972) Plant viruses with a divided genome. Annu Rev Phytopathol 10:125–150
15. Sainsbury F, Saxena P, Aljabali AA, Saunders K, Evans DJ, Lomonossoff GP (2014) Genetic engineering and characterization of cowpea mosaic virus empty virus-like particles. Methods Mol Biol 1108:139–153
16. Aljabali AA, Sainsbury F, Lomonossoff GP, Evans DJ (2010) Cowpea mosaic virus unmodified empty viruslike particles loaded with metal and metal oxide. Small 6(7):818–821
17. Butler PJ (1999) Self-assembly of tobacco mosaic virus: the role of an intermediate aggregate in generating both specificity and speed. Philos T Roy Soc B 354(1383):537–550
18. Adams MJ, Antoniw JF, Kreuze J (2009) Virgaviridae: a new family of rod-shaped plant viruses. Arch Virol 154(12):1967–1972
19. Namba K, Pattanayek R, Stubbs G (1989) Visualization of protein-nucleic acid interactions in a virus. Refined structure of intact tobacco mosaic virus at 2.9 a resolution by X-ray fiber diffraction. J Mol Biol 208(2):307–325
20. Adams MJ, Heinze C, Jackson AO, Kreuze JF, MacFarlane SA, Torrance L (2012) Virus taxonomy: ninth report of the international committee on taxonomy of viruses. In: King AMQ, Adams MJ, Carstens EB, Lefkowitz EJ (eds) Family virgaviridae. Elsevier, San Diego, pp 1139–1162
21. Hamamoto H, Sugiyama Y, Nakagawa N, Hashida E, Matsunaga Y, Takemoto S, Watanabe Y, Okada Y (1993) A new tobacco mosaic virus vector and its use for the systemic production of angiotensin-I-converting enzyme inhibitor in transgenic tobacco and tomato. Biotechnology 11(8):930–932
22. Turpen TH, Reinl SJ, Charoenvit Y, Hoffman SL, Fallarme V, Grill LK (1995) Malarial epitopes expressed on the surface of recombinant tobacco mosaic virus. Biotechnology 13(1):53–57
23. Dedeo MT, Duderstadt KE, Berger JM, Francis MB (2010) Nanoscale protein assemblies from a circular permutant of the tobacco mosaic virus. Nano Lett 10(1):181–186
24. Zhou K, Li F, Dai G, Meng C, Wang Q (2013) Disulfide bond: dramatically enhanced assembly capability and structural stability of tobacco mosaic virus nanorods. Biomacromolecules 14(8):2593–2600
25. Zhang J, Zhou K, Wang Q (2016) Tailoring the self-assembly behaviors of recombinant tobacco mosaic virus by rationally introducing

covalent bonding at the protein-protein interface. Small 12(36):4955–4959
26. Finbloom JA, Han K, Aanei IL, Hartman EC, Finley DT, Dedeo MT, Fishman M, Downing KH, Francis MB (2016) Stable disk assemblies of a tobacco mosaic virus mutant as nanoscale scaffolds for applications in drug delivery. Bioconjug Chem 27:2480.
27. Zahr OK, Blum AS (2012) Solution phase gold nanorings on a viral protein template. Nano Lett 12(2):629–633
28. Zhou K, Zhang J, Wang Q (2015) Site-selective nucleation and controlled growth of gold nanostructures in tobacco mosaic virus nanotubulars. Small 11(21):2505–2509
29. Endo M, Fujitsuka M, Majima T (2007) Porphyrin light-harvesting arrays constructed in the recombinant tobacco mosaic virus scaffold. Chemistry 13(31):8660–8666
30. Geiger FC, Eber FJ, Eiben S, Mueller A, Jeske H, Spatz JP, Wege C (2013) Tmv nanorods with programmed longitudinal domains of differently addressable coat proteins. Nanoscale 5(9):3808–3816
31. Altintoprak K, Seidenstücker A, Welle A, Eiben S, Atanasova P, Stitz N, Plettl A, Bill J, Gliemann H, Jeske H, Rothenstein D, Geiger F, Wege C (2015) Peptide-equipped tobacco mosaic virus templates for selective and controllable biomineral deposition. Beilstein J Nanotechnol 6:1399–1412
32. Uhde K, Fischer R, Commandeur U (2005) Expression of multiple foreign epitopes presented as synthetic antigens on the surface of potato virus X particles. Arch Virol 150(2):327–340
33. Eber FJ, Eiben S, Jeske H, Wege C (2015) Rna-controlled assembly of tobacco mosaic virus-derived complex structures: from nanoboomerangs to tetrapods. Nanoscale 7(1):344–355
34. Wu Z, Mueller A, Degenhard S, Ruff SE, Geiger F, Bittner AM, Wege C, Krill Iii CE (2010) Enhancing the magnetoviscosity of ferrofluids by the addition of biological nanotubes. ACS Nano 4(8):4531–4538
35. Adams MJ, Antoniw JF, Bar-Joseph M, Brunt AA, Candresse T, Foster GD, Martelli GP, Milne RG, Zavriev SK, Fauquet CM (2004) The new plant virus family flexiviridae and assessment of molecular criteria for species demarcation. Arch Virol 149(5):1045–1060
36. Kendall A, Bian W, Maris A, Azzo C, Groom J, Williams D, Shi J, Stewart PL, Wall JS, Stubbs G (2013) A common structure for the potexviruses. Virology 436(1):173–178
37. Shenton W, Douglas T, Young M, Stubbs G, Mann S (1999) Inorganic-organic nanotube composites from template mineralization of tobacco mosaic virus. Adv Mater 11:253–256
38. Fowler CE, Shenton W, Stubbs G, Mann S (2001) Tobacco mosaic virus liquid crystals as templates for the interior design of silica mesophases and nanoparticles. Adv Mater 13:1266–1269
39. Royston E, Lee SY, Culver JN, Harris MT (2006) Characterization of silica-coated tobacco mosaic virus. J Colloid Interface Sci 298:706–712
40. Royston ES, Brown AD, Harris MT, Culver JN (2009) Preparation of silica stabilized tobacco mosaic virus templates for the production of metal and layered nanoparticles. J Colloid Interface Sci 332:402–407
41. Kröger N, Deutzmann R, Sumper M (1999) Polycationic peptides from diatom biosilica that direct silica nanosphere formation. Science 286:1129–1132
42. Foo CW, Huang J, Kaplan DL (2004) Lessons from seashells: silica mineralization via protein templating. Trends Biotechnol 22(11):577–585
43. Bruckman MA, Randolph LN, Gulati NM, Stewart PL, Steinmetz NF (2015) Silica-coated Gd(Dota)-loaded protein nanoparticles enable magnetic resonance imaging of macrophages. J Mater Chem B Mater Biol Med 3(38):7503–7510
44. Lemloh M-L, Altintoprak K, Wege C, Weiss I, Rothenstein D (2017) Biogenic and synthetic peptides with oppositely charged amino acids as binding sites for mineralization. Materials 10(2):119
45. Stöber W, Fink A, Bohn E (1968) Controlled growth of monodisperse silica spheres in micron size range. J Colloid Interface Sci 26:62–69
46. Alonso JM, Gorzny ML, Bittner AM (2013) The physics of tobacco mosaic virus and virus-based devices in biotechnology. Trends Biotechnol 31:530–538
47. Pouget E, Dujardin E, Cavalier A, Moreac A, Valery C, Marchi-Artzner V, Weiss T, Renault A, Paternostre M, Artzner F (2007) Hierarchical architectures by synergy between dynamical template self-assembly and biomineralization. Nat Mater 6:434–439
48. Steinmetz NF, Shah SN, Barclay JE, Rallapalli G, Lomonossoff GP, Evans DJ (2009) Virus-templated silica nanoparticles. Small 5(7):813–816
49. Kumar K, Kumar Doddi S, Kalle Arunasree M, Paik P (2015) CPMV-induced synthesis of

hollow mesoporous SiO2 nanocapsules with excellent performance in drug delivery. Dalton Trans 44(9):4308–4317
50. van Rijn P, van Bezouwen LS, Fischer R, Boekema EJ, Böker A, Commandeur U (2015) Virus-SiO2 and virus-SiO2-au hybrid particles with tunable morphology. Part Part Syst Charact 32(1):43–47
51. Montague NP, Thuenemann EC, Saxena P, Saunders K, Lenzi P, Lomonossoff GP (2011) Recent advances of cowpea mosaic virus-based particle technology. Hum Vaccin 7(3):383–390
52. Lee KL, Uhde-Holzem K, Fischer R, Commandeur U, Steinmetz NF (2014) Genetic engineering and chemical conjugation of potato virus X. Methods Mol Biol 1108:3–21
53. McBride K, Summerfelt K (1990) Improved binary vectors for agrobacterium-mediated plant transformation. Plant Mol Biol 14(2):269–276
54. Hoekema A, Hirsch PR, Hooykaas PJJ, Schilperoort RA (1983) A binary plant vector strategy based on separation of vir- and T-region of the agrobacterium tumefaciens Ti-plasmid. Nature 303(5913):179–180
55. Sarikaya M, Tamerler C, Schwartz DT, Baneyx F (2004) Materials assembly and formation using engineered polypeptides. Annu Rev Mater Res 34(1):373–408
56. Sambrook J, Russell DW (2001) Molecular cloning: a laboratory manual, 3rd edn. Cold Spring Harbor Laboratory Press, New York
57. Porta C, Spall VE, Findlay KC, Gergerich RC, Farrance CE, Lomonossoff GP (2003) Cowpea mosaic virus-based chimaeras. Effects of inserted peptides on the phenotype, host range, and transmissibility of the modified viruses. Virology 310(1):50–63
58. Uhde-Holzem K, Schlosser V, Viazov S, Fischer R, Commandeur U (2010) Immunogenic properties of chimeric potato virus X particles displaying the hepatitis C virus hypervariable region I peptide R9. J Virol Methods 166(1–2):12–20
59. Lico C, Capuano F, Renzone G, Donini M, Marusic C, Scaloni A, Benvenuto E, Baschieri S (2006) Peptide display on potato virus X: molecular features of the coat protein-fused peptide affecting cell-to-cell and phloem movement of chimeric virus particles. J Gen Virol 87(Pt 10):3103–3112
60. Cruz SS, Chapman S, Roberts AG, Roberts IM, Prior DA, Oparka KJ (1996) Assembly and movement of a plant virus carrying a green fluorescent protein overcoat. Proc Natl Acad Sci U S A 93(13):6286–6290
61. Gooding GV Jr, Hebert TT (1967) A simple technique for purification of tobacco mosaic virus in large quantities. Phytopathology 57:1285
62. Raghavendra K, Adams ML, Schuster TM (1985) Tobacco mosaic virus protein aggregates in solution: structural comparison of 20s aggregates with those near conditions for disk crystallization. Biochemistry 24:3298–3304
63. Safancon H, Iwanami T, Karasev AV, van der Vlugt RAA, Wellink J, Wetzel T, Yoshikawa N (2012) Classification and nomenclature of viruses-Secoviridae. Ninth report of the International Committee on the Taxonomy of Viruses. Elsevier, San Diego, pp 881–899
64. Koenig R (1972) Anomalous behavior of the coat proteins of potato virus X and cactus virus X during electrophoresis in dodecyl sulfate-containing polyacrylamide gels. Virology 50(1):263–266
65. Goelet P, Lomonossoff GP, Butler PJ, Akam ME, Gait MJ, Karn J (1982) Nucleotide sequence of tobacco mosaic virus Rna. Proc Natl Acad Sci U S A 79(19):5818–5822
66. Laemmli UK (1970) Cleavage of structural proteins during the assembly of the head of bacteriophage T4. Nature 227(5259):680–685
67. Uhde-Holzem K, Fischer R, Commandeur U (2007) Genetic stability of recombinant potato virus X virus vectors presenting foreign epitopes. Arch Virol 152(4):805–811
68. Le DHT, Lee KL, Shukla S, Commandeur U, Steinmetz NF (2017) Potato virus X, a filamentous plant viral nanoparticle for doxorubicin delivery in cancer therapy. Nanoscale 9(6):2348–2357
69. Baratova LA, Grebenshchikov NI, Dobrov EN, Gedrovich AV, Kashirin IA, Shishkov AV, Efimov AV, Jarvekulg L, Radavsky YL, Saarma M (1992) The organization of potato virus X coat proteins in virus particles studied by tritium planigraphy and model building. Virology 188(1):175–180
70. Goodman RM, McDonald JG, Horne RW, Bancroft JB (1976) Assembly of flexuous plant viruses and their proteins. Philos Trans R Soc Lond Ser B Biol Sci 276(943):173–179
71. Voinnet O, Rivas S, Mestre P, Baulcombe D (2003) An enhanced transient expression system in plants based on suppression of gene silencing by the P19 protein of tomato bushy stunt virus. Plant J 33(5):949–956
72. Virudachalam R, Harrington M, Markley JL (1985) Thermal stability of cowpea mosaic virus components: differential scanning calorimetry studies. Virology 146(1):138–140
73. Altintoprak K, Seidenstücker A, Krolla-Sidenstein P, Plettl A, Jeske H, Gliemann H, Wege C (2017) Rna-stabilized protein nanorings: high-precision adapters for biohybrid design. Bioinsp Biomim Nan 6(4):208–223

Chapter 24

Propagation and Isolation of Tobacco Mosaic Virus That Surface Displays Metal Binding and Reducing Peptides for Generation of Gold Nanoparticles

Andrew J. Love and Michael E. Taliansky

Abstract

In this chapter we describe an approach for propagating, isolating and characterizing tobacco mosaic virus (TMV) genetically modified to surface display a metal binding and reducing peptide, and its utilization for the production of free gold metal nanoparticles from metal salt precursors.

Key words Plant virus, Virus purification, Virus modification, Surface display of functional peptides, Metal nanoparticle production

1 Introduction

Metal nanoparticles with a size range of 10–100 nm have a plethora of mechanical, dielectric, catalytic, and optical properties which are now being exploited in diverse sectors of the economy such as communications, materials science, cosmetics, energy, and medical industries. These nanoparticles are traditionally chemically synthesized by reducing metal salts in solution into agglomerates of metal atoms by addition of hazardous reductants such as the organophosphorus compound THPC (tetrakishydroxymethylphosphonium chloride), or by use of sodium borohydride [1]. Alternatively nanoparticles can be produced using physical approaches by vapourizing bulk metal substrates and precipitating them, or by laser ablating bulk metals into particles [2–4]. There is now an interest in alternative systems for nanoparticle production, which are potentially more environmentally friendly and may offer different opportunities for production of 3D nanomaterials into which metal nanoparticles are integrated. One such "green" approach that has been taken is to use peptides comprising amino acids with metal ion binding (histidine, aspartic and glutamic acids) and reducing

Christina Wege and George P. Lomonossoff (eds.), *Virus-Derived Nanoparticles for Advanced Technologies: Methods and Protocols*, Methods in Molecular Biology, vol. 1776, https://doi.org/10.1007/978-1-4939-7808-3_24,
© Springer Science+Business Media, LLC, part of Springer Nature 2018

(tyrosine, asparagine, glutamine, and tryptophan) activities which have been optimally configured to produce nanoparticles from metal salt precursors that are of suitable size (~40 nm) for nanotech purposes [5].

However, free peptides can be expensive to synthesize and may be unstable which may limit their utility. In order to overcome these limitations plant viruses such as tobacco mosaic virus (TMV) which can accumulate to high yields in plants (1–10 g/kg fresh plant material), may be used to generate bulk quantities of free peptide [6]. TMV is a rod of 300 nm length and 18 nm diameter with a central channel of 4 nm. It is composed of 2130 coat proteins (CP) surrounding a positive sense single-stranded RNA genome of 6395 nucleotides [7]. In addition peptide sequences can be introduced into specific regions of the TMV coat protein, such that once the virus assembles, the peptides will be densely presented on the particle surface in a manner which promotes enhanced stability and reactivity. The TMV CP is a 17.5 kDa protein consisting of 159 amino acids (aa) (including start codon), which has previously been modified to permit insertion/replacements of small (1–20 aa) sequences at defined sites, which will be surface displayed after virion assembly. Modifications have been performed at amino acids 155–156, 193–198, 208–213, 193–213, and C terminal regions of the CP [8–11]. A variety of full-length TMV clones with introduced restriction sites in the CP gene are available for convenient cloning of peptides into these surface display sites. In our previous publication [9] we introduced into the 193–213 region of the TMV CP an optimal metal-binding and reducing peptide (MBP; sequence: SEKLWWGASL) [5] which can produce ~40 nm metal nanoparticles from gold salts. We found that the MBP TMV assembled into network-like aggregates composed of structures that were longer and thinner than traditional TMV rods. Unlike TMV, the MBP TMV networks could produce free, stable, crystalline gold nanoparticles under shaking conditions after the addition of $K(AuCl_4)$, in the absence of exogenous reductants. We also demonstrated that the ~10–40 nm nanoparticles were predominantly colloidal with little association with the virus complexes. This is of potential utility as a bioreactor for free nanoparticle production or may enable spontaneous in situ metal nanoparticles to be formed in 3D structures (into which the virus complexes are integrated), rather than complete metallization of the 3D structures which would typically be achieved using hazardous chemicals. In this chapter we focus predominantly on the molecular methods used to produce the MBP TMV from clones, its isolation, characterization, and use for the generation of gold nanoparticles.

2 Materials

Unless indicated otherwise, standard reagents, chemicals and secondary antibodies were obtained from Sigma-Aldrich. Ultrapure water was produced using a Milli-Q integral system (EMD Millipore), and nuclease-free water was purchased from ThermoFisher Scientific.

2.1 Production of In Vitro Transcription Templates

1. pSNC004 and pSNC004 MBP: Plasmid pSNC004 contains the full length sequence of the RNA of TMV strain U1 flanked by an upstream T7 promoter and has been modified at nucleotides 193–198 and 208–213 of the coat protein to introduce *Ngo*MIV and *Bst*Z17I restriction sites for inserting small peptides for display on the surface of the virus [9]. We have previously digested at these sites and introduced a metal binding and reducing peptide (MBP; sequence: SEKLWWGASL) using traditional cloning methods [9].
2. *Kpn*I-*HF* restriction enzyme plus 10× CutSmart Buffer (New England Biolabs).
3. 10× DNA loading buffer: 65% (v/v) glycerol, 10 mM Tris—HCl (pH 7.5), 0.3% (w/v) bromophenol blue.
4. Low melting point agarose.
5. 5× TBE stock: 0.44 M Tris base, 0.44 M boric acid, 10 mM EDTA, adjust pH to 8.3 with HCl.
6. SYBR Safe DNA gel stain (ThermoFisher Scientific).
7. Combs that have different prong sizes to produce 200 and 50 μL wells.
8. 1 Kb Plus DNA ladder (ThermoFisher Scientific).
9. Safe Imager™ 2.0 Blue Light Transilluminator (ThermoFisher Scientific).
10. Razor blade for cutting out gel slices.
11. QIAquick Gel Extraction kit and associated QG, PE, and elution buffers (Qiagen).
12. 100% isopropanol.
13. NanoDrop 1000 spectrophotometer (Thermo Scientific).

2.2 Production of Infectious Transcripts

1. T7 mMessage mMachine in vitro transcription kit (Ambion, Life Technologies).
2. Nuclease-free water.

2.3 Plant Inoculations, Harvesting, and Sample Preparation

1. Nuclease-free water.
2. Carborundum powder (180 grit).
3. *Nicotiana benthamiana* plants grown under 16 h day-length at 23 °C to the five-leaf stage of development.

4. Glass rods.
5. Tin foil.
6. Liquid nitrogen.
7. Mortar and pestle.

2.4 15% Acrylamide SDS-PAGE

1. 2× Laemmli buffer: 4% (w/v) SDS, 20% (v/v) glycerol, 125 mM Tris—HCl (pH 6.8), 0.02% (w/v) bromophenol blue.
2. Mini-PROTEAN Electrophoresis System (Bio-Rad).
3. 1.5 mm Bio-Rad combs.
4. Resolving gel buffer: 1.5 M Tris—HCl pH 8.8.
5. Stacking gel buffer: 0.5 M Tris—HCl pH 6.8.
6. 40% (w/v) acrylamide–bisacrylamide solution 37.5:1.
7. 10% (w/v) ammonium persulfate (APS) in ultrapure water (*see* **Note 1**).
8. 10% and 0.1% (w/v) SDS.
9. TEMED (*N*,*N*,*N'*,*N'*-tetramethyl-ethylenediamine) solution, 99% purity.
10. Whatman 3MM paper.
11. SDS-PAGE running buffer: 0.025 M Tris—HCl, pH 8.3, 0.192 M glycine, 0.1% SDS (w/v).
12. Novex® Sharp Pre-stained Protein Standard (ThermoFisher Scientific).

2.5 Protein Transfer onto Membranes

1. Immobilon-P membrane (polyvinylidene fluoride, PVDF) (EMD Millipore).
2. 100% methanol.
3. Transfer buffer: 25 mM Tris base, 192 mM glycine, 10% (v/v) methanol.
4. Blotting cassette.
5. Sponges.
6. Whatman 3MM blotting paper.

2.6 Western Blot Analysis

1. Ponceau stain: 0.5% (w/v) Ponceau S Red, 1% (v/v) acetic acid.
2. 100% methanol.
3. Whatman 3MM blotting paper.
4. Anti-TMV antiserum raised in rabbits, produced at The James Hutton Institute.
5. Anti-rabbit IgG alkaline phosphatase conjugate (Sigma-Aldrich).

6. 1× phosphate buffered saline (PBS): 0.01 M Na_2HPO_4, 0.0018 M KH_2PO_4, 0.1370 M NaCl, 0.0027 M KCl, adjust pH to 7.4 with HCl.
7. Blocking buffer: 1% (w/v) bovine serum albumin, 0.05% (v/v) Tween 20 in 1× PBS.
8. NBT/BCIP substrate solution (ThermoFisher Scientific).

2.7 Immunotrapping Transmission Electron Microscopy

1. Anti-TMV antiserum raised in rabbits, produced at The James Hutton Institute.
2. 0.7 M sodium phosphate buffer pH 7.
3. Carbon-coated 200 mesh copper grids.
4. Miracloth.
5. Whatman 3MM blotting paper.
6. 0.44 μm filtered ultra-pure water.
7. 2% (w/v) uranyl acetate.
8. Transmission electron microscope (JEOL 1400 transmission electron microscope).

2.8 Virus Isolation

1. Plant virus extraction buffer: 0.1 M sodium phosphate buffer pH 7.8, 20 mM EDTA and 0.1% (v/v) β-mercaptoethanol. Cool to 4 °C.
2. Centrifugation: Sorvall RC6+ centrifuge with a Fiberolite F14-6x250y rotor (Thermo Scientific), 250 mL polypropylene copolymer Nalgene© centrifuge tubes (Sigma-Aldrich).
3. Ultracentrifugation: Optima L-80 XP Ultracentrifuge, an SW 41 Ti Rotor and Ultra-Clear open topped thin-walled tubes (Beckman Coulter).
4. Chloroform.
5. PEG 8000.
6. NaCl.
7. 25 mM Tris—HCl pH 7.8.
8. Sucrose cushion: 25 mM Tris–HCl pH 7.8, 20% (w/v) sucrose.
9. 10 mM and 0.01 M Tris–HCl buffer pH 7.8.
10. Sterile plastic or glass rods.
11. Micro balance (weighing range 110 g–0.1 mg).

2.9 Characterization of the Isolated Virus

1. 70 μL UV cuvettes (BrandTech Scientific).
2. Spectrophotometer.

2.10 Production of Gold Nanoparticles and UV-Vis Analysis

1. 3 mM potassium tetrachloroaurate in ultrapure water.
2. Purified WT TMV and MBP TMV (from Subheading 3.8, **step 17**).

3. Shaker set at 100 rpm and 20 °C.
4. 70 μL UV cuvettes (BrandTech Scientific).
5. Beckman DU640 spectrophotometer (Beckman Coulter).

2.11 TEM Analysis of Gold Nanoparticles

1. 0.44 μm filtered ultrapure water.
2. Carbon coated 200 mesh copper grids.
3. Whatman 3MM blotting paper.
4. Transmission electron microscope (e.g., JEOL 1400 transmission electron microscope) at 100 kV.

3 Methods

3.1 Production of In Vitro Transcription Templates

1. Combine 1 μg of pSNC004 MBP (or pSNC004 control) (*see* **Note 2**) with 10 μL 10× CutSmart Buffer, and then add ultrapure water to bring the volume to 97.5 μL and gently mix by pipetting.
2. Add 2.5 μL of *Kpn*I-*HF* to the reaction, then mix and split the reaction into five fresh Eppendorf tubes (20 μL per tube; this improves heat regulation in the sample). Incubate for 3 h in a 37 °C water bath.
3. Remove the digested pSNC004 MBP (or pSNC004 control) and combine the five 20 μL reactions into one Eppendorf tube. Add 11 μL of 10× DNA loading buffer. Put on ice. To check digestion, mix 200 ng of uncut plasmid with 3 μL of 10× DNA loading buffer and ultrapure water giving a final volume of 20 μL.
4. Mix 1 g of low melting point agarose with 100 mL of 1× TBE and heat in the microwave oven until the agarose has melted.
5. Let cool to ~50 °C and add 3 μL of SYBR safe DNA stain, mix and pour into horizontal casting trays, taking care to avoid bubbles.
6. Place combs that produce both ~200 μL and ~50 μL capacity wells into the molten gel and let set (*see* **Note 3**).
7. Once set, remove combs and place the gel into an electrophoresis tank filled with 1× TBE buffer. The gel should be covered with the buffer and the system checked to ensure that it can support an electric current. Turn off electricity.
8. To the 50 μL wells add 15 μL 1 Kb Plus DNA ladder, and the whole undigested plasmid control. To the 200 μL wells add 100 μL of digested sample (from **step 3**).
9. Run the gel at 100 V until the bands are resolved. Under a blue light transilluminator photograph the gel and then excise the linearized pSNC004 MBP (and linearized pSNC004 con-

trol) using a razor blade. The linearized fragment should be around 9064 bp in length.

10. Weigh the excised gel fragment, place in an Eppendorf tube and add 3 volumes (v/w) of QG buffer (from the QIAquick Gel Extraction kit) to the gel slice. Incubate at 50 °C for 10 min to dissolve the gel. Mix and then add 1 gel volume (v/w) of isopropanol and mix.
11. Add the solution to a QIAquick spin column in collection tube and centrifuge for 1 min at 16,000 × *g*. Remove flow through and add 500 μL QG buffer to the column and centrifuge again for 1 min at 16,000 × *g*. Remove flow through.
12. Add 750 μL of PE buffer to the column and incubate for 5 min prior to centrifuging for 1 min at 16,000 × *g*. Transfer column to fresh collection column and centrifuge for 3 min at 16,000 × *g* to remove ethanol traces.
13. Place the column in a fresh Eppendorf tube and add 15 μL of elution buffer. Incubate for 2 min at room temperature, centrifuge for 1 min at 16,000 × *g*. Add another 15 μL elution buffer to the column and incubate and centrifuge again.
14. Set a Nanodrop spectrophotometer to measure the OD 260 nm. Calibrate it against water, and then obtain baseline values for 1.5 μL elution buffer. Place 1.5 μL of the eluted linearized plasmid on the apparatus and take a measurement. After subtracting the baseline, the obtained OD 260 nm value will indicate the DNA concentration (i.e., 50 ng/μL dsDNA is equivalent to OD 260 nm= 1).

3.2 Production of Infectious Transcripts

1. Thaw components of the T7 mMessage mMachine in vitro transcription kit at room temperature, taking care to vortex the 10× Reaction Buffer and 2× NTP/CAP to resolubilize the precipitate.
2. At room temperature, add 750 ng purified *Kpn*I-linearized pSNC004 MBP (or linearized pSNC004 control) (from Subheading 3.1, **step 14**) to an Eppendorf tube, and then add 10 μL 2× NTP/CAP, 2 μL 10× Reaction Buffer, 2 μL Enzyme Mix, and make up to a final volume of 20 μL using nuclease-free water (*see* **Note 4**).
3. Mix the tube by flicking and then briefly centrifuge to collect the reaction mix in the bottom of the tube.
4. Incubate at 37 °C for 3 h (*see* **Note 5**).

3.3 Plant Inoculation, Harvesting, and Sample Preparation

1. Dilute infectious virus transcripts 1:50 (v/v) in RNA-free water (*see* **Note 6**) and add sterile carborundum abrasive to a final concentration of 1% (w/v). This should be enough transcript to inoculate ~80 *Nicotiana benthamiana* plants.

Fig. 1 *Nicotiana benthamiana* showing strongly stunted and distorted leaves at 16 dpi which were harvested for virus purification (harvested material indicated by a white circle). Large outer leaves with no symptoms were not harvested

2. On to the uppermost leaf of each plant, pipette 10 μL of transcript/abrasive and then immediately gently rub with a glass rod. This produces local minor damage that permits entry of the virus transcript into the plant.
3. Plants are left under 16 h day-length at 23 °C, and should be frequently observed for the development of systemic symptoms (*see* Fig. 1), which should be apparent by 16 days post-inoculation (dpi) (*see* **Note 7**).
4. Symptomatic leaf material should be carefully harvested (*see* Fig. 1). To avoid cross-contamination, immediately weigh, wrap in tin foil, and freeze in liquid nitrogen. Store if required at −70 °C (*see* **Note 8**).
5. Grind the plant material to a fine powder using a mortar and pestle in conjunction with liquid nitrogen. The majority of this material can be used for subsequent isolation of the virus (*see* Subheading 3.8); however ~30 mg should be retained for western blot analysis (*see* Subheadings 3.5 and 3.6), and another 30 mg can be used for electron microscopy (*see* Subheading 3.7).

3.4 15% Acrylamide SDS-PAGE

1. For protein analysis, add 30 mg of frozen crushed leaf material to 100 μL of 2× Laemmli loading buffer, mix by pipetting and then boil for 5 min at 100 °C.
2. Centrifuge the samples for 5 min at 16,000 × *g*, collect the supernatants and store at −20 °C until required for western blot analysis.

3. Place clean glass casting plates for 10.1 cm × 7.3 cm × 1.5 mm gels into the Mini-PROTEAN casting system (*see* **Note 9**).
4. In a Falcon tube combine 6.75 mL 40% (w/v) acrylamide with 4.5 mL resolving gel buffer, 6.4 mL ultrapure water, 180 μL 10% (w/v) APS, and 180 μL 10% (w/v) SDS.
5. Invert gently to mix and then add 12 μL of TEMED, followed by gentle mixing. Immediately pipette the solution between the glass plates in the casting rig until it is 75% full. On top of this, pipette 0.1% (w/v) SDS until the top of the glass plates is reached.
6. Once the gels are set (takes typically around 30 min), pour off the SDS and rinse the top of the gel with sterile water and then drain and blot with Whatman 3MM paper until dry (*see* **Note 10**).
7. Reinsert the gels into the casting apparatus.
8. Make the stacking gel by combining 700 μL 40% (w/v) acrylamide, 1.98 mL stacking gel buffer, 4.8 mL ultrapure water, 75 μL 10% (w/v) APS, and 75 μL 10% (w/v) SDS. After mixing by inversion, add 12 μL of TEMED, mix and immediately pipette on top of the resolving gel.
9. Carefully position the 1.5 mm Bio-Rad combs into the stacking gel and let set (takes around 30 min).
10. Remove the combs and position the gel plates into the running apparatus and fill the central cavity with SDS-PAGE running buffer and check for leaks. Assuming there are no leaks, put this assembly into a gel tank filled with SDS-PAGE running buffer and top up the central reservoir to cover the gel wells.
11. Wash out the wells briefly by pipetting in the running buffer (*see* **Note 11**).
12. Load 20 μL of each of the protein samples into the wells and also add 15 μL of Novex prestained ladder to one of the wells (*see* **Note 12**).
13. Run gels at 200 V until the dye front emerges from the bottom of the gels, and then turn off.
14. Remove the gels and cut off the stacking gel in preparation for blotting onto membranes.

3.5 Protein Transfer onto Membranes

1. Cut Immobilon-P membrane to the size of the gels and soak in 100% methanol for 15 s and then place in Milli-Q water for 2 min, prior to equilibration in Transfer buffer for at least 2 min.
2. Open the transfer cassettes and place a Transfer buffer soaked sponge on top of the black cassette lid followed by two soaked Whatman 3MM paper sheets. Onto this carefully place the gel with the stacking component removed, avoiding bubbles.

3. Next, layer the membrane over the gel followed by another two sheets of soaked Whatman 3MM and a sponge. Press the assembly from side to side to remove bubbles, then close the cassette.
4. Immediately insert the cassette into the tank filled with Transfer buffer, ensuring appropriate polarity of the terminals (black cassette lid next to black electrode).
5. Carry out the transfer overnight at 4 °C at 30 mA (*see* **Note 13**).

3.6 Western Blot Analysis

1. Check membranes for equal loading by immersing in Ponceau stain for 5 min and destaining in water. Photograph (*see* **Note 14**).
2. Place membranes in 100% methanol and then air-dry on Whatman 3MM paper.
3. Dilute antibodies against TMV (raised in rabbits) 1:10,000 in blocking buffer and add to the blots. Incubate with gentle shaking for 1 h at room temperature.
4. Wash by vigorously shaking in 1× PBS for 5 min; drain and repeat the wash step three times.
5. Add the anti-rabbit IgG alkaline phosphatase conjugate to the blots at a dilution of 1:1000 in blocking buffer (*see* **Note 15**).
6. Shake for 1 h at room temperature and then wash the blots in 1× PBS (*see* **step 4**).
7. Incubate the blots with BCIP/NBT until dark bands become visible (typically within 15 min).
8. Stop the reaction by washing the membrane with water. Photograph. No band should be visible in the uninfected plant material, whereas signals should be visible in the WT TMV and MBP TMV (*see* Fig. 2). The WT TMV coat protein should be visible as a band at around 17.5 kDa and the MBP TMV should have a slightly higher molecular weight than the WT TMV due to incorporation of the MBP peptide (*see* Fig. 2). This data also indicates that there is no appreciable difference in the level of accumulation between wild-type and modified viruses in plants.
9. Modification of the TMV CP can lead to formation of particles indistinguishable from the WT virus, but it has also been reported that different virus structures such as aggregates and nanoscale networks can form. In order to assess the production of these structures in plants, immunotrapping EM should be carried out using polyclonal antibodies raised against the virus coat protein, or alternatively antibodies specific for the amino acid sequence inserted into the virus structure can also be used (*see* Subheading 3.7).

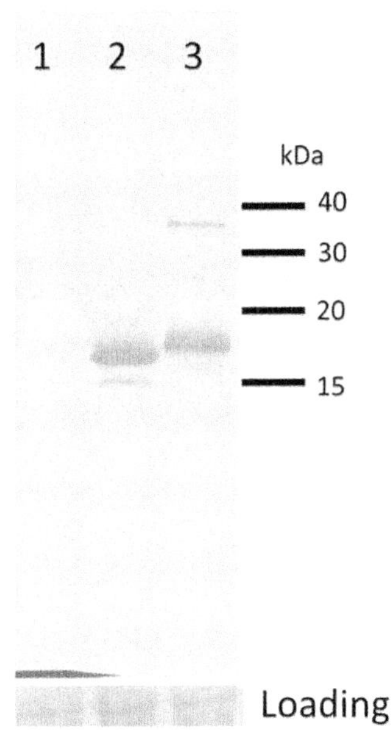

Fig. 2 Western blot of (1) uninfected plant material; (2) and of WT TMV and (3) MBP TMV infected plant material probed with a TMV antibody. The loading control below the blot is the Ponceau stained large Rubisco (ribulose-1,5-bisphosphate carboxylase/oxygenase) subunit, and the molecular weights are indicated on the right side. Figure taken from Frontiers in Plant Science with permission from the corresponding author and under the terms specified in the Creative Commons Attribution License (CC BY) [9]

3.7 Immunotrapping Transmission Electron Microscopy

1. Dilute TMV antiserum (raised in rabbits) 1:1000 in 0.7 M sodium phosphate buffer pH 7 and pipette 30 μL onto the shiny side of carbon-coated copper grids, prior to incubation at 37 °C for 60 min.
2. Cover the grids with 0.7 M sodium phosphate buffer pH 7 and agitate gently for 10 min, then take the grids and carefully drain by placing Whatman 3MM at the very edge of the grid and wicking off. Repeat the wash step. Keep the grids moist.
3. Take 30 mg of ground up MBP TMV, WT TMV, and uninfected plant material (from Subheading 3.3, **step 5**) and dilute 1:4 and 1:40 (w/v) in 0.7 M sodium phosphate buffer pH 7.0.
4. Filter this twice through Miracloth to remove large particles.
5. Pipette 20 μL of these samples onto the antiserum-coated grids and incubate overnight at 4 °C.
6. Wash by gently pipetting 50 μL sterile water on the grids and wicking off by placing Whatman 3MM at the very edge of the grid. Repeat this wash step.
7. To each grid add 20 μL of 2% (w/v) uranyl acetate and leave for 1 min. Wick off with Whatman 3MM as before and let air-dry.

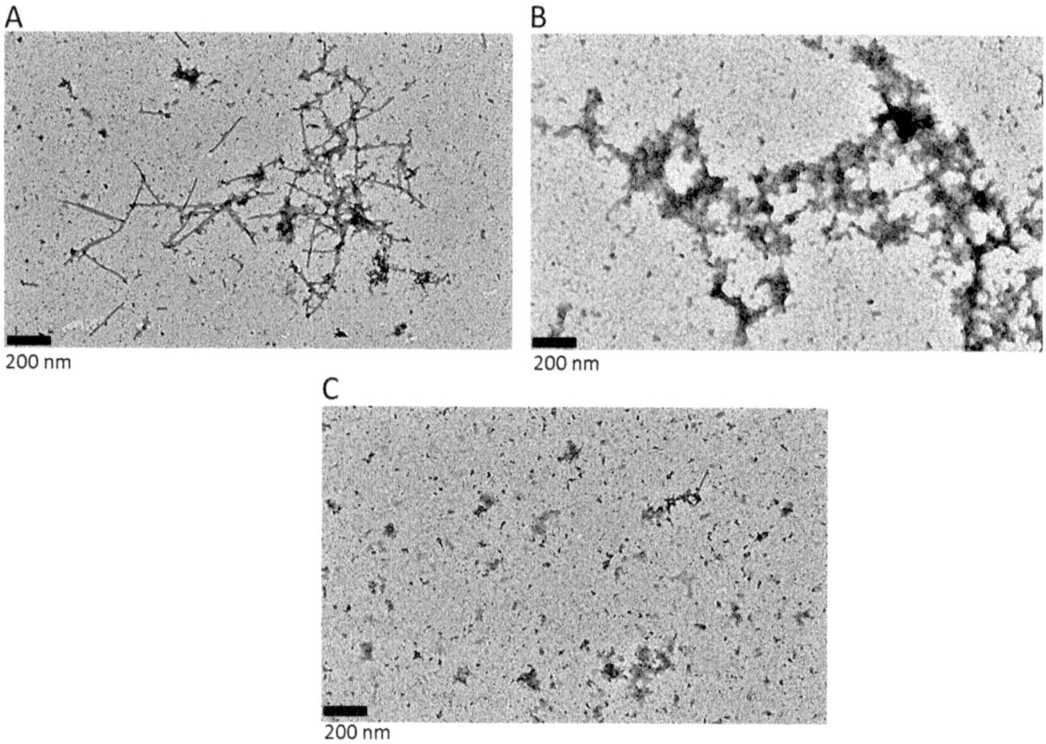

Fig. 3 TMV antibody immunotrapping EM performed on crude plant extracts of plants infected with (**a**) WT TMV and (**b**) MBP TMV, and (**c**) of uninfected plant material. Figure taken from Frontiers in Plant Science with permission from the corresponding author and under the terms specified in the Creative Commons Attribution License (CC BY) [9]

8. Load grids into a transmission electron microscope set at 80 kV, and obtain representative images. As indicated in Fig. 3, TMV rods (300 nm × 18 nm) would be expected in the WT TMV infected plant material (a), whereas no obvious uniform structures should be observed in uninfected plant material (c); which demonstrates the specificity of the immunotrapping method. The modification to the TMV CP may not change the virus structure; however in the case of the immunotrapped MBP TMV sample a network-like structure should be consistently observed (b). After confirming the detection of the virus, it should be purified from the plant material.

3.8 Virus Isolation

1. To the frozen plant powder (from Subheading 3.4), add 2 parts (w/v) plant virus extraction buffer (4 °C). These should be combined in a 250 mL polypropylene copolymer Nalgene© centrifuge tube in a chemical flow hood, then weighed to ensure even weight distribution between tubes prior to centrifugation (*see* **Note 16**).
2. Incubate at 4 °C for 1 h with gentle shaking (~50 rpm).

3. Centrifuge at 4 °C for 20 min at 10,000 × *g* in a Fiberolite F14-6x250y rotor (*see* **Note 17**), then decant the supernatant into fresh 250 mL centrifuge tubes, and add 1 part chloroform to six parts of the decanted supernatant (*see* **Note 18**).
4. Centrifuge at 10,000 × *g* for 20 min at 4 °C as before, then decant the supernatant into a fresh centrifuge tube and add NaCl and PEG 8000 to a final concentration of 1% and 2% (w/v), respectively, and mix until dissolved.
5. Incubate without shaking at 4 °C overnight to facilitate precipitation.
6. Pellet the virus by centrifuging the solution at 10,000 × *g* for 20 min at 4 °C.
7. Resuspend the pellet in 25 mM Tris–HCl pH 7.8 (*see* **Note 19**).
8. Centrifuge at 2300 × *g* at 4 °C for 2 min and decant the supernatant into a fresh tube.
9. Resuspend the pellet in 25 mM Tris–HCl pH 7.8 (*see* **Note 19**) and centrifuge at 2300 × *g* as before (*see* **step 8**), and decant and combine the supernatant with the previously collected supernatant (*see* **Note 20**). Store the clarified supernatants at 4 °C until required.
10. Add 2 mL of sucrose cushion to the bottom of an Ultra-Clear centrifuge tube (*see* **Note 21**). Cool to 4 °C.
11. Gently decant 8 mL of clarified supernatant drop-wise onto the 2 mL sucrose cushion.
12. Weigh the tubes in a microbalance to ensure weight uniformity prior to centrifugation in a SW 41 Beckman Coulter Optima L-80 XP rotor (Beckman Coulter) at 175,500 × *g* at 4 °C for 2 h.
13. Carefully discard the supernatant and resuspend the pellet in 10 mM Tris–HCl buffer (pH 7.8) by gently breaking up the pellet with sterile plastic or glass rod (*see* **Note 22**). Incubate overnight at 4 °C to ensure solubilization.
14. Add 0.01 M Tris–HCl buffer (pH 7.8) to the resuspended pellet to bring to 10 mL.
15. To a fresh Ultra-Clear tube, add 10 mL of resuspended pellet and then weigh on a microbalance to ensure weight uniformity prior to ultracentrifugation (2, 4, or 6 tubes should be prepared to balance the rotor).
16. Centrifuge at 175,500 × *g* in a SW41 rotor at 4 °C for 2 h.
17. Resuspend the pellet in one pellet volume of ultrapure water and store on ice in a 4 °C cold room until use.

3.9 Characterization of the Isolated Virus

1. To a 70 μL UV cuvette, add 70 μL of undiluted, 1:10 and 1:100 dilutions of the isolated virus (from Subheading 3.8, **step 17**) in water.
2. Insert a 70 μL UV cuvette containing ultrapure water into a spectrophotometer to set the baseline readings at OD 260 and 280.
3. Obtain OD 260 and 280 readings for the virus samples.
4. A OD_{260}/OD_{280} ratio of 1.25 indicates that the virus is highly pure and the $OD_{260}/3$ multiplied by the dilution factor will indicate the concentration of the virus (*see* **Note 23**).
5. Check the structural integrity of the isolated viruses by following **step 5** onward of the immunotrapping TEM procedure, except that purified virus diluted 1:20, 1:50, 1:100, and 1:1000 in water should be put directly on to carbon coated copper grids (non-antisera coated) for 1 min prior to following **steps 6–8** (*see* Subheading 3.7). The images obtained should be consistent with the structures observed in the immunotrapping TEM carried out on crude plant extracts (*see* Fig. 4) (*see* **Note 24**).

3.10 Production of Gold Nanoparticles and UV-Vis Analysis

1. In a 2 mL Eppendorf tube, combine 0.2 mg of pure MBP TMV or WT TMV (from Subheading 3.9, **step 4**) in ultrapure water with 3 mM potassium tetrachloroaurate to a final volume of 400 μL.
2. Shake at 100 rpm at 20 °C (*see* **Note 25**).
3. Set a spectrophotometer to obtain UV-Vis profiles (300–750 nm) and blank the spectrophotometer by calibrating against a 70 μL UV cuvette filled with ultrapure water.
4. To a 70 μL cuvette, add 70 μL of a 0.5 mg/mL suspension of MBP TMV or WT TMV and measure the UV-Vis spectrum using the spectrophotometer.
5. Add 70 μL of either the MBP TMV or WT TMV/potassium tetrachloroaurate reaction (from **step 2**) to a 70 μL cuvette and measure the UV-Vis spectra using the spectrophotometer. From these values subtract the measurements obtained in **step 4**—thus giving a nanoparticle only specific signal.

Gold nanoparticles typically produce a peak absorbance in the range of 400–600 nm wavelengths, which is determined by the size of the nanoparticles. An example of the expected UV-Vis spectral timecourse of gold nanoparticle formation in the reactions is shown in Fig. 5. No spectral change would be expected in reactions containing WT TMV mixed with the gold salt (not shown). The peak at 550 nm indicates nanoparticle formation, but the 500–600 nm absorption band indicates that a range of gold nanoparticle sizes were produced.

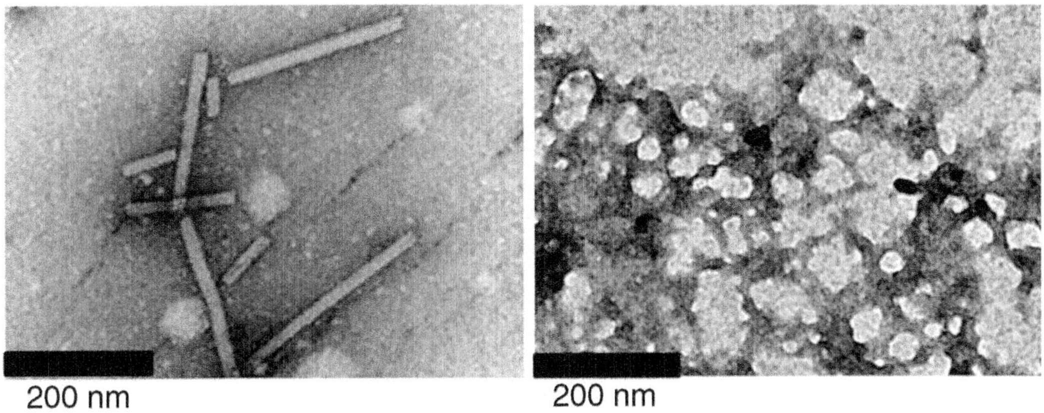

Fig. 4 TEM of purified viruses. The left panel shows purified and isolated WT TMV; the right panel shows a purified MBP TMV nanonet. Figure taken from Frontiers in Plant Science and adapted with permission from the corresponding author and under the terms specified in the Creative Commons Attribution License (CC BY) [9]

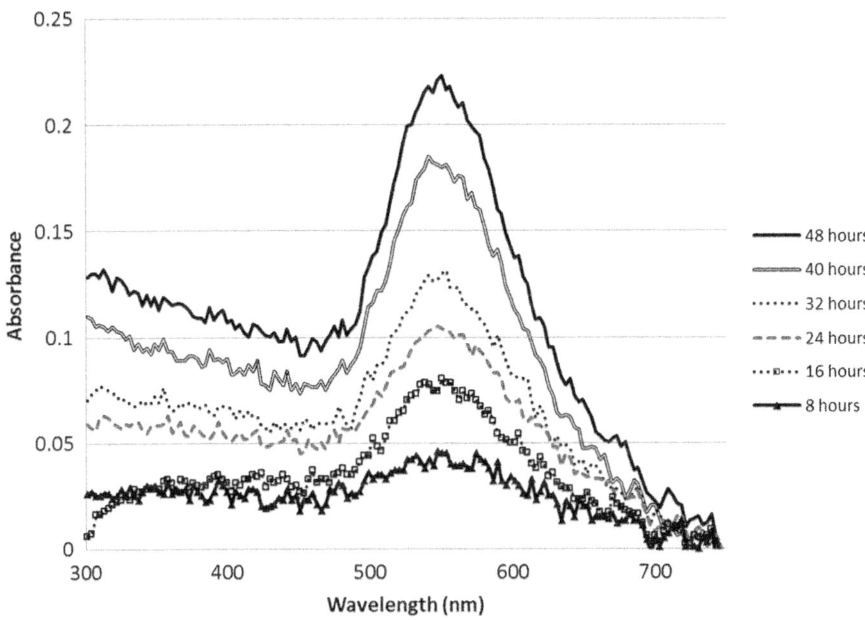

Fig. 5 UV-Vis analysis on 3 mM potassium tetrachloroaurate/MBP TMV reactions over time, with characteristic 550 nm peak indicating gold nanoparticle formation. Figure taken from Frontiers in Plant Science with permission from the corresponding author and under the terms specified in the Creative Commons Attribution License (CC BY) [9]

3.11 TEM Analysis of Gold Nanoparticles

1. From the concluded nanoparticle reaction (*see* Subheading 3.10, **step 5**), take 5 μL and dilute 1:10 and 1:100 in filtered ultrapure water.
2. Pipette 10 μL of the dilutions onto carbon coated copper grids and incubate for 30 s.

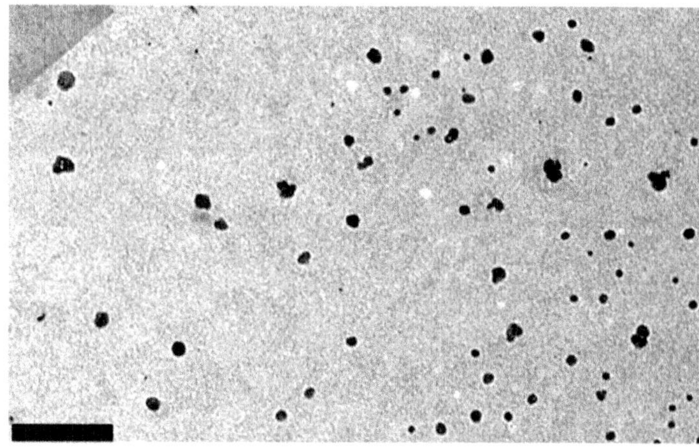

Fig. 6 TEM analysis of the gold nanoparticles produced by using the MBP TMV in conjunction with the gold salts

3. Wick the sample grids off with Whatman 3MM blotting paper by gently placing the paper to the very edge of the grid.
4. Wash by pipetting 20 μL of ultrapure water on to the grids and wicking with blotting paper.
5. Let samples completely air-dry prior to loading into a TEM operating at 100 kV for image capture. Gold nanoparticles should be observable as electron dense unattached particles of 10–40 nm in size (*see* Fig. 6) (*see* **Note 26**).

4 Notes

1. Store 10% (w/v) APS at 4 °C until use, and only use APS made within 4 days as it rapidly starts to lose reactivity.
2. For the purpose of appropriate controls later in the experiment we recommend that plasmid pSNC004 should also be prepared.
3. Wells of the correct size can be obtained by taking a standard comb and taping over the gaps with electrical tape.
4. If the linearized plasmid is too dilute for a 20 μL reaction, several reactions can be set up since a minimum of 100 ng of DNA can be used in the reactions. For example five reactions of 150 ng of DNA template would be comparable to a single reaction using 750 ng template.
5. This increased reaction time is required in order to maximize the amount of transcripts produced from such a large template (>5 kb).

6. If several reactions were used to produce the transcripts, this dilution factor can be adjusted accordingly. For example a single 20 μL 750 ng template reaction would be diluted to 1 mL, whereas 5× 20 μL reactions with 150 ng template would be combined and diluted to 1 mL.
7. Around 16 dpi it would be expected that systemic symptoms such as leaf wrinkling, chlorosis, stunting and mosaic patterns should be obvious (*see* Fig. 1).
8. For virus preparations in general, if the material has to be stored for any length of time we recommend flash-freezing the fresh leaf material prior to storage at −70 °C, and also to avoid, where possible repeated freeze–thaw cycles. This will limit disruption of the virus structure.
9. We recommend that plates are cleaned with Pyroneg detergent (ThermoFisher Scientific), washed with water, dipped in 100% ethanol and wiped dry with tissue in order to remove contaminants and promote uniform gel formation.
10. If the gel is not dried properly a thin liquid boundary may form between the upper stacking gel and resolving gel which may lead to poor sample resolution and potential sample mixing during electrophoresis.
11. Washing out the wells will remove pieces of gel and/or unpolymerized reagents which may interfere with sample migration.
12. We routinely fill empty wells with Laemmli loading buffer to facilitate uniform sample migration when the gel runs.
13. Alternatively, transfer can be carried out for 1 h at 100 V, by adding a Bio-Rad gel icepack to the running tank and applying agitation via a magnetic stir bar to facilitate uniform heat distribution.
14. Complete removal of Ponceau stain can be facilitated by addition of 0.1 M NaOH to the blots for 1 min, then thoroughly wash in water.
15. In plant material infected with viruses there may be a high level of host peroxidase activity, and this may result in false positive signals when horseradish peroxidase (HRP) conjugated antibodies are used in combination with the peroxidase substrates; consequently we urge some caution when using HRP conjugated antibodies.
16. To ensure optimal performance of the 250 mL tubes during centrifugation steps they should be 80% full. If the amount of crushed plant powder is less than this, 50 mL polypropylene copolymer Nalgene© Oak Ridge centrifuge tubes can be used (ThermoFisher Scientific).
17. If 50 mL polypropylene copolymer tubes are used, they can be centrifuged in an F21-8x50y rotor.

18. If the peptides inserted into the surface exposed regions of the virus are very hydrophobic (which is not the case for the MBP peptide), we recommend avoiding the use of chloroform and recommend proceeding to **step 4** but with a 30 min centrifugation.
19. Initially add a small volume of buffer and resuspend by pipetting, gradually adding more buffer until ~1 pellet volume is added.
20. This pellet resuspension and supernatant collection can be repeated multiple times to ensure optimal yields of virus in suspension, especially considering that **steps 10–12** require at least 8 mL of supernatant, or multiples of 8 mL (2, 4, or 6 sets of 8 mL) for balancing the ultracentrifuge.
21. For each 8 mL of clarified supernatant produced at **step 9**, 1 tube should be set up, and for later balancing and centrifugation stages it would be expected that pairs of tubes, namely 2, 4, or 6 tubes should be prepared.
22. Initially add a small volume of buffer and resuspend by pipetting, gradually adding more buffer until ~3 pellet volumes is added. Consider that 10 mL of resuspended pellet or multiples thereof are required for the next stages for balancing the ultracentrifuge.
23. The yield of the modified virus can also be double checked by making dilutions of the purified modified virus in Laemmli buffer (1:1, 1:10, 1:100) and running those on an SDS-PAGE with known concentrations of WT TMV, and carrying out western blot analysis using anti-TMV antiserum; as described in Subheadings 3.5–3.7.
24. If required, further physicochemical analysis such as atomic force microscopy, dynamic light scattering and zeta potential measurements can be carried out to provide more data on virus shape, size and surface charge respectively [9]. However, this requires access to more specialized equipment which may not be available in a standard molecular biology lab.
25. These reactions typically take 48 h to conclude, and the production of gold nanoparticles can be monitored using UV-Vis analysis. We would recommend measuring UV-Vis (300–750 nm) profiles of the samples every 8 h. Reactions are typically finished when there is no additional increase in absorbance at the 515 nm wavelength.
26. Further analysis of the nanoparticles can be carried out using zeta potential measurements, dynamic light scattering, selected area electron diffraction, energy-dispersive X-ray spectroscopy; which respectively determines the propensity of nanoparticles to aggregate, their particle size, crystallinity, and chemical composition [9]. Some of these parameters require specialist

equipment not normally found in a molecular biology lab, but the data they provide is useful for characterization for downstream applications of the nanoparticles.

Acknowledgments

A. J. L., and M. E. T. were supported by funding from The Genomia Fund and the Scottish Government Rural and Environmental Science and Analytical Services Division.

References

1. Narayanan KB, Sakthivel N (2011) Green synthesis of biogenic metal nanoparticles by terrestrial and aquatic phototrophic and heterotrophic eukaryotes and biocompatible agents. Adv Colloid Interface Sci 169:59–79
2. Kundu S, Panigrahi S, Praharaj S, Basu S, Ghosh SK, Pal A et al (2007) Anisotropic growth of gold clusters to gold nanocubes under UV irradiation. Nanotechnology 18:075712
3. Okitsu K, Yue A, Tanabe S, Matsumoto H, Yobiko Y (2001) Formation of colloidal gold nanoparticles in an ultrasonic field: control of rate of gold(III) reduction and size of formed gold particles. Langmuir 17:7720–7717
4. Tsuji T, Kakita T, Tsuji M (2003) Preparation of nano-size particles of silver with femtosecond laser ablation in water. Appl Surf Sci 206:314–320
5. Tan YN, Lee JY, Wang DIC (2010) Uncovering the design rules for peptide synthesis of metal nanoparticles. J Am Chem Soc 132:5677–5686
6. Yusibov V, Shivprasad S, Turpen TH, Dawson W, Koprowski H (1999) Plant viral vectors based on tobamoviruses. In: Hammond J, McGarvey P, Yusibov V (eds) Plant biotechnology. New products and applications. Springer, Berlin, pp 81–94
7. Clare DK, Orlova EV (2010) 4.6 angstrom Cryo-EM reconstruction of tobacco mosaic virus from images recorded at 300 keV on a 4k x 4k CCD camera. J Struct Biol 171:303–308
8. Haynes JR, Cunningham J, Vonseefried A, Lennick M, Garvin RT, Shen SH (1986) Development of a genetically-engineered, candidate polio vaccine employing the self-assembling properties of the tobacco mosaic-virus coat protein. Bio-Technology 4:637–641
9. Love AJ, Makarov VV, Sinitsyna OV, Shaw J, Yaminsky IV, Kalinina NO et al (2015) A genetically modified tobacco mosaic virus that can produce gold nanoparticles from a metal salt precursor. Front Plant Sci 6:984
10. Petukhova NV, Gasanova TV, Stepanova LA, Rusova OA, Potapchuk MV, Korotkov AV et al (2013) Immunogenicity and protective efficacy of candidate universal influenza a nanovaccines produced in plants by tobacco mosaic virus-based vectors. Curr Pharm Des 19:5587–5600
11. Turpen TH, Reinl SJ, Charoenvit Y, Hoffman SL, Fallarme V, Grill LK (1995) Malarial epitopes expressed on the surface of recombinant tobacco mosaic-virus. Bio-Technology 13:53–57

Chapter 25

TMV-Templated Formation of Metal and Polymer Nanotubes

Alexander M. Bittner

Abstract

Metals and polymers are probably the most important construction materials, but also have many more functions, e.g., for electronics. The interaction of metal ions with tobacco mosaic virus (TMV) was originally used for the preparation of heavy metal isomorphic replacement for structural analysis. Metal ions can also be the precursors for metal clusters, particles, and layers. Various strategies have been developed, which allow the creation of metal layers on the external surface of TMV. Such layers can be made as metal tubes, enveloping a complete virion. An alternative strategy is adsorption of metal nanoparticles. If a dense coating of TMV is achieved, again a tube results. Nanoscale tubes have various physical properties that depend on size, crystallinity, uniformity, but especially on the nature of the metal. Polymer coatings are as yet rarely investigated, though they can be prepared quite easily.

Here, a series of exemplary protocols is provided, which covers all of these different concepts.

Key words TMV, Adsorption, Reduction, Electroless deposition, Nanoparticle, Nanotube

1 Introduction

The TMV particle (virion) is a tube of about 2100 coat protein subunits, with a single RNA strand integrated in the protein helix. It possesses a 4 nm wide internal cylindrical cavity, usually filled with water and ions. If TMV is applied as a template for shaping metal deposits (*see also* Chapters 12, 14, and 24 for TMV and other viral templates), this channel can take part in most or even all metallization reactions; it was specifically used for the deposition of 3–5 nm wide metal wires [1–3]. Such ultrasmall structures will not be considered here, although they may in some cases form synchronously to a growing tube.

The adsorption of metal ions is driven by a delicate interplay between charges and chemical bonds both on the virion and in the metal complex (*see* also Chapters 23, 24, and 26). For most metals, one has to consider not just a metal ion, but a complex with ligands such as chloride. The influence of water as solvent or

Christina Wege and George P. Lomonossoff (eds.), *Virus-Derived Nanoparticles for Advanced Technologies: Methods and Protocols*, Methods in Molecular Biology, vol. 1776, https://doi.org/10.1007/978-1-4939-7808-3_25,
© Springer Science+Business Media, LLC, part of Springer Nature 2018

as ligand is absolutely crucial. The first experimental question to be addressed should be potential hydrolysis of the chosen metal complex—the charge, but also the chemical nature (polycondensation), can easily change by hydrolysis. A well-known example, and a very relevant one, is $PdCl_4^{2-}$. This complex is the standard way to introduce Pd(II). Its hydrolysis is slow (minutes to hours), but depends on pH. It can be followed by eye since the initially pale yellow solution darkens with time. Dark yellow solutions contain polycondensated Pd(II), which can attain diameters well above the virus size! The exact interaction mechanism between ions and TMV is unknown—charges, but also covalent (complex) bonds must be involved. This automatically entails a significant or even dominating influence of TMV coat protein variations, as long as these concern amino acids on the external nanotube surface. The influence of certain engineered mutations can be of much greater importance than that of the virus strain or even species used; this was specifically demonstrated for coat protein mutants with additional N-terminally exposed thiols (1-Cys, 2-Cys) [4, 5], or a C-terminal His_6 tag [6], elongating the natural chain of 158 amino acids. The naturally occurring TMV variant E50Q, however, also came out to exhibit substantially different metallization capacities than wild-type TMV [6], but most likely due to a reduced RNA content in combination with altered protein interactions, rather than due to the altered amino acid side chain. Hence, in the following, the chapter will focus on protein-induced effects.

The metal ions adsorbed to the TMV protein surfaces are then reduced by various chemicals (*see* Chapters 12 and 14). Borohydrides are the most popular reductants, due to simple handling, although they are rather strong, meaning their redox potentials are very negative. More noble metal ions require only mild reductants of more positive redox potentials. To avoid debris, one should remove all free metal ions (by centrifugation) before adding the reductant, else the dissolved metal ions will react first, and aggregate quickly to form precipitates of micron to millimeter size. After the reduction, the reductant should be removed, too, by washing. Ideally, only one layer of ions can adsorb and be reduced, such that at best a monolayer of metal on TMV can be obtained. For most intended uses of the resulting TMV-templated metal structures, the procedure has to be repeated several times to achieve coatings of few nanometers, or tens of nanometers, thickness.

Electroless deposition or metallization can be simpler to achieve. Electroless here refers exclusively to autocatalytic deposition of metals. This means that the deposited metal catalyzes its own deposition, and hence the process is continuous as long as sufficient amounts of reactants (metal ions and reductant) are provided. The initiation is a separate preparation step where a noble metal nanoparticle is either adsorbed, or produced in situ.

This particle has to be very small, usually <10 nm, in order to preserve the TMV structure. The in situ strategy appears to create particles that are partially <1 nm. The TMV surface is thus decorated with many similar nucleation seeds. When the electroless process is started, this leads to quick coalescence of simultaneously growing particles, which is the prerequisite for homogeneous layers [4, 6–8].

When TMV is itself adsorbed or immobilized on a macroscopic solid surface, the procedure of removing solutions is much simplified, compared to freely dispersed virions: Reagents can be replaced by simply rinsing with water.

Another way of producing a type of metal tubes is coating by a dense layer of nanoparticles (*see also* Chapters 14, 24, and 34). The particles do not coalesce, so the tubular shape is merely geometric. However, the simplicity of the procedure is highly attractive, and the result is very suitable for further modifications [9].

It is obvious that the strategies are not limited to pure metals, but work also for alloys [2]. Oxidation will often form hydroxides or oxides (*see* Chapter 12), which can be better produced by hydrolysis reactions [8, 10].

A substantially different strategy to produce stiff TMV-templated synthetic nanotubes is the deposition of organic polymer shells (*see also* Chapter 9). Coating with polymers can be achieved in many ways, but experimentally, only covalent linking of molecules and polymerization on TMV are explored in some detail. An especially simple and mild approach is based on oxidative polymerization of aniline, which creates a poly(aniline) layer on TMV, which at the same time aggregates end-to-end, yielding long poly(aniline)–TMV nanotubes [11]. The protocol reported here has the capability of permitting the easy application of a second coating of poly(sulfonated)styrene.

2 Materials

Prepare all solutions with deionized water (18.2 MΩ cm; e.g., purified by a MilliQ system, or other comparable water quality). Ideally the organic content should be measured (ppb range). All chemicals should be at least of analytical grade.

2.1 TMV Variant Preparations

TMV variants should be supplied dissolved in water or buffer, or immobilized on a solid scaffold. 0.2 mg/mL of the different purified TMV types, all in 0.1 M pH 7 phosphate buffer, are used as stock solutions here. The concentration in the reactions will be listed below, in case they are available. TMV wt (wild type) is available through standard purification procedures [6]. Different mutants can be based on distinct TMV strains, which are thought to be all similarly suitable, hence all are handled identically. The mutant TMV-1-

Cys can be produced as described in [4]. TMV-2-Cys is synthesized according to [5]. TMV-E50Q is accessible as described in [6], where TMV-His_6, a mutant in which each coat protein carries a His_6 tag at the carboxylate terminus, is also described. (In this book, *see* Chapters 23, 24, and 27 for TMV and TMV mutant preparation from plants.)

Purification procedures for reactants: All TMV-containing "solutions" are theoretically suspensions, not solutions, so in each mixing step they should be carefully agitated by vortexing a few seconds. The TMV preparations are normally stored in buffer, and first require careful dialysis. This is most conveniently done by dilution with water to the desired concentration, and >10 min dialysis against water in a Slide-A-Lyzer 10,000 MWCO mini dialysis unit (Thermo Scientific), repeated at least twice. The filling level of the unit should be checked before and after the dialysis. It may rise slightly, thus changing the concentration.

2.2 Substrate Surfaces

1. Gold-coated steel in shape of a CR2032 coin disk (for tests in a battery setup).
2. Polytetrafluorethylene (PTFE) substrate (Bytac Saint-Gobain Performance Plastics Poestenkill), masked by Parafilm® if required, for imaging. The substrate should be gold-coated stainless steel for application in a battery device.
3. Oxidized silicon wafer: Si(111) wafer (>1 Ω cm, n-type doped with P) cut with a dicer saw; alternatively, precut pieces of the wafer can be used. Rinse the wafer pieces with acetone, isopropanol, and water, dry with nitrogen, and treat with air plasma for several seconds (PlasmaPen, PVA TePla) to assure a hydrophilic silicon oxide layer (*see* **Note 1**).
4. Clean glassware thoroughly: Mechanically (by rubbing) with soap water, followed by the wafer cleaning procedure.
5. Mica: Freshly cleaved by removal of the top surface with adhesive tape.
6. Standard grids for TEM analysis: Carbon films, or carbon films on polymer (Pioloform, Formvar; Plano, Wetzlar).
7. In an SEM, any substrate can be used, but electrical conductivity is a plus (*see* **Note 2**).

2.3 Deposition Reactants

For purification, some cases require dialysis in 300,000 MWCO dialysis tubing (Pierce).

The pH can be adjusted with 1 M NaOH and 0.1 M HCl solutions.

1. Phosphate buffer pH 7.0.
2. 0.02 M Sodium tetrachloropalladate Na_2PdCl_4 (99.998%) in methanol (*see* **Note 3**).
3. Cobalt electroless plating solution: 0.033 M cobalt (II) sulfate ($CoSO_4 \cdot H_2O$), 0.008 M glycine (NH_2CH_2COOH), 0.5 M

sodium tetraborate ($Na_2B_4O_7$), 0.175 M dimethylamine borane ($(CH_3)_2NH{\cdot}BH_3$, 97%) (*see* **Note 4**).

4. Ethanol CH_3CH_2OH (puriss. p.a.) (*see* **Note 5**).
5. 1.36 mM Na_2PdCl_4 in 1 M NaCl at pH 7.0 (adjust with NaOH or HCl), prepare freshly.
6. Metallization solution: 223 mM nickel acetate ($Ni(CH_3COO)_2 \cdot 4H_2O$, 98%), 228 mM lactic acid (CH_3-CHOH-COOH, 85% in water), 42.4 mM $(CH_3)_2NH{\cdot}BH_3$, adjusted to pH 7.5 with NaOH.
7. Gold sol 6 nm (Aurion) (*see* **Note 6**).
8. 0.1 M citric acid ($(HOOC\text{-}CH_2)_2C(OH)\text{-}COOH$ (≥99.5%).
9. Aniline $C_6H_5NH_2$ (*see* **Note 7**).
10. Ammonium persulfate $(NH_4)_2S_2O_8$ (Sigma).
11. 0.01 M HCl.
12. 0.01 M NaOH.
13. Poly(sodium 4-styrene sulfonate) (also called poly(sulfonated styrene)) (Sigma).
14. Sodium chloride NaCl (p.a., ≥99.5%).

2.4 Analysis

1. Field emission SEM (FESEM) (FEI Quanta 250).
2. pH meter.
3. TEM (FEI Tecnai) (*see* **Note 8**).
4. UV-Vis spectrometer.

3 Methods

3.1 Immobilization Procedures

3.1.1 Immobilization of TMV-1-Cys

Prepare virus-assembled surfaces by immersing the Parafilm-patterned PTFE or Au-coated stainless steel discs in a solution of 0.2 mg/mL purified TMV-1-Cys in 0.1 M pH 7 phosphate buffer and incubate overnight. A standing-up configuration is obtained [4].

3.1.2 Immobilization of TMV wt

Dialyze TMV against pure water for 1 h, change the water at least twice. Dry a droplet of 0.001 mg/mL up to 1 mg/mL aqueous TMV on an oxidized Si wafer, cleaved mica, glass, or any other clean surface. A lying-down configuration is obtained.

3.2 Particle Metallization

3.2.1 Electroless Cobalt Deposition on Immobilized TMV-1-Cys [7]

1. Add 0.02 M of Na_2PdCl_4 in methanol to the immoblilized TMV to a final Na_2PdCl_4 concentration of 0.0015 M.
2. Incubate for 30 min on the solid surface. This will Pd-(II)-activate the virions.
3. Remove the solution, and replace it with a cobalt electroless plating solution.

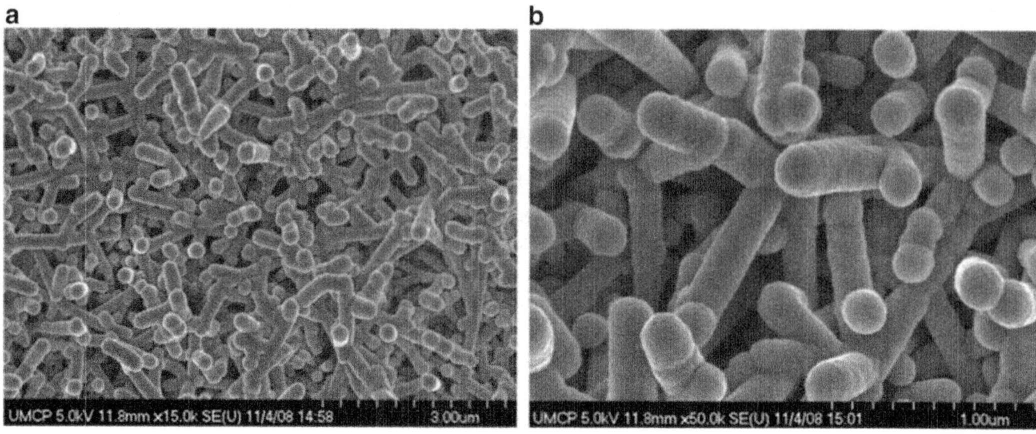

Fig. 1 SEM analysis of cobalt-coated TMV-1-Cys, assembled vertically on a PTFE surface at low (**a**) and high (**b**) magnification. Reproduced from Gosh et al. [7]

4. Incubate for 10 min on the TMV-loaded solid surface.
5. Remove the plating solution, wash with ethanol, and dry overnight under vacuum. Figure 1 shows SEM images of the produced structure. The cobalt surface is prone to oxidation when exposed to air.

3.2.2 Electroless Nickel Deposition on TMV Mutants [6]

This procedure gives different results on the various mutants. TMV-His_6 is metallized so readily that **step 1** can be left out (*see* Fig. 2).

1. Mix TMV suspension (concentration range 1 μg/mL to 1 mg/mL) with an equal volume of freshly prepared 1.36 mM Na_2PdCl_4 in 1 M NaCl at pH 7.0 (adjust with NaOH or HCl).
2. Centrifuge (18,000 × *g*, 10 min). Wash the resulting brownish pellet with water, sediment (18,000 × *g*, 5 min) and resuspend in 300 μL H_2O.
3. Mix the suspension with an equal volume of metallization solution.
4. Stop the ongoing metallization by taking aliquots and drying them rapidly on a solid or on a TEM grid.

3.2.3 Palladium Deposition on TMV-2-Cys [12]

1. Mix TMV-2-Cys in deionized water (unspecified concentration) with Na_2PdCl_4 (40–160 μg Pd/mL) and incubate for 20 min at 50 °C. Aggregates form, and the pH drops from 5 to 3. Pd tubes of ≈35 nm diameter are obtained.
2. Repeat as often as desired. Each cycle adds ≈3 nm in diameter.

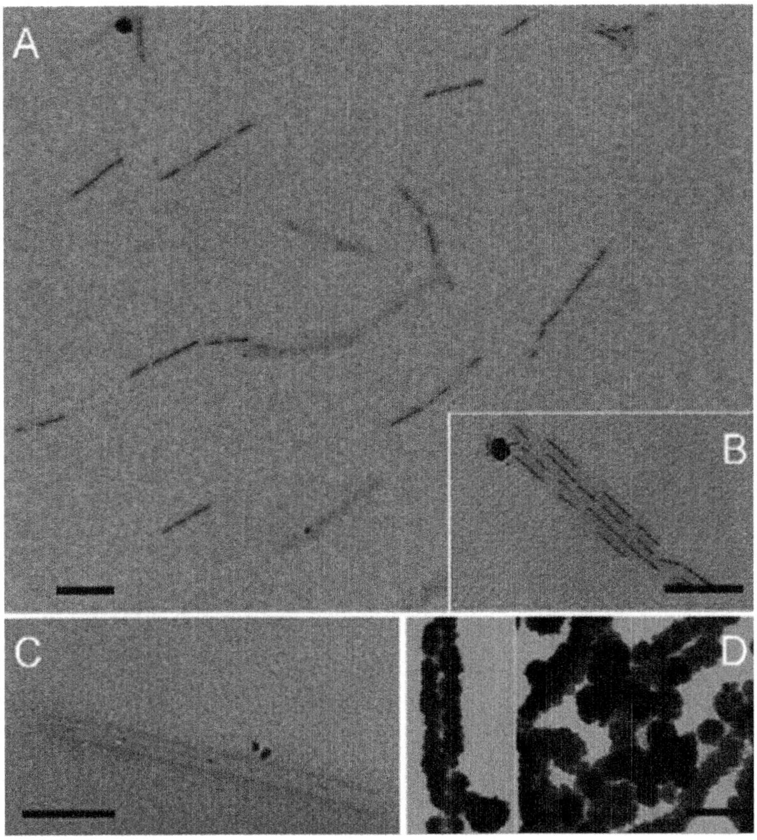

Fig. 2 TMV after nickel metallization, analyzed by TEM (no staining). The E50Q mutant shows metallization of the internal 4 nm wide channel (**a**, **b**). The wild type shows similar behavior (**c**). A mutant with a hexahistidine tag on its surface directs the metallization to the surface, even in absence of palladium sensitization (**d**). All scale bars are 100 nm. Reproduced from Kadri et al. [6]

3. The extent of palladium deposition onto the TMV and pH can be monitored after each coating cycle with UV-Vis spectroscopy, pH monitoring, and TEM (*see* Fig. 3).

3.2.4 Gold Nanoparticles in Dense Coating on TMV wt [9]

1. Sonicate the Au sol for 30 s in a sonication bath.
2. Mix 70 μL of Au sol with 25 μL of 0.1 M citric acid. Dialyze 1.5 h with a (10,000 MWCO) Slide-A-Lyzer mini dialysis unit against pure water (exchanged every 30 min).
3. Add 5 μL of 0.05 mg/mL TMV suspension. The measured pH should be 2.6–2.8. Mix thoroughly by vortexing for a few seconds. Incubate overnight at room temperature.

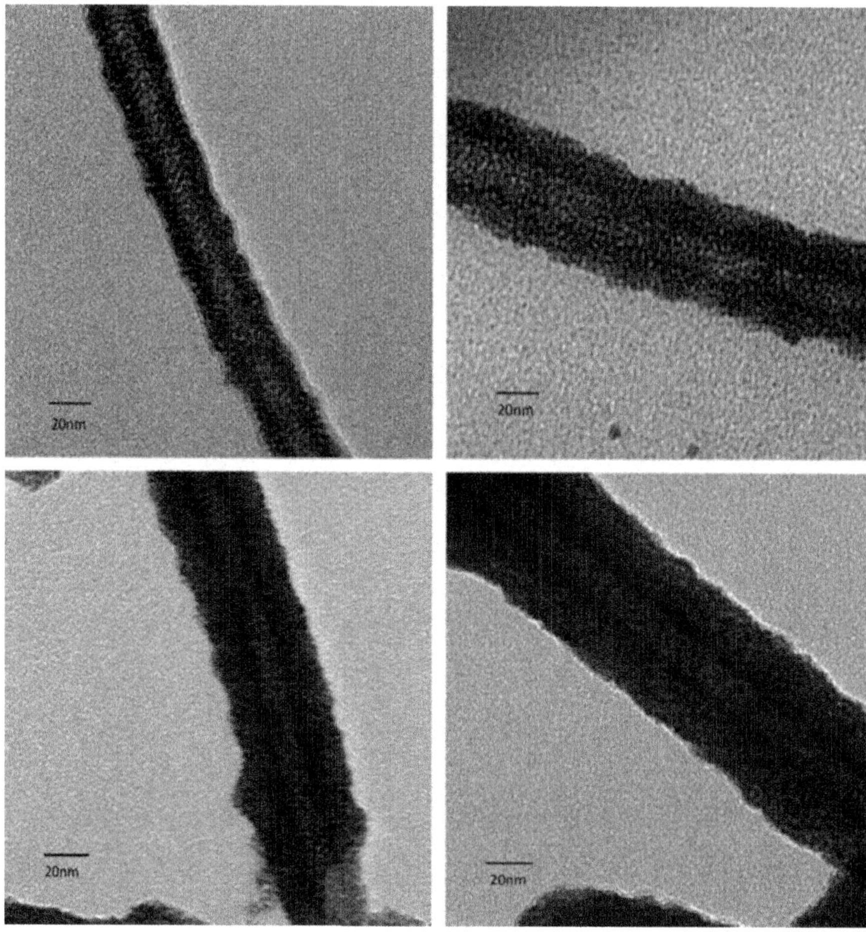

Fig. 3 TEM images of palladium-coated TMV after various coating cycles (each 20 min at 50 °C): 1 (top left), 4 (top right), 5 (bottom left), 6 (bottom right). The scale bar is 20 nm. Reprduced from Freer et al. [12]

4. Vortex for a few seconds and dialyze against water with a (10,000 MWCO) Slide-A-Lyzer mini dialysis unit for 60 min, with water exchange after 30 min. The structure can be visualized by STEM (*see* Fig. 4).

3.3 TMV wt Coating with Poly(Aniline) and Poly(Sulfonated Styrene) [11]

1. Add consecutively distilled aniline (10 mL) and ammonium persulfate solution (10 mg/mL, 1 mL) to TMV solution (1 mg/mL, 4 mL). Adjust the pH to 6.5 with 0.01 M HCl and 0.01 M NaOH. Incubate the reaction mixture at room temperature for 24 h.
2. Add poly(sulfonated)styrene (20 mg) and aniline (10 mL). Adjust the pH of the reaction solution to 4, and incubate at room temperature for 24 h.
3. Dialyze against water with a 300,000 MWCO dialysis tube. The structure can be visualized by TEM (*see* Fig. 5).

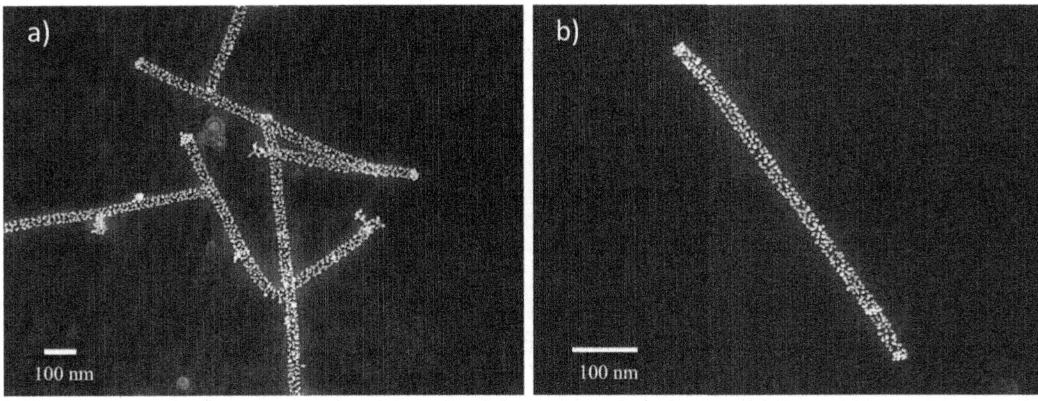

Fig. 4 TMV coated by a dense layer of 6 nm diameter gold nanoaparticles, imaged by bright field STEM at two zoom levels (**a**, **b**). Reproduced from Khan et al. [9]

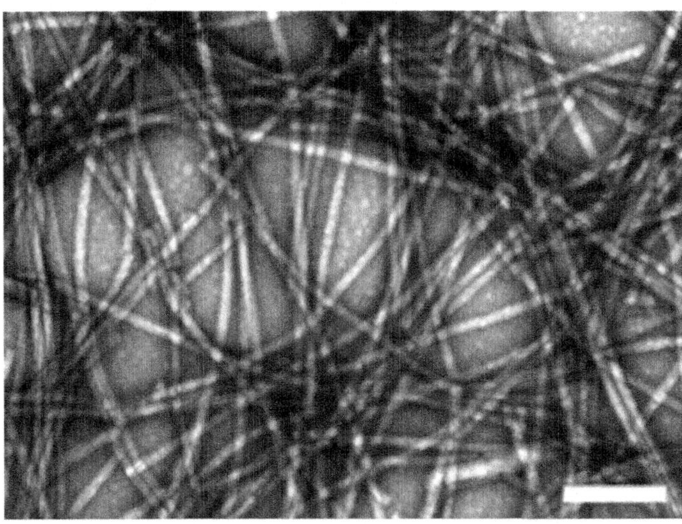

Fig. 5 Uranyl-stained poly(aniline)-coated TMV; scale bar 200 nm. Reproduced from Rong et al. [11]

3.4 Analysis

3.4.1 TEM

Dry small droplets (1–5 μL) of modified TMV solution on TEM grids, then analyze with a standard TEM, at 100–400 kV (*see* **Note 8**).

3.4.2 SEM

Dry small droplets (1–2 μL) of solution on a smooth surface, e.g., silicon wafer, then analyze by SEM (*see* **Note 9**).

4 Notes

1. The air plasma can be replaced by a 1 mbar oxygen plasma (Femto Plasma System, Diener).
2. Low voltage operation (≤3 kV) is preferred.

3. This salt hydrolyzes slowly in water, forming yellow aggregates. The color of the mM solutions used should be very slightly yellowish, and it should not be stored for more than some hours.
4. Do not store solutions, but prepare freshly.
5. Ethanol for washing samples should be absolute (100%).
6. The solution color should be slightly reddish, but never dark red and nontransparent, which would point to aggregation.
7. Aniline should be freshly distilled.
8. TEM can be replaced by STEM in a suitably equipped SEM (e.g., FEI Quanta 250, FEI Helios 450).
9. Generally the voltage should be below 3 kV.

References

1. Knez M, Bittner AM, Boes F, Wege C, Jeske H, Maiß E, Kern K (2003) Biotemplate synthesis of 3 nm nickel and cobalt nanowires. Nano Lett 3:1079–1082
2. Balci S, Hahn K, Kopold P, Kadri A, Wege C, Kern K, Bittner AM (2012) Electroless synthesis of 3 nm wide alloy nanowires inside Tobacco Mosaic Virus. Nanotechnology 23:045603
3. Kobayashi M, Onodera K, Watanabe Y, Yamashita I (2010) Fabrication of 3-nm platinum wires using a tobacco mosaic virus template. Chem Lett 39:616–618
4. Royston E, Ghosh A, Kofinas P, Harris MT, Culver JN (2008) Self-assembly of virus-structured high surface area nanomaterials and their application as battery electrodes. Langmuir 24:906–912
5. Lee SY, Royston E, Culver JN, Harris MT (2005) Improved metal cluster deposition on a genetically engineered tobacco mosaic virus template. Nanotechnology 16:S435–S441
6. Kadri A, Maiss E, Amsharov N, Bittner AM, Balci S, Kern K, Jeske H, Wege C (2011) Engineered tobacco mosaic virus mutants with distinct physical characteristics in planta and enhanced metallization properties. Virus Res 157:35–46
7. Ghosh A, Guo J, Brown AD, Royston E, Wang C, Kofinas P, Culver JN (2012) Virus-assembled flexible electrode-electrolyte interfaces for enhanced polymer-based battery applications. J Nanomater 2012:795892. 1–6
8. Shah SN, Khan AA, Espinosa A, Garcia MA, Nuansing W, Ungureanu M, Heddle JG, Chuvilin AL, Wege C, Bittner AM (2016) Virus-templated near-amorphous iron oxide nanotubes. Langmuir 32:5899–5908
9. Khan AA, Fox EK, Górzny ML, Nikulina E, Brougham DF, Wege C, Bittner AM (2013) pH control of the electrostatic binding of gold and iron oxide nanoparticles to Tobacco Mosaic Virus. Langmuir 29:2094–2098
10. Balci S, Bittner AM, Schirra M, Thonke K, Sauer R, Hahn K, Kadri A, Wege C, Jeske H, Kern K (2009) Catalytic coating of virus particles with zinc oxide. Electrochim Acta 54:5149–5154
11. Rong J, Oberbeck F, Wang X, Li X, Oxsher J, Niu Z, Wang Q (2009) Tobacco mosaic virus templated synthesis of one dimensional inorganic–polymer hybrid fibres. J Mater Chem 19:2841–2845
12. Freer AS, Guarnaccio L, Wafford K, Smith J, Steilberg J, Culver JN, Harris MT (2013) SAXS characterization of genetically engineered tobacco mosaic virus nanorods coated with palladium in the absence of external reducing agents. J Colloid Interface Sci 392:213–218

Chapter 26

Semiconducting Hybrid Layer Fabrication Scaffolded by Virus Shells

Petia Atanasova

Abstract

The formation of virus-based semiconducting hybrid thin films is a two-step process, which involves assembly of virus particles as a template layer and subsequent selective mineralization of the virus surface with inorganic nanoparticles to build a semiconducting organic–inorganic hybrid film. Here, we present the use of the convective assembly technique to obtain homogeneous and dense template monolayers of wild-type tobacco mosaic virus (wt-TMV) and the TMV mutant E50Q, of which most particles do not have detectable amounts of RNA in the protein tube. On the top of the aligned virus layer, zinc oxide (ZnO) is deposited to prepare virus–ZnO semiconducting hybrid films with controllable thickness under mild conditions of the chemical bath deposition (CBD).

Key words Semiconducting hybrid films, Tobacco mosaic virus, TMV mutants, ZnO, Convective assembly, Chemical bath deposition

1 Introduction

Owing to their unusual electrical, optical, and mechanical properties and potential application in nanotechnology, nanostructured materials have attracted considerable interest in recent years. However, due to limitations of conventional techniques to produce 1D and 2D nanostructures (e.g., microcontact printing, electron-beam lithography and photolithography), formation of more complex architectures using self-assembly of molecular components, also known as "bottom-up" method, is becoming increasingly important [1]. In addition, mild reaction conditions of the chemical bath deposition (CBD) for the synthesis of inorganic phases allow the mineralization of diverse biomolecules or even bigger biological entities, present in nature. Thereby, the biological phase can function as a template, scaffold, or stabilizer in the synthesis of bio–inorganic hybrid materials with novel tailored functionalities [2–4]. CBD is a simple method for the deposition of nanocrystalline semiconducting films from solution. Here, the inorganic phase is synthesized and

Christina Wege and George P. Lomonossoff (eds.), *Virus-Derived Nanoparticles for Advanced Technologies: Methods and Protocols*, Methods in Molecular Biology, vol. 1776, https://doi.org/10.1007/978-1-4939-7808-3_26,
© Springer Science+Business Media, LLC, part of Springer Nature 2018

deposited in parallel in the same bath [5]. The paper that presented this technique for the first time was published in 1868 by C. Puscher and describes the wonderful colors of sulfide layers formed from thiosulfate solution of copper, lead, and antimony salts [6]. Among the bio objects frequently used as templates, the rod-like tobacco mosaic virus (TMV) is of particular interest due to its high aspect ratio, precisely defined size as well as number and nature of its surface functional groups. TMV can be genetically or chemically modified in order to influence its biochemical properties [7, 8]. Furthermore, the virus is stable up to 90 °C, in a broad range of pH (2–9) and possesses chemical stability in several organic solvents [9]. In summary, TMV is an ideal scaffold for the deposition of inorganic materials such as the metals Au [10], Pt [11], Pd [12], Ni, Co [13], and Cu [14]; the metal sulfides CdS and PbS [15]; and the metal oxides ZnO [3, 16], SiO_2 [17, 18], and Co_3O_4 [19].

The virus templates can be organized into hierarchically ordered structures, which opens possibilities for their deployment in functional devices. Common techniques for the preparation of structured thin films are, for instance, drop casting, dip coating [20], spin coating [21], and Langmuir–Blodgett techniques [22]. Among them, convective assembly is an efficient method for depositing colloidal solutions in a microliter range over surface areas of several square centimeters [23, 24]. The colloidal particles in suspension are moved to the triple contact line formed between substrate, air, and solution and deposited on the substrate surface (*see* Fig. 1).

Their movement behavior depends on particle shape, particle–particle and particle–substrate interactions. Factors like capillary forces, substrate wettability, and surface tension control the particles' packing density on the substrate. The alignment of wild-type (wt) TMV particles in buffer on a substrate surface by convective assembly was systematically examined by Kuncicky et al. [23] and Wargacki et al. [24] It was shown that the thickness of the resulting TMV film depends on the deposition rate and the particle concentration. We have successfully used this technique to investigate the organizational behavior of wt-TMV and several TMV

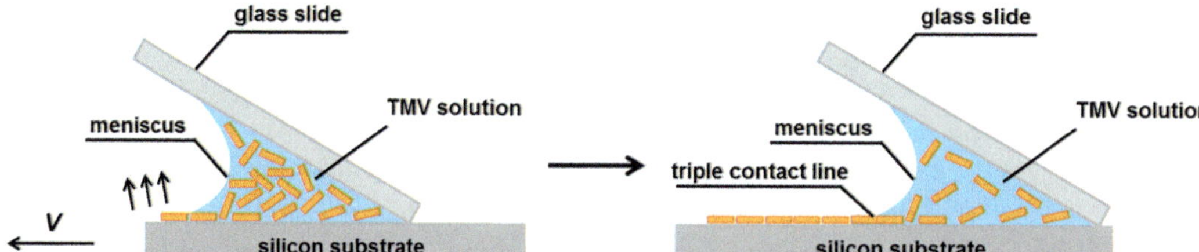

Fig. 1 Schematic representation of the convective assembly of suspended colloidal particles. A droplet from the suspension is placed between the substrate and the glass slide forming a triple contact line between substrate, air and glass slide. Moving the substrate horizontally with withdrawal speed *v*, the colloidal particles assemble on the substrate surface

mutants from buffer-free solutions [25]. Genetically modified Lys (T158 K), Cys (S3C) [26], 30%TMV-His_6 (30% His_6CP/70% wt CP) [7, 27, 28], and the RNA-free E50Q mutant [27, 28] (*see* Chapter 7) gave different surface arrangements under the same assembly conditions. The smoothest, almost defect-free monolayer was achieved with the E50Q mutant which has the same surface chemistry as the wt-TMV but higher flexibility due to the absence of continuous RNA in most of the particles [25, 27, 29]. Subsequent selective deposition of nanocrystalline ZnO particles on the virus monolayers resulted in the formation of bio–inorganic hybrid materials with an architecture governed by the predetermined template structure. The thickness of the semiconducting layer is controlled very precisely by the number of the deposition cycles. Such bio–inorganic hybrid films were deposited as semiconducting layers on substrates with prestructured field-effect transistors (FETs) [25]. It was shown that E50Q–ZnO hybrid films have an enhanced FET performance compared to wt-TMV–ZnO hybrid films based on wt-TMV homogeneous, dense but disordered template layers [25].

2 Materials

Prepare all virus solutions from buffer-free virus stock solutions using ultrapure water (resistivity 18.2 MΩ cm at 25 °C) and store at 4 °C. Dispose of virus-containing waste according to the disposal regulations for genetically modified organisms (GMOs) and specifically RNA-containing plant pathogens. For isolation and purification of virus material, *see* Chapters 23, 24, or 27 in this book.

2.1 Substrate Cleaning

1. Silicon wafers (100 p-doped polished wafers, Silchem) cut into 1.5 × 1.5 cm pieces.
2. Glass slides (1.4 × 4 cm).
3. Tweezers.
4. Ultrapure water (resistivity 18.2 MΩ cm at 25 °C).
5. 1:1 (v/v) ethanol–acetone mixture (p.a. grade).
6. Ethanol (p.a. grade).
7. Ultrasonic bath (e.g., Sonorex Super RK103H, 160/320 W).
8. Plasma cleaner, oxygen plasma (e.g., Femto low-pressure plasma system, Diener electronic GmbH).
9. Watch glass (diameter ~10 cm).
10. Beaker (100 mL volume).

2.2 Convective Assembly

1. wt-TMV solution: 5 mg/mL in ultrapure water or TMV E50Q solution: ~2 mg/mL in ultrapure water (1.9 mg/mL was used in our experiments).

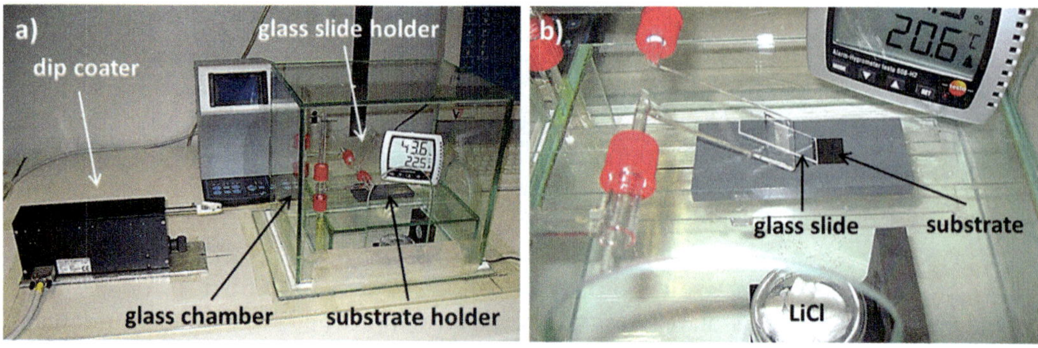

Fig. 2 (**a**) Image of a home-made convective assembly set-up consisting of horizontally positioned dip coater connected to a PC controller, glass chamber to maintain the humidity, substrate holder connected to the dip coater pinch, holder for the glass slide, hygrometer and LiCl container. (**b**) Closer view of the position of the fixed glass slide and silicon substrate prior to the start of the experiment

2. Computer controlled dip coater (KSV Instruments) positioned horizontally to provide continuous motion of the silicon substrate at a constant rate.
3. Glass chamber (*see* Fig. 2a) (*see* **Note 1**).
4. Substrate holder to fix the substrate (*see* Fig. 2a) (*see* **Note 2**).
5. Plasma-cleaned silicon substrate.
6. Holder to fix the glass slide (*see* Fig. 2a).
7. Plasma-cleaned glass slide.
8. LiCl (saturated solution) (*see* **Note 3**).
9. Hygrometer.
10. Thermometer.
11. Materials for AFM or SEM analysis.

2.3 ZnO Mineralization

Prepare all precursor stock solutions for ZnO mineralization using methanol (VLSI Grade) as a solvent (*see* **Note 4**). All storage vessels, dried before use at 80 °C for at least 2 h, should be closed tightly to avoid methanol evaporation.

1. 34 mM zinc acetate dihydrate ($ZnAc_2$) stock solution: Weigh 0.3734 g (1.7 mmol) $ZnAc_2$ (Sigma-Aldrich) and transfer it to a glass vessel. Add 50 mL methanol, close well, and sonicate in an ultrasonic bath (320 W) for 15 min.
2. 75 mM tetraethylammonium hydroxide (TEAOH) stock solution: Add 1.5 mL 1.5 M TEAOH in methanol solution (Aldrich Chemistry) to 28.5 mL methanol in a glass vessel and mix well.
3. 21.7 mM polyvinylpyrrolidone (PVP) stock solution: Weigh 12.855 g (1.286 mmol) PVP (*see* **Note 5**) and transfer to a glass vessel. Add 50 mL methanol, close well, and sonicate in an ultrasonic bath (320 W) for 15 min.

4. Peristaltic pump (e.g. ISM831A, ISMATEC, tube diameter 1.02 mm).
5. Glass vessels with lids (15 mL volume) (*see* **Note 6**).
6. Magnetic stirrer.
7. Magnetic stir bar.
8. Oil bath.
9. Thermocouple thermometer.
10. Watch glass (diameter around 10 cm).
11. Methanol (VLSI Grade).
12. Materials for AFM or SEM analysis.

3 Methods

3.1 Substrate Cleaning

1. Place the silicon wafers and the glass slides in a beaker. Add ultrapure water until the wafers and slides are fully covered. Cover the beaker with a watch glass and sonicate in an ultrasonic bath (320 W) for 10 min.
2. Wash the wafers and the glass slides ten times with ultrapure water.
3. Dry the wafers and the glass slides under an inert gas stream.
4. Place the wafers and the glass slides in a beaker. Add ethanol–acetone mixture until the wafers and slides are fully covered. Cover the beaker with a watch glass and sonicate in an ultrasonic bath (320 W) for 10 min.
5. Wash the wafers and the glass slides five times with ethanol.
6. Dry the wafers and the glass slides thoroughly under inert gas stream (*see* **Note 7**).
7. Place the wafers and the glass slides in a plasma cleaner and treat with oxygen-plasma (power 30 W, 0.75 Bars, gas flow 55 sccm) for 10 min (*see* **Note 8**).
8. Repeat **steps 1** and **2**.
9. Store in ultrapure water at room temperature (*see* **Note 9**).

3.2 Convective Assembly

1. Dry a plasma-treated silicon substrate and a glass slide thoroughly under inert gas stream.
2. Fix the substrate on the holder in the glass chamber and connect the holder to the horizontally positioned dip coater slide (*see* Fig. 2a, b). Fix the glass slide to the vertical holder in such way that the shorter slide edge is perpendicular to the withdrawal direction and touches the silicon substrate under 45° (*see* Fig. 2b). Move the substrate horizontally until the

slide edge stands around 1–2 mm from the substrate edge (*see* Fig. 2b).

3. Wait until humidity is stabilized around 35% ± 2% (*see* **Note 3**). The assembly process is conducted at 23 ± 2 °C.
4. When the required humidity is reached, spread the corresponding volume (*see* **Note 10**) of the virus solution with a micropipette along the contact line between the substrate and the glass slide (*see* Fig. 2b). Make sure that the complete volume of the virus solution builds a homogeneous and continuous film between the substrate and the glass slide forming a well-defined meniscus (*see* Fig. 1).
5. Set the required withdrawal speed (*see* **Note 10**).
6. Start the virus assembly and wait until the virus droplet is spread across the whole substrate surface. At the end, stop the withdrawal and wait until the rest of the virus solution completely evaporates (*see* **Note 11**), remove the glass slide and store the substrate with the immobilized viruses at room temperature (*see* **Note 12**).
7. Confirm the quality of the virus coverage via AFM or SEM analysis.

3.3 ZnO Mineralization

1. Heat an oil bath to 60 °C.
2. Prepare ZnO deposition solution from precursor stock solutions just prior to use: Mix 1 volume of $ZnAc_2$ stock solution with 1 volume of PVP stock solution. Then, add dropwise 1 volume of TEAOH stock solution to the $ZnAc_2$–PVP mixture using a peristaltic pump with a flow rate of 1.00 mL/min under continuous stirring.
3. Place a silicon substrate with immobilized virus monolayer (from Subheading 3.2, **step 6**) in a glass vessel (*see* **Note 13**). Add 1 mL from the deposition solution and close the vessel well (*see* **Note 6**).
4. Place the vessel in the preheated oil bath and heat at 60 °C for 1.5 h.
5. Remove the vessel from the oil bath and place the substrate in a watch glass filled with methanol. Wash the substrate five times with fresh methanol portions each. Dry the substrate surface under an inert gas stream. Store at room temperature.
6. Check the quality of the obtained hybrid films via AFM or SEM analysis (*see* Fig. 3).
7. To prepare a thicker ZnO semiconducting film on the virus template layer repeat **steps 3–5** (*see* **Note 14**).

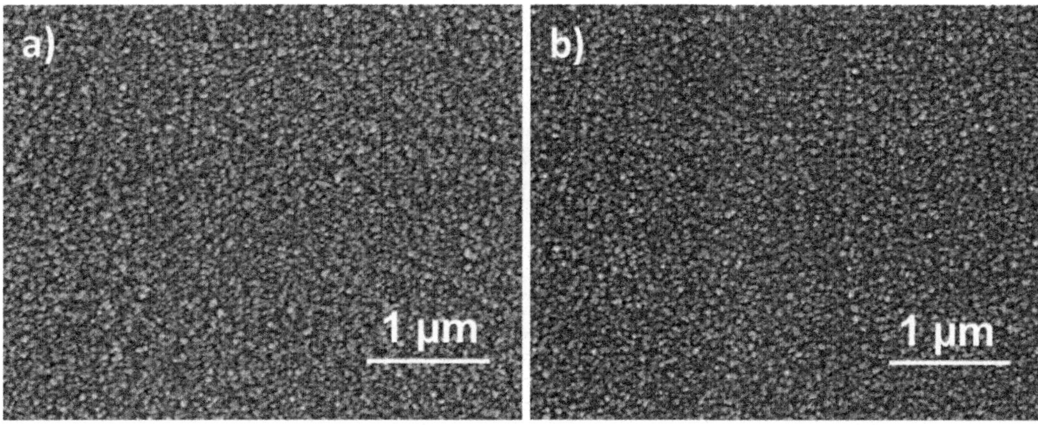

Fig. 3 SEM images (top view) of (**a**) wt-TMV–ZnO and (**b**) E50Q–ZnO hybrid films, respectively, prepared after 10 deposition cycles. Convective assembly conditions: (**a**) 5 μL 5 mg/mL wt-TMV in ultrapure water, withdrawal speed 1.2 mm/min; (**b**) 8 μL 1.9 mg/mL E50Q in ultrapure water, withdrawal speed 0.6 mm/min

4 Notes

1. The glass chamber should keep the assembly conditions (humidity and temperature) constant. Therefore, during the convective assembly the chamber should be well sealed.
2. The substrate holder is placed on the chamber bottom and connected also to the horizontally positioned dip coater (*see* Fig. 2a).
3. A small opening in the glass chamber is used to mount the substrate and the glass slide on the holders. Thereby, the humidity in the chamber adapts to ambient conditions. Therefore, prior to the virus assembly adjust the humidity, monitored with a hygrometer placed in the glass chamber, using a saturated LiCl solution or solid LiCl. The solid LiCl drops the humidity faster. If the humidity drops further during the assembly process, briefly open the chamber door or remove the LiCl solution.
4. Prepare and work with methanol solutions always in a hood, since methanol is highly toxic for humans. It can destroy the optic nerve, which leads to permanent blindness. Precursor stock solutions can be used for couple of days, while the deposition solution should be freshly prepared prior to the ZnO mineralization and used only within a day.
5. We found out that the quality of PVP is crucial for the success of the ZnO mineralization. Our experience showed that there is a difference in the mineralization rate using various PVP batches although bought from the same supplier. Therefore, in order to get good reproducibility, buy a large amount PVP of the same batch, which lasts for the whole series of experiments. If the mineralization is completely suppressed, the ratio

of the different stock solutions in the deposition solution has to be freshly determined. PVP used here: Sigma-Aldrich, Mw 10000, Lot # BCBJ4889V.

6. Glass vessels must be tightly lockable to prevent methanol evaporation. Since the deposition solution in the vessels is small, even slight loss of solvent will result in big differences in the reactant concentrations and hence will have an impact on the quality of the semiconductor. For good reproducibility, the vessels should be dried at 80 °C for several hours prior to use.
7. Silicon substrates should be free from organic solvent before placing in the plasma cleaner.
8. Plasma treatment removes the rest of organic impurities from the substrate and makes the silicon surface highly hydrophilic forming a thin SiO_2 layer on the top. The contact angle of the cleaned substrates tends to 0° and the roughness (Rms) is around 0.2 nm.
9. For good reproducibility, clean the substrates prior to the immobilization and mineralization of the viruses.
10. Depending on the required density of the template layer, different convective assembly conditions can be applied. As can be seen in Fig. 4, homogeneous, dense but disordered layers (*see* Fig. 4a) can be achieved with 5 μL, 5 mg/mL wt-TMV in ultrapure water aligned at a withdrawal speed of 1.2 mm/min. Homogeneous wt-TMV-based monolayers with some distances between TMV bundles (*see* Fig. 4b) can be prepared applying 8 μL, 1.9 mg/mL wt-TMV in ultrapure water, at a withdrawal speed of 0.6 mm/min. The E50Q mutant forms very dense and almost defect free monolayers (*see* Fig. 4d) by the use of 8 μL, 1.9 mg/mL E50Q in ultrapure water and a withdrawal speed of 0.6 mm/min [25]. The assembly process is conducted with buffer-free virus solutions to exclude a possible influence of salt residues from the buffer on the subsequent TMV mineralization. However, if presence of salt residues is not crucial for the mineralization, then homogeneous and dense monolayers of wt-TMV (*see* Fig. 4c) can be achieved with 7 μL, 5 mg/mL wt-TMV in 10 mM sodium–potassium phosphate buffer (SPP, pH 7.1) convectively assembled at a withdrawal speed of 0.8–0.9 mm/min. (At lower speed, a higher volume of virus solution is necessary, since the assembly takes longer time and the small droplet evaporates.)
11. Wait until the virus solution residue is completely evaporated, otherwise the rest of the solution will spread on the already immobilized virus layer and will change the template structure.
12. Take the substrate with immobilized viruses or already mineralized hybrid layers with tweezers holding both opposite edges of the substrate. Be sure not to place the tweezers on the

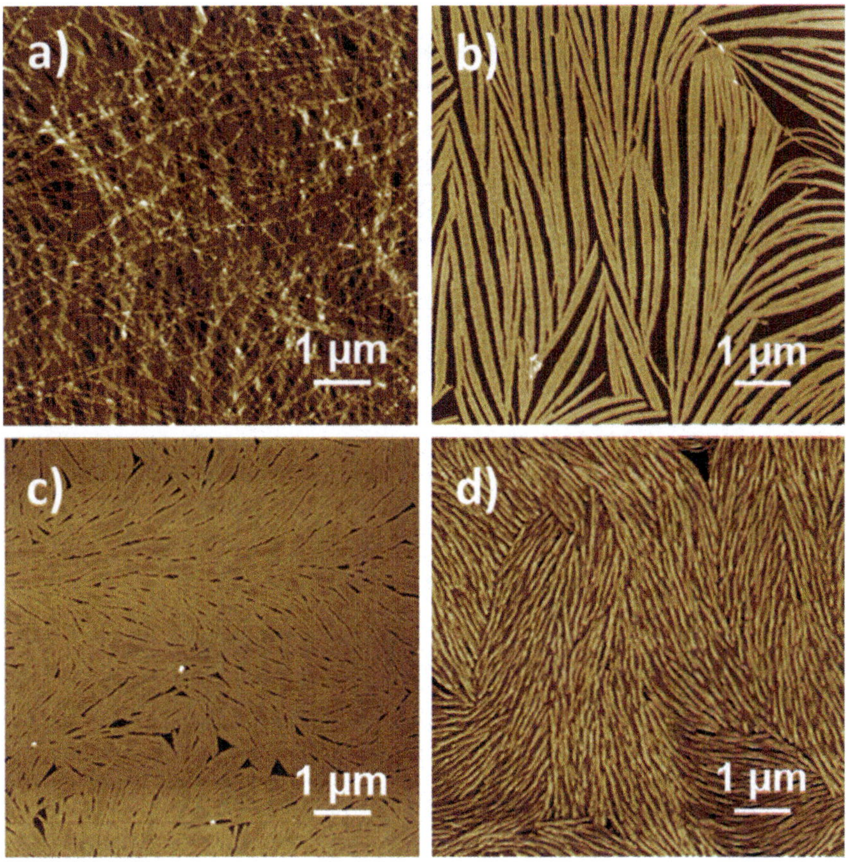

Fig. 4 AFM images of convectively assembled virus template layers with different density: (**a**) 5 μL, 5 mg/mL wt-TMV in ultrapure water, withdrawal speed 1.2 mm/min; (**b**) 8 μL, 1.9 mg/mL wt-TMV in ultrapure water, withdrawal speed 0.6 mm/min; (**c**) 7 μL, 5 mg/mL wt-TMV in 10 mM sodium–potassium phosphate buffer (SPP, pH 7.1), withdrawal speed 0.8 mm/min; (**d**) 8 μL, 1.9 mg/mL E50Q in ultrapure water, withdrawal speed 0.6 mm/min

TMV/mineralized side since the thin film on the substrate surface will be destroyed.

13. To ensure good reproducibility and to exclude possible virus drying to effect the deposition behavior, mineralization within the same day of the virus immobilization on the silicon wafer is recommended.

14. ZnO deposition conducted for 1.5 h is named 1 cycle. Usually, 1 cycle is not enough to completely cover the template layer and more deposition cycles are required. It is important to note that the amount of deposited ZnO is not constant at each mineralization cycle. At the beginning, the quantity of the ZnO nanoparticles attached on the template is lower and it accelerates with increasing numbers of deposition cycles. Therefore, if a certain semiconducting thickness is required, a series of samples with different numbers of deposition cycles

has to be prepared. Then, the thickness of the semiconducting film has to be determined. This can be done by SEM analysis of the sample's cross section.

Acknowledgments

The financial support through the priority program SPP 1569 of the DFG is gratefully acknowledged.

References

1. Thiruvengadathan R, Korampally V, Ghosh A, Chanda N, Gangopadhyay K, Gangopadhyay S (2013) Nanomaterial processing using self-assembly-bottom-up chemical and biological approaches. Rep Prog Phys 76(6):066501. https://doi.org/10.1088/0034-4885/76/6/066501
2. Atanasova P, Weitz RT, Gerstel P, Srot V, Kopold P, van Aken PA, Burghard M, Bill J (2009) DNA-templated synthesis of ZnO thin layers and nanowires. Nanotechnology 20(36):365302. https://doi.org/10.1088/0957-4484/20/36/365302
3. Atanasova P, Rothenstein D, Schneider JJ, Hoffmann RC, Dilfer S, Eiben S, Wege C, Jeske H, Bill J (2011) Virus-templated synthesis of ZnO nanostructures and formation of field-effect transistors. Adv Mater 23(42):4918–4922. https://doi.org/10.1002/adma.201102900
4. Zhou H, Fan TX, Han T, Li XF, Ding J, Zhang D, Guo QX, Ogawa H (2009) Bacteria-based controlled assembly of metal chalcogenide hollow nanostructures with enhanced light-harvesting and photocatalytic properties. Nanotechnology 20(8):085603. https://doi.org/10.1088/0957-4484/20/8/085603
5. Hodes G (2007) Semiconductor and ceramic nanoparticle films deposited by chemical bath deposition. Phys Chem Chem Phys 9(18):2181–2196. https://doi.org/10.1039/b616684a
6. Puscher C (1868) Ueber ein neues und billiges Verfahren, ohne Anwendung von Farben verschiedene Metalle (wie Gold, Silber, Kupfer, Argentan, Messing, Tombak, Eisen, Zink) mit prachtvollen Lüsterfarben zu überziehen. Polytech J 190(CVIII):3
7. Eiben S, Stitz N, Eber F, Wagner J, Atanasova P, Bill J, Wege C, Jeske H (2014) Tailoring the surface properties of tobacco mosaic virions by the integration of bacterially expressed mutant coat protein. Virus Res 180:92–96. https://doi.org/10.1016/j.virusres.2013.11.019. S0168-1702(13)00426-7 [pii]
8. Royston E, Ghosh A, Kofinas P, Harris MT, Culver JN (2008) Self-assembly of virus-structured high surface area nanomaterials and their application as battery electrodes. Langmuir 24(3):906–912. https://doi.org/10.1021/la7016424
9. Alonso JM, Gorzny ML, Bittner AM (2013) The physics of tobacco mosaic virus and virus-based devices in biotechnology. Trends Biotechnol 31(9):530–538. https://doi.org/10.1016/j.tibtech.2013.05.013
10. Bromley KM, Patil AJ, Perriman AW, Stubbs G, Mann S (2008) Preparation of high quality nanowires by tobacco mosaic virus templating of gold nanoparticles. J Mater Chem 18(40):4796–4801. https://doi.org/10.1039/b809585j
11. Lee SY, Choi JW, Royston E, Janes DB, Culver JN, Harris MT (2006) Deposition of platinum clusters on surface-modified tobacco mosaic virus. J Nanosci Nanotechnol 6(4):974–981. https://doi.org/10.1166/jnn.2006.146
12. Manocchi AK, Horelik NE, Lee B, Yi H (2010) Simple, readily controllable palladium nanoparticle formation on surface-assembled viral nanotemplates. Langmuir 26(5):3670–3677. https://doi.org/10.1021/la9031514
13. Knez M, Bittner AM, Boes F, Wege C, Jeske H, Maiss E, Kern K (2003) Biotemplate synthesis of 3-nm nickel and cobalt nanowires. Nano Lett 3(8):1079–1082
14. Balci S, Bittner AM, Hahn K, Scheu C, Knez M, Kadri A, Wege C, Jeske H, Kern K (2006) Copper nanowires within the central channel of tobacco mosaic virus particles. Electrochim Acta 51(28):6251–6257. https://doi.org/10.1016/j.electacta.2006.04.007
15. Shenton W, Douglas T, Young M, Stubbs G, Mann S (1999) Inorganic-organic nanotube composites from template mineralization

of tobacco mosaic virus. Adv Mater 11(3):253. https://doi.org/10.1002/(Sici)1521-4095(199903)11:3<253::Aid-Adma253>3.0.Co;2-7

16. Balci S, Bittner AM, Schirra M, Thonke K, Sauer R, Hahn K, Kadri A, Wege C, Jeske H, Kern K (2009) Catalytic coating of virus particles with zinc oxide. Electrochim Acta 54(22):5149–5154. https://doi.org/10.1016/j.electacta.2009.03.036
17. Royston ES, Brown AD, Harris MT, Culver JN (2009) Preparation of silica stabilized Tobacco mosaic virus templates for the production of metal and layered nanoparticles. J Colloid Interface Sci 332(2):402–407. https://doi.org/10.1016/j.jcis.2008.12.064. S0021-9797(08)01733-5 [pii]
18. Altintoprak K, Seidenstucker A, Welle A, Eiben S, Atanasova P, Stitz N, Plettl A, Bill J, Gliemann H, Jeske H, Rothenstein D, Geiger F, Wege C (2015) Peptide-equipped tobacco mosaic virus templates for selective and controllable biomineral deposition. Beilstein J Nanotech 6:1399–1412. https://doi.org/10.3762/bjnano.6.145
19. Schenk AS, Eiben S, Goll M, Reith L, Kulak AN, Meldrum FC, Jeske H, Wege C, Ludwigs S (2017) Virus-directed formation of electrocatalytically active nanoparticle-based Co3O4 tubes. Nanoscale 9:6334. https://doi.org/10.1039/c7nr00508c
20. Dimitrov AS, Nagayama K (1996) Continuous convective assembling of fine particles into two-dimensional arrays on solid surfaces. Langmuir 12(5):1303–1311. https://doi.org/10.1021/La9502251
21. Niwa D, Fujie T, Lang T, Goda N, Takeoka S (2012) Heterofunctional nanosheet controlling cell adhesion properties by collagen coating. J Biomater Appl 27(2):131–141. https://doi.org/10.1177/0885328210394470
22. Ariga K, Nakanishi T, Michinobu T (2006) Immobilization of biomaterials to nano-assembled films (self-assembled monolayers, langmuir-blodgett films, and layer-by-layer assemblies) and their related functions. J Nanosci Nanotechnol 6(8):2278–2301. https://doi.org/10.1166/jnn.2006.503
23. Kuncicky DM, Naik RR, Velev OD (2006) Rapid deposition and long-range alignment of nanocoatings and arrays of electrically conductive wires from tobacco mosaic virus. Small 2(12):1462–1466. https://doi.org/10.1002/smll.200600399
24. Wargacki SP, Pate B, Vaia RA (2008) Fabrication of 2D ordered films of tobacco mosaic virus (TMV): processing morphology correlations for convective assembly. Langmuir 24(10):5439–5444. https://doi.org/10.1021/La7040778
25. Atanasova P, Stitz N, Sanctis S, Maurer JH, Hoffmann RC, Eiben S, Jeske H, Schneider JJ, Bill J (2015) Genetically improved monolayer-forming tobacco mosaic viruses to generate nanostructured semiconducting bio/inorganic hybrids. Langmuir 31(13):3897–3903. https://doi.org/10.1021/acs.langmuir.5b00700
26. Geiger FC, Eber FJ, Eiben S, Mueller A, Jeske H, Spatz JP, Wege C (2013) TMV nanorods with programmed longitudinal domains of differently addressable coat proteins. Nanoscale 5(9):3808–3816. https://doi.org/10.1039/c3nr33724c
27. Mueller A, Kadri A, Jeske H, Wege C (2010) In vitro assembly of Tobacco mosaic virus coat protein variants derived from fission yeast expression clones or plants. J Virol Methods 166(1-2):77–85. https://doi.org/10.1016/j.jviromet.2010.02.026
28. Kadri A, Maiss E, Amsharov N, Bittner AM, Balci S, Kern K, Jeske H, Wege C (2011) Engineered Tobacco mosaic virus mutants with distinct physical characteristics in planta and enhanced metallization properties. Virus Res 157(1):35–46. https://doi.org/10.1016/j.virusres.2011.01.014
29. Culver JN, Dawson WO, Plonk K, Stubbs G (1995) Site-directed mutagenesis confirms the involvement of carboxylate groups in the disassembly of tobacco mosaic-virus. Virology 206(1):724–730. https://doi.org/10.1016/S0042-6822(95)80096-4

Chapter 27

Dual Functionalization of Rod-Shaped Viruses on Single Coat Protein Subunits

Christina Wege and Fania Geiger

Abstract

Plant viruses are emerging as versatile tools for nanotechnology applications since it is possible to modify their multivalent protein surfaces and thereby introduce and display new functionalities. In this chapter, we describe a tobacco mosaic virus (TMV) variant that exposes two selectively addressable amino acid moieties on each of its 2130 coat protein (CP) subunits. A lysine as well as a cysteine introduced at accessible sites of every CP can be modified with amino- and/or thiol-reactive chemistry such as *N*-hydroxysuccinimide esters (NHS ester) and maleimide containing reagents alone or simultaneously. This enables the pairwise immobilization of distinct molecules in close vicinity to each other on the TMV surface by simple standard conjugation protocols. We describe the generation of the mutations, the virus propagation and isolation as well as the dual functionalization of the TMV variant with two fluorescent dyes. The labeling is evaluated by SDS-PAGE and spectrophotometry and the degree of labeling (DOL) calculated.

Key words Tobacco mosaic virus, Fluorescent labeling, Coat protein dual functionalization, Surface modification, Degree of labeling

1 Introduction

Methods enabling surface modifications of plant viruses or virus-like particles have gained increasing importance in nanotechnology research since numerous plant virus derivatives have been developed into promising proof-of-concept tools for applications ranging from medical diagnostics via biosensor devices up to the fabrication of nanostructured battery electrodes [as reviewed for example in [1–5]]. The attractiveness of plant viruses is based on their regular and precise shapes on the nanometer scale, the possibility to genetically modify their genetic cargo including the genes encoding the own viral building blocks, high expression and purification yields, a high stability of several types of particles, and being noninfectious for humans and warm-blooded animals. These features even raise options to assemble artificial plant virus-derived structures and expanded three-dimensional materials [4, 6–9].

Christina Wege and George P. Lomonossoff (eds.), *Virus-Derived Nanoparticles for Advanced Technologies: Methods and Protocols*, Methods in Molecular Biology, vol. 1776, https://doi.org/10.1007/978-1-4939-7808-3_27,

© Springer Science+Business Media, LLC, part of Springer Nature 2018

Tobacco mosaic virus (TMV) is one of the most promising candidates in nanotechnology applications, as it combines all of these advantageous possibilities [10–12] and has been tested for the extent of genetic alterations that are tolerated by the virus over several decades [[13], and references therein]. The 2130 identical coat protein (CP) subunits are arranged in a helical manner incorporating the single-stranded RNA genome to form a rigid rod of 300 nm length with an outer diameter of 18 nm, and an inner channel of 4 nm diameter [14–16]. There have been studies using TMV as carrier scaffold or template in batteries [17], ferrofluids [18], biosensors [19–21], imaging agents [22, 23], vaccines [24, 25], cell culturing [26–28], and many more (reviewed in [1, 11, 12]). A number of TMV-like particles have been created to fulfill the different challenges of these applications. There have been described star-like, kinked, and branched structures [29–32] as well as rods with mixed CP variants [33] or serially ordered longitudinal domains of distinct CP types [34, 35]. Such TMV derivatives have been created through the use of mutant, natural, or engineered RNA molecules with TMV origin of assembly sequences nucleating nanorod growth in vitro. Different types of TMV particles have also been arranged on surfaces by either immobilizing fully [36] or partially assembled particles [37], or by bottom-up assembly of the rods with surface-bound viral RNA [38]. Under suitable conditions, TMV CPs multimerize into rod-like structures and disk aggregates also in the absence of RNA strands [reviewed in [39]]. Binding of molecules to all kinds of TMV-like assemblies can be either achieved by using the naturally occurring amino acids (aa) like glutamic acid or tyrosine [40–42], by genetically introduced amino acids with specific addressability and higher reactivity allowing for orthogonal coupling reactions also under mild conditions [43, 44], or by the incorporation of selectively addressable unnatural amino acids [45]. The TMV CP has been genetically modified with cysteines at different sites either exposed or buried inside the viral nucleoprotein helix. Such mutants were used to enhance rod assembly and stability or to bind fluorescent chromophores serving as light-harvesting system, respectively [46, 47]. Widely used are TMV variants with addressable amino or thiol groups originating from lysines [34, 48] or cysteines [34, 49] exposed on the outer nanorod surface, respectively. They have been employed by several groups for metal deposition [50–52] and for a number of applications requiring the immobilization of dyes [23, 49, 53], polymers [23], peptides [54], or enzymes [12, 19, 21]. This is because the corresponding conjugation chemistry based on amino-reactive succinimidyl (NHS) esters and thiol-reactive maleimides, respectively, is easy to employ and numerous reagents, ranging from linker and dye molecules up to stealth polymers and nanoparticle coatings, are commercially available.

So far, however, no TMV variant has been reported that is accessible to both these standard coupling reactions simultaneously on every CP subunit, thus enabling a pairwise installation of distinct functionalities with sub-3-nm distance on the outer TMV coat. We thus describe here the generation of TMV rods exposing a lysine and cysteine on the outer surface of each CP, called TMV_{DM} (double mutant) in the following. The TMV_{DM} is based on a previously established TMV_{Cys} [34] variant having a single amino acid substitution (S3C) close to the CP N-terminus. In addition to this exchange, the threonine at the C-terminal aa position 158 of the CP_{Cys} is replaced by a lysine (T158K), as described for a TMV_{Lys} [34] variant before. Hence, the new TMV_{DM} variant allows for a dual functionalization on single CPs with a maleimide and a NHS ester-containing molecule. Using the TMV_{DM} variant, a surface density higher than 2130 molecules per rod can be achieved for immobilized target molecules, and studies requiring a very close proximity of the investigated molecules can be conducted.

The generation of the plant-infectious plasmid $pTMV843_{DM}$ containing the genetic information for the TMV_{DM} variant is described only briefly since it is based on well-known molecular biology techniques and parallels the production of other mutant TMV CPs published earlier [55]. Mechanical inoculation of *Nicotiana benthamiana* and *Nicotiana tabacum* plants with plasmid $pTMV843_{DM}$ is explained in more detail. TMV_{DM} particles are isolated from successfully infected plants according to Gooding & Hebert [56]. Dual functionalization of the TMV_{DM} rods is then exemplified for a combination of two fluorescent dye derivatives, namely Atto 647 maleimide and Atto 488 NHS ester. Their covalent installation on TMV_{DM} CPs is verified through fluorescent bands generated upon SDS-PAGE, with the respective colors green for Atto 488, red for Atto 647, and yellow for CP subunits modified with both dyes (*see* Fig. 1a). The binding of the dyes also results in a shift of the respective CP protein bands in the SDS–polyacrylamide gels, due to the additional molecular weight of the coupled dyes (*see* Fig. 1a, b). The approximate degrees of labeling are calculated using the appropriate correction factors for the fluorescent dyes and the Lambert–Beer law (*see* Table 2). The novel TMV_{DM} carrier nanoparticles enable the pairwise immobilization of two distinct molecules. With a 50-fold excess of dye over CP-reactive sites, at least 50% of the CP subunits were modified under the labeling conditions applied in this study. This was achieved in a simple one-pot reaction and left the tobamoviral rods intact for subsequent uses. Coupling efficiencies might be further improved through serial application of the two dye derivatives and adjusting the pH for the second conjugation reaction. The protocol may also be transferred to other types of target molecules including bifunctional linkers, in order to display distinct types of functions on the tobamoviral carrier rods in so far unprecedented surface densities.

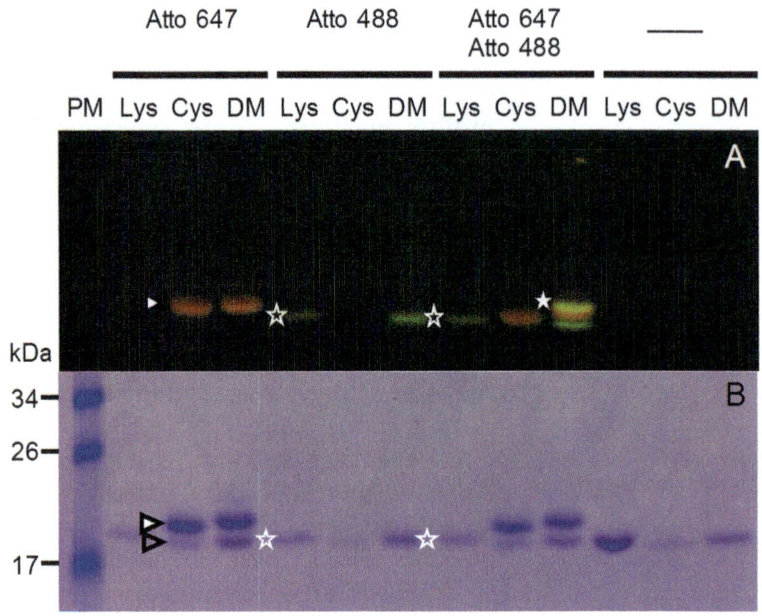

Fig. 1 15% SDS-PAGE of TMV_{DM} (DM), TMV_{Lys} (Lys), and TMV_{Cys} (Cys) particles unlabeled or labeled with Atto647 maleimide (Atto647) and/or Atto488 NHS ester (Atto488), respectively. PM: protein molecular weight marker. (**a**) Detection of the fluorescent dyes with a DR195M transilluminator. (**b**) Coomassie Blue staining (performed after fluorescence detection; *see* **Note 12**). Empty triangle: unlabeled TMV CP, filled triangle: TMV CP labeled with Atto 647, unfilled star: TMV CP labeled with Atto 488, filled star: TMV CP labeled with Atto 647 and Atto 488

2 Materials

All solutions, buffers, and media are prepared with deionized water (ddH_2O; 18.3 MΩ cm) if not stated otherwise.

2.1 CP Mutation

1. Plasmid p843TMV_{Cys} containing the infectious cDNA of TMV_{Cys}, a mutant of the wild-type TMV vulgare strain exposing a cysteine on every CP subunit on the outer virus surface due to an S3C amino acid exchange in the coat protein [34].
2. Taq- and ProofStart-DNA Polymerase (Qiagen).
3. 10× PCR buffer (Qiagen).
4. Q-solution (Qiagen).
5. dNTP mix (5 mM each).
6. PCR cycler.
7. Primers according to Table 1.
8. Equipment for horizontal agarose gel electrophoresis and DNA detection (for details *see* Chapter 2 in Molecular Cloning [57]).
9. GFX PCR, DNA, Gel Band purification Kit (GE Healthcare).

Table 1
Primer sequences used for an overlap extension PCR introducing a lysine codon into the TMV_{Cys} CP ORF at aa position 158

Name	Primer sequence (5′–3′)
F1	CCA ACC TCG AGG ATT ACA AAC GTG AGA GAC GGA GG
R1	CCA ACC TCG AGC GCG ATC CAA GAC ACA ACC CTT CG
RLys	TCT TGA CTA GCT CA***C T***TT GCA GGA
FLys	CCT CTA GTC CTG CAA ***AG***T GAG CTA G
CPF	CCG CTT TCT CTG GAG TTT GTG TCG
CPR	CGT GCC TGC GGA TGT ATA TGA ACC

Nucleotides inserting the mutation are shown in bold and italics

10. Restriction enzymes *Bsi*WI and *Nco*I and appropriate digestion buffers.
11. T_4-DNA ligase and appropriate buffer.
12. Competent *E. coli* NM522 (New England Biolabs).

2.2 Virus Propagation

1. *E. coli* NM522 transformed with $p843TMV_{DM}$ (mutated plasmid $p843TMV_{Cys}$) with both S3C and T158K mutations of TMV CP gene.
2. Lysogenic broth (LB) medium: 10 g/L tryptone, 5 g/L yeast extract, 10 g/L NaCl, adjust pH to 7.0 with HCl or NaOH; sterilize.
3. Ampicillin: stock solution 100 mg/mL in sterilized ddH_2O.
4. 37 °C shaking incubator for bacteria.
5. Maxi-Plasmid purification kit for large volumes.
6. 10 mM sodium–potassium phosphate (SPP) buffer pH 7.2. Mix 10 mM NaH_2PO_4 and 10 mM K_2HPO_4 till you get a pH of 7.2.
7. Six-leaf-stage *Nicotiana benthamiana* plants.
8. Five-to-six-leaf stage *Nicotiana tabacum* 'Samsun' nn plants.
9. Silicon carbide powder 320 mesh (Carborundum).

2.3 TMV Isolation from Plants

1. 20 g symptomatic leaves of infected *N. tabacum* 'Samsun' nn plants.
2. 0.5 M SPP buffer pH 7.2.

3. β-mercaptoethanol.
4. Electrical blender.
5. *n*-butanol.
6. Electrical stirrer/stirring bar.
7. 10 mM SPP buffer pH 7.2.
8. Polyethylene glycol 6000 (PEG 6000).
9. NaCl (solid).
10. Filter paper (e.g., Whatman Grade 595).
11. Miracloth (Merck).
12. Centrifuge and rotor for 50–100 mL volume and 10,000 × *g*.

To check virus integrity after isolation TEM can be used.

13. Transmission electron microscope (TEM, e.g., Tecnai G^2 Sphera, FEI).
14. TEM copper grids 400-mesh coated with formvar or pioloform (Plano).
15. 1% (w/v) uranyl acetate pH 4.3 (optionally supplemented with 250 μg/mL bacitracin as wetting agent).

2.4 Labeling with Dye Derivatives

1. Atto 488 NHS ester and Atto 647 maleimide (Sigma-Aldrich), or other fluorescent dyes as NHS ester or maleimide derivatives (*see* **Note 1**).
2. Dimethyl sulfoxide (DMSO), anhydrous >99.9%.
3. TMV_{DM} particles in 10 mM SPP-buffer pH 7.2 (as prepared in Subheading 3.3, see below).
4. Optima L-80 XP ultracentrifuge (Beckman Coulter) and rotor 42.4 Ti (Beckman Coulter) or equivalent ultracentrifuge with a fixed angle rotor holding small tubes (up to 1 mL) and capable of running at 120,000 × *g*.
5. Cellulose propionate centrifuge tubes (Beckman Coulter) or equivalent.
6. Glove box if available (*see* **Note 2**).

2.5 SDS-PAGE Detection of Fluorescent Protein Bands

1. Equipment and solutions for a standard SDS-PAGE with a 4% stacking and 15% separation gel.
2. Coomassie Brilliant Blue G-250 and destaining solution or any other standard staining solution for SDS-PAGE.
3. Aluminum foil or cardboard box for lightproof covering of the SDS-PAGE setup.
4. DR195M transilluminator (Clare Chemical Research) or other detection device matching the chosen fluorescent dyes with a camera capable of taking colored pictures (*see* **Note 3**).

2.6 Spectrophotometry and Calculation of Dye Labeling Efficiencies

1. 10 mM SPP buffer pH 7.2.
2. ddH_2O.
3. Suspensions of TMV_{DM} labeled with Atto 488 and/or Atto 647 in 10 mM SPP buffer pH 7.2.
4. Atto 647 solution in 10 mM SPP buffer pH 7.2 (e.g., 40 μM).
5. Atto 488 solution in 10 mM SPP buffer pH 7.2 (e.g., 100 μM).
6. NanoDrop ND-1000 spectrophotometer (Thermo Scientific) or equivalent spectrophotometer capable of measuring small volumes.
7. Extinction coefficient ε for TMV at 260 nm (i.e., $\varepsilon_{260} = 3\frac{\text{ml}}{\text{mg} \times \text{cm}}$), for Atto 488 at 501 nm (i.e., $\varepsilon_{501} = 90{,}000\frac{\text{l}}{\text{mol} \times \text{cm}}$) and for Atto 647 at 645 nm (i.e., $\varepsilon_{645} = 120{,}000\frac{\text{l}}{\text{mol} \times \text{cm}}$) or ε for the respective dyes at their maximum absorption wavelength.

3 Methods

3.1 CP Mutation

The point mutation T158K is inserted in the CP sequence of TMV_{Cys} by an overlap extension PCR according to Chapter 14 in Molecular Cloning [57]. Restriction enzymes are used to insert the mutated CP cDNA sequence into the plasmid backbone of $p843TMV_{Cys}$. Since these steps are all standard molecular techniques, they are only described briefly.

1. Set up two PCRs in parallel according to manufacturer's instruction using $p843TMV_{Cys}$ as template and primer pairs F1 and RLys, or R1 and FLys, respectively. Use the following program: 94 °C 10 min; 20 cycles of 94 °C, 45 s, 58 °C, 45 s, 72 °C 1 min; 72 °C 10 min.
2. Apply the PCR products to a horizontal agarose gel electrophoresis, purify the fragments with the GFX PCR, DNA, Gel Band purification Kit and use them as templates for the next PCR (**step 3**).
3. Set up a PCR with the gel-purified fragments as templates and primers F1 and R1. Use the following program: 94 °C 10 min; 5 cycles of 94 °C, 45 s, 58 °C, 45 s, 72 °C 1 min; 20 cycles of 94 °C, 45 s, 68 °C, 45 s, 72 °C 1 min; 72 °C 10 min.
4. Use the GFX PCR, DNA, Gel Band purification Kit to purify the PCR product.
5. Double-digest the purified PCR product of **step 3** and the plasmid pTMV843Cys both with the two restriction enzymes

*Bsi*WI and *Nco*I according to enzyme manufacturer's instructions.

6. Separate the resulting fragments on a horizontal agarose gel, cut out the two gel sections (as identified by ethidium bromide-staining of a parallel lane separated from the preparative part of the gel) containing the desired pTMV-backbone fragment or the mutated CP sequence, respectively and purify them using the GFX PCR, DNA, Gel Band purification Kit.
7. Ligate the fragments of the purified pTMV backbone and of the CP gene cDNA sequence (containing both the S3C mutation and the new T158K mutation) with T4-DNA ligase according to the manufacturer's instruction.
8. Transform the ligation products into competent *E. coli* NM522 according to standard procedures.
9. Isolate the plasmids of resulting clones and check by restriction enzyme assays followed by sequencing of the CP gene cDNA for clones containing the desired mutations. The correct resulting plasmid is called p843TMV$_{DM}$ in the following sections.

3.2 Virus Propagation

1. Culture *E. coli* NM 522 transformed with p843TMV$_{DM}$ overnight in LB-Medium with 100 μg/mL ampicillin.
2. Purify p843TMV$_{DM}$ with a Maxi-Plasmid purification kit.
3. Dilute p843TMV$_{DM}$ in 10 mM SPP buffer pH 7.2 to a concentration of 2.5 ng/μL.
4. Inoculate at least ten six-leaf-stage *N. benthamiana* plants by rubbing 25 μL of the plasmid dilution together with silicon carbide powder on the upper surface of each of the two upper leaves (*see* **Note 4**).
5. Harvest infected leaves 21 days post inoculation (dpi) (*see* **Note 5**).
6. Homogenize a small piece (approximately 1 cm^2) in 400 μL 10 mM SPP buffer pH 7.2.
7. Centrifuge for 1 min at 20,800 × *g* at room temperature.
8. Use the supernatant to inoculate two upper leaves of approximately five five-to-six-leaf stage *N. tabacum* 'Samsun' nn plants. For each leaf use 25 μL together with a breeze of carborundum.
9. Harvest the infected leaves 21 dpi and proceed directly with the TMV isolation or freeze the material at −20 °C (*see* Fig. 2).

3.3 TMV Isolation from Plants

TMV$_{DM}$ particles were isolated from infected *N. tabacum* leaves according to Gooding and Hebert [56]. Virus integrity was checked via TEM (*see* Fig. 3).

1. All steps are performed at room temperature.

Fig. 2 *N. benthamiana* and *N. tabacum* plants or leaves, respectively, infected with TMV_{DM} or TMVwt compared to mock-inoculated plants at 21 dpi

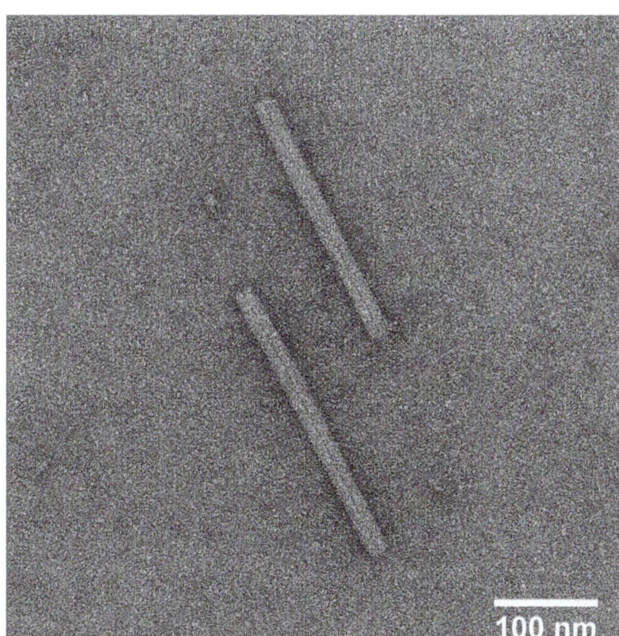

Fig. 3 Transmission electron micrographs of TMV_{DM} particles isolated from *N. tabacum* plants following UAc negative staining

2. Mix 20 g of TMV_{DM} infected leaf material with 40 mL 0.5 M SPP buffer pH 7.2 with 1% mercaptoethanol.
3. Homogenize in an electrical blender.
4. Filtrate homogenate through Miracloth and filter paper.
5. Add slowly *n*-butanol (drop by drop) while stirring (8 mL butanol/100 mL filtrate) to the green solution obtained after filtering.
6. After chloroplast precipitation stir for another 15 min.
7. Transfer sample to centrifuge tubes.
8. Centrifuge for 30 min at 10,000 × *g*.
9. Add 4 g/100 mL PEG 6000 to the supernatant while stirring.
10. After the PEG 6000 has dissolved, transfer sample to centrifuge tube.
11. Centrifuge for 15 min at 10,000 × *g*.
12. Resuspend the pellet in 20 mL 10 mM SPP buffer pH 7.2.
13. Add 0.4 g NaCl and 0.4 g PEG 6000 per 10 mL suspension.
14. After NaCl and PEG 6000 has dissolved, transfer sample to centrifuge tube.
15. Centrifuge for 15 min at 10,000 × *g*.
16. Resuspend pellet (containing TMV_{DM}) in 10 mL SPP buffer pH 7.2.
17. Determine the TMV concentration with UV spectrometry measuring the absorption at 260 nm and using the extinction coefficient for TMV: $\varepsilon_{260} = 3\frac{\text{ml}}{\text{mg} \times \text{cm}}$.

For TEM analysis to check virus integrity proceed with the following steps:

18. Prepare a TMV solution with 200 μg/mL in 10 mM SPP buffer pH 7.2.
19. Incubate one drop of TMV-solution on a formvar or pioloform coated 400 mesh copper grid for 5 min.
20. Stain the TMV with 1% uranyl acetate by placing the grid on top of a uranyl acetate (optionally supplemented with bacitracin) drop for 1 min.
21. Wash the grid dropwise with approximately 2 mL ddH_2O.
22. Observe the grid at 120 kV in a TEM.

3.4 Labeling with Dye Derivatives

1. Dilute TMV_{DM} in 10 mM SPP buffer pH 7.2 to a concentration of 200 μg/mL (*see* **Note 6**).
2. Dissolve 1 mg Atto 488 NHS ester in 60 μL and 1 mg Atto 647 maleimide in 70 μL anhydrous DMSO, respectively (each

final concentration: 17 mM), and split them into 5–10 μL aliquots, if possible in a glove box (*see* **Note 2**).

3. Mix 30 μL TMV_{DM} particles (200 μg/mL) in 10 mM SPP buffer pH 7.2 with 1 μL Atto 488 NHS ester solution and/or 1 μL Atto 647 maleimide solution (17 mM each). This corresponds to a 50-fold molar excess of dye molecules over reactive sites at the CPs. As a control, mix 30 μL of TMV_{DM} with 2 μL anhydrous DMSO (*see* **Note 7**).
4. Incubate for 3 h at room temperature. Keep the samples in the dark.
5. Add 70 μL of 10 mM SPP pH 7.2 per sample.
6. Transfer the samples to centrifuge tubes.
7. Centrifuge for 1.5 h at 4 °C, 120,000 × *g*.
8. Collect the supernatants in new 1.5 mL reaction tubes and resuspend the pellets in 30 μL 10 mM SPP pH 7.2.

3.5 SDS-PAGE Detection of Fluorescent Protein Bands

1. Prepare a SDS-PAGE with a 4% stacking and 15% separation gel according to standard procedures.
2. Mix 7.5 μL of each labeling reaction with 7.5 μL of 2× reducing protein sample buffer.
3. Heat the samples for 5 min at 95 °C.
4. Load 15 μL of the samples and the prestained protein molecular weight marker on the stacking gel.
5. Run the gel according to standard procedures, but keep the gel in the dark using a cardboard box or aluminum foil.
6. Directly after disassembling the SDS-PAGE setup, take a picture with the DR195M transilluminator (or equivalent) to see the fluorescent CP bands (*see* **Note 8**).
7. Fix and stain the gel with Coomassie Blue staining solution or any other standard protein staining solution.

3.6 Spectrophotometry and Degree of Labeling (DOL)

In order to calculate the labeling efficiencies, the TMV particle concentration as well as the dye concentrations have to be determined. In principle, all these concentrations can be calculated from spectrophotometric absorption values using the Lambert–Beer law, but since the dyes do not only absorb at their absorption maxima (A_{max}), but also at 260 nm and 320 nm (the values used to calculate the virus concentration), respectively, their contribution to the absorption at these wavelengths has to be taken into account. Therefore, correction factors (CFs) have to be determined first (*see* **Note 9**). Furthermore, Atto 647 is also absorbing at 501 nm, the absorption maximum of Atto 488. Hence, the correction factor CF_{488} has to be determined for Atto 647 in order to calculate the Atto 488 concentration correctly, for the double-labeled particles only. The correction factor for a specific wavelength λ is defined as $CF_\lambda = \frac{A_\lambda}{A_{max}}$.

1. Measure the absorption of each virus-containing sample at 260, 320, 501 and 645 nm.
2. Measure the absorption of an Atto 488 and an Atto 647 solution (their concentration does not matter as long as the absorption values are in the linear range; usually we diluted the dye stock solution (17 mM) 1:100–1:1000 for the measurement) at 260, 320, 501, and 645 nm in order to calculate concentrations and correction factors.

   Take the values of the spectrophotometry measurements of the dye solutions without virus and calculate the correction factors:

3. $CF_{260(\mathrm{Atto647})} = \frac{A_{260}}{A_{645}}$, $CF_{320(\mathrm{Atto647})} = \frac{A_{320}}{A_{645}}$ and $CF_{501(\mathrm{Atto\,647})} = \frac{A_{501}}{A_{645}}$ for the Atto 647 solution. Calculated values in Table 2.
4. $CF_{260(\mathrm{Atto488})} = \frac{A_{260}}{A_{501}}$ and $CF_{320(\mathrm{Atto488})} = \frac{A_{320}}{A_{501}}$ for the Atto 488 solution. Calculated values in Table 2.
5. Calculate for each labeled virus solution the Atto 647 concentration using the Lambert–Beer law $c_{647} = \frac{A_{645}}{\varepsilon_{645} \times 0.1\mathrm{cm}}$ (*see* **Note 10**). For example, in our case forthe originally 200 μg/mL TMV_{DM} labeled with Atto 647 and Atto 488:

$$c_{647} = \frac{0.053}{120{,}000\frac{1}{\mathrm{mol} \times \mathrm{cm}} \times 0.1\mathrm{cm}} = 4.42\mu\mathrm{M}.$$

6. Calculate for each labeled virus solution the Atto 488 concentration using the Lambert–Beer law corrected for the contribution of the Atto 647 at 501 nm: $c_{488} = \frac{A_{501} - \left(A_{645} \times CF_{501(\mathrm{Atto647})}\right)}{\varepsilon_{501} \times 0.1\mathrm{cm}}$.

   For example, in our case for the originally 200 μg/mL TMV_{DM} labeled with Atto 647 and Atto 488:

$$c_{488} = \frac{0.03 - (0.053 \times 0.022)}{90{,}000\frac{1}{\mathrm{mol} \times \mathrm{cm}} \times 0.1\mathrm{cm}} = 3.33\mu\mathrm{M}.$$

7. Calculate for each labeled virus solution the TMV concentration using the Lambert–Beer law corrected for the contribution

Table 2
UV/Vis absorption values of spectrophotometry measurements for solutions of the individual dyes, and for TMV_{DM} labeled with Atto 488 NHS ester (TMV_{DM} 488), or Atto 647 maleimide (TMV_{DM} 647), or both dye derivatives (TMV_{DM} 488/647), respectively, using a 50-fold excess per dye derivative over the corresponding CP-reactive site

Sample	A_{260}	A_{320}	A_{501}	A_{645}	c_{488} [µM]	c_{647} [µM]	c_{TMV} [mg/mL]	c_{CPTMV} [µM]	DOL_{488} [%]	DOL_{647} [%]
Atto 488 solution	0.232	0.116	0.95	0.004	105.56	0.33	–	–	–	–
Atto 647 solution	0.115	0.101	0.017	0.78	0.04	65.00	–	–	–	–
TMV_{DM} 488	0.047	0.013	0.032	0.006	3.54	0.50	0.10	5.41	65	9
TMV_{DM} 647	0.123	0.049	0.01	0.084	1.11	7.00	0.25	13.34	7	52
TMV_{DM} 488/647	0.074	0.027	0.03	0.053	3.33	4.42	0.14	7.66	42	58
TMV_{DM}										
Correction factors	$CF_{260\,(Atto488)}$	$CF_{320\,(Atto488)}$	$CF_{260\,(Atto647)}$	$CF_{320(Atto647)}$	$CF_{501(Atto647)}$					
	0.244	0.122	0.147	0.129	0.022					

Correction factors, concentrations and approximate degrees of labeling were calculated as explained in Subheading 3.6 (*see* **Note 13**)

A_{260}: Absorption at 260 nm, A_{320}: Absorption at 320 nm, A_{501}: Absorption at 501 nm, A_{645}: Absorption at 645 nm, c_{488}: concentration of Atto 488, c_{647}: concentration of Atto 647, c_{TMV}: concentration of TMV, c_{CPTMV}: concentration of TMV CPs, DOL_{488}: degree of labeling of TMV CPs with Atto 488, DOL_{647}: degree of labeling of TMV CPs with Atto 647, $CF_{260(Atto488)}$: correction factor at 260 nm for Atto 488, $CF_{320(Atto\,488)}$: correction factor at 320 nm for Atto 488, $CF_{260\,(Atto\,647)}$ correction factor at 320 nm for Atto 647, $CF_{320\,(Atto\,647)}$: correction factor at 320 nm for Atto 647, $CF_{501(Atto\,647)}$: correction factor at 501 nm for Atto 647

of the Atto 647 and Atto 488 at 260 and 320 nm, taking also into account that Atto 647 contributes to A_{501}:

$$c_{\mathrm{TMV}} = \frac{\begin{array}{l}(A_{260} - A_{645} \times \mathrm{CF}_{260(\mathrm{Atto647})} - (A_{501} - A_{645} \times \mathrm{CF}_{501(\mathrm{Atto647})}) \times \mathrm{CF}_{260(\mathrm{Atto488})}) - \\ (A_{320} - A_{645} \times \mathrm{CF}_{320(\mathrm{Atto647})} - (A_{501} - A_{645} \times \mathrm{CF}_{501(\mathrm{Atto647})}) \times \mathrm{CF}_{320(\mathrm{Atto488})})\end{array}}{\varepsilon_{\mathrm{TMV}} \times 0.1\mathrm{cm}}.$$

For our example of originally 200 μg/mL $\mathrm{TMV_{DM}}$ labeled with Atto 488 and Atto 647:

$$c_{\mathrm{TMV}} = \frac{\begin{array}{l}(0.074 - 0.053 \times 0.147 - (0.03 - 0.053 \times 0.022) \times 0.24) - \\ (0.027 - 0.053 \times 0.129 - (0.03 - 0.053 \times 0.022) \times 0.122)\end{array}}{3 \frac{\mathrm{ml}}{\mathrm{mg} \times \mathrm{cm}} \times 0.1\mathrm{cm}} = 0.14\mathrm{mg} / \mathrm{ml}.$$

8. Convert the TMV concentration from mg/mL into mol/L coat proteins using the molecular weight of a TMV particle (39.4 × 10^6 Da) and 2130 CPs per particle:

$$c_{\mathrm{CPTMV}}\left[\frac{\mathrm{mol}}{\mathrm{L}}\right] = \frac{C_{\mathrm{TMV}}\left[\frac{\mathrm{g}}{\mathrm{L}}\right]}{39.4 \times 10^6 \left[\frac{\mathrm{g}}{\mathrm{mol}}\right]} \times 2130.$$ In our example for $\mathrm{TMV_{DM}}$ labelled with Atto 488 and Atto 647:

$$c_{\mathrm{CPTMV}} = \frac{0.14\left[\frac{\mathrm{g}}{\mathrm{L}}\right]}{39.4 \times 10^6 \left[\frac{\mathrm{g}}{\mathrm{mol}}\right]} \times 2130 = 7.57\mu\mathrm{M}.$$

9. Calculate the degree of labeling for Atto 647: $\mathrm{DOL}_{647} = \frac{c_{647}}{c_{\mathrm{CPTMV}}}$ (*see* **Note 11**). In our example for $\mathrm{TMV_{DM}}$ labeled with Atto 647 and Atto 488: $\mathrm{DOL}_{647} = \frac{4.42\mu\mathrm{M}}{7.57\mu\mathrm{M}} = 58\%$.

10. Calculate the degree of labeling for Atto 488: $\mathrm{DOL}_{488} = \frac{c_{488}}{c_{\mathrm{CPTMV}}}$ (*see* **Note 11**) In our example for $\mathrm{TMV_{DM}}$ labeled with Atto 647 and Atto 488: $\mathrm{DOL}_{488} = \frac{3.33\mu\mathrm{M}}{7.57\mu\mathrm{M}} = 44\%$.

4 Notes

1. There is a great number of fluorescent dye conjugates available with maleimide or NHS ester as reactive group. In addition to conditions meeting your intended subsequent application, you should consider the following aspects upon choosing the dyes:

(1) You should be able to detect the dyes with your gel detection device. Ideally, the dyes are excited at their excitation maximum and detected at their emission maximum. Make sure that the fluorescence emission maxima can be distinguished with your device, otherwise you will not be able to discriminate between the dyes. You should be able to detect each dye individually so that you can create overlay pictures of both colors emitted from the protein bands. An alternative is a suboptimal excitation at a wavelength which can excite both dyes but not at their excitation maximum, however, with a detection device capable of color discrimination. This will allow for a qualitative, but not a quantitative evaluation of the labeling.

(2) Your spectrophotometer should be able to measure absorptions at the wavelengths of both absorption maxima of the dyes, and additionally UV absorption at 260 and 320 nm.

To (3) calculate the labeling efficiencies consistently, at least one of the dyes should not or only negligibly absorb at the absorption maxima of the other dye, otherwise it is not possible to determine the correction factors necessary for the calculations.

(4) Depending on the follow-up experiments planned, you might consider that some dyes introduce or change the charge of the virus surface. For most dyes you can find information about their net charge after coupling in the data sheets.

2. A glove box is not essential, but dissolving and aliquoting the dyes under inert gas prolongs their lifetime regarding coupling reactivity. This is due to the sensitivity of maleimides and NHS esters toward water resulting in hydrolysis. Keep the dyes as well as any sample containing dyes in the dark since they are light-sensitive. Use either amber or black reaction tubes or wrap them with aluminum foil.
3. Your detection system needs to be compatible with your chosen dyes. For details *see* **Note 1**.
4. Mechanical inoculation of the plants with p843TMV$_{DM}$ might be a little tricky. You have to find the right amount of silicon carbide powder and rubbing pressure to get the plasmid into the plant cells without killing them. Depending on your experience, you might have to use a higher number of plants to get sufficient infected ones. You might also try to get *N. tabacum* plants infected through mechanical inoculation directly, but it is easier with *N. benthamiana*.

5. You can freeze any infected plant material at −20 °C and use it later.
6. In this protocol we use 30 μL of virus particles with a concentration of 200 μg/mL. This volume can be scaled up, just make sure that you use the same excess of dye, and that the DMSO concentration originating from the dye solution does not exceed 50% since TMV starts to disassemble in high DMSO concentrations.
7. NHS ester coupling is most efficient at pH 8.0–9.0, but also hydrolysis is high in this pH range. Maleimide coupling has its optimum at pH 7.0–7.5. We chose pH 7.2 in combination with a mixture of both dye derivatives to keep hydrolysis low, but you can change the pH of the coupling solution to achieve different degrees of labeling. You might also start with one dye conjugate at its pH optimum and change the pH upon adding the second dye. Keep in mind that maleimides can also react with primary amines especially at a pH > 8.5. Do not use primary amine buffers such as Tris-(hydroxymethyl)-aminomethane (Tris) or glycine buffers.
8. Taking the fluorescent pictures of your unfixed SDS–polyacrylamide gel, make sure that you know the orientation of your gel on the picture, since you will not see the prestained protein marker bands. This is necessary to assign the bands and to compare with the corresponding subsequent Coomassie Blue stain.
9. For some dyes, there are correction factors available in the data sheets.
10. Make sure that you know which path length was used in the spectrophotometer you use. The NanoDrop instrument has a path length of 1 mm for UV–Vis measurement instead of the typical 1 cm. This is important for the concentration calculations. Check also that your measured value is in the linear range (can be checked by a calibration curve) otherwise you cannot use the Lambert–Beer law for calculating concentrations.
11. Viruses can back-mutate to their original sequence while passaged from plant to plant. In the case you recognize poor or no labeling at all, you should check the TMV particles for the desired mutations on the genomic level. To do so, isolate the RNA directly from the respective TMV particle preparation, do a reverse transcription PCR for a segment containing the CP gene (plus about 100 bp upstream and downstream) and screen the resulting DNA for the mutations by sequencing with primers annealing 50–100 bp upstream or downstream the CP gene borders, respectively.

12. In this experiment, TMV_{Lys} and TMV_{Cys} were also used as controls to show that only for the TMV_{DM} a yellow band can be seen if both dyes are used for labeling. For the same sample, there is also a green and a red band detected. These three bands show that some CP subunits are labeled with both dyes (yellow), others with a single dye only: either Atto 488 (green), or Atto 647 (red). Binding of both dyes to one CP also results in a larger electrophoretic mobility shift compared to unlabeled or only singly dye-labeled CPs. The band intensities recorded for this gel are not quantitative since the dyes are not equally excited at the excitation wavelength of the DR 195M transilluminator, and also fluorescence detection with the corresponding type of digital camera is not quantitative.
13. Calculating the degree of labeling is based on some assumptions pointed out in the following. It is assumed that the extinction coefficient is the same for the immobilized dye as for unbound dye, which might not be the case. The absorption of Atto 488 at the optimal excitation wavelength of 645 nm of Atto 647 is neglected, since it is very marginal. Otherwise a calculation would be impossible since no correction factor could be determined. This means that there is an error in the calculation which is for instance represented by the apparent DOL_{488} of 8.3% for TMV_{DM} labeled with Atto 647 only (Table 2). The measurement will therefore not provide you with exact values but with approximate values. They can effectively help to optimize the reaction conditions since changes in the DOL are reflected very well.

References

1. Bittner AM, Alonso JM, Górzny ML, Wege C (2013) Nanoscale science and technology with plant viruses and bacteriophages. In: Mateu MG (ed) Structure and physics of viruses: an integrated textbook, Subcellular biochemistry, vol 68. Springer Science+Business Media, Dordrecht, pp 667–702. https://doi.org/10.1007/978-94-007-6552-8_22. (ISSN: 0306-0225)
2. Wen AM, Steinmetz NF (2016) Design of virus-based nanomaterials for medicine, biotechnology, and energy. Chem Soc Rev 45(15):4074-4126. https://doi.org/10.1039/C5CS00287G
3. Mao C, Liu A, Cao B (2009) Virus-based chemical and biological sensing. Angew Chem Int Ed 48(37):6790–6810. https://doi.org/10.1002/anie.200900231
4. Lomonossoff GP, Evans DJ (2014) Applications of plant viruses in bionanotechnology. Plant Viral Vectors 375:61–87. https://doi.org/10.1007/82_2011_184
5. Lin B, Ratna B (2014) Virus hybrids as nanomaterials: methods and protocols. In: Methods in Molecular Biology, vol 1108. Humana Press, Springer, New York, NY
6. Singh P, Gonzalez MJ, Manchester M (2006) Viruses and their uses in nanotechnology. Drug Development Research 67(1):23–41. https://doi.org/10.1002/ddr.20064
7. Young M, Willits D, Uchida M, Douglas T (2008) Plant viruses as biotemplates for materials and their use in nanotechnology. Annual Review of Phytopathology 46:361–384. https://doi.org/10.1146/annurev.phyto.032508.131939
8. Luckanagul J, Lee LA, Nguyen QL, Sitasuwan P, Yang XM, Shazly T, Wang Q (2012) Porous alginate hydrogel functionalized with virus as three-dimensional scaffolds for bone differen-

tiation. Biomacromolecules 13(12):3949–3958. https://doi.org/10.1021/bm301180c

9. Bryksin AV, Brown AC, Baksh MM, Finn MG, Barker TH (2014) Learning from nature - novel synthetic biology approaches for biomaterial design. Acta Biomaterialia 10(4):1761–1769. https://doi.org/10.1016/j.actbio.2014.01.019
10. Culver JN, Brown AD, Zang F, Gnerlich M, Gerasopoulos K, Ghodssi R (2015) Plant virus directed fabrication of nanoscale materials and devices. Virology 479–480:200–212. https://doi.org/10.1016/j.virol.2015.03.008
11. Fan XZ, Pomerantseva E, Gnerlich M, Brown A, Gerasopoulos K, McCarthy M, Culver J, Ghodssi R (2013) Tobacco mosaic virus: a biological building block for micro/nano/biosystems. J Vac Sci Technol, A 31(5):050815. https://doi.org/10.1116/1.4816584
12. Koch C, Eber FJ, Azucena C, Förste A, Walheim S, Schimmel T, Bittner AM, Jeske H, Gliemann H, Eiben S, Geiger FC, Wege C (2016) Novel roles for well-known players: from tobacco mosaic virus pests to enzymatically active assemblies. Beilstein J Nanotechnol 7:605–612. https://doi.org/10.3762/bjnano.7.54
13. Culver JN (2002) Tobacco mosaic virus assembly and disassembly: determinants in pathogenicity and resistance. Annu Rev Phytopathol 40:287–308. https://doi.org/10.1146/annurev.phyto.40.120301.102400
14. Caspar DLD (1963) Assembly and stability of the tobacco mosaic virus particle. Advances in Protein Chemistry 18:37–121. https://doi.org/10.1016/S0065-3233(08)60268-5
15. Fromm SA, Bharat TAM, Jakobi AJ, Hagen WJH, Sachse C (2015) Seeing tobacco mosaic virus through direct electron detectors. Journal of Structural Biology 189(2):87–97. https://doi.org/10.1016/j.jsb.2014.12.002
16. Namba K, Pattanayek R, Stubbs G (1989) Visualization of protein-nucleic acid interactions in a virus - refined structure of intact tobacco mosaic-virus at 2.9-a resolution by X-Ray fiber diffraction. Journal of Molecular Biology 208(2):307–325. https://doi.org/10.1016/0022-2836(89)90391-4
17. Pomerantseva E, Gerasopoulos K, Chen XY, Rubloff G, Ghodssi R (2012) Electrochemical performance of the nanostructured biotemplated V2O5 cathode for lithium-ion batteries. Journal of Power Sources 206:282–287. https://doi.org/10.1016/j.jpowsour.2012.01.127
18. Wu ZY, Mueller A, Degenhard S, Ruff SE, Geiger F, Bittner AM, Wege C, Krill CE (2010) Enhancing the magnetoviscosity of ferrofluids by the addition of biological nanotubes. ACS Nano 4(8):4531–4538. https://doi.org/10.1021/nn100645e
19. Koch C, Wabbel K, Eber FJ, Krolla-Sidenstein P, Azucena C, Gliemann H, Eiben S, Geiger F, Wege C (2015) Modified TMV particles as beneficial scaffolds to present sensor enzymes. Front Plant Sci 6:1137. https://doi.org/10.3389/fpls.2015.01137
20. Fan XZ, Naves L, Siwak NP, Brown A, Culver J, Ghodssi R (2015) Integration of genetically modified virus-like-particles with an optical resonator for selective bio-detection. Nanotechnology 26(20):205501. https://doi.org/10.1088/0957-4484/26/20/205501
21. Baecker M, Koch C, Eiben S, Geiger F, Eber FJ, Gliemann H, Poghossian A, Wege C, Schoening MJ (2017) Tobacco mosaic virus as enzyme nanocarrier for electrochemical biosensors. Sens Actuators, B 238:716–722. https://doi.org/10.1016/j.snb.2016.07.096
22. Bruckman MA, Randolph LN, Gulati NM, Stewart PL, Steinmetz NF (2015) Silica-coated Gd(DOTA)-loaded protein nanoparticles enable magnetic resonance imaging of macrophages. J Mater Chem B 3(38):7503–7510. https://doi.org/10.1039/C5TB01014D
23. Shukla S, Eber FJ, Nagarajan AS, DiFranco NA, Schmidt N, Wen AM, Eiben S, Twyman RM, Wege C, Steinmetz NF (2015) The Impact of aspect ratio on the biodistribution and tumor homing of rigid soft-matter nanorods. Adv Health Mater 4(6):874–882. https://doi.org/10.1002/adhm.201400641
24. Smith ML, Lindbo JA, Dillard-Telm S, Brosio PM, Lasnik AB, McCormick AA, Nguyen LV, Palmer KE (2006) Modified tobacco mosaic virus particles as scaffolds for display of protein antigens for vaccine applications. Virology 348(2):475–488. https://doi.org/10.1016/j.virol.2005.12.039
25. Banik S, Mansour AA, Suresh RV, Wykoff-Clary S, Malik M, McCormick AA, Bakshi CS (2015) Development of a multivalent subunit vaccine against tularemia using tobacco mosaic virus (TMV) based delivery system. Plos One 10(6):e0130858. https://doi.org/10.1371/journal.pone.0130858
26. Kaur G, Wang C, Sun JA, Wang QA (2010) The synergistic effects of multivalent ligand display and nanotopography on osteogenic differentiation of rat bone marrow stem cells. Biomaterials 31(22):5813–5824. https://doi.org/10.1016/j.biomaterials.2010.04.017
27. Zan XJ, Feng S, Balizan E, Lin Y, Wang Q (2013) Facile method for large scale alignment of one dimensional nanoparticles and control over myoblast orientation and differentiation.

ACS Nano 7(10):8385–8396. https://doi.org/10.1021/nn403908k

28. Sitasuwan P, Lee LA, Bo P, Davis EN, Lin Y, Wang Q (2012) A plant virus substrate induces early upregulation of BMP2 for rapid bone formation. Integr Biol (Camb) 4(6):651–660. https://doi.org/10.1039/c2ib20041d
29. Eber FJ, Eiben S, Jeske H, Wege C (2013) Bottom-up-assembled nanostar colloids of gold cores and tubes derived from tobacco mosaic virus. Angewandte Chemie-International Edition 52(28):7203–7207. https://doi.org/10.1002/anie.201300834
30. Eber FJ, Eiben S, Jeske H, Wege C (2015) RNA-controlled assembly of tobacco mosaic virus-derived complex structures: from nanoboomerangs to tetrapods. Nanoscale 7(1):344–355. https://doi.org/10.1039/c4nr05434b
31. Taliansky ME, Kaplan IB, Yarvekulg LV, Atabekova TI, Agranovsky AA, Atabekov JG (1982) A study of TMV ts mutant Ni2519. II. Temperature-sensitive behavior of Ni2519 RNA upon reassembly. Virology 118(2):309–316. https://doi.org/10.1016/0042-6822(82)90350-6
32. Gallie DR, Plaskitt KA, Wilson TMA (1987) The effect of multiple dispersed copies of the origin-of-assembly sequence from TMV RNA on the morphology of pseudovirus particles assembled in vitro. Virology 158(2):473–476. https://doi.org/10.1016/0042-6822(82)90350-6
33. Eiben S, Stitz N, Eber F, Wagner J, Atanasova P, Bill J, Wege C, Jeske H (2014) Tailoring the surface properties of tobacco mosaic virions by the integration of bacterially expressed mutant coat protein. Virus Research 180:92–96. https://doi.org/10.1016/j.virusres.2013.11.019
34. Geiger FC, Eber FJ, Eiben S, Mueller A, Jeske H, Spatz JP, Wege C (2013) TMV nanorods with programmed longitudinal domains of differently addressable coat proteins. Nanoscale 5(9):3808–3816. https://doi.org/10.1039/c3nr33724c
35. Schneider A, Eber FJ, Wenz N, Altintoprak K, Jeske H, Eiben S, Wege C (2016) Dynamic DNA-controlled "stop-and-go" assembly of well-defined protein domains on RNA-scaffolded TMV-like nanotubes. Nanoscale 8:19853–19866. https://doi.org/10.1039/C6NR03897B
36. Royston E, Ghosh A, Kofinas P, Harris MT, Culver JN (2008) Self-assembly of virus-structured high surface area nanomaterials and their application as battery electrodes. Langmuir 24(3):906–912. https://doi.org/10.1021/la7016424
37. Yi H, Rubloff GW, Culver JN (2007) TMV microarrays: hybridization-based assembly of DNA-programmed viral nanotemplates. Langmuir 23(5):2663–2667. https://doi.org/10.1021/la062493c
38. Mueller A, Eber FJ, Azucena C, Petershans A, Bittner AM, Gliemann H, Jeske H, Wege C (2011) Inducible site-selective bottom-up assembly of virus-derived nanotube arrays on RNA-equipped wafers. ACS Nano 5(6):4512–4520. https://doi.org/10.1021/nn103557s
39. Butler PJ (1999) Self-assembly of tobacco mosaic virus: the role of an intermediate aggregate in generating both specificity and speed. Philos Trans R Soc Lond B Biol Sci 354(1383):537–550. https://doi.org/10.1098/rstb.1999.0405
40. Bruckman MA, Steinmetz NF (2014) Chemical modification of the inner and outer surfaces of tobacco mosaic virus (TMV). Virus Hybrids as Nanomaterials: Methods and Protocols 1108:173–185. https://doi.org/10.1007/978-1-62703-751-8_13
41. Bruckman MA, Kaur G, Lee LA, Xie F, Sepulveda J, Breitenkamp R, Zhang X, Joralemon M, Russell TP, Emrick T, Wang Q (2008) Surface modification of tobacco mosaic virus with "click" chemistry. Chembiochem 9(4):519–523. https://doi.org/10.1002/cbic.200700559
42. Wu LY, Zang JF, Lee LA, Niu ZW, Horvatha GC, Braxtona V, Wibowo AC, Bruckman MA, Ghoshroy S, zur Loye HC, Li XD, Wang Q (2011) Electrospinning fabrication, structural and mechanical characterization of rod-like virus-based composite nanofibers. Journal of Materials Chemistry 21(24):8550–8557. https://doi.org/10.1039/c1jm00078k
43. Witus LS, Francis MB (2011) Using synthetically modified proteins to make new materials. Acc Chem Res 44(9):774–783. https://doi.org/10.1098/rstb.1999.0405
44. McKay Craig S, Finn MG (2014) Click chemistry in complex mixtures: bioorthogonal bioconjugation. Chemistry & Biology 21(9):1075–1101. https://doi.org/10.1016/j.chembiol.2014.09.002
45. Wu FC, Zhang H, Zhou Q, Wu M, Ballard Z, Tian Y, Wang JY, Niu ZW, Huang Y (2014) Expanding the genetic code for site-specific labelling of tobacco mosaic virus coat protein and building biotin-functionalized virus-like particles. Chem Commun 50(30):4007–4009. https://doi.org/10.1039/c3cc49137d

46. Zhou K, Li F, Dai GL, Meng C, Wang QB (2013) Disulfide bond: dramatically enhanced assembly capability and structural stability of tobacco mosaic virus nanorods. Biomacromolecules 14(8):2593–2600. https://doi.org/10.1021/bm400445m
47. Miller RA, Presley AD, Francis MB (2007) Self-assembling light-harvesting systems from synthetically modified tobacco mosaic virus coat proteins. Journal of the American Chemical Society 129(11):3104–3109. https://doi.org/10.1021/ja063887t
48. Demir M, Stowell MHB (2002) A chemoselective biomolecular template for assembling diverse nanotubular materials. Nanotechnology 13(4):541–544. https://doi.org/10.1088/0957-4484/13/4/318. Pii S0957-4484(02)31463-6
49. Yi HM, Nisar S, Lee SY, Powers MA, Bentley WE, Payne GF, Ghodssi R, Rubloff GW, Harris MT, Culver JN (2005) Patterned assembly of genetically modified viral nanotemplates via nucleic acid hybridization. Nano Letters 5(10):1931–1936. https://doi.org/10.1021/nl051254r
50. Lee SY, Choi J, Royston E, Janes DB, Culver JN, Harris MT (2006) Deposition of platinum clusters on surface-modified tobacco mosaic virus. J Nanosci Nanotechnol 6(4):974–981. https://doi.org/10.1098/rstb.1999.0405
51. Lim JS, Kim SM, Lee SY, Stach EA, Culver JN, Harris MT (2010) Biotemplated aqueous-phase palladium crystallization in the absence of external reducing agents. Nano Lett 10(10):3863–3867. https://doi.org/10.1021/nl101375f
52. Yang C, Choi CH, Lee CS, Yi H (2013) A facile synthesis-fabrication strategy for integration of catalytically active viral-palladium nanostructures into polymeric hydrogel microparticles via replica molding. ACS Nano 7(6):5032–5044. https://doi.org/10.1021/nn4005582
53. Harder A, Dieding M, Walhorn V, Degenhard S, Brodehl A, Wege C, Milting H, Anselmetti D (2013) Apertureless scanning near-field optical microscopy of sparsely labeled tobacco mosaic viruses and the intermediate filament desmin. Beilstein J Nanotechnol 4:510–516. https://doi.org/10.3762/bjnano.4.60
54. Altintoprak K, Seidenstucker A, Welle A, Eiben S, Atanasova P, Stitz N, Plettl A, Bill J, Gliemann H, Jeske H, Rothenstein D, Geiger F, Wege C (2015) Peptide-equipped tobacco mosaic virus templates for selective and controllable biomineral deposition. Beilstein Journal of Nanotechnology 6:1399–1412. https://doi.org/10.3762/bjnano.6.145
55. Kadri A, Maiss E, Amsharov N, Bittner AM, Balci S, Kern K, Jeske H, Wege C (2011) Engineered tobacco mosaic virus mutants with distinct physical characteristics in planta and enhanced metallization properties. Virus Res 157(1):35–46. https://doi.org/10.1016/j.virusres.2011.01.014
56. Gooding GV Jr, Hebert TT (1967) A simple technique for purification of tobacco mosaic virus in large quantities. Phytopathology 57(11):1285
57. Green MR, Sambrook J (2012) In: Inglis J (ed) Molecular cloning a laboratory manual, vol 1, 4th edn. Cold Spring Harbor, New York, NY

Chapter 28

Drug-Loaded Plant-Virus Based Nanoparticles for Cancer Drug Delivery

Michael A. Bruckman, Anna E. Czapar, and Nicole F. Steinmetz

Abstract

Nature has designed nanosized particles, specifically viruses, equipped to deliver cargo to cells. We report the chemical bioconjugation and shape shifting of a hollow, rod-shaped tobacco mosaic virus (TMV) to dense spherical nanoparticles (SNPs). We describe methods to transform TMV rods to spheres, load TMV rods and spheres with the chemotherapeutic drug, doxorubicin (DOX), to deliver modified particles to breast cancer cells, and to determine the IC_{50} values of the plant virus-based drug delivery system.

Key words Tobacco mosaic virus (TMV), Viral nanoparticles (VNPs), Bioconjugation, Doxorubicin (DOX), Drug delivery, Breast cancer

1 Introduction

Cancer is among the leading causes of death in the USA and other developed countries [1]. Cancer treatment hinges on the use of harmful chemotherapeutic drugs that, while effective at eradicating tumor cells, also damage healthy cells and tissue. Nanotechnology offers the potential to improve therapeutic safety as well as efficacy, therefore improving quality of life and most importantly patient survival. Currently, FDA-approved nanoparticle-based drug delivery vehicles have been liposomal and micelle formulations [2]. Viral nanoparticles (VNPs) have been receiving increased interest for gene therapy and oncolytic virotherapy [3–5]. The VNPs described in this chapter, and studied and developed by our group, are those derived from plant viruses.

Plant VNPs offer a variety of advantages over other nanoparticle platforms—such as liposomes, inorganic, and polymeric systems—that include their general safety and bioavailability properties [6]. For example, the unmodified plant virus, tobacco mosaic virus (TMV), is cleared from the body quickly via the reticuloendothelial system and does not induce hemolysis or coagulation to occur

Christina Wege and George P. Lomonossoff (eds.), *Virus-Derived Nanoparticles for Advanced Technologies: Methods and Protocols*, Methods in Molecular Biology, vol. 1776, https://doi.org/10.1007/978-1-4939-7808-3_28,
© Springer Science+Business Media, LLC, part of Springer Nature 2018

[7]. Additionally, no systemic toxic response was noted in mice for doses up to 100 mg/kg of nanoparticles. However, because TMV and other VNPs are protein-based nanoparticles, they do generate an immune response. Strategies to overcome immune surveillance include polymer coatings, such as PEGylation [8]. Several sophisticated chemical and biological techniques [9, 10] are available to program the surface functionality and cargo delivery properties of VNPs. Some of these techniques will be described in this chapter.

Specifically, we will focus on TMV, a plant virus that has been studied for over 100 years. This particular rod-shaped plant virus, which forms a 300 × 18 nm-sized hollow nanotube, has been the focus for a number of breakthroughs in general virology and plant pathology throughout the years [11]. For example, TMV has received significant attention for its use as a nanosized building block for developing basic nanoparticle manipulation strategies such as self-assembly rules and bioconjugation protocols to finding application in a variety of scientific fields including electronics, sensing, light harvesting, cell growth, biomedical diagnostics, and drug delivery [12–15].

While shape-engineering is facilitated through bottom-up RNA-templated self-assembly, thermal shape-switching can also be applied to yield RNA-free spherical-shaped nanoparticles (SNPs). Because data indicate that shape is a critical design parameter to next-generation nanomedicine, studying TMV rods and SNPs side-by-side will help define design rules for drug delivery.

Development of SNPs for use is still in its infancy with new advancements in their bioconjugation and encapsulation properties only recently being made available [16]. A recent study, for example, identified methods to non-specifically bind proteins and epitopes to the SNP using electrostatic and hydrophobic interactions [16]. Additionally, we demonstrated how functional SNPs could be generated from functional TMV rods using a continuous flow nanomanufacturing system [17].

In this book chapter, we will present methods to load TMV rods and spheres with a common chemotherapeutic drug for treatment of breast cancer cells (MDA-MB-231). TMV and SNP nanoparticles were loaded with doxorubicin (DOX) using an EDC-mediated coupling strategy that reacts the free amine of DOX to the carboxylic acid side chains of glutamic acid and aspartic acid residues on TMV and SNPs. TMV has two surface functional carboxylic acids (GLU97 and GLU106) located along its interior channel and an additional 12 relatively unreactive carboxylic acids amino acids contained within its coat protein structure [18, 19]. Following transition to SNPs, it is expected (but not confirmed) that all available carboxylic acids participate in bioconjugation. A detailed description of protein-based nanoparticle characterization techniques is also included in this chapter. Following bioconjugation of TMV rods and SNPs with DOX, drug delivery studies targeting the breast cancer cell line, MB-MDA-231, is described. Overall this article provides an outline for the synthesis of drug-loaded VNPs and provides methods for characterization, including tissue culture evaluation.

2 Materials

All solutions are prepared using ultrapure water, referred to as H_2O. All buffers and TMV solutions are stored at 4 °C.

2.1 Thermal Transition of TMV Rods to Spheres (SNPs)

1. Tobacco mosaic virus (*see* **Note 1**).
2. Dialysis tubing (*see* **Note 2**).
3. Double distilled water.
4. 0.2 mL PCR tubes with caps (*see* **Note 3**).
5. Peltier thermal cycler.
6. Suggested centrifuges and rotors:
   (a) For large volumes (>1.5 mL): Beckman Coulter Optima L-90K Ultracentrifuge; Type 50.2 Ti rotor; Polypropylene tubes (Part # 41121703, Beckman).
   (b) For small volumes (<1.5 mL): Beckman Coulter Optima MAX-XP; TLA-110 rotor; Polypropylene tubes (Part # 357448, Beckman); Tube adapter (Part # 360951, Beckman).

2.2 Loading TMV Rods and Spherical Nanoparticles (SNPs) with DOX

All solutions are prepared freshly for each chemical reaction. Buffers are sterile filtered (using a 0.22 μm filter, e.g., Millipore) after being adjusted to the appropriate volume and pH. Buffers and solutions are stored at 4 °C (*see* **Note 4**).

2.2.1 EDC-Mediated Conjugation of DOX to Glutamic and Aspartic Acid Residues on TMV Rods and SNPs

1. 0.1 M (HEPES) buffer: Dissolve 23.8 g 4-(2-hydroxyethyl)-1-piperazineethanesulfonic acid (HEPES) in 800 mL H_2O. Adjust pH to 7.4 with 10 M NaOH, then dilute to 1 L with H_2O.
2. 0.1 M EDC stock solution: 15.5 mg/mL 1-ethyl-3-(3-dimethylaminopropyl)carbodiimide (EDC) in H_2O. Prepare immediately before use, EDC is very unstable in solution, EDC should be stored desiccated at −20 °C.
3. 0.1 M HOBt stock solution: 13.5 mg/mL hydroxybenzotriazole (HOBt) in DMSO. Prepared immediately before use, HOBt should be stored desiccated at −20 °C.
4. 0.1 M DOX stock solution: 58 mg/mL doxorubicin hydrochloride (DOX, Indofine Chemical Company) in DMSO.
5. 40% (w/v) sucrose in H_2O.
6. 0.01 M potassium phosphate buffer, pH 7.4 (referred to as 0.01 M KP buffer).

2.2.2 Encapsulation of DOX in SNPs

1. 0.2 mL PCR tubes with caps (*see* **Note 3**).
2. Peltier thermal cycler.
3. 0.01 M KP buffer.

2.3 Characterization of DOX Loaded VNPs

2.3.1 UV–Vis Absorbance Spectroscopy

1. A spectrophotometer capable of measuring absorbance at 260–570 nm.
   (a) We suggest the Thermo Scientific NanoDrop 2000 for individual sample measurements.
   (b) We suggest Corning™ Costar™ Black Clear-bottom plates (07200565) for 96-well plate absorbance measurements.

2.3.2 Scanning Electron Microscopy (SEM)

1. Silicon wafers (*see* **Note 5**).
2. Sputter coater with metal target (*see* **Note 6**).
3. Double-sided conductive tape.
4. Aluminum pin stub that fits the SEM.

2.3.3 SDS-PAGE

1. 4× SDS-PAGE sample buffer (Invitrogen).
2. 4–12% NuPAGE gels in SDS-PAGE gel doc station (Invitrogen).
3. 10× MOPS running buffer (Invitrogen).
4. Coomassie Blue gel staining solution: 1 g Coomassie R250, 100 mL glacial acetic acid, 400 mL methanol, 500 mL H_2O.
5. Gel destain solution: 200 mL methanol, 100 mL glacial acetic acid, 700 mL H_2O.
6. UV light (either be handheld or in a gel imaging unit).

2.4 Drug Delivery of DOX-VNPs to Cancer Cells

2.4.1 Cell Culture

1. MDA-MB-231 breast cancer cells.
2. Cell incubator maintained at 37 °C and 5% CO_2 humidified atmosphere.
3. RPMI-1640 growth medium (Corning).
4. Fetal bovine serum.
5. Penicillin–streptomycin, 10,000 Units/mL in 10 mM citrate buffer (ThermoFisher).

2.4.2 Cell Killing Assay of DOX-Loaded VNPs

1. Trypsin–EDTA (0.05% v/v), Phenol Red (10 mg/L) (ThermoFisher).
2. PBS for cell culture (pH 7.4, 300 mOsm/kg) (VWR).
3. 75 cm^2 cell culture polystyrene cell-culture-treated for optimal cell adherence with 0.2 μm filter tops (Corning).
4. 175 cm^2 cell culture treated flask polystyrene cell-culture-treated for optimal cell adherence with 0.2 μm filter tops (Corning).
5. Conical bottom 15 mL centrifuge tube (50 mL is also acceptable if required).
6. Hemocytometer.

7. Sterile, tissue culture treated for optimal cell adherence clear bottom polystyrene 96-well plate (Corning).
8. Multichannel pipette (10–100 μL).
9. MTT Cell Proliferation Assay (ATCC).
10. UV–Vis absorbance plate reader.

3 Methods

3.1 Thermal Transition of TMV Rods to Spheres (SNPs)

1. Dialyze TMV against water overnight at room temperature (*see* **Note 2**).
2. Measure concentration of TMV particles in solution using a UV spectrometer (*see* **Note 7**).
3. Dilute TMV in water to a concentration of 0.3 mg/mL.
4. Dispense 50 μL per PCR tube.
5. Place tubes in PCR heat cycler.
6. Run heat program once. Recommended heat program: 60 s at 25 °C, 90 s at 96 °C, 60 s at 25 °C.
7. Remove from heat cycler immediately and combine into a centrifuge tube.
8. Centrifuge at 160,000 × *g* for 2 h.
9. Discard supernatant.
10. Resuspend pellet in enough H_2O to achieve 1.5 mg/mL concentration, roughly 10 μL per PCR tube conversion.

3.2 Loading TMV Rods and SNPs with DOX

All reactions are done at 1 mg/mL TMV (or SNP) protein concentration at 1 mL total volume. Covalent immobilization of DOX to TMV rods and SNPs will be accomplished using an EDC coupling bioconjugation strategy.

3.2.1 EDC-Mediated Conjugation of DOX to TMV Rods and SNPs

DOX is conjugated to solvent exposed glutamic and aspartic acid residues on TMV rods and SNPs, using an EDC coupling bioconjugation strategy.

1. Add 100 μL TMV (at 10 mg/mL stock solution, *see* **Note 8**) to 840 μL 0.1 M HEPES buffer pH 7.4 at room temperature. Briefly mix solution by vortexing or by hand.
2. Add 34 μL DOX (60 eq per coat protein).
3. Add 8.5 μL HOBt (15 eq per coat protein) and mix.
4. Add 5.7 μL of EDC (10 eq) at three separate time intervals (total of 30 eq): when reaction starts, 6 h after starting, and 18 h after starting the reaction (*see* **Note 9**).
5. This reaction will proceed at room temperature for 24 h.

6. Purify the modified TMV samples by ultracentrifugation for 2.5 h at 160,000 × *g* over a 40% (w/v) sucrose cushion (*see* **Note 10**).
7. Resuspend the pellet in 0.01 M KP buffer and store at 4 °C.

3.2.2 Encapsulation of DOX Particles in SNPs

1. Add 30 μL dialyzed TMV (at 10 mg/mL stock solution) to 1445 μL water at room temperature.
2. Add 25 μL DOX (0.01 M, 10 eq per coat protein) and mix.
3. Dispense 50 μL per PCR tube.
4. Place in PCR heat cycler.
5. Run heat program once. Recommended heat program: 60 s at 25 °C, 90 s at 96 °C, 60 s at 25 °C.
6. Remove from heat cycler and combine into ultracentrifuge tube.
7. Purify the modified TMV samples by ultracentrifugation for 2 h at 4 °C at 160,000 × *g*.
8. Resuspend the pellet in enough 0.01 M KP buffer to achieve about 1 mg/mL final concentration and store at 4 °C (*see* **Note 11**).
9. Use immediately (within 6 h) for cell killing assay (*see* Subheading 3.4.2).

3.3 Characterization of DOX Loaded VNPs (See Fig. 1)

3.3.1 UV–Vis Absorbance Spectroscopy

1. Measure the absorbance of the solution (*from* Subheading 3.2.2, **step 8**) and note the readings at 260 nm (RNA), 280 nm (protein), and 480 nm (DOX) (*see* **Note 12**).
2. Use Beer's law ($A = \varepsilon c l$) to determine the DOX loading per TMV. Here, A is the absorbance, ε is the extinction coefficient [mL/(cm mg)], c is the concentration [mg/mL], and l is the path length [cm]. Expanding this equation gives:

$$\left(\frac{A(480)}{11{,}500}\right)\left(\frac{40{,}000{,}000}{\frac{A(260)}{6}}\right)=\frac{\#\text{DOX}}{\text{TMV}}.$$

Here, εDOX = ~11,500, MW TMV ~ 4.00 × 10^7 Da. Additionally, the ε TMV increases from 3 to 6 following modification (*see* **Note 13**).

3.3.2 Scanning Electron Microscopy

1. Dilute SNP sample (*from* Subheading 3.2.2, **step 8**) to 0.2 mg/mL (*see* **Note 14**).
2. Dry down 10 μL of sample onto a 5 × 5 mm diced silicon wafer (*see* **Note 15**).
3. When dry, flow 10 μL of water on the wafer for 30 s, then remove carefully with a pipette and allow the wafer to dry.

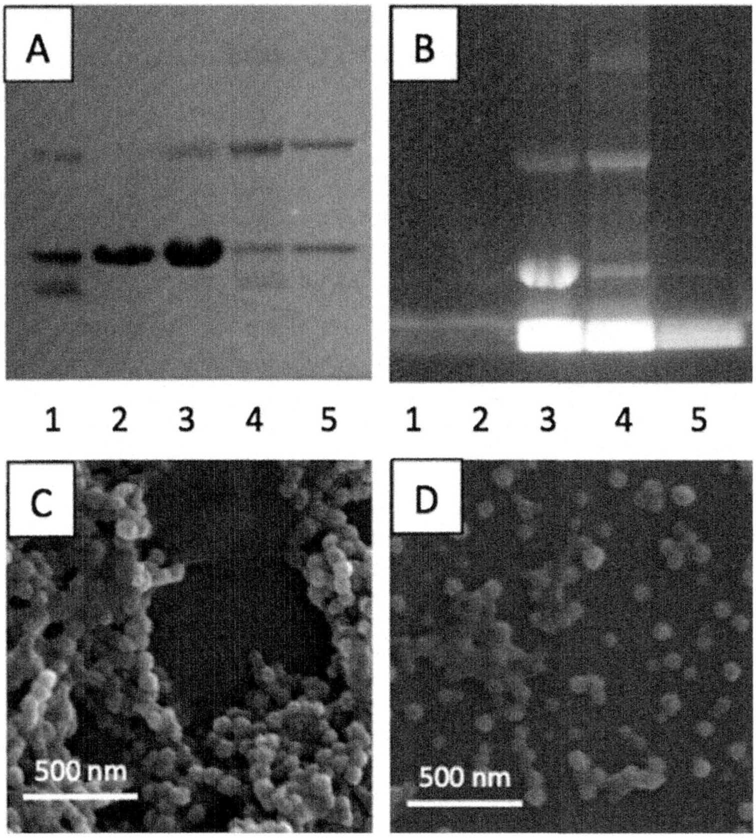

Fig. 1 (**a**) SDS-PAGE gel and (**b**) fluorescence image of: 1 SNPs; 2 TMV; 3 DOX-TMV; 4 DOX-SNP; 5 encapsulated DOX-SNP. Scanning electron images of (**c**) DOX-SNP and (**d**) encapsulated DOX-SNP

4. Sputter coat the silicon wafer with a thin layer of metal (30 s is sufficient).
5. Silicon wafer samples are mounted onto an aluminum pin stub using double-sided conductive tape.
6. Image samples using an available SEM. We obtained quality images operating at 5 kV.

3.3.3 SDS-PAGE

1. Dilute sample (*from* Subheading 3.2.2, **step 8**) to 1.5 mg/mL.
2. Mix 12 μL sample with 4 μL of 4× SDS PAGE sample buffer.
3. Heat to 100 °C for 5 min.
4. After allowing the sample to cool for 1 min, add 10 μL to an empty well in the denaturing 4–12% (w/v) NuPAGE gel.
5. Run gel in doc station at 200 V for 45 min in 1× MOPS running buffer.
6. When separation is complete, image gel under UV light to identify DOX locations.

7. Stain gel using Coomassie Blue gel stain for 1 h on rocker.
8. Destain gel using destaining solution until only protein bands remain. This may take 1–24 h depending on the gel being used. Destaining is best achieved on a rocker (*see* **Note 16**).

3.4 Drug Delivery of DOX-VNPs to Cancer Cells (See Fig. 2)

3.4.1 Cell Culture

1. MDA-MB-231 breast cancer cells are maintained at 37 °C in a 5% CO_2 humidified atmosphere. The cells are maintained in RPMI-1640 medium supplemented with 10% (v/v) heat-inactivated fetal bovine serum (FBS), and 1% (v/v) penicillin–streptomycin. All reagents were heated to 37 °C prior to use.
2. Cell growth medium was stored at 4 °C when not in use.
3. Cells were passaged every 2–3 days depending on confluency. Passage cells at a sub-cultivation ratio of 1:4.

3.4.2 Cell Killing Assay of DOX-Loaded VNPs (See Note 17)

Steps 1–12 describe cell seeding, **steps 13–17** describe particle incubation (drug delivery) and **steps 18–22** describe MTT cell proliferation Assay. Start experiment on a day where cells are confluent.

1. To carry out cell seeding discard growth medium and rinse cells with PBS.
2. Add 4–5 mL (for 175 mm^2 flask) trypsin–EDTA solution to flask and return flask to cell incubator for 5 min.

A)

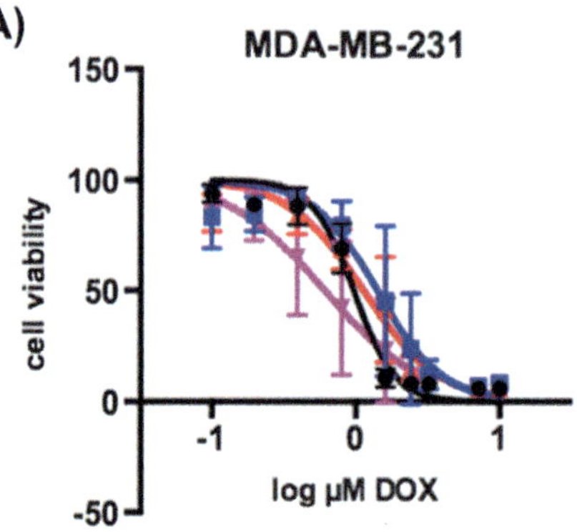

B)

MDA-MB-231	DOX	DOXTMV	DOXSNP	eDOXSNP
IC50 [μM]	0.97	1.41	1.13	0.61
STD	0.02	0.05	0.04	0.07

Fig. 2 (**a**) Cell viability of MDA-MB-231 cells after treatment with DOX (Black), DOX-TMV (blue), DOX-SNP (red), and encapsulated DOX-SNP (pink). Cell viability was determined after 72 h using MTT assay; dose dependency was evaluated testing a concentration range of 0.1–10 μM normalized to DOX. Data shown are averaged from three biological replicas (each performed in triplicate); error bars indicate the standard deviation. (**b**) IC50 values and standard deviation comparing DOX, DOX-TMV, DOX-SNP, and encapsulated DOX-SNP. Data were analyzed and graphed using Prism v6.0b software

3. Check cells under inverted microscope to confirm cell layer is detached.
4. Add 5 mL growth medium and aspirate cells by gently pipetting.
5. Collect cells and add 3 mL cells back into flask for culture and collect remaining cells in 15 mL conical bottom centrifuge tube.
6. Centrifuge cells at 500 × *g* for 5 min.
7. Collect and remove supernatant and replace with growth medium. Agitate cells to disperse.
8. Count cells using a hemocytometer (*see* **Note 18**).
9. Dilute cells to 100,000 cells/mL in growth medium.
10. Pipette 100 μL of cell suspension (10,000 cells total) to each well of a sterile, tissue culture-treated, 96-well clear bottom plate for planned experiment (*see* **Note 19**).
11. Place 96-well plate in cell incubator overnight.
12. After 24 h incubation, check wells with an inverted microscope to confirm cell attachment.
13. For nanoparticle incubation, remove growth medium using a pipette (we recommend a microchannel pipette).
14. Add 100 μL fresh growth medium and 10 μL of sample (from Subheading 3.2.2, **step 8**) to each well.
15. Place 96-well plate back into cell incubator for 24 h to allow particles to incubate with cells.
16. After 24 h incubation, remove cell growth medium from each well and rinse twice with PBS.
17. Add 100 μL fresh growth medium to each well and place plate back into cell incubator. Incubate cells for 24, 48, and/or 72 h.
18. Following cell incubation, add 10 μL MTT reagent to perform MTT cell proliferation assay.
19. Incubate for 4 h in cell incubator.
20. Add 100 μL Detergent reagent and leave overnight at room temperature.
21. Record absorbance at 570 nm using a plate reader.
22. Determine IC50 value by plotting results using software such as Excel or Prism.

4 Notes

1. Tobacco mosaic virus is obtained from *Nicotiana benthamiana* plants using a standard virus purification protocol. Briefly, this protocol relies on the following steps: organic

extraction, PEG/salt precipitation, and ultracentrifugation. In this book, *see* Chapters 23 and 27 for TMV purification.

2. This is a desalting step, therefore any MWCO is acceptable (3.5 kDa, 7 kDa, 10 kDa, or higher). We generally use 10 kDa. For small volumes (<0.5 mL) we suggest using Slide-A-Lyzer MINI Dialysis Devices. For large volumes (>0.5 mL) we suggest using SnakeSkin Dialysis tubing. Both products are available from ThermoFisher Scientific. Derivatives of these products are available from a variety of other commercial vendors.
3. We recommend tubes that are connected in strips of 8 for ease of pipetting and transferring. We also recommend using a multichannel pipette. Replace tubes after each transition of TMV to SNPs.
4. Buffers are kept until they appear to have particulates floating in solution.
5. We use precut 5 mm × 5 mm silicon wafers from Ted Pella.
6. Ensure that the sputter coater has one of the following targets installed: gold, palladium, or platinum.
7. The extinction coefficient for TMV is 3.0 mL/(mg·cm) at 260 nm.
8. We have found it difficult to achieve a high concentration of SNPs. Therefore, we suggest storing SNPs at a concentration of 5 mg/mL. Therefore, the amount of SNPs required for this reaction will be 200 μL and the amount of buffer drops by 100 μL.
9. If higher loading is desired, increase the amount of HOBt, EDC, and DOX while maintaining a molar ratio of 1:2:4 for HOBt–EDC–DOX.
10. A sucrose cushion is prepared in a 25 mL tube by adding 21 mL phosphate buffer, then carefully adding the reaction solution (1 mL) to the bottom of the tube with a glass pipette, then slowly and carefully adding 3 mL (roughly two full glass pipettes) 40% (w/v) solution of sucrose to the bottom of the centrifuge tube. Be careful not to disturb the cushion after it is made.
11. To speed up resuspension of DOX-SNPs, use the following techniques: sonication, aspiration with pipette, and extended mixing with a vortex. Sonication is the most effective method to quickly disperse SNPs.
12. The A_{260}/A_{280} for intact native TMV is 1.2. Modification with doxorubicin will increase this value. Therefore, a protein assay, such as the Modified Lowry's Protein Assay (ThermoFisher), may be required to determine TMV concentration.

13. The extinction coefficient of DOX modified TMV/SNP was calculated based on real experiments correlating the absorbance spectrum of TMV to protein concentration results obtained by a Modified Lowry Protein Assay (ThermoFisher).
14. SEM is most effective when imaging SNPs. TMV is difficult to image using SEM, however it can be achieved if a suitable TEM is not available. We suggest drying down a 0.1–0.5 mg/mL solution of SNPs, but higher or lower concentrations can be used if required.
15. Pay attention to the effect that drying conditions have on sample quality. We often flow a light stream of air over the top of the sample to reduce drying time. This can be achieved by holding a tube streaming air over the samples or placing on the sash of a chemical hood.
16. Destaining can be accelerated by adding Kimwipes® to the staining solution while on the rocker.
17. For statistically significant results, we suggest the following protocol; test cell killing in triplicates on three separate passages.
18. To obtain the most accurate cell count, aim to dilute cells to about 1,000,000 cells/mL for counting on the hemocytometer.
19. Initial tests for cell killing assay should be focused on optimizing the appropriate number of cells to obtain the best signal.

Acknowledgments

This work was supported in part by a grant from the National Science Foundation (DMR 1452257 to N.F.S.).

References

1. Group, U.S.C.S.W. 2015. vol. 2016 (U.S. Department of Health and Human Services, Centers for Disease Control and Prevention and National Cancer Institute)
2. Marchal S, Hor AE, Millard M, Gillon V, Bezdetnaya L (2015) Anticancer drug delivery: an update on clinically applied nanotherapeutics. Drugs 75:1601–1611
3. Yang Y et al (2015) Systemic delivery of an oncolytic adenovirus expressing decorin for the treatment of breast cancer bone metastases. Human gene therapy 26:813
4. Patil S, Rao RS, Majumdar B (2015) Clinical trials with oncolytic viruses: current and future prospects. The journal of contemporary dental practice 16:i–ii
5. Giacca M, Zacchigna S (2012) Virus-mediated gene delivery for human gene therapy. Journal of controlled release 161:377–388
6. Pokorski JK, Steinmetz NF (2011) The art of engineering viral nanoparticles. Mol Pharm 8:29–43
7. Bruckman MA et al (2014) Biodistribution, pharmacokinetics, and blood compatibility of

native and PEGylated tobacco mosaic virus nano-rods and -spheres in mice. Virology 449:163–173
8. Lee KL et al (2015) Stealth filaments: polymer chain length and conformation affect the in vivo fate of PEGylated potato virus X. Acta biomaterialia 19:166–179
9. Aljabali AA, Shukla S, Lomonossoff GP, Steinmetz NF, Evans DJ (2013) CPMV-DOX delivers. Molecular Pharmaceutics 10:3–10
10. Wen AM, Le N, Zhou X, Steinmetz NF, Popkin DL (2015) Tropism of CPMV to professional antigen presenting cells enables a platform to eliminate chronic infections. ACS Biomaterials Science & Engineering 1:1050–1054
11. Klug A (1999) The tobacco mosaic virus particle: structure and assembly. Philos Trans R Soc Lond B Biol Sci 354:531–535
12. Koch C et al (2015) Modified TMV particles as beneficial scaffolds to present sensor enzymes. Frontiers in Plant Science 6:1137
13. Bromley KM, Patil AJ, Perriman AW, Stubbs G, Mann S (2008) Preparation of high quality nanowires by tobacco mosaic virus templating of gold nanoparticles. Journal of Materials Chemistry 18:4796–4801
14. Miller RA, Presley AD, Francis MB (2007) Self-assembling light-harvesting systems from synthetically modified tobacco mosaic virus coat proteins. Journal of the American Chemical Society 129:3104–3109
15. Yildiz I, Shukla S, Steinmetz NF (2011) Applications of viral nanoparticles in medicine. Current opinion in biotechnology 22:901–908
16. Bruckman MA, Czapar AE, VanMeter A, Randolph LN, Steinmetz NF (2016) Tobacco mosaic virus-based protein nanoparticles and nanorods for chemotherapy delivery targeting breast cancer. J Control Release 231:103
17. Bruckman MA, VanMeter A, Steinmetz NF (2015) Nanomanufacturing of tobacco mosaic virus-based spherical biomaterials using a continuous flow method. ACS Biomater Sci Eng 1:13–18
18. Czapar AE, Steinmetz NF (2016) In: Lu Z-R, Sakuma S (eds) Nanomaterials in pharmacology. Springer, New York, NY, pp 65–85
19. Bruckman MA, Steinmetz NF (2014) Chemical modification of the inner and outer surfaces of Tobacco Mosaic Virus (TMV). Methods Mol Biol 1108:173–185

Chapter 29

Construction of Artificial Enzymes on a Virus Surface

Chunxi Hou and Junqiu Liu

Abstract

Combination of artificial enzyme design and self-assembly strategies leads to a novel way to construct supramolecular enzymes. To address this challenge, auxotrophic expression systems show great potential because they can introduce nonnatural catalytic groups into the subunits of protein assemblies. Among nonnatural amino acids, selenocysteine is the catalytic group of glutathione peroxidase (GPx). With the aid of computer simulation, we have incorporated selenocysteine into natural protein assemblies such as tobacco mosaic virus (TMV) and ferritin by cysteine auxotrophic technology, resulting in the conversion of TMV and ferritin into supramolecular enzymes.

Key words Artificial enzyme, Tobacco mosaic virus, Ferritin, Self-assembly, Glutathione peroxidase

1 Introduction

Engineering supramolecular enzymes is especially attractive in the fields of chemistry and materials science [1]. Recently, supramolecular enzymes have been developed successfully by using natural protein assemblies as enzyme scaffolds. Viruses that can self-assemble into well-defined nanoparticles of high order are perfect models. Once enzyme functions have been installed on the coat proteins, followed by naturally inherent self-assembly, these protein assemblies will become desirable supramolecular enzymes with numerous catalytic centers exposed on the periodic structures [2–5]. So far, auxotrophic expression technology is a widely applied technique for the development of supramolecular enzymes owing to the ability to incorporate catalytic groups into protein monomers [6].

Glutathione peroxidase (GPx) was the first selenium-containing enzyme found in mammals [7]. Based on its structure and catalysis, the catalytic selenocysteine of GPx has been introduced into a variety of scaffolds ranging from small molecules,

Christina Wege and George P. Lomonossoff (eds.), *Virus-Derived Nanoparticles for Advanced Technologies: Methods and Protocols*, Methods in Molecular Biology, vol. 1776, https://doi.org/10.1007/978-1-4939-7808-3_29,
© Springer Science+Business Media, LLC, part of Springer Nature 2018

peptides, proteins, and supramolecular polymers by using a variety of techniques. Here, we describe an approach for the construction of glutathione peroxidase (GPx) onto the nanodisk/nanotube structure of tobacco mosaic virus (TMV) [8–10]. Natural GPx has a catalytic selenocysteine moiety and a substrate binding site for glutathione. After modeling, docking, and analysis, we have chosen site142 of the TMV coat protein (TMVcp) to design and install the catalytic group, and site149 to incorporate the binding group arginine to enhance substrate binding. The selenium-containing TMVcp mutants are first dispersed in a basic solution of pH 8.0 in the presence of DTT, in an attempt to obtain protein monomers. In order to obtain disk/tube-like protein assemblies, we perform a buffer exchange to pH 7.0 and pH 5.5 solutions, respectively, and keep them at 4 °C for 2 days to ensure the growth of 20S disks and tubes, respectively. Eventually, self-assembling supramolecular enzymes with excellent catalytic ability to protect cells from oxidative damage have been prepared, as demonstrated successfully [11, 12].

2 Materials

The reagents for enzymatic analysis are all from Sigma. The reagents for the determination of GPx activity include: phenylmethylsulfonyl fluoride (PMSF), glutathione reductase (GR), reduced glutathione (GSH), and nicotinamide adenine dinucleotide phosphate (NADPH). All other reagents are of analytical grade and may be obtained for example from Beijing Chemical Plant. The ultrapure water in this study is prepared by a Millipore water purification system (Merck).

2.1 Construction of Mutant TMVcp for Mimicking a GPx Catalytic Center

Software for enzymatic mimicking: To find the catalytic center, we use the software AutoDock and PyMOL [13–15], followed by comparison with the crystal structure of TMVcp.

1. TMVcp gene: The gene is from plasmid pET20b-TMVcp [16], a gift from Professor M. B. Francis (University of California, Berkeley, USA).
2. Reagents for molecular biology: PrimerSTAR HS DNA polymerase, Agarose Gel DNA Purification Kit, and DNA Fragment Purification Kit (all from Takara), Dpn I (Fermentas).
3. All the primers are designed by software primer 5.0 [17] or DNAman [18]:
4. Material for insertion of TMVcp142Cys and TMVcp142 Cys149Arg into plasmids (of choice) via restriction and ligation.

Primers for the project	
Primers for C27G and C123G exchange:	
PU27	5′-AACCTGGGCACCAATGCGCT-3′
PD27	5′-GCATTGGTGCCCAGGTTAAT-3′
PU123	5′-TGGCGATTCGCGGTGCCATC-3′
PD123	5′-GGCACCGCGAATCGCCACGG-3′
Primers for installation of catalytic center (TMVcp142Cys):	
PU142	5′-GTTGCAGCTTTGAAAGCAG-3′
PD142	5′-TCAAAGCTGCAACGGTTATA-3′
Primers for installation of catalytic center with improved substrate binding TMVcp142Cys149Arg:	
PU149	5′-GCAGCCGCCTGGTGTGG-3′
PD149	5′-ACCAGGCGGCTGCTGCT-3′

5. Auxotrophic system: Strain BL21cysE51 (Professor A. Boeck, University of Munich, Germany; competent cells). The cysteine auxotrophic strain is constructed by insertion of the *cysE51* allele from JM39/2 into *E. coli* strain BL21 [19].
6. Material for transformation of BL21cysE51.

2.2 Expression of TMVcp Mutant Se-TMVcp 142Cys149Arg Mimicking GPx

1. Auxotrophic BL21cysE51 cells containing TMVcp142Cys149Arg gene-construct (*see* **Note 1**).
2. LB medium (10 g/L Bacto tryptone, 5 g/L yeast extract, 10 g/L NaCl) supplemented with 50 μg/mL ampicillin and 50 μg/mL kanamycin.
3. 1000× chloramphenicol storage solution: 10 mg/mL.
4. 1000×isopropy-β-D-thiogalactoside (IPTG) storage solution: 1 M.
5. 50 mM phosphate buffer (pH 7.4), 150 mM NaCl, cool to 4 °C.
6. M9 cultivation medium (inorganic salts 6.0 g/L Na_2HPO_4, 3.0 g/L KH_2PO_4, 0.5 g/L NaCl, 0.2 g/L NH_4Cl, 0.241 g/L $MgSO_4$, 20 normal amino acids (100 mg/L each), and 20 g/L glucose) supplemented with 50 μg/mL ampicillin and 50 μg/mL kanamycin (*see* **Notes 2** and **3**).
7. M9 production medium (inorganic salts 3.0 g/L K_2HPO_4, 1.0 g/L NH_4Cl, 2.0 g/L NaAc, 10.0 mg/L $CaCl_2$, 429.0 mg/L MgAc, and 3.6 g/L $FeCl_2$; 20 normal amino acids (100 mg/L each); 50 mg/L DL-selenocysteine; four kinds of bases (guanine, thymine, adenine, and cytosine, 50 mg/L each); 100 μg/L biotin; 4 ‰ (v/v) glycerol; 20 g/L

glucose), supplemented with 50 μg/mL ampicillin, 50 μg/mL kanamycin, and 400 μg/mL rifampicin.

8. Equipment for SDS-PAGE.
9. Size exclusion chromatography: Sephadex G25 column and G75 column (Amersham Pharmacia Biotech).
10. 50 mM PB buffer (storage solution) (pH 7.0, containing 1 mM EDTA): Take 0.305 g of $NaH_2PO_4 \cdot 2H_2O$ and 1.09 g of $Na_2HPO_4 \cdot 12H_2O$ into a 100-mL volumetric flask. After mixing them with 100 mL water, dissolve 3.72 mg EDTA into the solution.

2.3 Assembly and Identification of Se-TMVcp 142Cys149Arg Tubes

1. 20 mM Tris–HCl buffer (pH 8.0).
2. Dialysis tubes with 8000 Da MWCO (*see* **Note 4**).
3. HPLC: Equilibrate Superdex Gel Filtration 200 columns for 2 h, load and inject 20 μL samples onto these equilibrated size exclusion columns, set the flow rate at 1 mL/min, run each sample for 30 min and repeat three times, if necessary; collect the fractions at the retention time of nanotubes (*see* **Note 5**).
4. 200 mM PB buffer (pH 7.0): Weigh 1.22 g of $NaH_2PO_4 \cdot 2H_2O$ and 4.37 g of $Na_2HPO_4 \cdot 12H_2O$ in a 100-mL flask, adjust to this final volume with water and dissolve. Then put 3.72 mg EDTA into the prepared solution.
5. 200 mM PB buffer (pH 5.5): Weigh 2.92 g of $NaH_2PO_4 \cdot 2H_2O$ and 0.466 g of $Na_2HPO_4 \cdot 12H_2O$ in a 100-mL flask, adjust to this final volume with water and dissolve. Then put 3.72 mg EDTA into the prepared solution.
6. Silicon plate, pretreated: Cut a 6 inch silicon plate into small pieces (8 mm × 8 mm, square). Then treat the small silicon plates with hydrogen peroxide/vitriol as follows: Silicon plates are first immersed in a solution of potassium permanganate overnight. Then add the silicon plates to a solution of hydrogen peroxide (70 mL)/vitriol (30 mL) and heat at 80 °C for 20 min. Subsequently, put the silicon plates into isopropanol for sonication for 7 min, then put the silicon plates into toluene and sonicate for 7 min, then put them into methanol and sonicate for 7 min. Finally, put the silicon plates in ultrapurified water for sonication for 7 min and store at room temperature.
7. Atomic force microscopy (AFM): All AFM imaging of nanotube samples is carried out on a Nanoscope III controller (Veeco Metrology, Santa Barbara, CA, USA). Drop the samples (1 μM) onto freshly treated silicon plates, then wash by purified water and dry by N_2 blowing. The imaging is done by tapping mode AFM.

8. Transmission electron microscopy (TEM): All TEM images are recorded by a JEM-2100F instrument with an accelerating voltage of 120 kV. Drop 5 μL samples onto Formvar-carbon-coated copper grids and keep for 1 min. Then, wash specimens by using a drop of deionized water and stain by adding a drop of 4% phosphotungstate. After drying for more than 4 h, subject samples to TEM imaging.

2.4 Enzyme Analysis of GPx-Like TMVcp Mutant

1. 10× H_2O_2 storage solution (2 mL, 5 mM; *see* **Note 6**): Take 1.0 μL of 30% H_2O_2 and mix with water to make 2 mL of storage solution. You will also need 2.5 and 10 mM H_2O_2 solutions.
2. 10×GSH storage solution (2 mL, 10 mM): Weigh 6.146 mg GSH (reduced form, molecular weight of 307.3 g/mol) and dissolve it in 2 mL PB buffer (pH 7.0) (*see* Subheading 2.2, **item 10**). It is used for preparation of storage solution of 2.5 and 5 mM GSH.
3. 10×NADPH storage solution (2 mL, 2.5 mM): Dissolve 4.167 mg NADPH in 2 mL of 1% $NaHCO_3$ solution (β-nicotinamide adenine dinucleotide phosphate, reduced tetra(cyclohexylammonium) salt, NADPH).
4. 1% $NaHCO_3$ solution: Dissolve 0.05 g $NaHCO_3$ in 5 mL deionized water.
5. 10×glutathione reductase storage solution (1 mL): Add 22 μL glutathione reductase into 1 mL PB buffer (pH 7.0) (*see* Subheading 2.2, **item 10**).
6. UV–Vis spectrophotometer: A SHIMADZU UV-2450 spectrophotometer is used for determination of GPx activities according to the absorbance at 340 nm.

2.5 Antioxidative Analysis of GPx Mimic of Se-TMVcp 142Cys149Arg Disks or Tubes

1. Bovine heart: 20 g of bovine heart is freshly prepared with connective tissue removed.
2. HEPES buffer, pH 7.4: Weigh 23.8 g HEPES and dissolve it in 90 mL deionized water. Use 1 M NaOH to adjust the pH to 7.4. Add water to reach the final volume of 100 mL and filter it for sterilization. Keep at 4 °C or store at −20 °C for long-term storage.
3. Filter: 200 mesh copper sieve.
4. Homogenizer.
5. Mitochondria preparation: Clean bovine heart and chop it into small fragments. Suspend bovine heart in HEPES buffer supplemented with 0.25 M sucrose and put into homogenizer to homogenize for 15 min intermittently. The whole process is kept at 4 °C to protect bovine heart from oxidation. Filtrate the samples to remove debris and store at −20 °C.

6. 20× Mitochondria storage solution (10 mg/mL): Adjust the mitochondria concentration to 10 mg/mL. Store at 4 °C.
7. Bradford reagent: The Bradford method is used for determination of protein concentration by using BSA as a standard.
8. 50 mM Tris–HCl buffer (pH 7.4): Add 150 mL water to a 500 mL glass beaker. Weigh 1.21 g of Tris base and transfer to the beaker. Adjust the pH with HCl. Make up to 200 mL with water and store at 4 °C.
9. 2.5× KCl storage solution (0.75 M): Weigh 1 g of KCl and use 7 mL solution of 50 mM Tris–HCl buffer, pH 7.4 to dissolve. Store at 4 °C.
10. 10× GSH storage solution (10 mM): Weigh 6.14 mg of glutathione (GSH) and transfer to a centrifuge tube containing 2 mL of water. Store at 4 °C (about stability, *see* **Note 7**).
11. 10× GSSG storage solution (30 mM): Weigh 36.7 mg of oxidized glutathione (GSSG) and dissolve it in 2 mL water. Store at 4 °C for <1 week before use.
12. 10× Vitamin C (Vc) storage solution (5 mM): Weigh 2 mg of Vc and dissolve it in 2 mL water. Store at 4 °C.
13. 10× $FeCl_2$ storage solution (5 mM): Take 6.3 mg $FeCl_2$ into a 10-mL centrifuge tube and add 10 mL deionized water.
14. 700 g/L trichloroacetic acid (TCA) (TCA is a hazardous compound, please refer to **Note 8** for important safety precautions).
15. 5 g/L thiobarbituric acid (TBA).
16. 20 mM PB buffer (pH 8.0 and 400 mL): Dissolve 0.165 mg $NaH_2PO_4 \cdot H_2O$, 6.784 mg $Na_2HPO_4 \cdot 2H_2O$, 6.16 mg DTT in 400 mL H_2O_2. DTT should be added immediately before use. Store at 4 °C.
17. 200 mM PB buffer (pH 7.0): Dissolve 1.22 mg $NaH_2PO_4 \cdot H_2O$, 4.37 mg $Na_2HPO_4 \cdot 2H_2O$, 6.16 mg DTT in 400 mL H_2O.
18. 200 mM PB buffer (pH 5.5): Dissolve 2.92 mg $NaH_2PO_4 \cdot H_2O$, 0.466 mg $Na_2HPO_4 \cdot 2H_2O$, 6.16 mg DTT in 400 mL H_2O. If necessary, use 1 M HCl to adjust the pH value.
19. Spectrophotometer (to detect absorbance at 520 and 532 nm).

3 Methods

3.1 Construction of a Catalytic Center of GPx on TMVcp

1. Computer simulation for conversion of TMVcp to into a GPx mimic is conducted according to the method described by Seeliger et al. (*see* Figs. 1 and 2) [13]. In order to achieve

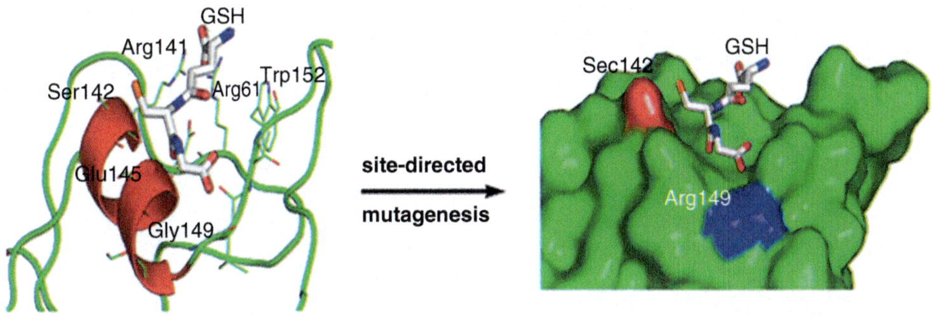

Fig. 1 Computer simulations for the construction of GPx mimicking Se-TMVcp142Cys149Arg. Site142 that locates in the α helix of the TMVcp is an ideal position to incorporate the catalytic selenocysteine of the GPx group. The native Arg141 and the introduced Arg149 (blue, point mutation) can form salt bridges with GSH to enhance substrate binding and stabilize the reactive intermediates. Several important surface amino acids nearby (Arg141, Arg61, Glu145, Gly149, and Trp152), involved in binding and catalysis, are shown. In the simulation, substrate GSH can bind to the shallow pocket and react with selenocysteine (red, point mutation). Adapted with permission from ref. 11; Copyright (2012) the American Chemical Society

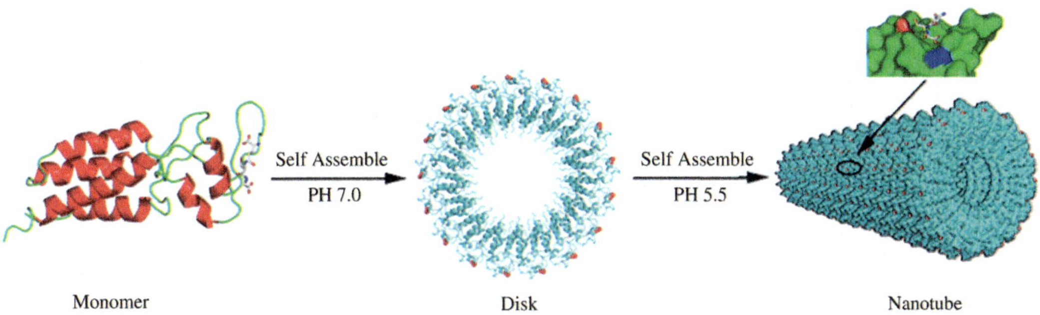

Fig. 2 Schematic representation of the assembly of an artificial supramolecular nanoenzyme from GPx mimicking Se-TMVcp142Cys149Arg. This is designed by computer simulation to function as an artificial GPx mimic after expression in a cysteine auxotrophic system. Self-assembly can be regulated by salt and pH values (pH 5.5 and pH 7.0), GPx active centers can be successfully exposed on the outer rim or surface of Se-TMVcp142Cys149Arg nanodisks or nanotubes, assembled from 34 or more than a thousand coat protein monomers, respectively. Adapted with permission from ref. 11; Copyright (2012) the American Chemical Society

cooperative catalysis, we also carried out simulation on ferritins (*see* Figs. 3 and 4).

2. Analyze the docking poses, hydrogen acceptors, hydrogen honors, and docking score, and confirm that they should be consistent with natural GPx.
3. C27 is the natural cysteine that is buried in wild type TMV, whereas C123 is the mutant RNA binding site artificially introduced by professor Francis (*see* Subheading 2.1, **item 1**). A necessary step is to remove both cysteine codons because cysteine codons could be substituted by selenocysteines in auxotropic systems, which may undermine the analysis of the activities of target selenocysteines.

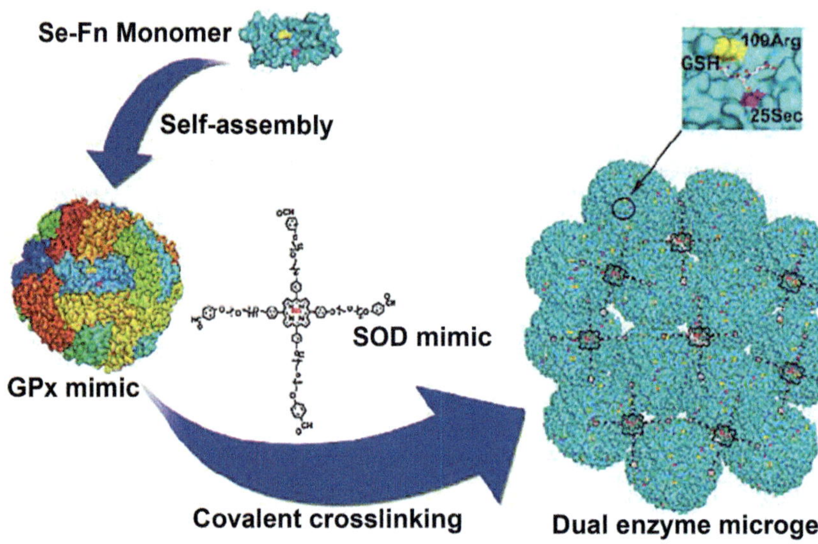

Fig. 3 Dual enzyme microgel with selenoferritin as GPx mimic and porphyrin derivative as SOD mimic. Adapted with permission from ref. 12; Copyright (2013) Wiley

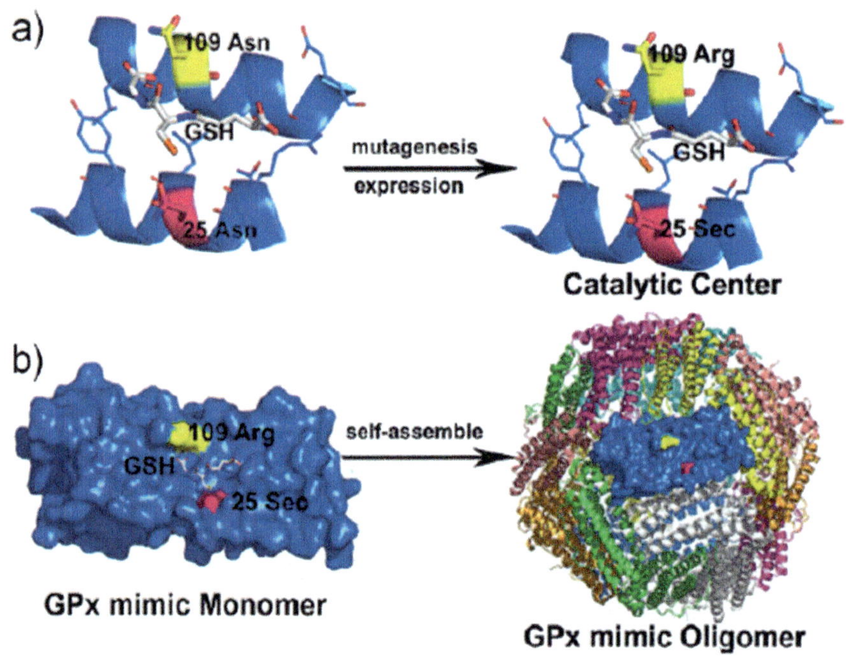

Fig. 4 (**a**, **b**) Schematic representation of the construction of Se-Fn and the assembly of the monomeric GPx mimic into an oligomer. Adapted with permission from ref. 12; Copyright (2013) Wiley

4. Substitute the two cysteines (C27 and C123) in the TMVcp variant used in this protocol (or the only C27 of wild-type TMVcp) with glycines by site-directed mutation PCR using the primers for C27G (PU27 and PD27) and for C123G (PU123 and PD123).

5. According to the results obtained by enzyme simulation, we choose site 142 as a position for the installation of the catalytic center. The TMVcp142Cys gene is obtained by site-directed mutagenesis using the primers PU142 and PD142 with plasmid pET-TMVcp as template. In order to improve substrate binding, another mutant TMVcp142Cys149Arg gene was obtained by site-directed mutagenesis using primers PU149 and PD149 with TMVcp142Cys as template.
6. Transfer two plasmids encoding TMVcp142Cys and TMVcp142Cys149Arg into BL21cysE51 competent cells and store stocks of correct clones at −80 °C. The procedure is done according to Green and Sambrook [20].

3.2 Expression of GPx-Like TMVcp Mutant (Se-TMVcp 142Cys149Arg)

A cysteine auxotrophic expression system is used for substituting cysteine with selenocysteine, thus converting TMVcp142Cys149Arg into Se-TMVcp142Cys149Arg.

1. Grow TMVcp142Cys149Arg gene-containing monoclonal BL21cysE51 cells in 20 mL LB medium supplemented with ampicillin (50 μg/mL) and kanamycin (50 μg/mL) at 37 °C overnight.
2. Inoculate the overnight culture into 1.0 L of M9 cultivation medium. Adjust the OD_{600} to the value of 0.1 and continue shaking at 37 °C.
3. As soon as an OD_{600} of 1.0 is reached, chloramphenicol is added to a final concentration of 10 μg/mL. In this step, chloramphenicol is used to inhibit translation of proteins owing to the absence of selenocysteine in medium.
4. Ten minutes later, add 1 mL of IPTG to a final concentration of 1 mM and incubate at 37 °C for 20 min.
5. Centrifuge the cells at 9200 × *g* and wash the pellet three times with cold 50 mM phosphate buffer pH 7.4 (containing 150 mM NaCl). Transfer the cleaned cells to M9 production medium (containing selenocysteine).
6. Agitate the culture flasks at a low velocity of 120 rpm at 37 °C. Low agitating velocity allows DL-selenocysteine to enter into cells slowly, thus could avoid adverse effects of DL-selenocysteine.
7. Monitor the expression level of Se-TMVcp142Cys149Arg and Se-TMVcp142Cys during the culture period by SDS-PAGE of samples taken at different time points (*see* Fig. 5).
8. After 7 h of culture, harvest cells by centrifugation of the culture at 9200 × *g* for 10 min.
9. Subject cells to ultrasonic crushing for 15 min. After centrifugation at 23,000 × *g* for 45 min, subject the supernatant to DEAE chromatography for protein purification. Use 50, 100

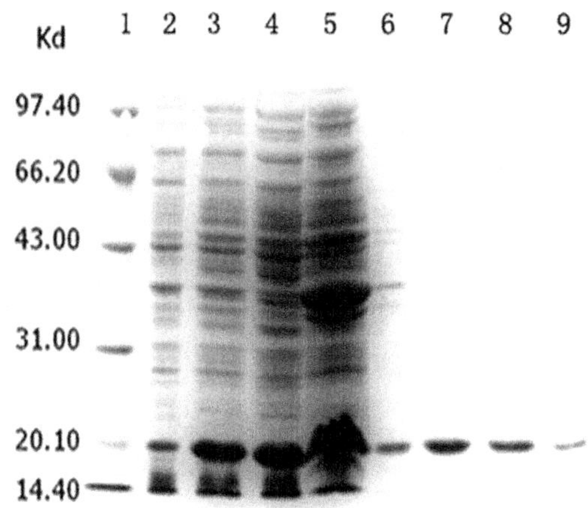

Fig. 5 SDS-PAGE after expression and purification of TMVcp142Cys149Arg. 1: Protein Marker. 2: Lysates before IPTG induction. 3: Lysates after expression of TMVcp. 4: Supernatant after expression of TMVcp. 5: Sediment after expression of TMVcp. 6 and 7: Purified TMVcp. 8 and 9: Purified TMVcp142Cys149Arg. Adapted with permission from ref. 11; Copyright (2012) the American Chemical Society

and 200 mM NaCl to elute, and collect the 200 mM eluted fractions identified according to the SDS-PAGE result.

10. Dialyse against 50 mM PB buffer (pH 7.0) and adjust to a final concentration of 0.15 mg/mL to store at −80 °C. Background spectra in the absence of protein are measured and subtracted.

3.3 Assembly and Identification of Se-TMVcp 142Cys149Arg Tubes

1. For dispersion, first exchange the Se-TMVcp142Cys149Arg solution (0.15 mg/mL, use 50 μL) into 20 mM PB buffer (pH 8.0) by dialysis overnight to obtain Se-TMVcp142Cys149Arg monomers. Generally, HPLC chromatography (KNAUER, Germany) is used to monitor all the distribution and assembly procedure.
2. In order to assemble disks, dialyze 50 μL of monomer solution against 200 mM PB buffer (pH 7.0) (*see* Subheading 2.5, **item 17**) at 4 °C for 1–2 days by using a 8000 Da cut-off membrane.
3. In order to assemble tubes, dialyze 50 μL of monomer solution against 200 mM PB buffer (pH 5.5) (*see* Subheading 2.5, **item 18**) by using a 8000 Da cutoff membrane at 4 °C for 4 days.
4. For structural determination, drop 5 μL of the sample onto the pretreated silicon plate to adsorb for 5 min. Then, rinse the plate three times with deionized water and dry under N_2.

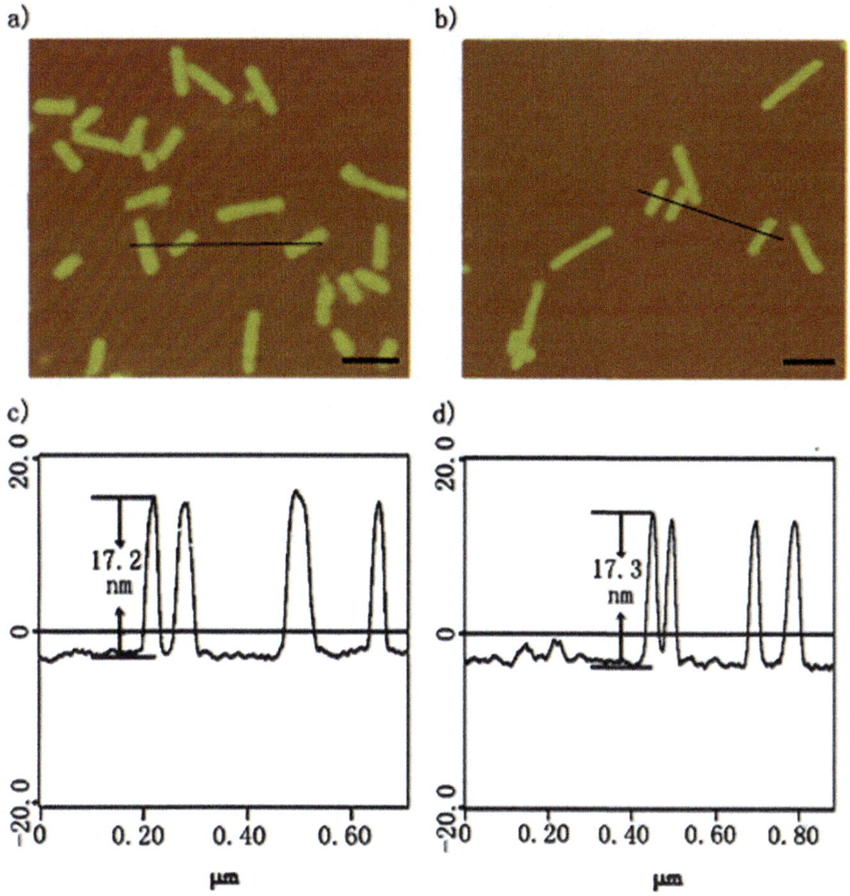

Fig. 6 AFM images of tubes assembled from Se-TMVcp142Cys149Arg and TMVcp. (**a**) Assembly of TMVcp at a concentration of 60 μM. (**b**) Assembly of Se-TMVcp142Cys149Arg at a concentration of 60 μM. (**c**, **d**) Height curves of assembled tubes corresponding to (**a**) and (**b**). Scale bars in (**a**) and (**b**) are 100 nm. Adapted with permission from ref. 11; Copyright (2012) the American Chemical Society

Analyze the samples by tapping mode atomic force microscopy (AFM) (*see* Fig. 6).

5. For statistical analysis, compute the lengths of tubes and disks by manual counting after five-time AFM measurements per dried drib samples. Collect data and make a bar chart for distribution analysis.
6. Drop disk and tube samples (30–150 μM) on Formvar-carbon-coated copper grids and let them dry for 1 min, rinse twice with ultrapure water, and dry by nitrogen blowing. Stain the samples on the copper grids negatively with sodium phosphotungstate prior to TEM detection according to a previously described method [21] (*see* Figs. 7 and 8).
7. Image all samples by using a FEI Tecnai 12 transmission electron microscope (TEM) with 200 kV accelerating voltage.

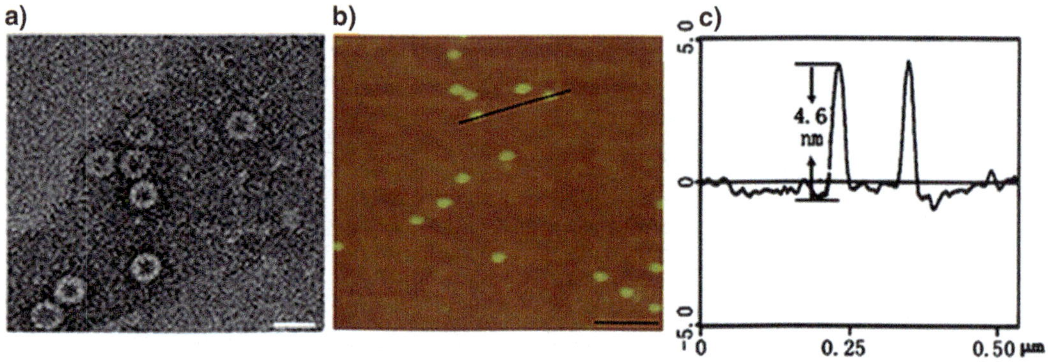

Fig. 7 TEM and AFM images of disks assembled from Se-TMVcp142Cys149Arg. (**a**) TEM image of Se-TMVcp142Cys149Arg disks which were assembled by exchanging buffers from pH 8.0 to pH 7.0. Samples were taken after equilibration for more than 40 h. (**b**) AFM image of Se-TMVcp142Cys149Arg disks. (**c**) Height curve of disk assemblies from (**b**). The scale bars in (**a**) and (**b**) are 20 nm and 100 nm, respectively. Adapted with permission from ref. 11; Copyright (2012) the American Chemical Society

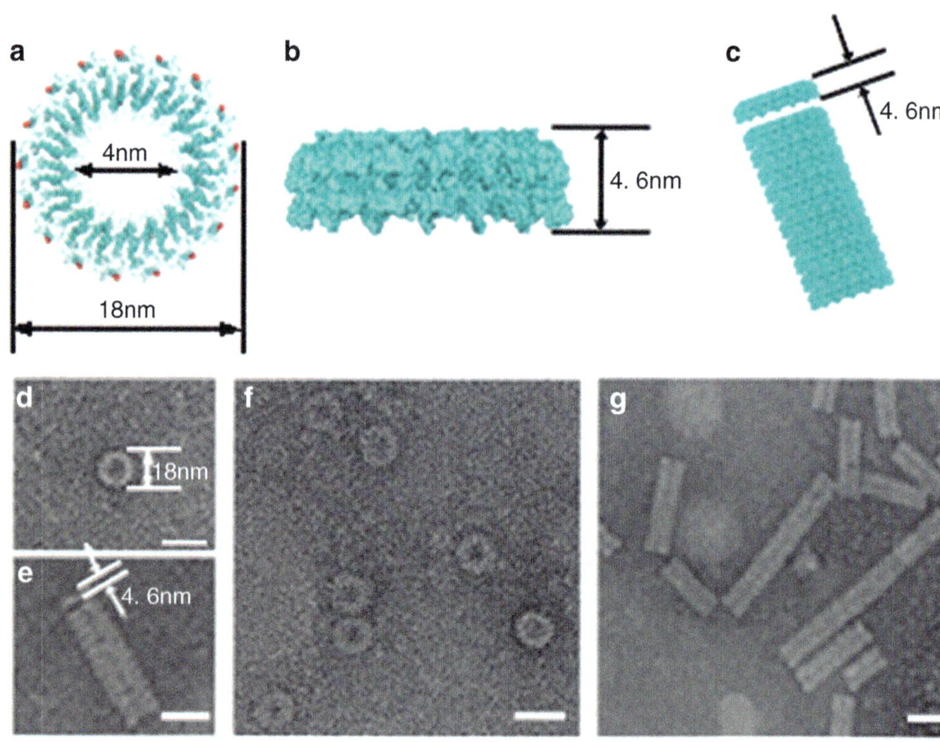

Fig. 8 Theoretical pictures made by software PyMOL and TEM images of Se-TMVcp142Cys149Arg disks and tubes purified by SEC-HPLC. (**a**) Vertical view of Se-TMVcp142Cys149Arg disks. (**b**) Side view of Se-TMVcp142Cys149Arg disks. (**c**) Transition of Se-TMVcp142Cys149Arg disks into Se-TMVcp142Cys149Arg tubes. (**d**, **f**) TEM images of purified disks. (**e**) TEM image of disk-to-tube transitions. (**g**) TEM image of purified Se-TMVcp142Cys149Arg tubes. Adapted with permission from ref. 11; Copyright (2012) the American Chemical Society

3.4 Enzyme Analysis of GPx-Like TMVcp Mutant

3.4.1 Determination of GPx Activity

GPx activity is determined by a coupled reductase method [22]. The spectrophotometer used is a UV/vis spectrophotometer.

1. Sample cells have a total volume of 500 μL for GPx determination. Make a mixture containing 50 μL GSH, 50 μL glutathione reductase, 50 μL NADPH (all from 10× stock solutions), 50 μL GPx-like enzyme sample solution (10–500 nM), and 250 μL PB buffer (50 mM, containing EDTA). Incubate them at 37 °C.
2. Three minutes later, add 50 μL H_2O_2 storage solution to initiate the reaction.
3. Record the absorbance at 340 nm (ε = 6220 1/(M cm), pH 7.0).
4. Use a sample cell without enzyme to measure the background substrate turnover rate. Subtract the background rate from the catalytic reaction.
5. The activity unit is defined as the amount of the enzyme that catalyzes the turnover of 1 μmol of NADPH per min. The specific activity is described as μmol/(min μmol) (Table 1).

3.4.2 Determination of Catalytic Kinetics of GPx

The kinetics determination for the reduction of H_2O_2 by GSH catalyzed by Se-TMVcp142Cys149Arg is according to the method described by Yu et al. [23, 24]

1. Make a variety of GSH solutions (2.5 mM, 5.0 mM, 10 mM) and H_2O_2 solutions (2.5 mM, 5.0 mM, 10 mM).
2. Preincubate 50 μL of Se-TMVcp142Cys149Arg with 50 μL GSH storage solution, 50 μL NADPH storage solution, and 50 μL glutathione reductase storage solution for 3 min.
3. Initiate the reaction by adding 50 μL H_2O_2 solution. The initial turnover rates can be obtained by changing the concentration of one substrate and keeping the concentration of the other constant.
4. Subsequently, the reaction rates are obtained at various concentrations of substrates following the subtraction of nonenzyme reaction rate (*see* Subheading 3.4.1, **steps 3–5**) (Table 2, Fig. 9).

3.5 Antioxidative Analysis of GPx Mimic of Se-TMVcp 142Cys149Arg Disks or Tubes

3.5.1 Extraction of Mitochondria of Bovine Heart

1. Weigh 20 g of bovine heart, chop, and wash three times with deionized water.
2. Add the chopped bovine heart to 2 volumes (w/v) of HEPES buffer, pH 7.4, and use a homogenizer at 4 °C to prepare a homogeneous suspension of bovine heart.
3. Centrifuge at 500 × *g* for 15 min and filter the supernatant carefully.

Table 1
GPx activities of Se-TMVcp142Cys149Arg and other GPx mimics

Catalyst	Substrate	Activity[a] (one disk or natural assemblies) (μmol/(min μmol))	Activity (one monomer) (μmol/(min μmol))	Selenium content (M/M per subunit)
TMVcp[b]	H_2O_2	ND	ND	0
TMVcp142Cys[c]	H_2O_2	3.4 ± 0.1	0.1 ± 0.03	0
Se-TMVcp142Cys[d]	H_2O_2	6320 ± 13.5	185.9 ± 0.41	0.95
Se-TMVcp142Cys149Arg[e]	H_2O_2	9356 ± 17.2	275.2 ± 0.50	0.96
Ebselen	H_2O_2	1.02[f]	1.02	1.0
Native GPx (rabbit liver)	H_2O_2	5780[g]	1445	1.0

Adapted with permission from ref. 11; Copyright (2012) the American Chemical Society
ND no detection
[a]Represents the activities of disk-like TMV assemblies
[b]TMVcp mutant without cysteine or selenocysteine
[c]TMVcp with a Cys at 142
[d]TMVcp mutant with a catalytic selenocysteine at 142
[e]TMVcp mutant with a catalytic selenocysteine at 142 and a recognition group Arg at 149
[f]Activity of a synthetic GPx mimic
[g]Activity of GPx from natural rabbit liver

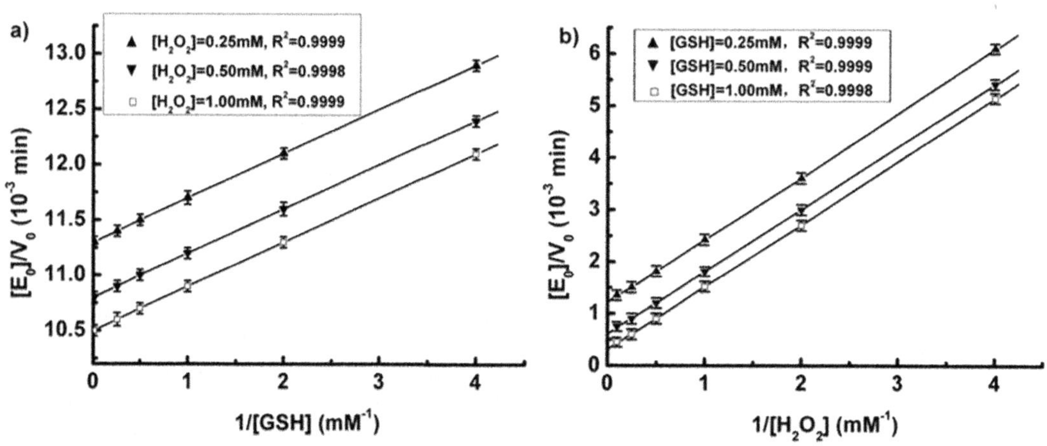

Fig. 9 Double reciprocal plots of H_2O_2 reduction by GSH in the presence of GPx mimic of monomer Se-TMVcp142Cys149Arg. (**a**) $[E_0]/V_0$ versus 1/[GSH] at $[H_2O_2]$ = 0.25 mM (filled triangle), 0.50 mM (inverted filled triangle) and 1.00 mM (open square). (**b**) $[E_0]/V_0$ versus $1/[H_2O_2]$ at [GSH] = 0.25 mM (filled triangle), 0.50 mM (inverted filled triangle) and 1.00 mM (open square). Adapted with permission from ref. 11; Copyright (2012) the American Chemical Society

4. Centrifuge the filtrate at 5000 × *g* for 15 min and remove the resulting supernatant.
5. Dissolve the sediment in 20 mL HEPES buffer, repeat **step 4**.
6. Resuspend the sediment in HEPES buffer, store at −80 °C.

Table 2
Apparent kinetic parameters[a] for H_2O_2 reduction by GSH catalyzed by Se-TMVcp142Cys149Arg (one active center)

$[H_2O_2]$ (mM)	k_{cat} (1/min)	$K_M^{H_2O_2}$ (×10^{-4} M)	$k_{cat}/K_M^{H_2O_2}$ (×10^6/M min)	[GSH] (mM)	k_{cat} (1/min)	K_M^{GSH} (×10^{-3} M)	k_{cat}/K_M^{GSH} (×10^5/M min)
0.25	91.54 ± 4.51	0.36 ± 0.02	2.54 ± 0.11	0.25	877.2 ± 40.22	1.09 ± 0.04	8.05 ± 0.37
0.50	95.37 ± 4.82	0.37 ± 0.05	2.58 ± 0.12	0.50	1851 ± 80.60	2.30 ± 0.01	8.05 ± 0.36
1.00	98.52 ± 5.71	0.38 ± 0.03	2.59 ± 0.15	1.00	2778 ± 130.5	3.33 ± 0.01	8.34 ± 0.41

[a]The data in the table were calculated from the plots in Fig. 4 with the following equation: $\frac{v_0}{[E_0]} = \frac{k_{cat}[GSH][H_2O_2]}{K_M^{GSH}[H_2O_2] + K_M^{H_2O_2}[GSH] + [GSH][H_2O_2]}$

3.5.2 Determine Concentration of Mitochondria

Use the Bradford method to determinate the concentration of mitochondria using BSA as standard.

3.5.3 Fe^{2+}/Vc Induced Damage to Mitochondria

1. Take appropriate amounts from stock solutions to prepare a 700 μL solution including: 400 μL of Tris–HCl buffer (pH 7.4), 100 μL of KCl (0.25 M), 100 μL of GSH (1 mM), 100 μL of GSSG (3 mM), and 50 μL of mitochondria (0.5 mg/mL).
2. Add 100 μL of Vc (0.5 mM), 100 μL of $FeCl_2$ (12.5 μM) to initiate oxidation reaction by rotating at 37 °C.
3. Add 50 μL of GPx-like TMV nanotubes to the above solution to protect mitochondria from oxidation.
4. Perform a series of experiments without GPx Se-TMVcp142Cys149Arg disks or tubes, Vc, or $FeSO_4$ to serve as control groups.
5. At different time points during culture, determine the swelling of mitochondria by measuring the absorbance at 520 nm and absorbance of malondialdehyde (MDA) at 532 nm (Fig. 10).

3.5.4 Determination of Lipid Peroxidation

1. Take 1 mL of analyte at different times, and mix with 1 mL of trichloroacetic acid (TCA, 700 g/L) and 1 mL of thiobarbituric acid (TBA, 5 g/L).
2. Incubate the sample solution at 80 °C for 40 min.
3. Centrifuge the solution at 200 × *g* for 10 min.
4. Determine the OD value of the supernatant at 532 nm.

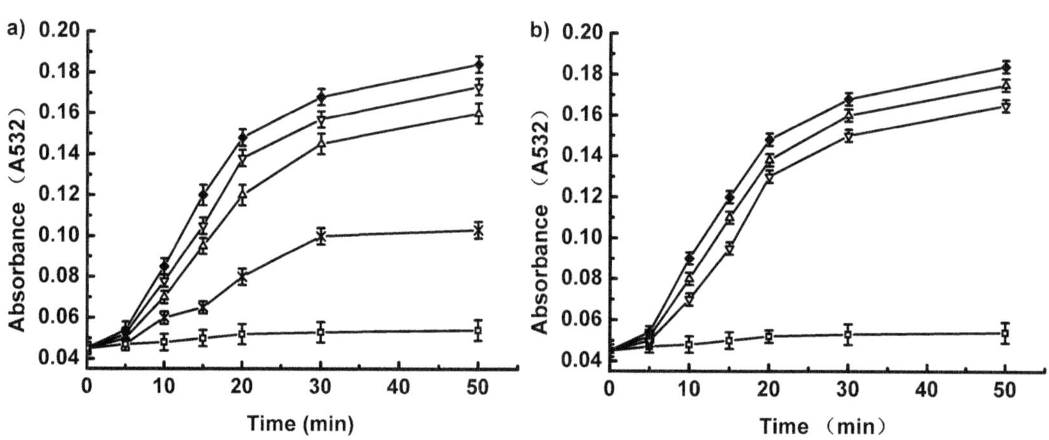

Fig. 10 Determination of lipid peroxidation in the presence of GPx mimics of Se-TMVcp142Cys149Arg disks and 2-SeCD. (**a**) Inhibition of MDA content over times by Se-TMVcp142Cys149Arg disks at various concentrations of 0 μM (filled diamond), 0.025 μM (inverted open triangle), 0.050 μM (open triangle), 0.100 μM (cross mark) and control of no oxidant (open square). (**b**) Inhibition comparisons of MDA content over times by Se-TMVcp142Cys149Arg disks (inverted open triangle) at 0.025 μM and 2-SeCD (open triangle) at 2 μM, no GPx mimics (filled diamond), control of no oxidant (open square)

4 Notes

1. Note that in auxotrophic cells, replacement of cysteine by selenocysteine depends on efficient charging of tRNAcys with Sec cysteine auxotrophic technology. For this purpose, cysteines in the medium should be removed totally by washing using pH 7.4 Tris–HCl buffer, when selenocysteine rather than cysteine should be added to the M9 medium for the second stage of culture of a cysteine auxotrophic stain.
2. A typical growth period for inoculated cells in M9 medium from OD value of 0.1–1.0 is about 5 h.
3. Lowering the rotation velocity to 180 rpm and extending the culture period to 7 h can improve the yield of selenium-containing protein.
4. Make sure that all the operations for Se-TMVcp142Cys149Arg should be at low temperature of 4 °C. Higher temperatures will greatly accelerate the oxidation of selenocysteine in the active site, leading to selenium-containing protein disks or tubes without GPx activity.
5. Without encapsulation by its intrinsic RNA, these selenium-containing GPx-like TMV assemblies seem to disassociate easily upon increasing temperature or centrifugation.
6. Purchased H_2O_2 (30%, v/v) has a density of 1.11 g/mL. Molecular weight of H_2O_2 is 34.01468 g/mol. 1 L of 30% H_2O_2 contains 1 L × 1.11 g/mL × 30% = 333 g H_2O_2. So 1 L of 30% H_2O_2 consists of 333 g/34.01 = 9.79 mol, and the molar concentration of 30% H_2O_2 is 9.79 M.
7. GSH solution could be stored at 4 °C and used in 2 days. GSH will spontaneously be oxidized in air so that freshly prepared GSH solution is preferred for each determination.
8. Precaution: Trichloroacetic acid (TCA) is a hazardous reagent, which can cause injury, disease and even death when inhalation, ingestion and skin contact with enough trichloroacetic acid happen. Operations should be undergone in a fume hood. 3M mask and gloves should be worn to protect against TCA contact.

Acknowledgment

We thank Prof. M.B. Francis for the expression plasmid containing the gene of tobacco mosaic virus coat protein, and we thank Prof. A. Boeck for providing the cysteine auxotrophic expression system. The work was supported by the Natural Science Foundation of China No. 21234004, 21420102007, 21574056, and 91527302 and the Chang Jiang Scholars Program of China.

References

1. Lee YJ, Yi H, Kim WJ, Kang K, Yun DS, Strano MS, Ceder G, Belcher AM (2009) Fabricating genetically engineered high-power lithium-ion batteries using multiple virus genes. Science 324:1051–1055
2. Carette N, Engelkamp H, Akpa E, Pierre SJ, Cameron NR, Christianen PC, Maan JC, Thies JC, Weberskirch R, Rowan AE (2007) A virus-based biocatalyst. Nat Nanotechnol 2:226–229
3. Rong JH, Oberbeck F, Wang XN, Li XD, Oxsher J, Niu ZW, Wang Q (2009) Tobacco mosaic virus templated synthesis of one dimensional inorganic-polymer hybrid fibres. J Mater Chem 19:2841–2845
4. Shenton W, Douglas T, Young M, Stubbs G, Mann S (1999) Inorganic-organic nanotube composites from template mineralization of tobacco mosaic virus. Adv Mater 11:253–256
5. Witus LS, Francis MB (2011) Using synthetically modified proteins to make new materials. Acc Chem Res 44:774–783
6. Stephanopoulos N, Carrico ZM, Francis MB (2009) Nanoscale integration of sensitizing chromophores and porphyrins with bacteriophage MS2. Angew Chem Int Ed 48:9498–9502
7. Schlick TL, Ding ZB, Kovacs EW, Francis MB (2005) Dual-surface modification of the tobacco mosaic virus. J Am Chem Soc 127:3718–3723
8. Bruckman MA, Liu J, Koley G, Yu L, Benicewicz B, Niu ZW, Wang Q (2010) Tobacco mosaic virus based thin film sensor for detection of volatile organic compounds. J Mater Chem 20:5715–5719
9. Gorzny ML, Walton AS, Evans SD (2010) Synthesis of high-surface-area platinum nanotubes using a viral template. Adv Funct Mater 20:1295–1300
10. Komatsu T, Terada H, Kobayashi N (2011) Protein nanotubes with an enzyme interior surface. Chem A Eur J 17:1849–1854
11. Hou CX, Luo Q, Liu JL, Miao L, Zhang CQ, Gao YZ, Zhang XY, Xu JY, Dong ZY, Liu JQ (2012) Construction of GPx active centers on natural protein nanodisk/nanotube: a new way to develop artificial nanoenzyme. ACS Nano 6:8692–8701
12. Gao YZ, Hou CX, Zhou LP, Zhang DM, Zhang CQ, Miao L, Wang L, Dong ZY, Luo Q, Liu JQ (2013) A dual enzyme microgel with high antioxidant ability based on engineered seleno-ferritin and artificial superoxide dismutase. Macromol Biosci 13:808–816
13. Seeliger D, de Groot BL (2010) Ligand docking and binding site analysis with pymol and autodock/vina. J Comput Aided Mol Des 24:417–422
14. DeLano, W.L. (2002) The PyMOL molecular graphics system. http://www.pymol.org
15. Dallakyan S (2012) A suite of automated docking tools. http://autodock.scripps.edu/
16. Miller RA, Presley AD, Francis MB (2007) Self-assembling light-harvesting systems from synthetically modified tobacco mosaic virus coat proteins. J Am Chem Soc 129:3104–3109
17. A PCR Primer Design software to design high amplification efficiency primers. http://www.premierbiosoft.com/
18. Lynnon DNAMAN is a one-for-all software package for molecular biology applications. http://www.lynnon.com
19. Mueller S, Senn H, Gsell B, Vetter W, Baron C, Boeck A (1994) The formation of diselenide bridges in proteins by incorporation of selenocysteine residues: biosynthesis and characterization of (Se)2-thioredoxin. Biochemistry 33:3404–3412
20. Green MR, Sambrook J (2012) Molecular cloning: a laboratory manual, 4th edn. Cold Spring Harbor Laboratory Press, Cold Spring Harbor, NY
21. Carter CB, Williams DB (2016) Transmission electron microscopy: diffraction, imaging, and spectrometry. Springer, Cham
22. Huang X, Liu Y, Liang K, Tang Y, Liu JQ (2008) Construction of the active site of glutathione peroxidase on polymer-based nanoparticles. Biomacromolecules 9:1467–1473
23. Yu HJ, Ge Y, Wang Y, Lin CT, Li J, Liu XM, Zang TZ, Xu JY, Liu JQ, Luo GM (2007) A fused selenium-containing protein with both Gpx and SOD activities. Biochem Biophys Res Commun 358:873–878
24. Jiang ZH, Arner ESJ, Mu Y, Johansson L, Shi JM, Zhao SQ, Liu SJ, Wang RY, Zhang TZ, Yan GL (2004) Expression of selenocysteine-containing glutathione s-transferase in *Escherichia Coli*. Biochem Biophys Res Commun 321:94–101

Chapter 30

Redox-Immunofunctionalized Potyvirus Nanoparticles for High-Resolution Imaging by AFM-SECM Correlative Microscopy

Agnès Anne, Arnaud Chovin, Christophe Demaille, and Thierry Michon

Abstract

We present in this chapter a new experimental approach allowing the high resolution imaging of immune complexes on virus particles. Combined atomic force—electrochemical microscopy (AFM-SECM) is used to image the presence of ferrocene functionalized specific antibodies on the surface of potyvirus particles. For this purpose, potyviruses, flexuous filamentous phytoviruses with a high aspect ratio, have been chosen. This technique allows analysis of the distribution of antibody labeling over the virus population. But, more importantly, it opens up the imaging of immune complexes decorating a single viral particle. Finally, its high resolution allows the characterization in situ of the ultrastructure of a single immune complex on the particle.

Key words Lettuce mosaic virus (LMV), Redox antibodies, Viral nanoparticles, Electrochemical atomic force microscopy, Single viral protein imaging, Potyvirus

1 Introduction

The antibody response is crucial for preventing many viral infections and may also contribute to the resolution of infection. Antibodies are produced against many epitopes, especially protein structures on the surface of virus particles. Antibodies can neutralize viral infectivity in a number of ways such as binding to virion receptors, blocking of uptake into cells, preventing the uncoating of genomes in endosomes, or through aggregation of virus particles [1–3]. Many structures of immune complexes between antibodies and their antigens have been solved. However, to date, the in situ distribution and topology of these complexes, for instance at the surface of cells or viruses, has mainly been analyzed using electron microscopy. Although a rather high resolution can be obtained by this method, it requires special treatments of the samples, especially removal of water that can damage some of the biological

Christina Wege and George P. Lomonossoff (eds.), *Virus-Derived Nanoparticles for Advanced Technologies: Methods and Protocols*, Methods in Molecular Biology, vol. 1776, https://doi.org/10.1007/978-1-4939-7808-3_30,
© Springer Science+Business Media, LLC, part of Springer Nature 2018

properties of the samples. Moreover, this technique does not give access to the surface contour of the objects.

We present here a high resolution correlative local probe microscopy technique combining atomic force (AFM) and scanning electrochemical microscopy (SECM) enabling: (1) resolution of the position of immune complexes on the virus surfaces, (2) probing selectively for redox functions linked to small immune complex clusters.

In addition, using genetic and/or chemical engineering approaches, the virus surface can be used as an experimental nanoplatform, for instance to organize enzymatically catalyzed electron transport. The biophysical properties of such biomimetic cascades can then be studied in detail by means of AFM-SECM correlative microscopy.

The use of this new high resolution imaging technique to resolve native biological structures at the nanoscale is described below using lettuce mosaic virus (LMV, genus potyvirus) as a model.

The protocol below will successively elaborate on:

- Purification of LMV particles from infected plants.
- Preparation of a redox-active ferrocene-PEG–antibody conjugate.
- Adsorption and redox immunomarking of LMV.
- AFM-SECM imaging of the redox immunomarked viral particles.

2 Material

Prepare all solutions using ultrapure deionized water (18 MΩ cm at 25 °C) and analytical grade reagents. Store buffers at 4 °C with no sodium azide added. When taking reagents out from the refrigerator or freezer let them equilibrate to room temperature before use unless otherwise stated. Filter all buffers through 0.22 μm units (cellulose acetate for LMV production and purification, or PVDF otherwise).

2.1 Production of Lettuce Mosaic Virus (LMV)

1. Viral material for inoculation: leaf material infected with lettuce mosaic virus, AF199 isolate [4] routinely maintained in *Nicotiana benthamiana* (*see* **Note 1**).
2. Plant material to be inoculated: 75 *Nicotiana benthamiana* plants (6 weeks, 10 cm high) (*see* **Note 2**).
3. Grinding agent: Carborundum (silicon carbide).
4. Grinding buffer (T1 buffer): 0.01 M phosphate buffer, pH 7.0. For 100 mL, mix 4.1 mL of 0.1 M KH_2PO_4, 5.9 mL of 0.1 M Na_2HPO_4, and 90 mL of water (*see* **Note 3**).

2.2 Extraction/Purification of LMV Particles

1. Homogenization buffer (T2 buffer): 0.1 M phosphate buffer pH 8.0 containing 0.02 M (0.15% (v/v)) 2-mercaptoethanol and 0.01 M EDTA. For 100 mL, dissolve 1.42 g Na_2HPO_4 in 70 mL of water and add 150 μL of 2-mercaptoethanol; under stirring, add 0.3 g of solid EDTA and adjust to pH 8.0 with 1 M NaOH, bring to a final volume of 100 mL with water.
2. Purification buffer (T3 buffer): 0.1 M phosphate buffer pH 8.0. For 100 mL, mix 3.7 mL of 0.1 M KH_2PO_4 and 96.3 mL of 0.1 M Na_2HPO_4 (*see* **Note 3**).
3. Blender (Waring brand) equipped with a 1 L stainless steel container.
4. Cheesecloth or diaper fabric, typically a cotton canvas 19 threads/m^2, 0.3 kg/m^2.
5. Triton X-100.
6. Polyethylene glycol 6000 (PEG_{6000}).
7. NaCl.
8. Sucrose cushion: 30% (w/v) in T3 buffer.
9. Sucrose gradient: Prepare 40%, 30%, 15%, and 5% (w/v) sucrose solutions in T2 buffer (*see* **Note 4**).
10. Refrigerated preparative centrifuge (Beckman, model Aventi JE or equivalent). Rotor JA14, 250 mL screw on cap polypropylene bottles (Subheading 3.2, **steps 2**, **4**, and **6**). Rotor JA 20, open-top 12 mL polypropylene tubes with thick-wall adapters or equivalent (Subheading 3.2, **step 10**).
11. Refrigerated preparative ultracentrifuge (Beckman, model XL70 or equivalent). Rotor 60 Ti (fixed angle), 22 mL screw on cap polycarbonate tubes (Subheading 3.2, **steps 8** and **11**). Rotor SW41 Ti with screw on cap swinging buckets, 12 mL open cap polypropylene tubes (Subheading 3.2, **step 12**).
12. UV–visible spectrophotometer.
13. Equipment for SDS-PAGE and western blotting, antibody for detection of coat protein of LMV, Coomassie Blue G250 staining solution.

2.3 Labeling Antibodies with Ferrocene-PEG-NHS

1. Antibody (Immunoglobulin G, IgG) to be labeled: polyclonal goat anti-rabbit IgG (Jackson ImmunoResearch Laboratories, Affinipure grade), used as an example in this protocol (*see* **Note 5**).
2. Redox polyethylene glycol (PEG) labeling reagent: Ferrocene-PEG succinimidyl ester, Fc-PEG-NHS (Mw ~3.8 kDa) [5, 6] (*see* **Note 6**). Store at 4 °C in a desiccator (*see* **Note 7**).
3. Buffer for the Fc-PEG-NHS/antibody coupling reaction (PBR buffer): 0.1 M phosphate buffer pH 8.0 (*see* **Note 8**). Dissolve 1.24 g KH_2PO_4 in 250 mL water, add 8.45 mL of 1 M NaOH.

4. Recovery buffer for the Fc-PEG antibody conjugate (PBS buffer): 10 mM phosphate saline buffer, 0.15 M NaCl, pH 7.4. Dissolve 0.34 g KH_2PO_4 and 2.25 g NaCl in 250 mL MQ water, add 2.1 mL of 1 M NaOH.
5. 0.1% (w/v) Sodium azide (NaN_3) in PBS. Dissolve 10 mg of NaN_3 in 10 mL of PBS.
6. Consumables: 1 mL sterile disposable syringe and needle (*see* **Note 9**); 0.2 μm PVDF syringe filter; Amicon ultra-4 centrifugal filter unit, 30 kDa cut-off (Millipore); 1.5 mL Eppendorf tubes; UV-transparent disposable plastic cuvettes, 1 cm path.
7. Equipment: Bench-top centrifuge; UV-visible spectrophotometer; desiccator.

2.4 Characterizing the Fc-PEG-IgG Conjugate (Optional)

1. Equipment for SDS-PAGE and staining with Coomassie brilliant blue G250.
2. Equipment for ICP-MS (inductively coupled plasma mass spectrometry).
3. Equipment for electrochemical evaluation of the recognition of a layer of antigens (primary antibody) adsorbed onto a carbon electrode surface [7] by Fc-PEG-IgG: Electrochemical workstation. Standard electrochemical cell equipped with a KCl-saturated calomel reference electrode (SCE) and a platinum counter electrode.

2.5 Adsorption of LMV and Blocking

1. LMV solution for adsorption. Just before use, dilute one aliquot of LMV viral particles (*from* Subheading 3.2, **step 15**) in PBA buffer to a concentration of 45 μg/mL.
2. PEG blocking reagent: Methoxy PEG_{2000} disulfide. Use as a 0.7 mM solution in water, split in 200 μL aliquots and store at −20 °C (*see* **Note 10**).
3. Bovine serum albumin (BSA) (IgG-free grade, Jackson ImmunoResearch Laboratories).
4. Buffer for PEG blocking of the surface (PBP buffer, pH 7.4): 100 mM phosphate buffer. Dissolve 0.310 g of $NaH_2PO_4 \cdot H_2O$ and 1.367 g of Na_2HPO_4 in 100 mL water.
5. Buffer for viral particles adsorption (PBA buffer, pH 7.4): 10 mM phosphate buffer. Dilute PBP buffer to one-tenth.
6. Ultraflat template stripped (TS) gold surfaces, commercially available (Platypus Technologies, USA), or produced in house at a much lower cost using a thermal gold evaporator as described in [8] (*see* **Note 11**).
7. O-ring fluid cell.
8. Glass or plastic container that can be flushed with water-saturated inert gas (nitrogen or argon), to be used to store the LMV bearing surfaces.

2.6 Redox Immunomarking of Surface Adsorbed LMV Particles

1. Primary antibodies: Rabbit polyclonal anti-LMV antibodies directed against either (1) the capsid protein (CP) or (2) the genome-linked viral protein (VPg) (*see* **Note 12**) (5 µg/mL in PBI).
2. Secondary antibody: Fc-PEGylated anti-rabbit antibody conjugate (*from* Subheading 3.3) (20 µg/mL in PBI).
3. Buffer for immunomarking of viral particles (PBI buffer): PBA buffer containing 1 mg/mL BSA, 0.1% (v/v) Tween 20, and 0.01% (w/v) NaN_3.
4. Buffer for viral particles adsorption (PBA buffer, pH 7.4) (*see* Subheading 2.5, **item 5**).

2.7 In Situ AFM-SECM Imaging of the Redox ImmunoMarked Viral Particles

1. Imaging buffer (PBA buffer, pH 7.4): 10 mM phosphate buffer.
2. 1% (v/v) glutaraldehyde (EM grade) solution in PBA buffer.
3. AFM setup endowed with scanning electrochemical (SECM) imaging capabilities, AFM-SECM probes (tips). A description of such a setup and tips can be found in the literature [9].

3 Methods

3.1 Production of Lettuce Mosaic Virus (LMV)

1. Preparation for inoculation: Grind in a mortar leaf material (1 g) from previously infected plants (LMV, AF199 isolate) in 5 mL of T1 buffer (tissue-to-maceration ratio of 1/5 (w/v)) and 0.1 g of carborundum.
2. Plant inoculation: Gently rub leaves from 75 young *Nicotiana benthamiana* plants with the preparation from **step 1**. Carborundum particles induce small wounds on the leaf surface and facilitate virus penetration.
3. Maintain the inoculated plants in the greenhouse until vein-clearing and mosaic symptoms appear on systemically infected leaves (*see* **Note 13**).
4. After 25 days *harvest the symptomatic leaves and weigh them. Then freeze the leaves at −20 °C before extraction* (*see* **Note 14**).

3.2 Extraction/Purification of LMV Particles

The whole extraction/purification process requires 3-days. All operations must be performed at 4 °C.

1. Day 1: Homogenize 300 g of the harvested leaves in a blender with 600 mL of 4 °CT2 buffer (*see* **Note 15**). This breaks up cell walls and releases viral particles and cell debris into the solution.
2. Spin down plant debris at 7000 × *g* for 20 min at 4 °C.
3. After filtration through two layers of cheesecloth, add 3% (v/v) Triton X-100 to the supernatant and gently stir for 3 h at 4 °C.

4. Spin down the remaining insoluble material at 12,000 × *g* for 10 min at 4 °C, then collect the supernatant.
5. Add 40 g/L PEG_{6000} and 12 g/L (0.2 M) NaCl to the supernatant. Let the viral material precipitate overnight under gentle stirring at 4 °C in the dark (*see* **Note 16**).
6. Day 2: Recover the precipitate as a pellet by centrifugation at 12,000 × *g* for 20 min.
7. Suspend the pellet in 60 mL (1/10th of the starting volume) of 4 °C T3 buffer containing 1% (v/v) Triton X-100 for 2 h.
8. Distribute the extract between four tubes and equilibrate weight by adding the appropriate volume of T3 buffer. Tubes should be filled up to 80% of their total volume. Centrifuge at 160,000 × *g* for 3 h at 4 °C (*see* **Note 17**).
9. Discard the supernatant and gently suspend the virus pellet in 15 mL of T2 buffer (5 mL for 100 g of harvested leaves) overnight on a shaker in the cold room.
10. Day 3 (optional): Centrifuge at 3000 × *g* for 10 min to pellet large debris. Save supernatant and suspend pellet again in 5 mL of T2 buffer. Centrifuge at 3000 × *g* for 10 min. Recover the supernatant and pool it with the saved 15 mL (final sample volume, 20 mL). Discard pellet (*see* **Note 18**).
11. Distribute the supernatant to 5 mL fractions and add each to the top of a 30% sucrose cushion (14 mL) in T3 buffer (for 15–20 mL supernatant, use between 3 and 4 centrifugation tubes). Centrifuge at 90,000 × *g* at 4 °C for 3 h. Suspend pellet in 2 mL of T3 buffer.
12. Purify the viral suspension through a 5–40% discontinuous sucrose gradient as follows: Fill centrifugation tubes with sucrose solutions prepared in T2 buffer: from bottom to top, 1.5 mL of 40%, 3 mL of 30%, 3 mL of 15%, and 1.5 mL of 5% sucrose solution. Carefully lay a volume of 0.5–1 mL of the virus suspension on the top of the gradient. Centrifuge at 80,000 × *g* at 4 °C for 1 h.
13. Remove the supernatant and suspend the pellet in the minimum amount of T3 buffer, typically 1 mL buffer for 100 g of starting plant material.
14. Quantify the concentration of viral particles in the suspension by UV absorption spectrometry: Measure absorbance at 260 nm, A_{260}. By using an extinction coefficient of 2.6 mg/(mL cm) at 260 nm, the concentration of viral particles C (mg/mL) is given by the formula: $C = A_{260}/2.6 \times l$, where l is the cell path length in cm.
15. Distribute the viral particle suspension (typically at 5–6 mg/mL) in 200 μL aliquots and store at −80 °C.

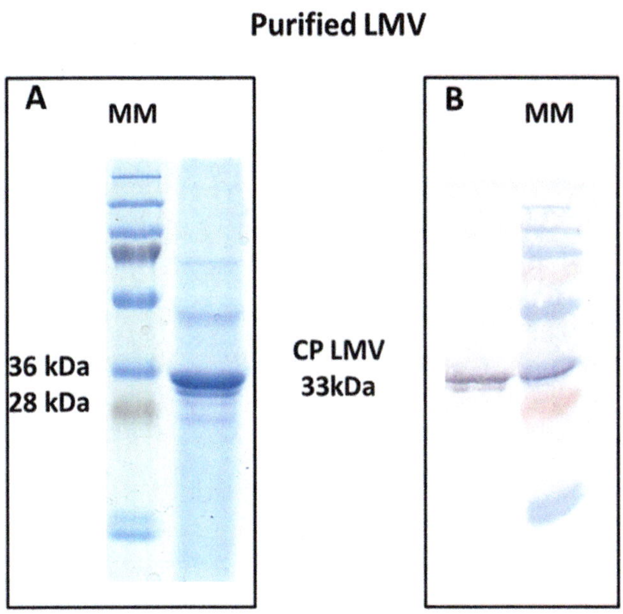

Fig. 1 An aliquot of the preparation was heat-denatured and its protein composition analyzed by SDS-PAGE (left panel). A major molecular species, migrating at 33 kDa, the expected position for the LMV coat protein (CP), was observed. The proteins were transferred from the gel to a nitrocellulose membrane (right panel). The presence of the LMV coat protein (CP) was revealed by immunodetection with the rabbit polyclonal antibody raised against LMV whole particles used in this study. Immunodetection was obtained by association of the LMV primary antibody with an anti-rabbit goat secondary antibody coupled to alkaline phosphatase

16. LMV characterization: The presence of LMV particles in the sample is ascertained by SDS-PAGE (sodium dodecyl sulfate–polyacrylamide gel electrophoresis) and western blotting. As shown in Fig. 1, the gel should show a major band at 33 kDa, corresponding to the molar mass of the coat protein of LMV (*see* **Note 19**).

3.3 Labeling Antibodies with Ferrocene-PEG-NHS

This protocol details the preparation of the ferrocene (Fc)-PEG-IgG conjugate to be used as a secondary antibody for redox immunodetection of viruses. Ferrocene-PEG chains are conjugated to the IgG by reacting the activated ester NHS function of the Fc-PEG-NHS reagent with the primary amines of the lysine residues exposed at the surface of the IgG (*see* Fig. 2). The amide bond thus formed is extremely robust, ensuring long-term stability of the Fc-PEG-IgG conjugate. The protocol is optimized for labeling 500 μg of antibody, utilizing a 100× fold molar excess of Fc-PEG-NHS to maximize the labeling yield. Typically, depending on the host species of the antibodies, 5–10 Fc-PEG chains are found to be attached to each IgG molecule [7].

1. Add 500 μg of the antibody to be labeled to 2 mL PBR buffer (*see* **Note 20**). Filter the resulting solution using a 0.22 μm

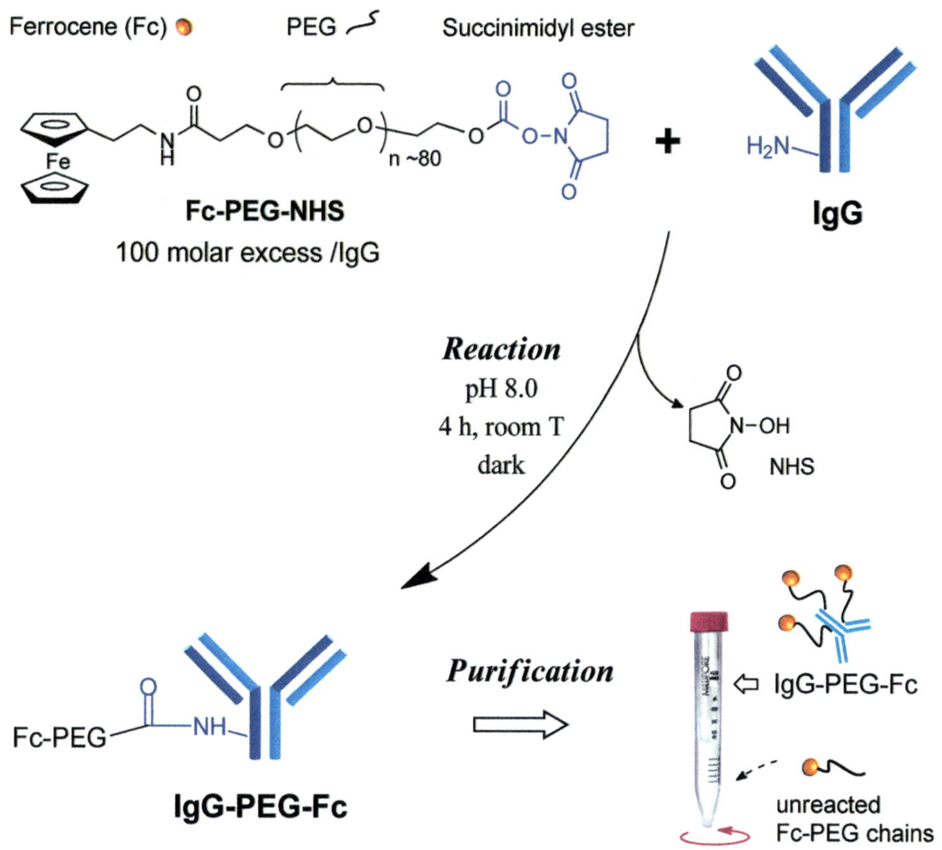

Fig. 2 Schematic representation of the covalent modification of the secondary antibody with the amine (-NH_2) reactive Fc-PEG-NHS derivative, followed by purification

PVDF filter. Transfer the antibody solution into an Amicon ultra-4 centrifugal filter unit (30 kDa) (*see* **Note 21**).

2. Concentrate the antibody solution by centrifugation at 2500 × *g* for about 10 min in order to bring the solution volume down to ~100 μL (*see* **Note 22**).
3. Pipette the ~100 μL concentrate and transfer it into a preweighed Eppendorf tube in order to determine its exact volume by weighing. Adjust the volume to 150 μL by adding PBR buffer. The antibody concentration at this stage is 3.3 mg/mL (23 μM).
4. Transfer the 150 μL antibody solution into the Eppendorf tube containing 1 mg of the PEG labeling reagent, pipette a few times to fully dissolve the Fc-PEG-NHS solid.
5. Incubate at room temperature for 4 h in the dark.
6. Purify the Fc-PEG labeled antibody by ultrafiltration in order to separate it from unconjugated Fc-PEG chains. Add ~850 μL of PBS buffer to the antibody solution. Transfer the ~1 mL

solution to a new, rinsed, Amicon ultra-4 centrifugal unit. (a) Adjust volume to 2 mL with PBS and (b) spin at 2500 × *g* for 10 min in order to bring the antibody solution (concentrate) volume down to ~100 μL. Discard the eluate or store it for subsequent UV-vis control analysis (optional, *see* **Note 23**). Repeat steps (a) and (b) five times (*see* **Note 24**).

7. Pipette the final ~100 μL concentrate and transfer it into a pre-weighted Eppendorf tube. Measure the exact volume of the concentrate by weighing the tube. Bring the volume to 500 μL by adding PBS buffer, the approximate labeled antibody concentration should then be about 1 mg/mL.
8. Add ~50 μL of PBS buffer containing 0.1% (w/v) NaN_3 to the purified labeled antibody solution then store in the fridge (*see* **Note 25**).

3.4 Characterizing the Fc-PEG-IgG Conjugate (Optional)

1. Efficient labeling of the IgG by the Fc-PEG chains can be qualitatively assessed by SDS-PAGE. The gel (*see* Fig. 3) typically shows bands in the high molecular weight region, ascribable to IgGs conjugated to a variable number of Fc-PEG chains. Bands corresponding to unlabeled IgGs are absent.
2. The labeling ratio (i.e. the average number of Fc-PEG chains per antibody) of the conjugate can be determined by ICP-MS (inductively coupled plasma mass spectrometry).

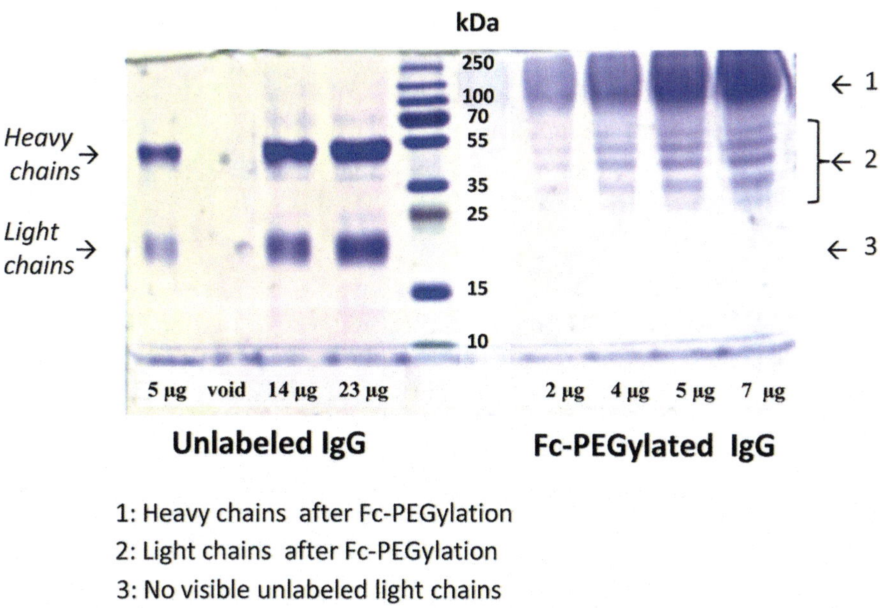

Fig. 3 SDS-PAGE analysis of the Fc-PEG-IgG conjugate. Efficient Fc-PEG labeling of the IgG is ascertained by the observation of bands in the high molecular weight region, ascribable to IgGs conjugated to a variable number of Fc-PEG chains

3. The molecular recognition (immunomarking) capability of the Fc-PEG-IgG antibody can be evaluated by electrochemically monitoring the recognition of a layer of antigens (primary antibody) adsorbed onto a carbon electrode surface [7]. Using this approach we verified that antibodies decorated by Fc-PEG chains fully preserve their molecular recognition capabilities.

3.5 Adsorption and Redox Immunomarking of LMV on Ultraflat Gold Surfaces

1. Mount a template-stripped (TS) gold surface as the bottom of an O-ring fluid cell, leaving ~1 cm^2 of the gold surface exposed.
2. Fill the cell with ~200 μL of a 45 μg/mL LMV solution in PBA buffer and let spontaneous virus adsorption occur for 10–20 min depending on the virus coverage targeted. Typically this should result in a surface coverage in adsorbed virus of 1–2 viral particles/μm^2 (*see* Fig. 4).
3. Pipette most of the LMV solution from the surface (be careful not to let the surface dry) and fill the cell with PBA buffer. Repeat this pipetting/buffer filling cycle five times. Leave the PBA buffer in contact with the surface for 10 min to desorb any loosely bound material.
4. If the viral protein to be marked is the VPg protein jump directly to **step 6**. If the capsid protein (CP) is targeted do the following: Add 20 μL of PBP buffer to the 200 μL methoxy PEG_{2000} disulfide aliquot and fill the cell with the resulting solution. Allow self-assembly of PEG-disulfide on gold layer for 2 h. From this moment, and during all of the next steps of virus labeling, the surface should be kept under a water-saturated inert atmosphere (nitrogen or argon) in order to protect the PEG disulfide layer from oxidation.
5. Rinse the surface thoroughly with PBA and soak it in PBA for 10 min.
6. Cover the surface with a 2 mg/mL BSA solution in PBA buffer and let BSA adsorb onto unoccupied surface sites for 1 h.
7. Rinse the surface thoroughly with PBA and soak it for 10 min.

3.6 Redox Immunomarking of Surface Adsorbed LMV Particles

1. Deposit on the surface ~200 μL of a 5 μg/mL solution of either the anti-CP or anti-VPg antibody prepared in PBI buffer. Allow the immunological recognition of the adsorbed virus particles by the antibodies to occur for 1 h.
2. Rinse the surface thoroughly with PBA and soak it for 10 min. Repeat this step twice.
3. Cover the surface with a 20 μg/mL solution of Fc-PEG-labeled secondary antibody prepared in PBI buffer. Allow the immunological recognition of the virus bound primary antibodies by the Fc-PEG labeled antibodies to proceed overnight (~15 h).

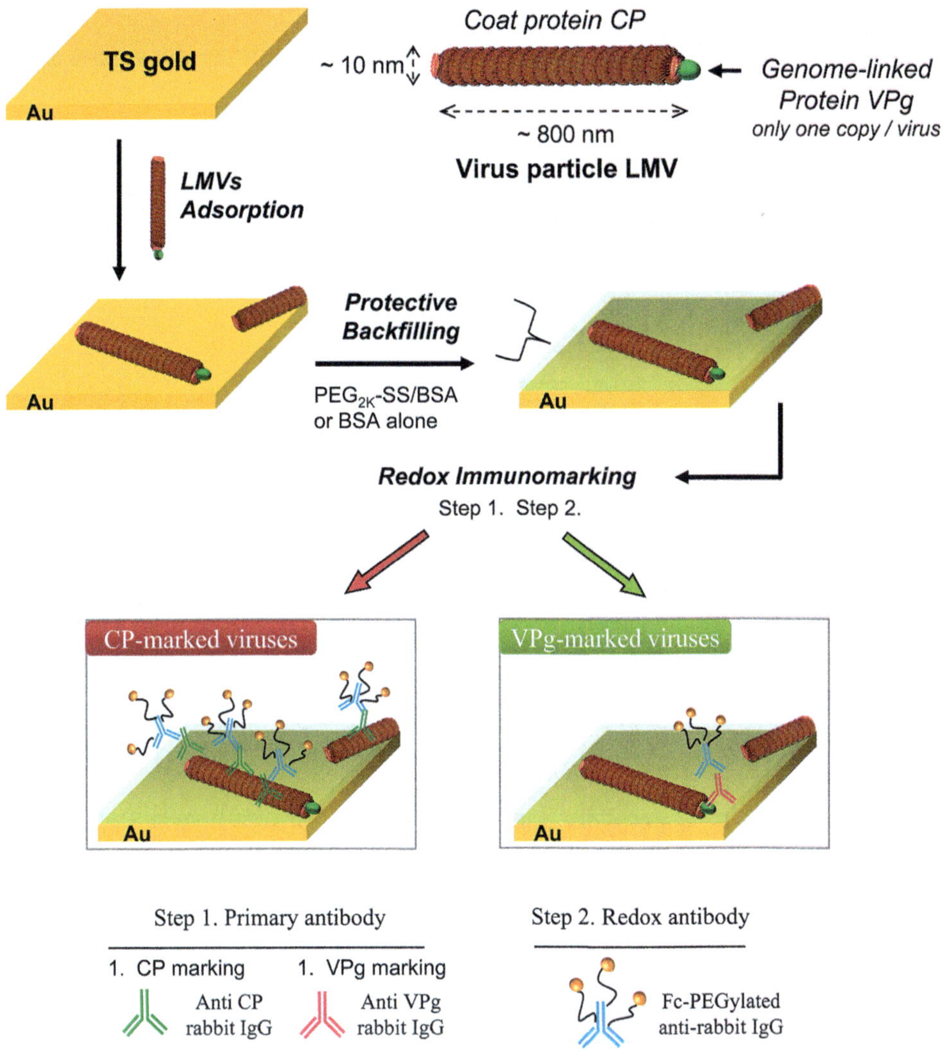

Fig. 4 Scheme for the immobilization of LMV particles on a gold surface and subsequent redox immune-marking (Reprinted with permission from ACS Nano 2015, 9:4911–4924. Copyright 2015 American Chemical Society)

4. Rinse the surface thoroughly with PBA and soak it in PBA for 10 min. Repeat this step twice.

3.7 In Situ AFM-SECM Imaging of the Redox Immunomarked LMV Particles

1. Cover the surface with ~200 μL of a 1% (v/v) glutaraldehyde solution in PBA buffer and leave it to react for 15 min. This treatment hardens the viral structure and is used to facilitate AFM-based imaging of viruses in liquid [10].
2. Rinse the surface thoroughly with PBA buffer.
3. The surface is now ready for in situ AFM-SECM imaging carried out in PBA buffer. Bias the AFM-SECM tip at +0.3 V/SCE and the LMV bearing surface at 0.0 V/SCE. Excite the AFM-SECM tip acoustically at its fundamental flexural reso-

nance frequency, and allow the AFM feedback mechanism to approach the tip toward the surface. Use a damping set-point of ~10–20% for both approach and imaging as described in [11]. The topography images should reveal isolated virus particles while the simultaneously acquired current images should show current "spots" decorating the virus particles (*see* Fig. 5), thus allowing the redox function of individual viral particles to

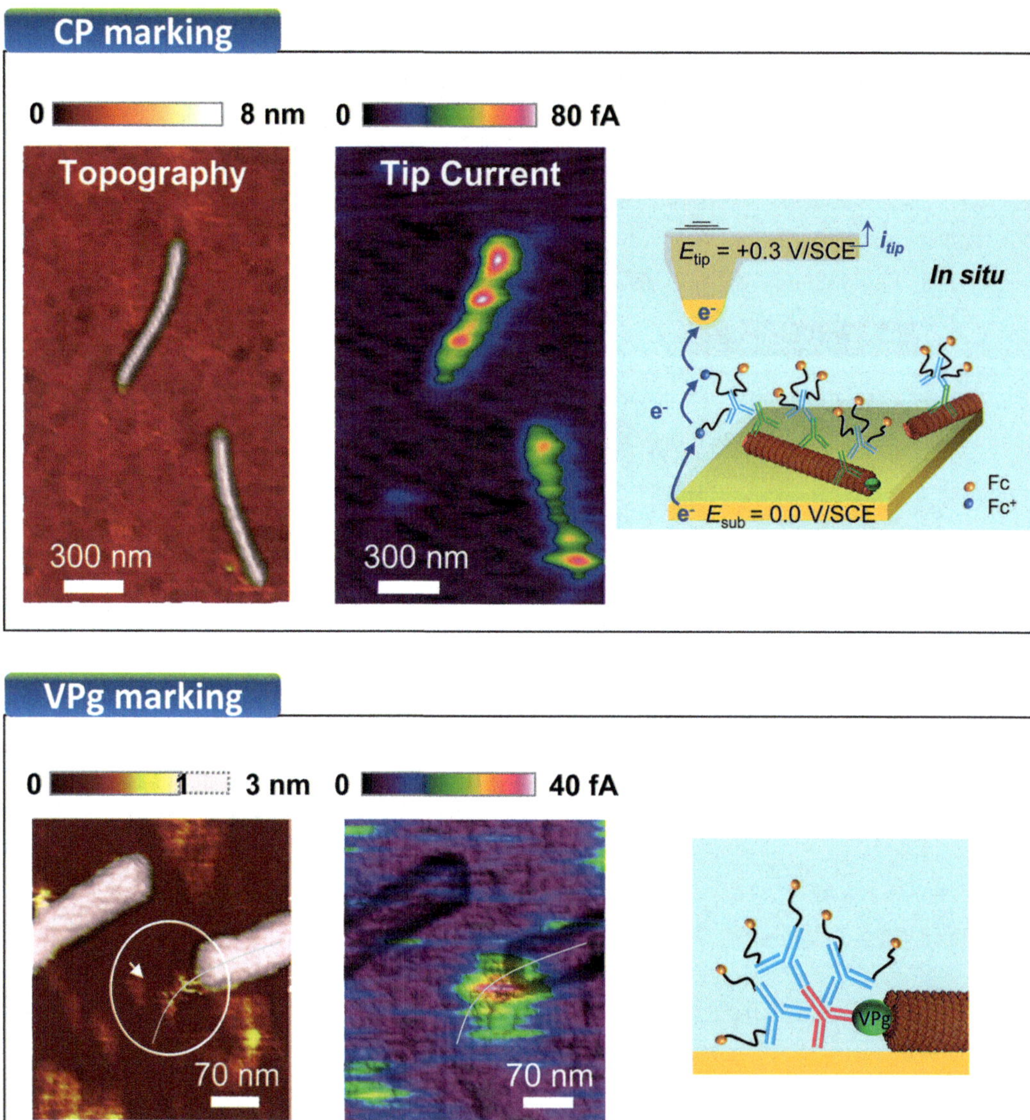

Fig. 5 Simultaneously acquired in situ topography and tip current AFM-SECM images of LMV particles immobilized on a gold surface and either CP- (Panel **a**) or VPg- (Panel **b**) redox immunomarked. The scheme in (**a**) depicts the electrochemical detection at the tip of the Fc heads borne by the Fc-PEGylated antibodies. The scheme in (**b**) represents the structure of an immunocomplex formed between VpG, one primary antibody, and three Fc-PEG labeled IgGs. Imaging medium: 10 mM pH 7.4 phosphate buffer (Reprinted with permission from ACS Nano 2015, 9:4911–4924. Copyright 2015 American Chemical Society)

be mapped. If CP proteins were targeted, a series of current spots should be visible all along the viral capsid (*see* Fig. 5a), whereas marked VPg should appear as an isolated current spot located at one of the extremities of viral particles (*see* Fig. 5b).

4 Notes

1. Infected leave material can be stored desiccated at 4 °C for a few years and used for inoculation according to needs.
2. All plants before and after inoculation are grown in confined greenhouse facilities under controlled light, moisture and temperature conditions: LED tubes Samsung 100 W (IRC white-warm-3000 K) covering a continuous 400–800 nm spectral range with two main peaks at 450 and 605 nm, absolute irradiance 25 and 43 μW/cm^2. Turn on tubes for a solar irradiance below 4.5 K.Lux. Set the day–night light cycle to 15 h and 9 h, respectively. Adjust the day-night temperature cycle to 22 °C and 19 °C, respectively. Maintain a relative humidity of 50%.
3. Preparation of 0.1 M KH_2PO_4: dissolve 13.6 g KH_2PO_4 in 1 L water. Preparation of 0.1 M Na_2HPO_4: dissolve 17.7 g Na_2HPO_4 in 1 L water.
4. Dissolve 40 g sucrose in 100 mL of T2 buffer. Use this stock 40% (w/v) solution to prepare 10 mL of each of the other sucrose solutions (30%, 15%, 5% (w/v) in T2).
5. Importantly, the antibody should be free of amines, carrier proteins and preservatives, such as sodium azide, as these species would interfere with the conjugation reaction.
6. The ferrocene-PEGylating NHS ester reagent is a specialized high-grade redox PEG product synthesized in house [5, 6], the PEG chain has a molar mass of 3500 Da, corresponding to ~80 (CH_2-CH_2O) monomer units.
7. Under these appropriate storage conditions (protection from humidity, oxygen and light), Fc-PEG-NHS reactants are stable for at least 1 year. The desiccator has to be taken out of the refrigerator the day before the conjugation reaction is carried out, and left to thermally equilibrate for several hours prior to opening. The required amount of Fc-PEG-NHS should then be weighed in an Eppendorf tube and stored overnight in a desiccator at room temperature before use.
8. pH 8 buffer offers the optimal trade-off between efficient coupling of the un-protonated form of the amines of the antibody with the succinimidyl ester NHS of the PEG chains and the competing hydrolysis of this reactive ester. The coupling buffer should be free of amines.

9. Preferably use all-plastic syringes with no rubber plunger seals, nor silicone lubricants.
10. Methoxy PEG_{2000} disulfide can be custom-synthesized as described elsewhere [12]. Reduced form, methoxy PEG thiol, available from different commercial sources may be used as an alternative. When possible, we purchase our PEGs from Jenkem Technology, a trustful manufacturer which offers high quality PEG products.
11. Template-stripped TS gold surfaces are prepared from mica substrates coated with a layer of gold. Glass slides are then glued to the gold surface. The resulting mica–gold–glass "sandwiches" can be stored for at least 1 month under ambient conditions before use. The ultraflat gold surface for adsorption is prepared just before use by detaching the glass slide from the mica substrate, either mechanically or with the help of tetrahydrofuran.
12. These antibodies are typically produced in house [13]. However commercially available anti-LMV capsid protein antibodies (Agdia-Biofords or Bioreba) may also be used, keeping in mind that commercial antibodies are often optimized for ELISA tests and may not be suitable for particle imaging.
13. Under the growth conditions described, a systemic infection spreads on the entire plant with symptoms present from lower to upper leaves. Take care not to wait until the appearance of tissue necrosis as it was observed that the yield of viral particles tends to decrease from this stage on.
14. Use the frozen leaves for immediate extraction or store them at −20 °C for extraction at a later date.
15. T2 buffer contains 2-mercaptoethanol. Subheading 3.2, **step 1** must be performed under the fume hood.
16. Light enhances tissue oxidation which results in a decrease of purification yield.
17. CAUTION: For all high-speed centrifugation steps, take care to prevent rotor unbalance by adjusting the weight of preparative tubes to maximal differences of about ±0.5 g.
18. This procedure can be optionally performed once or twice for washing pellets containing small darker aggregates. It was observed that this can increase the yield of viral particles at least by factor 2.
19. A western blot using antibodies specific for the major coat protein is highly recommended to check that the purified particles are truly LMV capsids and not from a contaminating virus.
20. Commercial antibodies (IgGs) are provided as sterile solutions in bottles closed by a septum, so one should use a sterile syringe and needle to withdraw the required volume of antibody.

21. Rinse the membrane of the Amicon unit by filling it with ~1 mL PBR buffer and spinning at 2500 × *g* for 5 min. Empty the unit just before use.
22. The spinning time and final volume are given for a fixed angle (38°) centrifuge rotor. For other types of rotors refer to the Amicon user's manual for the appropriate spinning time/conditions yielding a final volume of ~100 μL after about 10 min.
23. Removal of the unreacted chains from the antibody solution can be monitored by recording UV spectra of the successive eluates. The spectrum of the first eluate should show a strong band in the 260 nm region ascribable to unreacted Fc-PEG-NHS (and hydrolyzed NHS containing species, typical absorbance at a wavelength of 260 nm, A_{260} ~ 1.2 in a 1 cm path cell). This band should rapidly decrease in intensity from one eluate to the next. Purification is considered to be complete when the band becomes undetectable (typically after 5 spins).
24. These repeated centrifugation/reconstitution steps reduce the concentration of unconjugated Fc-PEG chains by a factor in excess of ~$1/10^6$. More precisely, the concentration of free Fc-PEG chains in the final concentrate is lowered to the nanomolar range.
25. The solution can be stored in the fridge for up to 6 months without significant loss of the recognition capability of the Fc-PEG-IgG conjugate.

Acknowledgments

This work was supported by the French Agence Nationale de la Recherche (ANR) under reference ANR-09-PIRI-0005 ("Cascade" project), and the mission for interdisciplinary of CNRS ("Nano" challenge program). Drs Cecilia Taofifenua and Laurent Nault are warmly thanked for their decisive contribution to the development of AFM-SECM imaging of viral particles. We thank Amandine Barra for SDS-PAGE and western blot analysis of LMV preparations.

References

1. Watkinson RE, McEwan WA, Tam JCH, Vaysburd M, Jame LC (2015) TRIM21 promotes cGAS and RIG-I sensing of viral genomes during infection by antibody-opsonized virus. PLoS Pathog 11:e1005253. https://doi.org/10.1371/journal.ppat.1005253
2. Marschall ALJ, Frenzel A, Schirrmann T, Schüngel M, Dübel S (2011) Targeting antibodies to the cytoplasm. MAbs 3:3–16
3. Stoermer KA, Morrison TE (2011) Complement and viral pathogenesis. Virology 411:362–373
4. Krause-Sakate R, Redondo E, Richard-Forget F, Salomão Jadão A, Houvenaghel MC, German-Retana S, Agenor Pavan M, Candresse T, Murilo Zerbini F, Le Gall O (2005) Molecular mapping of the viral determinants of systemic wilting induced by a Lettuce mosaic

virus (LMV) isolate in some lettuce cultivars. Virus Res 109:175–180
5. Anne A, Moiroux J (1999) Quantitative characterization of the flexibility of poly(ethylene glycol) chains attached to a glassy carbon electrode. Macromolecules 32:5829–5835
6. Abbou A, Anne A, Demaille C (2004) Probing the structure and dynamics of end-grafted flexible polymer chain layers by combined atomic force - electrochemical microscopy. Cyclic voltammetry within nanometer-thick macromolecular poly(ethylene glycol) layers. J Am Chem Soc 126:10095–10108
7. Anne A, Demaille C, Moiroux J (1999) Elastic bounded diffusion dynamics of ferrocene-labeled poly(ethylene glycol) chains terminally attached to the outermost monolayer of successively self-assembled monolayers of immunoglobulins. J Am Chem Soc 121:10379–10388
8. Hegner M, Wagner P, Semenza G (1993) Ultralarge atomically flat template-stripped Au surfaces for scanning probe microscopy. Surf Sci 291:39–46
9. Abbou J, Demaille C, Druet M, Moiroux J (2002) Fabrication of submicrometer-sized gold electrodes of controlled geometry for scanning electrochemical-atomic force microscopy. Anal Chem 74:6355–6363
10. Kuznetsov YG, McPherson A (2011) Atomic force microscopy in imaging of viruses and virus-infected cells. Microbiol Mol Biol Rev 75:268–285
11. Nault L, Taofifenua C, Anne A, Chovin A, Demaille C, Besong-Ndika J, Cardinale D, Carette N, Michon T, Walter J (2015) AFM-SECM imaging of redox-immunomarked proteins on native potyviruses: from subparticle to single-protein resolution. ACS Nano 9:4911–4924
12. Anne A, Chovin A, Demaille C, Lafouresse M (2011) High-resolution mapping of redox-immunomarked proteins using electrochemical-atomic force microscopy in molecule touching mode. Anal Chem 83:7924–7932
13. Roudet-Tavert G, German-Retana S, Delaunay T, Delécolle B, Candresse T, Le Gall O (2002) Interaction between potyvirus helper component-proteinase and capsid protein in infected plants. J Gen Virol 83:1765–1770

Chapter 31

Presenting Peptides at the Surface of Potyviruses In Planta

Flora Sánchez and Fernando Ponz

Abstract

Potyviruses are plant viruses with elongated, flexuous virions amenable to modifications in the only viral structural protein, the coat protein (CP). Out of the several theoretically possible modifications to the CP, the one most exploited for peptide presentation is the genetic fusion of the peptide-to-be-expressed, to the CP N-terminus. Successful high-level expression of the modified CP has been achieved this way. The purified recombinant viral particles incorporate most, if not all, the properties of the expressed peptides. For many purposes, the recombinant virus particles present in extracts of infected plants should be purified for further use. Procedures for carrying out the whole process, from cloning to purification are described in the chapter.

Key words Potyvirus, Recombinant CP, Peptide presentation, Virion purification

1 Introduction

Potyviruses form the most abundant group of plant viruses described to date. They encapsidate a genomic single-stranded (+) RNA molecule of ~10 kb in a membrane-less elongated and flexuous capsid about 700–750 nm long, and 15–18 nm wide. This particle is formed by approximately 2000 copies of a single viral protein (the coat protein, CP), which covers the viral RNA. The mature CP is proteolytically released from a polyprotein precursor which is cleaved by viral-encoded proteinases to produce the mature viral functional proteins [1]. Recently, some new potyviral proteins have been described, which are produced by alternative mechanisms [2–4].

The overall architecture of potyviral virions makes them especially well-suited for peptide presentation on their external surface. No potyvirus crystals have ever been described, so no X-ray diffraction structure is known. However, there is an appreciable degree of structural information on the particle from other approaches, thus paving the way for rational biotechnological developments based

Christina Wege and George P. Lomonossoff (eds.), *Virus-Derived Nanoparticles for Advanced Technologies: Methods and Protocols*, Methods in Molecular Biology, vol. 1776, https://doi.org/10.1007/978-1-4939-7808-3_31,

© Springer Science+Business Media, LLC, part of Springer Nature 2018

upon it [5]. Based on biochemical and immunological experimental approaches, it is long known that the CP N-terminal and C-terminal regions project toward the exterior of the flexuous tubular core of the virion [6]. Available structural models do not predict these externally projected regions of the CP to be particularly structured [7]. These regions, especially the N-terminal one, also represent the most immunodominant antigenic stretches of the CP when injected into animals [8], and the epitopes predominantly recognized by polyclonal antibodies obtained against the whole CP [6]. These properties derive directly from the spatial position of these regions.

It is logical that most, if not all, efforts to present foreign peptides on potyvirus particles have concentrated on the N-terminal region of the protein, both for antigenicity and for antibody recognition. Plum pox virus was the first potyvirus to be used for these developments [9], in this case searching for increases in the antigenic potential of certain peptides. The work with this potyvirus already identified the issue of the potential impairment of virus infectivity due to the insertions. This issue has not yet been properly solved, mostly because the underlying reasons are not well identified. Some researchers have pointed to peptide length and peptide overall charge as the most critical aspects, and even proposed some rules for expression success preserving virus infectivity [10]. However, this aspect still needs clarification. The zucchini yellow mosaic virus CP N-terminal region was eliminated and replaced with a nonviral peptide, a different approach illustrating the potency and flexibility of the potyviral platform for viral nanoparticle (VNP) peptide expression [11]. Our own work has focused on another potyvirus, turnip mosaic virus (TuMV). The production of peptides, inserted in the N-terminal region of the viral CP, was a second-generation development of the exploitation of infectious cDNA clones of the virus for the in planta expression and production of heterologous proteins [12]. The expression of a peptide derived from the human vascular endothelial growth factor receptor-3 (VEGFR-3), a protein involved in angiogenesis, lymphangiogenesis, and neurogenesis, showed the suitability of the virus for peptide expression with the double purpose of increasing antigenicity and antibody sensing [13].

This chapter will describe in detail the experimental approach to generating TuMV virions presenting a foreign peptide, using a VEGFR-3 peptide as an example, producing them in plants, and purifying them. Further improvements for the generation, production and purification procedures of potyviral VLPs in planta have been recently published [14] will not be covered in the chapter.

2 Materials

2.1 Media, Buffers, Reagents, and Solutions

1. Luria–Bertani (LB) medium: 10 g/L Bacto tryptone, 10 g/L, NaCl, 5 g/L yeast extract, pH 7.0 with NaOH. Add 0.1 mg/mL ampicillin.
2. LB agar with ampicillin: LB medium with 10 g/L agar added. Add 0.1 mg/mL ampicillin.
3. 0.5 M EDTA, pH 8.0, adjust pH with NaOH.
4. 0.5 M potassium phosphate buffer, pH 7.5: 0.5 M monobasic dihydrogen phosphate (KH_2PO_4)–0.5 M dibasic monohydrogen phosphate (K_2HPO_4), 16:84. Dilute this buffer for lower molarities (50 mM).
5. 0.5 M potassium phosphate buffer, pH 7.5, 0.01 M EDTA.
6. 0.25 M potassium phosphate buffer, pH 7.5, 0.01 M EDTA.
7. 0.5 M Tris–HCl, pH 8.2 and pH 7.5: 60.55 g Tris in 800 mL of water. Adjust pH to 8.2 or 7.5, respectively, with HCl. Make up to 1 L with water. Store at 4 °C.
8. 10 mM Tris–HCl pH 7.5, 10 mM EDTA.
9. Extraction buffer: 0.5 M Tris–HCl, pH 8.2, 2% (w/v) PVP-40, 1% (w/v) PEG 6000, 0.8% (w/v) NaCl, 0.05% (v/v) Tween 20.
10. 0.05 M carbonate buffer, pH 9.6. Na_2CO_3 1.59 g/L, $NaHCO_3$ 2.93 g/L.
11. 10× PBS: 80 g/L NaCl, 2 g/L KCl, 11.5 g/L Na_2HPO_4, 2 g/L KH_2PO_4.
12. PBS-Tween: 0.05% (v/v) Tween 20 in 1× PBS.
13. 0.5 M ammonium acetate.
14. Chloroform.
15. CsCl.
16. Mineral oil.
17. NaCl.
18. PEG 6000.
19. Carborundum.
20. Vector: p35Tunos-vec01-Nat1 [12].
21. Vector: pCR-Blunt II-TOPO vector (Invitrogen) (or any other vector prepared to clone blunt-ended PCR fragments).
22. Ampicillin stock solution 50 mg/mL in dH_2O, kanamycin stock solution 50 mg/mL in in dH_2O.
23. Primers (*see* Table 1).
24. dH_2O.
25. 100% glycerol.

Table 1
Primers used in construction and sequence determination

Primer	Sequence
TuMlu	ACGCGTGGATGATTGAACAAG
Tu130U	CAAGCAATCTTTGAGGATTATG
Tu340L	GTCGCGTTTTCCCTCTTC

2.2 Antibodies, Enzymes, and Kits

1. Monoclonal antibody anti-potyvirus group (anti-POTY) (Agdia, SRA 27200), diluted 1:200.
2. Geneclean Turbo kit (MP Biomedicals) to eliminate primers, buffer changes and DNA concentration.
3. Qiagen Plasmid extraction kit, Qiagen HiSpeed plasmid Midi kit, Qiagen plasmid mini kit (Qiagen).
4. Restriction enzymes: *Mlu*I, *Nae*I, *Afe*I with appropriate reaction buffers + blunt end restriction enzyme of choice.
5. Shrimp alkaline phosphatase.
6. T4 DNA ligase and reaction buffer.
7. Zero Blunt® TOPO® PCR Cloning Kit (Invitrogen).
8. Proofreading DNA polymerase.

2.3 Other

1. 3–4-week-old *Nicotiana benthamiana* plants grown at 21 °C (day, 16 h) and 18 °C (night, 8 h) at the four-leaf stage.
2. 3-week-old *Brassica juncea* plants grown at 25 °C maximum and 19 °C minimum at the four-leaf stage.
3. *E. coli* competent cells.
4. Blender.
5. Centrifuge, ultracentrifuge, fixed-angle rotors for ultracentrifuge, ultracentrifuge tubes.
6. Glass rods, cotton sticks.
7. Syringe, needle.
8. Miracloth, gauze.
9. Gene-specific primers for PCR with appropriate restriction sites. *See* **Note 1**.
10. Agarose gels and materials for gel electrophoresis.
11. Reagents for IC-RT-PCR and PCR, including buffers.
12. Equipment for electron microscopy.
13. Equipment for MALDI-TOF analysis.
14. Reagents for SDS-PAGE and Western Blot (with anti-POTY and antibody against the foreign peptide).

15. Polyacrylamide gel electrophoresis: 5% (w/v) gel, gel loading buffer containing Bromophenol Blue and Xylene Cyanol, staining solution containing 0.02% (v/v) Methylene Blue, gel extraction buffer (pH 8.0) containing 0.5 M ammonium acetate and 1 mM EDTA.
16. Plant growth chamber and greenhouse.
17. Spectrophotometer.

3 Methods

3.1 Potyvirus Vectors with Unique Restriction Sites Close to the 5′ End of the CP Cistron

Vector ZYMV-AGII has a polylinker (PstI, ScaI, SpeI, NheI, and SalI) close to the 5′ end of the CP cistron [15]. Fusions of different peptides to the amino terminus of CP, either by addition or substitution, have been described in the ZYMV-AGII vector [11].

Vector PPV-NATMluI has a unique *Mlu*I site close to the amino terminus of the CP, and expression of several peptides fused to the N-terminus of the CP has been described [9].

p35Tunos-vec01-Nat1, with a unique *Nae*I site coincident with the first two codons of the CP cistron has been described [12] and used for the fusion of peptides to the CP [13]. In the following methods, we will refer only to this particular vector.

3.2 Insertion of the Peptide Coding Sequence

The insertion of the peptide coding sequence is carried out by replacing the wild type *Mlu*I-*Nae*I fragment (191 bp) in the vector p35Tunos-vec01-Nat1 by a modified fragment that includes the peptide coding sequence at the 3′ terminal part. This *Mlu*I-*Nae*I fragment codes for a partial NIb gene and the first amino acid of the CP. Figure 1 illustrates the strategy.

1. Design the lower primer. This primer contains, from 5′ to 3′: (a) a restriction site for a blunt restriction enzyme (*see* **Note 2**), (b) the complementary sequence of the peptide-encoding sequence, and (c) a sequence of 17 nucleotides matching the 3′ end of the wild type *Mlu*I-*Nae*I fragment (GGCCTGGTGATAAACAC). An example with VEGFR-3 aa 1279–1298 peptide is given in **Note 1**.
2. Perform a PCR with template vector p35Tunos-vec01-Nat1 using a proofreading polymerase (e.g., *Pfu*). The oligonucleotide TuMlu is the upper primer. Reaction conditions: a denaturation step of 10 min at 94 °C, 35 cycles of 30 s at 94 °C, 1 min at 52 °C, and 1 min at 72 °C, and a final elongation step of 72 °C for 15 min. Check the result of the PCR by 2% agarose gel electrophoresis.
3. Clone the fragment into pCR-Blunt II-TOPO vector. Verify the integrity of the fragment by sequencing.

(a)

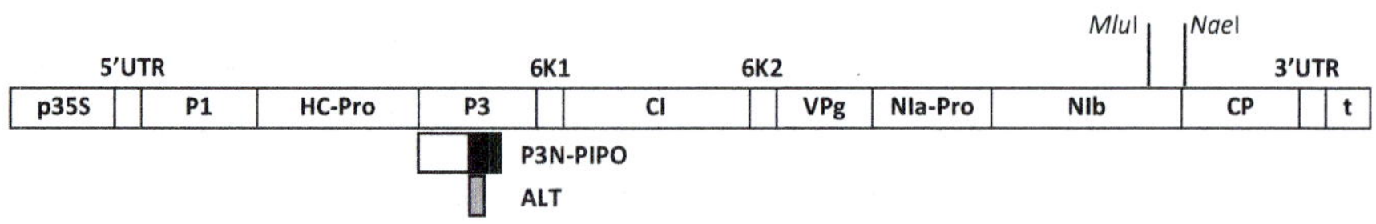

(b)

```
1         10        20        30        40        50
TAC GCG TGG ATG ATT GAA CAA GCT CCA TTT AGC TCC CTA GCA CAA GAA GGG AAA GCT CCT
ATG CGC ACC TAC TAA CTT GTT CGA GGT AAA TCG AGG GAT CGT GTT CTT CCC TTT CGA GGA
 Y   A   W   M   I   E   Q   A   P   F   S   S   L   A   Q   E   G   K   A   P>
   MluI
60        70        80        90        100       110
TAC ATA GCG GAA ACA GCG CTG AGG AAG CTC TAC CTT GAT AAG GAA CCA GCT CAA GAG GAT
ATG TAT CGC CTT TGT CGC GAC TCC TTC GAG ATG GAA CTA TTC CTT GGT CGA GTT CTC CTA
 Y   I   A   E   T   A   L   R   K   L   Y   L   D   K   E   P   A   Q   E   D>
                    AfeI
120       130       140       150       160       170
CTC ACC CAT TAT TTG CAA GCA ATC TTT GAG GAT TAT GAA GAT GGT GCT GAG GCT TGT GTT
GAG TGG GTA ATA AAC GTT CGT TAG AAA CTC CTA ATA CTT CTA CCA CGA CTC CGA ACA CAA
 L   T   H   Y   L   Q   A   I   F   E   D   Y   E   D   G   A   E   A   C   V>

180       190       200       210       220       230
TAT CAC CAG GCC GGC GAA ACG CTT GAT GCA GAG TTG ACA GAC GAG CAA AAG CAG GCA GAG
ATA GTG GTC CGG CCG CTT TGC GAA CTA CGT CTC AAC TGT CTG CTC GTT TTC GTC CGT CTC
 Y   H   Q   A   G   E   T   L   D   A   E   L   T   D   E   Q   K   Q   A   E>
                NaeI
240       250       260       270       280       290
AAG GAG AAG AAG GAG AGA GAG AAG GCA GAA AAG GAA CGA GAG AGG CAA AAG CAG TTG GCA
TTC CTC TTC TTC CTC TCT CTC TTC CGT CTT TTC CTT GCT CTC TCC GTT TTC GTC AAC CGT
 K   E   K   K   E   R   E   K   A   E   K   E   R   E   R   Q   K   Q   L   A>

300       310       320       330       340
CTC AAG AAA GGC AAG GAT GTT GCA CAA GAA GAG GGA AAA CGC GAC
GAG TTC TTT CCG TTC CTA CAA CGT GTT CTT CTC CCT TTT GCG CTG
 L   K   K   G   K   D   V   A   Q   E   E   G   K   R   D>
```

(c)

Vector	Construct	aa sequence Nt-CP
p35Tunos-vec01-Nat1	-	A GETLDAELTDEQ
p35Tunos-vec01-Nat1	With inserted peptide	A- inserted peptide -GETLDAELTDEQ

Fig. 1 Representations of chimeric clones. (**a**) Schematic illustration of p35Tunos-vec01-Nat1 indicating the positions of *MluI* and *NaeI* restriction sites for insertion of the foreign peptide. The insertions are actually performed through the replacement of a *MluI*-*NaeI* fragment containing the peptide sequence. Abbreviations: generic names of potyvirus cistrons (P3N-PIPO and P3N-ALT are generated through transcriptional slippage of the viral RNA polymerase); viral UTRs; 35S promoter from cauliflower mosaic virus (p35S); *nos* terminator (t). (**b**) Nucleotide and amino acid sequences of p35Tunos-vec01-Nat1 in a region of 340 bp starting at the *MluI* restriction site. Restriction sites for *MluI*, *NaeI*, and *AfeI* are underlined and indicated. In bold are shown the primers TuMlu, Tu130U, Tu340L, and the part of "lower primer" that matches the TuMV sequence, at their corresponding sites positions. The amino acid sequence of NIb is highlighted in light grey and the amino acid sequence of CP highlighted in blue (darker grey in the paper version). (**c**) Comparison of the amino acid sequences of p35Tunos-vec01-Nat1 and the resulting construction with peptide inserted at the site of insertion in the N-terminal region (Nt) of TuMV CP. The insertion of the coding sequence for the peptide is between the first and the second amino acid of the TuMV CP (alanine, glycine)

4. Prepare 5 μg of the p35Tunos-Vec01-Natl vector for the ligation reaction by double digestion with *Mlu*I and *Nae*I restriction enzymes (*see* **Note 3**). Change the restriction enzymes buffer and concentrate by Geneclean Turbo kit. Treat the digested plasmid with 2 U of shrimp alkaline phosphatase at 37 °C for 30 min. Inactivate the phosphatase by incubating the sample at 65 °C for 15 min. Quantify the concentration of the prepared vector by agarose gel electrophoresis.
5. Obtain the modified DNA fragment by double digestion of the pCR-Blunt II-TOPO vector with insert sequence-confirmed clone with *Mlu*I and selected restriction enzyme corresponding to the restriction site introduced into the 'lower' primer. Digest 5 μg of plasmid DNA.
6. Purify the insert from polyacrylamide gel electrophoresis (SDS-PAGE) by the crush and soak method, follow the described protocol [16] with the modifications indicated later.
7. Load the digestion in a 5% polyacrylamide gel, the gel loading buffer should contain Bromophenol Blue (BPB) and Xylene Cyanol (XC). Run the electrophoresis at 7 V/cm until the BPB reaches the bottom of the gel. The dye XC migrates at a similar position as a DNA fragment of 260 bp. The modified fragment (larger than 191 bp) should migrate faster than the XC.
8. Detach the glass plates and stain the gel with 0.02% Methylene Blue for 15 min or until the DNA bands are visible.
9. Cut the selected band with a scalpel, transfer to a microcentrifuge tube, crush the polyacrylamide with a disposable pipette tip, and add 300 μL of 0.5 M ammonium acetate, 1 mM EDTA pH 8.0 solution. Incubate at 30 °C or 37 °C on a rotary wheel.
10. Purify and concentrate the eluted DNA by Geneclean Turbo kit. Quantify the concentration of the insert by agarose gel electrophoresis.
11. Ligate the double-digested phosphatase-treated vector (from **step 4**) with the insert. To 800 ng of double-digested phosphatase-treated vector, add 20 ng of the insert, 1 μL of 10× Ligase buffer and 200 U of T4 DNA ligase, in a final volume of 10 μL. Incubate at 14 °C for 8–12 h.
12. Transform *E. coli* competent cells by the heat-shock method with the ligation reaction products. Plate 1/10 and the rest on LB agar plates containing 0.1 mg/mL ampicillin. Incubate the plates at 30 °C overnight (*see* **Note 4**).
13. Prepare plasmids for 4–6 different clones (minipreparation scale). Grow overnight in liquid medium containing ampicillin at 30 °C, prepare the plasmids by the Qiagen plasmid mini kit. Analyze the presence of recombinants by cutting

the plasmids with the appropriate enzymes (*Afe*I, *Mlu*I, and *Nae*I). *See* **Note 5**.

14. Confirm integrity of the putative chimeric viruses by sequencing.
15. Purify the confirmed plasmids at large scale by HiSpeed Qiagen plasmid Midi kit (*see* **Note 6**).

3.3 Infectivity and Stability Assays of the Recombinant Virus

1. Dust the two younger fully expanded leaves of *Nicotiana benthamiana* plants in the four true-leaf stage with carborundum and inoculate with 5 μL of purified plasmid DNA (passage 0) by rubbing with a glass rod or a pipette tip.
2. Symptoms of virus infection (younger, noninoculated leaves will appear smaller and more wrinkled than the mock inoculated) should appear at 7–15 days after inoculation. Figure 2 shows mock inoculated and infected leaves of *N. benthamiana*.
3. Amplify a fragment of the virus progeny by immunocapture RT-PCR (IC-RT-PCR). Coat PCR tubes with anti-POTY diluted 1:200 in 0.05 M carbonate buffer pH 9.6 by incubating for 2 h at 37 °C. After two washings with PBS-Tween, add plant samples prepared by grinding leaf material in extraction buffer and incubate overnight at 4 °C. After three washings with PBS–Tween, RT-PCR is performed with primers Tu130U and Tu340L. Reaction conditions are: reverse transcription of 30 min at 42 °C, a denaturation step of 4 min at 95 °C, 35

Fig. 2 Detached leaves of *N. benthamiana*. The left panel shows a leaf from a noninfected plant. The right panel shows leaves from different plants infected with the recombinant potyvirus (TuMV expressing the VER3 peptide) exhibiting different symptoms of virus infection

cycles of 20 s at 94 °C, 30 s at 53.5 °C, and 1 min at 72 °C, and a final elongation step of 72 °C for 5 min. Clean the fragment from primers by Geneclean Turbo kit. Send to sequence with primer Tu340L. Check the resulting sequence for integrity.

4. Prepare a crude sap from infected plants (from **step 2**) by grinding symptomatic leaf material in 1:10 (w/v) 50 mM potassium phosphate buffer pH 7.5. Inoculate *Nicotiana benthamiana* by rubbing the crude sap (passage 1) on carborundum-dusted leaves with cotton sticks.
5. Repeat **step 3** to check the stability of the recombinant virus in passage 1.

3.4 Recombinant Virus Purification

1. Inoculate 25 *Brassica juncea* plants in the four true-leaf stage by dusting the two younger expanded leaves with carborundum and rubbing with crude sap from *Nicotiana benthamiana* passage 1 (from Subheading 3.3, **step 4**). Symptoms of virus infection (mosaic with pale green and dark green areas) will appear at 6–8 days after inoculation. Keep the inoculated plants for 4 weeks before virus purification.
2. Virus purification is carried out according to [17] with minor modifications, in three main steps: Infected tissue grinding and clarification (Subheading 3.4.1), virus concentration (Subheading 3.4.2) and virus isopycnic centrifugation (Subheading 3.4.3). All the steps are performed at 4 °C.

3.4.1 Infected Tissue Grinding and Clarification

1. Collect the infected leaves and determine the exact weight (we typically harvest about 250 g from 25 infected plants).
2. Grind the material in a blender adding 0.5 M potassium phosphate buffer, pH 7.5. Use 100 mL of buffer per 50 g of leaves.
3. Filter the purée through gauze. Add 50 mL of chloroform per 50 g of leaves. Stir in a magnetic stirrer for 10 min at 4 °C.
4. Transfer to 250 mL centrifuge bottles and separate the phases by centrifuging at 500 × *g* for 10 min at 4 °C.
5. Collect the upper aqueous phase and filter through Miracloth.
6. Centrifuge the filtrate at 5000 × *g* for 10 min at 4 °C.
7. Filter the supernatant through Miracloth, measure the volume in a graduated cylinder and transfer to an Erlenmeyer flask on ice.

3.4.2 Virus Concentration

1. Add NaCl to a final concentration of 4% (w/v), and when NaCl is dissolved, add PEG 6000 to a final concentration of 6% (w/v) and stir for 90 min at 4 °C.
2. Centrifuge at 12,000 × *g* for 10 min at 4 °C.

3. Discard the supernatant. Add 0.5 M potassium phosphate buffer, pH 7.5 and 0.01 M EDTA pH 8.0 to the pellet (use ~10 mL per 50 g of leaves). Resuspend the pellet by stirring overnight at 4 °C on a magnetic stirrer.
4. Clarify the suspension by centrifuging at 9000 × *g* for 10 min at 4 °C.
5. Filter the supernatant through Miracloth.
6. Pellet the virus by ultracentrifugation at 80,000 × *g* for 2 h at 4 °C.
7. Discard the supernatant. Add 8 mL (50–500 g of collected leaves, Subheading 3.4.1, **step 1**) of 0.25 M potassium phosphate buffer, pH 7.5, 0.01 M EDTA to the pellet and resuspend the pellet stirring for 2 h at 4 °C on a magnetic stirrer.

3.4.3 Isopycnic Centrifugation

1. Transfer the preparation to a graduated cylinder and adjust the volume to 9 mL with 0.25 M potassium phosphate buffer, pH 7.5, 0.01 M EDTA at 4 °C.
2. Transfer the preparation to a 10 mL plastic tube containing 3.6 g of CsCl. Dissolve by inverting the tube. Final volume is 10 mL and the refractive index is 1.3596.
3. Transfer to an ultracentrifuge disposable tube (total volume of the tube should be approximately 14 mL). Fill the tube completely with mineral oil and close it according to the instructions of the manufacturer. If even tubes are used they will be balanced. If the gradient tubes are odd, balance with another tube containing saturated $ZnSO_4$ and mineral oil.
4. Centrifuge at 150,000 × *g* for 18 h at 4 °C in a fixed-angle rotor without brake.
5. The virus band should be visible approximately at the middle of the tube (*see* Fig. 3). Open the tube or puncture the top of the tube with a hypodermic needle in order to remove vacuum. Remove the virus band by puncturing the tube just below the band with a needle attached to a 5 mL syringe.

3.4.4 Determining Virus Concentration and Eliminating CsCl

1. Transfer the contents of the syringe to a clean tube. Measure the absorbance of the solution, diluted 1:10 in dH_2O, at 260 and 280 nm. A ratio A_{260}/A_{280} of 1.14–1.25 will indicate purity of the viral preparation [18]. Calculate the concentration of the virus by the formula of the Lambert Beer law $c = A/\varepsilon$, where c is the concentration in mg/mL, A is the absorbance at 260 nm ($A_{0.1\%,\,1\text{cm}}$ at 260 nm), and ε is the TuMV absorption coefficient (2.65 M^{-1} cm^{-1}) (*see* **Note 7**).
2. Eliminate the CsCl by diluting the preparation at least ten times with 0.25 M potassium phosphate buffer, pH 7.5, 0.01 M EDTA at 4 °C (up to 30 mL total volume) and centrifuging at 80,000 × *g* for 2 h at 4 °C.

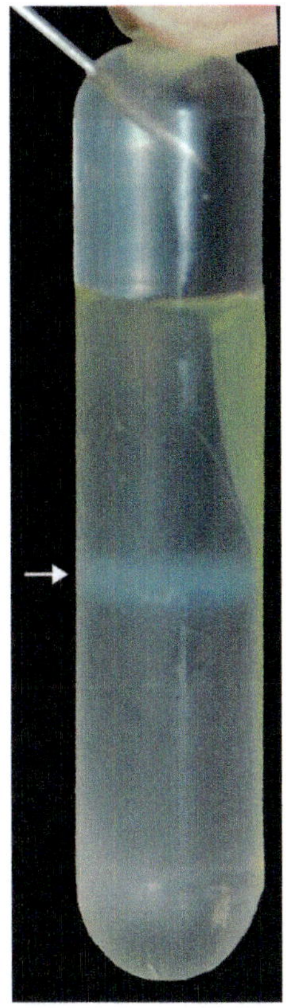

Fig. 3 Ultracentrifuge tube after potyvirus centrifugation through a cesium chloride gradient. The white arrow marks the whitish band containing the virus

3. Discard the supernatant and resuspend the pellet in 1 mL 10 mM Tris–HCl pH 7.5, 10 mM EDTA precooled at 4 °C by stirring on a magnetic stirrer for 2 h at 4 °C.
4. Determine the concentration of the preparation (*see* **step 1** and **Note 8**). Add an equal volume of 100% glycerol, mix and store at −20 °C.

3.5 Quality Control of the Purified Recombinant Virions

The purified recombinant virus particles should be assessed for expected traits, like the presence of the intact foreign peptide and its biological functionality. The recombinant viral coat protein (CP) should exhibit a slower electrophoretic mobility in Western blot assays in comparison with the wild-type CP. Optimally this analysis should be performed separately with two different antibodies. The anti-POTY mAb (Agdia) will recognize both types of CPs and reveal the slower mobility of the peptide-carrying one. A second antibody recognizing epitopes present in the foreign peptide should react only with the recombinant CP. An example of this analysis is shown in Fig. 4.

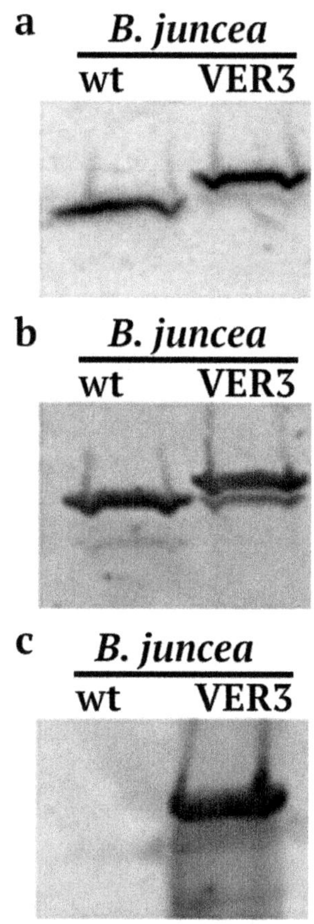

Fig. 4 Analysis of the recombinant potyvirus coat protein by SDS-PAGE and western blot. TuMV particles purified from infected Indian mustard plants (*B. juncea*) were disassembled in Laemmli loading buffer and electrophoretically separated. In each panel the left lane corresponds to wild-type virus and the right lane to the recombinant one, expressing the VER3 peptide. (**a**) Coomassie Blue-stained gel, (**b**) Western blot detected with the anti-POTY MAb, (**c**) Western blot using a polyclonal antibody against the VER3 peptide (reproduced from [13] with permission from Elsevier)

The integrity of the foreign peptide can easily be assessed if the recombinant CP is subjected to MALDI-TOF analysis, which provides the exact Mw of the protein. Again both wild-type and recombinant CP should be assayed in order to obtain the peptide Mw in the recombinant protein and compare with the expected one. The methodological steps for both these analyses have been thoroughly described in the Methods in Molecular Biology Series (*see* for instance, [19, 20]).

The assessment of peptide biological functionality will depend on the specific cases. Almost universally, however, the peptide should be detected on the surface of the potyvirus particle by an antibody recognizing it (with any immunological technique). If the recombinant virus particles immunodecorated with the specific peptide antibody are inspected under the electron microscope,

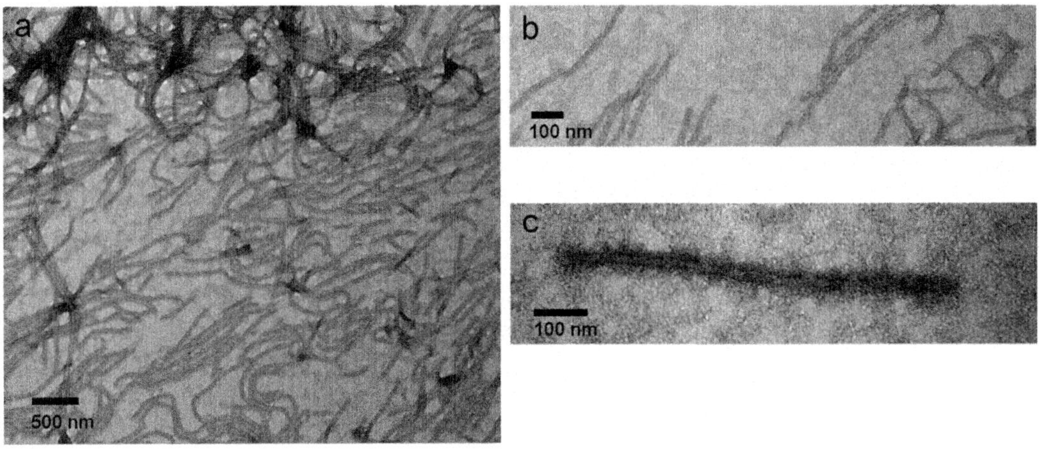

Fig. 5 Electron micrographs of negatively UAc-stained nonrecombinant and recombinant TuMV VNP preparations. (**a**) A low magnification (×8000) shows that the recombinant has the typical morphology of potyvirus particles. Higher magnifications show that nonrecombinant TuMV virions are not decorated with a polyclonal antibody against the foreign peptide expressed (**b**, ×120,000), whereas recombinant TuMVs are heavily decorated (**c**, ×200,000), revealing the presence of the peptide on the VNP surface (reproduced from [13] with permission from Elsevier)

images of wider and darker particles should be obtained, revealing antibody attachment (*see* Fig. 5). Other functionalities will depend on the specific cases. Features such as increased peptide immunogenicity, affinity for ligands, attachments to specific cell types, etc. require specifically designed assays.

4 Notes

1. For inserting the VEGFR-3 aa 1279–1298 peptide (LASEEFEQIESRHRQESGFR), the designed primer was the following: AGGCCTGAAGCCGCTTTCTTG TCTATGTCTAGACTCGATCTGCTCGAATTC TTCTGAAGCAAG**GGCCTGGTGATAAACAC** (underlined is the restriction site for *Stu*I, followed by the complementary sequence of 19 amino acids of the peptide-encoding sequence, and in bold 17 nucleotides matching the 3′ end of the wild type *Mlu*I-*Nae*I fragment). The clone of the PCR product will be digested with *Mlu*I and *Stu*I. Cloning the *Mlu*I-*Stu*I fragment in p35Tunos-vec01-Nat1 (*Mlu*I-*Nae*I digested) will reconstruct the *Mlu*I site but not the *Nae*I site.
2. The target sequence of the selected restriction enzyme should neither be present in the wild-type *Mlu*I-*Nae*I fragment, nor in the added foreign peptide.
3. Check that the digestion is completed by running an aliquot of the digestion reaction in a 0.5% (w/v) agarose gel compared with undigested plasmid DNA.

4. Incubation at 30 °C is mandatory, as this plasmid is unstable at 37 °C.
5. The correct recombinant plasmids should have a single *Mlu*I site and two *Afe*I sites (fragment sizes: 7.7, 5.5 kbp), one of them within the *Mlu*I-*Nae*I fragment (*see* Fig. 1), and one site for *Nae*I only if the recognition site for this enzyme was chosen for the 5′ end of the primer. A plasmid without insert will have zero sites for *Mlu*I and *Nae*I and one site for *Afe*I.
6. The usual concentration of the recombinant plasmids with this kit is 200–300 ng/μL.
7. Mean of the absorption coefficient determined for a few potyvirus species ranges from 2.4 to 2.9 [18].
8. The usual yield is 2.4 mg of virus per 100 g of infected *B. juncea* leaves.

Acknowledgments

The authors thank Ivonne González-Gamboa, Carmen Yuste-Calvo, Papaiah Sardaru, and Silvia López-González for their generous gift of graphic material. The technical support of Lucía Zurita is deeply acknowledged. Work about nanobiotechnological developments of TuMV has been funded over the years by several INIA grants.

References

1. Ivanov KI, Eskelin K, Lohmus A et al (2014) Molecular and cellular mechanisms underlying potyvirus infection. J Gen Virol 95:1415–1429. https://doi.org/10.1099/vir.0.064220-0
2. Chung BY, Miller WA, Atkins JF et al (2008) An overlapping essential gene in the Potyviridae. Proc Natl Acad Sci U S A 105:5897–5902. https://doi.org/10.1073/pnas.0800468105
3. Mingot A, Valli A, Rodamilans B et al (2016) The P1N-PISPO trans-frame gene of Sweet potato feathery mottle potyvirus is produced during virus infection and functions as RNA silencing suppressor. J Virol 90:3543–3557. https://doi.org/10.1128/JVI.02360-15
4. Hagiwara-Komoda Y, Choi SH, Sato M et al (2016) Truncated yet functional viral protein produced via RNA polymerase slippage implies underestimated coding capacity of RNA viruses. Sci Rep 6:21411. https://doi.org/10.1038/srep21411
5. Kendall A, McDonald M, Bian W et al (2008) Structure of flexible filamentous plant viruses. J Virol 82:9546–9554. https://doi.org/10.1128/JVI.00895-08
6. Shukla DD, Strike PM, Tracy SL et al (1988) The N and C-termini of the coat proteins of potyviruses are surface-located and the N-terminus contains the major virus-specific epitopes. J Gen Virol 69:1497–1508. https://doi.org/10.1099/0022-1317-69-7-1497
7. Baratova LA, Efimov AV, Dobrov EN et al (2001) In situ spatial organization of potato virus A coat protein subunits as assessed by tritium bombardment. J Virol 75:9696–9702. https://doi.org/10.1128/JVI.75.20.9696-9702.2001
8. Shukla DD, Tribbick G, Mason TJ et al (1989) Localization of virus-specific and group-specific epitopes of plant potyviruses by systematic immunochemical analysis of overlapping peptide fragments. Proc Natl Acad Sci U S A 86:8192–8196
9. Fernández-Fernández MR, Martínez-Torrecuadrada JL, Casal JI et al (1998) Development of an antigen presentation system based on plum pox potyvirus. FEBS Lett 427:229–235. https://doi.org/10.1016/S0014-5793(98)00429-3

10. Kimalov B, Gal-On A, Stav R et al (2004) Maintenance of coat protein N-terminal net charge and not primary sequence is essential for zucchini yellow mosaic virus systemic infectivity. J Gen Virol 85:3421–3430. https://doi.org/10.1099/vir.0.80417-0
11. Arazi T, Shiboleth YM, Gal-On A (2001) A nonviral peptide can replace the entire N terminus of zucchini yellow mosaic potyvirus coat protein and permits viral systemic infection. J Virol 75:6329–6336. https://doi.org/10.1128/JVI.75.14.6329-6336.2001
12. Touriño A, Sánchez F, Fereres A et al (2008) High expression of foreign proteins from a biosafe viral vector derived from *Turnip mosaic virus*. Span J Agric Res 6:48–58. https://doi.org/10.5424/sjar/200806S1-373
13. Sánchez F, Sáez M, Lunello P et al (2013) Plant viral elongated nanoparticles modified for log-increases of foreign peptide immunogenicity and specific antibody detection. J Biotechnol 168:409–415. https://doi.org/10.1016/j.jbiotec.2013.09.002
14. González-Gamboa I, Manrique P, Sánchez F et al (2017) Plant-made potyvirus-like particles used for log-increasing antibody sensing capacity. J Biotechnol 254:17. https://doi.org/10.1016/j.jbiotec.2017.06.014
15. Arazi T, Slutsky SG, Shiboleth YM et al (2001) Engineering zucchini yellow mosaic potyvirus as a non-pathogenic vector for expression of heterologous proteins in cucurbits. J Biotechnol 87:67–82. https://doi.org/10.1016/S0168-1656(01)00229-2
16. Sambrook J, Russell DW (2006) Isolation of DNA fragments from polyacrylamide gels by the crush and soak method. CSH Protoc 2006. https://doi.org/10.1101/pdb.prot2936
17. Spence NJ (1992) The identification, distribution and ecology of bean common mosaic virus in Africa. PhD Thesis, University of Birmingham, UK., Birmingham
18. Hollings M, Brunt AA (1981) Potyvirus group. CMI/AAB descriptions of plant viruses No 245
19. Kurien BT, Scofield RH (2015) Western blotting: an introduction. Methods Mol Biol 1312:17–30. https://doi.org/10.1007/978-1-4939-2694-7_5
20. Savary BJ, Vasu P (2012) Routine identity confirmation of recombinant proteins by MALDI-TOF mass spectrometry. Methods Mol Biol 824:37–50. https://doi.org/10.1007/978-1-61779-433-9_2

Chapter 32

Engineering of M13 Bacteriophage for Development of Tissue Engineering Materials

Hyo-Eon Jin and Seung-Wuk Lee

Abstract

M13 bacteriophages have several qualities that make them attractive candidates as building blocks for tissue regenerating scaffold materials. Through genetic engineering, a high density of functional peptides and proteins can be simultaneously displayed on the M13 bacteriophage's outer coat proteins. The resulting phage can self-assemble into nanofibrous network structures and can guide the tissue morphogenesis through proliferation, differentiation and apoptosis. In this manuscript, we will describe methods to develop major coat-engineered M13 phages as a basic building block and aligned tissue-like matrices to develop regenerative nanomaterials.

Key words M13 bacteriophage, Nanofiber, Phage engineering, Tissue regenerating materials, Self-assembly

1 Introduction

The ability to design biomimetic matrices with precise control to provide chemical, physical, and mechanical cues for regulating cell behavior is critical for the development of tissue regenerating materials [1–4]. The design of tissue engineering materials takes the in vivo cellular microenvironment into consideration [5–11]. In vivo, cells are in close contact both with each other and with the extra cellular matrix (ECM) that contains many diffusible factors. The ECM consists of an interconnected network of fibrous proteins that form self-organized structures to provide biochemical signals, physical support, and topographical cues for regulating cellular behavior (i.e., proliferation, differentiation, polarization, and migration). Recently, top-down and bottom-up material synthesis approaches [6, 8–15] have been developed to approach a level of complexity observed in the natural tissue environment. Nanofibrous materials offer increased surface area for cell engagement [10, 13], thereby improving cell interaction and material integration, as well as cell proliferation and differentiation [6, 12, 13]. Controlling the

Christina Wege and George P. Lomonossoff (eds.), *Virus-Derived Nanoparticles for Advanced Technologies: Methods and Protocols*, Methods in Molecular Biology, vol. 1776, https://doi.org/10.1007/978-1-4939-7808-3_32,
© Springer Science+Business Media, LLC, part of Springer Nature 2018

density of displayed signaling molecules has permitted quantitative studies of cell behavior [9, 11, 12, 16–18]. Characterization of cell growth on aligned topographical surfaces has shown that cells can respond to ordered textures with polarization and coalignment [6, 13, 19]. Recently, mechanical properties of the materials could regulate the cellular behaviors such as proliferation or differentiation of stem cells (*see* Chapter 39) [20, 21]. Furthermore, by controlling the mechanical stiffness in stimuli-responsive manners, we can more closely mimic in vivo microcellular environments. These various engineering approaches provide different combinations of advantageous material attributes. However, no single technique has been able to simultaneously satisfy the multiple requirements of tissue regenerating scaffolds: to provide a cell-conducive environment, display signaling molecules in a controlled manner, form self-organized nanofilamentous scaffolds, control mechanical stiffness in dynamic and responsive manners, and control macroscopic cellular behavior [6, 10].

1.1 Phages as Novel Bionanomaterials

Phage engineering provides unprecedented opportunities for creating novel nanomaterials by integrating the principles of biology, chemistry, and materials sciences. M13 is a bacteriophage (virus) composed of a single-stranded DNA encapsulated by various major and minor coat proteins. It has a long-rod filamentous shape that is approximately 880 nm long and 6.6 nm wide [22, 23]. Through genetic modification, functional peptides can be simultaneously displayed on the pIII and pIX minor coat proteins and the pVIII major coat protein [23–25]. M13 phages have several features that make them very attractive, compared with conventional materials, for constructing nanoscale materials and devices [26–30]. M13 replicates through bacterial amplification and is nonlytic, thus producing little cell debris during amplification and thereby simplifying the purification processes [31, 32]. Therefore, mass scale production of the virus can be easily realized through infection of *E. coli* cells, resulting in a monodisperse population of phages [26, 31]. Due to their monodispersity and long-rod shape, the phages have the ability to form higher order structures and have been extensively studied as highly organized liquid crystalline systems [26, 33, 34]. The concentration of the viral suspension, the ionic strength of the solution, and the externally applied force fields have been used to modulate viral organization in these systems and have previously been optimized for the construction of one-, two-, and three-dimensional phage-based materials [26, 35–37]. Furthermore, through the insertion of random gene sequences into the phage genome, a large combinatorial library can be displayed on the phage's major and minor coat proteins. Smith et al. developed the process of phage display to identify phages that exhibit various biologically active peptide motifs and to select desired properties through high-throughput screening processes [23, 24, 38].

Recently, we developed novel phage-based biomimetic nanofiber matrices that could easily be engineered to display various biochemical cues and to form self-assembled nanostructures for regulating cellular behaviors. Using this phage-based tissue matrix design system, we investigated various protein and cellular interfaces and discovered new findings on the interactions between proteins and cells. In our recent work [39], we examined the effects of a collagen-derived biochemical cue (DGEA) on bone-associated cells. We engineered M13 phage to display the DGEA-peptide in high densities on their major coat proteins and studied their effects on mouse-derived bone stem cells (preosteoblasts, MC3T3-E1). Through our study, we showed that the DGEA-peptides could stimulate bone stem cells to outgrow, an event that is linked to osteogenic differentiation. We expanded our protein-cell interface investigation by using other cell types and biochemical cues. Our recent papers [40, 41] reported that phages engineered to express RGD (integrin binding peptide) and IKVAV (neural cell stimulating peptide) peptides formed two- and three-dimensional matrices that regulated the directional growth of neural cells in a chemical and physical cue-specific manner. In addition, our recent paper [39] presented a facile strategy for immobilizing growth factors on genetically engineered phage matrices for tissue regeneration. We modified M13 phages to express streptavidin-binding peptides (HPQ) and/or integrin binding peptides (RGD) on their major and minor coat proteins. The resulting phages formed nanofibrous matrices that could easily immobilize the streptavidin-conjugated growth factors FGF-b and NGF for neural cell proliferation and differentiation. We verified that the immobilized growth factors possessed a prolonged ability to stimulate the target cells. We demonstrated the synergistic roles of the growth factors and integrin binding peptides in controlling cell morphologies and the growth of neural cells. Our phage matrices, which could be easily functionalized with various ligands and growth factors, can be used as convenient test beds for investigating the functions of various biochemical stimulants on numerous cell types. In our recent work, we combined self-assembly ability of phage with the RGD- and EEEE- (four glutamates) engineered phage to form hierarchically organized soft and hard tissue to enhanced the mechanical properties of the phage through the biomineralization of calcium phosphates which closely mimicked the structure of dental enamels [28]. We also grew calcium carbonate based compositely materials which could support the fibroblast cell growth in a similar approach [42]. By expanding similar approaches, we expect that novel tissue engineering materials can be further developed and that we can study the matrices and cellular interactions on the molecular level. In addition, we can fabricate the functional matrices to deliver therapeutic gene or protein information [43, 44]. Therefore, in this manuscript, we describe a general method to engineer the phage with a specific ligand in the major coat protein to create aligned tissue-like structures, and use them for tissue regenerating materials.

2 Materials

Prepare all solutions using autoclaved ultrapure water and analytical grade reagents. Prepare and store all reagents and buffers at room temperature, unless indicated otherwise.

2.1 Genetic Engineering of Phage

2.1.1 Genetic Engineering

1. Oligonucleotide primers (Table 1) (Integrated DNA Technologies) [39, 40, 45].
2. Phusion™ High-Fidelity DNA Polymerase (New England Biolabs).
3. M13KE phage vector (New England Biolabs) containing a *Pst*I restriction site in pVIII gene (*see* **Notes 1** and **2**).
4. Agarose for electrophoresis.
5. 1 kb DNA ladder (New England Biolabs).
6. Material for SYBR® Safe DNA Gel staining (ThermoFisher).
7. QIAquick Gel extraction kit (Qiagen).
8. Essential restriction enzymes: *Pst*I.
9. 10× NEB Buffer 3.1 (New England Biolabs).
10. QIAquick® PCR Purification Kit.
11. T4 DNA ligase with ligase buffer (New England Biolabs).

2.1.2 Electroporation

1. Electrocompetent XL1-Blue (Agilent) or other F^+ strain (freezer stock + mid-log culture).
2. Two sterile electroporation cuvettes (0.1 cm gap).
3. 20% (w/v) glucose solution. Filter-sterilize.
4. SOC medium (Invitrogen): 5 g/L yeast extract, 20 g/L tryptone, 0.584 g/L NaCl, 0.186 g/L KCl, 2.4 g/L $MgSO_4$. Autoclave, cool to <50 °C, add 20 mL 20% glucose solution per liter.
5. MicroPulser™ Electroporator (Bio-Rad).
6. pUC18 (control plasmid).
7. 14 mL Falcon polypropylene round-bottom tubes (BD Biosciences).
8. 50 mL Falcon conical tubes (BD Biosciences).
9. Microcentrifuge tubes (VWR).
10. LB broth Miller (EMD Millipore Corporation).
11. LB medium: 25 g/L LB broth Miller, autoclave.

Table 1
Primer sequences for pVIII major coat protein engineering of M13 phage

Primer name	Oligonucleotide primer sequence[a]	Insert peptide sequence[b]	Function
IKVAV primer	5′-ATATAT**CTGCAG**NK(NNK)$_2$ *CGTGGTGAC*(NNK)$_2$GATCCCGCAAAAGCGGCCTTTAACTCCC-3′	A*XXXIKVAVXX*DP[c]	Neural cell stimulation
RGD primer	5′-ATATAT**CTGCAG**NK(NNK)$_2$ *CGTGGTGAC*(NNK)$_2$GATCCCGCAAAAGCGGCCTTTAACTCCC-3′	A*XXXRGDXX*DP[c]	Integrin binding
DGEA primer	5′-ATATAT**CTGCA***GATGGTGAGG CT*GATCCCGCAAAAGCGGCC-3′	A*DGEA*DP (*see* **Note 11**)	Bone cell stimulation
HPQ primer	5′-TATAT **CTGCAG**NK(NNK)$_2$*CAT CCGCAG*(NNK)$_2$GATCCCGCAAAAGCGGCCTTTAACTCCC-3′	A*XXXHPQXX*DP[c]	Streptavidin binding
Linearize primer	5′-CCT**CTGCAG**CGAAAGACAGC ATCGG-3′		M13KE backbone for PCR reaction

[a]For primer oligonucleotide sequences, the restriction sites (*PstI*) are shown in bold, and the insert is underlined and in italics
[b]For the resulting peptide sequence, the insert is underlined and in italics
[c]Constructed from partial library approach, selected sequence indicated

12. Bacto Agar (BD Biosciences).
13. Top agar: Dissolve 25 g/L of LB broth Miller, 7 g/L of agar, and 1 g of $MgCl_2 \cdot 6H_2O$ in deionized water and autoclave. Divide the solution into 25 mL aliquots in 50 mL Falcon conical tubes, store solid at room temperature. Melt the top agar completely using a microwave and hold at 45 °C before use.
14. IPTG/X-gal stock solution: Mix 0.5 g IPTG (isopropyl-β-D-thiogalactopyranoside) and 0.4 g X-gal (5-bromo-4-chloro-3-indolyl-β-D-galactopyranoside) in 10 mL DMF (dimethyl formamide). Solution can be stored at −20 °C in the dark.
15. LB/IPTG/X-gal plates: Dissolve 25 g/L of LB broth Miller, 15 g/L agar in deionized water. Autoclave, cool to <70 °C, add 1 mL IPTG/X-gal stock per liter and pour. Store plates at 4 °C in the dark.
16. Tetracycline (Calbiochem) stock solution: 15 mg/mL in 1:1 ethanol–water (v/v). Store at −20 °C. Vortex before using.
17. LB/Tet plates: 25 g/L of LB broth Miller, 15 g/L agar. Autoclave, cool to <70 °C, add 1 mL tetracycline stock solution per liter and pour. Store plates at 4 °C in the dark.
18. XL1-Blue streak plate: Streak an LB/Tet plate with XL1-Blue to make colonies.
19. Mid-log culture of bacteria: Inoculate a single colony of XL1-Blue from a plate into 3 mL of LB with 5 μg/mL tetracycline solution in mL Falcon conical centrifuge tubes and incubate at 37 °C for 6–8 h by shaking at 225 rpm (OD_{600} ~0.5).

2.1.3 Preparation of Phage Stock

1. 14 mL Falcon polypropylene round-bottom tubes (BD Biosciences).
2. Mid-log phase XL1-Blue.
3. LB Broth MILLER (EMD Millipore Corporation).
4. LB medium: 25 g/L LB broth Miller, autoclave.
5. Miniprep Kit for DNA sequencing (QIAPrep® QIAGEN group).
6. Sequencing primer (Integrated DNA Technologies): 5′-CCCTCATAGTTAGCGTAACG-3′

2.2 Phage Mass Amplification and Purification

1. LB Broth MILLER (EMD Millipore Corporation).
2. Erlenmeyer flasks, 250 mL and 4 L.
3. Miniprep Kit for DNA sequencing (QIAPrep® QIAGEN group).
4. PEG (Fisher Scientific)/NaCl (Crystal Macron Finest Chemicals): 20% (w/v) polyethylene glycol 8000, 2.5 M NaCl. Autoclave, mix well to combine separated solution layers while still warm. Store at room temperature.

5. 50 mM TBS buffer: 50 mM Tris–HCl (pH 7.5) (Cellgro), 150 mM NaCl. Autoclave.
6. 10 mL syringe.
7. 0.2 μm PES syringe filter (Fisher).
8. 1 L centrifuge bottles (Beckman Coulter): UV irradiation for at least 2 h before use.
9. 50 mL centrifuge bottles (Beckman Coulter): Autoclave and dry before use.
10. Centrifuge (Beckman Coulter, Avanti J-26 XP).
11. Fixed-angle centrifuge rotor for 1 L (Beckman Coulter, JLA-8.1000).
12. Fixed-angle centrifuge rotor for 50 mL (Beckman Coulter, JA-20).

2.3 Phage Film Fabrication

1. Glass substrate (slide glass or cover glass), 1 × 1 cm or 1 × 2.5 cm.
2. Piranha solution: 1:4 (v/v) H_2O_2–H_2SO_4.
3. DI water (NANOpure Diamond™ Barnstead).
4. Nitrogen stream.
5. 1% (v/v) 3-aminopropyl triethoxysilane (APTES) in absolute ethanol.
6. Phosphate Buffered Saline (PBS): 8 g/L NaCl, 0.2 g/L KCl, 1.44 g/L Na_2HPO_4, 0.24 g/L KH_2PO_4, adjust pH to 7.4 with HCl. Autoclave.
7. Humidified incubator (37 °C).

3 Methods

3.1 Genetic Engineering of Phage

The M13 phage is engineered in its pVIII major coat protein.

The following procedure is specific for the M13KE cloning vector, but could easily be adapted for other phage cloning vectors: Engineer M13 phage major coat protein (pVIII) to display RGD or IKVAV peptide motif by using a partial library cloning approach [39, 40]. Insert an octapeptide at the N-terminus of the major coat protein (pVIII) of M13 phage. The insert is positioned between the first and the fifth amino acids of the wild-type pVIII, replacing residues 2–4 (Ala-Glu-Gly-Asp-Asp to Ala-(Insert)-Asp) (Fig. 1) Different alternative peptides can be added based on this protocol.

3.1.1 Genetic Engineering

1. Design oligonucleotide primers following the convention in Table 1 (*see* **Notes 3** and **4**). Five primers are designed and ordered to insert the desired four peptides (e.g., IKVAV, RGD, DGEA, HPQ). Use two primers for one desired peptide insertion/display on the pVIII major coat protein of M13 phage.

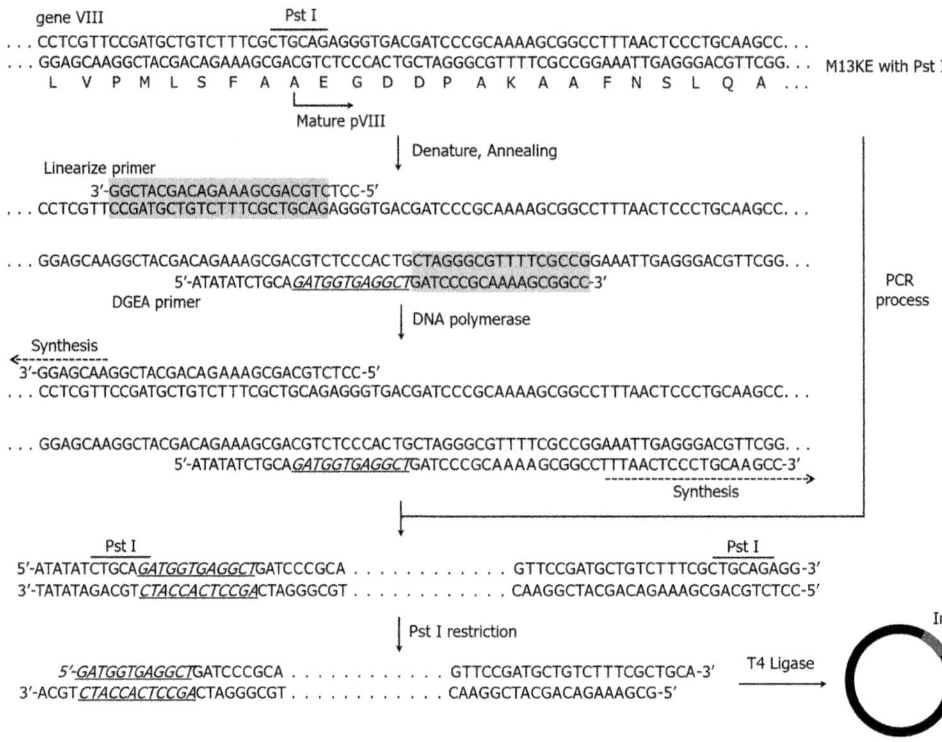

Fig. 1 Construction of a desired peptide on pVIII gene in M13KE. Schematic shows the sequences of the peptide cloning site as well as the designing and cloning a peptide (e.g., DGEA; underlined and in italics) into M13KE

2. To incorporate the gene sequences, perform a polymerase chain reaction (PCR) using Phusion Polymerase, two primers (insert and linearization), and the M13KE vector with an engineered *Pst*I site as the template. PCR ingredients and conditions are shown in Table 2.
3. Gel-purify the PCR product on a 0.6% (w/v) agarose gel in TAE buffer (*see* **Note 5**), including a 1 kb DNA ladder. Visualize by SYBR® Safe DNA Gel staining.
4. Excise the PCR product from the gel. Extract the PCR product by a QIAquick® Gel Extraction Kit (*see* **Note 6**).
5. The PCR product is then cut using *Pst*I to create sticky ends. We favor digestion as follows: 50 μL gel extracted DNA (~10 μg), 37 μL H_2O, 10 μL 10× NEB Buffer 3.1 and 3 μL *Pst*I (60 U). Incubate at 37 °C for 3–5 h.
6. Purify the DNA by QIAquick® PCR Purification Kit.
7. Ligate the construct. Suggested starting parameters for 10 μL of ligation mixture: 100 ng of restricted PCR DNA; 1 μL of 10× ligase buffer; and 200 U of T4 DNA Ligase (*see* **Note 7**). Incubate 18 h at 16 °C.

Table 2
Phage cloning PCR Conditions

PCR ingredients	pVIII PCR conditions	
~50 ng dsDNA template[a]	98 °C 1 min	
2.5 μL 10 μM forward primer 2.5 μL 10 μM reverse primer 1 μL dNTP (10 mM mixed bases) 1 μL DMSO	98 °C 15 s 58 °C[b] 20 s 72 °C 3 min 30 s	× 25 cycles
10 μL 5× HF Phusion Polymerase Buffer adjust with sterile H_2O to 50 μL	72 °C 10 min 4 °C ∞	
0.5 μL Phusion Polymerase Enzyme		

[a]~1 μL, use any template that has a *Pst*I site for the pVIII M13 engineering
[b]Primer annealing temperature = Primer T_m (lower of the two primers) − 2

3.1.2 Electroporation

Electroporate 1 μL of ligated DNA into 50 μL of electrocompetent *E. coli* XL1-Blue or other F^+ strain (*see* **Note 8**).

1. Prechill two sterile electroporation cuvettes (0.1 cm gap) and preheat sterile SOC medium to 37 °C. Set the electroporator to a voltage of 1700 V.
2. Thaw the electroporation-competent cells on ice. After mixing the cells gently, aliquot 50 μL of cells into each of the two prechilled tubes (one tube for the experimental transformation and one tube for the pUC18 control transformation).
3. Keep the tubes on ice.
4. Add either the 1 μL of ligated DNA or 1 μL pUC18 to the cells with gentle mixing.
5. Transfer the cell–DNA mixture to a chilled electroporation cuvette, tapping the cuvette until the mixture settles evenly on the bottom.
6. Slide the cuvette into the electroporation chamber until the cuvette sits flush against the electrical contacts.
7. Pulse the sample once, then quickly remove the cuvette. Immediately add 950 μL of SOC medium to resuspend the cells.
8. Transfer the cells to a sterile 14 mL Falcon polypropylene round-bottom tube. Incubate the tube at 37 °C for 1 h with shaking at 225 rpm.
9. Prepare 10, 100, and 1000-fold dilutions of the outgrowth in LB. Transfer 10 μL of each dilution to a microcentrifuge tube with 200 μL of a mid-log culture of XL1-Blue and mix with 3 mL of top agar (held at 45 °C).
10. Vortex briefly and spread on LB/IPTG/X-gal plates. Incubate overnight at 37 °C and check for blue plaques the next day.

3.1.3 Preparation of Phage Stock

1. Next day, prepare several 14 mL round bottom flasks, and add 1 mL LB medium and 10 μL mid-log culture of XL1-Blue to each 14 mL flask. The volume ratio between LB medium and mid-log culture *E. coli* should be 1:100.
2. Collect several blue plaques of phage-infected *E. coli* from petri dish culture (LB/IPTG/Xgal plates), and culture in each of the 14 mL round bottom flask. Place the tube for 6–8 h incubation at 37 °C with shaking.
3. Transfer the content to a microcentrifuge tube (1.5 mL) in a sterile environment. Centrifuge at 15,000 × *g* for 10 min.
4. Transfer 80% of supernatant (phage suspension) to a new microcentrifuge tube and stock at 4 °C until DNA sequence analysis. Phage stock can be stored up to 2 month without further purification.
5. Remove remaining culture medium and use the pellet (*E. coli*) for DNA extraction. Extract the phage vector by QIAprep® Spin Miniprep Kit (Qiagen).
6. Analyze phage DNA sequences using a sequencing primer by DNA sequencing. If DNA sequencing result has correct DNA sequences for a desired peptide sequence, phage stock can be used for next step (Subheading 3.2).

3.2 Phage Mass Amplification and Purification

1. Prepare LB medium in two Erlenmeyer flasks (250 mL) with a final volume of 20 mL in each flask and autoclave them. The medium should not be hot for incubation.
2. Add 200 μL of mid-log culture XL1-Blue to both Erlenmeyer flasks and add 200 μL of the phage stock solution to the second Erlenmeyer flask.
3. Incubate at 37 °C for 6–8 h with shaking at 225 rpm.
4. Transfer 20 mL of phage culture media to a centrifuge bottle and centrifuge at 12,000 × *g* at 4 °C for 20 min. Transfer the upper 80% of supernatant (phage solution) to a 50 mL Falcon conical tube (keep it at 4 °C up to 1 month, or use immediately).
5. Prepare an Erlenmeyer flask with 1.5 L of LB medium and autoclave. Add 15 mL of mid-log culture XL1-Blue from **step 3** and phage solution into the Erlenmeyer flask.
6. Incubate at 37 °C overnight with shaking at 225 rpm.
7. Collect a 1 mL sample from the Erlenmeyer flask and proceed with the miniprep protocol for DNA sequencing to check eventual contamination.
8. Transfer the culture media to 1 L centrifuge tubes and spin for 20 min at 15,000 × *g* at 4 °C.

9. Transfer the supernatant to new 1 L centrifuge tubes and add 1/5 volume of PEG/NaCl. Allow the phages to precipitate at 4 °C overnight.
10. Centrifuge the tubes for 20 min at 15,000 × *g* at 4 °C, then discard the supernatant. Centrifuge for 10 min again. The remaining phages on the tube walls will spin down to the pellet.
11. Resuspend the phage pellet using 10 mL of 50 mM TBS buffer and transfer each tube content to a new 50 mL centrifuge tube.
12. Spin for 20 min at 12,000 × *g* at 4 °C.
13. Transfer the supernatant to a new 50 mL centrifuge tube, by filtering with a 10 mL syringe and a 0.2 μm filter. After filtration, add 1/6 volume (v/v) of PEG/NaCl to the 50 mL centrifuge tube. Allow the phage to precipitate at 4 °C for at least 4–6 h, preferably overnight.
14. Spin the PEG precipitation at 12,000 × *g* for 20 min at 4 °C. Decant and discard the supernatant, respin the tube briefly, and remove residual supernatant with a pipette.
15. Suspend the pellet in 1 mL of 50 mM TBS buffer. Transfer the suspension to a microcentrifuge tube and spin at 15,000 × *g* for 10 min at 4 °C to pellet residual cells.
16. Transfer the supernatant to a fresh microcentrifuge tube and reprecipitate by adding 1/6 volume (v/v) of PEG/NaCl. Incubate at 4 °C for 30–60 min.
17. Centrifuge at 15,000 × *g* for 10 min at 4 °C. Discard the supernatant, respin briefly, and remove residual supernatant with a micropipette (*see* **Note 9**).
18. Suspend the pellet in 250 μL of TBS. Centrifuge at 15,000 × *g* for 10 min to pellet any remaining insoluble materials. Transfer the supernatant to a fresh tube. This supernatant is the amplified engineered phage.

3.3 Phage Film Fabrication

1. Place glass slide substrates in piranha solution for 10 min, thoroughly rinse with DI water, and dry in a stream of nitrogen.
2. Treat the substrates with 1% APTES solution and anneal at 100 °C for 10 min.
3. For the preparation of isotropic phage films, place a phage solution (50 μL of 10^{12} viruses/mL) in PBS (*see* **Note 10**); Using a micropipette, spread onto the APTES-treated glass surface (1 × 1 cm); and allow the droplet to dry overnight at 37 °C in a humidified incubator.

   For the preparation of aligned phage films with anisotropic topography, place a droplet (5 μL) of the phage solution (10–30 mg/mL) on one edge of the APTES-treated glass substrate

(1 × 2.5 cm); drag the droplet along using a glass slide parallel to the long axis of the glass substrate to apply a shear force in a single direction (*see* Fig. 2).

4. Dry the film samples overnight at room temperature and rinse gently with PBS prior to cell culture experiments.
5. Resulting film morphologies can be characterized using polarizing optical microscopy and atomic force microscopy imaging.

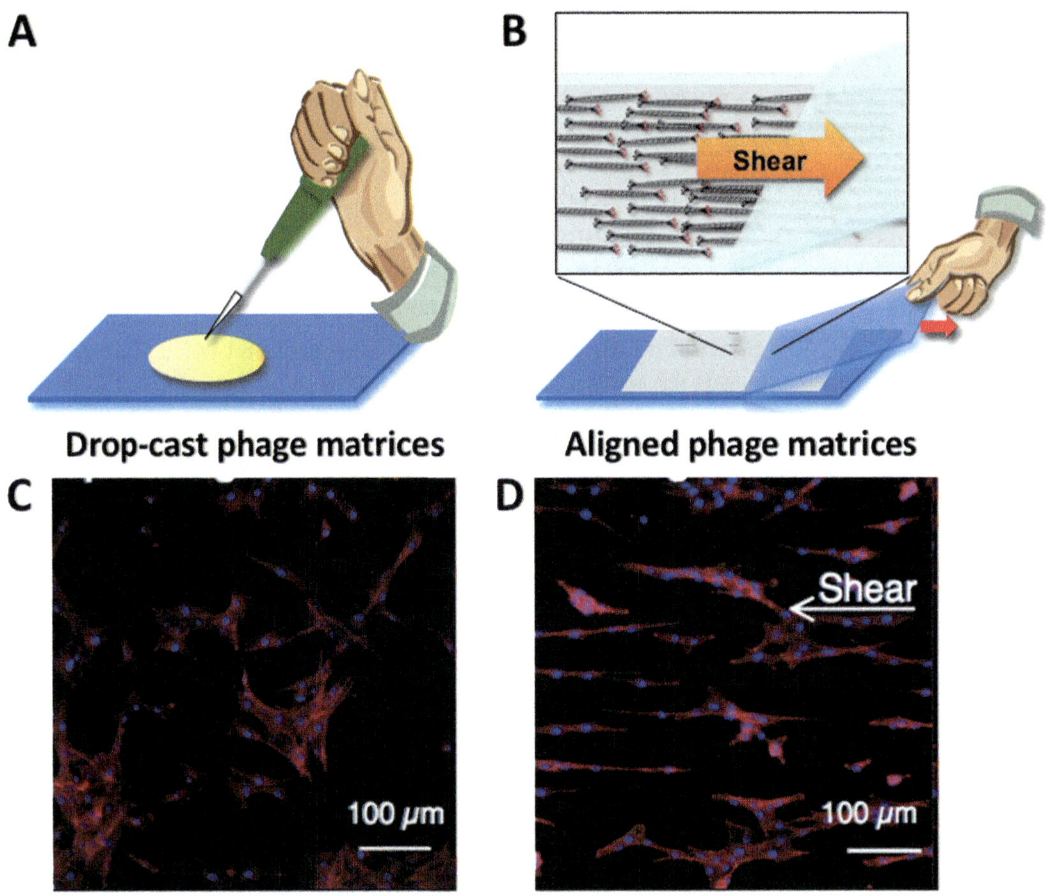

Fig. 2 Control of cell growth morphologies using phage tissue engineering materials. Schematic diagram of the drop cast phage film (**a**) and shear induced directionally aligned phage matrices (**b**). Composite images of fibroblast cells (NIH-3T3) cultured on top of drop cast phage matrices (**c**) and shear induced aligned phage matrices (**d**). Actin filaments stained with phalloidin (red), and nuclei counterstained with DAPI (blue)

4 Notes

1. M13KE vector (NEB) has a *PstI* restriction site (CTGCAG) in LacZα. Modify the *PstI* restriction site (CTGCAG) by mutating the nucleotide change T6246 to A6246 in M13KE vector using the QuikChange® Kit (Stratagene).
2. To facilitate the recircularization of the engineered plasmid, a *Pst*I restriction site (CTGCAG) is created by mutating the nucleotide base at position 1372 of M13KE vector (NEB) from T to A by site directed mutagenesis using methods described in the QuikChange® Kit (Stratagene).
3. Bear in mind that the sequence SFA (Ser-Phe-Ala) preceding the leader peptidase cleavage site is part of the pVIII signal sequence. The first residue of the displayed peptide immediately follows this sequence.
4. For randomized positions, the relative representations of each amino acid can be improved by limiting the third position of each codon to G or T (= A or C on the synthetic library oligonucleotide).
5. Gel purification is most efficient with lower % agarose gels, staying in the 0.6–0.8% range if possible.
6. To minimize the risk of DNA damage, it is best to limit the UV exposure of the DNA. Therefore, use long-wavelength UV for as short a time as possible to get the bands cut out.
7. These are starting parameters. You may have to optimize the ligation conditions.
8. We use *E. coli* host strain XL1-Blue (recA1 endA1 gyrA96 thi-1 hsdR17 supE44 relA1 lac [F′ proAB lacI^{q}ZΔM15 Tn10 (Tet^r)]). Other *E. coli* host strains such as ER2738 and DH5αF′ can be used. Any strains are probably used if they have a robust F$^+$ genotype.
9. The phage pellet is a white pellet on the side of the tube.
10. Concentration of phage particles is calculated its absorption at 269 and 320 nm: Dilute the phage solution 100 times with TBS buffer. In this case, dilution factor is 100. Measure the concentration of phage particles with UV-visible spectroscopy using this equation.

$$\text{Phages / ml} = \frac{(A_{269} - A_{320}) \times 6 \times 10^{16}}{\text{number of phage nucleotide bases}} \times \text{dilution factor.}$$

11. The phage prefers having a net negative charge after the genetic engineering of the major coat proteins. Therefore, we prefer a partial library approach for oligonucleotide primers design.

Nucleotides NNK (N = A, T, G, C; K = G/T) for 20 amino acids are added around a desired peptide sequence. In the case of DGEA peptide engineering on the major coat protein, a partial library approach is not necessary due to the highly negative charge of the DGEA peptide sequence.

5 Summary

M13 bacteriophages are excellent candidate materials for the development of tissue regenerating materials. Through genetic engineering, a high density of functional peptides and proteins can be simultaneously displayed on the M13 bacteriophage's outer major coat proteins. The resulting phage can self-assemble into nanofibrous network structures and can regulate the tissue morphogenesis including proliferation, differentiation and apoptosis. In this manuscript, we describe methods to develop major coat-engineered M13 phages as a basic building block and to regulate tissue-like matrices to control the growth of fibroblast cells.

Acknowledgment

This work was supported under the framework of international cooperation program managed by the National Research Foundation of Korea (NRF-2016K2A9A1A01951919). H.-E.J. was supported by the National Research Foundation of Korea (NRF) grant funded by the Korea government (MSIP) (No. 2016R1C1B1008824). We acknowledge funding support from the Tsinghua-Berkeley Shenzhen Institute.

References

1. Huber AB, Kolodkin AL, Ginty DD, Cloutier JF (2003) Signaling at the growth cone: ligand-receptor complexes and the control of axon growth and guidance. Annu Rev Neurosci 26:509–563
2. Milner R, Campbell IL (2002) The integrin family of cell adhesion molecules has multiple functions within the CNS. J Neurosci Res 69:286–291
3. Rutka JT, Apodaca G, Stern R, Rosenblum M (1988) The extracellular-matrix of the central and peripheral nervous systems - structure and function. J Neurosurg 69:155–170
4. Oster SF, Deiner A, Birgbauer E, Sretavan DW (2004) Ganglion cell axon pathfinding in the retina and optic nerve. Semin Cell Dev Biol 15:125–136
5. Tirrell M, Kokkoli E, Biesalski M (2002) The role of surface science in bioengineered materials. Surf Sci 500:61–83
6. Engel E, Michiardi A, Navarro M, Lacroix D, Planell JA (2008) Nanotechnology in regenerative medicine: the materials side. Trends Biotechnol 26:39–47
7. Griffith LG, Swartz MA (2006) Capturing complex 3D tissue physiology in vitro. Nat Rev Mol Cell Biol 7:211–224
8. Langer R, Vacanti JP (1993) Tissue engineering. Science 260:920–926
9. Schense JC, Bloch J, Aebischer P, Hubbell JA (2000) Enzymatic incorporation of bioactive peptides into fibrin matrices enhances neurite extension. Nat Biotechnol 18:415–419

10. Stevens MM, George JH (2005) Exploring and engineering the cell surface interface. Science 310:1135–1138
11. Liu WF, Chen CS (2005) Engineering biomaterials to control cell function. Mater Today 8:8
12. Silva GA et al (2004) Selective differentiation of neural progenitor cells by high-epitope density nanofibers. Science 303:1352–1355
13. Yang F, Murugan R, Wang S, Ramakrishna S (2005) Electrospinning of nano/micro scale poly(L-lactic acid) aligned fibers and their potential in neural tissue engineering. Biomaterials 26:2603–2610
14. Teng YD et al (2002) Functional recovery following traumatic spinal cord injury mediated by a unique polymer scaffold seeded with neural stem cells. Proc Natl Acad Sci U S A 99:3024–3029
15. Zhang SG (2003) Fabrication of novel biomaterials through molecular self-assembly. Nat Biotechnol 21:1171–1178
16. Saha K, Irwin EF, Kozhukh J, Schaffer DV, Healy KE (2007) Biomimetic interfacial interpenetrating polymer networks control neural stem cell behavior. J Biomed Mater Res A 81:240–249
17. Ellis-Behnke RG et al (2006) Nano neuro knitting: peptide nanofiber scaffold for brain repair and axon regeneration with functional return of vision. Proc Natl Acad Sci U S A 103:5054–5059
18. Stroumpoulis D, Zhang HN, Rubalcava L, Gliem J, Tirrell M (2007) Cell adhesion and growth to peptide-patterned supported lipid membranes. Langmuir 23:3849–3856
19. Gomez N, Chen SC, Schmidt CE (2007) Polarization of hippocampal neurons with competitive surface stimuli: contact guidance cues are preferred over chemical ligands. J R Soc Interf 4:223–233
20. Saha K et al (2008) Substrate modulus directs neural stem cell behavior. Biophys J 95:4426–4438
21. Ivanovska IL, Shin JW, Swift J, Discher DE (2015) Stem cell mechanobiology: diverse lessons from bone marrow. Trends Cell Biol 25:523–532
22. Dogic Z, Fraden S (1997) Smectic phase in a colloidal suspension of semiflexible virus particles. Phys Rev Lett 78:2417–2420
23. Smith GP, Petrenko VA (1997) Phage display. Chem Rev 97:391–410
24. Petrenko VA, Smith GP, Gong X, Quinn T (1996) A library of organic landscapes on filamentous phage. Protein Eng 9:797–801
25. Huang Y et al (2005) Programmable assembly of nanoarchitectures using genetically engineered viruses. Nano Lett 5:1429–1434
26. Lee SW, Mao CB, Flynn CE, Belcher AM (2002) Ordering of quantum dots using genetically engineered viruses. Science 296:892–895
27. Nam KT et al (2006) Virus-enabled synthesis and assembly of nanowires for lithium ion battery electrodes. Science 312:885–888
28. Chung WJ et al (2011) Biomimetic self-templating supramolecular structures. Nature 478:364–368
29. Lee BY et al (2012) Virus-based piezoelectric energy generation. Nat Nanotechnol 7:351–356
30. Lee JH et al (2017) Phage-based structural color sensors and their pattern recognition sensing system. ACS Nano 11:3632–3641
31. Webster R (1996) Biology of the filamentous bacteriophage. In: Phage display of peptides and proteins. Academic Press, San Diego, CA, pp 1–20
32. Russel M (1995) Moving through the membrane with filamentous phages. Trends Microbiol 3:223–228
33. Onsager L (1949) The effects of shape on the interaction of colloidal particles. Ann N Y Acad Sci 51:627–659
34. Dogic Z, Fraden S (2006) Ordered phases of filamentous viruses. Curr Opin Colloid Interface Sci 11:47–55
35. Merzlyak A, Lee SW (2006) Phage as templates for hybrid materials and mediators for nanomaterial synthesis. Curr Opin Chem Biol 10:246–252
36. Flynn CE, Lee SW, Peelle BR, Belcher AM (2003) Viruses as vehicles for growth, organization and assembly of materials. Acta Mater 51:5867–5880
37. Lee S-W, Belcher AM (2004) Virus-based fabrication of micro- and nanofibers using electrospinning. Nano Lett 4:387–390
38. Chung W-J, Kwon K-Y, Song J, Lee S-W (2011) Evolutionary screening of collagen-like peptides that nucleate hydroxyapatite crystals. Langmuir 27:7620–7628
39. Yoo SY, Kobayashi M, Lee PP, Lee SW (2011) Early osteogenic differentiation of mouse preosteoblasts induced by collagen-derived DGEA-peptide on nanofibrous phage tissue matrices. Biomacromolecules 12:987–996
40. Merzlyak A, Indrakanti S, Lee SW (2009) Genetically engineered nanofiber-like viruses for tissue regenerating materials. Nano Lett 9:846–852

41. Chung WJ, Merzlyak A, Yoo SY, Lee SW (2010) Genetically engineered liquid-crystalline viral films for directing neural cell growth. Langmuir 26:9885–9890
42. Tom S, Jin H-E, Heo K, Lee S-W (2016) Engineered phage films as scaffolds for CaCO 3 biomineralization. Nanoscale 8:15696–15701
43. Jin HE, Farr R, Lee SW (2014) Collagen mimetic peptide engineered M13 bacteriophage for collagen targeting and imaging in cancer. Biomaterials 35:9236–9245
44. Yoo SY et al (2016) M13 bacteriophage and adeno-associated virus hybrid for novel tissue engineering material with gene delivery functions. Adv Healthc Mater 5:88–93
45. YoungáYoo S (2011) Facile growth factor immobilization platform based on engineered phage matrices. Soft Matter 7:1660–1666

Chapter 33

Displaying Whole-Chain Proteins on Hepatitis B Virus Capsid-Like Particles

Julia Heger-Stevic, Philipp Kolb, Andreas Walker, and Michael Nassal

Abstract

The highly immunogenic icosahedral capsid of hepatitis B virus (HBV) can be exploited as a nanoparticulate display platform for heterologous molecules. Its constituent core protein (HBc) of only ~180 amino acids spontaneously forms capsid-like particles (CLPs) even in *E. coli*. The immunodominant c/e1 epitope in the center of the HBc primary sequence comprises a solvent-exposed loop that tolerates insertions of flexible peptide sequences yet also of selected whole proteins as long as their 3D structures fit into the two acceptor sites. This constraint is largely overcome in the SplitCore system, where the sequences flanking the loop are expressed as two separate but self-complementing entities, with the foreign sequence fixed to the carrier at one end only. Both the contiguous and the split type of CLP strongly enhance immunogenicity of the displayed sequence but also nonvaccine applications can easily be envisaged. After a brief survey of the basic features of the two HBc carrier forms, we provide conceptual guidelines concerning which foreign proteins are likely to be presentable, or not, on either carrier type. We describe generally applicable protocols for CLP expression in *E. coli*, cell lysis and CLP enrichment by sucrose gradient velocity sedimentation, plus a simple but meaningful gel electrophoretic assay to assess proper particle formation.

Key words Capsid-like particles, CLPs, CLP-vaccines, HBc, Nanoparticulate antigen carrier, Native agarose gel electrophoresis, Protein-display, SplitCore system, Velocity sedimentation

1 Introduction

1.1 HBV CLPs, and Their Use as Nanoparticulate Protein-Display Platform

Hepatitis B virus (HBV) is a small DNA virus that replicates through protein-primed reverse transcription [1]. Infectious virions consist of a lipid-based outer envelope into which three envelope proteins are embedded, collectively known as hepatitis B surface antigen (HBsAg). The smallest form produced in yeast is the active ingredient of the most widely used anti-HBV vaccines. The inner nucleocapsid, also called core particle or hepatitis B core antigen (HBcAg), is an icosahedrally symmetric protein-only container for the viral

---

Julia Heger-Stevic, Philipp Kolb, and Andreas Walker contributed equally to this work.

Christina Wege and George P. Lomonossoff (eds.), *Virus-Derived Nanoparticles for Advanced Technologies: Methods and Protocols*, Methods in Molecular Biology, vol. 1776, https://doi.org/10.1007/978-1-4939-7808-3_33,

© Springer Science+Business Media, LLC, part of Springer Nature 2018

genome, formed by multiple copies of a single species of core protein (HBc) of 183–185 amino acids, depending on HBV subtype. HBc can spontaneously (i.e., in the absence of any further HBV gene products) assemble into genome-less noninfectious capsid-like particles (CLPs) that structurally resemble authentic virus capsids and share their exceptional immunogenicity [2]. Recombinant expression of such CLPs in *E. coli* has enabled mutational and structural analyses which revealed an N terminal assembly domain encompassing the first ~140 amino acids (*see* Fig. 1a, b), plus a C terminal arginine-rich domain (CTD; starting at amino acid (aa) position 150) that binds nucleic acids but is dispensable for particle formation [3, 4]. Specific packaging of the HBV pregenomic RNA (the template for reverse transcription into DNA) strictly requires the viral polymerase protein [5]. In its absence, HBc CLPs carrying the CTD encapsidate host RNA non-sequence specifically [3, 6, 7]. Based on electron cryomicroscopy (cryo EM) and X-ray crystallography [8, 9], the assembly-domain adopts an all-α-helical fold with a shape resembling the letter “L” (*see* Fig. 1e). Strong interactions between the central, antiparallel α-helices α3 and α4 from two monomers create a four-helix bundle, resulting in a stable dimer (*see* Fig. 1d). Multiple weak interactions [10] between the C terminal

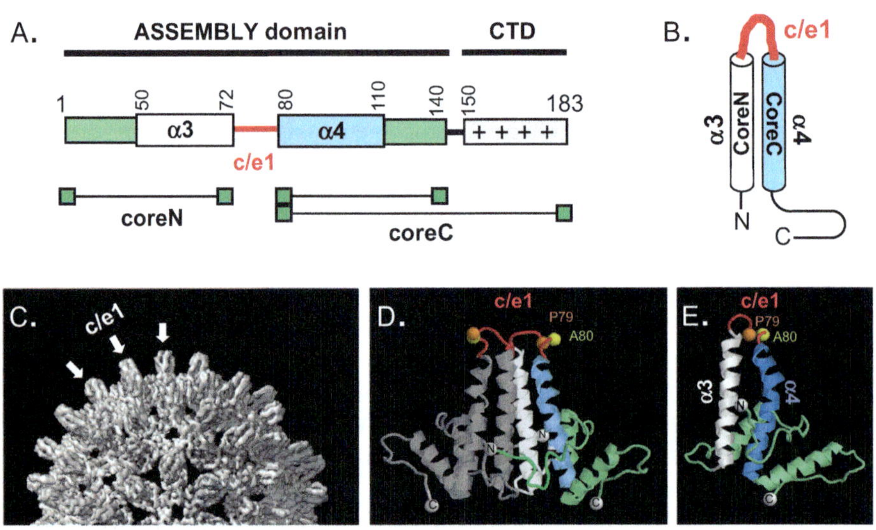

Fig. 1 Structural features of the HBc carrier protein. (**a**) Domain organization. The sequences comprising the assembly domain and the C terminal domain (CTD) are indicated on the top; + signs symbolize clusters of Arg residues. The borders of the central helices α3 and α4 are depicted as white and light-blue boxes, the c/e1 epitope is marked in red. The sequence stretches referred to as coreN and coreC in the text (irrespective of how far they extend beyond the end of the assembly domain) are indicated at the bottom. (**b**) Schematic view of the structure of an HBc monomer. The representation highlights the helix-loop-helix scaffold onto which foreign sequences can be grafted by insertion into the c/e1 loop. (**c**) Wild-type HBc T = 4 CLP. The top half of a 3D reconstruction derived by cryo-EM is shown (B. Böttcher and M. Nassal, unpublished). (**d**, **e**) 3D structures of the HBc dimer (**d**) and monomer (**e**). Structures are based on X-ray crystallography of CLPs from CTD-less HBc (PDB accession number 1QGT). Color coding of the central helices and the c/e1 loop as in (**b**)

regions of the assembly domain mediate assembly of predominantly 120 dimers into an icosahedral particle shell (triangulation number T = 4; ~34 nm diameter). A minor form (T = 3; ~30 nm diameter) consists of only 90 dimers [11]. The four-helix bundles appear as prominent spikes on the CLP surface (*see* Fig. 1c), with the loops connecting α3 and α4 forming the most solvent-exposed parts. The loop comprises the immunodominant c/e1 B cell epitope formed by aa residues around position 78. Exposure and tolerance to insertion of short and/or structurally flexible foreign sequences have made this loop the preferred site for engineering HBV CLPs into an immune-enhancing particulate carrier platform [12, 13]. Even selected whole-chain proteins could successfully be inserted in their native form into the c/e1 loop without compromising CLP formation [14–16]; the latter is crucial for the strong immunogenicity demonstrated for various such foreign protein-bearing CLPs [17]. A common feature of well-presentable foreign proteins is their propensity to adopt a stable three-dimensional (3D) fold in which N and C terminus are in close spatial proximity. This allows the stressless fixation of both termini to the two acceptor sites in the carrier protein (*see* Fig. 2a). This limitation is largely overcome in the SplitCore system (*see* Fig. 2b) where the foreign protein is linked to the carrier over one end only. This is enabled by the strong propensity of the N proximal (coreN) and C proximal half (coreC) of the HBc assembly domain (*see* Fig. 1a) to self-complement into a split monomer that remains competent for dimerization and subsequent multimerization [18, 19].

Here we provide conceptual guidance for the proper design of foreign protein-bearing HBc CLPs, both for the contiguous and the SplitCore system, and describe generally applicable protocols for the expression, purification, and quality control of the respective CLPs in *E. coli*. Its fast growth without need for expensive media and equipment often allows easy access to milligram amounts of a recombinant protein per liter of culture. However, its prokaryotic nature also poses restrictions on the types of heterologous proteins, including fusions with HBc, that can be expressed in a soluble form (*see* Subheading 1.2.1). In this regard it is important to note that the sucrose gradient separation and gel electrophoretic quality control protocols described below should easily be adaptable to HBc-CLPs expressed in eukaryotic hosts.

1.2 Conceptual Considerations for Protein Display on HBc CLPs

Proteins are not inert solids but form their 3D structures by numerous individually weak interactions that are largely defined by the primary sequence but still are difficult to predict. HBc combines many advantageous features of a universal carrier protein, e.g., the strong independent folding propensity of the coreN and coreC segments. However, this may be very different for to-be-displayed foreign proteins, and further (including nonproductive) folding

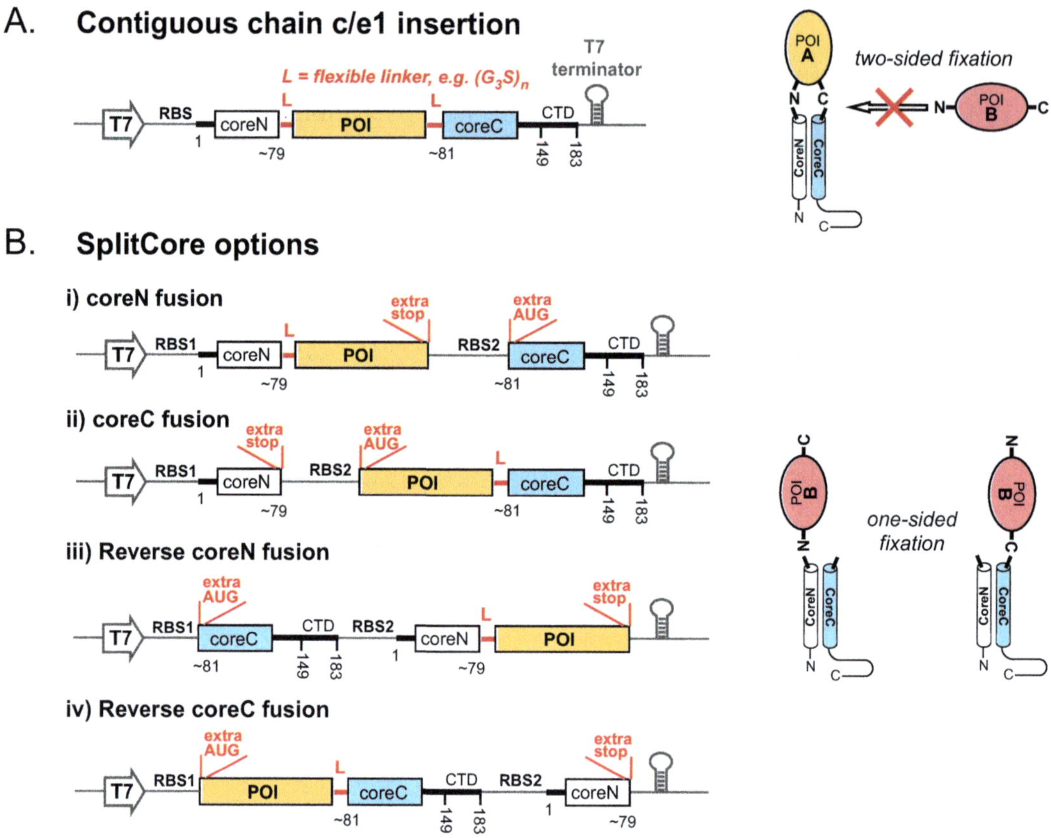

Fig. 2 General expression vector designs for protein display on HBc CLPs. (**a**) Contiguous chain system. Typical vectors feature a T7 promoter (T7), ribosome binding site (RBS), the coding region for the HBc fusion with the protein of interest (POI), and a T7 terminator. The fusion protein carries the POI fused via flexible linkers (L) to the coreN and the coreC parts; coreC may comprise the CTD or not. CLPs with CTD package up to 20-fold more *E. coli* RNA. (**b**) SplitCore system design options. The POI can be fused to coreN (i) or coreC [(ii); with our without CTD]. Translation of the two segments can be mediated by two RBSs on one bicistronic mRNA. The cistron order can also be reversed (iii, iv) to counteract formation of translation initiation inhibiting RNA secondary structures that occur in the other order. Extra stop and start codons required for the split approach are indicated in red. The schemes on the right illustrate how a POI A with closely juxtaposed termini fits into the two carrier acceptor sites of the contiguous chain system, while a POI B with distant termini can only be accomodated in the SplitCore system thanks to the one-sided fixation

options can be introduced by the artificial fusion between carrier and insert protein. Hence the guidelines outlined below will reduce but not necessarily obviate experimental testing of more than one expression strategy.

1.2.1 To-Be-Displayed Protein

Proteins come in vastly differing sizes and folds, making it difficult to reliably predict whether or not a given protein can successfully be displayed on HBc CLPs (or any other particulate carrier protein). However, a few general considerations can save time and effort in the design phase of a protein display project. Altogether, it is highly worthwhile to gather as much information on the display candidate protein; in particular, structural knowledge will aid in choosing the most promising HBc CLP platform.

Primary Sequence Length

The surface area of HBc CLPs is finite. Hence if the size of a foreign entity exceeds a certain limit, 180 (T = 3) or 240 (T = 4) copies of it will interfere with CLP assembly. The space requirements of a protein depend on primary sequence length and also on shape and relative orientation to the radial CLP surface; especially for nonglobular proteins the SplitCore system with its one-sided carrier fixation offers additional spatial freedom. The upper size limit has not been determined, but we have successfully displayed complex fusion proteins, e.g., consisting of the green fluorescent protein (GFP) plus bacterial protein antigens, with molecular mass of >60 kDa [18]. CLP surface crowding may also be reduced by reducing the number of heterologous proteins per carrier subunit. One option is the coexpression of insert-bearing with nonmodified HBc subunits to generate mosaic particles [20], another is the use of covalently linked HBc "tandem" dimers in which only one c/e1 loop per dimer carries the insert [21] (*see* Chapter 7).

Quaternary Structure

Higher-order quaternary structures can interfere with CLP assembly. A typical example is the antiparallel dimer formed by many variants of GFP. Here, the termini of each monomer protrude from opposite ends of the dimer, incompatible with both being joined to the surface of one CLP [22]. In addition, inter-insert interactions may lead to undesired aggregation. This can be prevented by engineering monomeric variants of the insert protein [22] which may, however, differ in antigenicity. A potential alternative is to covalently link the two monomers of the to-be-displayed protein into a single chain that is then fused over one end to the SplitCore carrier. Notably, dimerization per se is not prohibitive for proper CLP formation [15], as shown for the borrelial OspC protein which forms a stable parallel dimer [23]. However, most higher-order quaternary structures are likely incompatible with the icosahedral carrier CLP structure. Hence monomeric proteins, or independently folding protein domains, with molecular masses <50 kDa are the most promising candidates for successful HBc CLP display.

Even then, the propensity to adopt a stable fold under the conditions in an *E. coli* cell remains a crucial criterion. A most useful predictor is that the heterologous protein as such can be expressed in a soluble form in *E. coli*. Conversely, a protein that is difficult to express in *E. coli* will generally be a poor candidate for CLP display. The circumsporozoite protein (CSP) from the malaria agent *Plasmodium falciparum* represents a notable exception where fixation to the carrier appears to reduce the aggregation tendency of free CSP [18].

Posttranslational Modifications

Proteins can undergo a multitude of posttranslational modifications (PTMs) that may affect overall structure and/or specific conformations. Here we focus only on oxidation of Cys-SH groups to disulfide bonds and glycosylation. Owing to the reducing environment in their cytoplasm and lower folding potential compared to eukaryotes [24], heterologous expression of Cys-rich proteins in *E. coli* is, per se, challenging. Proteins that do not stably fold under reducing conditions tend to form improper intramolecular and intermolecular disulfide bonds upon exposure to oxidizing conditions (e.g., upon cell lysis), leading to nonspecific aggregation and insolubility. To be able to display such proteins on HBc CLPs, two options should be considered. The first is using one of the various *E. coli* expression strains that are engineered to have a more oxidizing cytoplasm, e.g., the Rosetta-gami (Novagen/Merck-Millipore) or SHuffle strains (New England Biolabs). For a Cys-rich tick protein we found that total yields were lower than in the standard BL21 strains (*see* below) lacking a modified redox system; however, the fraction of soluble fusion protein was significantly increased [25]. Another option is to mutationally replace from one to all of the endogenous Cys-residues. This allowed the successful preparation of soluble tick protein-displaying SplitCore CLPs. Although the Cys-free tick protein as such was less stably structured than its wild-type counterpart [25], the corresponding CLPs induced strong antibody responses against the wild-type tick protein [16]; however, this may be different for other to-be-displayed protein antigens.

Glycosylation can profoundly affect protein stability and antigenicity [26]. *E. coli* is not equipped to perform mammalian-like glycosylation and, despite recent progress, engineered *E. coli* strains harboring mammalian glycosylation pathways [27] are not yet readily available. Hence before embarking on attempts to display a naturally glycosylated protein on CLPs it should be tested for the free protein whether the lack of glycosylation negatively affects folding or stability, or interferes with the intended downstream application of the respective CLPs.

1.2.2 HBc CLP Display Platform

Full-Length Versus CTD-Less HBc as Carrier

The two-domain structure of HBc (*see* Fig. 1a) offers two basic options for its use as carrier platform (*see* Fig. 2). CLPs from full-length HBc including the CTD package up to ~4000 nt of *E. coli* RNA [6, 7], whereas CLPs from HBc lacking the CTD, e.g., HBc149, contain 10- to 20-fold less RNA [3] (*see* also Fig. 5). Most early studies employed HBc149 because of its much higher expression in simple *E. coli* expression strains. A major reason is the accumulation in the HBc183 CTD of Arg codons that are rarely used in *E. coli*. This is overcome by using *E. coli* expression strains that provide extra copies of the rare tRNAs on a separate plasmid, such as BL21 CodonPlus cells (Agilent Technologies). Alternatively, codon usage in the HBc coding sequence, especially for the CTD, can be adapted to *E. coli* by translationally silent mutations (J. Heger-Stevic and

M. Nassal, unpublished data); this can also be useful for the to-be-displayed protein. Hence the choice of HBc183 vs. a CTD-less variant depends on the intended downstream application. The RNA-CTD interactions in HBc183-based CLPs further stabilize the particles, and in vaccine applications the packaged RNA can act as TLR7/8 agonist that promotes IgG2a induction and reduces T_H cell dependence of the IgG response [28]. If transduction of foreign genetic material is not desired, CTD-less HBc is the preferred choice. As examples for the application of the protocols given in Subheading 3, we will use constructs carrying GFP as heterologous c/e1 insert in the contexts of HBc149 und HBc183 as carrier.

Contiguous Chain Versus SplitCore CLPs as Carrier

While the SplitCore system offers much wider applicability than the contiguous chain HBV CLP system (*see* Fig. 2b and below), the latter, where applicable, requires just one well-designed fusion protein. Successful CLP formation depends primarily on the spatial distance between the insert protein's termini which will both be fused to the c/e1 acceptor sites in the carrier protein (*see* Fig. 2a). For instance, replacement of Ala80 in the c/e1 loop leaves the carboxyl end of Pro79 and the amino terminus of Ser81 as fusion sites that are about 4 Å apart. Larger distances between the insert protein's termini may be balanced, within limits, by the inclusion of flexible hydrophilic linker sequences, e.g., by an increased number of $(Gly)_3$Ser repeats. However, for the borrelial OspA protein with its termini >50 Å apart [29], even very long linkers only partially rescued CLP formation [30]. Moreover, such long linkers may have an often undesirable immunogenic potential of their own. Hence, whenever it is known or suspected that the candidate display protein has termini that are far apart, the SplitCore system is more promising, as exemplified by successful CLP formation with proteins that failed to properly assemble in the contiguous chain system, such as OspA, CSP, or several tick proteins [16, 18, 19].

Additional Options Specific for the SplitCore System

The SplitCore system also offers additional design options (beyond full-length vs. CTD-less split HBc as building block) that may be exploited for improved CLP formation and/or tailoring a CLP's specific properties. Most relevant is the ability of both carrier protein segments to act as acceptors (*see* Fig. 2b). Hence a heterologous protein may be fused via its N terminus to coreN, or via its C terminus to coreC (both may also be combined). As in simple reporter protein fusions, N-terminal versus C-terminal fusion can markedly affect the folding and function of the protein of interest. Hence if a CLP display candidate protein is known to have a buried N-terminus but an exposed C-terminus, fusion to coreC is certainly a better choice, and vice versa. Furthermore, for a cotranslationally folding two-domain protein, the order of appearance from the ribosome can strongly affect the folding pathway [31]. Hence if a specific heterologous protein aggregates when fused to coreC, fusion to coreN is a viable alternative. Not least, a protein fused to coreN will mostly expose its C-terminus to the solvent, while the reverse is true for a fusion to coreC. For vaccine applications, these opposing orientations can profoundly impact the induced antibody repertoire [18]. Hence if the location of a neutralizing epitope in a protein antigen is known, SplitCore CLPs designed to expose this part of the protein are more likely to evoke relevant antibodies.

A complication of the SplitCore system is the requirement to express both segments in, ideally, identical molar amounts. The most universal solution, in our hands, is the generation of a bicistronic mRNA from a single expression vector [18, 19] (*see* Fig. 2b). Polycistronic mRNAs are rare in eukaryotes but common in prokaryotes where ribosomes can recognize internal translation initiation sites by binding of the 3′ proximal 16S rRNA residues to a complementary ribosome binding site (RBS) shortly upstream of the initiator AUG codon (for a recent review, *see* [32]). Hence translation of the downstream cistron can be achieved by simply introducing a second RBS (RBS2 in Fig. 2b) in front of the second open reading frame. However, efficient translation also requires that the RBS and initiator AUG not be buried within RNA secondary structures [33]. These will be affected by the base-pairing potential of the surrounding RNA sequence and hence vary between different coding sequences. If one of the two SplitCore protein segments is poorly expressed from one vector, this can often be alleviated by reversing the cistron order on the bicistronic mRNA (*see* Fig. 2b, iii and iv) [19, 25]; as an example we will use a reverse cistron order SplitCore construct carrying a split GFP, with the smaller GFPβ11 segment on HBc149-based coreC, and the larger GFP segment β1-10 on coreN. Alternatively, RNA secondary structure prediction programs may be consulted to design mutations that will destabilize interfering structures.

1.2.3 Protein-Displaying CLP Expression: Vectors and Expression Strains

There are no specific limitations regarding the vector system for expression of contiguous chain or SplitCore CLPs in *E. coli*. However, systems based on the bacteriophage T7 RNA polymerase for mRNA production are among the most efficient and also in our hands commonly gave the highest CLP yields. Our standard vector pET28a2 [22] is based on the commercial pET28 plasmid (Novagen/EMD-Millipore) except the original kanamycin resistance gene has been replaced by an ampicillin resistance cassette; this avoids the recovery period after transformation and before plating on antibiotic-containing agar plates which is mandatory for antibiotics targeting translation. Like other pET vectors, it features a T7 promoter followed by lac operator and RBS sequences upstream, plus a T7 terminator downstream of the intended expression cassette (*see* Fig. 2); a medium-copy pBR322 origin; and the *lacI* gene encoding the lac repressor (useful general information can be found in the pET system manual, freely available at the merckmillipore.com home page). However, we have also successfully used other T7 promoter-based vectors, e.g., pRSF-Duet (Novagen/EMD-Millipore) which features a different, high-copy replication origin and two T7 promoters; the second T7 promoter, intended for simultaneous expression of two proteins, can easily be removed during cloning, as in the SplitCore split GFP construct used as one example (*see* Fig. 3a).

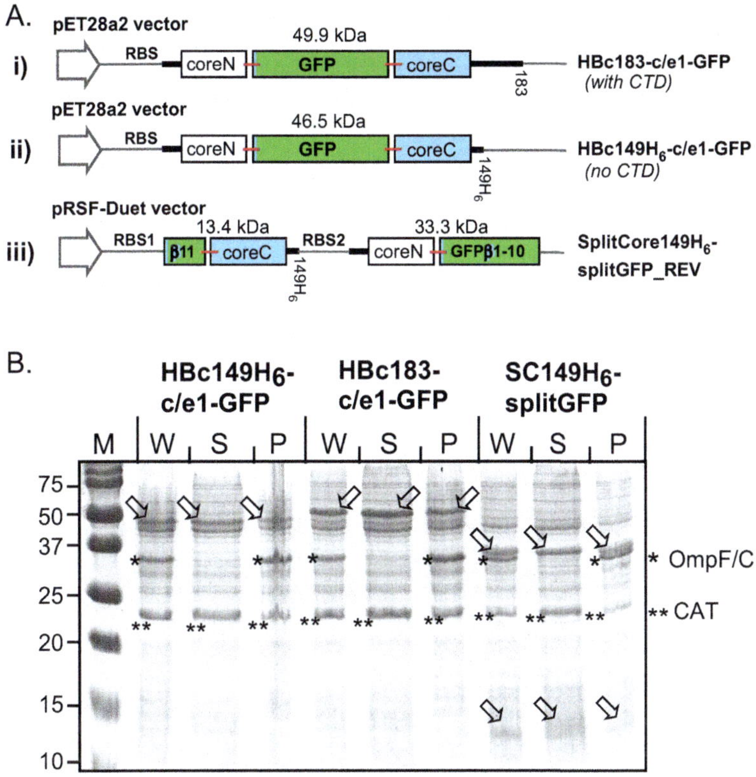

Fig. 3 Assessing expression and solubility of fusion proteins intended to form protein-presenting HBc CLPs. (**a**) Schematic representation of three representative GFP-presenting HBc constructs. GFP was chosen as example for a to-be-displayed protein as its color and fluorescence allow direct visualization of the respective fusion proteins upon velocity sedimentation (*see* Fig. 4) and native agarose gel electrophoresis (*see* Fig. 5). The top two contiguous chain constructs (i, ii) differ only in the coreC part which contains (i) or does not contain (ii) the CTD; both are encoded by ampicillin resistance conferring pET28a2 plasmids [22]. The SplitCore construct (iii) carries a split GFP, with the small GFPβ11 fragment fused to coreC and the large GFPβ1–10 fragment fused to coreN as described [18]. However, the cistron order is reversed and the vector backbone is based on the kanamycin resistance conferring plasmid pRSF-Duet. (**b**) SDS-PAGE analysis of soluble and insoluble lysate components. 200 ml cultures of *E. coli* BL21*Cp cells transformed with the plasmid constructs shown in (**a**) and induced as described in Subheading 3.1 were lysed in 3 ml TN300-based lysis buffer each as described in Subheading 3.2. Then aliquots of the whole lysate before centrifugation (W; 2 μl), and of the soluble supernatant (S; 5 μl) and insoluble pellet (P; 2 μl) after centrifugation were analyzed by SDS-PAGE in 15% polyacrylamide gels. M, marker proteins having the molecular masses indicated on the left (in kDa). *E. coli* OmpF and/or OmpC membrane proteins (marked by an asterisk) typically abound in the insoluble fraction, chloramphenicol acetyltransferase (CAT; marked by double asterisks) from the Codonplus plasmid in the soluble fraction. The positions of the GFP-HBc fusion proteins are marked by arrowheads; the signals in lanes S indicate that substantial fractions of all three fusion proteins are soluble

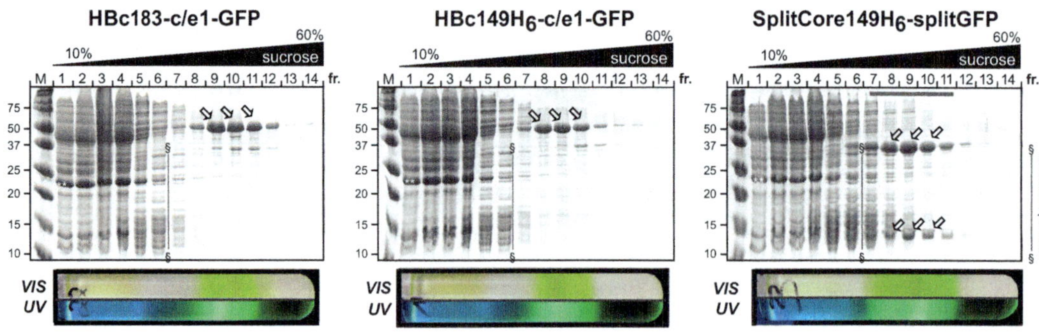

Fig. 4 Assessing protein chemical purity of HBc fusion proteins after sucrose gradient sedimentation. Cleared lysates containing the GFP-HBc fusion proteins shown in Fig. 3b (3 ml each) were sedimented through 10–60% sucrose step gradients in a TST41-14 rotor as described in Subheading 3.3 and collected in 14 fractions. 3.5 μl of each fraction were analyzed by SDS-PAGE as in Fig. 3. The positions of the relevant fusion protein bands are indicated by arrowheads. Their accumulation in fractions 7–11 already suggests CLP formation. The slightly faster sedimentation of the HBc183-based sample is due to the much higher RNA content (*see* Subheading "Full-Length Versus CTD-Less HBc as Carrier"; and Fig. 5). The bulk of *E. coli* proteins and CAT (~24 kDa, marked with double asterisks) typically remain in the top five fractions. The numerous small proteins cosedimenting into fractions 5–7 are largely derived from ribosomal proteins (vertical line bordered by § signs). The bottom panels show the original gradient tubes observed at daylight (VIS) or on a UV transilluminator (UV)

Our standard HBc carrier protein is based on the core protein from a genotype D, subtype ayw HBV isolate (GenBank accession no. CAA24706, but without the 29 aa precore sequence), encoded by a synthetic gene providing numerous restriction sites for easy mutagenesis [34]. However, the successful use of rodent hepadnavirus core proteins [2, 18] as display platform implies that HBc from most other HBV isolates as well as new, engineered variants [35] will perform equally well. For Cp149 based vectors, addition of a C terminal His_6-tag may increase the efficiency of CLP formation [22].

T7 promoter-controlled constructs are not expressed in common *E. coli* cloning strains (e.g., DH5α, TOP10, and various others) which lack T7 RNA polymerase; however, such strains are strongly recommended for cloning and plasmid stock generation. Assuming that readers are familiar with basic cloning and transformation procedures we will not further allude to these issues. The most commonly used *E. coli* strains providing RNA T7 polymerase, and well suited for HBc CLP expression, are BL21(DE3) cells and their derivatives. Carrying a genomic copy of the T7 RNA polymerase gene under control of the lacUV5 promoter, expression of the RNA polymerase is induced by isopropyl-β-D-1-thiogalactopyranoside (IPTG) which relieves repression by the expression vector-encoded lac repressor. Useful derivatives are BL21 Star (DE3) cells (ThermoFisher) which feature enhanced mRNA stability, and the already mentioned CodonPlus strains which provide rare tRNAs from an

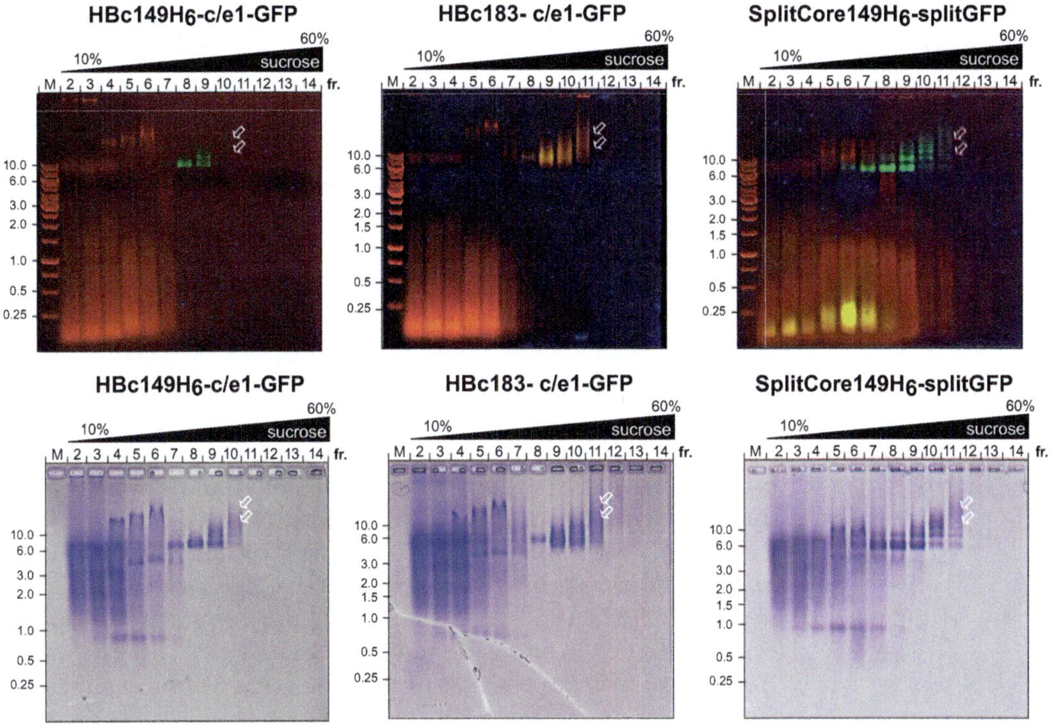

Fig. 5 Assessing purity and CLP formation by native agarose gel electrophoresis (NAGE). Equal aliquots (3.5 μl) of the sucrose gradient fractions shown in Fig. 4 were subjected to NAGE in 1% agarose gels containing Ethidium Bromide as described in Subheading 3.4 and inspected for fluorescence and protein content. (**a**) Fluorescence detection. Gels were photographed on a UV transilluminator, revealing the distribution of RNA (orange) and GFP (green). The yellowish color in fractions 8–11 of the HBc183-based sample derives from the simultaneous green GFP fluorescence plus Ethidium Bromide fluorescence of packaged RNA. The arrowheads indicate slower migrating oligomeric/multimeric particles accumulating in the lower gradient fractions. Note that the fragmented RNA especially in the upper gradient fractions would not be detected by SDS-PAGE inspection (Fig. 4). (**b**) Protein detection. The gels shown in (**a**) were subsequently stained with Coomassie Blue. The distinct appearance of the main bands in fractions 7–9 indicates the presence of proper CLPs.

extra plasmid that confers resistance to chloramphenicol. For routine purposes, we often use BL21 Star (DE3) cells that are pretransformed with the CodonPlus plasmid (referred to as BL21*Cp cells) to then introduce the specific CLP expression vector. Transformants are then selected by using chloramphenicol (for the CodonPlus plasmid) plus the appropriate antibiotic for the expression vector, i.e., ampicillin (or carbenicillin; see below) for pET28a2 vectors.

As outlined in Subheading "Posttranslational Modifications," in some cases the use of special purpose *E. coli* strains such as Rosetta-gami of SHuffle Express (or T7 SHuffle Express for T7 RNA polymerase-based expression) may be advantageous. However, BL21*Cp cells are most universally applicable and therefore are the first choice for all pilot experiments.

1.2.4 CLP Purification: General Considerations

A key feature of CLPs is their large size compared to nonassembled proteins. Hence separation methods discriminating molecules on the basis of their hydrodynamic diameters are the first choice for CLP purification; in turn, they can yield additional information on the particle status of a given preparation. A simple and robust method is velocity sedimentation in sucrose gradients, described in detail below. An alternative is size exclusion chromatography in an appropriate matrix such as Sepharose CL4B or Superose 6 [6]. For CLPs bearing a net surface charge at near-physiological pH, ion exchange chromatography may also be applied; for instance, we have successfully used anion-exchange chromatography as an orthogonal follow-up purification for wild-type HBV CLPs (bearing a negative surface charge at neutral pH; *see* Subheading 1.2.5) after sucrose gradient enrichment (A. Walker, J. Heger, M. Nassal, unpublished data). Affinity purification using antibodies against the CLP-displayed protein should be possible but is expensive and the harsh conditions required for release from the affinity matrix may jeopardize particle integrity. A preenrichment of the CLPs may also be achieved by fractionated ammonium sulfate precipitation, as used early on for nonmodified HBc CLPs [3]. However, for foreign protein-presenting CLPs the optimal ammonium sulfate concentration would have to be determined in advance, and the high ionic strength may affect particle integrity. Hence we usually do not include this step.

CLP Enrichment via Velocity Sedimentation in Sucrose Gradients

Beyond substantial separation power, one of the advantages of sucrose gradient sedimentation is the single-use nature of the consumables, i.e., rotor tubes and gradient media. Hence technical concerns typical for column chromatography, e.g., contamination with material from previous runs or system clogging, are no issue.

A theoretical treatment of velocity sedimentation is beyond the scope of this chapter; some fundamental principles as well as practical guidance can be found in [36], and on the websites of centrifuge manufacturers. In brief, velocity sedimentation is a nonequilibrium method that separates particles according to the rate at which they travel through a gradient medium under a given centrifugal force. This rate is indicated in Svedberg units (with $1S = 10^{-13}$ s, i.e., a unit of time), and is a characteristic property of a specific particle, affected by multiple parameters such as particle mass, density, and shape, as well as medium viscosity which itself depends on temperature. In general, sedimentation speed increases with particle size and density. Nonmultimeric proteins often have sedimentation coefficients of <5S, intact prokaryotic ribosomes have 70S, their subunits have 50S and 30S. Reported sedimentation coefficients for unmodified HBV CLPs are in the range of 70S–100S [3]. As material constants, s-values are independent of rotor speed or geometry. If the same rotor is run at lower speed, it will just take longer until a given particle reaches the same position in the tube. Run time and speed can therefore be adjusted over a wide range, and conditions working well for one rotor-type can be adjusted to different rotors and centrifuges. Useful online calculation tools are publicly available at several sites, e.g., www.sciencegateway.org/tools/rotor.htm or www.beckmancoulter.com/wsrportal/wsr/research-and-discovery/products-and-services/centrifugation/rotors/index.htm?t=3.

1.2.5 CLP Integrity Assessment: General Considerations

The usual aim for a CLP preparation is to obtain protein-chemically pure intact CLPs. Protein-chemical purity is assessed by SDS-PAGE analysis and Coomassie Blue staining (or other protein staining method). Final proof for proper CLP formation requires electron microscopy, with image reconstruction after electron cryomicroscopy ("cryo-EM") offering the highest resolution [37]. As appropriate EM facilities are not routinely available, simpler techniques can already give highly useful hints whether a specific fusion construct forms CLPs or not. A distribution of the protein(s) of interest in a sucrose gradient as in Fig. 4 already suggests particle formation yet additional information is easily obtained by native agarose gel electrophoresis (NAGE) as shown in Fig. 5.

The larger pore size of agarose vs. polyacrylamide gels allows intact large particles, such as viruses or CLPs, to enter the matrix and migrate in an electric field according to their surface charge [38]; due to their low diffusion coefficients, homogeneous particles will appear as distinct bands (nonassembled proteins diffuse rapidly in agarose gels, giving a smeary appearance). The separation range of the agarose gels can be adjusted via the agarose concentration. For instance, T = 4 (diameter 34 nm) and T = 3 (diameter 30 nm) wild-type HBV CLPs do not separate in 1% (w/v) agarose gels but do so at 2.5% (w/v) agarose [18]. Agarose gels containing Ethidium Bromide, as commonly used in DNA fragment separation, provide information on the distribution of contaminating nucleic acids (mostly RNA) in the sucrose gradient fractions, yet also on the RNA content of HBc CLPs, as the nucleic acid stain can penetrate the CLP shell and bind to the packaged RNA [3]. The typical procedure outlined in Subheading 3.4 involves first inspection of Ethidium Bromide-stained nucleic acids, followed by Coomassie Blue detection of protein. Intact CLPs with packaged RNA will give one distinct band that stains with both.

Note that particles without a net surface charge at the pH of the electrophoresis buffer do not move from the loading slot. Their negative surface charge at neutral pH causes wild-type HBV CLPs to migrate toward the anode; however, surface charge of CLPs displaying a heterologous protein will largely depend on that protein's solvent-accessible positively vs. negatively charged side chains. Alternatively, a protein that does not move may have formed aggregates too large to enter the gel. To distinguish between these possibilities, the pH of the electrophoresis buffer can be varied, or the CLP surface charge can be chemically modified; for instance, succinylation has successfully been used to reverse the positive charge on Lys side-chains [39]. A lack of impact on mobility suggests protein or CLP aggregation.

NAGE can also be combined with immunoblotting to achieve higher sensitivity and/or specificity than general in-gel staining methods. To this end, the gel contents can be blotted to a polyvinylidenfluoride (PVDF) membrane by capillary transfer [18]. Further processing of the membrane can then be performed exactly as in Western blotting after SDS-PAGE.

1.2.6 Considerations for Further CLP Purification

The enrichment achieved by a single sucrose gradient sedimentation may be sufficient for some applications but not for others. If higher purity is required, we here provide some aspects to be considered.

Insufficient CLP purity may originate from overloading the first sucrose gradient (too much total protein), or too large a sample volume relative to the volume of the gradient. Subjecting the appropriate CLP-containing fractions from the first gradient to a second round of velocity sedimentation, possibly using a smaller sample volume, helps to improve resolution. The sucrose concentration in the sample must be lower than that of the least concentrated sucrose solution of the second gradient which is easily achieved by dialysis against sucrose-free TN buffer (size exclusion chromatography is a viable alternative); given the large size of CLPs, the molecular weight cut-off of the dialysis tubing can be as high as 100 kDa. To account for the accompanying increase in sample volume and decrease in protein concentration, the dialyzed sample can be concentrated, e.g., by a centrifugal ultrafiltration device. However, reducing the sample to a very small volume, even if only transiently during centrifugation, should be avoided because at a concentration of several milligrams per milliliter, even highly soluble CLPs (like most proteins in general) may start to aggregate and precipitate on the ultrafiltration membrane. Figure 6 shows the further enrichment by this procedure of the SplitCore-split GFP CLP preparation which contained substantial amounts of contaminating proteins (*see* Fig. 4) and fragmented RNA (*see* Fig. 5a) after the first gradient separation. Any of the alternative purification methods mentioned in Subheading 1.2.4 may also be used as a second purification step if compatible with the specific properties of the displayed protein.

Lastly, it should be noted that bacterial lipopolysaccharide (LPS), a biologically highly active polymer detected by neither common protein stains nor Ethidium Bromide, does not completely separate from CLPs during sucrose gradient sedimentation. A simple yet efficient way to remove LPS is based on Triton X-114 phase partitioning, during which LPS and other *E. coli* membrane components, but not the CLPs, accumulate in the

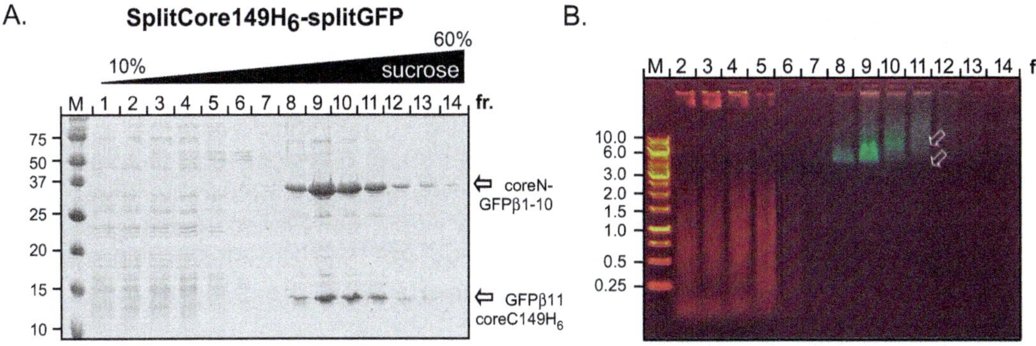

Fig. 6 Further enrichment of CLPs by a second sucrose gradient sedimentation. The CLP fractions from the SplitCore149H6-splitGFP example preparation still contained substantial amounts of *E. coli* proteins (*see* Fig. 4) and RNA (*see* Fig. 5) after the first sucrose gradient run. Fractions 7–11 (horizontal bar in Fig. 4) were therefore subjected to a second sucrose gradient enrichment. The pooled fractions were dialyzed and concentrated as described briefly in Subheading 1.2.6; then the concentrate was equally distributed to two TST41.14 rotor tubes and resedimented under identical conditions. Gradient fractions were then analyzed as before. (**a**) SDS-PAGE. Protein-chemical purity of the split fusion protein (intensity of the Coomassie Blue stained specific bands vs. that of unrelated bands in the rest of the lane) increased from ~70% in fraction 9 of the first gradient to >85% in fraction 9 of the second gradient. (**b**) NAGE. The bulk of contaminating bacterial RNA remained in the top fractions, as shown by Ethidium Bromide staining

detergent-rich phase [18]. Of note, the sucrose concentration must be lowered to 5% (w/v) or less, otherwise the phases will not separate well.

2 Materials

2.1 HBc Fusion Protein Expression

1. A petri dish with Luria–Bertani (LB) medium-based agar (10 g/l Bacto tryptone, 5 g/l yeast extract, 10 g/l NaCl, 15 g/l agar) containing the appropriate selection markers, with easily visible colonies of the desired *E. coli* expression strain transformed with the proper expression vector (*see* **Note 1**).

   For the example preparations, *E. coli* BL21*Cp cells transformed with pET28a2 based vectors for contiguous chain GFP insertions or with a pRSF-Duet vector for the SplitCore construct were used; hence ampicillin (50 μg/ml) plus chloramphenicol (34 μg/ml) were appropriate for the pET28a2 vectors and kanamycin (50 μg/ml) plus chloramphenicol (34 μg/ml) for the pRSF-Duet vector. (*see* **Note 2**)
2. Sterile LB medium (*see* **item 1** without agar) containing the appropriate antibiotics (*see* **item 1**).
3. 1 M IPTG stock solution in pure sterile water.
4. TN50 buffer: 25 mM Tris–HCl, pH 7.5, 50 mM NaCl.
5. Sterile 50 ml (for the preculture) and 1 l (for the induction culture) baffled Erlenmeyer flasks.
6. Orbital shaker(s) operating at 37 °C and at 20 °C.

7. A UV-VIS spectrometer and matching disposable plastic cuvettes (to monitor bacterial growth via optical density at 600 nm [OD_{600}]).
8. A clinical centrifuge with temperature control plus rotors suited for 10–15 ml and 50 ml tubes (*see* **item 9**), capable of generating ≥4000 × *g*; alternatively, a preparative high-speed centrifuge with a rotor suited for 250 ml centrifuge bottles (e.g., Sorvall Evolution RC Superspeed/SLA1500 rotor; Sorvall Lynx 4000 Superspeed/F14-6x250y rotor) can be used (*see* **Note 3**).
9. Centrifuge tubes/bottles approved for the RCF to be applied. For example:for *15 ml*: Greiner polypropylene centrifuge tubes 17 × 120 mm, conical bottom, maximal RCF 3500 × *g*;

   for *50 ml*: Greiner polypropylene centrifuge tubes 30 × 150 mm, conical bottom, maximal RCF 3200 × *g*;

   for *250 ml*: Nalgene wide-mouth centrifuge bottles with sealing caps, Style 3141, maximal RCF 27,500 × *g*.

2.2 Sonication-Mediated Lysis of *E. coli* Cells

1. TN300 buffer: 25 mM Tris–HCl, pH 7.5, 300 mM NaCl (*see* **Note 11**). If disulfide bridge formation is of concern (*see* **Note 12**), adding DTT (1–5 mM final concentration; from a 1 M DTT stock solution in water) and EDTA (1 mM final concentration; from a 0.5 M EDTA stock solution, pH 8.0) to the respective TN buffer maintains reducing conditions and prevents heavy metal ion catalyzed air oxidation [4].
2. 10 mg/ml Hen egg white lysozyme stock solution in water (to digest the bacterial cell wall).

   Nonionic detergent: 10% (v/v) Triton X-100 or NP40 stock solutions in water.
3. 25× Protease inhibitor cocktail: For Roche cOmplete Protease inhibitor cocktail tablets (Roche) dissolve 1 tablet in 2 ml of water to obtain a 25× stock solution; store at −20 °C until use. A potential alternative is the trypsin-like and thrombin-like protease inhibitor 4-(2-aminoethyl)-benzenesulfonyl fluoride hydrochloride (AEBSF; available from various suppliers) which, at 0.1 mM final concentration, seems to efficiently block *E. coli* protease-mediated degradation of HBc's vulnerable arginine-rich CTD. If using this product, consult the manufacturer's recommendations for proper application.
4. 25 U/μl benzonase stock solution (an engineered DNA and RNA degrading endonuclease from *Serratia marcescens*, available for example from Merck-Millipore) (*see* **Note 13**).
5. A rotary shaker capable of holding 15 ml Greiner centrifuge tubes.
6. 30 ml centrifuge tubes or 20 ml glass beaker.

7. A sonifier with timed operation and pulse mode capabilities suited for sample volumes of several milliters, e.g., Branson B12 or newer equivalent, with an appropriate tip (e.g., micro-tip) (*see* **Note 14**).
8. Screw-cap centrifuge tubes approved for use at >4000 × *g*; e.g., Sarstedt polypropylene tubes 107 × 25 mm, conical bottom, maximal RCF 8800 × *g*.
9. Equipment for SDS-PAGE: Gels, Laemmli buffer, SDS-PAGE sample buffer, Coomassie Blue staining and destaining solution.
10. Optional: 0.22 μm filter units.

2.3 CLP Enrichment by Sucrose Gradient Velocity Sedimentation

1. The cleared bacterial lysate from Subheading 3.2, **step 9** or optional **step 11**.
2. 80% (w/v) sucrose stock solution: 800 g sucrose per 1 l of TN300 buffer (*see* Subheading 2.2, **item 1**) (or TN50 or TN150 buffer; *see* **Note 11**).
3. 10%, 20%, 30%, 40%, 50%, and 60% (w/v) Sucrose solutions: Prepare by filling the following volumes of the 80% (w/v) sucrose stock solution (**item 2**) up to 100 ml with the same TN buffer as used for preparing item 2: 10% = 13.0 ml; 20% = 27.1 ml; 30% = 42.4 ml; 40% = 59.1 ml; 50% = 77.3 ml; and 60% = 97.1 ml. If desired, add DTT and EDTA to 1–5 mM and 1 mM final concentration (*see* **Note 12**).
4. A temperature-controlled preparative ultracentrifuge (e.g., Beckman-Coulter Optima LE-80K or equivalent) suited for medium to large volume swing-out rotors, e.g., SW40Ti (Beckman-Coulter) or TST41.14 (Kontron) holding six tubes of 14 ml volume; or SW28 rotor (Beckman-Coulter) holding six tubes of 38.5 ml volume.
5. Matching centrifuge tubes (e.g., Beckman-Coulter polyallomer tubes 14 ml, 14 × 95 mm, thin-wall, cat. no. 331374; and 38.5 ml, 25 × 89 mm, thin-wall, cat. no. 344058). The usable volume of both types of tubes is slightly less than the nominal volume.
6. Equipment for SDS-PAGE: Gels, Laemmli buffer, SDS-PAGE sample buffer, Coomassie Blue R-250 staining and destaining solution.

2.4 CLP Quality Control by Native Agarose Gel Electrophoresis (NAGE)

1. Sucrose gradient fractions from Subheading 3.3, **step 7**.
2. Electrophoresis grade agarose.
3. 0.5 M EDTA, pH 8.0.
4. 10× TAE electrophoresis buffer: For 1 l, dissolve 48.4 g Tris base in ~600 ml pure water, add 20 ml of 0.5 M EDTA, pH 8.0, and slowly add 11.44 ml glacial acetic acid; then fill up with water to 1 l (*see* **Note 30**).

5. 10 mg/ml Ethidium Bromide stock solution in water.
6. DNA size marker in a loading buffer containing Bromophenol Blue and/or Xylene Cyanol.
7. Coomassie Blue staining and destaining solution (as used in SDS-PAGE analysis).
8. A submerged horizontal electrophoresis cell, matching gel tray and power supply, all as commonly used for DNA fragment separation; the size of the gel does not matter. Minigels (e.g., 7 × 10 cm, 50 ml gel volume) with combs for 8–15 loading slots work well.
9. A gel documentation system equipped with a 300 nm UV transilluminator, or a LASER scanner capable of recording Ethidium Bromide fluorescence.

3 Methods

All handling of the recombinant bacteria should be performed according to good laboratory practice and in line with the respective biosafety regulations. The following protocols are based on CLP expression in a 200 ml culture and do not differ for contiguous chain versus SplitCore CLPs. While easily scaled up or down, for well-expressed and well-folding HBc fusion proteins this can yield milligram amounts of CLPs. The example preparations of the GFP-presenting CLPs shown in Fig. 3 were all generated this way. They comprise a contiguous chain construct based on HBc183 (including the CTD) and one based on HBc149H$_6$ (without the CTD), plus a reverse cistron order SplitCore construct carrying a split GFP [18], with the small segment GFPβ11 on coreC149H$_6$ and the large segment GFPβ1-10 on coreN.

3.1 HBc Fusion Protein Expression

1. Ino.culate a single colony (*see* **Note 4**) from the LB agar plate into 5–10 ml LB medium with antibiotics in a 50 ml Erlenmeyer flask and shake overnight at 37 °C. The suspension should become visibly turbid.
2. The next morning, collect the preculture in a 15 ml or 50 ml screw-cap centrifuge tube and pellet the cells at low speed (3200–3500 × *g* for 10 min) in the clinical centrifuge at 4 °C. Discard the supernatant and resuspend the cell pellet in 5–10 ml of fresh LB medium with antibiotics. This washing step removes β-lactamase secreted into the medium by ampicillin/carbenicillin-resistant bacteria (*see* **Note 5**).
3. Inoculate the cell suspension into 200 ml LB medium with appropriate antibiotics in the 1 l Erlenmeyer flask and shake at 37 °C. Once the suspension becomes visibly turbid, start spectrophotometrically monitoring the OD_{600nm} until it

reaches a value between 0.6 and 1.0 (*see* **Note 6**); this correlates with exponential growth which is important for maximal protein expression.

4. Remove the flask from the shaker and let it cool down to room temperature; then add 1/1000 volume of 1 M IPTG (*see* **Note 7**).
5. Continue shaking for 12–16 h (e.g., overnight) at 20 °C (*see* **Note 8**).
6. Harvest the cells by low speed centrifugation (4000–5000 × *g*, 10–15 min, 4 °C) in a 250 ml centrifuge bottle. Alternatively, collect the cells by sequential centrifugation of 50 ml aliquots in a 50 ml Falcon tube using the clinical centrifuge (3200 × *g*, 10–15 min, 4 °C). Under the culture conditions described, ~3 g cells (wet weight) per 200 ml culture can be expected.
7. Resuspend the cell pellet in 10–20 ml of TN50 buffer, followed by repelleting of the cells in a 50 ml screw cap tube (3200 × *g*, 10–15 min, 4 °C); discard the supernatant (*see* **Note 9**).
8. Freeze the cell pellet at −80 °C until further workup (*see* **Note 10**).

3.2 Sonication-Mediated Lysis of *E. coli* Cells

Efficient yet gentle cell lysis is the key for high yields of soluble CLPs. Properly conducted, the mild detergent/sonication based method described below works well for preparative laboratory scale work. If other lysis methods are to be used (e.g., a French pressure cell for larger volumes), always make sure that the cells are actually broken up, and that the conditions applied do not impose too much thermal or other stress on your protein of interest.

1. Resuspend the frozen cell pellet from Subheading 3.1, **step 8** in 3.0 ml (*see* **Note 15**) of TN300 buffer (or TN50 or TN150 buffer; *see* **Note 11**), containing 1–5 mM DTT and 1 mM EDTA if desired (*see* **Note 12**).
2. Add 1/20 volume of the nonionic detergent stock solution (0.5% final concentration).
3. Add 1/100 volume of the 10 mg/ml lysozyme stock solution (0.1 mg/ml final concentration).
4. Add 1/25 volume of the 25× protease inhibitor cocktail solution (*see* **Note 16**).
5. Vortex for about 20 s to mix all components and incubate on ice for 10–15 min until the solution becomes viscous due to the release of *E. coli* genomic DNA (*see* **Note 17**).
6. Add 1–2 μl (25–50 U) of Benzonase, vortex briefly, and incubate at room temperature on a rotary shaker or equivalent mixing device until viscosity drops, due to fragmentation of genomic DNA (*see* **Note 18**).

7. Transfer the suspension into the 30 ml centrifuge tube for the subsequent sonication step and cool for 5 min on ice. Alternatively, a small glass beaker (20 ml volume) may be used. This allows better heat dissipation and observability of the position of the sonicator tip in the solution; however, directly touching with the sonicator tip can cause the glass to break.
8. Sonify the sample to completely disrupt the bacterial cell wall and membranes that are already weakened by the lysozyme and detergent treatment. Also, genomic DNA will now be fully released and sheared into small pieces. The following conditions work well with a Branson B12 sonifier equipped with a standard microtip: Position the tip a few mm underneath the surface of the precooled sample and sonicate for 10 s at 50% maximal power, then let the sample cool down for 50 s. Repeat this sonication/cooling cycle up to ten times. Larger sample volumes may require more cycles. Vessels with a thick plastic well may require longer cooling periods to allow the heat to dissipate. (*see* **Note 19**)
9. Set aside a small aliquot (20 μl) of the suspension (to be used for assessing total expression level of the protein of interest) and centrifuge the rest of the sample at 8000 × *g* for 30 min at 4 °C to pellet insoluble cell debris and protein aggregates. Transfer the supernatant ("cleared lysate") into a fresh 15 ml screw cap tube and store at 4 °C until the subsequent ultracentrifugation step (*see* **Note 20**). Resuspend the insoluble pellet in a volume of fresh TN buffer corresponding to that of the original lysate.
10. To 20 μl each of the supernatant, the resuspended pellet and the set-aside sample of unfractionated lysate, add an appropriate SDS-PAGE sample buffer and boil for 5 min. Analyze 5–10 μl of each sample side-by-side by SDS-PAGE followed by Coomassie Blue (or other protein) staining. The relative amounts of the protein of interest in the whole lysate, the soluble fraction, and the pellet (labeled W, S, P in Fig. 3b) give an indication as to how well the protein is expressed, and which fraction of it is soluble. The pellet typically shows a prominent band of ~35 kDa (marked with an asterisk in Fig. 3b) which corresponds to the *E. coli* outer membrane proteins OmpF and/or OmpC [40]. A prominent band of ~24 kDa in the soluble fraction (marked with a double-asterisk in Fig. 3b) is due to the well soluble chloramphenicol acetyltransferase (CAT) encoded by the Codonplus plasmid (when using Codonplus cells). If the protein bands are too faint or too strong, repeat with larger or smaller amounts. In case of doubt, the identity of desired band(s) can be verified by immunoblotting using antibodies against HBV core protein, or against the heterologous protein of interest (*see* **Note 21**).

11. *Optional:* Pass the cleared lysate through a sterile 0.22 μm filter unit. This can increase shelf-life of the lysate by preventing microbial growth (e.g., of bacteria that survived the lysis procedure) and by removal of macroscopic aggregates that might reduce resolution during the subsequent sucrose gradient centrifugation step.

3.3 CLP Enrichment by Sucrose Gradient Velocity Sedimentation

The procedure uses 10–60% (w/v) sucrose step gradients as these are easily prepared. If a gradient mixer is available, continuous 10–60% (w/v) sucrose gradients may be used analogously. However, the step gradient procedure will give satisfactory results in most instances. The indicated centrifugation conditions separate intact CLPs well from the bulk of soluble *E. coli* proteins (remaining in the top gradient region), and from aggregates (sedimenting to the bottom region). A similar distribution should be aimed at if a different rotor and/or modified running conditions are to be used. Example gradients for the two contiguous chain HBc and the SplitCore GFP constructs (*see* Fig. 3) are shown in Fig. 4.

Before using an ultracentrifuge, it is mandatory to become familiar with proper operation of the specific model, rotor and tubes by consulting the user manual, and/or the manufacturer's pertinent web-based information. As a nonequilibrium technique, velocity sedimentation in sucrose gradients has a limited focusing effect, caused by the increasing viscosity of more concentrated sucrose solutions. However, molecules in the top part of the loaded sample will have a longer distance to travel to a specific gradient position than those in the bottom part. Hence smaller sample volumes achieve higher resolution. For separating CLPs using a TST41.14 or SW40Ti rotor, the sample volume should be 3 ml or less, for an SW28 8 ml or less. Volumes given below are for a TST41.14 or equivalent rotor, those for the SW28 rotor are also provided separately thereafter.

1. Determine the volume of the cleared lysate (from Subheading 3.2, **step 9** or **11**) to be subjected to sucrose gradient centrifugation. This information is used to determine the appropriate rotor tube size, and for adjusting the volume of the 10% (w/v) sucrose solution at the gradient top.
2. Prepare the sucrose step gradient by pipetting 1.8 ml of the 60% sucrose solution into two appropriate TST41.14 rotor tubes (one for the sample, the other as a counterweight in the centrifuge). Overlay this bottom layer with the same volume of the 50% sucrose solution by slowly letting the overlay solution run out of a disposable pipette whose outlet is kept just above the surface of the underlying solution (*see* **Note 22**). Repeat this with the lower concentration sucrose solutions; this will give a gradient volume of 6 × 1.8 ml = 10.8 ml, leaving ~2.5 ml

for the sample. For slightly larger or smaller sample volumes, use the 10% (w/v) sucrose solution to adjust the gradient volume such that the remaining free space at the top of the tube is just sufficient to accommodate the actual sample volume. Slight volume variations of the 10% (w/v) sucrose layer will only have a minor impact on the sedimentation behavior of fully formed CLPs. For an SW28 or equivalent rotor, use 5 ml of each of the sucrose solutions and apply using a disposable 5 or 10 ml pipette and a pipettor (e.g., Pipetboy or similar); this will leave ~8.0 ml volume for your sample.

3. Carefully load the sample on top of one of the gradient tubes (*see* **Note 23**); use TN buffer to adjust the counterweight tube to exactly the same weight.
4. Place the tubes appropriately in the rotor and run the TST41.14/SW40Ti rotor at 41,000 RPM (maximum RCF 300,000 × *g*; average RCF in the middle of the tube at r = 11.3 cm 213,000 × *g*) for 2 h at 20 °C (*see* **Note 24**). For an SW28 rotor use 28,000 RPM for 3.5 h at 20 °C (maximum RCF 141,000 × *g*, average RCF at r = 11.8 cm 103,000 × *g*) (*see* **Note 24**).
5. Harvest the gradient (*see* **Note 25**) in 14 equally sized fractions (TST41.14: 14 × 0.9 ml; SW28: 14 × 2.7 ml) from the top (*see* **Note 26**) by using a micropipette with a 1 ml disposable tip held just beneath the actual liquid surface (*see* **Note 27**). Transfer the respective solutions into appropriately sized and labeled tubes and store at 4 °C (*see* **Note 28**).
6. Monitor the protein distribution in the gradient by analyzing 5–10 μl of each fraction using SDS-PAGE and Coomassie Blue staining (*see* Subheading 3.2, **step 10**). For well-expressed soluble CLPs this will typically reveal a peak of the desired fusion protein(s) slightly below the center of the gradient, i.e., around fractions 7–10 out of the 14 fractions collected from top to bottom, as shown in the example gradients in Fig. 4 (*see* **Note 29**).
7. Store the relevant fractions at 4 °C (*see* **Note 28**) or alternatively at −20 °C or −80 °C; the sucrose acts as cryoprotectant.

3.4 CLP Quality Control by Native Agarose Gel Electrophoresis (NAGE)

1. Prepare a 1%(w/v) agarose gel (*see* **Note 31**) in 1× TAE buffer containing a final concentration of 0.5 μg/ml of Ethidium Bromide.
2. Run 2–10 μl aliquots of each gradient fraction on the agarose gel (*see* **Note 32**) plus, in a separate lane, a DNA size marker in a loading buffer containing Bromophenol Blue and/or Xylene Cyanol to visualize progress of electrophoresis. The positions of the DNA marker fragments after electrophoresis can serve as a reference when comparing different samples on separately run gels. For a 7 × 10 cm minigel a constant voltage of 80–100 V is appropriate (*see* **Note 33**).

3. Inspect Ethidium Bromide staining on a UV transilluminator and take a picture for future reference to band positions. Alternatively use an appropriate LASER scanner. Fragmented *E. coli* RNAs will appear as a smear, RNA in CLPs as a distinct band. The example gels in Fig. 5a represent a special case, as GFP on CTD-less CLPs is directly visible by its green fluorescence. The full-length HBc-based GFP CLPs appear yellowish, due to the simultaneous emission of orange fluorescence from the Ethidium Bromide-stained packaged RNA and of green fluorescence from GFP.
4. Transfer the agarose gel into a tray containing Coomassie Blue staining solution and incubate with shaking for 30–60 min; do not stain overnight as destaining of agarose gels is very slow.
5. Transfer the stained gel into a tray containing destaining solution and incubate with shaking until the background staining is acceptably low. Depending on staining intensity and gel thickness, this can take several days. Paper towel immersed in the destaining solution speeds up destaining.
6. Document Coomassie Blue staining using appropriate video/camera equipment (*see* Fig. 5b).
7. Compare Ethidium Bromide vs. Coomassie Blue staining patterns (*see* **Note 34**). Colocalization of both stains in one band indicates the presence of intact CLPs. Protein-presenting as well as a nonmodified HBc CLPs may partly form higher order structures with distinct electrophoretic mobility which typically accumulate in the lower gradient fractions (marked with arrowheads in Fig. 5). Though not formally proven, together with the ladder-like appearance in NAGE this implies they represent stable dimers, trimers and so on of intact CLPs. The NAGE assay helps to include or exclude this material from the intended downstream application.
8. Pool the gradient fractions meeting your quality criteria and store at 4 °C (mid-term) or frozen at −20 °C or −80 °C (long-term).
9. *Optional*: If further enrichment is required, subject the pooled CLP fractions to a second sucrose gradient and validate by SDS-PAGE and NAGE (*see* Fig. 6). Alternatively, one of the other purification methods discussed in Subheading 1.2.6 may be used.

4 Notes

1. *E. coli* cells on agar plates have a limited life span; cells should have been plated no longer than 1 or at most 2 weeks prior to the experiment. Frozen glycerol stocks may be an alternative;

however, in our hands their use has often resulted in lower expression yields.

2. *E. coli* cells transformed with pET vectors encoding ampicillin resistance may tolerate carbenicillin better than ampicillin although both antibiotics share the same mechanism-of-action. Hence when experiencing unexpectedly low numbers of colonies after transformation, replacing ampicillin by the same concentration of carbenicillin may help.
3. If no equipment for 250 ml bottles is available, cells from the 200 ml induction culture in Subheading 3.1, **step 6**, may be collected using 50 ml tubes in the clinical centrifuge.
4. A multiple colony inoculum speeds up the schedule but bears some risk that not all bacteria are producers.
5. Ampicillin and carbenicillin kill only actively dividing bacteria. Transformed cells secrete the resistance factor, β-lactamase, into the medium, reducing the effective antibiotic concentration with time. Hence prolonged growth can promote overgrowth of the culture by nontransformed cells or cells that have lost the ampicillin-resistance conferring plasmid, resulting in low yields. Vectors conferring resistance to translation-inhibiting antibiotics, e.g., kanamycin, avoid this potential problem.
6. The time required depends on the number of bacteria in the preculture inoculum, richness of the medium (LB is a rich medium), temperature, and oxygen supply. *E. coli* grow much faster under aerobic conditions (doubling time around 30 min in rich medium at 37 °C). The baffled flask and an appropriate shaking speed ensure maximal aeration.
7. The final IPTG concentration of 1 mM ensures maximal induction; however, for proteins folding poorly slower expression rates may be advantageous as they prevent accumulation of aggregation-prone folding intermediates. The same holds for induction temperature. Induction at 20 °C is usually a good compromise between expression and folding rates. Temperatures above 25 °C commonly yield more protein, yet also a higher proportion of aggregates. The reverse is true for lower induction temperatures. IPTG concentration, induction temperature and duration are therefore among the first parameters for optimization of soluble CLP yields.
8. Overnight induction at 20 °C will often give the highest CLP yield. If for some reason a higher induction temperature is desired, induction time should be reduced.
9. Discarding the supernatant removes undesired medium components and/or metabolites.
10. Proteins frozen in intact cells usually retain functionality and proper folding over months or years unless subjected to

repeated freeze–thaw cycles. Conversely, freezing will facilitate breaking up the cells in the subsequent lysis step.

11. Assembly and stability of wild-type HBV CLPs are promoted by high salt conditions [6]. However, for composite CLPs the to-be-displayed protein may also affect these parameters, often in a poorly predictable way. While TN300 is a good starting point buffers containing less NaCl, e.g., TN50 or TN150 (with 50 mM and 150 mM NaCl, respectively) may be used instead.
12. The Cys61 residues in the HBc dimer interface are optimally juxtaposed to form a disulfide bridge [41, 42]. The two C-terminal Cys183 residues per HBc dimer can disulfide-crosslink most or all of the CLP constituent dimers [4]. As neither type of disulfide bridge affects the general structure of the CLPs, their presence or absence is often not relevant. However, for special applications or when trying to display Cys-rich heterologous proteins, maintaining reducing conditions through the entire workup procedure may be advantageous.
13. Benzonase may be replaced by DNase I plus RNase A, each at a final concentration of 0.1 mg/ml lysate.
14. A French pressure cell may be used instead, with advantages especially for larger cell pellets than used here. However, proper conditions will have to be optimized for complete cell lysis.
15. Larger volumes may improve lysis efficiency but also inappropriately increase sample volume for the subsequent ultracentrifugation step, especially when only a medium-sized rotor is available. For a usual cell pellet from a 200 ml culture, use at least 2 ml and not more than 6 ml of TN buffer to resuspend the cells.
16. The inhibitor stock solution can only be kept for limited time; consult the manufacturer's website for details.
17. A lack of visible viscosity increase suggests poor cell lysis; in this case add another 1/100 volume of the 10 mg/ml lysozyme stock solution and incubate for another 10–15 min.
18. Undigested high molecular weight DNA may interfere with subsequent steps; sufficiently digested samples should be easily pipettable using a 1 ml disposable tip on a micropipette.
19. The most important issues, regardless of the specific sonifier, are: (1) before use, always consult the instruction manual for safe and proper operation; (2) do not let the microtip come out of the solution as this will cause excessive foaming and potentially protein denaturation at the air-liquid interface; moreover, the microtip may be damaged; (3) do not let the microtip touch the walls or bottom of the vessel, especially when using a glass beaker, as this will cause it to break; (4) combine short sonication with long cooling phases, as indicated, to ensure the sample

does not heat up which might cause thermal denaturation; (5) make sure the microtip is intact, as even hairline cracks caused by inappropriate operation will drastically reduce energy transfer from the tip into the solution. Inefficient lysis will result in unbroken cells that retain the protein of interest. If uncertain, microscopically inspect a small aliquot of the sample for the presence of intact *E. coli* cells. Protein in unbroken cells as well as protein aggregates caused by oversonification will be both present in the insoluble fraction of the lysate. Apparent insolubility may therefore not always be an intrinsic property of a given fusion protein.

20. The cleared lysate still contains degradative *E. coli* enzymes, including proteases that might attack your protein of interest when the protease inhibitors become exhausted. Overnight storage at 4 °C is usually safe (unless the to-be-displayed protein is particularly sensitive) yet extended storage should be avoided. Conversely, overnight storage often leads to formation of insoluble material, likely derived from *E. coli* membrane components that can be removed by an additional centrifugation step at 8000 × *g*. This can help to improve resolution during the subsequent sucrose gradient purification.
21. The percentage of acrylamide in the gel should match the expected molecular mass of the fusion protein. For SplitCore constructs comprising only one modified carrier segment (*see* Subheading "Contiguous Chain Versus SplitCore CLPs as Carrier") detection of the small nonmodified other segment will require 15% polyacrylamide gels, or use of an alternative gel system, e.g., Tris-Tricine gels [43].
22. The higher-concentrated sucrose solutions are very viscous, and hence the micropipette should be operated slowly.
23. Two sequential sucrose gradients can increase CLP purity (*see* Subheading 3.4, **step 9**). If using this approach, a sample from the center of the first gradient must first be brought to less than 10% sucrose (e.g., by dialysis) before loading on the second gradient so as not to exceed the density of the top gradient layer.
24. Running the gradients at lower temperature will decrease sedimentation velocity due to the increased viscosity of the sucrose solution. This will require longer run times.
25. There are various techniques to harvest a gradient; the one described is simple and does not require specialized equipment.
26. A larger number of fractions gives higher resolution yet is also more laborious. In our hands, 14 equally sized fractions are a good compromise for both SW28 and TST41.14 gradients.
27. Aspirating liquid from too deep under the surface will cause material from the lighter top layers to mix with that originally present in the denser fractions.

28. In the absence of microbial growth, wild-type HBV CLPs as well as various foreign protein presenting CLPs can be kept for weeks or months at 4 °C without decomposition. However, different to-be-displayed heterologous proteins may differ in this respect.
29. While the bulk of soluble *E. coli* proteins remains in the top four to five fractions, ribosomal proteins (mostly <50 kDa) usually accumulate around fraction 6, i.e., shortly above the typical CLP peak.
30. 0.5× TBE can as well be used as electrophoresis buffer.
31. To separate T = 4 from T = 3 CLPs, use 2.5% (w/v) agarose instead.
32. The sucrose content of the gradient fractions obviates the need to add a loading buffer containing sucrose or glycerol to increase sample density. If a DNA loading buffer is added it must not contain SDS which is present in some commercial products as this will disintegrate CLPs and/or denature their constituent proteins.
33. Like any separation system, agarose gels can be overloaded; this depends on the thickness of the gel and the protein concentration in the sample. Approximately 1–3 μg of CLPs per lane usually gives good results. If necessary, repeat the NAGE using more or less sample per lane.
34. Due to their lower RNA content a given amount of CLPs from CTD-less HBc will stain more weakly with Ethidium Bromide than the same amount of CLPs from CTD-containing HBc.

Acknowledgments

Work in the authors' laboratory pertinent to this chapter has been supported by the Deutsche Forschungsgemeinschaft (DFG), most recently through grants NA154/9-4 and NA154/11-1.

We thank Jolanta Vorreiter and Andrea Pfister for excellent technical assistance.

References

1. Nassal M (2008) Hepatitis B viruses: reverse transcription a different way. Virus Res 134(1–2):235–249
2. Whitacre DC, Lee BO, Milich DR (2009) Use of hepadnavirus core proteins as vaccine platforms. Expert Rev Vaccines 8(11):1565–1573. https://doi.org/10.1586/erv.09.121
3. Birnbaum F, Nassal M (1990) Hepatitis B virus nucleocapsid assembly: primary structure requirements in the core protein. J Virol 64(7):3319–3330
4. Nassal M, Rieger A, Steinau O (1992) Topological analysis of the hepatitis B virus core particle by cysteine-cysteine cross-linking. J Mol Biol 225(4):1013–1025. [pii] 0022-2836(92)90101-O
5. Beck J, Nassal M (2007) Hepatitis B virus replication. World J Gastroenterol 13(1):48–64

6. Porterfield JZ, Dhason MS, Loeb DD et al (2010) Full-length hepatitis B virus core protein packages viral and heterologous RNA with similarly high levels of cooperativity. J Virol 84(14):7174–7184. https://doi.org/10.1128/JVI.00586-10. [pii] JVI.00586-10
7. Strods A, Ose V, Bogans J et al (2015) Preparation by alkaline treatment and detailed characterisation of empty hepatitis B virus core particles for vaccine and gene therapy applications. Sci Rep 5:11639. https://doi.org/10.1038/srep11639
8. Böttcher B, Wynne SA, Crowther RA (1997) Determination of the fold of the core protein of hepatitis B virus by electron cryomicroscopy. Nature 386(6620):88–91. https://doi.org/10.1038/386088a0
9. Wynne SA, Crowther RA, Leslie AG (1999) The crystal structure of the human hepatitis B virus capsid. Mol Cell 3(6):771–780. [pii] S1097-2765(01)80009-5
10. Ceres P, Zlotnick A (2002) Weak protein-protein interactions are sufficient to drive assembly of hepatitis B virus capsids. Biochemistry 41(39):11525–11531
11. Kenney JM, von Bonsdorff CH, Nassal M et al (1995) Evolutionary conservation in the hepatitis B virus core structure: comparison of human and duck cores. Structure 3(10):1009–1019
12. Pumpens P, Grens E (2001) HBV core particles as a carrier for B cell/T cell epitopes. Intervirology 44(2–3):98–114
13. Ulrich R, Nassal M, Meisel H et al (1998) Core particles of hepatitis B virus as carrier for foreign epitopes. Adv Virus Res 50:141–182
14. Kratz PA, Böttcher B, Nassal M (1999) Native display of complete foreign protein domains on the surface of hepatitis B virus capsids. Proc Natl Acad Sci U S A 96(5):1915–1920
15. Skamel C, Ploss M, Böttcher B et al (2006) Hepatitis B virus capsid-like particles can display the complete, dimeric outer surface protein C and stimulate production of protective antibody responses against Borrelia burgdorferi infection. J Biol Chem 281(25):17474–17481. https://doi.org/10.1074/jbc.M513571200. [pii]: M513571200
16. Kolb P, Wallich R, Nassal M (2015) Whole-chain tick saliva proteins presented on hepatitis B virus capsid-like particles induce high-titered antibodies with neutralizing potential. PLoS One 10(9):e0136180. https://doi.org/10.1371/journal.pone.0136180. [pii]: PONE-D-15-04395
17. Nassal M, Skamel C, Vogel M et al (2008) Development of hepatitis B virus capsids into a whole-chain protein antigen display platform: new particulate Lyme disease vaccines. Int J Med Microbiol 298(1-2):135–142. https://doi.org/10.1016/j.ijmm.2007.08.002. [pii] S1438-4221(07)00112-9
18. Walker A, Skamel C, Nassal M (2011) SplitCore: an exceptionally versatile viral nanoparticle for native whole protein display regardless of 3D structure. Sci Rep 1:5. https://doi.org/10.1038/srep00005
19. Kolb P, Nguyen TTA, Walker A et al (2015) SplitCore: advanced nanoparticulate molecular presentation platform based on the hepatitis B virus capsid. In: Khudyakov Y, Pumpens P (eds) Viral nanotechnology. CRC Press, Boca Raton, FL, pp 187–208
20. Vogel M, Diez M, Eisfeld J et al (2005) In vitro assembly of mosaic hepatitis B virus capsid-like particles (CLPs): rescue into CLPs of assembly-deficient core protein fusions and FRET-suited CLPs. FEBS Lett 579(23):5211–5216. https://doi.org/10.1016/j.febslet.2005.08.044. [pii] S0014-5793(05)01042-2
21. Peyret H, Gehin A, Thuenemann EC et al (2015) Tandem fusion of hepatitis B core antigen allows assembly of virus-like particles in bacteria and plants with enhanced capacity to accommodate foreign proteins. PLoS One 10(4):e0120751. https://doi.org/10.1371/journal.pone.0120751. [pii] PONE-D-14-49828
22. Vogel M, Vorreiter J, Nassal M (2005) Quaternary structure is critical for protein display on capsid-like particles (CLPs): efficient generation of hepatitis B virus CLPs presenting monomeric but not dimeric and tetrameric fluorescent proteins. Proteins 58(2):478–488. https://doi.org/10.1002/prot.20312
23. Kumaran D, Eswaramoorthy S, Luft BJ et al (2001) Crystal structure of outer surface protein C (OspC) from the Lyme disease spirochete, Borrelia burgdorferi. EMBO J 20(5):971–978. https://doi.org/10.1093/emboj/20.5.971
24. Rosano GL, Ceccarelli EA (2014) Recombinant protein expression in Escherichia coli: advances and challenges. Front Microbiol 5:172. https://doi.org/10.3389/fmicb.2014.00172
25. Kolb P, Vorreiter J, Habicht J et al (2015) Soluble cysteine-rich tick saliva proteins Salp15 and Iric-1 from E. coli. FEBS Open Bio 5:42–55. https://doi.org/10.1016/j.fob.2014.12.002
26. Shental-Bechor D, Levy Y (2009) Folding of glycoproteins: toward understanding the biophysics of the glycosylation code. Curr Opin Struct Biol 19(5):524–533. https://doi.org/10.1016/j.sbi.2009.07.002. S0959-440X(09)00098-0

27. Valderrama-Rincon JD, Fisher AC, Merritt JH et al (2012) An engineered eukaryotic protein glycosylation pathway in Escherichia coli. Nat Chem Biol 8(5):434–436. https://doi.org/10.1038/nchembio.921. [pii] nchembio.921
28. Bachmann MF, Zabel F (2015) Immunology of virus-like particles. In: Khudyakov Y, Pumpens P (eds) Viral nanotechnology. CRC Press, Boca Raton, FL, pp 121–128
29. Li H, Dunn JJ, Luft BJ et al (1997) Crystal structure of Lyme disease antigen outer surface protein A complexed with an Fab. Proc Natl Acad Sci U S A 94(8):3584–3589
30. Nassal M, Skamel C, Kratz PA et al (2005) A fusion product of the complete Borrelia burgdorferi outer surface protein A (OspA) and the hepatitis B virus capsid protein is highly immunogenic and induces protective immunity similar to that seen with an effective lipidated OspA vaccine formula. Eur J Immunol 35(2):655–665. https://doi.org/10.1002/eji.200425449
31. Gloge F, Becker AH, Kramer G et al (2014) Co-translational mechanisms of protein maturation. Curr Opin Struct Biol 24:24–33. https://doi.org/10.1016/j.sbi.2013.11.004. [pii] S0959-440X(13)00198-X
32. Gualerzi CO, Pon CL (2015) Initiation of mRNA translation in bacteria: structural and dynamic aspects. Cell Mol Life Sci 72(22):4341–4367. https://doi.org/10.1007/s00018-015-2010-3
33. Tuller T, Waldman YY, Kupiec M et al (2010) Translation efficiency is determined by both codon bias and folding energy. Proc Natl Acad Sci U S A 107(8):3645–3650. https://doi.org/10.1073/pnas.0909910107. [pii] 0909910107
34. Nassal M (1988) Total chemical synthesis of a gene for hepatitis B virus core protein and its functional characterization. Gene 66(2): 279–294
35. Lu Y, Chan W, Ko BY et al (2015) Assessing sequence plasticity of a virus-like nanoparticle by evolution toward a versatile scaffold for vaccines and drug delivery. Proc Natl Acad Sci U S A 112(40):12360–12365. https://doi.org/10.1073/pnas.1510533112. [pii] 1510533112
36. Rickwood D (ed) (1984) Centrifugation (2nd edition)—a practical approach, Practical approaches in biochemistry. IRL Press, Oxford
37. Böttcher B (2015) Electron cryomicroscopy and image reconstruction of viral nanoparticles. In: Khudyakov Y, Pumpens P (eds) Viral Nanotechnology. CRC Press, Boca Raton, FL, pp 27–60
38. Serwer P, Griess GA (1999) Advances in the separation of bacteriophages and related particles. J Chromatogr B Biomed Sci Appl 722(1–2):179–190
39. Walker A, Skamel C, Vorreiter J et al (2008) Internal core protein cleavage leaves the hepatitis B virus capsid intact and enhances its capacity for surface display of heterologous whole chain proteins. J Biol Chem 283(48):33508–33515. https://doi.org/10.1074/jbc.M805211200. [pii] M805211200
40. Silhavy TJ, Kahne D, Walker S (2010) The bacterial cell envelope. Cold Spring Harb Perspect Biol 2(5):a000414. https://doi.org/10.1101/cshperspect.a000414. [pii] cshperspect.a000414
41. Nassal M (1992) The arginine-rich domain of the hepatitis B virus core protein is required for pregenome encapsidation and productive viral positive-strand DNA synthesis but not for virus assembly. J Virol 66(7):4107–4116
42. Selzer L, Kant R, Wang JC et al (2015) Hepatitis B virus core protein phosphorylation sites affect capsid stability and transient exposure of the C-terminal domain. J Biol Chem 290(47):28584–28593. https://doi.org/10.1074/jbc.M115.678441. [pii] M115.678441
43. Schägger H (2006) Tricine-SDS-PAGE. Nat Protoc 1(1):16–22. https://doi.org/10.1038/nprot.2006.4. [pii] nprot.2006.4

Chapter 34

Dual-Functionalized Virus–Gold Nanoparticle Clusters for Biosensing

Carissa M. Soto and Walter J. Dressick

Abstract

Metallic nanoscale 3D architectures concentrate electromagnetic energy at precise spatial locations to enable sensing and photocatalysis applications. We have developed solution-based methods to reproducibly fabricate 3D gold nanostructures useful as efficient surface-enhanced Raman spectroscopy (SERS) biosensors. Virus capsids were recruited as templates to assemble gold nanoparticles on their surfaces at well-defined locations to prepare the nanoscale 3D structures. Cowpea mosaic virus (CPMV) and its variants were selected as specific templates due to their high symmetry, scalability, and stability, which have proven useful in materials science applications. While the methods described herein were optimized for the CPMV capsids, they also provide a useful starting point for researchers who are working toward the nanoassembly of metal nanoparticles on other protein scaffolds.

Key words Gold nanoparticles, Cowpea mosaic virus (CPMV), Optical metamaterials, Surface-enhanced Raman spectroscopy (SERS), Biosensors, Virus-like particles (VLP), Plasmonics, Nanomaterials

1 Introduction

Optical metamaterials are composites engineered to exhibit technologically useful optical properties not normally found in natural materials, such as negative indices of refraction, superlensing, spatially controlled electromagnetic fields, and photonic bandgaps [1–3]. The unusual properties of these materials originate from the combinations of metals, semiconductors, and/or insulator materials arranged in precisely repeating patterns at scales smaller than the wavelength of light they affect. For metamaterials interacting with visible or near-IR light, fabrication of composites featuring repetitive nanoscale noble metal and insulator components precisely positioned within the material are typically prepared. Such materials often utilize shape-controlled Au, Ag, or, to a lesser extent, Cu nanoparticle (NP) components whose natural plasmons span the visible–IR spectral region [4].

Christina Wege and George P. Lomonossoff (eds.), *Virus-Derived Nanoparticles for Advanced Technologies: Methods and Protocols*, Methods in Molecular Biology, vol. 1776, https://doi.org/10.1007/978-1-4939-7808-3_34,
© Springer Science+Business Media, LLC, part of Springer Nature 2018

We have been investigating metamaterial architectures [5–11], with a recent interest in those capable of reproducibly concentrating electromagnetic energy at precise spatial locations for the development of efficient surface-enhanced Raman spectroscopy (SERS) biosensors [12]. Excitation of the noble metal plasmons with visible or near-IR light leads to concentration of the electromagnetic field in the regions between the NPs to form "hot spots," which enhance Raman signals from any molecules present in the interparticle gaps.

In general, the fabrication of previous SERS-active architectures has been demonstrated using both "top-down" and "bottom-up" approaches. Top-down approaches utilize techniques such as lithography [13], oblique angle deposition [14], layer-by-layer deposition [15], metallization [16], and etching [17], alone or in various combinations, to arrange noble metal NPs on surfaces at precisely fixed positions, separations, and orientations. Although top-down fabrication approaches are highly scalable for the preparation of two dimensional (2D) structures, they are not efficient means for preparation of high symmetry three dimensional (3D) architectures exhibiting nanoscale positional control [1]. In contrast to 2D structures, high symmetry 3D structures exhibit orientation-independent SERS signals [18, 19] required for use as mobile taggants and probes. Such 3D structures are best prepared using bottom-up approaches [1, 20] in which biomolecules possessing the desired symmetry are utilized to self-assemble bound metal NPs to form the requisite SERS-active biocompatible composite. Typical biotemplates include DNA [21, 22] and viral protein capsids [5–12, 23]. However, while available DNA template architectures are limited primarily by the imagination of the researcher, scale-up remains an issue that currently limits this technique to laboratory use. In contrast, viral capsids, which are typically available in larger quantities via conventional isolation or genetic engineering techniques, can provide a variety of high symmetry templates for nanoparticle binding and organization [5–12, 23]. A critical issue in the use of any template method is the ability to reproducibly utilize it to fabricate the required pure composite material.

In this chapter we describe a bottom-up approach using a virus capsid as a template to build 3D nanocomposites for biosensing applications. We describe in detail preparation and purification steps for the self-assembly and binding of Au NPs on the faces of cowpea mosaic virus (CPMV) capsid proteins to create biocompatible composite metamaterials useful as SERS substrates. The resulting composites comprise 6–12 Au NPs attached to the capsid surface (in which the Au NPs exhibit icosahedral symmetry with respect to the capsid) and are denoted as nanoclusters (NC) due to the 3D nature and size of the structure (*see* Fig. 1). Plasmonic interactions between adjacent Au NPs on the NCs concentrate the

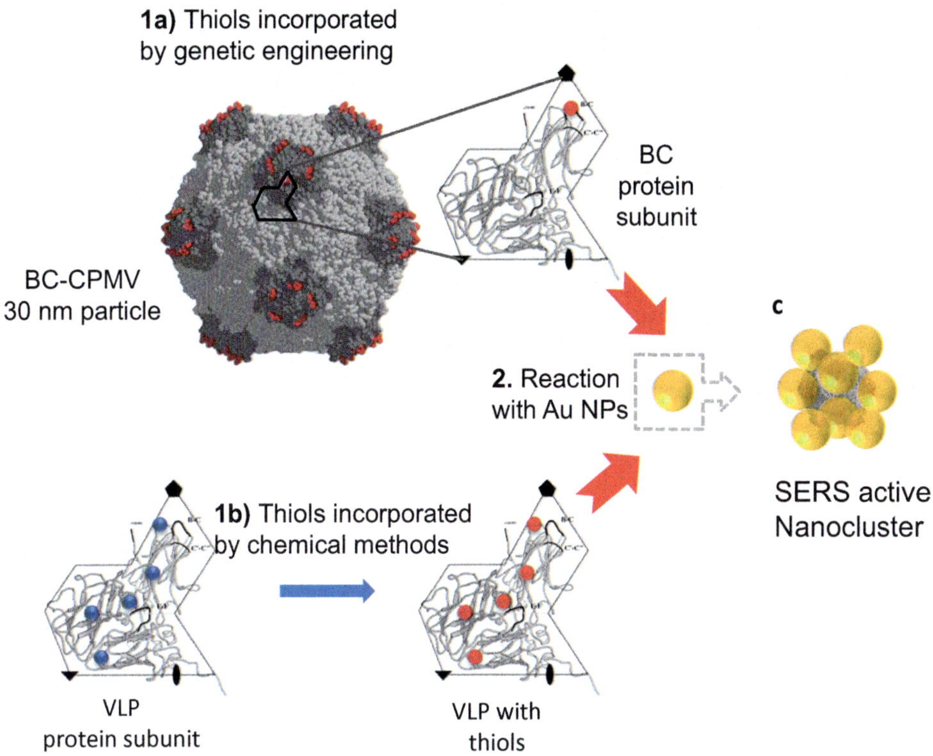

Fig. 1 General concept of Nanoclusters (NC) preparation. On *step 1* thiol groups (cysteine amino acid, red circle) are incorporated via (**a**) genetic engineering to the BC loop or (**b**) via chemical methods to the lysines (blue circles) on the VLP. On *step 2* Au NPs are bound to the thiol-containing protein capsid to produce (**c**) the NCs

electromagnetic field inside the interparticle gaps, creating hot spots leading to orientation-independent SERS signals for molecules as large as DNA localized there via solution treatments [12].

We describe the preparation of two kinds of NCs by using two CPMV variants, the BC-CPMV [5, 24] and virus-like-particles (VLPs) [25, 26]. Both capsids are readily scaled for production in multigram quantities. For the purpose of the current chapter, the first functionalization constitutes the incorporation of thiol groups on the virus capsid while the second is the binding of Au NP at the thiol groups. These doubled-functionalized virus capsids (BC-NC and VLP-NC) contain Au NP strategically positioned in 3D for its SERS activity (*see* Fig. 1). The BC-CPMV or simply BC is produced by generically engineering the wild-type (WT) CPMV to express a cysteine group (i.e., thiols) at each BC loop [24], resulting in a total of 60 thiols pentagonally organized in groups of 5 at each of the 12 capsid vertices [5]. The BC contains viral RNA inside the capsid cavity. In contrast the VLP-CPMV comprises the WT-CPMV protein capsid synthesized in the absence of infection, leaving an "empty" (RNA-free) capsid [25]. The native lysines [27] of the VLP are chemically modified to contain solvent-accessible thiol groups for Au NPs binding [12, 28, 29]. One advantage of the VLP-NC is its noninfectious nature due to the absence of RNA while exhibiting identical behavior to the BC-NC system with respect to Au NP self-assembly, structure, and SERS activity.

Herein, we focus on describing in detail the most successful protocols and presenting notes regarding critical variables to successfully produce the doubled-labeled virus particles, which ultimately are nanoclusters with SERS biosensor capabilities. The biosensing experiments are beyond the scope of the current chapter and can be found elsewhere [12].

2 Materials

Deionized water (18.2 MΩ·cm) is prepared by passing water through an Elix® 5 Milli-Q Plus Ultra-Pure Water System (Millipore Corp.). All buffers are filtered sterilized using a Nalgene filter 0.2 μm pore size for sterilization with 500 ml capacity or syringe filters (0.2 μm PES: polyethersulfone; e.g., Thermo Fisher Scientific) when filtering samples in the ml range. All buffers are used at room temperature (22 ±1 °C) unless otherwise noted. All the glassware used for Au NPs synthesis and long-term storage are cleaned with aqua regia prior to use.

1. BC-CPMV (John E. Johnson's laboratory at The Scripps Research Institute, USA).
2. VLP-CPMV (George P. Lomonossoff's laboratory at the John Innes Centre, UK; *see also* Chapter 23).

2.1 Clean Equipment with Aqua Regia (See Note 1)

Aqua regia (*see* **Note 1**): 300 ml 12 M HCl (ACS grade), 100 ml (70% (w/w; ACS plus grade) HNO_3. First pour HCl into a 1 l glass beaker, then slowly add HNO_3.

2.2 Gold Nanoparticles Synthesis

2.2.1 60 mM Sodium Citrate (SC) Solution

Mix 0.45 g sodium citrate tribasic dihydrate (ACS reagent, Sigma-Aldrich) with 25 ml of Milli-Q water. Prepare a day prior to the synthesis of Au NPs (*see* **Note 2**) in a plastic Falcon tube (50 ml capacity) and let sit overnight at room temperature. Measure the resultant pH.

2.2.2 1% (w/v) Hydrogen Tetrachloroaurate (III) ($HAuCl_4$) Solution

1. Mix 1 g of hydrogen tetrachloroaurate (III) ($HAuCl_4{\cdot}3H_2O$, 99.9%, Sigma-Aldrich) with 100 ml of Milli-Q water following **items 2** and **3** to ensure complete transfer of the solid (*see* **Note 3**).
2. Add 5 ml of water to the 1 g $HAuCl_4$ bottle and transfer the solution using a 1 ml pipette a 100 ml volumetric flask (previously cleaned with aqua regia; *see* **Note 1**).

3. Rinse the $HAuCl_4$ bottle with 5 ml of water making sure the bottle is rinsed well and transfer the solution to the volumetric flask each time, repeat 3×.
4. Add water to the volumetric flask until the mark to complete 100 ml.
5. Close tightly with a glass stopper [cleaned with aqua regia (*see* **Note 1**)] and mix by hand. This is a 1% (w/v) solution, which is equivalent to 25 mM.
6. Cover with aluminum foil and store at room temperature for at least 1 month prior to use (*see* **Note 3**).

2.2.3 AuNP Standard Solution

30 nm diameter Au NP dispersion of known concentration from a commercial source (e.g., Ted Pella Inc., CA) for calibration curve (Table 1).

2.2.4 Setup for Au NP synthesis

All glassware previously cleaned with aqua regia (*see* **Note 4**), for setup (*see* Fig. 2).

1. 250 ml two-neck round bottom flask (neck joint of 24/40), matching glass stopper for round bottom flask.
2. Condenser (neck joint 24/40).
3. Thermometer.
4. Teflon stir bar and magnetic stirrer.
5. Heating mantle.
6. Staco Variable Autotransformer or variable AC transformer (Variac).
7. Laboratory support jack (Lab-Jack).

Table 1
Au NP dispersion dilution series for calibration curve preparation

Standard Au NP dispersion ID[a]	Resultant concentration (particles/ml) × 10^{11}	Measured absorbance at 525 nm
Stock	2.00	1.02
Dilution 1	1.50	0.76
Dilution 2	1.13	0.62
Dilution 3	0.84	0.43
Dilution 4	0.63	0.35
Dilution 5	0.48	0.28
Dilution 6	0.36	0.18
Dilution 7	0.27	0.06

[a]Dilution series prepared by mixing 500 μl of previous dispersion with 500 μl of filtered Milli-Q water

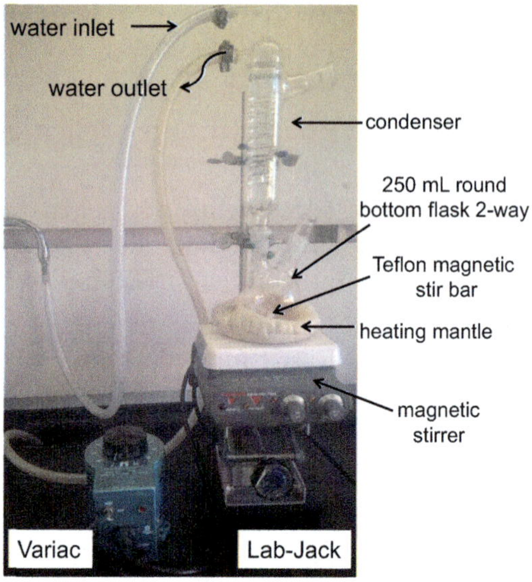

Fig. 2 Experimental setup for Au NP synthesis

2.3 Artificial Thiols Incorporation on VLP

1. 1 mg/ml N-succinimidyl S-acetylthiopropionate (Pierce™ SATP, Thermo Fisher Scientific): 1 mg/ml SATP in dimethyl sulfoxide (DMSO). Prepare freshly, vortex, use immediately.
2. 100 mM potassium phosphate buffer, pH 7.5: 16 ml of 0.2 M of KH_2PO_4 and 84 ml of 0.2 M of K_2HPO_4 diluted with Milli-Q water up to 200 ml (adjust pH if needed prior to volume completion).
3. 0.5 M hydroxylamine solution: 250 μl 100 mM EDTA pH 7.5, 34 mg hydroxylamine–HCl (Thermo Scientific Pierce), 750 μl 100 mM potassium phosphate buffer, pH 7.5; use immediately.
4. 100 mM and 10 mM potassium phosphate buffer, pH 6.0: 13 ml of 0.2 M of KH_2PO_4 and 87 ml of 0.2 M of K_2HPO_4 diluted with Milli-Q to 200 ml (adjust pH if needed prior to volume completion) for 100 mM buffer. Dilute the 100 mM buffer 10× to prepare a 10 mM buffer.
5. 500 mM EDTA, pH 6.0: Mix 50 ml of 1 M EDTA pH 8.0 and 25 ml of water. Adjust pH to 6.0 using 1 M HCl and complete the volume to 100 ml with Milli-Q water.

2.4 Purification of NCs

1. Basic Water: 1 μl 2 M KOH, 400 ml fresh Milli-Q water, resultant pH: 8.5–9.0. Use a pH meter to measure the pH. Continue adding microliter amounts of the base until the solution reaches the optimum pH range; use immediately.
2. 20 mM thioctic acid (DL-TA, 99%, Thermo Fisher Scientific): 800 μl ethanol (200 proof, Warner Graham Company), 200 μl Milli-Q water, 4 mg of DL-thioctic acid. Keep the solution tightly capped at 4 °C for long-term storage (1 year).

3. 50% (v/v) Glycerol (molecular biology grade) in Milli-Q water.
4. 10× TBE (Tris-borate-EDTA) buffer: 108 g/l Tris(hydroxymethyl) aminomethane (molecular biology grade), 55 g/l boric acid, 40 ml of 0.5 M EDTA pH 8. Mix and store at room temperature.
5. 1% (w/v) Low Melting Agarose in TBE: Weigh 2 g of low melting agarose (SeaPlaque™ GTG™ Agarose, genetic technology grade, Lonza), complete the volume to 200 ml with 1× TBE. Heat in a microwave until the solid is dissolved, mix well and pour over the horizontal electrophoresis apparatus (8 cm × 8 cm) using a 6-well comb (1 cm length, 2 mm depth), let the mixture cool down until the gel hardens. Add 1× TBE to cover the electrophoresis apparatus and remove the comb gently.
6. Ice packs.
7. 10 mM Bis-Tris, 1 mM EDTA: pH 6.5 buffer. Buffer is prepared by standard methods.
8. 1000 units/ml β-agarase I (New England Biolabs) in 10 mM Bis-Tris, 1 mM EDTA: pH 6.5 buffer.

2.5 Devices

1. Spectrophotometer for UV-Vis spectroscopy (525 nm, 280 nm, 260 nm).
2. Particle Size Analyzer for Dynamic Light Scattering (DLS) measurements.
3. 5 ml prepacked Hi-Trap Desalting Column (GE Healthcare).
4. 15 ml 100 k molecular weight cutoff (MWCO) concentrators (Amicon® Ultra-15, Fisher Scientific).
5. Syringe filter (0.2 μm, PES: polyethersulfone).
6. Eppendorf Centrifuge 5810 R equipped with a swing-bucket rotor.
7. Teflon magnetic stir bar.
8. 4 ml capacity concentrator (100 k MWCO; Amicon® Ultra-4, Fisher Scientific).
9. Scalpel/razor blade for cutting out gel bands.
10. Heating block.
11. Incubator at 42 °C.
12. Optional: equipment for TEM and AFM.

3 Methods

3.1 Clean Equipment with Aqua Regia (See Note 1)

1. Place air-dried glassware and magnetic stir bars (previously cleaned with water, soap, and ethanol) inside a glass container to be used as a primary or secondary container. Fill up the 250 ml two neck round bottom flask with active aqua regia

and cover the ends of the two necks with a 30 ml beaker each. Likewise, fill up the primary containers containing small items with aqua regia and cover with a watch glass or petri dish (*see* **Note 4**).

2. Next day carefully transfer the aqua regia to a 1 l beaker. Rinse the clean glassware with water. Collect the rinse water in a waste bottle. Add water to the clean glassware and let stand overnight.
3. Remove the water, rinse the clean glassware with water, and air dry. Store at room temperature.

3.2 Gold Nanoparticles Synthesis, Based on the Puntes Method [30], (See Note 5)

1. Using previously cleaned glassware and magnetic stir bar, assemble the equipment inside a chemical hood as shown in Fig. 2.
2. At least a day prior to the synthesis, validate the equipment for boiling water (100 °C) and 90 °C (*see* **Note 6**).
3. Pour 145 ml of water (Milli-Q ~pH 6) into the round bottom flask. Add 5.5 ml of 60 mM SC solution (prepared a day before) and stir using the stirring bar. Cover the round bottom flask with aluminum foil. Turn on the Variac to optimum setting for boiling water (80% 140 V in our setting); do not run water through condenser and let boil (*see* **Note** 7).
4. Immediately after boiling begins add 1 ml of 1% (w/v) $HAuCl_4$ and continue to boil for 10 min (*see* **Note 8**).
5. Remove the heating mantle by adjusting the Lab-Jack and start running tap water through the condenser; remove the aluminum foil. Let the round bottom flask cool for at maximum 5 min, (*see* **Note** 7). After 5 min the temperature is typically 85 °C. Turn off the water from the condenser. Using the Lab-Jack move the mantle back to touch the round bottom flask and set the Variac to the corresponding setting to maintain the temperature to 90 °C (50% 140 V in our setting). Wait for 10 min.
6. Confirm that the temperature of the pink solution is 90 °C. Add 1 ml of 60 mM SC. Wait for 2 min.
7. Add 1 ml of 1% $HAuCl_4$, after which the solution will turn darker red (burgundy). Continue to heat at 90 °C for 30 min with constant stirring. Check the temperature every 5 min to make sure that is kept constant at 90 °C (*see* **Notes 6** and **9**).
8. Remove the round bottom flask from heat and start running water through the condenser. The solution typically cools down to 70 °C in 7 min, after which the condenser water can be turned off and let cool down until the red Au NP dispersion reaches room temperature.

9. Transfer the Au NP dispersion to a 250 ml bottle (previously cleaned with aqua regia), cap, and cover the bottle with aluminum foil and store at 4 °C for up to 3 months.
10. Let the Au NP dispersion age for 1 week prior to usage (*see* **Note 10**).
11. As described in the following, characterize the Au NP dispersion by UV-Vis spectroscopy and determine the Au NP concentration. Use DLS for size determination.

3.3 UV-Vis Characterization

1. Place 500 μl of Au NP dispersion into a 1 ml cuvette and complete the volume with 500 μl of water. Measure the absorbance at 525 nm using a UV-Vis spectrophotometer (*see* **Notes 11** and **12**) (*see* Fig. 3a).

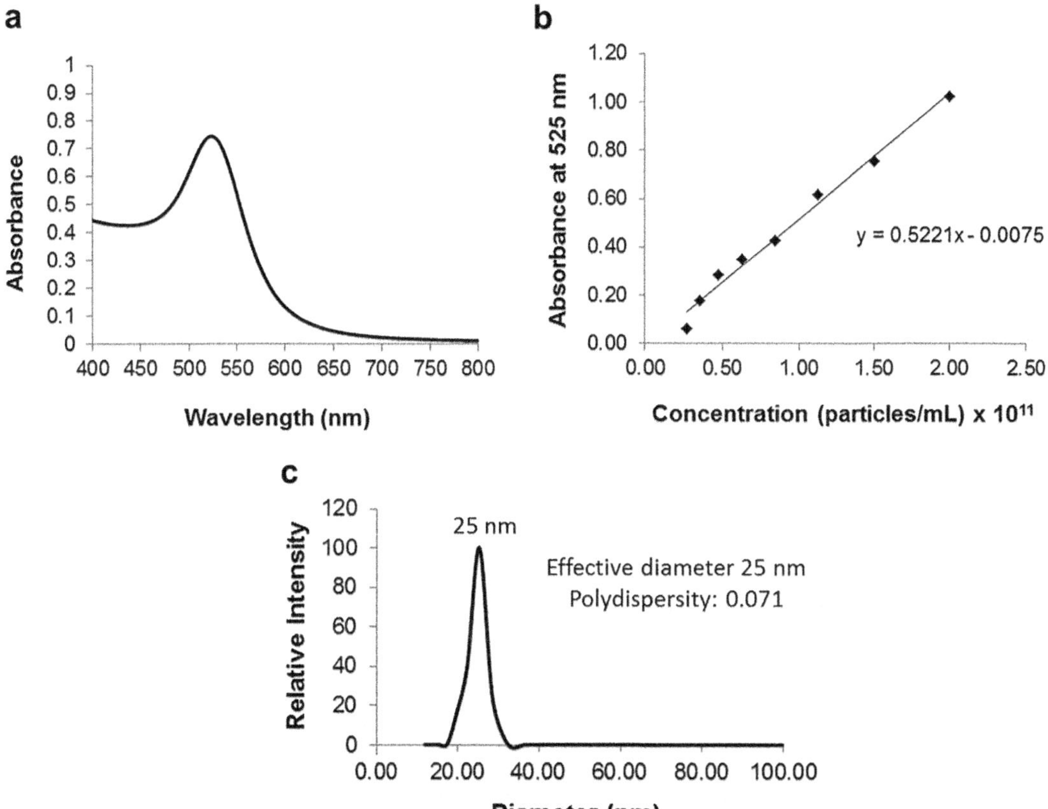

Fig. 3 Characterization and quantification of Au NPs. (**a**) Absorbance of Au NP as prepared in our laboratory. Au NP dispersion was diluted 50% in water prior to UV-Vis measurement, $\lambda_{max} = 525$ nm. (**b**) Calibration curve for Au NP quantification. Dispersions were diluted according to Table 1 and absorbance at 525 nm measured. Standard Au NP dispersion corresponds to commercially available 30 nm Au NPs (Ted Pella Inc. Redding, CA). The equation is the linear regression of "*Absorbance at 525 nm vs. Concentration*" data; it corresponds to the format $y = mx + b$ where m is the slope and b is the intercept in y (*see* **Note 14**). (**c**) Dynamic Light Scattering, DLS of Au NPs

3.4 Calibration Curve Preparation for Quantification of Au NPs

1. Prepare a dilution series of an Au NP dispersion of known source and concentration (*see* **Note 13**) as shown in Table 1 using filtered Milli-Q water as the solvent.
2. Determine the absorbance at 525 nm (corresponding λ max) of each dispersion (*see* Table 1).
3. Graph absorbance (y axis) vs. # particles/ml (*x* axis), Fig. 3b.
4. Using the resultant linear regression of the calibration curve determine the concentration of the newly synthesized Au NP (*see* Subheading 3.3, **Note 14**).

3.5 Size Determination via DLS (See Note 15)

1. Wash the plastic cuvettes (four clear optical windows, maximum capacity of 4.5 ml) with filtered Milli-Q water (filtration via a 0.45 μm syringe filter).
2. Use a minimum of 2 ml of Au NPs for the DLS.
3. Determine the optimum parameters in your DLS. If sample dilution is necessary make sure to use filtered Milli-Q water for the dilutions (*see* **Note 16**).
4. Report the data as the relative intensity of the species vs. diameter in nm (*see* Fig. 3c).

3.6 Removal of Small Molecules and Buffer Exchange via Hi-Trap Desalting Columns

1. Equilibrate a 5 ml prepacked Hi-Trap Desalting Column with 25 ml of the buffer required in each protocol (*see* Subheading 3.7).
2. Dilute the virus particles in the equilibration buffer (buffer is dictated by the following step in a procedure) up to 1.5 ml (*see* **Note 17**).
3. Inject the 1.5 ml virus samples manually using a sterile 5 ml plastic syringe.
4. The first 1.5 ml is the void volume of the column. Collect the following fractions and analyze by UV-Vis (*see* **Note 18**).

3.7 NC Assembly

3.7.1 Incorporation of Artificial Thiols into VLPs [11, 28] (See Note 19)

1. Mix 25 μl of VLP as received (in 10 mg/ml in 50 mM potassium phosphate pH 7.0) with 1.5 ml of 100 mM KP pH 7.5 in an Eppendorf tube and vortex.
2. Filter the solution using a syringe filter (0.2 μm, PES). Add 100 mM KP pH 7.5 to complete to 1.5 ml, using the tube marking as a volume reference.
3. Inject the 1.5 ml solution from **step 2** to a Hi-Trap column previously equilibrated in 100 mM KP pH 7.5 (*see* Subheading 3.6) and collect 1.5 ml fractions.
4. Measure the UV-Vis of each fraction. The VLP-containing sample shows a peak centered at 280 nm (*see* **Note 18**).
5. Mix 1.5 ml of 0.1 mg/ml of VLP with 4 μl of 1 mg/ml SATP in DMSO (*see* **Note 20**). Vortex and incubate overnight in the dark at room temperature.

6. Use a Hi-Trap column preequilibrated in 100 mM KP pH 7.5 to remove the excess of SATP using same methods as used previously in **steps 3** and **4**. The particles recovered are VLP-SHR (*see* **Note 21**).

3.7.2 Deprotection of VLP-SHR to Produce VLP-SH (See Note 21)

1. Mix 1.5 ml of 38 μg/ml VLP-SHR with 5 μl of 0.5 M hydroxylamine, vortex, and incubate at room temperature for 2 h in the dark.
2. Remove the excess of hydroxylamine using a Hi-Trap column (preequilibrate with 10 mM potassium phosphate, pH 6.0). The VLP-SH containing fraction is identified by UV-Vis (λ_{max} at 280 nm) and used immediately for Au NP coupling (*see* **Note 18**).

3.7.3 BC or VLP-SH Reaction with Au NPs

1. Concentrate 40 ml of Au NP 3×10^{11} particles/ml using two 15 ml 100 k molecular weight cutoff (MWCO) concentrators). Distribute 30 ml of Au NP (3×10^{11} particles/ml) in two tubes, spin for 3 min at 3200 × *g* at room temperature using an Eppendorf Centrifuge 5810 R equipped with a swing-bucket rotor.
2. Transfer the filtrate to a 50 ml Falcon tube. Keep the concentrated Au NPs in the filters and add 5 ml of Au NP to each filter and spin again for 3 min at 3200 × *g* at room temperature.
3. The initial 40 ml Au NP will be concentrated down to 600 μl. Transfer the Au NPs to an Eppendorf tube (*see* **Note 22**).
4. In a 15 ml sterile plastic tube, mix 1171 μl of the Au NPs filtrate (recovered after concentration, **step 2**) with 200 μl of 100 mM potassium phosphate buffer pH 6.0, 4 μl of 500 mM EDTA pH 6.0, and 25 μl of 0.02 mg/ml BC or VLP-SH (from Hi-Trap column in 10 mM potassium phosphate buffer pH 6.0) in that order (*see* **Notes 23**) and vortex the solution.
5. Add the Au NPs slowly to the virus containing mixture (from **step 4**) in 100 μl increments until the 600 μl of the concentrated Au NPs are all added to the solution from **step 4**. Mix gently by hand in between Au NP additions. Vortex and incubate at room temperature in the dark for 36 h (*see* **Notes 10**, **24**, and **25**).

3.7.4 Coating of Au NPs with Stabilizer and Agarose Electrophoresis (See Note 26)

1. In a 20 ml glass vial containing a Teflon magnetic stir bar add 20 ml of basic water (pH 8.5–9.0) and 10 μl of 20 mM TA and stir vigorously.
2. Add the reaction mixture from Subheading 3.7.3 to the TA in basic water in 100 μl increments, let it mix in between additions. Mix for 4 h at room temperature. Store at 4 °C overnight.
3. Add 200 μl of 10× TBE and vortex. Concentrate using a 5 ml capacity concentrator (100 k MWCO) by centrifuging at 3200 × *g* for 3 min. The final volume will be ~500 μl.

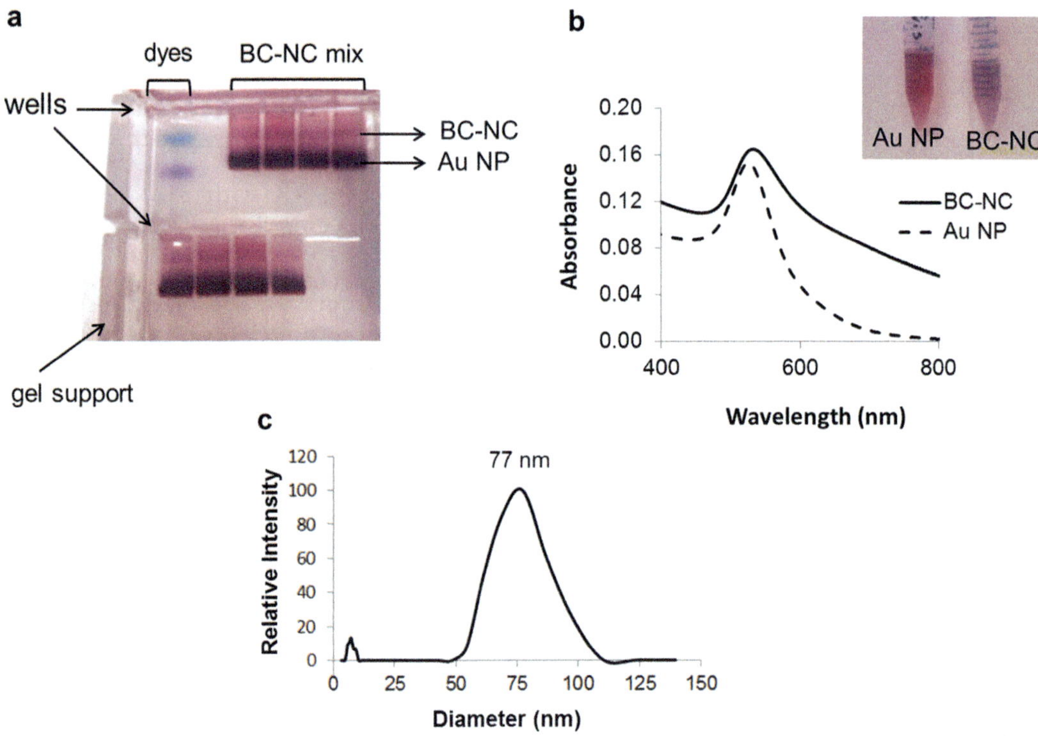

Fig. 4 Characterization of purified NCs. (**a**) Photograph of an agarose gel after it was run. The dyes xylene cyanol (light blue) and bromophenol blue (purple) are used to track the gel. After separation, the NCs and Au NPs bands are easily distinguishable. The BC-NC runs parallel to the xylene cyanol while the free Au NPs run farther as expected for smaller nanoparticles. One reaction is run in a single gel distributed in a total of eight wells (*see* Subheading 3.7.4). (**b**) Absorbance of purified NCs (solid line) in comparison to the negative control, free Au NPs (dashed line). The NCs plasmon peak is red shifted and a broader band emerges around 600 nm in comparison to a single peak at 525 nm for the free Au NPs. The inset shows a photograph of the isolated NCs (purple) and Au NPs (red), demonstrating that the methods produce nanoparticles in bulk. (**c**) DLS of NC, a peak centered at 77 nm is representative of the NCs

4. Add 50 μl of the 50% glycerol (v/v) to the 500 μl reaction mix and vortex. Load 60 μl of the mixture per well into a 1% (w/v) low melting agarose/TBE and run at 100 V for 30 min or until the NC and free Au NP bands are well resolved (note intense red bands Fig. 4a). Put ice packs on top of the gel apparatus to keep the system cool (*see* Fig. 4a).

3.7.5 Recovery of the BC-NC or VLP-NC from Agarose Gels

1. Cut the bands corresponding to the NCs (*see* Fig. 4a, ~200 mg of gel per band) and transfer to a 2 ml Eppendorf tube. Cut the bands into little pieces using a razor blade, which facilitates the melting of the gel pieces. Weigh the Eppendorf tubes containing the bands to make sure to have less than 200 mg per tube. If a larger amount of gel is used per tube the extraction is less efficient. Redistribute the gel pieces into the tubes as needed.

2. Add 400 μl of 10 mM Bis-Tris, 1 mM EDTA: pH 6.5 buffer to each tube and incubate at room temperature for 30 min. Remove the buffer gently, avoiding retrieval of gel pieces, and add 400 μl of fresh buffer. Incubate at room temperature for 30 min. Discard the buffer (*see* **Note 27**).
3. Melt the agarose pieces by incubating the gel-containing tubes 5 min at 70 °C in a heating block. Check the tubes after 2.5 min of incubation. Mix them gently by hand; heat for 2.5 min. Ensure that the gel is completely melted; one can heat for additional time, up to a total of 7 min if necessary. Transfer the tubes immediately to a 42 °C heating block.
4. Equilibrate at 42 °C for 10 min. Add 2 μl of β-agarase enzyme (1000 units/ml) per tube. Mix gently and incubate at 42 °C for 4 h. Incubate at room temperature overnight (*see* **Note 28**).
5. Wash the samples twice to remove the neoagaro-oligosaccharides. Add the samples into the filters (4 ml capacity) and complete the volume with 10 mM Bis-Tris, 1 mM EDTA: pH 6.5 buffer. Centrifuge 3 min, 3200 × *g*. Add 5 ml of buffer (10 mM Bis-Tris, 1 mM EDTA: pH 6.5). Centrifuge 3 min, 3200 × *g*. Reconstitute the volume to 2 ml (10 mM Bis-Tris: 1 mM EDTA, pH 6.5; *see* **Note 29**).

3.8 NCs Characterization: UV-Vis and DLS (See Fig. 4)

1. Use 1 ml of the NC dispersion from the previous section (*see* Subheading 3.7.5) to determine the absorbance in the visible range (*see* Subheading 3.3). Compare its absorbance with a corresponding negative control that contains only Au NPs recovered in the same way as the NCs (*see* **Note 30**).
2. Use 2 ml of the NC dispersion to determine its size by DLS (*see* Subheading 3.5)
3. For details about TEM and AFM characterization *see* our previous publication [12].

4 Notes

1. Aqua regia is a corrosive acid that delivers chlorine gas upon preparation (highly toxic, do not inhale). Use proper protective equipment and perform all the manipulations in a chemical fume hood. Never use metal tools since they will degrade in aqua regia; use only glass or Teflon tools. Use secondary containers when cleaning glassware to prevent any spill. Once the aqua regia solution is mixed it is active for several days until it changes from an orange color that evolves bubbles (active) to a light yellow solution without any bubbles (not active). After the solution is not usable, dilute with water and dispose of as a Corrosive Acid following the proper regulations. It is

not recommended to neutralize and dispose of in any drain since after all the cleaning steps the solution may be contaminated with heavy metals. Collect spent aqua regia in glass bottles and dispose of as a hazardous waste following your institution and government regulations. Do not cover tightly while the solution is still active since the gas evolution will increase the pressure of the container; cover lightly with a petri dish or a watch glass while in use.

2. Milli-Q water was previously stored in a glass bottle for more than a week resulting in a measured pH ~5.5–6.1. The optimum SC solution was the one prepared one day before Au NP synthesis with Milli-Q water pH ~6. Upon dilution of the 60 mM SC solution (pH 8.3) to 2.2 mM with Milli-Q water (pH ~6), the pH of the resulting SC solution was measured as 7.3.

3. $HAuCl_4$ is hygroscopic, and therefore to facilitate complete transfer of 1 g accurately without the need of weighing 1 g of the salt, it is purchased in 1 g bottles and used completely in each preparation. We determined that the 1% $HAuCl_4$ solution is usable up to 1 year; if the solution is older than 1 year the Au NP synthesis becomes irreproducible. The solution must be aged for at least one month prior to use.

4. Use a 1 l beaker as a secondary container to hold the 250 ml two neck-round bottom flask and a small boiling dish (300 ml capacity, 10 cm diameter) as a primary container to submerge magnetic stir bars, glass stoppers, and quartz/glass cuvettes. Use a glass test tube as a primary container (15 cm long, 2 cm diameter) to submerge the thermometer in aqua regia.

5. It is important to note that we synthesize Au NP in our laboratories since commercial Au NP was not reactive enough for the purpose of the biosensor preparation (the requirement of six 30 nm in diameter Au NP per virus capsid for an active biosensor). While the Au NP synthesis was done using published protocols pioneered by Puntes and coworkers [30] here we present details on how we prepared the Au NP in our laboratories with the resources available to us at the time for this particular application.

6. We recommend measuring the temperature of the dispersion as opposed to measuring the vapor temperature in the Fig. 2 apparatus, especially when determining the temperature of the mixture at 90 °C. Small variations in the temperature will promote differences in the Au NP size. One may encounter a slight variation of the Variac setting for keeping Au NP dispersion at 90 °C. Make sure to monitor the temperature continuously during the entire synthesis.

7. In our setup (80% 140 V), it takes ~15 min to boil; it is not recommended to have a system that will take much longer to boil. It is known that longer heating times will alter the concentration

of intermediates causing changes in the size of the Au NP [30]. For the growth step, it is not advisable to cool longer than 5 min since it will take longer to rewarm up to 90 °C. Once the system is validated (Variac setting for 100 °C and 90 °C) keep using the same setting as needed for future preparations.

8. Typical color changes, 1 min: light pink; 2 min: slightly darker pink; at 10 min the solution is red. According to Puntes et al. [30] this step produces the "seeds".
9. After 15 min a temperature spike may occur; if this happens, decrease the temperature by removing the heating mantle slightly from contact with the round bottom flask. If faster cooling is desired, flow water through the condenser for 2 min. A range of 88–90 °C is optimum during the reaction.
10. This Au NP preparation in our laboratory setting produces Au NPs of 25–30 nm in diameter. The Au NPs are aged 1 week prior to any reaction with thiol-containing-CPMV scaffolds. The Au NPs stay active for up to three months for reactions with BC or VLP-SH.
11. We used a Cary5000 (Agilent Technologies, Santa Clara, CA) spectrophotometer equipped with an NIR detector and disposable plastic cuvettes of 1.5 ml capacity volume, path length 1 cm, spectral range 220–900 nm (BrandTech™ BRAND™ UV-Cuvette, Fisher Scientific).
12. A reaction results in a 150 ml solution of an absorbance of 1.2–1.6 at 525 nm. Based on the Beer-Lambert Law the direct relationship of absorbance to concentration is accurate at absorbance ranges (0.1–1). Make dilutions accordingly to fulfill that requirement, Fig. 3a.
13. We used a 30 nm diameter Au NP dispersion of 2×10^{11} particles/ml from Ted Pella Inc. (Redding, CA). Any 30 nm Au NP dispersion of known concentration from a commercial source can be used since it is well documented that the plasmon peak intensity is directly related to the concentration of Au NPs of similar sizes. The calibration curve is valid as long as the Au NPs synthesized is close in size (within ~5 nm) to the Au NPs used as the standard.
14. To determine the concentration of an Au NP in a dispersion of unknown concentration use the corresponding absorbance at 525 nm (y) and solve for x in the equation; $x = (y + 0.0075)/0.5221$ in our case.
15. DLS measurements were performed in a Brookhaven Instrument (NanoBrook Omni, Particle Solutions software v. 3.1) using disposable plastic cuvettes of 4.5 ml capacity, four optical windows, spectral range: 340–800 nm, 1 cm path length (standard cuvette). At least 2 ml of the Au NPs were used for the measurements.

16. The objective is to have a small polydispersity (<0.1) and an effective diameter close to the diameter of the most abundant particle (relative intensity of 100, Fig. 3c). Our instrument parameters consist of data acquisition time 150 s, laser wavelength of 640 nm, scattering angle of 173°, and temperature of 25 °C. Some instruments with lower sensitivity may require longer acquisition time or are set up to average measurements. If the setup averages individually, then select to average 10 measurements at 1 min per measurement.
17. The genetic engineering [31] and production of BC [11] was done at John E. Johnson's laboratory by methods previously published. We received the samples from their laboratories and kept them at 4 °C for long-term storage.
18. Calculate the concentration of VLP (or any of its variants: VLP-SHR and VLP-SH) using the Beer–Lambert law, $A = \mathcal{E}bc$, where A is the absorbance at 280 nm, $\mathcal{E}_{280\text{nm}}$ = 1.28 ml/(mg cm) (extinction coefficient), b = 1 cm (cuvette path length), and c is the concentration in mg/ml. To calculate the concentration of BC, measure the absorbance at 260 nm using $\mathcal{E}_{260\text{nm}}$ = 8 ml/(mg cm).
19. The incorporation of SH to VLP was done as previously published for the WT-CPMV [29] and it was not further optimized since the resulting VLP-NC was of premium quality. Therefore, we did not have a need to optimize further. The VLP was kept at 4 °C for long-term storage. For more details on all the steps followed to optimize the Au NP coupling to BC refer to our previous publication and corresponding supporting information [11], where we described the importance of Au NP age, type of buffer, ionic strength of the buffer ethylenediaminetetraacetic acid (EDTA) concentration, Au NP to virus ratio, Au NP concentration, and reaction time.
20. SATP can be stored at −20 °C as previously weighed in 1 mg fractions. Use this protocol to prepare the fractions. Let the solid equilibrate to room temperature and weigh individual fractions in Eppendorf tubes. Label the tubes with the amount in mg and when ready to use take a tube and add the corresponding amount of DMSO to obtain a 1 mg/ml solution.
21. The VLP-SHR fraction can be stored at 4 °C up to a month until ready to use. The resultant VLP-SHR solution (750 μl of 76 μg/ml) was diluted 1:1 with 750 μl of 100 mM KP pH 7.5. Make sure to have the Au NPs ready before starting the deprotection since 8 h after deprotection the VLP-SH will start to cross-link. While the deprotection is taking place make sure to concentrate the Au NP prior to the Au NP reaction (*see* Subheading 3.7.3).

22. When concentrating Au NPs, do not spin down the concentrators more than twice because the filters may leak. If needed use a clean filter for further concentration. If a swing-bucket rotor is not available consult the filters manufacturer's guidelines for alternative rotors and corresponding speeds. Once the Au NP dispersion is concentrated one may have around ~200 μl of Au NP in each filter. Transfer the Au NP to a clean 1.5 ml Eppendorf tube gently, use a 200 μl pipette. Add 100 μl of the filtrate (recovered from the centrifugation of Au NPs) to each filter to rinse the filter and transfer to the Eppendorf tube where the Au NPs are being collected. At the end the Au NPs will be in a final ~600 μl volume.
23. The final volume of the reaction is set to 2 ml to have the Au NP 20 times concentrated from its initial concentration (since the initial volume of Au NP as synthesized was 40 ml). The final concentration of EDTA is 1 mM and the concentration of phosphate buffer is 10 mM. Variations in buffer type, pH, and EDTA concentrations will change the efficiency of the reaction. When completing the volume to 2 ml use the filtrate recovered after concentrating Au NP. Do not resuspend the AuNPs in water or in any other buffer; the filtrate is the best choice for completing the volume for highest reaction efficiency with the CPMV capsids. After the Au NP is concentrated it is important to begin the reaction promptly (within 15–30 min) to avoid aggregation. If the Au NPs are kept concentrated more than 20× for longer periods of time they may aggregate; once the Au NPs are set at 20× concentration they are stable for at least 3 days.
24. During our optimization experiments, optimum conditions were found to be a 20-fold excess of Au NP particles relative to virus particles in the reaction mixture. Based on the geometrical constraints of an icosahedral particle (virus, 30 nm) the maximum number of Au NP of similar size (30 nm) that can be accommodated on the virus surface is 12 Au NPs. Therefore, a 20-fold excess of particles corresponds to a virus:Au NP particle ratio of 1:240. Calculations are done by using the molecular weight of the VLP (3.9×10^6 g/mol) or BC (5.6×10^6 g/mol) and the Avogadro's number to the determine the number of virus particles in the virus solutions. The Au NP concentration in particles/ml is determined from the calibration curve (*see* Fig. 3b).
25. We determined that 36 h was the optimum reaction time. Longer reaction times generate larger aggregates that are not useful for the biosensor application. We determined the optimum reaction conditions by following the increase of the particles size over time by DLS.

26. This method should be followed exactly as described herein. If the TA is added directly to the reaction mixture, the NCs aggregate irreversibly. The use of basic water and prior dilution of the TA is critical.
27. Even if the color of the solution is red, discard it. It is not necessary to collect the buffer since most of the NCs stay inside the gel. On the contrary, free Au NP diffuses from the gel more readily. When having a negative control (free Au NP band) we kept only the fraction that we purified from the gel to have a viable negative control.
28. Stability at each stage. The reaction mixture post TA treatment (for coating the unreacted Au NPs) can be stored at 4 °C for a week prior to purification. The gel pieces before the β-agarase treatment can be stored at 4 °C for 1 week. The solution after β-agarase digestion can be stored at 4 °C for 1 week.
29. To remove the neoagaro-oligosaccharides the post β-agarase digestion solution is washed twice with water or buffer (depending on NC application) using 4 ml 100 k MWCO filters. If water is necessary (example: immobilization on TEM grids for imaging) make sure to immobilize the sample immediately since the NCs are less stable in water; they tend to aggregate in water (within ~2 h). For the biosensor work we used 10 mM Bis-Tris, 1 mM EDTA: pH 6.5 buffer; the NCs in buffer were stable during the entire experiment (~1 week). Once the NCs are dried on a surface they are stable for weeks, but the container must be sealed to avoid any contamination from the environment. In summary, as in any new sensitive characterization technique it is recommended to prepare the NCs fresh the first time the experiment is done. From the buffers we tested the 10 mM Bis-Tris, 1 mM EDTA: pH 6.5 was the most successful to keep the NCs from aggregating in solution.
30. It is important to have a valid negative control to demonstrate that differences in UV-Vis, DLS, and SERS activity corresponds to the NCs as opposed to the treatment of the individual Au NPs during the methods. It is well documented that changes in buffer conditions and aggregation may change the spectral characteristics of Au NP dispersion. Therefore, a negative control is crucial for further experimentation. The negative control is recovered from the agarose gels by cutting out the free Au NP containing bands and following the same procedures as used for the NC purification.

Acknowledgments

This work was supported by the Office of Naval Research 6.1 base funds. We thank J. Phelps and J. E. Johnson for BC-CPMV samples and Y. Meshcheriakova and G. Lomonossoff for VLP samples.

References

1. Muehlig S, Cunningham A, Dintinger J, Scharf T, Burgi T, Lederer F, Rockstuhl C (2013) Self-assembled plasmonic metamaterials. Nanophotonics 2(3):211–240. https://doi.org/10.1515/nanoph-2012-0036
2. Hedayati MK, Faupel F, Elbahri M (2014) Review of plasmonic nanocomposite metamaterial absorber. Materials 7(2):1221–1248. https://doi.org/10.3390/ma7021221
3. Yao K, Liu YM (2014) Plasmonic metamaterials. Nanotechnol Rev 3(2):177–210. https://doi.org/10.1515/ntrev-2012-0071
4. Kretschmer F, Muhlig S, Hoeppener S, Winter A, Hager MD, Rockstuhl C, Pertsch T, Schubert US (2014) Survey of plasmonic nanoparticles: from synthesis to application. Part Part Syst Charact 31(7):721–744. https://doi.org/10.1002/ppsc.201300309
5. Blum AS, Soto CM, Wilson CD, Cole JD, Kim M, Gnade B, Chatterji A, Ochoa WF, Lin TW, Johnson JE, Ratna BR (2004) Cowpea mosaic virus as a scaffold for 3-D patterning of gold nanoparticles. Nano Lett 4(5):867–870. https://doi.org/10.1021/nl0497474
6. Blum AS, Soto CM, Wilson CD, Brower TL, Pollack SK, Schull TL, Chatterji A, Lin TW, Johnson JE, Amsinck C, Franzon P, Shashidhar R, Ratna BR (2005) An engineered virus as a scaffold for three-dimensional self-assembly on the nanoscale. Small 1(7):702–706. https://doi.org/10.1002/smll.200500021
7. Blum AS, Soto CM, Wilson CD, Whitley JL, Moore MH, Sapsford KE, Lin TW, Chatterji A, Johnson JE, Ratna BR (2006) Templated self-assembly of quantum dots from aqueous solution using protein scaffolds. Nanotechnology 17(20):5073–5079. https://doi.org/10.1088/0957-4484/17/20/006
8. Blum AS, Soto CM, Wilson CD, Amsinck C, Franzon P, Ratna BR (2007) Electronic properties of molecular memory circuits on a nanoscale scaffold. IEEE Trans Nanobioscience 6(4):270–274. https://doi.org/10.1109/tnb.2007.908978
9. Blum AS, Soto CM, Sapsford KE, Wilson CD, Moore MH, Ratna BR (2011) Molecular electronics based nanosensors on a viral scaffold. Biosens Bioelectron 26(6):2852–2857. https://doi.org/10.1016/j.bios.2010.11.021
10. Zhou JC, Soto CM, Chen MS, Bruckman MA, Moore MH, Barry E, Ratna BR, Pehrsson PE, Spies BR, Confer TS (2012) Biotemplating rod-like viruses for the synthesis of copper nanorods and nanowires. J Nanobiotechnol 10:Article No. 18. https://doi.org/10.1186/1477-3155-10-18
11. Fontana J, Dressick WJ, Phelps J, Johnson JE, Rendell RW, Sampson T, Ratna BR, Soto CM (2014) virus-templated plasmonic nanoclusters with icosahedral symmetry via directed self-assembly. Small 10(15):3058–3063. https://doi.org/10.1002/smll.201400470
12. Lebedev N, Griva I, Dressick WJ, Phelps J, Johnson JE, Meshcheriakova Y, Lomonossoff GP, Soto CM (2016) A virus-based nanoplasmonic structure as a surface-enhanced Raman biosensor. Biosens Bioelectron 77:306–314. https://doi.org/10.1016/j.bios.2015.09.032
13. Bahns JT, Guo Q, Montgomery JM, Gray SK, Jaeger HM, Chen L (2009) High-fidelity nano-hole-enhanced Raman spectroscopy. J Phys Chem C 113(26):11190–11197. https://doi.org/10.1021/jp900764a
14. Malvadkar NA, Demirel G, Poss M, Javed A, Dressick WJ, Demirel MC (2010) Fabrication and use of electroless plated polymer surface-enhanced Raman spectroscopy substrates for viral gene detection. J Phys Chem C 114(24):10730–10738. https://doi.org/10.1021/jp101542j
15. Mühlig S, Cialla D, Cunningham A, März A, Weber K, Bürgi T, Lederer F, Rockstuhl C (2014) Stacked and tunable large-scale plasmonic nanoparticle arrays for surface-enhanced Raman spectroscopy. J Phys Chem C 118(19):10230–10237. https://doi.org/10.1021/jp409688p
16. Mu C, Zhang J-P, Xu D (2010) Au nanoparticle arrays with tunable particle gaps by template-assisted electroless deposition for high performance surface-enhanced Raman scattering. Nanotechnology 21(1):Article No. 015604. https://doi.org/10.1088/0957-4484/21/1/015604
17. Becker M, Stelzner T, Steinbrück A, Berger A, Liu J, Lerose D, Gösele U, Christiansen S (2009) Selectively deposited silver coatings on gold-capped silicon nanowires for surface-enhanced Raman spectroscopy. ChemPhysChem 10(8):1219–1224. https://doi.org/10.1002/cphc.200800809
18. Alù A, Engheta N (2009) The quest for magnetic plasmons at optical frequencies. Opt Express 17(7):5723–5730. https://doi.org/10.1364/oe.17.005723
19. Urban AS, Shen X, Wang Y, Large N, Wang H, Knight MW, Nordlander P, Chen H, Halas NJ (2013) Three-dimensional plasmonic nanoclusters. Nano Lett 13(9):4399–4403. https://doi.org/10.1021/nl402231z
20. Zhao X (2012) Bottom-up fabrication methods of optical metamaterials. J Mater Chem 22(19):9439–9449. https://doi.org/10.1039/C2JM15979A
21. Barrow SJ, Wei X, Baldauf JS, Funston AM, Mulvaney P (2012) The surface plasmon

modes of self-assembled gold nanocrystals. Nat Commun 3:Article No. 1275. https://doi.org/10.1038/ncomms2289

22. Zhao Y, Xu L, Liz-Marzán LM, Kuang H, Ma W, Asenjo-García A, García de Abajo FJ, Kotov NA, Wang L, Xu C (2013) Alternating plasmonic nanoparticle heterochains made by polymerase chain reaction and their optical properties. J Phys Chem Lett 4(4):641–647. https://doi.org/10.1021/jz400045s
23. Zahr OK, Blum AS (2012) Solution phase gold nanorings on a viral protein template. Nano Lett 12(2):629–633. https://doi.org/10.1021/nl203368v
24. Johnson J, Lin T, Lomonossoff G (1997) Presentation of heterologous peptides on plant viruses: genetics, structure, and function. Annu Rev Phytopathol 35:67–86. https://doi.org/10.1146/annurev.phyto.35.1.67
25. Sainsbury F, Saxena P, Aljabali AAA, Saunders K, Evans DJ, Lomonossoff GP (2014) Genetic engineering and characterization of cowpea mosaic virus empty virus-like particles. In: Lin B, Ratna B (eds) Virus hybrids as nanomaterials, Methods mol biol, vol 1108. Humana Press, Clifton, pp 139–153. https://doi.org/10.1007/978-1-62703-751-8_11
26. Aljabali AAA, Sainsbury F, Lomonossoff GP, Evans DJ (2010) Cowpea mosaic virus unmodified empty virus-like particles loaded with metal and metal oxide. Small 6(7):818–821. https://doi.org/10.1002/smll.200902135
27. Chatterji A, Ochoa WF, Paine M, Ratna BR, Johnson JE, Lin T (2004) New addresses on an addressable virus nanoblock: uniquely reactive lys residues on cowpea mosaic virus. Chem Biol 11:855–863. https://doi.org/10.1016/j.chembiol.2004.04.011
28. Steinmetz NF, Evans DJ, Lomonossoff GP (2007) Chemical introduction of reactive thiols into a viral nanoscaffold: a method that avoids virus aggregation. Chembiochem 8(10):1131–1136. https://doi.org/10.1002/cbic.200700126
29. Soto CM (2014) A programmable fluorescent viral nanoblock: sensing made easy in a single step. In: Lin B, Ratna B (eds) Virus hybrids as nanomaterials, Methods mol. biol, vol 1108. Humana Press, Clifton, pp 155–172. https://doi.org/10.1007/978-1-62703-751-8_12
30. Bastús NG, Comenge J, Puntes V (2011) Kinetically controlled seeded growth synthesis of citrate-stabilized gold nanoparticles of up to 200 nm: size focusing versus ostwald ripening. Langmuir 27:11098–11105. https://doi.org/10.1021/la201938u
31. Wang Q, Lin T, Johnson JE, Finn MG (2002) Natural supramolecular building blocks: cysteine-added mutants of cowpea mosaic virus. Chem Biol 9:813–819. https://doi.org/10.1016/S1074-5521(02)00166-7

Chapter 35

TMV-Based Adapter Templates for Enhanced Enzyme Loading in Biosensor Applications

Claudia Koch, Arshak Poghossian, Christina Wege, and Michael J. Schöning

Abstract

Nanotubular tobacco mosaic virus (TMV) particles and RNA-free lower-order coat protein (CP) aggregates have been employed as enzyme carriers in different diagnostic layouts and compared for their influence on biosensor performance. In the following, we describe a label-free electrochemical biosensor for improved glucose detection by use of TMV adapters and the enzyme glucose oxidase (GOD). A specific and efficient immobilization of streptavidin-conjugated GOD ([SA]-GOD) complexes on biotinylated TMV nanotubes or CP aggregates was achieved via bioaffinity binding. Glucose sensors with adsorptively immobilized [SA]-GOD, and with [SA]-GOD cross-linked with glutardialdehyde, respectively, were tested in parallel on the same sensor chip. Comparison of these sensors revealed that TMV adapters enhanced the amperometric glucose detection remarkably, conveying highest sensitivity, an extended linear detection range and fastest response times. These results underline a great potential of an integration of virus/biomolecule hybrids with electronic transducers for applications in biosensorics and biochips. Here, we describe the fabrication and use of amperometric sensor chips combining an array of circular Pt electrodes, their loading with GOD-modified TMV nanotubes (and other GOD immobilization methods), and the subsequent investigations of the sensor performance.

Key words Tobacco mosaic virus (TMV), Coat protein, Enzyme nanocarrier, Glucose biosensor, Glucose oxidase, Amperometric sensor, Pt electrode array, Streptavidin–biotin

1 Introduction

Enzyme activity may be affected to a critical extent upon immobilization on solid supports, which is, however, advantageous or necessary for numerous applications, including many biosensor layouts. The choice of the right immobilization strategy is therefore essential for the generation of biosensors with good and reliable performance, storage stability, and reusability. The efficiency of enzyme immobilization is of equal importance, since a high level of immobilization results in a good cost–benefit ratio. The immobilization method of choice should further allow strong attach-

Christina Wege and George P. Lomonossoff (eds.), *Virus-Derived Nanoparticles for Advanced Technologies: Methods and Protocols*, Methods in Molecular Biology, vol. 1776, https://doi.org/10.1007/978-1-4939-7808-3_35,
© Springer Science+Business Media, LLC, part of Springer Nature 2018

ment to the detector surfaces, to avoid unwanted detachment during measurements. Concomitantly, however, the surface-coupled enzymes need to maintain their structure and thus catalytic activity, which must not be hindered by the coupling process. To reach these goals, a broad spectrum of different immobilization techniques has been reported in the literature, including cross-linking, covalent or affinity binding, physical adsorption, entrapment, and various linker-based strategies (refer to reviews such as [1–5] for further details).

Recently, the coupling of (bio)chemical recognition elements with multivalent nanoscale biological building blocks, such as virus particles, has been found to be a very promising strategy for the creation of biohybrids; these open up novel opportunities for the high-density presentation of enzymes [6–12]. One of the most promising viral backbones is the plant virus tobacco mosaic virus (TMV) [13–15]. TMV is a rigid nanotube with a length of 300 nm and an outer diameter of 18 nm. Native TMV consists of 2130 identical coat protein (CP) subunits which self-assemble into a helical structure embedding a single-stranded RNA genome [16, 17]. Numerous studies have demonstrated the huge potential of the robust TMV particles as versatile building blocks for functional materials and as high-yield nanotemplates, allowing the presentation of active molecules or synthetic compounds (*see* Fig. 1) [18].

Recently, TMV has successfully been introduced as enzyme carrier scaffold in a new approach for the improvement of label-free glucose biosensors [6]. In this study, enzyme coupling to the TMV capsid was achieved by a cysteine-modified TMV CP variant (TMV_{Cys}) [19], yielding versatile carrier rods with more than 2000 selectively addressable thiols (also described in Chapter 27). The thiol group of a cysteine exposed on the surface of every CP subunit allowed the coupling of bifunctional maleimide-PEG-biotin linkers, resulting in biotinylated TMV-based nanoparticles (TMV_{Cys}/Bio). Streptavidin-conjugated glucose oxidase ([SA]-GOD) was immobilized via biotin–streptavidin affinity interaction on the surface of the TMV_{Cys}/Bio rods, or of RNA-free CP_{Cys}/Bio aggregates (ring-shaped "disks" and lower order oligomers) produced from the biotinylated rods, respectively (*see* also [7]). The affinity binding method allowed control over the bioactive molecules' orientation, in order to avoid enzyme deactivation and/or blocking of their active sites [20]. The TMV-based adapter templates were then adsorbed on sensor surfaces, thereby allowing the immobilization of enzymes at high surface densities in the detector devices. For comparison, glucose sensors with adsorptively immobilized [SA]-GOD, as well as [SA]-GOD cross-linked with glutardialdehyde (techniques often used for enzyme immobilization on sensors) were tested in parallel. Physical adsorption is a simple way to immobilize enzymes without substantial loss of their activity. However, a weak bond between the sensor surface and the enzyme

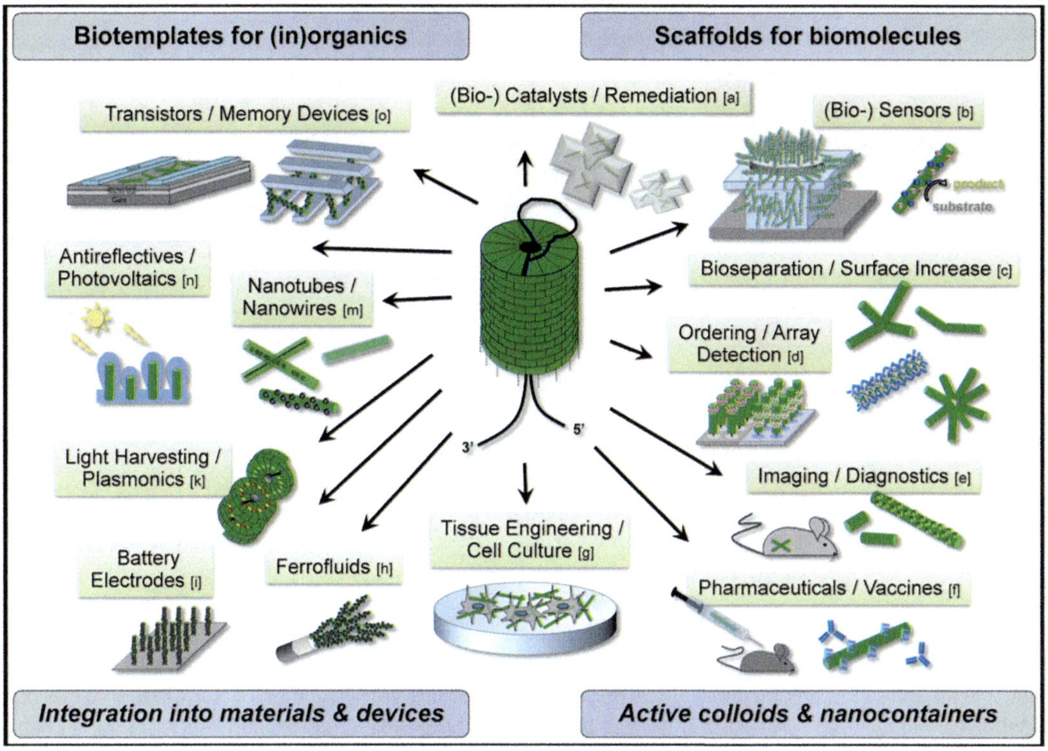

Fig. 1 TMV as a versatile multivalent soft-matter template: TMV and related tobamoviruses are used as templates for the construction of various biohybrid nanoobjects and nanostructured materials. For references (**a**–**o**) ***see*** original work [18]. Reproduced according to the Creative Commons Attribution 2.0 International Public License from [18]

can lead to a fast leakage of the enzymes. An immobilization by cross-linking with glutardialdehyde or other bifunctional agents is attractive due to its simplicity and the strong chemical binding between biomolecules. The main drawback is the possible loss of enzyme activity due to distortion of the enzyme conformation and chemical alteration of the active sites during cross-linking [1]. Each of these immobilization strategies was used to immobilize [SA]-GOD conjugates on one out of four single Pt electrodes on the same sensor chip, followed by amperometric detection of varying glucose concentrations. An exemplary distribution of electrodes (glucose sensors) on the chip with the differently immobilized [SA]-GOD conjugates is summarized in Fig. 2.

Comparable studies in ELISA-plate wells showed that by using these viral adapters, considerably higher amounts of enzymes could be installed from commercial [SA]-GOD preparations than through direct or biotin linker-mediated coupling. This resulted in up to 45 times higher turnover rates, increased reusability ($t_{50\%}$ = 13 days instead of 4–6 days) as well as enhanced stability ($t_{50\%}$ = 3 weeks instead of 1 week) compared to conventionally

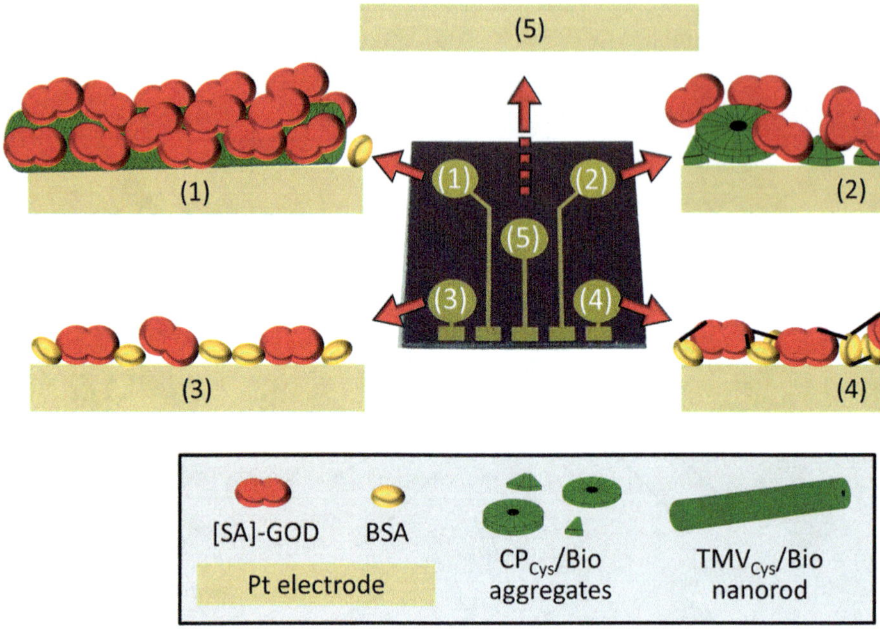

Fig. 2 Example of electrode loading with [SA]-GOD from a BSA-containing stock solution in different layouts. Shown is an array of Pt electrodes loaded with [SA]-GOD-modified TMV nanotubes (1) or CP aggregates (2), adsorptively immobilized [SA]-GOD (3) as well as [SA]-GOD cross-linked with glutardialdehyde (4), all subjected to enzyme loading in the presence of BSA. A bare electrode (5) serves for electrochemical functionality testing of the as-produced electrodes

bound enzymes, indicating a stabilizing effect of the TMV environment over weeks [7]. In the amperometric detection setup, Pt electrodes of electrochemical sensor chips loaded with [SA]-GOD-modified TMV sticks exhibited faster response times ($t_{90\%}$ = 5 s), higher glucose sensitivities (4.9 nA/mM mm^2), an extended nearly linear detection range (0.1–7.4 mM glucose concentration), and less noise signal compared to sensors using other immobilization methods (for typical response curves *see* Fig. 3) [6] (*see* **Note 1**).

Taken together, these results underline a great potential for the integration of TMV as enzyme carrier in sensing devices. Despite the observed benefits (higher enzyme loading, better sensor stability and faster response times), there are up to now only few reports on using TMV particles as scaffolds in sensing applications (*see* e.g. [15, 21] and references therein). In combination with inorganic materials, TMV nanoparticles coated with palladium as a sensing material have promoted the development of a surface acoustic wave (SAW) hydrogen sensor [22]. Using organically modified TMV nanorods, it was possible to create a thin film sensor for the detection of volatile organic compounds (VOC) such as ethanol and methanol, after deposition of oligo-aniline-grafted TMV particles onto a glass substrate [23]. Another biosensor approach is based on TMV-like particles modified with 2,4,6-trinitrotoluene-

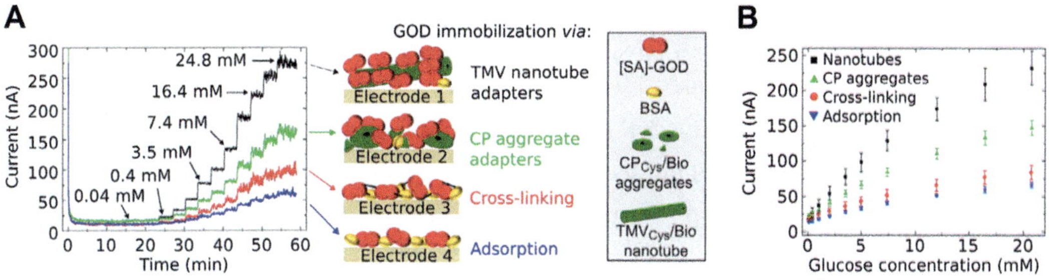

Fig. 3 Dynamic responses (**a**) and mean calibration curves (**b**) of four glucose sensors with variously immobilized [SA]-GOD conjugates on the same chip, recorded in solutions with different glucose concentrations. Reproduced according to the Creative Commons Attribution 2.0 International Public License from [6]

(TNT)-binding peptides for sensing in solution [24]. Recently, it was shown that the combination of a novel TMV-VLP receptor layer for antibody binding with an optical microdisk resonator can be used as transducer for biosensing applications [25].

The following protocol will focus on a recently established label-free biosensor layout profiting from biotinylated TMV-based adapter templates for the efficient and durable installation of the glucose sensor enzyme GOD on the surface of sensor electrodes for amperometric readout [6]. It describes the preparation of biotin-fashioned TMV-based adapters (both nanosticks and CP aggregates) and the fabrication of appropriate sensor electrodes, TMV adapter immobilization on their surfaces, their equipment with [SA]-GOD via bioaffinity coupling and, finally, proof-of-concept measurements demonstrating the glucose detection capabilities of the system in comparison to control layouts.

2 Materials

Prepare all buffers and solutions with deionized water (ddH_2O, 18.3 MΩ cm; e.g., purified by a membraPure system (Aquintus) or other comparable water quality) if not stated otherwise. Prepare all virus dilutions in 10 mM sodium potassium phosphate (SPP) pH 7.0, if not otherwise recommended. Dissolve all virus-containing pellets in 10 mM SPP pH 7.0.

2.1 Biotin Linker Coupling to TMV_{Cys} Nanosticks [7]

1. 10 mM sodium potassium phosphate (SPP) buffer pH 7.0: Mix 10 mM NaH_2PO_4 and 10 mM K_2HPO_4 until a pH of 7.0 is obtained.
2. 15 mg TMV_{Cys} [19] in SPP: Use a virus particle preparation of typically 5 mg/ml isolated from about 20 g leaves of systemically infected *Nicotiana tabacum* 'Samsun' nn plants (*see* for example Subheading 2.3 in Chapter 27).

3. Maleimide-activated biotin linker to enable high-affinity binding of [SA]-GOD conjugates (maleimide-PEG_{11}-biotin EZ-Link®, Thermo Scientific) (*see* **Note 2**).
4. Dimethyl sulfoxide (DMSO), anhydrous, >99.9%.
5. Highly accurate balance.
6. Ultracentrifuge with fixed-angle rotor holding small tubes (up to 1.5 ml) and capable of running at 120,000 × *g* [e.g., Optima L-80 XP Ultracentrifuge (Beckmann Coulter) and rotor 45 Ti (Beckmann Coulter)].
7. 1.5 ml tubes suitable for ultracentrifugation (UC).
8. UV–Vis spectrophotometer [e.g., NanoDrop ND-1000 spectrophotometer (Thermo Scientific)] or equivalent instrument capable of measuring the UV absorption of small volumes (e.g., 1–2 μl).
9. SDS–polyacrylamide electrophoresis (SDS-PAGE) equipment and solutions for discontinuous Laemmli gels with a 4% stacking gel and 15% separation gel, as for example listed in Subheading 2.5 of Chapter 23.
10. Coomassie Brilliant Blue R-250 and destaining solution or similar standard staining solution for SDS-PAGE.
11. ImageJ by Rasband, W.S., US National Institutes of Health, Bethesda, Maryland, USA; software can be downloaded from https://imagej.nih.gov/ij/download.html.
12. Standard molecular biology instrumentation and consumables such as microliter pipettes and microcentrifuge tubes.

2.2 CP_{Cys}/Bio Preparation

1. 1 ml of a 5 mg/ml TMV_{Cys}/Bio solution (produced in Subheading 3.1).
2. Ice-cold glacial acetic acid.
3. Centrifuge with fixed angle rotor holding small tubes (up 2 ml) and capable of running at 20,000 × *g* and 4 °C.
4. Dialysis tubing: 8 kDa molecular weight cutoff (MWCO) (e.g., Spectra/Por® 7 Dialysis Membrane Pretreated RC Tubing, Spectrum Laboratories, Inc., Rancho Dominguez, CA, USA).
5. Two clips, suitable to block the tubing.
6. 3 × 5 l ddH_2O at 4 °C.
7. 10 mM SPP, pH 7.0.
8. UV–Vis spectrophotometer [e.g., NanoDrop ND-1000 spectrophotometer (Thermo Scientific)] or equivalent instrument capable of measuring the UV absorption of small volumes (e.g., 1–2 μl).

2.3 Sensor Chip Fabrication [26]

1. UNIVEX 350 Evaporator System (Leybold/Oerlikon) for electron-beam evaporation.
2. Photoresist (AZ 5214, MicroChemicals GmbH); a light-sensitive material used in photolithography to form a patterned coating on a surface.
3. Ultrasonic wedge bonder (Kulicke & Soffa Pte Ltd).
4. Silicone rubber (Momentive Performance Materials GmbH, Germany) for encapsulation of bond pads and bond wires.
5. Printed circuit boards: homemade (FH Aachen, Germany).
6. Oxidation furnace (Tempress, Netherlands) for thermal oxidation of Si wafer.
7. p-type Si wafer (Topsil GlobalWafers, thickness: 356–406 μm; specific resistivity: ~1000 Ω cm; polished).
8. Photomask for opening windows for deposition of metal layers.
9. Mask Aligner for lithography (MJB3, Suss Microtec, Germany); UV-radiation: 10 mW/cm^2, $\lambda = 365$ nm.
10. Titanium (Umicore, granules).
11. Platinum (Umicore, granules).
12. Acetone (Technic France).
13. Liquid glue to connect chip and board.

2.4 Sensor Chip Loading with Viral Adapters and Immobilization of [SA]-GOD

1. 10 mM SPP, pH 7.0.
2. 0.33 μg/μl TMV_{Cys}/Bio and 0.33 μg/μl CP_{Cys}/Bio in 10 mM SPP pH 7.0: dilute prepared TMV_{Cys}/Bio (Subheading 3.1) and CP_{Cys}/Bio solutions (Subheading 3.2) in 10 mM SPP pH 7.0.
3. Humid chamber to avoid evaporation of protein solutions.
4. Streptavidin-conjugated glucose oxidase ([SA]-GOD; 1 μg/μl in storage buffer (0.1 M K-phosphate buffer pH 6.0, supplemented with 10 μg/μl BSA (bovine serum albumin) and 1200 ppm 5-nitro-5-bromo-1,3-dioxane)) with 1–2 [SA] partners per GOD (provided by the supplier SDT, Baesweiler, Germany).
5. 2% (w/v) glutardialdehyde [diluted from 25% (w/v) glutardialdehyde solution in water (Merck)].

2.5 Electrochemical Characterization of Sensor Electrodes

1. 250 mM glucose (glucose stock solution).
2. Measuring vessel with holder able to bear two sensor chips boards in parallel.
3. 10 mM SPP, pH 7.0.

4. Potentiostat (PalmSens, Palm Instruments BV) equipped with an eight-channel multiplexer (CH8, Palm Instruments BV) to switch between the individual channels.
5. Liquid-junction Ag/AgCl reference electrode (3 M KCl, Metrohm).
6. Pt counter electrode with a surface area of approximately 1 cm.
7. Dark Faraday cage of approximately 60 × 40 × 30 cm: homemade (FH Aachen, Germany).
8. Software: PSTrace (PalmSens, PalmSens BV).

3 Methods

Carry out all processes and preparation steps at room temperature (RT), unless otherwise specified.

3.1 TMV_{Cys} Isolation and Biotin Linker Coupling

1. Dilute a stock solution of TMV_{Cys} particles from infected *N. tabacum* leaves (isolated according to Gooding and Herbert [27] as described in detail in Chapter 23 or 27) to yield a total of 3 ml TMV_{Cys} of 5 mg/ml in 10 mM SPP buffer pH 7.0 (*see* **Note 3**).
2. Divide the maleimide-PEG_{11}-biotin linker in small packages of around 5 mg each and freshly dissolve one dose prior to the experiment in DMSO (f.c. 250 nmol/μl). Store the dissolved linker solution at 4 °C.
3. Mix 10 mg TMV_{Cys} (i.e., 2 ml of **step 1**) with 12 μmol (i.e., 5 μl stock solution) linker and incubate under gentle agitation at RT for 16 h (*see* **Note 4**).
4. Split the TMV_{Cys} and linker containing solution in equal parts and transfer them into tubes suitable for ultracentrifugation (UC).
5. Remove unbound maleimide-biotin linker molecules by UC (120,000 × *g*, 4 °C, 2 h): collect the supernatants with nonreacted linkers by use of a pipette and discard them. Resuspend UC pellets containing the linker-fashioned virus particles (TMV_{Cys}/Bio) in the initial amount of the TMV solution (here: 1 ml for each tube) in 10 mM SPP (pH 7.0) and combine both portions.
6. Determine the TMV concentration using a spectrophotometer and the extinction coefficient for TMV at 260 nm (ε_{260} = 3 ml/(mg•cm) [28]) (*see* **Note 5**).
7. Analyze the samples on a discontinuous Laemmli SDS-PA gel according to standard methods [29]. Prepare a 15% separation gel with a 4% stacking gel. Mix 1.5 μg of the TMV_{Cys}/Bio sample with SDS-gel loading buffer freshly supplemented with

reducing agent (dithiothreitol, β-mercaptoethanol or similar) to obtain a final volume of 10 μl. Heat for 5 min at 95 °C, load samples and a protein molecular weight marker into the slots and perform the electrophoresis. After disassembly of the SDS-PAGE setup, fix and stain the gel with Coomassie Blue R250 or a similar colloidal dye according to standard methods (for details on SDS-PAGE, you may also refer for example to Chapter 23, Subheading 3.3.1). After destaining of the gel background and the appearance of clear bands, document the image and use ImageJ (or a different suitable image analysis software) to assess the percentage of TMV CPs carrying a linker molecule (*see* Fig. 4 and **Note 6**).

3.2 CP_{Cys}/Bio Preparation from TMV_{Cys}/Bio Particles

1. Mix 1 Vol. TMV_{Cys}/Bio suspension (from Subheading 3.1, **step 6**; typically 600 μl) with 2× Vol. ice-cold glacial acetic acid [30]. Incubate for 20 min on ice. To separate RNA and CPs, centrifuge at 20,000 × *g* at 4 °C for 20 min. Collect the supernatant containing the CPs.
2. Incubate a ~10 cm long piece of dialysis tubing (8 kDa MWCO) for 2 min in ultrapure water. Block one end of the tubing with a clip. Transfer the supernatant into the dialysis tubing and block the second end with a clip.
3. Dialyze against ultrapure water (~ 5 l ddH_2O) for 24–48 h at 4 °C until the tubing content seems cloudy or turbid. Change water every few hours.

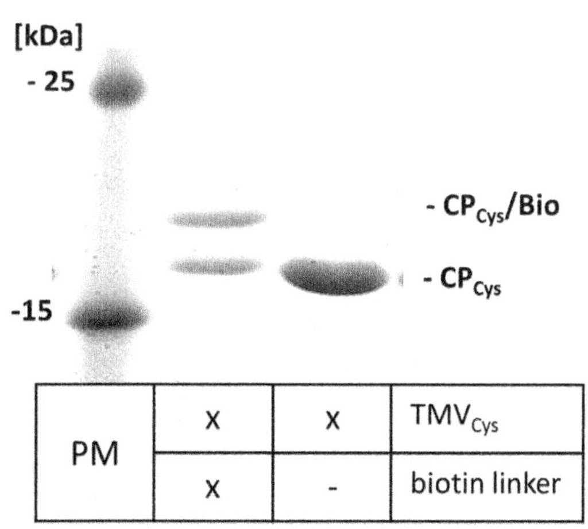

Fig. 4 Biotinylation of TMV_{Cys} behind Cyscoupling of maleimide-PEG_{11}-biotin (biotin linker) to the thiol groups of TMV_{Cys} (CP_{Cys}) results in a shift of the CP signal (CP_{Cys}/Bio). Samples were analyzed using a 15% SDS-PA gel. CPs were stained with Coomassie Brilliant Blue R250. Image evaluation using ImageJ revealed a coupling efficiency of approximately 50%

4. Transfer content into a 15 ml tube and subsequently divide the solution into 1.5 ml microcentrifuge tubes and centrifuge at 20,000 × *g* at 4 °C for 20 min.
5. Discard the supernatant and dissolve the CP_{Cys}/Bio-containing pellet in 10 mM SPP, pH 7.0 [total volume of all pellets: ½ Vol. of the initial TMV solution (here: 300 μl)]. Combine the resuspended pellets. Store at RT (or at 4 °C for prolonged storage).
6. Determine the approximate protein concentration of the CP_{Cys}/Bio solution using a spectrophotometer and the extinction coefficient for TMV CPs at 280 nm (ε_{280} = 1.3 ml/(mg•cm) [28]) (*see* **Note 7**).

3.3 Amperometric Sensor Chips Fabrication

The used chips consist of a p-Si-SiO_2-Ti-Pt layer structure. The process steps of sensor chip fabrication are described in detail in [26] and are shown in Fig. 5. Briefly:

1. Grow 500 nm SiO_2 by thermal wet oxidation on a p-Si wafer at 1100 °C for 45 min.
2. For structuring the electrode distribution on the sensor surface, use spin-coating of photoresist at 4000 min^{-1} onto the Si wafer (*see* **Note 8**).

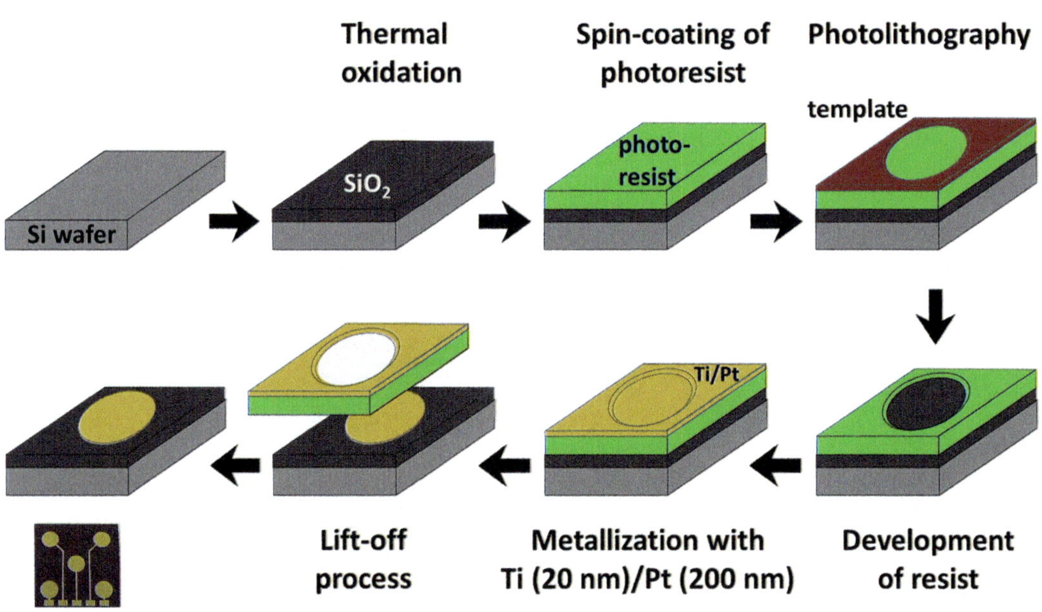

Fig. 5 Flowchart showing the fabrication steps of an amperometric sensor chip; thermal wet oxidation for insulation → deposition of photoresist layer and patterning through template (photomask) → development of resist by UV → deposition of electrode material → lift-off process of photoresist and metal layers

3. Use a template (photomask) to obtain a desired pattern on the photoresist: expose sensor with photomask to UV radiation ($t = 5$ s, $I = 10$ mW/cm^2, $\lambda = 365$ nm).
4. Deposit 20 nm titanium by electron-beam evaporation as adhesion layer on top of the SiO_2.
5. Deposit a 200 nm layer of Pt as electrode material.
6. Use acetone as solvent in the lift-off process (*see* **Note 9**).
7. Separate the processed wafer into chips of 10 mm × 10 mm. After cleaning, glue the sensor chips onto printed circuit boards.
8. Use an ultrasonic wedge bonder to provide electrical connection between sensor chip electrodes and the printed circuit boards.
9. Use silicone rubber to encapsulate the bond pads and bond wires (*see* **Note 10**).

A schematic layer structure (a), layout of Pt electrodes on a chip (b) and a photograph of a bonded and encapsulated sensor chip on a printed circuit board (c) are depicted in Fig. 6. Every chip comprises five circular working electrodes, each having a diameter of 2 mm (an area of 3.14 mm^2) and a distance to the next Pt electrode of ~2 mm to avoid cross-talk.

3.4 Sensor Chip Loading with Viral Adapters and Immobilization of [SA]-GOD

Recently published data revealed that 1 μg TMV_{Cys}/Bio sticks are sufficient to saturate an area of 2 mm diameter on the Pt surface of the electrodes [6]. To investigate the influence of the viral adapters on the biosensor performance, glucose sensors with adsorptively immobilized [SA]-GOD as well as [SA]-GOD cross-linked on the electrodes with glutardialdehyde can be tested in parallel, which enables comparative studies. An example for the possible distribution of differently immobilized [SA]-GOD conjugates is shown in Fig. 2. Since the commercial [SA]-GOD preparation contains 10 μg/μl BSA, the BSA molecules can adsorb nonspecifically and will reside in areas not exposing biotin linkers (where high-affinity binding of [SA]-GOD will replace BSA). BSA is thus also available

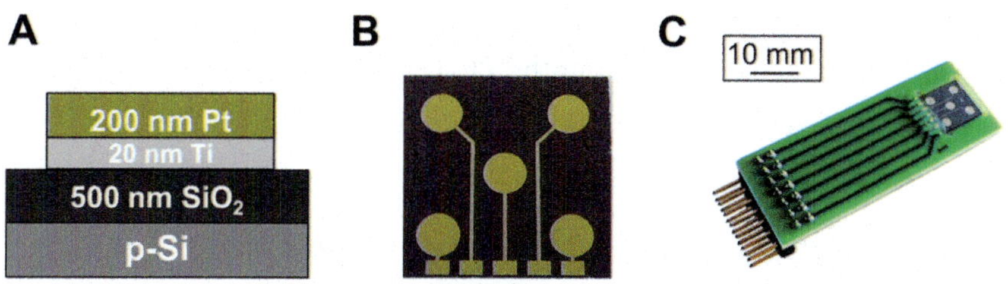

Fig. 6 Schematic layer structure (**a**), layout of Pt electrodes on a chip (**b**) and photograph of a bonded and encapsulated sensor chip on a printed circuit board (**c**)

as bridging molecule upon glutardialdehyde-mediated cross-linking.

1. Prepare for electrode (1) of the sensor chip 3 μl of TMV_{Cys}/Bio, and for electrode (2) 3 μl CP_{Cys}/Bio in 10 mM SPP buffer, respectively, with a concentration of 0.33 μg/μl each.
2. Incubate the 3 μl of TMV_{Cys}/Bio on electrode (1) and the 3 μl of CP_{Cys}/Bio on electrode (2) of one and the same sensor chip at RT for 1 h in a humid chamber (*see* **Note 11**).
3. Rinse the sensor chip thoroughly by repeated dipping in deionized (DI) water to remove nonadsorbed virus particles.
4. Mix 9 μl of [SA]-GOD solution with 6 μl 10 mM SPP buffer. Apply 5 μl of this mixture onto electrode (1) (functionalized with TMV_{Cys}/Bio), 5 μl on electrode (2) (functionalized with CP_{Cys}/Bio), and 5 μl on the so far unmodified electrode (3), respectively. Equip electrode (4) with [SA]-GOD undergoing cross-linking with both [SA]-GOD molecules and BSA by glutardialdehyde during the immobilization. To this end, mix 3 μl of [SA]-GOD solution with 2 μl cross-linker (2% glutardialdehyde; total volume 5 μl) (*see* **Note 12**). Incubate the enzyme solution on each electrode surface overnight at 4 °C in a humid chamber to allow specific bioaffinity binding of streptavidin and biotin. (The bare electrode (5) serves for electrochemical functionality testing of the as-produced electrodes.)
5. Rinse the sensor chip thoroughly by repeated dipping in deionized (DI) water to remove nonattached [SA]-GOD conjugates, BSA and unstably adsorbed virus particles or CP aggregates serving as adapter templates.

3.5 Electrochemical Characterization of Sensor Electrodes

Integrate the sensor chip loaded with variously immobilized [SA]-GOD conjugates (*see* Subheading 3.4) into the measurement setup (*see* Fig. 7).

1. Fill the measuring vessel with 10 mM SPP pH 7.0, add a magnetic stir bar and install holder on a magnetic stirrer.
2. Set the sensor chip into the holder so that the electrodes are immersed within the 10 mM SPP pH 7.0 buffer and connect the chip with the potentiostat (*see* **Note 13**).
3. Install the conventional liquid-junction Ag/AgCl reference electrode and the Pt counter electrode.
4. Apply a constant potential of +0.6 V vs. Ag/AgCl to the working electrodes (1) to (4).
5. Measure the background current of each sensor electrode in the absence of glucose for 10 min.
6. Add glucose from a 250 mM glucose stock solution to the buffer to reach a glucose concentration of 2 μM. Use the magnetic

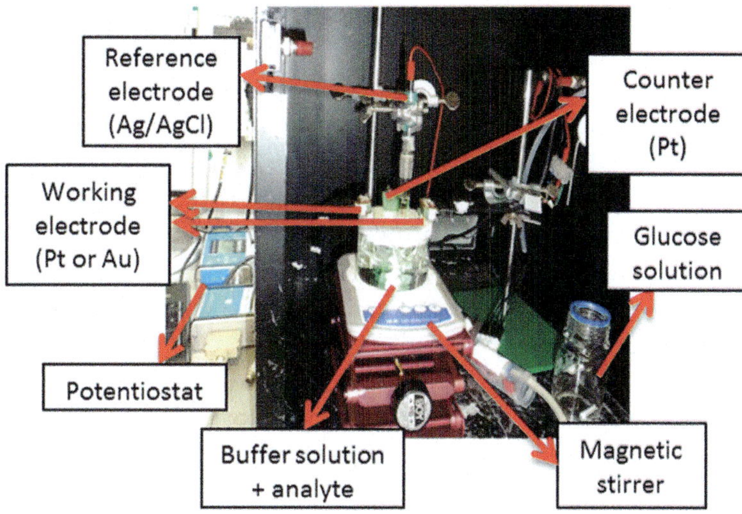

Fig. 7 Experimental setting for electrochemical characterization of amperometric glucose biosensors

stirrer to obtain a homogeneous mixture of the glucose solution.

7. Record the sensor performance of the working electrodes for 3–4 min.
8. Increase the glucose concentration in the measuring vessel sequentially by addition of more glucose from the stock solution in appropriate steps up to 25 mM glucose to the constantly stirred buffer solution. Measure the electrodes' performance after each step for 3–4 min, as described before.

As described in more detail recently [6], the presence of TMV-based adapter templates is expected to improve the performance of the respective sensor electrodes substantially, with the beneficial effects of nanostick templates surpassing those of plain CP coatings. Typical sensor response curves and the corresponding currents are shown in Fig. 3.

4 Notes

1. Explanation of some essential terms: reusability = the ability of a sensor to maintain its performance characteristics over a certain period of time; the reproducibility of output readings is determined at constant conditions; $t_{50\%}$ = the time point at which the sensor's initial performance is decreased to 50% of the original values (under constant conditions); response time = the time period during which the sensor signal reaches a specified percentage (e.g., 90%) of its final value in response to a stepwise change of the analyte concentration; sensitiv-

ity = the ratio of change in the sensor output signal to the change in the analyte concentration ≈ slope of calibration curve; nearly linear concentration range = best fit between the calibration curve and a specified straight line.

2. It is expected that these long bifunctional linkers with a total length of 6 nm provide all degrees of freedom required for a dense immobilization of enzymes on the adapters, resulting in full coverage of the TMV carrier sticks [7].
3. Store all TMV-containing solutions at 4 °C in 10 mM SPP buffer pH 7.0.
4. 10 mg TMV_{Cys} with 12 μmol linker corresponds to a 22-fold molar excess of maleimide-PEG_{11}-biotin linkers (in relation to the 2130 attachment sites on every TMV_{Cys} particle). This is sufficient to obtain TMV-based nanoscaffolds with 50–90% of the CP_{Cys} attachment sites biotinylated (in dependence on linker reactivity and coupling conditions), allowing for efficient and simple high-affinity binding of [SA]-GOD conjugates.
5. In principle, TMV particle concentration can be calculated from spectrophotometric absorption values using the Lambert-Beer-Law, but it has to be taken into account that the protein content (CP content) of TMV is only 95% (w/w).
6. The CP_{Cys} of TMV_{Cys} has a molecular weight of 17.6 kDa, a maleimide-PEG_{11}-biotin linker molecule ~0.922 kDa. A successful coupling event of a linker molecule to a thiol group of a CP_{Cys} (resulting in CP_{Cys}/Bio) leads to a mass increase up to ~ 18.5 kDa, which results in a distinguishable upward shift of the CP signal upon SDS-PAGE. Therefore, two bands should be visible in these samples. The biotinylation efficiency can be determined by densitometry using ImageJ, given that the amount of CP per band in combination with a colloidal (e.g., Coomassie Blue) stain is in the linear dynamic range of the pixel intensities. The pixel intensity sum of both bands must be set to 100% in order to then calculate the percentage of each fraction (CP_{Cys} or CP_{Cys}/Bio, i.e., faster vs. retarded band).
7. The CP-aggregates devoid of RNA are heterogeneous. After incubation at RT in the buffer applied, they will predominantly contain two-layer (34mer) or four-layer (68mer) disks and oligomers thereof [31]. Before starting the experiment, dilute the CP_{Cys}/Bio-containing solution with 10 mM SPP (pH 7.0) to a concentration of 0.33 μg/μl.
8. Prior to photoresist coating, heat wafer to 180 °C for 5 min to remove adsorbed water from the substrate surface, to allow a better adhesion of the hydrophobic photoresist. The typical thickness of the resist should be about 1.4 μm. Use a soft-bake

step (90 °C for 5 min) to reduce the remaining content of solvent in the resist.

9. Expose the whole wafer to acetone. During this incubation the residual photoresist will be stripped off, together with parts of the metallic thin films covering it. As a result, only the metal layers in direct contact with the underlying SiO_2 surface will remain and form the pattern defined by the applied photomask.
10. A good encapsulation of all bond pads and bond wires is necessary in order to prevent undesired short circuits.
11. Prepare more than one sensor chip in parallel. You can store the sensor chips in 10 mM SPP, pH 7, at 4 °C for several hours/days without bacteriostatic preservatives.
12. The [SA]-GOD solution applied contains 1 μg/μl [SA]-GOD in 0.1 MK-phosphate buffer, pH 6.0, and is supplemented with 10 μg/μl BSA and 1200 ppm 5-nitro-5-bromo-1,3-dioxan. The BSA amount contained in the enzyme solution allows the glutardialdehyde mediated cross-linking without further addition of BSA.
13. While using a potentiostat equipped with eight-channel multiplexer, two sensor chips (in summary eight electrodes) can be characterized in parallel.

References

1. Sassolas A, Blum LJ, Leca-Bouvier BD (2012) Immobilization strategies to develop enzymatic biosensors. Biotechnol Adv 30(3):489–511
2. Brena B, Gonzalez-Pombo P, Batista-Viera F (2013) Immobilization of enzymes: a literature survey. Methods Mol Biol 1051:15–31
3. DiCosimo R, McAuliffe J, Poulose AJ, Bohlmann G (2013) Industrial use of immobilized enzymes. Chem Soc Rev 42(15):6437–6474
4. Singh RK, Tiwari MK, Singh R, Lee JK (2013) From protein engineering to immobilization: promising strategies for the upgrade of industrial enzymes. Int J Mol Sci 14(1):1232–1277
5. Mateo C, Palomo JM, Fernandez-Lorente G, Guisan JM, Fernandez-Lafuente R (2007) Improvement of enzyme activity, stability and selectivity via immobilization techniques. Enzyme Microb Technol 40(6):1451–1463
6. Bäcker M, Koch C, Eiben S, Geiger F, Eber F, Gliemann H et al (2017) Tobacco mosaic virus as enzyme nanocarrier for electrochemical biosensors. Sensors Actuators B Chem 238:716–722
7. Koch C, Wabbel K, Eber FJ, Krolla-Sidenstein P, Azucena C, Gliemann H et al (2015) Modified TMV particles as beneficial scaffolds to present sensor enzymes. Front Plant Sci 6, Article 1137
8. Carette N, Engelkamp H, Akpa E, Pierre SJ, Cameron NR, Christianen PC et al (2007) A virus-based biocatalyst. Nat Nanotechnol 2(4):226–229
9. Chatterji A, Ochoa W, Shamieh L, Salakian SP, Wong SM, Clinton G et al (2004) Chemical conjugation of heterologous proteins on the surface of cowpea mosaic virus. Bioconjug Chem 15(4):807–813
10. Aljabali AA, Barclay JE, Steinmetz NF, Lomonossoff GP, Evans DJ (2012) Controlled immobilisation of active enzymes on the cowpea mosaic virus capsid. Nanoscale 4(18):5640–5645
11. Pille J, Cardinale D, Carette N, Di Primo C, Besong-Ndika J, Walter J et al (2013) General strategy for ordered noncovalent protein assembly on well-defined nanoscaffolds. Biomacromolecules 14(12):4351–4359

12. Cardinale D, Michon T (2016) Enzyme nanocarriers. Pan Standford Publishing, CRC Press, Taylor and Francis Group, Boca Raton, FL
13. Bittner AM, Alonso JM, Gorzny ML, Wege C (2013) Nanoscale science and technology with plant viruses and bacteriophages. Subcell Biochem 68:667–702
14. Culver JN, Brown AD, Zang F, Gnerlich M, Gerasopoulos K, Ghodssi R (2015) Plant virus directed fabrication of nanoscale materials and devices. Virology 479-480:200–212
15. Fan XZ, Pomerantseva E, Gnerlich M, Brown A, Gerasopoulos K, McCarthy M et al (2013) Tobacco mosaic virus: a biological building block for micro/nano/bio systems. J Vac Sci Technol A 31(5) 050815
16. Caspar DL (1963) Assembly and stability of the tobacco mosaic virus particle. Adv Protein Chem 18:37–121
17. Butler PJ (2009) Assembly of tobacco mosaic virus. R Soc Lond B Biol Sci 276(943): 151–163
18. Koch C, Eber FJ, Azucena C, Forste A, Walheim S, Schimmel T et al (2016) Novel roles for well-known players: from tobacco mosaic virus pests to enzymatically active assemblies. Beilstein J Nanotechnol 7:613–629
19. Geiger FC, Eber FJ, Eiben S, Mueller A, Jeske H, Spatz JP et al (2013) TMV nanorods with programmed longitudinal domains of differently addressable coat proteins. Nanoscale 5(9):3808–3816
20. Chen C, Xie Q, Yang D, Xiao H, Fu Y, Tan Y et al (2013) Recent advances in electrochemical glucose biosensors: a review. RSC Adv 3(14):4473–4491
21. Mao C, Liu A, Cao B (2009) Virus-based chemical and biological sensing. Angew Chem 48(37):6790–6810
22. Srinivasan K, Cular S, Bhethanabotla VR, Lee SY, Harris MT, Culver JN (2005) AIChE annual meeting, Cincinnati, USA, p 74d/1
23. Bruckman MA, Liu J, Koley G, Li Y, Benicewicz B, Niu Z et al (2010) Tobacco mosaic virus based thin film sensor for detection of volatile organic compounds. J Mater Chem 20(27):5715–5719
24. Zang F, Gerasopoulos K, Fan XZ, Brown AD, Culver JN, Ghodssi R (2014) An electrochemical sensor for selective TNT sensing based on Tobacco mosaic virus-like particle binding agents. Chem Commun 50(85):12977–12980
25. Fan XZ, Naves L, Siwak NP, Brown A, Culver J, Ghodssi R (2015) Integration of genetically modified virus-like-particles with an optical resonator for selective bio-detection. Nanotechnology 26(20):205501
26. Bäcker M, Delle L, Poghossian A, Biselli M, Zang W, Wagner P et al (2011) Electrochemical sensor array for bioprocess monitoring. Electrochim Acta 56(26):9673–9678
27. Gooding G Jr, Hebert T (1967) A simple technique for purification of tobacco mosaic virus in large quantities. Phytopathology 57(11):1285
28. Zaitlin M, Israel H (1975) Tobacco mosaic virus (type strain). CMI/AAB descriptions of plant viruses. No. 151. http://www.dpvweb.net/
29. Green MR, Sambrook J (2012) Molecular cloning: a laboratory manual, 4th edn. Cold Spring Harbor Laboratory Press, New York
30. Fraenkel-Conrat H (1957) Degradation of tobacco mosaic virus with acetic acid. Virology 4(1):1–4
31. Butler PJ (1984) The current picture of the structure and assembly of tobacco mosaic virus. J Gen Virol 65(Pt 2):253–279

Chapter 36

Integrated Methods to Manufacture Hydrogel Microparticles Containing Viral–Metal Nanocomplexes with High Catalytic Activity

Cuixian Yang, Eunae Kang, and Hyunmin Yi

Abstract

Controlled synthesis of small and catalytically active noble metal nanoparticles under mild aqueous conditions is an unmet challenge. Genetically modified tobacco mosaic virus (TMV) can serve as a preferential precursor adsorption and growth sites for the controlled synthesis of palladium (Pd) nanoparticles with high catalytic activity. Here we describe detailed methods for the synthesis of Pd-TMV nanocomplexes as well as their integration into polymeric hydrogel microparticle platforms with controlled dimensions via a simple replica molding process. Such Pd-TMV-containing hydrogel particles may be useful in environmental remediation of toxic chemicals such as carcinogenic dichromate ions.

Key words *Tobacco mosaic virus* (TMV), Palladium (Pd) nanoparticles, Nanocatalysts, Hydrogel microparticles, Polyethylene glycol (PEG), Dichromate reduction

1 Introduction

Palladium (Pd) is an essential noble metal for catalysis in a large range of reactions, from hydrogenation [1], oxidation [2] to C–C coupling reactions such as Heck and Suzuki Coupling [3–5]. Controlled synthesis of small, uniform, and catalytically active Pd nanoparticles under mild conditions is thus an important topic for many applications, including environmental cleanup [6], chemical synthesis [7], and hydrogen storage [8]. Due to their precisely controlled dimensions and structures, as well as the ability to install additional functionalities via genetic modification in a programmable manner, viral particles can serve as suitable supports for metal or metal oxide nanoparticle synthesis [5, 9]. In addition to icosahedral and filamentous viruses, the rigid rod-shaped tobacco mosaic virus (TMV) has been extensively utilized owing to its additional advantages such as a robust structure and high stability (e.g., pH 2–10, various organic solvents, temperature up to 90 °C) [10].

Christina Wege and George P. Lomonossoff (eds.), *Virus-Derived Nanoparticles for Advanced Technologies: Methods and Protocols*, Methods in Molecular Biology, vol. 1776, https://doi.org/10.1007/978-1-4939-7808-3_36,
© Springer Science+Business Media, LLC, part of Springer Nature 2018

Here we report detailed methods for the synthesis of small (~1.2 nm diameter) and catalytically active Pd nanoparticles on genetically modified TMV particles (TMV1cys) [12] under mild aqueous conditions. This is achieved by simply mixing the particles with Pd precursors (Na_2PdCl_4) and incubating at elevated temperature (50 °C), as shown in the schematic diagram of Fig. 1a [11]. The TMV1cys templates have one cysteine genetically substituted on the outer surface of each of over 2000 identical coat proteins [12]. These modified particles provide enhanced adsorption of Pd precursors that leads to a high local precursor concentration for nucleation and preferential Pd nanoparticle formation exclusively

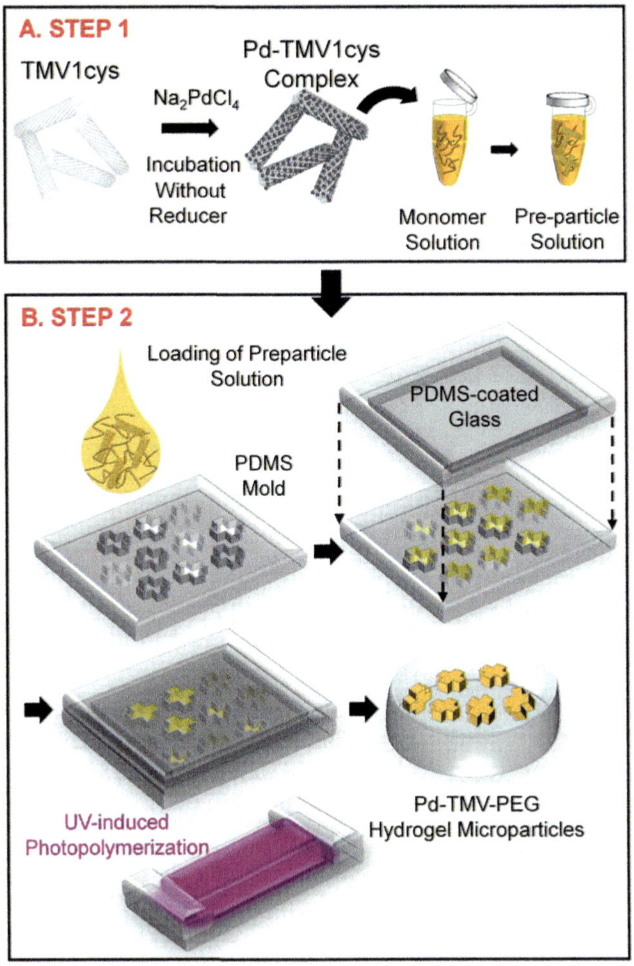

Fig. 1 Schematic diagram for the synthesis-fabrication methods of polymeric hydrogel microparticles containing viral-templated Pd nanocatalysts. (**a**) Step 1: Simple mixing and incubation of TMV1cys and Pd precursor leads to the exclusive formation of small Pd nanoparticles on the TMV1cys. (**b**) Step 2: Replica molding of the preparticle solution containing Pd-TMV complexes, PEGDA and photoinitiator Darocur 1173 via UV-induced photopolymerization yields uniform hydrogel microparticles containing Pd-TMV complexes with controlled shapes, high catalytic activity and stability. (Reproduced from [11] with permission from American Chemical Society)

on their tubular structures, a process aided by several electron-donating amino acids [11, 13]. We also report a simple method for embedding the resulting stable Pd-TMV nanocomplexes in a polymeric hydrogel microparticle format as shown in the schematic diagram of Fig. 1b. The complexes are mixed with polymerizable prepolymer polyethylene-glycol diacrylate (PEGDA) and a photoinitiator (Darocur 1173) and filled into microwells composed of polydimethyl siloxane (PDMS, silicone rubber) [11]. Irradiation with UV light emitted by a simple handheld lamp (365 nm, 8 W) triggers photoinduced radical polymerization, encapsulating the nanocomplexes in a stable, cross-linked hydrogel in a microparticle format of controlled 2D shape. These microparticles are characterized by high catalytic activity and stability as well as easy reuse, without any regeneration treatment, for dichromate (Cr(VI)) reduction. Therefore these particles provide an important Pd surface-catalyzed electron transfer reaction for environmental cleanup [11, 14].

2 Materials

All solutions are prepared in deionized water unless otherwise indicated.

2.1 Synthesis of Pd Nanoparticles on TMV Biotemplates

1. Genetically modified tobacco mosaic virus (TMV1cys). The first generation was from Professor James N. Culver's laboratory at the Institute for Bioscience and Biotechnology Research of University of Maryland Biotechnology Institute, USA. The TMV1cys utilized in this study was extracted from infected tobacco leaves using phosphate buffer, followed by chloroform phase separation, PEG8000 sedimentation and sucrose gradient for TMV purification as previously described [12] (*see* Chapter 37).
2. 100 mM sodium phosphate buffer (pH 7.0).
3. 20 mM sodium tetrachloropalladate (II) (Na_2PdCl_4) (*see* **Note 1**), prepare freshly.
4. Analog dry block heater.
5. Microfuge 22R centrifuge (Beckman Coulter).
6. Vortexer.
7. UV Spectrophotometer (optional).

2.2 Replica Molding

1. Polyethylene-glycol diacrylate (PEGDA, average M_n 700 Da, Sigma-Aldrich).
2. 2-Hydroxy-2-methylpropiophenone (photoinitiator, Sigma-Aldrich).
3. Polyethylene glycol (PEG, M.W. 200 Da; PEG200, Sigma-Aldrich).

4. Polydimethylsiloxane (PDMS) elastomer kit (Sylgard 184, Dow Corning Corporation).
5. Silicon master mold with the microwell patterns, fabricated via standard photolithography [11].
6. PDMS-coated glass slide (for covering PDMS mold upon UV cross-linking the Pd-TMV-hydrogel microparticles, *see* Fig. 1b): Prepare a thin film of Sylgard 184 mixture (10% w/w cross linker agent in base) on a cleaned 1″ × 1″ glass slide by spin-coating with 2000 rpm for 30 s. Cure overnight at 65 °C, and peel off the PDMS layer at the center area of the glass slide that will correspond to the microwell section in the PDMS mold, in order to leave a small gap between mold and glass coverage preventing the glass slide from touching the filled microwells of the PDMS mold.
7. Handheld UV lamp (8 W, 365 nm, e.g., Spectronics Corp., Westbury, NY, USA).
8. Aluminum mirror (e.g., Thorlabs, Newton, NJ, USA).
9. Humidity chamber (e.g., Thermo Fisher Scientific, Waltham, MA, USA).
10. Deionized water containing 0.5% (v/v) Tween 20 (TW20, Thermo Fisher Scientific).
11. Vacuum chamber.

2.3 Dichromate Reduction

1. 100 mM potassium dichromate ($K_2Cr_2O_7$) (purity: 99.5%).
2. 100 mM sodium formate (HCOONa) (purity: 99%).
3. Sulfuric acid (H_2SO_4) (purity: 99.9%).
4. Quartz cuvette (path length 1 cm).
5. Vortexer.
6. UV–Vis spectrophotometer (e.g., Evolution™ 300, Thermo Scientific).

3 Methods

3.1 Synthesis of Pd Nanoparticles on TMV1cys Templates

1. Dilute purified TMV stock solution to 1.2 mg/ml by using 100 mM sodium phosphate buffer (pH 7.0).
2. Add 100 μl of 1.2 mg/ml TMV in sodium phosphate buffer solution into 100 μl of 20 mM Na_2PdCl_4 aqueous solution to obtain 200 μl of a mixture containing 0.6 mg/ml TMV and 10 mM Na_2PdCl_4.
3. Incubate the mixture from **step 3** at 50 °C for 30 min in a dry heating block. The color of the mixture should change from yellowish brown to brown (*see* Fig. 2a), and the conversion can be tracked with UV-vis absorbance (*see* Fig. 2b, c).

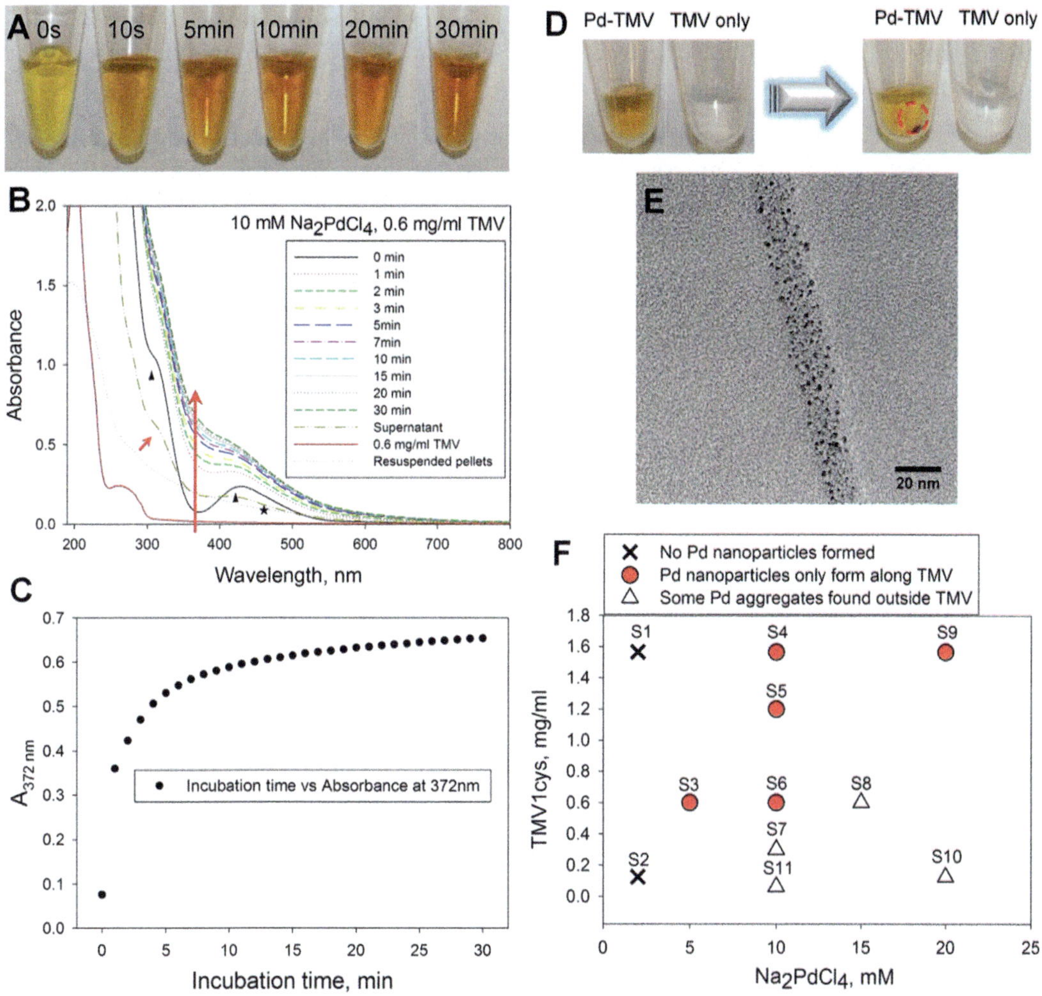

Fig. 2 Synthesis of Pd-TMV nanocomplexes. (**a**) The color of the mixture of TMV1cys and $NaPdCl_4$ turns dark over time upon incubation at 50 °C. (**b**) UV–Vis absorbance spectra increase over a broad wavelength range over time. (**c**) Absorbance at 372 nm rapidly increases and then plateaus upon incubation. (**d**) The as-synthesized Pd-TMV complexes are very stable, and can be readily separated by centrifugation and resuspension. (**e**) Transmission electron microscopy (TEM) can be utilized to examine the formation of TMV-templated Pd nanoparticles. (**f**) Phase diagram of TMV1cys *vs.* Na_2PdCl_4 concentrations showing synthesis conditions suitable for the exclusive formation of Pd nanoparticles on TMV. (Reproduced from [11] with permission from American Chemical Society)

4. After the incubation, centrifuge the solution at 9000 ×*g* for 5 min.
5. Discard the supernatant, and collect the brown pellets (Pd-TMV complex nanostructures) at the bottom of the Eppendorf tube (*see* Fig. 2d). Resuspend the pellets in 110 μl of deionized water.
6. Put the Pd-TMV solution on a vortexer for ~1 min to form a clear suspension for microparticle fabrication or TEM measurement.

3.2 Fabrication of Microparticles Containing Pd-TMV

The fabrication of the polymeric microparticles is micromolding-based. The microparticles contain Pd-TMV nanocomplexes from Subheading 3.1.

1. Prepare a polydimethylsiloxane (PDMS) mold containing 2080 microwells by using commercially available Sylgard® 184: mix base with 10% (w/w) crosslinking agent, pour the mixture on a silicon master mold (*see* **Note 2**) and remove air bubbles by incubating in a vacuum chamber followed by overnight curing at 65 °C.
2. Mix 110 μl of Pd-TMV solution (Subheading 3.1, **step 6**) with 30 μl of PEG200 to make solution **1**. Vortex the mixture vigorously until a clear suspension is formed (*see* **Note 3**).
3. To make solution **2**, mix 50 μl of PEGDA with 10 μl of photoinitiator. Vortex the mixture vigorously until a clear suspension is formed.
4. Add solution **1** into solution **2** to have 200 μl of preparticle solution with a TMV final concentration of 0.6 mg/ml (*see* **Note 3**). The resulting volume ratio of Pd-TMV solution–PEGDA–PEG 200–photoinitiator is 55:25:15:5.
5. To make preparticle solutions with TMV final concentrations of 0.3, 0.6, 1.2, 2.4, and 4.8 mg/ml, centrifuge varying volumes (100, 200, 400, 800, and 1600 μl) of the Pd-TMV solution (0.6 mg/ml TMV, 10 mM Na_2PdCl_4 from Subheading 3.1, **step 6**) at 9000 × *g* for 5 min. Resuspend the collected pellets in 110 μl of deionized water, mixed with 30 μl PEG200, then add to solution **2** to obtain 200 μl total volume of preparticle solutions (*see* Fig. 3b).
6. Fill the microwells (2080 wells per 2 cm × 2 cm mold) with 200 μl of the preparticle solution by scratching the solution on the PDMS mold with a disposable pipet tip (*see* **Note 4**).
7. Remove excess preparticle solution on the mold by suctioning up with a pipet, and seal the filled mold with a PDMS-coated glass slide except for the microwell area (2 cm × 2 cm) (again, *see* **Note 4**).
8. Place the sealed micromolds on an aluminum mirror and expose to 365 nm UV light with an 8 W handheld UV lamp for 15 min (*see* **Note 5**).
9. The polymerized PEG microparticles with encapsulated Pd-TMV complexes (Pd-TMV-PEG microparticles) are collected from the microwells by first physically bending and squeezing the mold, then placing water containing 0.5% (v/v) Tween 20 on the mold surface.
10. The microparticles are collected by pipetting up and down the Tween 20/water solution on the microwells a few times before transferring into a storage vial.

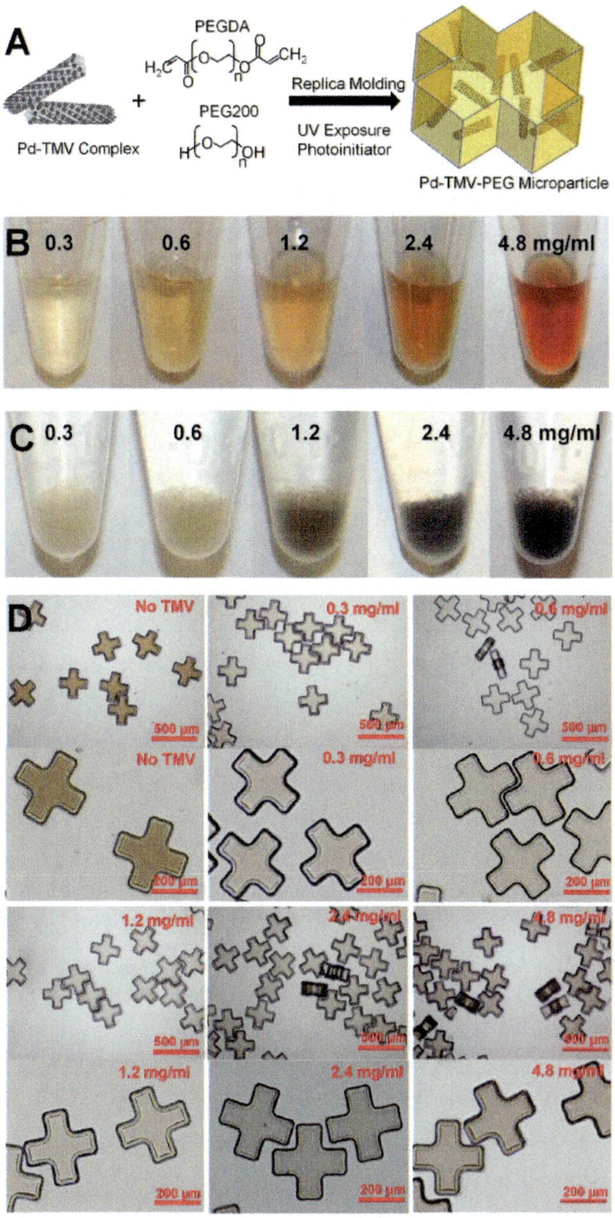

Fig. 3 Fabrication of PEG-based hydrogel microparticles containing Pd-TMV complexes. (**a**) Schematic diagram shows that UV-induced polymerization of prepolymer solutions prepared by simple mixing of Pd-TMV complexes with PEGDA, PEG 200 (inert short chain porogen) and photoinitiator leads to the formation of the hydrogel microparticles with controlled shapes. (**b**) Preparticle solutions containing various amounts of Pd-TMV complexes. (**c**) Microparticles containing higher Pd-TMV content show darker color. (**d**) Brightfield micrographs of the as-prepared microparticles show reliable fabrication with varying Pd-TMV contents. (Reproduced from [11] with permission from American Chemical Society)

11. This particle removal with Tween 20–water solution is repeated several times for the complete collection of the microparticles (*see* Fig. 3c) which should show uniform and consistent shapes and color (*see* Fig. 3d).
12. The prepared Pd-TMV–PEG microparticles are stored in 0.5% (v/v) Tween 20 solution at room temperature until catalytic reaction studies. Up to 2 week storage under this condition has shown to not affect the catalytic activity negatively [11].

3.3 Dichromate Reduction Reaction to Examine Catalytic Activity

This reaction is performed to examine the catalytic activity of the Pd-TMV complexes in the polymer microparticles.

1. Add 20 μl of 100 mM potassium dichromate into 20 ml of 100 mM sodium formate to obtain 20 ml reaction solution with final concentrations of 0.1 mM potassium dichromate and 100 mM sodium formate.
2. Adjust the reaction solution to pH 3.0 with H_2SO_4 solution.

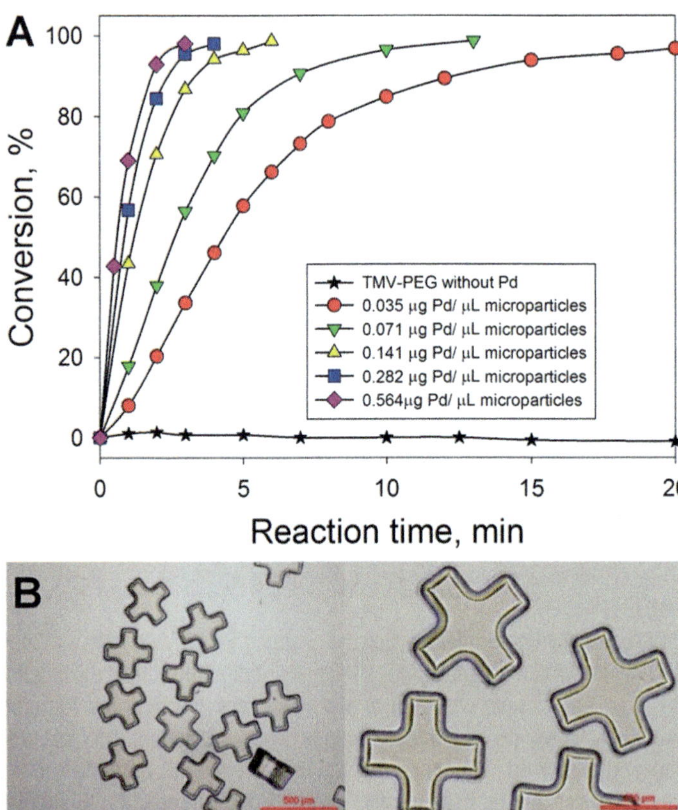

Fig. 4 Dichromate reduction reaction results. (**a**) Conversion plot from changes in UV-vis absorbance at 350 nm, dichromate ion's characteristic absorbance peak. Higher Pd-TMV content leads to more rapid conversion. (**b**) Brightfield micrographs of the microparticles after five consecutive batch reactions show high stability. (Reproduced from [11] with permission from American Chemical Society)

3. Add Pd-TMV-PEG microparticles (from Subheading 3.2, **step 12**) to 1 ml of reaction solution prepared as described in **steps 1** and **2** in a 2 ml microcentrifuge tube.
4. Shake the tube vigorously using a vortexer to carry out the reaction.
5. Sample the supernatant periodically and measure the reaction progress by UV/vis spectrophotometry every minute for 20 min between 200 and 600 nm wavelength (*see* Fig. 4a for representative conversion, measured by UV absorbance at 350 nm). The microparticles are stable upon five repeated reaction cycles of 20 min each without any reactivation treatment (*see* Fig. 4b).

4 Notes

1. The Pd precursor power (Na_2PdCl_4) should be protected from air and moisture to avoid oxidation during storage.
2. The silicon master mold is prepared by following standard photolithography procedures [15].
3. Murky, nontransparent appearance of the mixture of solutions **1** and **2** indicates aggregation of Pd-TMV complexes. Repeat vortexing and sonication for uniform dispersion and mixing.
4. **Step 6** and 7 must be carried out in a humidity chamber with relative humidity above 90% to prevent rapid evaporation of the preparticle solution in the microwells for uniform particle and mesh sizes [16].
5. For the safety, direct exposure of the UV light to eyes and skin should be avoided.

References

1. Stanislaus A, Cooper BH (1994) Aromatic hydrogenation catalysis: a review. Catal Rev 36(1):75–123
2. Stahl SS (2005) Palladium-catalyzed oxidation of organic chemicals with O2. Science 309(5742):1824–1826
3. Phan NTS, Van Der Sluys M, Jones CW (2006) On the nature of the active species in palladium catalyzed Mizoroki–Heck and Suzuki–Miyaura Couplings—homogeneous or heterogeneous catalysis, a critical review. Adv Synth Catal 348(6):609–679
4. Miyaura N, Suzuki A (1995) Palladium-catalyzed cross-coupling reactions of organoboron compounds. Chem Rev 95(7):2457–2483
5. Yang C, Yi H (2015) Viral templated palladium nanocatalysts. ChemCatChem 7(14):2015–2024
6. Yang C, Manocchi AK, Lee B, Yi H (2010) Viral templated palladium nanocatalysts for dichromate reduction. Appl Catal B Environ 93(3–4):282–291
7. Yang C, Manocchi AK, Lee B, Yi H (2011) Viral-templated palladium nanocatalysts for Suzuki coupling reaction. J Mater Chem 21(1):187–194
8. Yamauchi M, Kobayashi H, Kitagawa H (2009) Hydrogen storage mediated by Pd and Pt nanoparticles. ChemPhysChem 10(15):2566–2576
9. Thakkar KN, Mhatre SS, Parikh RY (2010) Biological synthesis of metallic nanoparticles. Nanomedicine 6(2):257–262. https://doi.org/10.1016/j.nano.2009.07.002
10. Young M, Willits D, Uchida M, Douglas T (2008) Plant viruses as biotemplates for materials and their use in nanotechnology. Annu Rev Phytopathol 46:361–384

11. Yang C, Choi C-H, Lee C-S, Yi H (2013) A facile synthesis–fabrication strategy for integration of catalytically active viral-palladium nanostructures into polymeric hydrogel microparticles via replica molding. ACS Nano 7(6):5032–5044
12. Yi H, Nisar S, Lee S-Y, Powers MA, Bentley WE, Payne GF, Ghodssi R, Rubloff GW, Harris MT, Culver JN (2005) Patterned assembly of genetically modified viral nanotemplates via nucleic acid hybridization. Nano Lett 5(10):1931–1936
13. Manocchi AK, Seifert S, Lee B, Yi H (2011) In situ small-angle X-ray scattering analysis of palladium nanoparticle growth on tobacco mosaic virus nanotemplates. Langmuir 27(11):7052–7058
14. Yang C, Meldon JH, Lee B, Yi H (2014) Investigation on the catalytic reduction kinetics of hexavalent chromium by viral-templated palladium nanocatalysts. Catal Today 233:108–116
15. Qin D, Xia Y, Whitesides GM (2010) Soft lithography for micro- and nanoscale patterning. Nat Protoc 5(3):491–502
16. Lewis CL, Choi CH, Lin Y, Lee CS, Yi H (2010) Fabrication of uniform DNA-conjugated hydrogel microparticles *via* replica molding for facile nucleic acid hybridization assays. Anal Chem 82(13):5851–5858

Chapter 37

Integrated Methods to Manufacture Hydrogel Microparticles with High Protein Conjugation Capacity and Binding Kinetics via Viral Nanotemplate Display

Sukwon Jung, Christina L. Lewis, and Hyunmin Yi

Abstract

Genetically modified tobacco mosaic virus (TMV) can serve as a potent nanotemplate for high capacity protein conjugation through covalent coupling to its coat proteins with precise nanoscale spacing. TMV's own genomic RNA can also be exploited for orientationally controlled assembly onto various platforms with sequence and spatial selectivity via nucleic acid hybridization. Here we describe detailed methods for fabrication of hydrogel microparticles with capture DNA sequences, chemical activation and programming of TMV templates, TMV assembly with the microparticles and protein conjugation via bio-orthogonal click reactions.

Key words Tobacco mosaic virus (TMV), Nucleic acid hybridization, Bio-orthogonal click reactions, Protein conjugation, Hydrogel microparticles, Chitosan

1 Introduction

Polymeric hydrogel microparticle suspension arrays offer attractive alternatives to planar substrates for various biosensing applications due to enhanced probe capacity in 3D networks, rapid solution-like kinetics, a hydrophilic environment for favorable probe–target interactions, and small sample volumes [1–5]. However, several hurdles remain with hydrogel microparticle fabrication strategies for reliable and rapid protein sensing. These challenges include limited mass transfer through polymeric networks with small mesh sizes [6, 7], potential oxidative damage to probe biomolecules during copolymerization procedures [8], and limited enhancement of probe capacity [4, 9]. Tobacco mosaic virus (TMV) is a biological nanotube with exquisitely controlled dimensions of 300 nm length and 18 nm diameter that can serve as potent template for high capacity protein conjugation through covalent conjugation on over 2000 precisely spaced identical coat proteins (CP)

Christina Wege and George P. Lomonossoff (eds.), *Virus-Derived Nanoparticles for Advanced Technologies: Methods and Protocols*, Methods in Molecular Biology, vol. 1776, https://doi.org/10.1007/978-1-4939-7808-3_37,
© Springer Science+Business Media, LLC, part of Springer Nature 2018

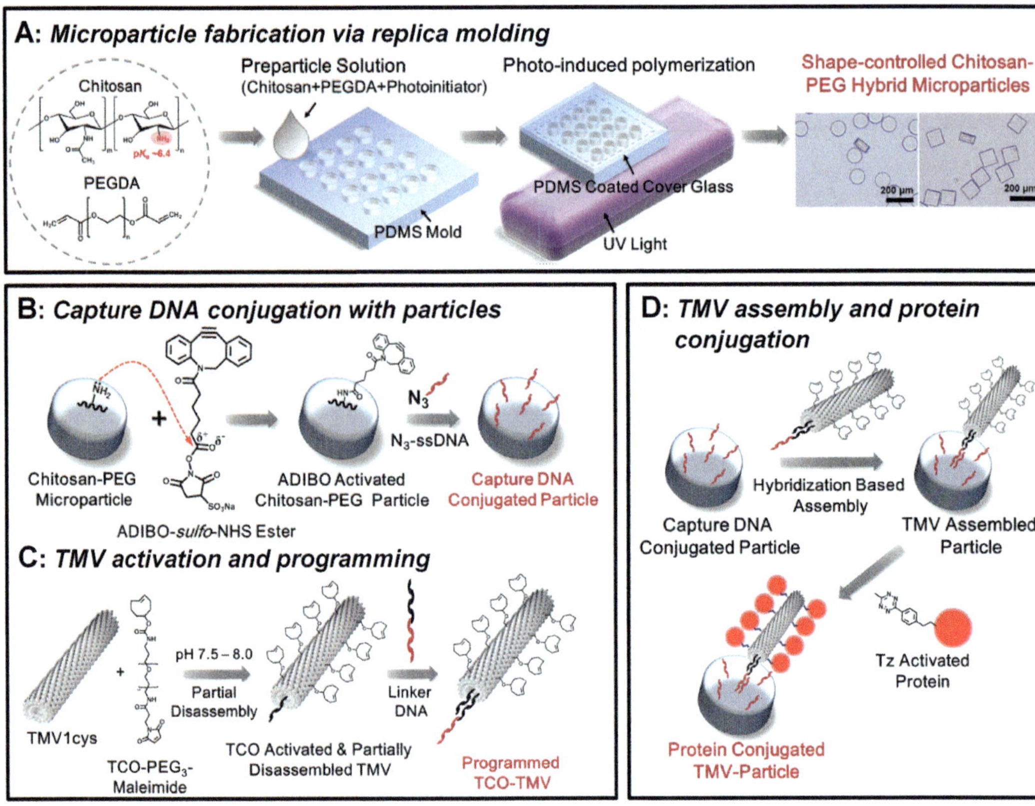

Fig. 1 Scheme for the fabrication of TMV-assembled microparticles. (Reproduced from [9] and [12] with permission from American Chemical Society)

[10]. Nucleic acid hybridization of TMV templates via its genomic RNA (6.4 kb), with a 5′-terminal portion of the viral omega sequence exposed through partial disassembly, permits programmable and orientationally controlled assembly to capture the TMV particles on single-stranded (ss) DNA-functionalized sites [10, 11].

Here we describe detailed methods to utilize genetically modified TMV templates (TMV1cys; one cysteine displayed on the outer edge of each CP by genetic exchange at position 1) for high capacity protein conjugation via high yield bio-orthogonal click reactions [9]. First, poly(ethylene glycol) (PEG)-based polymeric hydrogel microparticles with controlled shapes containing capture DNAs are fabricated from PEG diacrylate (PEGDA) by simple replica molding (*see* Fig. 1a) via copolymerization with chitosan. Chitosan is an aminopolysaccharide providing abundant primary amine groups with low pK_a (~6.4) [10] for activation with amine-reactive *N*-hydroxysuccinimidyl (NHS) ester form of azadibenzocyclooctyne (ADIBO), which then reacts with azide-modified ssDNA to form stable triazole linkages in the strain-promoted alkyne–azide cyclization (SPAAC) reaction scheme (*see* Fig. 1b)

Table 1
Single-stranded DNA sequences for TMV assembly [9]

	Sequence
Capture DNA[a]	5′-ATGATGATGATGATGATG-3′
Linker DNA[b]	5′-GTTTGTTGTTGTTGGTAATTGTTG *TTTTTT* CATCATCATCATCATCAT-3′

[a]Capture DNA sequences are all modified with azide at the 5′-end
[b]TMV 5′-end-complementary sequence–*Spacer*–Capture DNA-complementary sequence

[12]. Second, the TMV templates are activated with thiol-reactive maleimide forms of ADIBO for a SPAAC reaction or trans-cyclooctene (TCO) for a rapid tetrazine (Tz)–TCO ligation reaction [13]. These activated TMV templates are partially disassembled to expose the 5′-end of their genomic RNA via ultracentrifugation at high pH (7.5–8) before they can be programmed with linker DNAs via RNA–DNA hybridization (*see* Table 1 and Fig. 1c). These activated and programmed TMV templates are then assembled on the capture DNA-conjugated microparticles via hybridization, then exposed to proteins that are preactivated with the counterparts for covalent conjugation via SPAAC (azide) or Tz–TCO (Tetrazine) reactions (*see* Fig. 1d). The TMV–microparticles prepared as described show substantially improved protein conjugation capacity (e.g., ~2400-fold over planar substrates) and kinetics due to the absence of slow mass transfer of proteins through polymer networks [9].

2 Materials

Prepare all buffer solutions at room temperature using deionized water unless indicated otherwise. Store all the buffer solutions at room temperature after filtration using a disposable sterile filter with 0.22 μm pore size. Prepare and store all reagents using analytical grade chemicals without further purification at room temperature, unless indicated otherwise. Follow waste disposal regulations when disposing of waste materials.

2.1 Microparticle Fabrication via Replica Molding

1. Silicon master mold with 1600 microwells (~100 μm diameter, 100 pl volume each).
2. Vacuum chamber.
3. 2% (w/v) chitosan oligosaccharide lactate (average M_n 5 kDa, >90% deacetylation, Sigma-Aldrich) (*see* **Note 1**).
4. Polyethylene glycol diacrylate (PEGDA) (average M_n 700 Da, Sigma-Aldrich) (*see* **Note 2**).

5. 2-hydroxy-2-methylpropiophenone (photoinitiator, Sigma-Aldrich) (*see* **Note 2**).
6. Polydimethylsiloxane (PDMS) elastomer kit (Sylgard 184, Thermo Fisher Scientific).
7. PDMS-coated glass slide (for covering PDMS mold upon UV cross-linking chitosan-PEG-hybrid hydrogel microparticles, *see* Fig. 1a): Prepare a thin film of Sylgard 184 mixture (10% w/w cross-linker agent in base, i.e., non-cross-linked PDMS) on a cleaned 1″ × 1″ glass slide (*see* **Note 3**) by spin coating with 2000 rpm for 30 s (*see* **Note 4**). Cure overnight at 65 °C, and peel off the PDMS layer at the center area of the glass slide that will correspond to the microwell section (0.7 cm × 0.7 cm) in the PDMS mold in order to leave a small gap between PDMS mold and glass coverage, preventing the glass slide from touching the filled microwells of the PDMS mold.
8. Handheld UV lamp (8 W, 365 nm, e.g., Spectronics Corp.).
9. Aluminum mirror (e.g., Thorlabs).
10. Humidity chamber with humidity above 90% (*see* **Note 5**).
11. Deionized water containing 0.5% (v/v) Tween 20 (*see* **Note 6**).
12. 5× saline sodium citrate (SSC) buffer solution containing 0.05% (v/v) Tween 20: Dilute 20× SSC (175.3 g/l NaCl, 88.2 g/l trisodium citrate (NaCit); e.g., molecular biology grade 20× SSC from Sigma-Aldrich) with water, and add Tween 20.

2.2 Capture DNA Conjugation to Microparticles via a SPAAC Reaction

Capture DNA conjugation to CS–PEG Microparticles is achieved via a Strain-promoted Alkyne–Azide Cycloaddition (SPAAC) Reaction.

1. 100 μM azide-terminated capture DNA (*see* Table 1) (e.g., from Integrated DNA Technologies, Coralville, IA) in TE buffer solution. Store the DNA solution at −20 °C (*see* **Note 7**).
2. Azadibenzocyclooctyne (ADIBO)–*sulfo*–NHS ester (Click Chemistry Tools): 10 mg/ml in extra dry dimethyl sulfoxide (DMSO, Thermo Fisher Scientific). Store the solution at −20 °C (*see* **Note 8**).
3. 5× SSC buffer solution containing 0.05% (v/v) Tween 20 (*see* Subheading 2.1, **item 10**).

2.3 TMV Activation and Programming

1. Tris–EDTA (TE) buffer: Mix 10 ml of 1 M Tris–HCl pH 8.0 and 2 ml of 0.5 M EDTA pH 8.0. Add water to make a 1 l solution.
2. Genetically modified TMV1cys [10, 14]: Extract and purify TMV1cys from infected tobacco leaves by following purification procedures described in ref. 14.

3. 0.1 M sodium phosphate buffer pH 7.0: Prepare 1 M of NaH_2PO_4 (monobasic) and Na_2HPO_4 (dibasic) by dissolving 138 g of $NaH_2PO_4 \cdot H_2O$ (M.W. 138) and 142 g of Na_2HPO_4 (M.W. 142) to make a final volume of 1 l, respectively. Mix 42.3 ml of 1 M NaH_2PO_4 and 57.7 ml of 1 M Na_2HPO_4 and dilute this mixture to 1 l with deionized water.
4. 0.1 M Tris buffer pH 7.5: Open a foil pouch of Trizma® Preset crystals (Sigma-Aldrich) and dissolve contents in deionized water to make a final volume of 1 l.
5. 10–40% (w/v) sucrose gradient: Dissolve 25 g of sucrose (molecular biology grade) in 0.1 M Tris buffer pH 7.5 to make a final volume of 100 ml. Put the solution into 5 ml centrifuge tubes and freeze at −20 °C. The sucrose gradient is formed when the frozen solution is thawed (*see* **Note 9**).
6. Ultracentrifuge, rotors (SW-55Ti for sucrose gradient, 70-Ti for pelleting), tubes.
7. Syringe (to remove the TMV band after ultracentrifugation).
8. TCO–PEG_3–Maleimide or ADIBO–maleimide (Click Chemistry Tools): 10 mg/ml in extra dry dimethyl sulfoxide (DMSO, Thermo Fisher Scientific). Store the solutions at −20 °C.
9. Diethylpyrocarbonate (DEPC)-treated water (i.e., RNase-free water): Mix 1 ml of DEPC and 1 l of deionized water. Autoclave the mixture for 20 min, on liquid cycle, to sterilize and deactivate the DEPC.
10. 5× SSC buffer: Dilute 20× SSC buffer solution with DEPC-treated water.
11. 5× SSC buffer solution containing 0.05% (v/v) Tween 20 (*see* Subheading 2.1, **item 10**).
12. 100 μM Linker DNA (*see* Table 1) in TE buffer solution (*see* Subheading 2.2, **item 1**). Store the DNA solution at −20 °C (*see* **Note 7**).
13. Centrifugal filter unit (Amicon Ultra 0.5, 100 kDa cutoff).

2.4 Protein Assembly with Microparticles

The assembly is performed via nucleic acid hybridization.

5× SSC buffer solution containing 0.05% (v/v) Tween 20 (*see* Subheading 2.1, **item 10**).

2.5 Protein Activation and Conjugation with TMV

1. Protein to be conjugated (e.g., antibody or R-phycoerythrin (R-PE) from Anaspec).
2. Borate-buffered saline: 0.3 M NaCl, 50 mM borate, pH 8.5 (50 mM borate, pH 8.5, is prepared from Thermo Scientific™ Pierce™ 20× concentrate, Thermo Fisher Scientific).

3. Phosphate-buffered saline (PBS): 0.137 M sodium chloride, 0.01 M phosphate, 0.0027 M potassium chloride, pH 7.4 (open a PBS packet (Sigma-Aldrich) and dissolve the contents to make 1 l solution).
4. Tetrazine (Tz)–PEG_5–NHS ester (Click Chemistry Tools) or NHS–PEG_{12}–Azide (Thermo Fisher Scientific): 10 mg/ml in DMSO. Store the solutions at −20 °C (*see* **Note 8**).
5. Centrifugal filter unit (Amicon Ultra 0.5, 100 kDa cutoff).
6. Spectrophotometer.
7. 5× SSC buffer solution containing 0.05% (v/v) Tween 20 (*see* Subheading 2.1, **item 10**).

3 Methods

Carry out all procedures at room temperature unless otherwise specified.

3.1 Chitosan-PEG (CS-PEG) Microparticle Fabrication via Replica Molding

1. Prepare a polydimethylsiloxane (PDMS) mold containing 1600 microwells by using commercially available Sylgard® 184: mix base with 10% (w/w) cross-linking agent, pour the mixture on a silicon master mold (*see* **Note 10**) and remove air bubbles by incubating in a vacuum chamber followed by overnight heat treatment at 65 °C.
2. Prepare preparticle solution by simply mixing chitosan solution, PEGDA, photoinitiator and deionized water (*see* **Note 11**) to 0.8% (w/v) chitosan, 40% (v/v) PEGDA, 2% (v/v) photoinitiator (final concentrations).

   The following **steps 3** and **4** must be carried out in a humidity chamber with relative humidity above 90% to prevent rapid evaporation of the preparticle solution in the microwells for uniform particle and mesh sizes [4].
3. Fill the microwells of the mold with the preparticle solution by scratching it on the mold with a disposable pipet tip (*see* Fig. 1a).
4. Remove excess preparticle solution on the mold by suctioning up with a pipet, and seal the filled mold with a glass slide coated with PDMS except for the area covering the microwell area (0.7 cm × 0.7 cm) (*see* Subheading 2.1, **item 7**).
5. Place the sealed mold on an aluminum mirror outside the humidity chamber, and expose to 365 nm UV light with an 8 W handheld UV lamp to the sealed mold for 15 min to cross-link the preparticle solution and thus to form chitosan–PEG microparticles (CS–PEG particles).

6. Place deionized water containing 0.5% (v/v) Tween 20 on top of the mold and wait for less than 1 min to hydrate and swell microparticles (*see* **Note 12**).
7. Upon removal of the water by pipetting, bend the mold carefully to remove microparticles from the mold.
8. Place deionized water containing 0.5% (v/v) Tween 20 on top of the mold again and pipet the microparticles into a 1.5 ml microcentrifuge tube.
9. Rinse the particles with washing procedures at least 5 times: mix the particles with 200 μl of 5× SSC containing 0.05% (v/v) Tween 20 by pipetting in a microcentrifuge tube, then exchange the buffer solution after the particles settle on the tube bottom.
10. Store the particles in 5× SSC at room temperature or 4 °C, at which condition the particles are stable for 6 months at minimum.

3.2 Capture DNA Conjugation to CS–PEG Microparticles via SPAAC Reaction

Capture DNA conjugation to CS–PEG microparticles is achieved via Strain-Promoted Alkyne–Azide Cycloaddition (SPAAC) reaction.

1. Activate the as-prepared CS–PEG particles with ADIBO molecules by incubating with 500 μM ADIBO–*sulfo*–NHS ester in 5× SSC containing 0.05% (v/v) Tween 20 for 1 h at room temperature on a nutating mixer (*see* Fig. 1b).
2. Rinse the ADIBO-activated particles with 5× SSC containing 0.05% (v/v) Tween 20 by following the washing procedures at least five times (*see* Subheading 3.1, **step 10**).
3. Incubate the ADIBO-activated particles with 10 μM azide-terminated capture DNA in 5× SSC containing 0.05% (v/v) Tween 20 for 24 h at room temperature on a nutating mixer.
4. Rinse the capture DNA-conjugated particles with 5× SSC containing 0.05% (v/v) Tween 20 by following the washing procedures at least 5 times (*see* Subheading 3.1, **step 10**).
5. Store the particles in 5× SSC at room temperature or 4 °C.

3.3 TMV TCO Activation and DNA-Linker-Mediated Programming

1. Incubate 0.1 ml of 3 mg/ml genetically modified TMV rods (TMV1cys, 0.3 mg total) with 10-times molar excess of TCO–PEG3–maleimide over the cysteines displayed on the TMV surfaces in 0.1 M sodium phosphate buffer (pH 7.0) for 2 h at room temperature on a nutating mixer (*see* Fig. 1c).
2. Purify and partially disassemble the TCO-activated TMVs by ultracentrifugation in a 10–40% (w/v) sucrose gradient in 0.1 M Tris buffer pH 7.5 at 4 °C at 48,000 × *g* for 2 h in order to expose TMVs' 5′-end RNA sequence. A white band of partially disassembled TMV with a protruding 5′-RNA portion

will be present in the middle of the centrifuge tube upon ultracentrifugation.

3. Remove the white band of TMV with a syringe and mix with 5× SSC buffer solution prepared with DEPC treated water (*see* **Note 13**).
4. Pellet the TMVs with ultracentrifugation at 90,000 × *g* for 1 h, and resuspend the pelleted TMVs in the 5× SSC buffer solution prepared with DEPC-treated water.
5. Incubate the TMV solution with 10-times molar excess of linker DNA overnight at room temperature on a nutating mixer in order to "program" them for assembly with the capture DNA-conjugated particles via nucleic acid hybridization (*see* **Note 14**).
6. Separate the excess linker DNAs from the TMV solution via centrifugal filtration (Amicon Ultra, 14,000 × *g* for 20 min) at 4 °C with 5× SSC containing 0.05% (v/v) Tween 20.

3.4 TMV Assembly with Microparticles via Nucleic Acid Hybridization

The assembly is performed via nucleic acid hybridization between the free portion of the DNA linkers partially bound to the TMV genomic RNA, and the capture DNA conjugated to CS–PEG microparticles.

1. Incubate the chemically modified and programmed TMV rods with the capture DNA-conjugated particles in 5× SSC containing 0.05% (v/v) Tween 20 overnight at 30 °C (*see* **Note 15**) on a nutating mixer (*see* Fig. 1d).
2. Gently rinse the particles according to the wash procedure (*see* Subheading 3.1, **step 10**).
3. The density of the assembled TMV can be modulated by varying the concentration of TMV rods (*see* Fig. 2a).

3.5 Protein Activation with Tetrazine and Conjugation with TMV-Assembled Particles

1. Exchange buffer solution of a 500 μl aliquot of 1 mg/ml R-PE solution, *see* **Note 16**) for borate-buffered saline via centrifugal filtration (Amicon Ultra, 14,000 × *g* for 15 min) at 4 °C to 250 μl of 2 mg/ml final concentration (*see* **Note 17**).
2. Incubate 250 μl of 2 mg/ml R-PE with a 20-fold molar excess of Tz–PEG_5–NHS ester or NHS–PEG_{12}–Azide for 30 min at room temperature on a nutating mixer (*see* **Note 18**).
3. Separate unreacted chemicals from the R-PE solution via centrifugal filtration (Amicon Ultra, 14,000 × *g* for 15 min) with PBS (pH 7.4).
4. Measure the concentration of the final R-PE solution via UV–Vis spectroscopy with characteristic absorbance peaks and molar extinction coefficient of the R-PE (1.96×10^4 1/(M cm) at 565 nm).

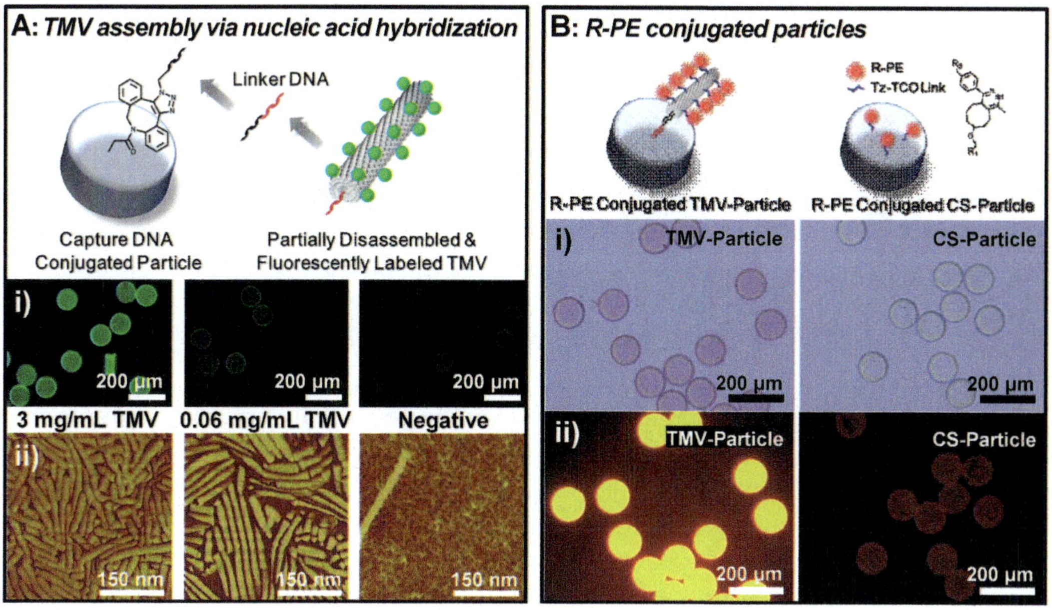

Fig. 2 Fluorescence and atomic force microscopy (AFM) results. (**a**) (i) Epifluorescence images of microparticles with fluorescently labeled TMV assembled on the microparticle surfaces using varying TMV concentrations. (ii) AFM images of the TMV-assembled microparticle surfaces showing densely packed TMV rods. (**b**) (i) Brightfield and (ii) epifluorescence micrographs of TMV-assembled and CS-microparticles after R-PE conjugation. (Reproduced from [9] and [12] with permission from American Chemical Society)

5. Incubate the previously prepared TCO (or ADIBO)-activated TMV particles with 2 μM Tz (or azide)-activated R-PEs in 5× SSC buffer solution containing 0.05% (v/v) Tween 20 for 2 h at room temperature on a nutating mixer (*see* Fig. 1d).
6. Gently rinse the microparticles according to the wash procedure (*see* Subheading 3.1, **step 10**) in order to remove the unconjugated R-PEs from the particle solution (*see* **Note 19**).
7. Protein conjugation capacity is significantly improved by surface-displayed TMV rods on the microparticles (roughly 53-fold compared to the microparticles without TMV, *see* Fig. 2b).

4 Notes

1. A small amount of chitosan oligosaccharide lactate can be dissolved in water without pH adjustment unlike long chain chitosan. Incubation in an ultrasonic bath for 30 min is required to fully dissolve the 2% (w/v) chitosan oligosaccharide lactate in deionized water.
2. Light sensitive materials. Store these materials in the dark for consistent results.

3. Cut a plain microscope glass slide in three pieces (1″ × 1″) using a glass cutter, wash the glass sections with acetone, 2-propanol, and deionized water sequentially. Dry them by blowing clean nitrogen gas followed by incubation at 90 °C in a convection oven.
4. Degassing of Sylgard 184 mixture is required before spin coating.
5. The humidity chamber is constructed with a commercial humidifier and a clear acrylic box.
6. Surfactant Tween 20 prevents the particles from sticking to the pipette tips and tube walls.
7. To minimize DNA damage by repeated freezing and thawing cycles, aliquot the DNA solution for two to four times use.
8. Handle the chemicals containing NHS ester functional group, which is moisture-sensitive, carefully. Reactivity of NHS ester toward amines is dramatically reduced by hydrolysis when exposed to moisture. Equilibrate vials containing the lyophilized chemicals at room temperature to prevent moisture condensation, and immediately add extra dry DMSO. We recommend preparing frozen aliquots of the stock solution in order to minimize their freeze–thaw cycles to two to four times.
9. Do not allow the tubes to touch one another during thawing in order to prevent disruption of the sucrose gradient.
10. The silicon master mold is prepared by standard photolithography procedures [15].
11. Chitosan solution and PEGDA are both viscous. To prepare a homogeneous mixture, incubate in an ultrasonic bath for 15 min followed by vigorous vortex mixing (<15 s).
12. The hydration step allows the microparticles to be removed from the mold by physical bending.
13. Carefully handle the partially disassembled TMVs with RNase-free solutions. Degradation of the exposed TMV's 5′-end RNA will lead to inconsistent TMV assembly results.
14. Upon hybridization, the programmed TMVs are stable.
15. We have found that optimal TMV assembly occurs at a hybridization temperature of 30 °C.
16. Other proteins such as antibodies can be conjugated using the procedure described in Subheading 3.5.
17. The amidation reaction with the primary amines of lysines on proteins is favored at alkaline conditions due to lysine's high pK_a value (10.5) [16].

18. High concentration of the reagents is required since hydrolysis of NHS ester is promoted at alkaline conditions.
19. Save the unconjugated proteins as these can be reused upon purification via centrifugal filtration.

Acknowledgment

We extend our special gratitude to the collaborators Professor Chang-Soo Lee at Chungnam National University and Dr. Chang-Hyung Choi at Harvard University, both of who continue to provide much inspiration and support.

References

1. Appleyard DC, Chapin SC, Doyle PS (2011) Multiplexed protein quantification with barcoded hydrogel microparticles. Anal Chem 83(1):193–199
2. Choi NW, Kim J, Chapin SC et al (2012) Multiplexed detection of mRNA using porosity-tuned hydrogel microparticles. Anal Chem 84(21):9370–9378
3. Le Goff GC, Srinivas RL, Hill WA et al (2015) Hydrogel microparticles for biosensing. Eur Polym J 72:386–412
4. Lewis CL, Choi CH, Lin Y et al (2010) Fabrication of uniform DNA-conjugated hydrogel microparticles via replica molding for facile nucleic acid hybridization assays. Anal Chem 82(13):5851–5858
5. Pregibon DC, Doyle PS (2009) Optimization of encoded hydrogel particles for nucleic acid quantification. Anal Chem 81(12):4873–4881
6. Jung S, Yi H (2013) Facile strategy for protein conjugation with chitosan-poly(ethylene glycol) hybrid microparticle platforms via strain-promoted alkyne-azide cycloaddition (SPAAC) reaction. Biomacromolecules 14(11):3892–3902
7. Lee AG, Arena CP, Beebe DJ et al (2010) Development of macroporous poly(ethylene glycol) hydrogel arrays within microfluidic channels. Biomacromolecules 11(12):3316–3324
8. Floyd RA, Carney JM (1992) Free radical damage to protein and DNA: mechanisms involved and relevant observations on brain undergoing oxidative stress. Ann Neurol 32(Suppl):S22–S27
9. Jung S, Yi H (2014) An integrated approach for enhanced protein conjugation and capture with viral nanotemplates and hydrogel microparticle platforms via rapid bioorthogonal reactions. Langmuir 30(26):7762–7770
10. Yi H, Nisar S, Lee SY et al (2005) Patterned assembly of genetically modified viral nanotemplates via nucleic acid hybridization. Nano Lett 5(10):1931–1936
11. Yi H, Rubloff GW, Culver JN (2007) TMV microarrays: hybridization-based assembly of DNA-programmed viral nanotemplates. Langmuir 23(5):2663–2667
12. Jung S, Yi H (2012) Fabrication of chitosan-poly(ethylene glycol) hybrid hydrogel microparticles via replica molding and its application toward facile conjugation of biomolecules. Langmuir 28(49):17061–17070
13. Karver MR, Weissleder R, Hilderbrand SA (2011) Synthesis and evaluation of a series of 1,2,4,5-tetrazines for bioorthogonal conjugation. Bioconjug Chem 22(11):2263–2270
14. Gooding GV Jr, Hebert TT (1967) A simple technique for purification of tobacco mosaic virus in large quantities. Phytopathology 57(11):1285
15. Qin D, Xia Y, Whitesides GM (2010) Soft lithography for micro- and nanoscale patterning. Nat Protoc 5(3):491–502
16. Nelson DL, Lehninger AL, Cox MM (2008) Lehninger principles of biochemistry. W. H. Freeman, New York

Chapter 38

Interactions Between Plant Viral Nanoparticles (VNPs) and Blood Plasma Proteins, and Their Impact on the VNP In Vivo Fates

Andrzej S. Pitek, Frank A. Veliz, Slater A. Jameson, and Nicole F. Steinmetz

Abstract

Plant viral nanoparticles (VNPs) are currently being developed as novel vessels for delivery of diagnostic and therapeutic cargos to sites of disease. With a rapid increase in the number of VNP variants and their potential applications in nanomedicine, the properties they acquire in the bloodstream need to be investigated. Biomolecules present in plasma are known to adsorb onto the surface of nanomaterials (including VNPs), forming a biointerface called the protein corona, which is capable of reprogramming the properties of VNPs. Here we describe a few general methods to isolate and study the VNP–protein corona complexes, in order to evaluate the impact of protein corona on molecular recognition of VNPs by target cells, and clearance by phagocytes. We outline procedures for in vivo screening of VNP fates in a mouse model, which may be useful for evaluation of efficacy and biocompatibility of different VNP based formulations.

Key words Viral nanoparticles (VNPs), Tobacco mosaic virus (TMV), Protein corona, Opsonization, Nanomedicine, Bio–nano interactions

1 Introduction

Plant viral nanoparticles (VNPs) have emerged in nanomedicine as a novel type of nanomaterials with applications in diagnostics, imaging and drug delivery [1–6]. Similarly to synthetic nanoparticles (NPs), VNPs permit encapsulation of cargo and selective delivery to pathogenic tissue [7], allowing for a potential decrease of dose and side effects while maintaining the same magnitude of therapeutic effect or contrast. In addition, VNPs are biocompatible, biodegradable, and structurally highly monodisperse [1], thus rendering them a promising platform technology.

To achieve tissue-specific delivery, two concepts are being studied: (1) Passive targeting, in which tissue specificity is achieved by utilizing the structure–function relationship. In cancer

Christina Wege and George P. Lomonossoff (eds.), *Virus-Derived Nanoparticles for Advanced Technologies: Methods and Protocols*, Methods in Molecular Biology, vol. 1776, https://doi.org/10.1007/978-1-4939-7808-3_38,
© Springer Science+Business Media, LLC, part of Springer Nature 2018

treatment, nanoparticles take advantage of the leaky vasculature of the tumor to enter and remain in its microenvironment (enhanced permeability and retention effect; EPR) [8, 9]. (2) Active targeting, in which the surface of VNPs is decorated by small molecules (such as peptides, proteins, antibodies, and nucleic acids) with high affinity toward the proteins present on the cells of targeted tissue. These moieties "address" VNPs to interact specifically with their molecular targets [8, 10]. Two main ways of modifying the VNP surface are:

- Genetic engineering allowing for introduction of amino acids serving as ligation handles for subsequent chemical conjugation; or for encoding peptide sequences such as targeting peptides, epitopes, or other functionalities [1];
- Chemical modification allowing covalent binding of any peptide and non-peptide-based molecule, for example PEG to induce stealth properties [1, 11].

Nevertheless, when introduced into the bloodstream, all NPs (including VNPs) are rapidly coated with plasma proteins [12, 13]. This phenomenon, known as the formation of a protein corona, changes the surface, and therefore the biological characteristics of (V)NPs, potentially pushing them off their target [14]. The protein corona is composed of proteins strongly bound to the NPs surface (hard corona) and proteins that are weakly bound to either the NPs surface or to the proteins of the hard corona, and are constantly exchanging with the surrounding proteins (soft corona) [12]. Together hard corona and soft corona compose full corona.

We have recently shown that plasma protein adsorption on VNPs is greatly reduced compared to synthetic particles, attesting to the biocompatible nature of VNPs. For example, around six times less hard corona amount per unit of surface was measured for tobacco mosaic virus (TMV) compared to silica NPs [15]. Nevertheless, the proteins that were bound to TMV could be classified as players of the innate immune system. The majority of corona proteins found on TMV were classified as complement proteins and antibodies capable of activating recognition and clearance by the sentinel cells of mononuclear phagocyte system (MPS) [15], which is in line with the immunogenic properties of VNPs. We also found that tumor cell- and thrombus-specific targeting moieties grafted to the VNPs surface can nonspecifically interact with plasma components, leading to effects such as aggregation, which limits VNP affinity toward the target and causes accumulation in capillary-rich organs such as lungs; or nonspecific uptake by tissues, leading to disadvantageous biodistribution [15]. Altogether, our data highlight that in vitro characterization of the protein corona can provide insights into the in vivo fates of (V)NPs; here we describe the methods to do so in detail.

Due to a vast number of plant viruses and methods of their modification, the number of variations of VNP surface types is overwhelming and rapidly increasing. The interactions between plasma proteins and all types of VNPs intended for biomedical applications should be carefully considered. To address this, we extend the methods for protein corona analysis previously described for synthetic NPs [16] with methods specifically optimized for the analysis of the protein corona formed on VNPs (TMV is used as an example).

In the first section of this chapter we outline:

- The method used to prepare the VNP–hard protein corona complexes and to isolate them from excess plasma using ultracentrifugation for downstream analysis [15, 16].
- The protocol used to characterize the amount and composition of hard protein corona on the surface of VNPs using sodium dodecyl sulfate–polyacrylamide gel electrophoresis (SDS-PAGE) [15].

In the second section, we describe methods to characterize the influence of protein corona on interactions between VNPs and proteins or cells:

- Dot blots—providing insight into the affinity between the surface of VNPs (or surface of VNP–protein corona complexes) and chosen antibodies, or other proteins of interest [14, 17]. This method is also useful for investigating the antibody-blocking properties of VNP "stealth" coatings [18].
- Flow cytometry (FACS) to quantify the influence of protein corona on interactions between the VNPs and the cells (e.g., targeting when using cancerous cell line, or clearance when using phagocytes) [15].

In the third and final section we describe in vivo procedures to assay the pharmacokinetic profiles and biodistribution of VNPs after intravenous injection into BALB/c mouse model [9, 18, 19].

2 Materials

2.1 Preparation and Analysis of VNP–Hard Protein Corona Complexes

2.1.1 VNP/Plasma Incubation and Separation

1. Blood plasma from human or other species of interest (*see* **Note 1**).
2. Purified plant viral nanoparticles (VNPs) (*see* **Note 2**).
3. Phosphate buffered saline (PBS): 0.01 M Na_2HPO_4, 0.0018 M KH_2PO_4, 0.0027 M KCl, and 0.137 M NaCl, pH adjusted to 7.4 with HCl (PBS is also available as tablets; Fisher Scientific or Sigma-Aldrich).
4. Sucrose cushion: 40% (w/v) sucrose in PBS.

5. Pipette or syringe + needle (to add the sucrose cushion to the bottom of the tube).
6. Ultracentrifuge (e.g., Beckman Coulter Optima L-90 K Ultracentrifuge with rotor type 50.2 Ti and tubes).
7. Benchtop centrifuge.
8. Sample rotator or shaker.
9. Vortex mixer.

2.1.2 SDS-PAGE

For analysis of hard corona protein composition and quantity.

1. 4× NuPAGE SDS sample buffer (Invitrogen).
2. β-mercaptoethanol.
3. NuPAGE 4–12% (w/v) Bis-Tris Gel (Invitrogen).
4. 1× NuPAGE MOPS SDS running buffer (Invitrogen).
5. Novex SeeBlue Plus2 Pre-Stained Protein Standard (Invitrogen).
6. Fixing solution: 40% (v/v) methanol, 10% (v/v) acetic acid, 50% (v/v) H_2O.
7. Staining solution: 1% (w/v) Coomassie Brilliant Blue powder in 50% (v/v) methanol, 10% (v/v) glacial acetic acid, 40% (v/v) H_2O.
8. Destaining solution: 5% (v/v) methanol, 10% (v/v) acetic acid, 85% (v/v) H_2O.
9. Imaging cabinet (e.g., Alpha Imager; Biosciences).

2.2 *In Vitro* Characterization of Protein Coronal Influence on Interactions Between VNPs, Proteins, and Cells

2.2.1 Dot Blot (VNP–Protein Interactions Analysis)

Dot blots are performed to analyze the VNP–protein interactions.

1. Nitrocellulose membrane.
2. Blotting paper.
3. PBS: 0.01 M Na_2HPO_4, 0.0018 M KH_2PO_4, 0.0027 M KCl, and 0.137 M NaCl, pH adjusted to 7.4 with HCl (PBS is also available as tablets; Fisher Scientific or Sigma-Aldrich).
4. Blocking solution: 5% (w/v) skim milk powder in PBS.
5. 35 × 10 mm petri dishes (or larger if necessary).
6. Fluorescently labeled VNPs; "bare" or hard corona complexes produced as described in Subheading 3.1.1 (*see* **Note 3** and Chapters 4, 19, and 30).
7. Antibody/protein of interest, present in the blood plasma.
8. Fluorescent imaging cabinet (e.g., MAESTRO Imager) capable of quantitative fluorescence intensity measurements.

2.2.2 FACS (VNP–Cell Interactions Analysis)

FACS analysis is performed for VNPs in serum-free medium ("bare" VNPs) and complete medium (where VNP–protein corona complexes form) to analyze the influence of the protein corona on VNP–cell interactions.

1. Fluorescently labeled VNPs (*see* **Note 3** and Chapters 4, 19, and 30).
2. Cell line of interest (e.g., HeLa cells to model interactions with cancer or RAW264.7 macrophages to model clearance; ATCC Manassas, VA, USA).
3. Cell culture medium (serum-free) appropriate for the used cell line (e.g., Gibco MEM or DMEM; Thermo Fisher Scientific).
4. Fetal bovine serum (FBS; or other species serum) for preparation of complete cell culture medium (serum-free medium + FBS = complete medium).
5. Penicillin/Streptomycin (e.g., Gibco by Thermo Fisher Scientific).
6. Cell incubator with 5% CO_2.
7. Tissue culture grade PBS, pH 7.0.
8. Enzyme-free Hanks'-based Cell Dissociation Buffer.
9. Hemocytometer.
10. 96-well v-bottom, nontreated, polystyrene, sterile plates (Corning).
11. Centrifuge with inserts for 96-well plates.
12. FACS buffer: 1 mM EDTA, 25 mM HEPES, and 1% (v/v) FBS in tissue culture grade PBS, pH 7.0.
13. Cell fixing solution: 2% (v/v) formaldehyde (prepared from paraformaldehyde) in FACS buffer.
14. Flow cytometer (e.g., BD LSR II or Accuri C6 system equipped with lasers/filters appropriate for fluorophore used to label VNPs).

2.3 In Vivo VNP Pharmacokinetics and Biodistribution

1. PBS: 0.01 M Na2HPO4, 0.0018 M KH_2PO_4, 0.0027 M KCl, and 0.137 M NaCl, pH adjusted to 7.4 with HCl (PBS is also available as tablets; Fisher Scientific or Sigma-Aldrich).
2. ~7-8-week-old Balb/C mice (Jackson Laboratories or Charles River; *see* **Note 4**).
3. Fluorescently labeled VNPs (*see* **Note 3** and Chapter 4, 19, and 30).
4. Sterile 300–500 μL syringes and 28–30 gauge needles (e.g., ½ in. needles).
5. Isoflurane (anesthetic) vaporizer connected to induction chamber and/or nose cone.

2.3.1 Pharmacokinetics (VNP Circulation Half-Life Analysis)

Pharmacokinetic experiments are performed to analyze the VNP circulation half-life.

6. Heparin-coated capillary tubes (e.g., Microhematocrit Capillary Tubes).
7. Heparin-coated microcentrifuge tubes (e.g., Fisher Scientific).
8. Fluorescent signal spectroscope or plate-reader (e.g., Tecan Infinite M200).
9. Low volume cuvette or 384-well flat bottom black plate.

2.3.2 Biodistribution (VNP Accumulation in Organs Analysis)

Biodistribution experiments are performed to analyze the VNP accumulation in organs.

10. Fluorescent imaging cabinet (e.g., MAESTRO Imager) capable of quantitative fluorescence intensity measurements.

3 Methods

3.1 Preparation and Analysis of VNP–Hard Protein Corona Complexes

3.1.1 VNP/Plasma Incubation and Separation (See Fig. 1)

1. Completely defrost plasma and allow it to equilibrate at the incubation temperature of 21 °C, or 37 °C (the choice of temperature might affect the kinetics of protein adsorption on VNP surface, however, with this consideration in mind, either temperature can be used depending on downstream experimental setup; 37 °C is considered most physiologically relevant). Plasma may be briefly vortexed in order to homogenize it.
2. Spin plasma at 16,000 × *g* for 3 min to remove possible protein aggregates that may have formed in the plasma. Carefully transfer the supernatant to a clean microcentrifuge tube (for incubation), discarding the protein pellet that may have formed.
3. Add VNPs to plasma at concentration of 0.2–0.5 mg/mL, ensuring homogeneity through immediate repeated aspiration with a pipette and vortexing (*see* **Note 5**).
4. Place sample on a rotator or shaker and incubate for 1 h (or amount of time required by experiment; *see* **Note 6**) under gentle agitation at the temperature chosen in **step 1**.
5. Immediately after incubation place sample into 25 mL ultracentrifuge tube (when using Beckman Coulter type 50.2 Ti rotor) and add PBS in the amount required to fill the tube (leaving ~3 mL empty space for sucrose cushion). Carefully add 3 mL of the sucrose cushion to the bottom of the tube using a pipette or syringe equipped with a needle (*see* **Notes 7** and **8**).
6. Spin down the VNP–protein complexes using ultracentrifugation. For most kinds of VNPs (e.g., TMV, CPMV, PVX)

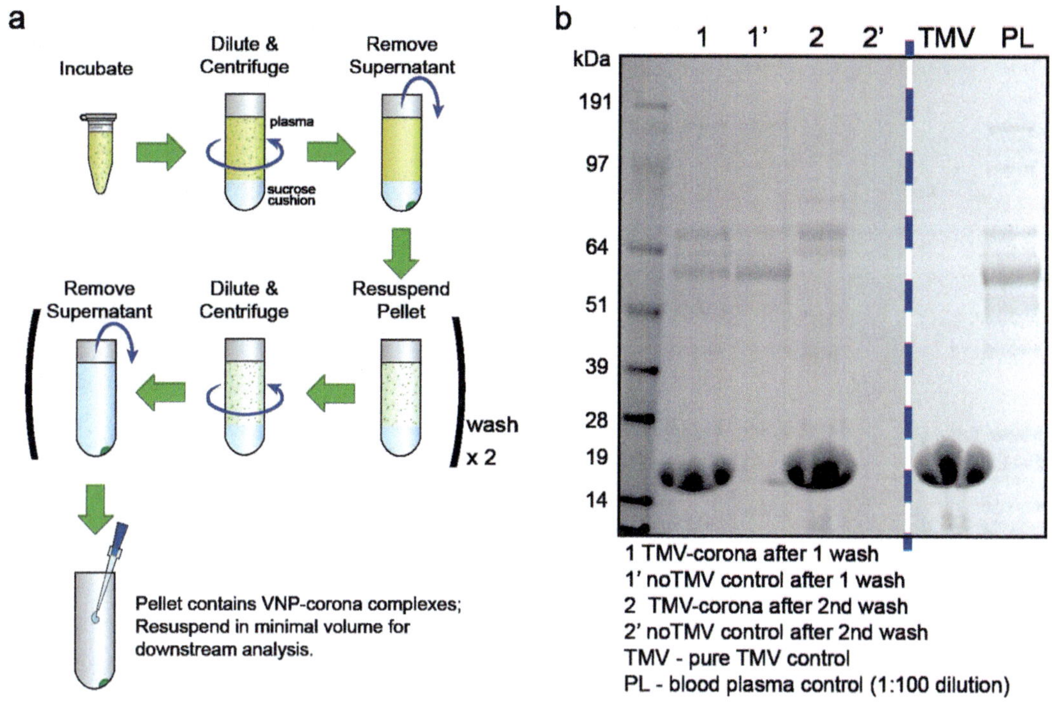

Fig. 1 (**a**) Schematic representation of the experimental steps involved in preparation of VNP–hard protein corona complexes. (**b**) SDS-PAGE analysis of TMV–corona samples prepared with different number of washes with PBS

centrifugation for 3 h under 160,000 × *g* will suffice (42,000 rpm when using Beckman Coulter type 50.2 Ti rotor). For particularly small or large particles the centrifugal force and time may need to be optimized (*see* **Note 9**).

7. After centrifugation, the VNP–protein complex should have formed a pellet at the bottom of the tube. Gently remove the supernatant, taking care not to disturb the pellet (immediate removal of supernatant is necessary to prevent the resuspension of VNPs in plasma).
8. Resuspend the pellet in small volume of PBS (e.g., 0.5 mL; a smaller volume allows more efficient resuspension) through repeated aspiration with a pipette. Add amount of PBS and sucrose cushion required to fill the ultracentrifuge tube (as in **step 5**).
9. Place VNP–protein complex sample in centrifuge and spin with the same time and velocity used in **step 6** (wash 1).
10. Repeat **steps 7–9** one more time (wash 2; *see* **Note 10**).
11. The resulting pellet contains VNP–hard corona complexes and is ready for downstream characterization [15, 16].

12. Resuspend the pellet in minimal volume of PBS (e.g., 50–100 μL; *see* **Note 11**). VNP–protein corona samples should be prepared fresh prior to each characterization to prevent proteolysis and ensure reproducibility.

3.1.2 SDS-PAGE

For analysis of hard corona protein composition and quantity.

1. Prepare samples by mixing: 10 μL of ~4 mg/mL VNP or VNP–protein corona sample, 4 μL of 4× NuPage LDS sample buffer and 2 μL β-mercaptoethanol (disulfate bond reducing agent).
2. Boil samples for 5–7 min at 100 °C to denature the proteins.
3. Prepare the NuPAGE Bis-Tris gel of appropriate acrylamide percentage (typically 4–12%). Discard the plastic strip and comb securing the gel and place it in the tank. Completely fill the inner chamber of gel module and partially fill the outside chamber of the tank with 1× NuPAGE MOPS SDS running buffer.
4. Load 10 μL of the Novex SeeBlue Plus2 Pre-Stained Protein Standard into a chosen well of the NuPAGE Bis-Tris gel.
5. Load the VNP control samples and VNP–protein corona samples side-by-side into the gel to allow easy comparison (*see* **Note 10**).
6. Run the gel electrophoresis for ~45 min at 200 V.
7. Remove the gel from the plastic casing. Use the following protocol for staining with Coomassie Brilliant Blue (ensure that the gel is completely immersed at each step; *see* **Note 12**). First, place the gel in fixing solution for 30 min. Then discard the fixing solution and place the gel in staining solution for 2 h to overnight. Decant the staining solution (which can be reused) and place the gel in destaining solution for ~4 h; change destaining solution at least two times or until desired band visibility is acquired.
8. Acquire the gel image using an imaging cabinet with transilluminating white light source (*see* **Note 13**) [15].

3.2 In Vitro Characterization of Protein Coronal Influence on Interactions Between VNPs, Proteins, and Cells

3.2.1 Dot Blot (See Fig. 2 and Note 14)

Dot blots are performed to analyze the VNP–protein interactions.

1. Prepare the dot blots in a number corresponding to the number of tested VNP samples (one per sample). Draw a grid and a blot outline on a nitrocellulose membrane using a conventional pencil (allow minimum 5 × 5 mm square per each protein spot on each blot); avoid touching the membrane with bare skin; cut the rectangular blots of appropriate size (e.g., 1.5 cm × 2 cm for six spot grid with labels; *see* Fig. 2 for an example).
2. Activate grid-labelled nitrocellulose membrane by soaking in PBS for 5 min.

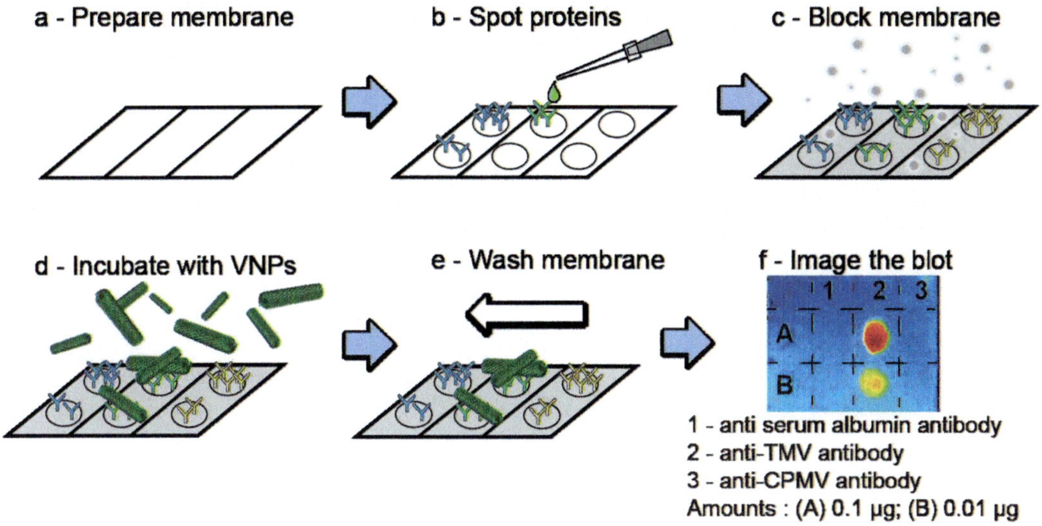

Fig. 2 Schematic representation of the experimental steps involved in screening for interaction between VNPs and antibodies using dot-blot analysis

3. Place the membrane on a blotting-paper support soaked with PBS; the support should be thoroughly moistened, but no "pools" of PBS should be visible on the surface; remove eventual air bubbles trapped between the membrane and the support.
4. Wait until there are no traces of droplets/liquid on the membrane surface; the surface should be liquid-free before one proceeds to spot the proteins (however, the membrane should be constantly moistened by the blotting-paper support).
5. Spot 1 μL droplets of antibody/protein solutions of interest (*see* **Note 15**) on the membrane in appropriate segments of the grid.
6. Leave the membrane for ~5 min to allow protein spots to absorb.
7. Place the arrays in blocking solution for 1 h at room temperature (21 °C). Ensure the blot is not attached to the bottom of the container and that the surface of the blot is completely covered in blocking solution.
8. Wash the arrays three times in PBS for 5 min.
9. Place the arrays in individual small petri dishes (e.g., 35 × 10 mm dishes for small blots) and pour the samples in, ensuring the surface of the blots is completely immersed in sample. Approximately 2 mL of 40 μg/mL VNPs or VNP–hard corona sample (*see* Subheading 3.1.1 for preparation protocol) in PBS or diluted plasma is needed when using 35 × 10 mm petri dishes. Incubate for 1–2.5 h at room

temperature gently agitating on the lab shaker. Wrap petri dishes in aluminum foil to prevent photobleaching.

10. Remove sample and wash arrays three to four times for 5 min in PBS.
11. Dry the dot blots on a fresh piece of blotting paper for ~2 h and image [14, 17, 18] using for example MAESTRO fluorescence imaging cabinet.
12. Perform densitometric analysis.

3.2.2 FACS (VNP–Cell Interactions Analysis)

FACS analysis is performed for VNPs in serum-free medium ("bare" VNPs) and complete medium (where VNP–protein corona complexes form) to analyze the influence of the protein corona on VNP–cell interactions.

1. Grow cells of choice (*see* **Note 16**) to confluence with the appropriate medium supplemented with 10% (v/v) FBS and 1% (w/v) penicillin/streptomycin at 37 °C and 5% CO_2.
2. Rinse confluent cells twice with PBS.
3. Incubate in enzyme-free Hanks'-based cell dissociation buffer for 15 min. If the cells do not fully dissociate shake and firmly tap the tissue culture flask to dislodge the cells.
4. Collect dissociation buffer with cells; divide in two equal volumes and spin under 100–500 × *g* for ~5 min.
5. Discard the supernatant; resuspend the cells in a minimal amount of serum-free and complete medium (*see* **Note 17**).
6. Count the cells using hemocytometer and adjust the medium volume to achieve a concentration of 1×10^6 cells/mL.
7. Transfer 200 μL of cell solution in serum-free and complete medium into 96-well v-bottom plate (i.e., 200,000 cells per well) in triplicates for each sample and cell-only control.
8. To each well add 1×10^5 to 1×10^6 VNPs per cell (e.g., add 13.5 μL of 1 mg/mL solution of TMV for 1×10^6 particles per cell) and incubate for 3–6 h in the dark at 37 °C and 5% CO_2. Remember to keep a triplicate of control cells without VNPs added.
9. Spin down the cells at 500 × *g* for 5 min and remove supernatant by placing the reversed 96-well plate on the lab paper to soak the medium. Do not tap the plate or aspirate the supernatant using a pipette to avoid the loss of cell pellet.
10. Add 220–240 μL of FACS buffer per well and aspirate up and down to resuspend the cells. Spin at 500 × *g* for ~5 min and remove supernatant using lab paper towel (wash 1).
11. Repeat the centrifugation one more time (wash 2).

12. Add 220–240 μL of cell fixing solution per well and incubate for 10 min at room temperature.
13. Spin down the cells at 500 × *g* for 5 min and remove supernatant by placing the reversed 96-well plate on the lab paper towel to soak the fixing solution.
14. Wash the cells in FACS buffer twice (*see* **step 10**).
15. Resuspend the final pellet in 300 μL of FACS buffer and store at 4 °C.
16. Analyze cells using a flow cytometer (*see* **Note 18**) [15].

3.3 In Vivo VNP Pharmacokinetics and Biodistribution

3.3.1 Pharmacokinetics

Pharmacokinetic experiments are performed to analyze the VNP circulation half-life.

1. Design the experiment. Depending on the number of time points necessary for the pharmacokinetic curve, divide the animals into $G = N_t/2$ groups (G = number of groups, N_t = number of time points; 2 = maximum number of bleeds per mouse). Use a minimum 3 animals in each group (*see* **Note 19**). An exemplary pharmacokinetic experiment design is shown in Fig. 3.
2. Prepare VNP solution in PBS at concentration of 2–4 mg/mL (*see* **Note 20**).

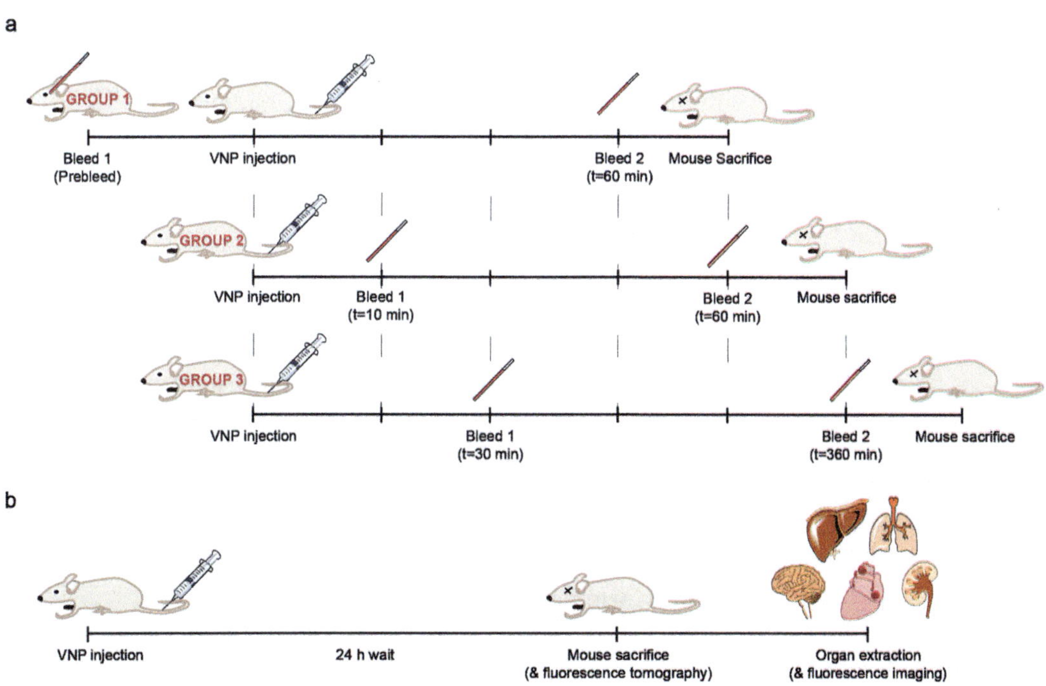

Fig. 3 (**a**) Schematic representation of the design of a pharmacokinetics experiment. (**b**) Schematic representation of the design of a biodistribution experiment

3. For the first group (*see* Fig. 3): Weigh the mice. Place the mice in the induction chamber connected to isoflurane vaporizer (set isoflurane gas flow to 2–2.5% and oxygen gas flow to ~2–3 L/min) until anesthetized. Prebleed the mice from orbital sinus using heparin-coated capillary (bleed 1; use nose cone for continuous administration of anesthesia while bleeding). Collect ~100–150 μL of blood and store it on ice in the dark in heparin-coated tubes. Inject VNPs at a dose of 200–400 μg per mouse (100 μL of VNP solution) by tail vein injection using for example a 500 μL insulin syringe with 29 gauge ½ in. needle and release mice. After the appropriate amount of time (*see* Fig. 3), collect blood from orbital sinus under anesthesia using capillary tubes (bleed 2). Collect ~100–150 μL of blood at each bleed and store it on ice in the dark in heparin-coated tubes. Sacrifice the animals immediately after second bleed. To minimize suffering do not allow the animal to wake from anesthesia.
4. For the second and third group (*see* Fig. 3): Weigh the mice. Place the mice in the induction chamber connected to isoflurane vaporizer (set isoflurane gas flow to 2–2.5% and oxygen gas flow to ~2–3 L/min) until anesthetized. Inject VNPs at a dose of 200–400 μg per mouse (100 μL of VNP solution) by tail vein injection and release mice. At the appropriate times, collect blood from orbital sinus under anesthesia using capillary tubes (bleed 1 and 2). Collect ~100–150 μL of blood at each bleed and store it on ice in the dark in heparin-coated tubes. Sacrifice the animals immediately after second bleed. To minimize suffering do not allow the animal to wake from anesthesia.
5. Centrifuge the blood samples at 2000 × *g* for 10 min at 4 °C to pellet the red blood cells as soon as possible after collection. Avoid excessive centrifugal forces, as they might cause sedimentation of VNPs.
6. Carefully recover the supernatant (blood plasma) and use at least 50 μL to measure fluorescence, using a fluorescent spectroscope or plate reader (*see* **Note 21**). For a calibration curve, use plasma collected before VNP injection spiked with known concentrations of VNPs.
7. Calculate the theoretical initial VNP concentration in plasma (C_{t0}), based on the mass of injected VNPs (m_{VNP}) and the total volume of the plasma in the mouse body (V_{tot}), using the following formula:

$$C_{t0} = \frac{m_{VNP}}{V_{tot}}$$

Both blood volume per gram of body mass and fraction of plasma in total blood can be found in the literature for each mouse type. The total volume of plasma can be calculated using the measured mass of the mouse body and values found in the literature.

8. Based on fluorescent measurements and the standard curve calculate the concentrations of VNPs at each time point (C_t). Use the obtained values and initial VNP concentration (C_{t0}) to calculate the % injected dose (% ID) of VNPs at different time points [18, 19].

$$\%\text{ID} = \frac{C_t}{C_{t0}} \times 100\%$$

3.3.2 Biodistribution (See Note 4)

Biodistribution experiments are performed to analyze the VNP accumulation in organs.

1. Prepare VNP solution in PBS at a concentration of 2–4 mg/mL (*see* **Note 20**).
2. Inject particles at a dose of 200–400 μg per mouse (100 μL of VNP solution) by tail vein injection (*see* Fig. 3); use a minimum of three animals per group/particle type (*see* **Note 19**); include control group of mice injected with PBS as background for imaging.
3. After 12–24 h sacrifice mice and collect heart, lungs, brain, kidneys, spleen, and liver (and other organs as needed); image and analyze particle accumulation using for example a MAESTRO imaging system with the appropriate excitation and emission filters, according to the fluorescent VNP label [9, 15].

4 Notes

1. Either commercially available plasma or plasma drawn from donors can be used. When handling the blood plasma, avoid freeze–thaw cycles and follow the guidelines of HUPO Proteome Project [20]. For blood draw procedure and plasma preparation from full blood refer to literature [16, 20].
2. The protocols for production and extraction of VNPs from plants are available in literature, including several chapters of this book. For example TMV can be propagated in *Nicotiana benthamiana* by mechanical inoculation using 5–10 μg of virus per leaf, and purified in high yields using established procedures involving PEG precipitation [11], as described in Chapters 23, 24, and 27.

3. The VNPs can be fluorescently labeled using a number of chemical strategies, depending on the availability of functional groups on the surface or in the interior of the capsid. For example, the EDC chemistry can be used to conjugate alkyne molecules to glutamic acid residues present in the internal channel of TMV. Subsequently, the alkyne–azide cycloaddition (click) reaction can be utilized to conjugate the azide-labeled dyes [11].
4. For any animal procedure, approval must be obtained from the Institutional Animal Care and Use Committee. Any mice designated for study of fluorescent VNPs biodistribution should be put on a "low fluorescence diet" for at least 2 weeks prior the study to decrease the background autofluorescence signal of the tissues.
5. To ensure that only well-dispersed VNPs are incubated in plasma, the VNP stock solution should be briefly spun down at 16,000 × *g*, 3–5 min to sediment aggregates. Well-dispersed VNPs will not pellet down under such conditions.
6. In general, protein corona forms rapidly. 1 h incubation is considered a sufficient time for corona formation, however if corona evolution over the time is investigated [21, 22], different incubation times may be considered.
7. If using smaller or larger tubes, the volumes of PBS and sucrose cushion should be adjusted.
8. Since prolonged centrifugation at high speeds is necessary to pellet-down the VNPs, the sucrose cushion is necessary to prevent the collapse of viral capsids after reaching the bottom of the tube. Additionally, the sucrose cushion prevents the unbound plasma proteins from sedimentation to the bottom of the tube altogether with VNPs.
9. If using large ultracentrifuge tubes (e.g., 25 mL) and relatively low volume of pure plasma sample (1–2 mL), two washes/centrifugations were determined to be sufficient in the removal of excess soft protein corona. When using small (1.5–2.0 mL) tubes, at minimum three washes/centrifugations are required to isolate the VNP–hard corona complexes. The number of washes necessary to obtain pure and stable hard corona samples can be experimentally determined (e.g., by using SDS-PAGE) [16].
10. When using low volume tubes, changing the tubes after each spin/wash (and/or use of low protein binding tubes) is recommended to minimize the carryover of plasma proteins that are not part of VNP–protein corona complexes. For large tubes, at least one tube change is required. As a control, it is recommended to prepare one sample containing plasma without VNPs and process it in exactly the same manner as all the

other samples (including compositional SDS-PAGE analysis) to identify protein carryover.

11. UV-VIS spectroscopy (e.g., using a Nanodrop instrument) may be used to roughly estimate the concentration of VNPs in hard corona samples. To do so, measure the absorbance of diluted solution of bare VNP and the VNP–protein corona at 260 nm (RNA maximum absorption) and 280 nm (protein maximum absorption). Use the Lambert–Beer law to determine the concentration of VNPs with their known extinction coefficient:

$$A = \varepsilon c l$$

where A is the absorbance, ε is the VNP extinction coefficient in mL cm^{-1} mg^{-1} (e.g., $\varepsilon_{TMV(260)} = 3$), c is the concentration in mg mL^{-1}, and l is the path length in centimeters.

    Compare the A_{260}/A_{280} ratio of bare VNP to the VNP–protein coron.a sample. For example, the $A_{260}/A_{280} = 1.2$ is indicative of intact and pure TMV VNPs; its decreased value for a hard corona sample may be indicative of an increased fraction of protein component over TMV nucleic acid.

12. For detection of low-intensity protein bands, alternative staining methods with higher sensitivity (e.g., silver staining) may be required.

13. The protein bands of interest can be cut out, trypsin-digested, extracted from the gel, and analyzed using mass spectrometry (MS) to determine the identities of the proteins [15]. One should ensure that the method used to stain the gel is compatible with MS.

14. Dot blots are a useful tool for identification of interactions between surface of VNPs and proteins. This is of particular significance when comparing different VNP stealth coatings and their impact on recognition by antibodies, but can also be applied for screening interactions between VNPs and other proteins, or between coronal proteins and other plasma/membrane proteins. In short, the protein of interest is spotted on the membrane support and incubated with sample containing fluorescently labeled VNPs. After a series of washes, the blots can be imaged for the fluorescent signal, which corresponds to VNPs binding to proteins of interest. Since some proteins (e.g., transmembrane proteins) can change their structure when adsorbed on a flat surface, the results obtained by dot-blot analysis should always be validated by another method (e.g., TEM immunogold recognition [23]).

15. It is advised to initially spot three concentrations of antibodies (e.g., 100, 20, and 10 μg/mL) to identify the amount necessary for high signal-to-noise ratio during imaging. In subse-

quent (repeat) experiments, it is advisable to pick one concentration of antibody/protein and spot it in triplicates.

16. Different cell lines should be selected depending on whether cellular targeting, nonspecific uptake or MPS clearance is investigated, e.g., HeLa cells to model interactions with cancer or RAW264.7 macrophages to model clearance.
17. Different plasma species can be used depending on the experiment and cell line used. For example, a medium supplemented with 10% human serum (instead of bovine) may be considered when using human-derived cell lines.
18. Difference in VNP uptake by cells in serum-free and complete media correspond to differences in cellular processing of "bare" VNPs and VNP–full corona [12] complexes, respectively. The FACS method can also be used to investigate differences in cellular uptake, in serum-free conditions, of "bare" VNP and preformed VNP–hard corona [12] complexes produced as described in Subheading 3.1.1. Confocal microscopy can be used to extend the flow cytometry studies with deeper insight into subcellular distribution of VNPs and VNP–corona complexes inside the cells [14, 24].
19. Three animals per group is regarded as a minimal number to test reproducibility of the result. However, in models where biological variance or experimental error is high (e.g., in most mouse cancer models, where the size of tumor and other organs may vary significantly from mouse to mouse), a group size >3 may be required to reduce the standard variation of obtained results, and ensure statistical significance.
20. The dispersion properties of VNPs in plasma may have dramatic consequences for biodistribution and clearance of VNPs in vivo, such as rapid clearance and accumulation of VNP aggregates in capillary vasculature of mouse lungs [15]. Thus, Dynamic Light Scattering (DLS) analysis may be employed to characterize the hydrodynamic diameters D_h of VNPs [16]. The D_h values obtained by DLS for well-dispersed VNPs in situ in plasma may not be accurate (especially for small particles), due to high "background" scattering of plasma proteins. However, due to exponential correlation between particle size and the scattering intensity, the aggregation of VNPs may still be detected.
21. If the volume of collected plasma is not sufficient, samples may be diluted with PBS prior to fluorescent measurement under the condition that the same dilution coefficient is applied to standards of plasma spiked with VNPs.

Acknowledgment

This work was supported in part by a grant from the National Science Foundation CAREER DMR 1452257 (to NFS) and grants from the National Institute of Health (NIH): NHLBI R21 HL121130 (to NFS) and a pilot grant from Case-Coulter Translational Research Partnership and the Harrington Heart & Vascular Institute.

References

1. Koudelka KJ, Pitek AS, Manchester M, Steinmetz NF (2015) Virus-based nanoparticles as versatile nanomachines. Annu Rev Virol 2:379–401. https://doi.org/10.1146/annurev-virology-100114-055141
2. Steinmetz NF (2013) Viral nanoparticles in drug delivery and imaging. Mol Pharm 10:1–2. https://doi.org/10.1021/mp300658j
3. Cho C-F, Shukla S, Simpson EJ, Steinmetz NF, Luyt LG, Lewis JD (2014) Molecular targeted viral nanoparticles as tools for imaging cancer. Methods Mol Biol 1108:211–230. https://doi.org/10.1007/978-1-62703-751-8_16
4. Brasch M, la Escosura de A, Ma Y, Uetrecht C, Heck AJR, Torres T, Cornelissen JJLM (2011) Encapsulation of phthalocyanine supramolecular stacks into virus-like particles. J Am Chem Soc 133:6878–6881. https://doi.org/10.1021/ja110752u
5. Douglas T, Young M (1998) Host–guest encapsulation of materials by assembled virus protein cages. Nature 393:152–155. https://doi.org/10.1038/30211
6. Steinmetz NF, Hong V, Spoerke ED, Lu P, Breitenkamp K, Finn MG, Manchester M (2009) Buckyballs meet viral nanoparticles: candidates for biomedicine. J Am Chem Soc 131:17093–17095. https://doi.org/10.1021/ja902293w
7. Wen AM, Wang Y, Jiang K, Hsu GC, Gao H, Lee KL, Yang AC, Yu X, Simon DI, Steinmetz NF (2015) Shaping bio-inspired nanotechnologies to target thrombosis for dual optical-magnetic resonance imaging. J Mater Chem B Mater Biol Med 3:6037–6045. https://doi.org/10.1039/C5TB00879D
8. Peer D, Karp JM, Hong S, Farokhzad OC, Margalit R, Langer R (2007) Nanocarriers as an emerging platform for cancer therapy. Nat Nanotechnol 2:751–760. https://doi.org/10.1038/nnano.2007.387
9. Shukla S, Wen AM, Ayat NR, Commandeur U, Gopalkrishnan R, Broome A-M, Lozada KW, Keri RA, Steinmetz NF (2014) Biodistribution and clearance of a filamentous plant virus in healthy and tumor-bearing mice. Nanomedicine (Lond) 9:221–235. https://doi.org/10.2217/nnm.13.100
10. Cole JT, Holland NB (2015) Multifunctional nanoparticles for use in theranostic applications. Drug Deliv Transl Res 5:295–309. https://doi.org/10.1007/s13346-015-0218-2
11. Bruckman MA, Steinmetz NF (2014) Chemical modification of the inner and outer surfaces of tobacco mosaic virus (TMV). Methods Mol Biol 1108:173–185. https://doi.org/10.1007/978-1-62703-751-8_13
12. Walczyk D, Baldelli Bombelli F, Monopoli MP, Lynch I, Dawson KA (2010) What the cell "sees" in bionanoscience. J Am Chem Soc 132:5761–5768. https://doi.org/10.1021/ja910675v
13. Monopoli MP, Baldelli Bombelli F, Dawson KA (2011) Nanobiotechnology: nanoparticle coronas take shape. Nat Nanotechnol 6:11–12. https://doi.org/10.1038/nnano.2011.267
14. Salvati A, Pitek AS, Monopoli MP, Prapainop K, Baldelli Bombelli F, Hristov DR, Kelly PM, Åberg C, Mahon E, Dawson KA (2013) Transferrin-functionalized nanoparticles lose their targeting capabilities when a biomolecule corona adsorbs on the surface. Nat Nanotechnol 8:137–143. https://doi.org/10.1038/nnano.2012.237
15. Pitek AS, Wen AM, Shukla S, Steinmetz NF (2016) The protein corona of plant virus nanoparticles influences their dispersion properties, cellular interactions, and in vivo fates. Small 12:1758–1769. https://doi.org/10.1002/smll.201502458
16. Monopoli MP, Pitek AS, Lynch I, Dawson KA (2013) Formation and characterization of the nanoparticle-protein corona. Methods Mol Biol 1025:137–155
17. Pitek AS, O'Connell D, Mahon E, Monopoli MP, Baldelli Bombelli F, Dawson KA (2012)

Transferrin coated nanoparticles: study of the bionano interface in human plasma. PLoS One 7:e40685. https://doi.org/10.1371/journal.pone.0040685

18. Pitek AS, Jameson SA, Veliz FA, Shukla S, Steinmetz NF (2016) Serum albumin "camouflage" of plant virus based nanoparticles prevents their antibody recognition and enhances pharmacokinetics. Biomaterials 89:89–97. https://doi.org/10.1016/j.biomaterials.2016.02.032
19. Lee KL, Shukla S, Wu M, Ayat NR, Sanadi El CE, Wen AM, Edelbrock JF, Pokorski JK, Commandeur U, Dubyak GR, Steinmetz NF (2015) Stealth filaments: polymer chain length and conformation affect the in vivo fate of PEGylated potato virus X. Acta Biomater 19:166–179. https://doi.org/10.1016/j.actbio.2015.03.001
20. Rai AJ, Gelfand CA, Haywood BC, Warunek DJ, Yi J, Schuchard MD, Mehigh RJ, Cockrill SL, Scott GBI, Tammen H, Schulz-Knappe P, Speicher DW, Vitzthum F, Haab BB, Siest G, Chan DW (2005) HUPO Plasma Proteome Project specimen collection and handling: towards the standardization of parameters for plasma proteome samples. Proteomics 5:3262–3277. https://doi.org/10.1002/pmic.200401245
21. Dell'Orco D, Lundqvist M, Oslakovic C, Cedervall T, Linse S (2010) Modeling the time evolution of the nanoparticle-protein corona in a body fluid. PLoS One 5:e10949. https://doi.org/10.1371/journal.pone.0010949
22. Casals E, Pfaller T, Duschl A, Oostingh GJ, Puntes VF (2011) Hardening of the nanoparticle-protein corona in metal (Au, Ag) and oxide (Fe_3O_4, CoO, and CeO_2) nanoparticles. Small 7:3479–3486. https://doi.org/10.1002/smll.201101511
23. Kelly PM, Åberg C, Polo E, O'Connell A, Cookman J, Fallon J, Krpetić Ž, Dawson KA (2015) Mapping protein binding sites on the biomolecular corona of nanoparticles. Nat Nanotechnol 10:472–479. https://doi.org/10.1038/nnano.2015.47
24. Lesniak A, Fenaroli F, Monopoli MP, Åberg C, Dawson KA, Salvati A (2012) Effects of the presence or absence of a protein corona on silica nanoparticle uptake and impact on cells. ACS Nano 6:5845–5857. https://doi.org/10.1021/nn300223w

Chapter 39

Fabrication of Plant Virus-Based Thin Films to Modulate the Osteogenic Differentiation of Mesenchymal Stem Cells

Kamolrat Metavarayuth, Huong Giang Nguyen, and Qian Wang

Abstract

Stem cells can interact and respond to the extracellular nanoscale environment. Viral nanoparticles have been utilized as building blocks to control cell growth and differentiation. By integrating stem cell research and virus nanoparticle chemistry together, a systematic analysis of the effects of nanotopography on stem cell differentiation can be accomplished. The fabrication of thin films of the viral nanoparticles is particularly valuable for such studies. Here, we describe two methods to fabricate plant virus-based thin films and procedures to study the osteogenic differentiation of mesenchymal stem cells on plant virus-based substrates. The method makes use of wild-type tobacco mosaic virus (wt-TMV), RGD-modified TMV (TMV-RGD), turnip yellow mosaic virus (TYMV), cowpea mosaic virus (CPMV), turnip vein clearing virus (TVCV), and potato virus X (PVX) for development of bone tissue engineering biomaterials.

Key words Viral nanoparticles, Plant virus, Thin film, Osteogenic differentiation, Mesenchymal stem cells

1 Introduction

The development of biomimetic scaffolding materials that can guide the proliferation, self-renewal and differentiation of multipotent stem cells into specific lineages is a major aim of tissue engineering. Since many cellular functions are dictated by the interactions between cells and biomaterials [1–3], the surface chemistry and topography of support materials play a pivotal role in modulating cell behaviors [3–9]. Therefore, the understanding of the various topographical cues that are responsible for cellular behaviors is a key to this application.

Recently, viral nanoparticles have been utilized as building blocks for targeted cell growth and differentiation [10–14]. The symmetrical arrangement of the viral coat proteins makes virus particles an ideal scaffold for displaying identical copies of functional groups for applications where multivalent ligand display is important. Furthermore, chemical and genetic modifications of virus

Christina Wege and George P. Lomonossoff (eds.), *Virus-Derived Nanoparticles for Advanced Technologies: Methods and Protocols*, Methods in Molecular Biology, vol. 1776, https://doi.org/10.1007/978-1-4939-7808-3_39,
© Springer Science+Business Media, LLC, part of Springer Nature 2018

surfaces to incorporate new functional groups have been extensively studied, providing a library of viruses with different surface properties [10–12, 15–17]. Our group has demonstrated that two-dimensional (2D) virus thin film-coated substrates can promote the osteogenic differentiation of bone-derived mesenchymal stem cells (BMSCs) [18–23]. The virus-based thin films can be achieved using two methods: direct deposition via electrostatic interactions and layer-by-layer (LbL) deposition. Figure 1 illustrates the processes of LbL-assisted deposition of virus nanoparticles on a glass substrate and the characterization of the resulting virus thin film using AFM imaging. This shows the height profile of wt-TMV on the top layer of the substrate and suggests that the viral particles are mostly intact and lie flat on the surface. The detailed procedures for the two methods for the fabrication of virus thin films coated on various substrates are outlined in this chapter.

In addition, this chapter describes the steps necessary to comprehensively study BMSC osteogenesis, including BMSC isolation and expansion from bone marrow of Wistar rats. The respective procedures have to follow the ethics, standards, and procedures of animal experimentation and typically demand for approval according to the protection of animals acts defined by the country hosting the research. The typical spindle-shaped morphology and highly oriented cytoskeleton structure of bone-derived mesenchymal stem cells obtained from the procedure are shown in the images in Fig. 2. We also describe the general protocol for performing quantitative real-time RT-PCR analysis (RT-qPCR) of the osteogenic markers listed in Table 1. Our RT-PCR analyses mainly emphasize that cells cultured on virus substrates undergo osteogenic differentiation faster than cells cultured on control flat substrates (*see* Fig. 3). For example, osteogenic markers are highly upregulated in cells grown on a turnip yellow mosaic virus (TYMV) substrate by day 14, which is 7 days earlier than upregulation of osteogenic markers in cells on control substrates (day 21). Moreover, a CellTiter Blue assay, immunofluorescence assays and cytochemical staining and quantification of alkaline phosphatase and calcium mineralization of common osteogenic markers were conducted to confirm the successful bone formation of rBMSCs (rat BMSCs) on virus-based thin film scaffolds. Our results show that cells on virus-based scaffolds have higher alkaline phosphatase activity, as well as higher calcium mineralization, compared to cells cultured on a control flat surface or cells cultured with virus particles suspended in culture medium (*see* Fig. 4). In addition, immunofluorescence staining of the commonly used osteogenic markers, osteocalcin and osteopontin, shows higher expression of these two proteins in bone-like nodules of cells on virus substrates (*see* Fig. 5). Quantification of bone morphogenetic protein-2 (BMP-2) by ELISA demonstrates the rapid upregulation of this protein in cells

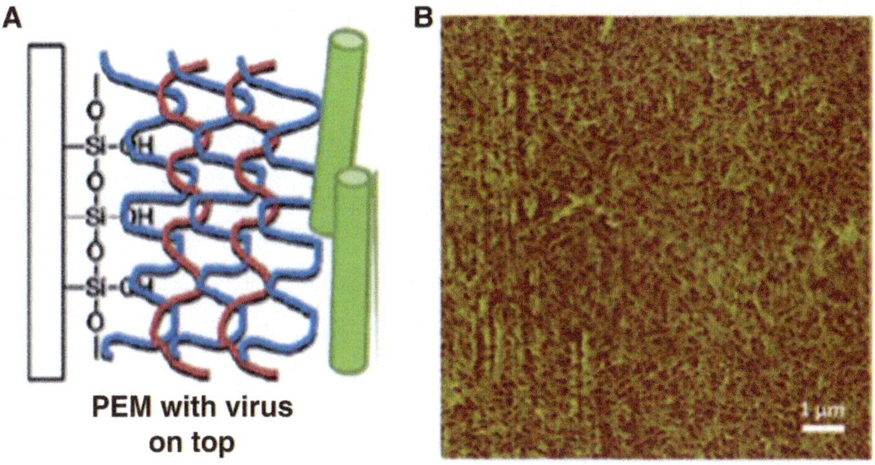

Fig. 1 Virus thin film fabrication. (**a**) Schematic illustration of LbL-assisted deposition of virus nanoparticles on a glass substrate. (**b**) AFM height image of wt-TMV on the top layer indicates the viral particles are mostly intact and laid flat on the surface

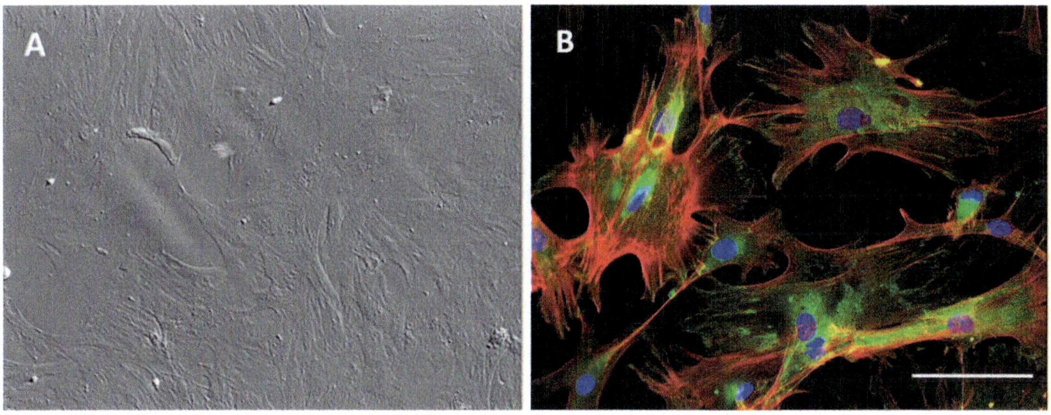

Fig. 2 Primary BMSCs isolated from the bone marrow of male Wistar rats. (**a**) Bright field image of rBMSCs cultured in primary media for 7 days. (**b**) Immunofluorescence images of rBMSCs on silane coated glass slide for 24 h. Color representation: nucleus (blue), vinculin (green), phalloidin (red). Scale bar is 100 μm

on virus substrates as early as 8 h after osteoinduction. Thus, it can be exploited as an early marker of osteogenesis in cells cultured on virus substrates (*see* Fig. 6). This chapter also provides the detailed protocols for these osteogenic studies.

2 Materials

Prepare all solutions using nanopure water (prepared by purifying deionized water to attain a sensitivity of 18 MΩ cm at 25 °C) and analytical grade reagents. Unless indicated otherwise, prepare and store all reagents at room temperature. Plant viruses that have been

Table 1
Primers used for RT-qPCR to measure gene expression levels

Gene	Sequence (5′-3′)
GAPDH	Forward: ACTAAAGGGCATCCTGGGCTACACTGA
	Reverse: TGGGTGGTCCAGGGTTTCTTACTCCTT
Arbp	Forward: CGACCTGGAAGTCCAACTAC
	Reverse: ATCTGCTGCATCTGCTTG
BGLAP	Forward: AAAGCCCAGCGACTCT
	Reverse: CTAAACGGTGGTGCCATAGAT
BMP-2	Forward: ACCAACCATGGGTTTGTGGTGGAAGT
	Reverse: TCCGCTGTTTGTGTTTGGCTTGACG
Col1a	Forward: TCCTGCCGATGTCGCTATC
	Reverse: CAAGTTCCGGTGTGACTCGTG
Sparc	Forward: ACAAGCTCCACCTGGACTACA
	Reverse: TCTTCTTCACACGCAGTTT
Spp1	Forward: GACGGCCGAGGTGATAGCTT
	Reverse: CATGGCTGGTCTTCCCGTTGC
Runx2	Forward: GCTTCTCCAACCCACGAATG
	Reverse: GAACTGATAGGACGCTGACGA

GAPDH glyceraldehyde 3-phosphate dehydrogenase, *Arbp* attachment region binding protein, *BGLAP* bone-gamma-carboxyglutamate protein (osteocalcin), *BMP2* bone morphogenetic protein 2, *Col1a* colagen-1-alpha, *Sparc* secreted protein acidic and rich in cysteine (Osteonectin), *SPP1* secreted phosphoprotein 1 (Osteopontin), *Runx2* runt-related transcription factor 2. The primers were synthesized commercially (Integrated DNA Technologies, Inc.), and evaluated at an annealing temperature of 58 °C

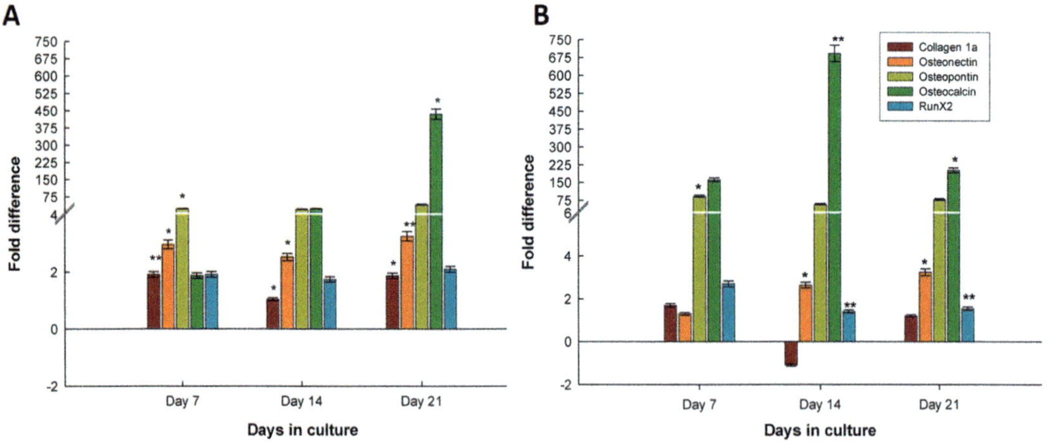

Fig. 3 RT-qPCR analysis for the gene expression in rBMSCs seeded on (**a**) tissue culture petri dish and (**b**) TYMV-coated substrate for 21 days. Cells were cultured under osteogenic media condition. (**) represents $p < 0.005$ and (*) for $p < 0.05$, respectively. Figure modified with permission from ref. 18. Copyright 2008 Elsevier

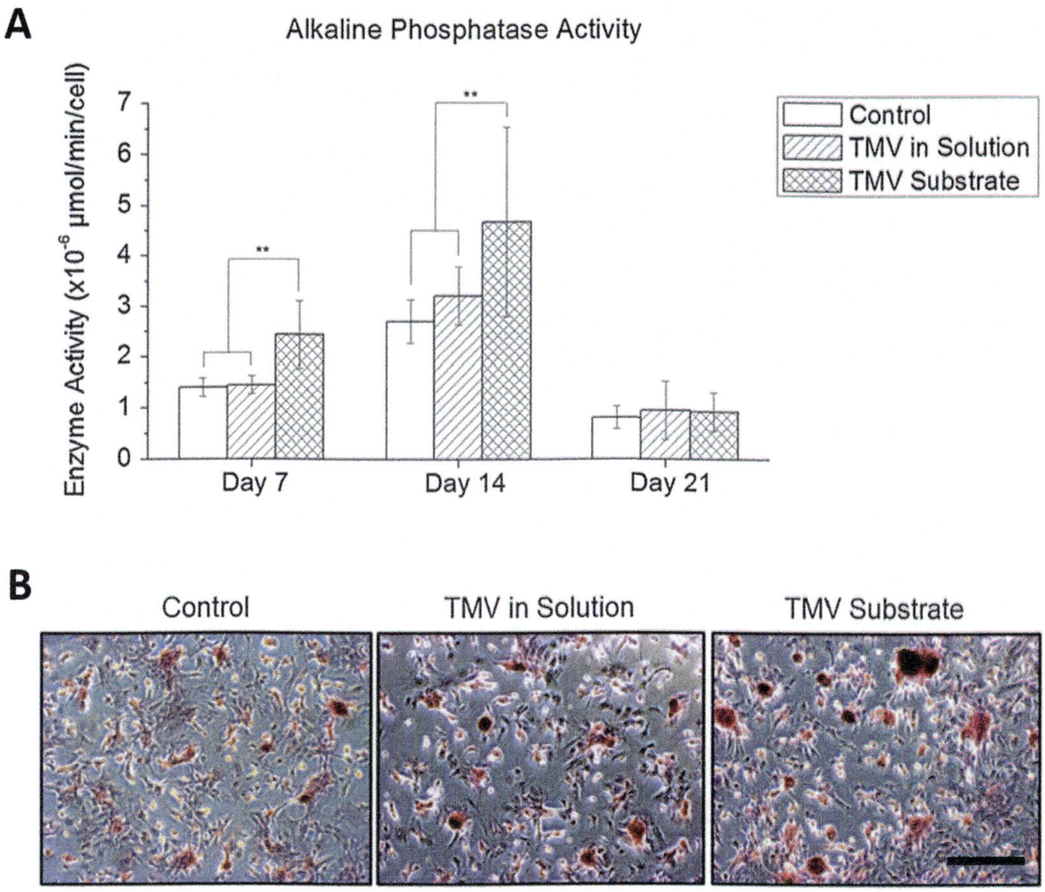

Fig. 4 Cytochemical analysis of the bone differentiation process of rBMSCs. (**a**) Alkaline phosphatase activity of cells under three different conditions (control, TMV in solution, and cells on TMV coated substrate). (**b**) Alizarin Red staining of each rBMSCs for calcium mineralization analysis. (**$p < 0.05$ based ANOVA). The error bars denote ±S.D. Scale bar is 200 μm. Figure modified with permission from ref. 21. Copyright 2012 The Royal Society of Chemistry

used to fabricate virus thin films for osteogenic induction of mesenchymal stem cells (MSCs) include wild-type tobacco mosaic virus (wt-TMV), RGD-modified TMV (TMV-RGD), turnip yellow mosaic virus (TYMV), cowpea mosaic virus (CPMV), turnip vein clearing virus (TVCV), and potato virus X (PVX). These plant viruses can be readily isolated and purified according to a previously reported protocol [23]. The virus coverage on the wafers is characterized using tapping-mode AFM images obtained from a NanoScope IIIA MultiMode AFM (Veeco). Si tips with a resonance frequency of approximately 300 kHz, a spring constant of about 40 N/m, and a scan rate of 0.5 Hz are used.

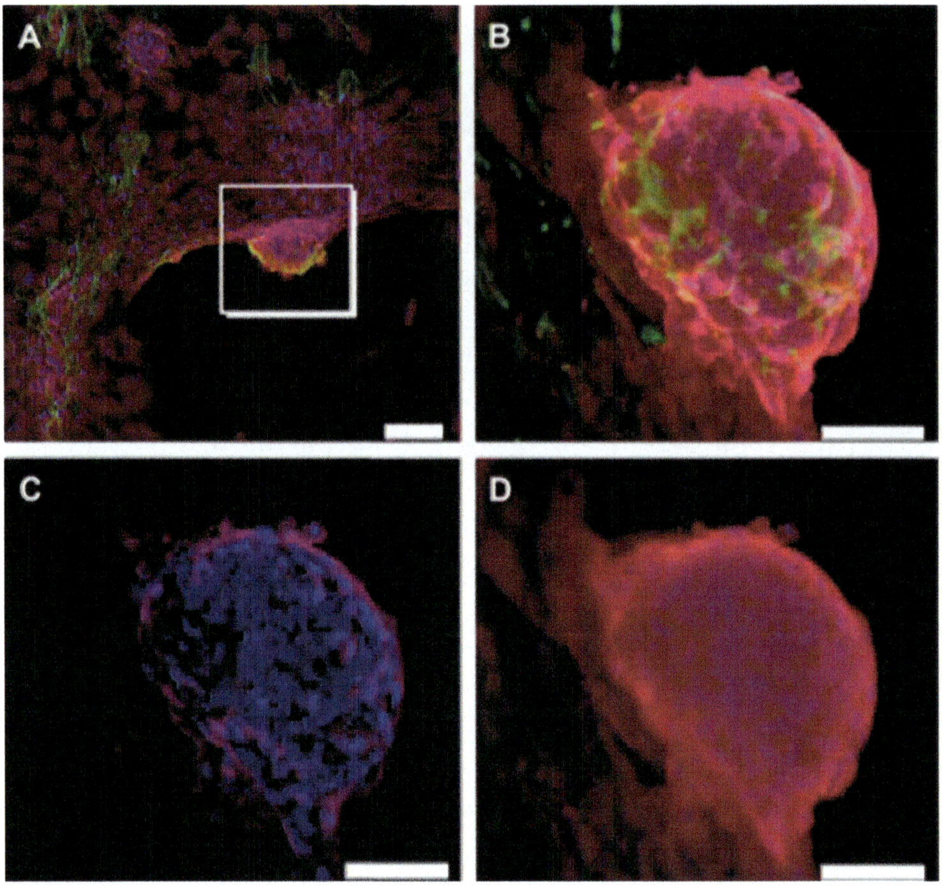

Fig. 5 Immunofluorescence imaging of rBMSCs that undergo osteogenesis and form bone-like nodule structure. (**a**) The nodules formed by the cells seeded on TYMV coated substrates after 14 days of induction in osteogenic medium at lower magnification (10×, scale bar: 100 mm). (**b**) Nodule structures at 40×, (**c**) expression of osteopontin in the cells arranged at the periphery of the nodule, and (**d**) osteocalcin expression in the cells constituting the nodule structure and in the neighboring cells. Scale bars: 50 mm; color representation: DAPI (blue), phalloidin (green), osteocalcin (red), osteopontin (pink). Reproduced with permission from ref. 18. Copyright 2008 Elsevier

2.1 Direct Deposition of Viral Particles Via Electrostatic Interactions

1. 95% v/v ethanol (EtOH).
2. Nitrogen gas.
3. 0.2 mg/mL and 1 mg/mL wt-TMV, TMV-RGD, TYMV, CPMV, TVCV, and/or PVX solution in nanopure water or virus storage buffer.
4. Virus storage buffer: 10 mM potassium phosphate buffer, pH 7.0 (prepared from potassium phosphate monobasic, KH_2PO_4 and potassium phosphate dibasic, K_2HPO_4).
5. Sterile tissue culture hood.
6. (3-aminopropyl)triethoxysilane (APTES)-coated glass slides.
7. Poly-D-lysine (PDL)-coated circle glass coverslips (18 mm diameter) or PDL coated 12-well plates (Corning®).

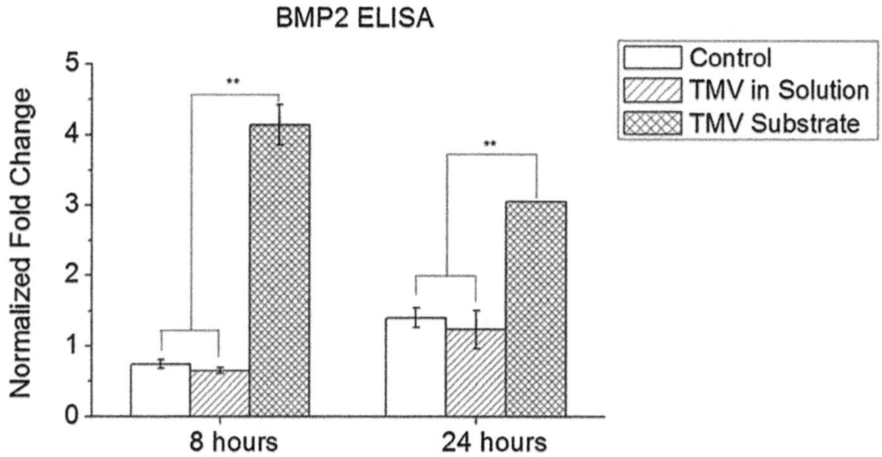

Fig. 6 Quantification of BMP2 protein expression at 8 and 24 h normalized to cell number by ELISA. The values are expressed as fold change compared to cells on TCP before osteoinduction. The error bars denote ±1 S.D. **p* < 0.01 based on equal variance two-tailed Student's ***t***-test

2.2 Fabrication of Virus Thin Films by LbL Deposition Method

1. APTES coated glass slides.
2. 1.0 mg/mL poly(allylamine hydrochloride) (PAH) in water.
3. 1.0 mg/mL poly(acrylic acid) (PAA) in water.
4. 1.0 mg/mL poly(styrenesulfonate) (PSS) in water.
5. 0.1 mg/mL wt-TMV, TMV-RGD, TYMV, CPMV, TVCV, and/or PVX solution in nanopure water or virus storage buffer, pH close to isoelectric point of virus (*see* Table 2).

2.3 rBMSC Isolation and Expansion (Cells Can Also Come from Commercial Sources)

1. Laminar air hood for surgery.
2. Blue cloth/incontinence sheet.
3. 70% v/v ethanol.
4. 100 mL beaker or tray for soaking instruments.
5. Rat: young adult 80 g male Wistar rats.
6. Death chamber and CO_2 (gas or dry ice).
7. 5-in. stout/toothed forceps.
8. Clippers (for shaving the rat).
9. 3 (and 4) in. gauze pad.
10. Betadine or iodine scrub.
11. #3 scalpel handle and #3 blades.
12. Hemostatic forceps.
13. Scissors.
14. Bone shears or hospital cut-all shears.
15. Two petri dishes (100 mm).
16. 5 mL and 20 mL syringes.

Table 2
Isoelectric point (pI) of some viruses which can be used for this study

Virus	pI
wt-TMV and TMV-RGD	3.5
TYMV	3.77–3.97
CPMV	5.5
TVCV	3.55
PVX	4.4

17. 16 g, 19½ g, 20 g, 21½ g needles.
18. 50 mL conical tubes.
19. Three T-75 Flasks (Corning®).
20. 4 mL of 0.25% trypsin–2.21 mM EDTA in HBSS without sodium bicarbonate, calcium, and magnesium; porcine parvovirus-tested (Corning®) per one T-75 Flask.
21. Two 50 mL beakers for storing surgical tools in 70% ethanol during cell extraction process.
22. Primary medium: DMEM with 4500 mg/L glucose and 4.00 mM L-glutamine, without sodium pyruvate (HyClone™). Prepared as instructed by the manufacturer.
23. Growth medium: Primary medium supplemented with 10% fetal bovine serum (FBS), 100 U/mL penicillin, 100 mg/mL streptomycin, and 250 ng/mL amphotericin B.
24. Osteogenic medium: Growth medium supplemented with 10 mM sodium β-glycerol phosphate, 50 mg/mL L-ascorbic acid 2-phosphate, and 10^{-8} M dexamethasone.
25. Incubator (37 °C, 5% CO_2, 80% humidity).
26. Mosconas (136.8 mM NaCl, 28.6 mM KCl, 11.9 mM $NaHCO_3$, 9.4 mM glucose, 0.08 NaH_2PO_4, pH 7.4, Sigma) or 1× PBS (10× PBS, *see* Subheading 2.5, **step 5a**).

2.4 RT-qPCR of Osteogenic Markers

1. Substrate: Virus coated 12-well plate (*see* Subheading 3.1.2).
2. Growth medium (*see* Subheading 2.3, **item 23**).
3. Osteogenic medium (*see* Subheading 2.3, **item 24**).
4. RNA extraction kit: RNeasy mini purification kit (Qiagen).
5. RNA quantity and quality analysis: Bio-Rad Experion (Bio-Rad Laboratories).
6. RNA reverse transcription: qScript cDNA SuperMix (Quanta Biosciences).
7. iQ5 real-time PCR detection system(Bio-Rad Laboratories).

8. Incubator (37 °C, 5% CO_2, 80% humidity).
9. Primers (*see* Table 1).

2.5 CellTiter Blue Assay

1. CellTiter Blue assay (Promega).
2. 10% CellTiter Blue in growth medium (*see* Subheading 2.3, **item 23**).
3. Incubator (37 °C, 5% CO_2, 80% humidity).
4. Microplates for fluorescence-based assays.

2.6 Staining and Quantification of AP and Ca Mineralization

1. Substrate: Virus coated glass coverslip (*see* Subheading 3.1.2).
2. Incubator (37 °C, 5% CO_2, 80% humidity).
3. Growth medium (*see* Subheading 2.3, **item 23**).
4. Osteogenic medium (*see* Subheading 2.3, **item 24**).
5. 4% (w/v) Formaldehyde: For 1 L of 4% formaldehyde, add 800 mL of 1× PBS to a glass beaker on a stir plate in a ventilated hood. Heat while stirring to approximately 60 °C. Prevent the solution from boiling. Add 40 g of paraformaldehyde powder to the heated PBS solution. The powder will not immediately dissolve. Slowly add 1 M NaOH dropwise until the solution clears. Once the paraformaldehyde is completely dissolved, cool and filter the solution. Then adjust the volume of the solution to 1 L with 1× PBS. Adjust pH with dilute HCl to approximately 6.9. The solution can be aliquoted and frozen or stored at 2–8 °C for up to 1 month.
6. 1-Step *p*-nitrophenyl phosphate (1Step PNPP) solution.
7. 0.1% (w/v) Alizarin Red, pH 4.1–4.5.
8. Aluminum foil.
9. 0.1 M and 2 M sodium hydroxide (NaOH).
10. Spectrophotometer.

2.7 Immunofluorescence Assay and Analysis

1. Substrate: Virus coated glass coverslip (*see* Subheading 3.1.2).
2. Incubator (37 °C, 5% CO_2, 80% humidity).
3. Growth medium (*see* Subheading 2.3, **item 23**).
4. Osteogenic medium (*see* Subheading 2.3, **item 24**).
5. 4% (w/v) formaldehyde (*see* Subheading 2.6, **item 5**).
6. 0.1% (v/v) Triton-X 100.
7. 10× PBS: 80 g/L NaCl, 2 g/L KCl, 14.4 g/L Na_2HPO_4, 2.4 g/L KH_2PO_4, adjust pH to 7.4 with HCl.
8. Blocking buffer: 1.5% (w/v) bovine serum albumin (BSA) in 1× PBS.
9. Primary antibody: 0.5 mg/mL mouse monoclonal antibody targeting BMP-2 (R&D systems).

10. Secondary antibody: 0.2 mg/mL goat anti-mouse antibody conjugated with horseradish peroxidase (Cayman Chemical).
11. Secondary antibody: 2 mg/mL goat anti-mouse antibody conjugated with Fluorescein (Chemicon).
12. Primary antibody: 0.2 mg/mL anti-vinculin mouse monoclonal antibody (Neomarkers).
13. 3,3′,5,5′-tetramethylbenzidine (TMB) solution: TMB ready-to-use tablet (Amresco). Dissolve one TMB tablet in 10 mL of 0.05 M phosphate–citrate buffer, pH 5.0. Add 2 μL of fresh 30% hydrogen peroxide immediately prior to use.
14. 2 M sulfuric acid (H_2SO_4).
15. Rhodamine Phalloidin (14 μM in 1× PBS).
16. 1 mg/mL DAPI (4,6-diamidino-2-phenylindole).
17. Glass slides for fluorescence microscopy.
18. Clear nail polish.
19. Olympus IX81 fluorescence microscope.
20. SlideBook 5, 3i (Intelligent Imaging Innovations).

3 Methods

Carry out all procedures at room temperature unless otherwise specified. The procedures involving animal models have to follow the ethics, standards, and procedures of animal experimentation and typically demand for approval according to the protection of animals acts defined by the country hosting the research.

3.1 Direct Deposition Via Electrostatic Interactions

3.1.1 Direct Deposition on Silane-Coated Glass Slides

1. Cut silane-coated glass slides into 1.5 cm^2 wafers, sterilize with 95% ethanol, then wash three times with nanopure water (*see* **Note 1**) and dry the wafers with nitrogen.
2. Drop an aqueous solution of virus (0.20 mL, 0.2 mg/mL) onto the dried wafer and let the wafers dry overnight in an incubator at 37 °C in the presence of CO_2 and a 65% relative humidity or at ambient temperature in a sterile tissue culture hood (*see* **Note 2**).

3.1.2 Direct Deposition on Poly-D-Lysine (PDL) Coated Substrates

1. If using glass coverslips, sterilize them by immersing them in 95%v/v ethanol for a few minutes. Then wash the coverslips three times with nanopure water to remove ethanol and dry them with nitrogen.
2. Drop a solution of virus in water or virus storage buffer (0.50 mL, 1 mg/mL) onto a cleaned glass coverslip or into a well of a 12-well plate. Incubate overnight in a sterile tissue culture hood with the lid closed to prevent evaporation.

3. After incubation, aspirate any remaining virus solution and wash the substrate three times by adding and aspirating about 1 mL nanopure water per wash, to remove loosely attached virus particles. Then dry the substrate in a sterile tissue culture hood (*see* **Note 3**).

3.2 Fabrication of Virus Thin Films by LbL Deposition Method

1. Alternately dip a cleaned glass (*see* Subheading 3.1.2, **step 1**) in a positively charged polymer such as PAH and a negatively charged polymer such as PAA or PSS for 15 min each for five times (two times for PAH and three times for PAA or PSS) until it reaches $(PAH/PAA$ or $PAH/PSS)_5$, where the subscript 5 indicates the number of layers. After the 5th coating, a positive polymer will be the outermost layer which will retain the negatively charged virus particles by electrostatic interaction in the next step.
2. Alternately immerse the $(PAH/PAA)_5$ coated substrate in the virus solution (0.1 mg/mL, pH close to isoelectric point of virus, *see* Table 2) for 20 min and PAH solution (1.0 mg/mL, pH 7.0) for 15 min until the desired number of layers is obtained. Use nanopure water to rinse in between each deposition step. Our results demonstrated that a minimum of three layers of virus particles deposited on the substrate with this method produced nearly 100% coverage as confirmed by AFM. QCM can also be used to track the coating procedure by monitoring the frequency shift corresponding to the mass of virus particles' increment on the substrate after each deposition.

3.3 rBMSC Isolation and Expansion

1. In the laminar hood for surgery, put down an absorbent blue cloth/incontinence sheet.
2. Put approximately 50–60 mL of 70%v/v ethanol in a beaker and soak the instruments.
3. Euthanize rat in a closed chamber supplied with CO_2.
4. Once all movement has stopped for approximately 3 min, perform cervical dislocation by placing a pair of stout forceps (toothed forceps) behind the head while holding the rat down. Pull on the rat tail until the neck gives way.
5. Move rat to laminar air hood.
6. Dip rat in 70% v/v ethanol or spray the rat with 70%v/v ethanol.
7. Shave lower half of rat using clippers. Make sure to get as much hair off the legs as possible.
8. Using a sterile 3-in. gauze pad, scrub the shaved area with Betadine or iodine scrub. Make sure to scrub all areas including feet and tail. Squeeze some betadine or iodine from the gauze onto the area around the anus and testicles to make sure you cover this area thoroughly.

9. Using a scalpel and forceps (or hemostatic forceps) make an incision along the dorsal aspect (back/top) of the leg.
10. Skin the leg, then find the femur (top leg bone) and make an incision on both sides of the bone. Then grab the bone with the toothed forceps or hemostatic forceps. Slice around the proximal (top) end of the bone and remove it from the socket.
11. Remove the foot and separate the femur from the tibia at the knee.
12. Use scalpel and gauze to remove any remaining tissue from exterior of bone (*see* **Note 4**). Be sure to remove any cartilage and metaphyseal ends from bones. The metaphyseal ends are like caps on the ends of the bones.
13. Place the excised long bones in Primary medium with 5× antibiotics for temporary storage (before the bone marrow is flushed out).
14. Change the gloves and then wipe the working surface with 70% (v/v) ethanol.
15. Prepare two 100 mm petri dishes. One is for primary medium stock and the other is for the aspirated cells. Put 23 mL of primary medium into each dish.
16. One by one, cut the proximal (close/top) end of femur and distal (far/bottom) end of tibia; then using a 5 mL syringe with 5 mL primary medium and a 16 g needle, bore a hole in the end of bone and flush the marrow into sterile petri dish (*see* **Note 5**).
17. Once all long bones (tibia and femur) have been flushed, break up the cell clumps by repeatedly aspirating with a 19½ g needle and then a 21½ g needle mount on 20 mL syringe (*see* **Note 6**).
18. Using a 21½ g needle, transfer the cells into a 50 mL conical tube.
19. Centrifuge at 200 × g for 5 min.
20. Resuspend cell pellet in 6 mL of primary medium. Be careful to not get any bubbles into the solution.
21. Transfer the resuspended pellet into three T-75 Flasks, 2 mL per flask.
22. Add 18 mL primary medium to each flask, bringing the final volume in each flask up to 20 mL.
23. Store cells in incubator at 37 °C and 5% CO_2.
24. At day 3, aspirate off medium and replace with 20 mL fresh medium.
25. At day 7, aspirate off medium and wash cultures twice with 12 mL Mosconas (or PBS) and replace with 20 mL fresh

medium to remove any hematopoetic cells and any other unattached elements (*see* **Note 7**).

26. At day 10, when cells have reached sub-confluent layer, rinse twice with Mosconas and lift cells from flask by adding 4 mL of 0.25% EDTA–trypsin solution (*see* **Note 8**).
27. After 4 mL of 0.25% EDTA–trypsin is incubated with cells in incubator for 3–5 min. Pipette cell suspension and transfer to sterile 15 mL centrifuge tube to pellet the cells by centrifugation at 233 × *g* for 5 min. After centrifugation, aspirate the supernatant and resuspend cell pellet in 1 mL fresh medium to break down the cell aggregates. Then add 2 mL of fresh medium to make total of 3 mL cell suspension.
28. Prepare three new flasks containing 14 mL of fresh medium, then aliquot 1 mL of cell suspension from **step 27** to each flask.
29. To maintain cells, replenish with fresh medium every 3–4 days (*see* **steps 26–28**).
30. rBMSC is best to be used for osteogenesis studies up to passage five.

3.4 RT-qPCR of Osteogenic Markers

1. Seed rBMSC at 4.5 × 10^4 cells per substrate (determined by counting cells with a hemocytometer under a microscope) and allow cells to attach overnight in growth medium. Grow cells in triplicates for three sets of cells per experiment. Each set is for one of three time points (8 h, 24 h, and 14 days). In addition, pellet 13 × 10^4 cells to be used as control undifferentiated cells.
2. Replace the growth medium with osteogenic medium and culture the three sets of cells for 8 h, 24 h, and 14 days, respectively.
3. Terminate triplicates of the cell cultures at 8 h and 24 h for BMP-2 gene expression analysis and at 14 days for analysis of later osteogenic markers by adding lysis buffer as instructed by RNA extraction kit of choice.
4. Extract total RNA of cells from each set of experiments and analyze the quality and quantity of the extracted RNA using the RNA extraction kit and BioRad Experion following the manufacturer's instructions.
5. Reverse-transcribe the extracted RNA using qScript cDNA SuperMix following the manufacturer's instructions.
6. Perform RT-qPCR using the following method: 60 cycles of PCR (95 °C for 20 s, 58 °C for 15 s, and 72 °C for 15 s), after an initial denaturation step of 5 min at 95 °C. Use the gene of glyceraldehyde 3-phosphate dehydrogenase (GAPDH) as the housekeeping gene. The primers used for RT-qPCR in this study are shown in Table 1. Each cell culture at 8 h and 24 h is

analyzed for GAPDH and BMP-2. The results showed that cells grown on virus substrates (except CPMV) had an upregulation of BMP-2 compared to cells grown on non-virus coated substrates and CPMV coated substrates. Each cell culture at 14 days is analyzed for GAPDH, BGLAP, and SPP1. Our experiments showed a result corresponding to that obtained for BMP-2: Cells grown on virus substrates (except CPMV) had an upregulation of the osteogenic markers BGLAP and SPP1.

3.5 CellTiter Blue Assay

1. Prepare 10% CellTiter Blue in growth medium.
2. Discard the medium from cells growing in 12-well plates and add 1 mL of 10% CellTiter Blue medium (a live cell compatible assay) to each plate well.
3. Incubate the plates in an incubator at 37 °C, 5% CO_2, and 80% humidity for 1 h.
4. Pipet the medium in each plate up and down to make sure that medium in each well is homogenous after *CellTiter Blue* reduction by cells and pipet 100 μL of the medium to a 96-well plate (Triplicate) (*see* **Note 9**).
5. Pipet 100 μL nanopure water to three wells to be used as blanks.
6. Pipet blank 10% *CellTiter Blue* medium to three wells to be used as control.
7. Measure fluorescence at $OD_{560ex}/OD_{590em.}$ The higher this value, the greater numbers of cells are in the well.

3.6 Staining and Quantification of AP and Ca Mineralization

1. Seed rBMSC at 4.5×10^4 cells per substrate and allow cells to attach overnight in growth medium. Grow cells in triplicates for three different time points (7, 14, and 21 days)
2. Replace the growth medium after 24 h with osteogenic medium and incubate for an additional 7, 14, and 21 days.
3. Determine the number of cells in each sample 1 h prior to cell fixation by *CellTiter Blue* assay (*see* Subheading 3.5).
4. Fix cells on the substrates with 4% w/v formaldehyde for 15 min at room temperature (*see* **Note 10**).
5. To determine AP activity, incubate each fixed sample in 500 μL of 1Step PNPP for 15 min at room temperature. Then transfer the solution to a new microfuge tube with 250 μL of 2 M NaOH to stop the reaction and measure the absorbance at 405 nm wavelength.
6. To compare calcium deposition, stain the fixed samples at day 14 (after the *CellTiter Blue* assay) with Alizarin Red solution for 30 min. Wrap the substrates in aluminum foil to protect them from light (*see* **Note 11**). After washing with ultrapure

water, add 200 μL of 0.1 M NaOH to each sample to extract the dye from the sample.

7. Quantify the amount of Alizarin Red dye by measuring the absorbance at 548 nm wavelength.
8. Normalize both absorbance values at 405 nm (PNPP) and 548 nm (Alizarin Red) against the cell number from *CellTiter Blue*. The ratio of absorbance value at 405 nm and the cell number obtained with *CellTiter Blue* represents the ALP activity. The higher the number, the greater the enzyme activity. The ratio of the absorbance value at 548 nm and the cell number determined with *CellTiter Blue* implies the calcium mineralization quantity. A higher number means a higher mineralization. Higher numbers of both values mean higher osteogenesis.

3.7 Immuno-fluorescence Assay and Analysis

This general protocol is specified for different uses in Subheadings 3.7.1–3.7.3.

1. Seed 2 sets of rBMSC at 1×10^5 cells per substrate for two time points (8 and 24 h) analysis and allow cells to attach overnight in growth medium.
2. Replace the growth medium after 24 h with osteogenic medium and incubate for additional 8 h and 24 h.
3. Determine the number of cells in each sample 1 h prior to cell fixation by *CellTiter Blue* assay (*see* Subheading 3.5).
4. Fix cells with 4% w/v formaldehyde for 30 min at room temperature.
5. Permeabilize cells with 0.1% v/v Triton X-100 for 15 min, after that block the cells in blocking buffer for 1 h at room temperature.
6. After blocking, incubate the cells overnight at 4 °C with primary antibody in blocking buffer. Then incubate with secondary antibody for 2 h at room temperature (*see* **Note 12**).
7. Wash the cells with blocking buffer.
8. Process cells as described in Subheading 3.7.1, 3.7.2, or 3.7.3.

3.7.1 Enzyme-Linked Immunosorbent Assay (ELISA)

1. Follow general protocol for immunofluorescence assays (*see* Subheading 3.7) using mouse monoclonal antibody targeting BMP-2 at 1:100 dilution in blocking buffer as primary antibody and goat anti-mouse antibody conjugated with horseradish peroxidase at 1:1000 dilution in blocking buffer as secondary antibody.
2. Add 1 mL of TMB solution to each sample and incubate on a rocker for 30 min at room temperature.
3. Stop the reaction by adding 500 μL of 2.0 M H_2SO_4 and measure absorbance at 450 nm wavelength.

4. Normalize the absorbance value against the cell number obtained from *CellTiter Blue* analysis done prior to this assay. This value quantifies the protein of interest per cell in each sample.

3.7.2 BMP2 Immunostaining

1. Follow general protocol for immunofluorescence assays (*see* Subheading 3.7) using mouse monoclonal antibody targeting BMP-2 at 1:100 dilution in blocking buffer as primary antibody and goat anti-mouse antibody conjugated with Fluorescein at 1:100 dilution in blocking buffer as secondary antibody.
2. Stain filamentous actin and nuclei by incubating cells with 5 μL rhodamine phalloidin and 1 μL DAPI in 1 mL in blocking buffer for 10 min at room temperature.
3. Wash the cells with blocking buffer. Then mount and seal sample onto glass slide using clear nail polish.
4. Image the stained substrates with fluorescence microscope.

3.7.3 Vinculin Immunostaining

1. Follow general protocol for immunofluorescence assays (*see* Subheading 3.7) using anti-vinculin mouse monoclonal antibody at 1:200 dilution in blocking buffer as primary antibody and goat anti-mouse antibody conjugated with Fluorescein at 1:100 dilution in blocking buffer as secondary antibody.
2. Stain filamentous actin and nuclei (*see* Subheading 3.7.2, **step 2**).
3. Wash the cells with blocking buffer. Then mount and seal sample onto a glass slide using clear nail polish.
4. Image the stained substrates with fluorescence microscope.
5. Analyze immunofluorescence images of vinculin with SlideBook 5 by setting the criteria to select vinculin spots with XY shape factor larger than 1.5 and area size between 0.5–15 mm^2. Then calculate the average size of vinculin for each image. After that calculate average vinculin size of cells on each substrate. The average vinculin size correlates to focal adhesion size of cells grew on different substrate which implies the strength of cell adhesion on the substrate. The smaller focal adhesion (FA) size might increase cellular motility and facilitate the formation of cell aggregates which supports osteogenic differentiation.

4 Notes

1. Washing the silane-coated glass wafer with nanopure water is necessary to remove residual glass after cutting, which can interfere during the subsequent virus-coating step.

2. Make sure to drag the virus solution to the edge of the glass wafer by using 200 μL pipette tip to spread virus solution onto the wafer while pulling the virus solution to the edge of the glass wafer to prevent uneven coating. While the virus solution is drying, it will tend to be drawn to the center of the wafer if there is no surface tension created by the interaction of virus solution and the edge of the wafer.
3. Washing the virus-coated PDL coverslip with nanopure water after coating also removes salt crystals from the virus storage buffer that remain on the surface with virus particles after coating.
4. The scalpel can also be used to scrape tissue from the bone. It also works well to help scrape the cartilage from the ends.
5. The size of the needle used is determined by the size of the bones. The younger the rat, the smaller the needle (larger needle number) is needed. A good needle size to start with is typically a 16 g or a 19½ g. Pay particular attention when selecting the needle size to the tibia (lower leg bone). If this bone breaks when the first size of needle is stuck into it, immediately switch to the next smaller needle size and put the needle into the same hole. Make sure to push the needle further down into the bone to prevent back flushing of cells out of the opening in the bone. Also be careful not to squirt the cells and medium out of the anterior (front) aspect of the proximal (top) end of the bone. This area is easily torn when the muscle and cartilage are removed. The area can be distinguished by its redder color than the surrounding bone. Slowly depressing the syringe plunger first to confirm that the cells will not squirt out of this location, then push the needle a little deeper into the bone and push out 2.5 mL (half of the syringe) and give it a few quick pushes. Sometimes marrow tends to hang onto the end of the bone, 1–1.5 mL of medium in the petri dish can be used to quickly squirt out these cells.
6. Do not let the cells sit stagnant for any longer than 5 min. Cells will begin to attach to a culture dish as soon as they get the chance.
7. If PBS is used, make sure to not leave it on the cells for too long. In contrast to Mosconas, PBS does not have glucose in it which acts as a food source for the cells.
8. After 4 mL of 0.25% EDTA–trypsin is incubated with cells in incubator for 3–5 min, check the dishes under the microscope. If many cells are still attached, tapping the bottom of the dishes on the counter top can dislodge some of them. Not all of the cells will come off the dish.

9. Be careful not to carry any suspended cells to the 96-well plate. They can continue the reaction and the result obtained will have higher cell activity.
10. Cells are briefly fixed here to preserve the alkaline phosphatase activity. Fixing cells for longer than 15 min can reduce the enzyme activity.
11. Since the reaction is highly light-sensitive, the substrates must be wrapped in aluminum foil during the Alizarin Red staining.
12. To prevent fluorescence bleaching, all incubation steps that involve fluorescence-conjugated antibodies need to be performed in a dark room.

Acknowledgment

This work was partially supported by USA NSF CHE-0748690 and USA SF CHE-1307319 and the University of South Carolina. We followed the Institutional Animal Care and Use Committee, School of Medicine, University of South Carolina for the rBMSC extraction guideline.

References

1. Lee CH, Shin HJ, Cho IH, Kang YM, Kim IA, Park KD, Shin JW (2005) Nanofiber alignment and direction of mechanical strain affect the ECM production of human ACL fibroblast. Biomaterials 26(11):1261–1270. https://doi.org/10.1016/j.biomaterials.2004.04.037
2. Andersson AS, Brink J, Lidberg U, Sutherland DS (2003) Influence of systematically varied nanoscale topography on the morphology of epithelial cells. IEEE Trans Nanobioscience 2(2):49–57
3. Fukushima N, Ohkawa H (1995) Hematopoietic stem cells and microenvironment: the proliferation and differentiation of stromal cells. Crit Rev Oncol Hematol 20(3):255–270. https://doi.org/10.1016/1040-8428(94)00163-N
4. Arnold M, Cavalcanti-Adam EA, Glass R, Blummel J, Eck W, Kantlehner M, Kessler H, Spatz JP (2004) Activation of integrin function by nanopatterned adhesive interfaces. Chemphyschem 5(3):383–388. https://doi.org/10.1002/cphc.200301014
5. He J, Zhou W, Zhou X, Zhong X, Zhang X, Wan P, Zhu B, Chen W (2008) The anatase phase of nanotopography titania plays an important role on osteoblast cell morphology and proliferation. J Mater Sci Mater Med 19(11):3465–3472. https://doi.org/10.1007/s10856-008-3505-3
6. Mendonca G, Mendonca DB, Aragao FJ, Cooper LF (2008) Advancing dental implant surface technology – from micron- to nanotopography. Biomaterials 29(28):3822–3835. https://doi.org/10.1016/j.biomaterials.2008.05.012
7. Yim EK, Reano RM, Pang SW, Yee AF, Chen CS, Leong KW (2005) Nanopattern-induced changes in morphology and motility of smooth muscle cells. Biomaterials 26(26):5405–5413. https://doi.org/10.1016/j.biomaterials.2005.01.058
8. Wan Y, Wang Y, Liu Z, Qu X, Han B, Bei J, Wang S (2005) Adhesion and proliferation of OCT-1 osteoblast-like cells on micro- and nano-scale topography structured poly(L-lactide). Biomaterials 26(21):4453–4459. https://doi.org/10.1016/j.biomaterials.2004.11.016
9. Liu Q, Cen L, Yin S, Chen L, Liu G, Chang J, Cui L (2008) A comparative study of proliferation and osteogenic differentiation of adipose-derived stem cells on akermanite and beta-TCP ceramics. Biomaterials 29(36):4792–4799.

https://doi.org/10.1016/j.biomaterials.2008.08.039

10. Lee LA, Nguyen QL, Wu L, Horvath G, Nelson RS, Wang Q (2012) Mutant plant viruses with cell binding motifs provide differential adhesion strengths and morphologies. Biomacromolecules 13(2):422–431. https://doi.org/10.1021/bm2014558
11. Pokorski JK, Steinmetz NF (2011) The art of engineering viral nanoparticles. Mol Pharm 8(1):29–43. https://doi.org/10.1021/mp100225y
12. Chung W-J, Merzlyak A, Lee S-W (2010) Fabrication of engineered M13 bacteriophages into liquid crystalline films and fibers for directional growth and encapsulation of fibroblasts. Soft Matter 6(18):4454–4459. https://doi.org/10.1039/C0SM00199F
13. Wang J, Wang L, Li X, Mao C (2013) Virus activated artificial ECM induces the osteoblastic differentiation of mesenchymal stem cells without osteogenic supplements. Sci Rep 3:1242. https://doi.org/10.1038/srep01242
14. Merzlyak A, Indrakanti S, Lee SW (2009) Genetically engineered nanofiber-like viruses for tissue regenerating materials. Nano Lett 9(2):846–852. https://doi.org/10.1021/nl8036728
15. Zhu H, Cao B, Zhen Z, Laxmi AA, Li D, Liu S, Mao C (2011) Controlled growth and differentiation of MSCs on grooved films assembled from monodisperse biological nanofibers with genetically tunable surface chemistries. Biomaterials 32(21):4744–4752. https://doi.org/10.1016/j.biomaterials.2011.03.030
16. Wang Q, Lin T, Tang L, Johnson JE, Finn MG (2002) Icosahedral virus particles as addressable nanoscale building blocks. Angew Chem Int Ed 41(3):459–462. https://doi.org/10.1002/1521-3773(20020201)41:3<459::AID-ANIE459>3.0.CO;2-O
17. Destito G, Schneemann A, Manchester M (2009) Biomedical nanotechnology using virus-based nanoparticles. In: Manchester M, Steinmetz N (eds) Viruses and nanotechnology, Current topics in microbiology and immunology, vol 327. Springer, Berlin, pp 95–122. https://doi.org/10.1007/978-3-540-69379-6_5
18. Kaur G, Valarmathi MT, Potts JD, Wang Q (2008) The promotion of osteoblastic differentiation of rat bone marrow stromal cells by a polyvalent plant mosaic virus. Biomaterials 29(30):4074–4081. https://doi.org/10.1016/j.biomaterials.2008.06.029
19. Kaur G, Valarmathi MT, Potts JD, Jabbari E, Sabo-Attwood T, Wang Q (2010) Regulation of osteogenic differentiation of rat bone marrow stromal cells on 2D nanorod substrates. Biomaterials 31(7):1732–1741. https://doi.org/10.1016/j.biomaterials.2009.11.041
20. Kaur G, Wang C, Sun J, Wang Q (2010) The synergistic effects of multivalent ligand display and nanotopography on osteogenic differentiation of rat bone marrow stem cells. Biomaterials 31(22):5813–5824. https://doi.org/10.1016/j.biomaterials.2010.04.017
21. Sitasuwan P, Lee LA, Bo P, Davis EN, Lin Y, Wang Q (2012) A plant virus substrate induces early upregulation of BMP2 for rapid bone formation. Integr Biol (Camb) 4(6):651–660. https://doi.org/10.1039/c2ib20041d
22. Sitasuwan P, Lee LA, Li K, Nguyen HG, Wang Q (2014) RGD-conjugated rod-like viral nanoparticles on 2D scaffold improve bone differentiation of mesenchymal stem cells. Front Chem 2:31. https://doi.org/10.3389/fchem.2014.00031
23. Metavarayuth K, Sitasuwan P, Luckanagul JA, Feng S, Wang Q (2015) Virus nanoparticles mediated osteogenic differentiation of bone derived mesenchymal stem cells. Adv Sci 2(10):150026–150034. https://doi.org/10.1002/advs.201500026

Chapter 40

Dual Surface Modification of Genome-Free MS2 Capsids for Delivery Applications

Ioana L. Aanei and Matthew B. Francis

Abstract

One of the hallmarks of virus-like particles (VLPs) is the fact that they possess distinguishable interior and exterior surfaces. Taking advantage of our knowledge of the amino acid location from X-ray crystal structures, we have developed a series of synthetic modifications of the MS2 bacteriophage viral capsid, including small molecule and polymer attachment, as well as conjugation with peptides, DNA and other proteins. These constructs have found applications in nanomaterial fabrication and as delivery vehicles with therapeutic potential. Importantly, the dual-modification strategies described herein could be extended to other VLP systems.

Key words MS2 bacteriophage, Site-specific modification, Virus-like particles, Delivery vehicle, Bioconjugation, Drugs, Polymers, Aptamers, Antibodies

1 Introduction

Virus-like particles are biodegradable, stable in a range of environmental conditions, and stand out as the most homogeneous particles available in their size regime. Building on these useful properties, scientists have worked toward augmenting the native properties of the viral capsid proteins by covalently attaching small molecules, polymers, inorganic particles, and other macromolecules. These modifications require reactions that can take place in aqueous solution at near-neutral pH, ambient temperatures, and low substrate concentrations, without the use of any protecting groups. In addition, when the design of the supramolecular assembly requires multiple modifications, the chemical groups often have to be installed in distinct locations on the protein scaffold using sequential, orthogonal steps. As such, the construction of chemically modified viral capsids lies at the experimental vanguard of both the organic and materials chemistry fields.

Christina Wege and George P. Lomonossoff (eds.), *Virus-Derived Nanoparticles for Advanced Technologies: Methods and Protocols*, Methods in Molecular Biology, vol. 1776, https://doi.org/10.1007/978-1-4939-7808-3_40,

© Springer Science+Business Media, LLC, part of Springer Nature 2018

VLPs offer the opportunity to attach numerous copies of a given molecule onto the same nanoscale scaffold. By conjugating imaging or therapeutic agents to the external surface of the carrier, delivery vehicles with the potential for the detection and treatment of disease can be constructed. If these modifications are made on the interior surface of the VLP, an additional benefit is the protection of the active molecule (e.g., a drug) from external factors (such as enzymes and other serum proteins) during delivery. With careful design of the attachment linkages, the drug molecules can also be engineered for release from the carrier upon endocytosis, leading to a controlled-release delivery system [1]. By attaching targeting groups to the outside of the VLP, there is potential to increase the binding avidity through multivalency effects to achieve a higher level of specificity in binding to receptors of interest [2, 3]. This dual-modification strategy can be used for constructing a wide array of agents that combine the benefits of nanoscale carriers with the advantages of small molecule active molecules and macromolecular targeting groups.

Two of the classical methods for appending new functional groups to proteins involve lysine and cysteine modification reactions [4, 5]. Lysines are usually prevalent and easily accessible on the surface of proteins, but the chemical methods available to modify these residues lack control over the number and location of the modifications. They also require a large excess of reagents to achieve high yields [6]. Although there are many commercially available cysteine-reactive reagents, there remain many instances [6] in which they cannot accomplish the desired modification, and alternative chemistries are needed. Our lab has focused on modifying amino acids that are not commonly targeted for modification, such as tryptophans (with metallocarbenoid reagents [7] or phenylene diamines [8, 9]) and tyrosines (with diazonium salts [10], Mannich reactions [11], pi-allyl palladium complexes [12], or anisidine derivatives [8, 9]). Unique positions in the protein sequence, such as the N-terminus, can also be modified through transamination reactions [13–17]. In addition to these, several oxidative coupling reactions have been developed in our lab to interact with unnatural amino acid residues inserted into a sequence at precise locations [18–21].

One system our lab has investigated extensively is the bacteriophage MS2 viral capsid, consisting of 180 sequence-identical subunits ($T = 3$) that can be expressed recombinantly in *E. coli* [18]. The 27 nm capsids self-assemble inside the cell without the native RNA genome. A useful feature of this capsid is the presence of 32 pores (~2 nm in diameter) that allow the access of small organic molecules to the interior of the protein shell without requiring disassembly. We aimed to design a set of synthetic protocols that could be used to label the interior and exterior surfaces differentially. We identified two cysteine residues in the sequence of the MS2 monomer protein, but upon further experimentation, we observed that they were not solvent accessible and could not be readily modified. Therefore, we introduced a new cysteine in position 87, which proved to be very reactive, allowing for up to 180 modifications on the interior surface [1]. We have attached numerous cargo molecules by reacting maleimides with the Cys87, including fluorophores [22–24], porphyrins for singlet-oxygen generation [25], gadolinium complexes for magnetic resonance imaging (MRI) contrast enhancement [26, 27], cryptophane cages for ^{129}Xe nuclear magnetic resonance [28], chelators for positron emission tomography (PET) [29] and taxol molecules [1] for chemotherapeutic purposes. We have also found that a native tyrosine residue (Tyr85) can be modified selectively using diazonium salts [30], providing a useful alternative for interior labeling.

There is a wide array of choices for targeting moieties that cannot be appended to viral capsids through genetic means (antibody fragments, proteins, nucleic acids, and peptoids). Initial attempts to attach these moieties onto the MS2 capsid were based on NHS ester chemistry to modify two lysines (Lys106 and Lys113) on the exterior surface [31]. This method did not lead to homogeneously labeled MS2 capsids, the modification levels were fairly low, and smaller reagents were seen to modify the lysine residues on the interior surface as well. Unfortunately, in the case of MS2, the N-terminus is located at important contact regions between the monomers, so N-terminal modification [13, 15, 16] was not a viable option. Our lab set out to explore new coupling reactions that could circumvent several of these issues. In previous work, we had found that anilines react rapidly with *N*-acyl phenylene diamines in the presence of sodium periodate [10]. This "oxidative coupling" strategy could be used in the presence of the other amino acids and nucleic acids without interference. In order to use this reaction for MS2 capsids, we needed the introduction of aniline groups into the capsid protein, which we achieved using the amber codon suppression technology [32]. Introducing the aniline as the *p*-aminophenylalanine (*p*AF) residue at position 19 proved to be optimal to achieve an even distribution of this reactive handle on the exterior surface of the capsid [18]. By modifying the *p*AF residues, we have decorated the MS2 capsid with DNA aptamers [22] that bind to tyrosine kinase 7 (PTK-7), a receptor found on Jurkat T-lymphocytes and many other cancer cell lines. By attaching photodynamic therapy agents on the inside of the targeted capsids and irradiating with 420 nm light, we were able to show selective killing of a PTK7-positive cell line in the presence of other cells [22]. Recent work in our lab has developed protocols for attaching full-length antibodies to the same *p*AF19 residue [24]. In this work, an antibody targeting the epidermal growth factor receptor (EGFR) overexpressed by numerous types cancer cells was used to achieve specificity to EGFR-positive cells.

As a result of these studies, the chemistry developed for the dual-surface modification of MS2 capsids is versatile enough to join a myriad of available targeting groups with numerous payloads. This general, two-step approach to modifying the two surfaces of the virus can now be extended to a variety of targeted vehicles and materials applications.

2 Materials

Prepare all solutions using ultrapure water (18 MΩ cm at 25 °C) and analytical grade reagents. Diligently follow all waste disposal regulations when disposing waste materials.

2.1 Buffers and Reagents

1. 3-(4-hydroxy-3-nitrophenyl)propanoic acid.
2. *N*-hydroxysuccinimide.
3. Dichloromethane (DCM).
4. 1-ethyl-3-(3-dimethylaminopropyl)carbodiimide (EDC).
5. 50 mM porphyrin maleimide in DMSO.
6. Dimethylformamide (DMF).
7. Phosphate buffers: Potassium phosphate monobasic at different ionic strengths and pHs: 200 mM, pH 6.5; 50 mM, pH 7.0; 50 mM + 150 mM NaCl, pH 7.0; 10 mM, pH 7.2; 50 mM, pH 7.2; 25 mM, pH 8.0; 50 mM, pH 8.0; 100 mM pH 8.0.
8. 100 mM sodium dithionite ($Na_2S_2O_4$ stock solution) in 200 mM phosphate buffer pH 6.5 (*see* **Note 1**), make fresh before use.
9. 50 mM sodium periodate solution ($NaIO_4$ stock solution) in water, make fresh before use.
10. 0.5 M Tris(2-carboxyethyl)phosphine hydrochloride (TCEP, pH 7.0 solution).
11. 10 mM dye maleimide in DMSO.
12. 10 mM chelator of interest with a maleimide functionality in DMSO.
13. 1× PBS: 137 mM NaCl, 12 mM phosphate, 2.7 mM KCl, pH 7.4.
14. 10 mM maleimide derivative of doxorubicin in 10 mM phosphate buffer pH 7.2.
15. Triethylamine.
16. 50 mM potassium ferricyanide ($K_3[Fe(CN)_6]$, stock solution) in water, can be stored at room temperature.
17. 40 mM NHS-anisidine (2,5-dioxopyrrolidin-1-yl 6-(4-(dimethylamino)phenoxy) hexanoate) in DMF.
18. *N,N*-diisopropylethylamine (DIPEA).

19. 25 mM cerium (IV) ammonium nitrate (stock solution) in water, make fresh before use.
20. 50 mM Bis-Tris buffer in water, adjust to pH 6.0 with HCl.

2.2 Other Materials

1. Sodium sulfate, solid (Na_2SO_4).
2. 200 μM Bacteriophage MS2 mutants with a *p*-aminophenylalanine (*p*AF) residue at position 19 and a Cys at position 87 (T19*p*AF N87C) in 10 mM phosphate buffer, pH 7.2 [18].
3. 200 μM Bacteriophage MS2 mutants with Tyr or Trp at positions 15 or 19, in 10 mM phosphate buffer, pH 7.2 [8, 9].
4. Eluent for silica gel chromatography: hexane–ethyl acetate 90:10, 80:20, 70:30, 60:40 v/v.
5. NAP-5 gel filtration desalting columns (GE Healthcare).
6. Spin concentrators, e.g., Amicon Ultra centrifugal filters with 3, 10, 30 and 100 kDa molecular weight cutoff (Millipore).
7. Equipment for electrospray ionization mass spectrometry (ESI-MS, e.g., Agilent 1260 series LC pump outfitted with an Agilent 6224 Time-of-Flight (TOF) LC/MS system).
8. UV spectrophotometer (e.g., NanoDrop 1000, Thermo).
9. 300 μM DNA of interest containing a 5′-end primary amine.
10. Equipment for SDS-PAGE analysis (e.g., 4–12% bis-Tris NuPage gels and Mini Gel Tank apparatus, Thermo).
11. 100 μM peptide of interest in 10 mM phosphate buffer, pH 7.2.
12. Reverse-phase HPLC instrumentation (e.g., 1100 Series HPLC Systems with an inline diode array detector (DAD) and inline fluorescence detector (FLD), Agilent).
13. Analytical and preparative reverse-phase HPLC columns (C18 stationary phase and $MeCN/H_2O$ with 0.1% TFA gradient mobile phase).
14. Sep-Pak C18 reverse phase chromatography column (Waters).
15. ^{64}Cu radioactive isotope, in HCl (University of Wisconsin): $^{64}CuCl_2$ diluted in 100 mM ammonium acetate, pH 6.2 (100 μL of ~30 mCi/mL).
16. HPLC SEC with an inline radioactivity detector (e.g., 590 HPLC pump (Waters), UV detector (Linear Systems), Model 105S-1 high-sensitivity radiation detector with 1 cm^3 CsI scintillating crystal coupled to a 1 cm^2 Si PIN photodiode/low-noise preamplifier (Carroll-Ramsey Associates), and fluorescence detector (Spectra system FL3000, Thermo). Chromatography traces were collected using PeakSimple data system and software (SRI Instruments), and analyzed using the Gaussian multipeak fitting feature of the OriginPro software v. 8.6.0 (OriginLab).

17. PolySep GFC-P5000 size exclusion chromatography column (300 × 7.8 mm, 5 μm particle size, 500 Å pore size, Phenomenex). Column flow rate 1.5 mL/min of 10 mM KH_2PO_4 containing 1 mM disodium ethylene diamine triacetic acid (EDTA), pH 7.2.
18. Methoxy-polyethyleneglycol-amine (molecular weight 5000 Da, Laysan Bio).
19. 0.45 μm Spin filter (nylon membrane, PALL).

3 Methods

For expression and purification of genome-free MS2 viral capsids, please *see* Chapter 21, Subheading 3.1.

3.1 Synthesis of Nitrophenol-NHS Conjugate

1. Activate 3-(4-hydroxy-3-nitrophenyl)propanoic acid (1 equiv) with *N*-hydroxysuccinimide (1.1 equiv) by stirring in DCM at room temperature in the presence of 1-ethyl-3-(3-dimethylaminopropyl)carbodiimide (EDC, 1.1 equiv) (*see* **Note 2**). A typical reaction scale is ~300 mg acid (1.5 mmoles). Allow to react at room temperature for 3 h. An orange color should be observed.
2. Pour the crude mixture into water and extract with 10 mL DCM twice. Dry the combined organic extracts with Na_2SO_4, then perform gravity filtration using filter paper, and concentrate via rotary evaporation.
3. Additional purification by silica gel chromatography with hexane–ethyl acetate mobile phase will yield the pure product as a bright yellow solid after rotary evaporation. Use 100 mL each of 90:10, 80:20, 70:30, 60:40 v/v mixtures. The product should elute with the 70:30 solvent.

3.2 Synthesis of MS2-DNA Conjugates for Delivery of Singlet Oxygen Generators

To prevent oxidation of the internal Cys residues by $NaIO_4$, the internal cargo has to be attached before the external modification takes place.

3.2.1 Internal Modification with Porphyrin

1. Mix a solution of MS2 T19*p*AF N87C in 10 mM phosphate buffer, pH 7.2 (final concentration 50–100 μM) with 10–20 equiv of porphyrin (or other singlet oxygen generator) maleimide as a 50 mM solution in DMSO. Allow the reaction to proceed for 2 h in the dark at room temperature (*see* **Note 3**).
2. Purify using a NAP-5 column equilibrated with 10 mM phosphate buffer, pH 7.2 to remove the excess porphyrin. Concentrate the conjugates using a spin concentrator with a 100 kDa molecular weight cutoff filter.
3. Characterize the extent of modification by electrospray ionization mass spectrometry (ESI-MS) and absorbance measurements (*see* **Note 4**).

3.2.2 DNA Modification with Nitrophenol

1. React the DNA of interest containing a 5′-end primary amine (300 μM final concentration, typical reaction scale 1 mg) with nitrophenol–NHS (*see* Subheading 3.1, 60–120 equiv) in 1:1 solution of DMF and 50 mM phosphate buffer, pH 8.0. Allow to react at room temperature for 1.5 h.
2. Purify the modified DNA by using NAP-5 gel filtration column. The product should be yellow due to the nitrophenol moiety.
3. Reduce the installed nitrophenol functionality by using a 100 mM stock solution of $Na_2S_2O_4$ in 200 mM phosphate buffer, pH 6.5 for a final $Na_2S_2O_4$ concentration of 10 mM. Allow to react for 20 min at room temperature. A change of color from pale yellow to colorless should be observed. Purify using NAP-5 desalting columns, then lyophilize for prolonged storage. When needed, resuspend in 50 mM phosphate buffer, pH 7.0 to a final concentration of 1 mM (determine the concentration by measuring the absorbance at 260 nm).

3.2.3 Aminophenol Conjugation to MS2

1. React MS2 capsids with the *p*AF unnatural amino acid (from Subheading 3.2.1) at a final concentration of 20 μM with aminophenol-DNA conjugate (from Subheading 3.2.2) at a final concentration of 200 μM in 50 mM phosphate buffer, pH 7.0, containing 150 mM NaCl. A typical reaction scale is 1 mg protein. Add $NaIO_4$ to a final concentration of 1 mM and allow to react at room temperature for 5 min. The solution can be quenched by addition of 10 equiv of TCEP solution (pH 7.0) per mole of oxidant.
2. Purify the MS2-DNA conjugate by using a NAP-5 gel filtration column and spin-concentrate using 100 kDa molecular weight cutoff filters.
3. Verify conjugation using SDS-PAGE analysis. A mass shift due to the DNA attachment should be observed (most likely a band will be observed for the MS2 monomer and a second band will have the additive molecular weight of the MS2 monomer and the DNA segment).

3.3 Synthesis of Fluorescently Labeled MS2-Peptide Conjugates for Imaging Applications

To prevent oxidation of the internal Cys residues by $NaIO_4$, the internal cargo has to be attached before the external modification takes place.

3.3.1 Internal Modification with Fluorescent Dyes

1. Mix a solution of MS2 T19*p*AF N87C in 10 mM phosphate buffer, pH 7.2 (final concentration 50–100 μM) with 10–20 equiv of 10 mM dye maleimide. Allow the reaction to proceed for 2 h in the dark at room temperature.
2. Purify using a NAP-5 column equilibrated with 10 mM phosphate buffer, pH 7.2 to remove the excess dye. Concentrate the conjugates using a spin concentrator with 100 kDa molecular weight cutoff filter.

3. Characterize the extent of modification by electrospray ionization mass spectrometry (ESI-MS) and absorbance measurements (*see* **Notes 5** and **6**).

3.3.2 Peptide Modification with Aminophenol

1. Synthesize the peptide of interest by using Solid Phase Peptide Synthesis (SPPS) methods [33] or purchase from commercial vendors.
2. Attach the nitrophenol moiety by reacting a solution of the peptide (100 μM final concentration in 10 mM phosphate buffer, pH 7.2) with nitrophenol-NHS (10 equiv from DMSO stock, *see* Subheading 3.1) overnight at room temperature (*see* **Note 7**).
3. Purify the nitrophenol–peptide using reverse-phase HPLC. The product should be yellow due to the nitrophenol moiety.
4. Reduce the installed nitrophenol functionality by using a 100 mM solution of $Na_2S_2O_4$ in 200 mM phosphate buffer, pH 6.5 for a final concentration of 10 mM. Allow to react for 20 min. Purify using Sep-Pak C18 columns, then lyophilize for prolonged storage.

3.3.3 Decorating the MS2 Surface with Peptides

1. React MS2 capsids with the *p*AF unnatural amino acid at a final concentration of 20 μM with aminophenol-peptides (*see* Subheading 3.3.2) at a final concentration of 200 μM in 50 mM phosphate buffer, pH 7.2. Add $NaIO_4$ to a final concentration of 1 mM and allow to react at room temperature for 5 min.
2. Purify the MS2-peptide conjugate by using a NAP-5 gel filtration column and spin-concentrate using 100 kDa molecular weight cutoff filters. Wash three times with 10 mM phosphate buffer, pH 7.2.
3. Verify conjugation using SDS-PAGE analysis or ESI-MS.

3.4 Synthesis of Radiolabeled MS2-Antibody Conjugates for Positron Emission Tomography

3.4.1 Internal Modification with Chelators

To prevent oxidation of the internal Cys residues by the $NaIO_4$, the internal cargo has to be attached before the external modification takes place.

1. Mix a solution of MS2 T19*p*AF N87C (final concentration 50–100 μM) with 10 equiv of the maleimide derivative of the chelator of interest as a 10 mM solution in DMSO (*see* **Note 3**). A typical reaction scale is 1 mg protein. Add a ¼ volume of the reaction solution of 100 mM phosphate buffer pH 8.0 to ensure optimal pH for the reaction (*see* **Note 8**). Allow the reaction to proceed for 2 h at room temperature.
2. Purify using a NAP-5 column equilibrated with 10 mM phosphate buffer, pH 7.2 to remove the excess chelator. Concentrate the conjugates using a spin-concentrator with 100 kDa molecular weight cutoff filter. Wash three times with 10 mM phosphate buffer, pH 7.2.
3. Characterize the extent of modification by ESI-MS.

3.4.2 Synthesis of Aminophenol-Antibody Conjugates

1. Buffer-exchange an antibody solution (usually 1 mL of 1 mg/mL solution) to 25 mM phosphate buffer pH 8.0 by using 30 kDa molecular weight cutoff spin concentrators.
2. Add 5 equiv of NHS-nitrophenol from a DMSO stock (*see* **Note 3**, *see* Subheading 3.1). Incubate at room temperature for 1 h 15 min.
3. Reduce the nitrophenol groups to aminophenol by adding 100 mM $Na_2S_2O_4$ to reach a final concentration of 10 mM. Allow the reaction to proceed at room temperature for 20 min.
4. Remove the excess small molecules using spin concentrators with a molecular weight cutoff of 30 kDa and wash three times with 10 mM pH 7.2 phosphate buffer. The concentration of the final product can be measured by UV absorbance at 280 nm (using $\varepsilon = 212{,}000\ M^{-1}\ cm^{-1}$).

3.4.3 Generation of MS2-Antibody Conjugates

1. Mix a solution of MS2 with the *p*AF unnatural amino acid on the external surface (from Subheading 3.4.1) with a solution of aminophenol-antibody (from Subheading 3.4.2). The ratio of MS2 capsid and antibody can be varied to generate conjugates with different numbers of antibodies attached to the VLP. A good starting point for the final concentration of the capsid is 50 μM MS2 monomer (~300 nM capsid), and ratios of 1–10 equiv of antibody per capsid.
2. Add $NaIO_4$ to reach a final concentration of 500 μM, and allow the reaction to proceed at room temperature for 5 min.
3. Remove the excess $NaIO_4$ by using NAP-5 size exclusion columns equilibrated with 10 mM pH 7.2 phosphate buffer. If needed, concentrate the product by using a 100 kDa MWCO spin concentrator. Wash three times with 10 mM phosphate buffer pH 7.2.
4. Analyze the extent of modification by using SDS-PAGE and subsequent optical densitometry. The light and heavy chains modified with the MS2 monomer protein (~14 kDa gel shift), as well as the unmodified denatured chains will be present.

3.4.4 Radiolabeling of MS2-Antibody Conjugates

1. Mix a solution of MS2-antibody-chelator conjugates (50–100 μL of 200 μM in MS2 monomer, Subheading 3.4.3) with excess $^{64}CuCl_2$ diluted in 100 mM ammonium acetate, pH 6.2 (100 μL of ~30 mCi/mL). Incubate at room temperature for 2 h.
2. Remove excess radioisotope by using NAP-5 desalting columns and 100 kDa spin filters, Wash three times with sterile PBS.
3. Verify the purity of the sample by using HPLC SEC with an in-line radioactivity detector. Use the UV trace to determine the concentration of the protein against a calibration curve.

3.5 Synthesis of PEGylated MS2 Viral Capsids for Delivery of Doxorubicin via $K_3[Fe(CN)_6]$ Oxidation

To prevent oxidation of the doxorubicin sugar moiety, a milder catalyst, potassium ferricyanide, $K_3[Fe(CN)_6]$, can be used. It is still recommended to attach the internal cargo before proceeding with the external surface modification.

3.5.1 Internal Modification with Chemotherapeutic Cargo

1. React a solution of the MS2 mutant containing the internal Cys residue (1 equiv, 50 μM final concentration) with a maleimide derivative of doxorubicin (0.75 equiv) in 10 mM phosphate buffer pH 7.2 for 3 h at room temperature [34], *see* **Note 9**. A typical reaction scale is 1 mg protein.
2. Remove the excess doxorubicin by using NAP-5 size exclusion columns equilibrated with 10 mM phosphate buffer pH 7.2. If needed, concentrate the product by using a 100 kDa MWCO spin concentrator. Wash three times with 10 mM pH 7.2 phosphate buffer.
3. Analyze the extent of modification by using ESI-MS, SDS-PAGE, and UV-Vis analysis (*see* **Note 10**).

3.5.2 Synthesis of Methoxy-PEG_{5k}-Aminophenol

1. Dissolve methoxy-polyethyleneglycol-amine with a molecular weight of 5000 Da (1 equiv) in DCM and add 5 equiv of nitrophenol-NHS ester (stock solution in DMSO, *see* Subheading 3.1). A typical reaction scale is 10 mg polymer. Add 10 equiv of triethylamine and stir the reaction at room temperature overnight. Note that the reaction mixture turns orange upon addition of triethylamine.
2. Evaporate the solvent under a gentle stream of nitrogen, and remove the residual DMSO under high vacuum.
3. Remove the excess nitrophenol-NHS by precipitating it with ~500 μL of water. Filter the crude reaction mixture through a 0.45 μm spin filter.
4. Concentrate the nitrophenol-PEG_{5k} using 3 kDa MWCO spin concentrators. Additional purification by RP-HPLC can be performed.
5. Lyophilize the nitrophenol-PEG_{5k}-OMe product for prolonged storage.
6. The nitrophenol moiety can be reduced using a 100 mM $Na_2S_2O_4$ solution with a final concentration of 10 mM. The reaction can be performed at room temperature for 20 min until the loss of yellow color (due to the nitrophenol reduction) can be observed.
7. Remove excess reducing agent by using 3 kDa MWCO spin concentrators and washing five times with water. Lyophilize the aminophenol-PEG_{5k}-OMe product for prolonged storage.
8. Dissolve the aminophenol-PEG_{5k}-OMe in water and adjust the solution to 1 mM concentration.

3.5.3 Attachment of PEG Chains onto the MS2 VLP

1. Mix the MS2 *p*AF mutant solution (1 equiv, 20–50 μM final concentration) with the PEG solution (2–5 equiv). Add the $K_3[Fe(CN)_6]$ catalyst at a final concentration of 5 mM. Allow the reaction to proceed for 30 min to 1 h at room temperature.
2. Remove the excess small molecules by using NAP-5 size exclusion columns equilibrated with 10 mM phosphate buffer pH 7.2. If needed, concentrate the product by using a 100 kDa MWCO spin concentrator. Wash three times with 10 mM phosphate buffer pH 7.2.
3. Analyze the extent of modification by using SDS-PAGE and optical densitometry. A mass shift should be observed. Usually, the PEG-modified MS2 monomers run higher than expected (~23 kDa versus 19 kDa expected).

3.6 Alternative Strategies for MS2 Surface Modification: Tyr and Trp Modification with CAN

Wild-type MS2 includes a number of native tyrosine and tryptophan residues that are not very solvent-accessible and not very reactive. A tyrosine residue on the interior surface of MS2 (Y85) has been successfully modified using tyrosine modification strategies developed in our group [11]. However, the location of this residue prevents large moieties from being attached, since the diffusion through the pores would limit the access of these reagents to the site of interest.

To circumvent these issues, our group has genetically engineered a mutant of MS2 that contains tyrosine (Tyr) and tryptophan (Trp) residues in solvent-exposed positions [15, 19] on the surface of the capsid. These residues can be modified in the presence of cerium(IV) ammonium nitrate (CAN) with phenylene diamine (both Tyr and Trp, [8, 9]) or anisidine derivatives (only Tyr, [8]). The protocol presented herein describes the modification of MS2 mutants with PEG_{5k} chains, although other modifications (PEG_{2k}, peptides, DNA, dye molecules) are also available through similar strategies [9].

3.6.1 Synthesis of Phenylene Diamine- and Anisidine-PEG_{5k} Derivatives

1. Synthesize the NHS esters of phenylene diamine and an anisidine derivative as described in [9, 18].
2. An anisidine-PEG_{5k} derivative can be synthesized by reacting a solution of methoxy-polyethyleneglycol-amine (5000 Da molecular weight, 1 equiv, 12.5 mM in H_2O) with a solution of 2,5-dioxopyrrolidin-1-yl 6-(4-(dimethylamino)phenoxy) hexanoate (NHS-anisidine derivative, 10 equiv, 40 mM in DMF). Add 2 equiv of *N,N*-diisopropylethylamine (DIPEA) and stir for 2 h at room temperature. A typical reaction scale is 10 mg polymer.
3. Remove the solvents in vacuo and resuspend in H_2O. Filter using a 0.45 μm spin filter.
4. Purify the resulting PEG derivative by using a NAP-5 size exclusion column. Elute in water and lyophilize for prolonged storage. A similar protocol can be followed for the phenylene diamine derivatives.

3.6.2 Attachment of PEG Chains to Tyr or Trp on the External Surface of MS2

1. Dissolve the MS2 mutants with Tyr or Trp at positions 15 or 19 at a final concentration of 25 μM (1 equiv) in 50 mM bis-Tris, pH 6.0. A typical reaction scale is 1 mg protein. Add the PEG-phenylene diamine or anisidine derivative to a final concentration of 500 μM (20 equiv). Add a fresh 25 mM solution of cerium(IV) ammonium nitrate (CAN) in dd-H_2O to reach a final concentration of 1.5 mM. Allow the mixture to react at room temperature for 1 h.
2. Quench the reaction by adding a solution of TCEP in dd-H_2O to a final concentration of 3 mM.
3. Analyze the extent of modification by SDS-PAGE and optical densitometry.

4 Notes

1. It is very important to store the dithionite tightly closed in a dry and well-ventilated place. Handle and store under inert gas and never allow the dithionite to get in contact with water during storage since the compound is air-, heat-, and moisture-sensitive. Please note that, if stored at 4 °C, the dithionite should be allowed to equilibrate to room temperature before opening the container.
2. A typical reaction scale setup is as follows: Add 3-(4-hydroxy-3-nitrophenyl)propanoic acid (325 mg, 1.54 mmol, 1.0 equiv) and *N*-hydroxysuccinimide (195 mg, 1.69 mmol, 1.1 equiv) to a scintillation vial and dissolve in 15 mL of DCM with stirring. Add EDC (325 mg, 1.69 mmol, 1.1 equiv), and the reaction mixture will immediately turn orange. Upon purification by silica gel chromatography, the product can be isolated as a bright yellow solid using rotary evaporation. Usual yields are ~60%.
3. Ensure that the final concentration of organic solvent is not higher than 5% in order to prevent denaturation of the protein structure.
4. To calculate the conjugation efficiency of the porphyrin to the MS2 capsids, measure the UV spectrum using a NanoDrop or similar devices. Use the absorbance of the porphyrin measured at the Soret band (ε = 266,000 M^{-1} cm^{-1} at 410 nm) and the absorbance of the protein at 260 nm (ε = 172,000 M^{-1} cm^{-1}) to determine the molar ratio (assume negligible porphyrin absorbance at 260 nm).
5. Use the extinction coefficient provided by the supplier and the absorbance measured at the maximum absorbance peak.
6. A high degree of modification can lead to fluorophore quenching due to the proximity of the Cys residues on the internal surface of the MS2 capsids. Subequimolar amounts of

dyes and shorter reaction times have been shown to modify the capsids sufficiently for flow cytometry and fluorescence microscopy experiments.

7. Alternatively, the nitrophenol can be attached while the peptide is still on beads, with 10 equiv nitrophenol-NHS, 10 equiv 2-(6-chloro-1*H*-benzotriazole-1-yl)-1,1,3,3-tetramethylaminium hexafluorophosphate (HCTU) and 20 equiv *N,N*-diisopropylethylamine (DIPEA) in DMF overnight at room temperature. Another approach is to convert a tyrosine residue into an oxidative coupling partner, as described in [3]). If necessary, cleave the peptide off the beads using appropriate conditions based on the nature of the resin.
8. The chelators may contain counterions that are fairly acidic, and thus the pH of the reaction mixture could be affected.
9. Extreme caution should be taken when working with doxorubicin due to its toxic effects on humans.
10. Doxorubicin absorbs at 490 nm with an extinction coefficient of $\varepsilon = 11{,}000\ M^{-1}\ cm^{-1}$ and has a weak fluorescence at 512 nm.

Acknowledgments

This work was supported by the Genentech Fellowship, the DOD Breast Cancer Research Program grant W81XWH-14-0400, and the NIH grant 5R21EB018055-02.

References

1. Wu W, Hsiao SC, Carrico ZM, Francis MB (2009) Genome-free viral capsids as multivalent carriers for taxol delivery. Angew Chem Int Ed 48:9493–9497
2. Ashley CE, Carnes EC, Phillips GK, Durfee PN, Buley MD, Lino CA, Padilla DP, Phillips B, Carter MB, Willman CL, Brinker CJ, Caldeira J do C, Chackerian B, Wharton W, Peabody DS (2011) Cell-specific delivery of diverse cargos by bacteriophage MS2 virus-like particles. ACS Nano 5:5729–5745
3. Obermeyer AC, Capehart SL, Jarman JB, Francis MB (2014) Multivalent viral capsids with internal cargo for fibrin imaging. PLoS One 9(6):e100678. https://doi.org/10.1371/journal.pone.0100678
4. Hermanson GT (2008) Bioconjugate techniques. Academic Press, Cambridge
5. Tilley SD, Joshi NS, Francis MB (2009) Proteins: chemistry and chemical reactivity. In: Wiley encyclopedia of chemical biology, vol 4. John Wiley & Sons, Inc., San Francisco, pp 158–174
6. Stephanopoulos N, Francis MB (2011) Choosing an effective protein bioconjugation strategy. Nat Chem Biol 7:876–884
7. Antos JM, Francis MB (2004) Selective tryptophan modification with rhodium carbenoids in aqueous solution. J Am Chem Soc 126:10256–10257
8. Seim KL, Obermeyer AC, Francis MB (2011) Oxidative modification of native protein residues using cerium(IV) ammonium nitrate. J Am Chem Soc 7:876–884
9. Seim KL (2013) Development of oxidative bioconjugation methodology for the site-selective modification of the electron rich aromatic amino acids. Dissertation, University of California, Berkeley
10. Hooker JM, Esser-Kahn AP, Francis MB (2006) Modification of aniline containing proteins using an oxidative coupling strategy. J Am Chem Soc 128:15558–15559
11. Joshi NS, Whitaker LR, Francis MB (2004) Three-component Mannich type reaction for

selective tyrosine bioconjugation. J Am Chem Soc 126:15942–15943

12. Tilley SD, Francis MB (2006) Tyrosine-selective protein alkylation using p-allylpalladium complexes. J Am Chem Soc 128:1080–1081
13. Witus LS, Netirojjanakul C, Palla KS, Muehl EM, Weng CH, Iavarone AT, Francis MB (2013) Site-specific protein transamination using *N*-methylpyridinium-4-carboxaldehyde. J Am Chem Soc 135:17223–17229
14. Dixon HBF, Fields R (1972) Specific modification of NH_2-terminal residues by transamination. Methods Enzymol 25:409–419
15. Palla KS, Witus LS, Mackenzie KJ, Netirojjanakul C, Francis MB (2014) Optimization and expansion of a site selective *N*-methylpyridinium-4-carboxaldehyde mediated transamination for bacterially expressed proteins. J Am Chem Soc 137:1123–1129
16. MacDonald JI, Munch HK, Moore T, Francis MB (2015) One-step site-specific modification of native proteins with 2-pyridinecarboxaldehydes. Nat Chem Biol 11:326–331
17. Obermeyer AC, Jarman JB, Francis MB (2014) N-terminal modification of proteins with o-aminophenols. J Am Chem Soc 136:9572–9579
18. Carrico ZM, Romanini DW, Mehl RA, Francis MB (2008) Oxidative coupling of peptides to a virus capsid containing unnatural amino acids. Chem Commun 10:1205–1207
19. Behrens CR, Hooker JM, Obermeyer AC, Romanini DW, Katz EM, Francis MB (2011) Rapid chemoselective bioconjugation through the oxidative coupling of anilines and aminophenols. J Am Chem Soc 133:16398–16401
20. Obermeyer AC, Jarman JB, Netirojjanakul C, El Muslemany K, Francis MB (2013) Mild bioconjugation through the oxidative coupling of ortho-aminophenols and anilines with ferricyanide. Angew Chem Int Ed 53:1057–1061
21. ElSohly AM, Francis MB (2015) Development of oxidative coupling strategies for site-selective protein modification. Acc Chem Res 48:1971–1978
22. Tong GJ, Hsiao SC, Carrico ZM, Francis MB (2009) Viral capsid DNA aptamer conjugates as multivalent cell-targeting vehicles. J Am Chem Soc 131:11174–11178
23. Stephanopoulos N, Carrico ZM, Francis MB (2009) Nanoscale integration of sensitizing chromophores and porphyrins with bacteriophage MS2. Angew Chem Int Ed 48:9498–9502
24. ElSohly AM, Netirojjanakul C, Aanei IL, Jager A, Bendall S, Farkas ME, Nolan GP, Francis MB (2015) Synthetically modified viral capsids as versatile carriers for use in antibody-based cell targeting. Bioconjug Chem 26:1590–1596
25. Stephanopoulos N, Tong GJ, Hsiao SC, Francis MB (2010) Dual-surface modified virus capsids for targeted delivery of photodynamic agents to cancer cells. ACS Nano 4:6014–6020
26. Datta A, Hooker JM, Botta M, Francis MB, Aime S, Raymond KN (2008) High relaxivity gadolinium hydroxypyridonate-viral capsid conjugates: Nanosized MRI contrast agents. J Am Chem Soc 130:2546–2552
27. Garimella PD, Datta A, Romanini DW, Raymond KN, Francis MB (2011) Multivalent, high-relaxivity MRI contrast agents using rigid cysteine-reactive gadolinium complexes. J Am Chem Soc 133:14704–14709
28. Meldrum T, Seim KL, Bajaj VS, Palaniappan KK, Wu W, Francis MB, Wemmer DE, Pines A (2010) A xenon-based molecular sensor assembled on an MS2 viral capsid scaffold. J Am Chem Soc 132:5936–5937
29. Farkas ME, Aanei IL, Behrens CR, Tong GJ, Murphy ST, O'Neil JP, Francis MB (2013) PET imaging and biodistribution of chemically modified bacteriophage MS2. Mol Pharm 10:69–76
30. Hooker JM, Kovacs EW, Francis MB (2004) Interior surface modification of bacteriophage MS2. J Am Chem Soc 126:3718–3719
31. Kovacs EW, Hooker JM, Romanini DW, Holder PG, Berry KE, Francis MB (2007) Dual-surface-modified bacteriophage MS2 as an ideal scaffold for a viral capsid-based drug delivery system. Bioconjug Chem 18:1140–1147
32. Mehl RA, Anderson JC, Santoro SW, Wan L, Martin AB, King DS, Horn DM, Schultz PG (2003) Generation of a bacterium with a 21 amino acid genetic code. J Am Chem Soc 125:935–939
33. Amblard M, Fehrentz J-A, Subra G (2006) Methods and protocols of modern solid phase peptide synthesis. Mol Biotechnol 33:239–254
34. Finbloom JA, Han K, Aanei IL, Hartman EC, Finley DT, Dedeo MT, Fishman M, Downing KH, Francis MB (2016) Stable disk assemblies of a tobacco mosaic virus mutant as nanoscale scaffolds for applications in drug delivery. Bioconjug Chem 27:2480–2485

Chapter 41

Virus-Based Cancer Therapeutics for Targeted Photodynamic Therapy

Binrui Cao, Hong Xu, Mingying Yang, and Chuanbin Mao

Abstract

Cancer photodynamic therapy (PDT) involves the absorption of light by photosensitizers (PSs) to generate cytotoxic singlet oxygen for killing cancer cells. The success of this method is usually limited by the lack of selective accumulation of the PS at cancer cells. Bioengineered viruses with cancer cell-targeting peptides fused on their surfaces are great drug carriers that can guide the PS to cancer cells for targeted cancer treatment. Here, we use cell-targeting fd bacteriophages (phages) as an example to describe how to chemically conjugate PSs (e.g., pyropheophorbide-a (PPa)) onto a phage particle to achieve targeted PDT.

Key words Bacteriophage, Bioconjugation, Photosensitizer (PS), Cancer cell targeting peptide, Photodynamic therapy (PDT)

1 Introduction

Photodynamic therapy (PDT) involves the absorption of light by a photoactive chemical (known as photosensitizer or PS) to enable the generation of cytotoxic singlet oxygen ($^1O_2$) from endogenous oxygen [1–3]. For cancer treatment, a PS needs to accumulate in a tumor and be irradiated with an appropriate wavelength of light to generate singlet oxygen [4]. The cytotoxic singlet oxygen induces immediate death of cancer cells or destruction of tumor tissues via apoptosis and necrosis [5]. Although PDT has been used for treating several types of cancer [6], this method still faces the problem of lacking selective accumulation of PSs at cancer cells.

The filamentous bacteriophage fd (phage in the following) is a bionanofiber (~7 nm wide and 1 μm long) that specifically infects bacteria [7]. Each phage particle is composed of five genetically modifiable coat proteins surrounding a circular single-stranded (ss) DNA that encodes the coat proteins. Specifically, ~3900 copies of the major coat protein (pVIII) form the main body of the phage while five copies of two out of the four minor coat proteins (pIII

Christina Wege and George P. Lomonossoff (eds.), *Virus-Derived Nanoparticles for Advanced Technologies: Methods and Protocols*, Methods in Molecular Biology, vol. 1776, https://doi.org/10.1007/978-1-4939-7808-3_41,
© Springer Science+Business Media, LLC, part of Springer Nature 2018

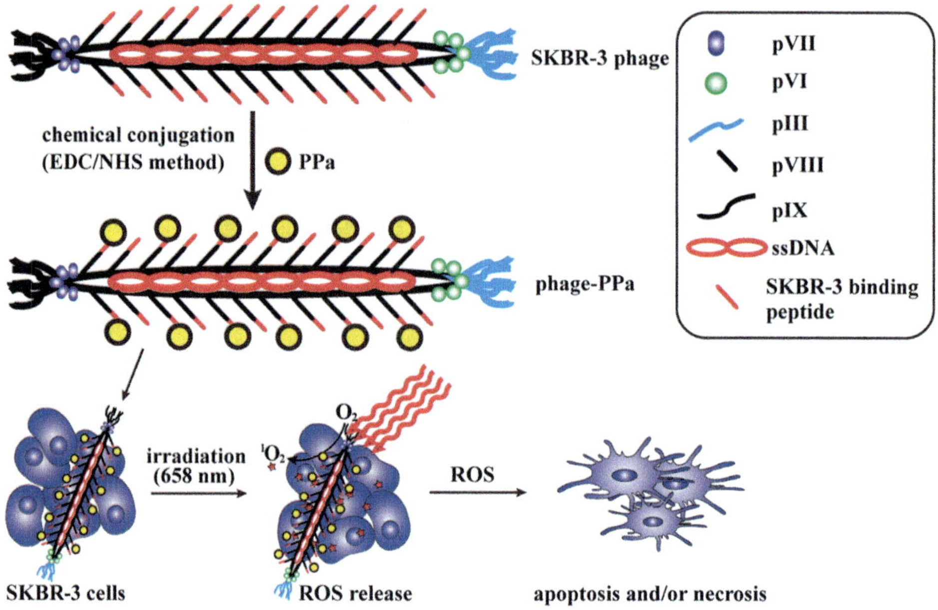

Fig. 1 Scheme of photosensitizer conjugation onto phage for photodynamic therapy (PDT)

and pVI together, and pVII and pIX together) constitute the two ends, respectively [8]. The coat protein surface of the phage can be genetically modified by fusing a foreign peptide sequence to the N-terminal (the solvent exposed) end of one or more of the coat proteins, a technique also known as phage display [9]. We have previously applied phage display to obtain a unique SKBR-3 breast cancer cell-specific phage [10]. The breast cancer cell-targeting phage has great potential as a novel photosensitizer carrier to achieve targeted PDT against cancer cells.

In this chapter, we describe the use of cancer cell-targeting phage as a novel PS carrier to achieve targeted cancer PDT; the SKRB-3 breast cancer cell is used as the target cell. Our previously identified SKBR-3-specific phage that surface displays a SKBR-3-targeting peptide, VSSTQDFP, on each copy of its pVIII major coat protein is employed as the PS carrier. Since the detailed protocol of the affinity selection of a cell-targeting phage has been previously described by us [11], this chapter is focused on how to use selected cell-targeting phages for cancer PDT. Basically, the procedure includes two main steps. *Step One: Partial modification of cancer cell-targeting phages with PS* (*see* Fig. 1). Our previously identified SKBR-3 breast cancer cell-targeting phages (termed as SKBR-3 phage) are amplified by infecting host bacteria and purified through double polyethylene glycol (PEG) precipitation. Then, the SKBR-3 phages are partially (~50%) modified with a PS, pyropheophorbide-a (PPa) [12], to form the phage-PPa complex (termed PPPa) through a well-established EDC/NHS (EDC: 1-ethyl-3-(-3-dimethylaminopropyl) carbodiimide hydrochloride; NHS: N-hydroxysuccinimide) chemistry [13]. After that, the photophysical properties and stability of the PPPa are tested before further use. *Step Two: Selective killing of cancer cells by PPPa through PDT* (*see* Fig. 1). The specific binding of PPPa to SKBR-3 cells and the selective killing of SKBR-3 breast cancer cells by PPPa upon radiation with a red light at 658 nm (in vitro PDT) are evaluated respectively. Through the two steps, we successfully fabricated PPPa and demonstrated its ability to selectively kill breast cancer cells [14]. Although we only show the protocol of phage-based cancer PDT, the same strategy can be extended to other genetically modifiable viruses, such as T7 phage and tobacco mosaic virus.

2 Materials

Prepare all solutions and gels using ultrapure water and analytical grade reagents. All the test tubes, flasks, cell culture plates, centrifuge tubes, and bottles should be sterile or autoclaved before use.

2.1 SKBR-3 Phage Amplification, Purification, and Quantification

1. *Escherichia coli* (*E. coli*) strain K91 Blukan cells (from Dr. George Smith at the University of Missouri) (*see* **Note 1**).
2. SKBR-3 phages (displaying a sequence of VSSTQDFP [10]) with a concentration of about 1×10^{13} virions/ml (*see* **Note 2**).
3. Luria broth (LB) medium: Mix 25 g of LB powder (10 g/l Bacto tryptone, 5 g/l yeast extract, 10 g/l NaCl) in 1 l of distilled water. Autoclave for 30–60 min at 121 °C.
4. 70 mg/ml kanamycin (stock solution). Filter through a sterile 0.22 μm filter and store at −20 °C. Use at 1:1000 dilution in LB broth.
5. 20 mg/ml tetracycline (stock solution) in 100% ethanol, aliquot and store at −20 °C. Use at 1:1000 dilution in LB broth.
6. PEG (stock solution): 2.5 M NaCl, 20% (w/v) PEG-8000 (*see* **Note 3**).
7. 1× TBS buffer: 8 g/l NaCl, 0.2 g/l KCl, 3 g/l Tris base, adjust pH to 7.4 with 1 M HCl, autoclave for 30 min at 121 °C to sterilize. Store at room temperature.
8. UV-Vis spectrometer.

2.2 Chemical Conjugation of PPa with SKBR-3 Phages to Form PPPa

1. Dimethyl sulfoxide (DMSO).
2. Pyropheophorbide-a (PPa).
3. 1-ethyl-3-(-3-dimethylaminopropyl) carbodiimide hydrochloride (EDC).
4. 2-mercaptoethanol.
5. N-hydroxysuccinimide (NHS).
6. Stock SKBR3 phage solution (1×10^{13} virions/ml) (from Subheading 3.1).
7. Dialysis tubing with a cut-off of 8–10 kD MWCO.

2.3 Tests of PPPa

1. PPPa (from Subheading 3.2).
2. 1% (w/v) uranyl acetate staining solution. Filter through 0.45 μm filter and store at 4 °C.
3. 400 mesh formvar-coated copper TEM grid (e.g., Ted Pella, 01822-F).
4. Low voltage transmission electron microscopy (TEM).
5. 50 nM 1, 3-diphenylisobenzofuran (DPBF) in DMSO.
6. Dimethyl sulfoxide (DMSO).
7. 100 μl quartz cuvettes.
8. Spectrophotometer.
9. UV-Vis spectrometer.

2.4 In Vitro PDT Studies

1. SKBR-3 breast cancer cells.
2. MCF 10A human breast epithelial cells.
3. McCoy's 5A Medium.
4. Eagle's Minimum Essential Medium (EMEM) containing Earle's Balanced Salt Solution, nonessential amino acids, 2 mM L-glutamine, 1 mM sodium pyruvate, and 1500 mg/l sodium bicarbonate.
5. Fetal bovine serum (FBS).
6. Fluorescence-based live/dead cell viability kit (L3224, ThermoFisher Scientific).
7. 5 mg/ml 3-(4, 5-dimethylthiazolyl-2)-2, 5-diphenyltetrazolium bromide (MTT). Prepare freshly.
8. 10 μM PPPa solution in McCoy's 5A Medium supplemented with 10% FBS.
9. 10 μM PPPa solution in EMEM supplemented with 10% FBS.
10. CO_2-humidified incubator.
11. Red laser light source (600–800 nm, Modulight, Inc.).
12. Cell culture flasks and plates.
13. Microplate reader.
14. Fluorescence microscope.
15. Laser power meter.
16. 1× PBS buffer: 8 g/l NaCl, 0.2 g/l KCl, 1.44 g/l Na_2HPO_4, 0.24 g/l KH_2PO_4, adjust pH to 7.4 with 1 M HCl, autoclave for 30 min at 121 °C to sterilize. Store at room temperature.

3 Methods

Bacteria are always cultured in a shaking incubator at 37 °C with a speed of 220 rpm unless otherwise indicated. Mammalian cells are always cultured in a CO_2 incubator at 37 °C in 5% CO_2-humidified atmosphere in cell culture flasks. When using commercial kits, we follow the manufacturer's recommended protocol unless otherwise stated.

3.1 SKBR-3 Phage Amplification, Purification, and Quantification

1. Inoculate 5 ml of LB media with 2 μl of stored K91 Blukan *E. coli* in a test tube.
2. Add 5 μl of kanamycin stock solution into the tube and incubate overnight.
3. Transfer 200 μl of the overnight culture into 40 ml of LB medium in a 250-ml flask and incubate until the OD_{600} reaches 0.4–0.6.
4. Mix 200 μl of the mid-log culture with 10 μl of the stocked SKBR-3 phage solution in an Eppendorf microcentrifuge tube and allow the mixture to stay at room temperature for 15 min without shaking (*see* **Note 4**).
5. Transfer all of the mixtures into 1 l of LB medium in a 4-l flask.
6. Add 1 ml of the tetracycline stock solution into the flask and incubate for 16 h (*see* **Note 5**).
7. Purify and enrich the amplified phage particles, as documented in our previously published protocol [15].
8. Determine the concentration of the amplified phage solution by using a UV-Vis spectrum. (*see* **Note 6**) $OD_{269} = 1$ is approximately equal to 7.2×10^{12} virions/ml (*see* **Note 7**).
9. Dilute the phage solution to 1×10^{13} virions/ml.

3.2 Chemical Conjugation of PPa with SKBR-3 Phages to Form PPPa (See Fig. 2)

1. Dissolve 1 mg of PPa in 1 ml of DMSO.
2. Add 0.4 mg of EDC to the PPa solution and incubate at 37 °C for 1 h with gentle shaking to activate the carboxylic group in PPa (*see* **Note 8**).
3. After incubation, add 1.2 μl of 2-mercaptoethanol (*see* **Note 9**).
4. Mix 22.5 μl of the EDC-activated PPa solution (*see* **Note 10**) and 0.6 mg of NHS with 1.28 ml of the stocked SKBR3 phage solution (from Subheading 3.1) (*see* **Notes 11** and **12**) and incubate for 2 h at room temperature to prepare ~50% partially modified PPPa (*see* **Note 13**).
5. Dialyze the mixture against water.
6. Store the produced PPPa solution at −20 °C.

3.3 Tests of PPPa

1. The morphology of PPPa is observed by transmission electron microscopy (TEM) (*see* Fig. 3). PPPa is negatively stained with 1% (w/v) uranyl acetate staining solution on a TEM grid and then visualized under a low-voltage TEM (80 kV). The detailed procedure is well documented in our previously published protocol [15] (*see* **Notes 14**).
2. The successful conjugation of phage with PPa is verified by UV-Vis absorption spectroscopy and fluorescence spectroscopy.

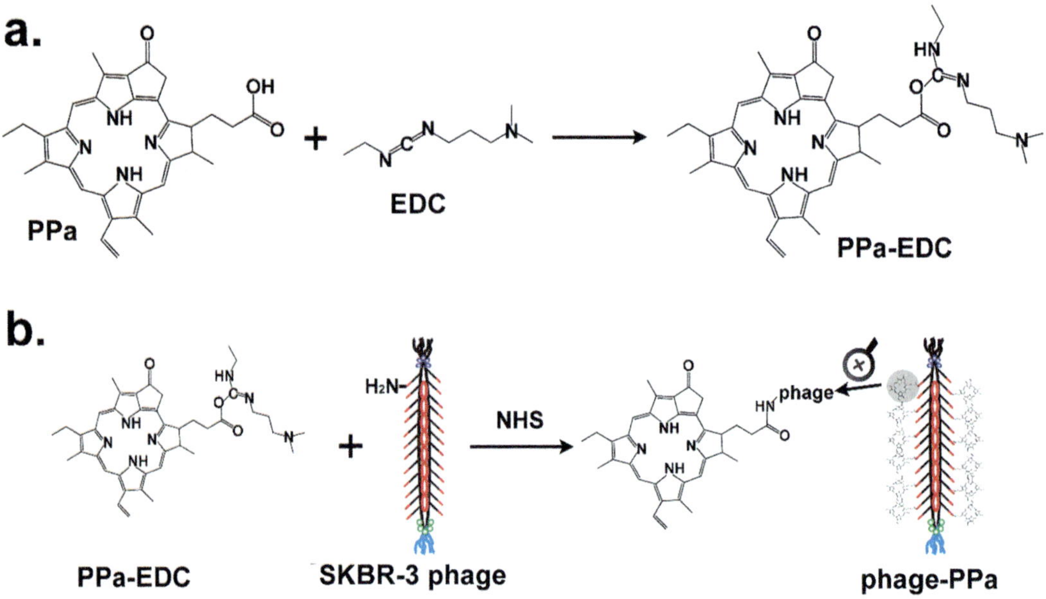

Fig. 2 (**a**) The carboxyl group of PPa is activated by EDC to form EDC-activated PPa. (**b**) EDC-activated PPa reacts with the amino groups on SKBR-3 phage in the presence of NHS to form phage-PPa complex

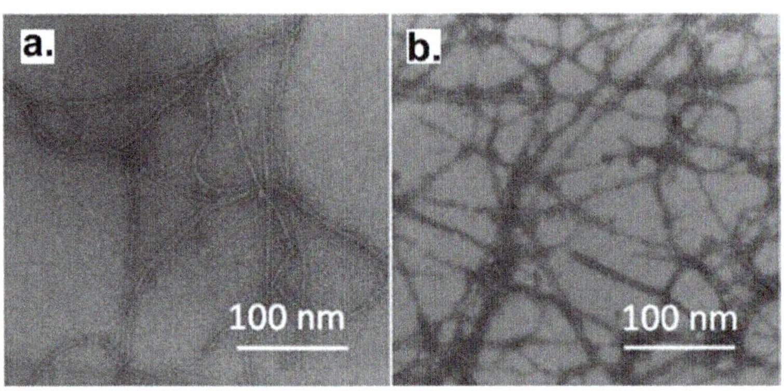

Fig. 3 TEM images of unmodified phage (**a**) and phage-PPa (**b**). Reproduced from [14] with permission

The UV-Vis spectra of phage and PPa are recorded in a 100 μl quartz cell in a mixture of 50% water and 50% DMSO (v/v) solvent at pH 7.0. The fluorescence spectra of PPPa are measured by a spectrophotometer in a 100 μl quartz cell in DMSO (*see* **Note 15**).

3. The generation of singlet oxygen from PPPa is tested by a steady state method using DPBF: 2.3 ml of DPBF (50 nM) solution in DMSO is mixed with 10 μl of PPPa solution in a quartz cell saturated with oxygen. The cell is placed into a UV-Vis spectrometer. The decay of the DPBF absorption is monitored at 0, 10, 30, 60, 90, 120, 150, 180 s at OD_{410} (*see* **Note 16**).

3.4 In Vitro PDT

SKBR-3 breast cancer cells and MCF 10A human breast epithelial cells are cultured in cell culture 24-well microplates in McCoy's 5A medium and EMEM, respectively, supplemented with 10% FBS at 37 °C in 5% CO_2-humidified incubator. Three wells of SKBR-3 breast cancer cells and three wells of MCF 10A human breast epithelial cells are cultured. Culture media are changed every 2 days until cells reach 90% confluence.

1. Incubate SKBR-3 and MCF 10A cells separately in the cell culture medium containing 10 μM PPPa for 1, 3, and 5 h (*see* **Note 17**).
2. Wash the cells three times with PBS buffer and fill the plates with their regular medium with 10% FBS (PPPa-free).
3. Apply a dose of red laser light (658 nm) at a power density of 50 mW/cm^2 to cells for a total of 2 min of exposure. An interval of 30 s is applied between every 10 s of laser exposure (*see* **Note 18**).
4. Following the PDT, incubate cells for additional 3 h.
5. A live/dead cell viability kit is used to evaluate cell death using fluorescence microscopy according to the manufacturer's protocol (*see* **Note 19**).
6. A MTT assay is applied to cells for quantitative estimation of cell viability. Cells are washed with PBS and treated with 15 μl of fresh MTT reagent (5 mg/ml) and incubated in fresh medium for 5 h. After incubation, the medium is removed and 100 μl of DMSO is added. The absorbance at 490 nm of the DMSO solution is measured using a microplate reader [16]. With more cells alive in one well, higher absorbance should be measured.

4 Notes

1. The K91 Blukan strain can be requested from Dr. Smith through this link: http://www.biosci.missouri.edu/SmithGP/PhageDisplayWebsite/PhageDisplayWebsiteIndex.html.
2. The SKBR-3 phage is a bioengineered fd-tet phage that surface displays a SKBR-3-targeting peptide, VSSTQDFP, on each of its 3900 copies of pVIII major coat protein. The SKBR-3-targeting peptide/phage was discovered by our group through a well-known biotechnique called phage biopanning [10]. The detailed protocols of the affinity selection of a cell targeting phage/peptide and amplification of the SKBR-3 phage can be found in our previous publications [10, 11].
3. Heat the mixture while stirring to allow the temperature to increase slowly from room temperature to 50 °C. The PEG

solution will become homogeneous and transparent around 35 °C. Stop heating and store the solution at room temperature for further use.

4. This step allows phage to infect K91 Blukan cells with higher efficiency.
5. If the yield of phage is lower than 10^{14} virions per liter of LB medium, increase the incubation time from 16 to 20 h. The bioengineered phage-infected bacteria grow more slowly than those infected with wild-type phage, so a longer incubation time may lead to a higher yield. If possible, the phage can also be amplified in a fermenter with the reported parameters (pH 7.4, speed 800 rpm, bubbled with oxygen gas, 28 °C) to maximize the yield [17].
6. Using a NanoPhotometer (IMPLEN Inc.) can save phage solution sample. Only 5 μl of phage solution is needed for each measurement.
7. $A = \varepsilon cl$ With A = absorbance (optical density, OD), unitless; ε = absorption coefficient or absorptivity, [ml/(mg cm)]; c = concentration of the substance, [mg/ml]; l = length of the path of light through the thickness of the cuvette, [cm].

   For fd phage: M_w of fd phage = ~2.17×10^7 g/mol, ε of phage = 3.84 ml/(mg cm) [18], when A (OD_{269}) = 1,

   $A = 1 = \varepsilon cl = 3.84$ ml/(mg cm) × c × 1 cm,

   c = 0.26 mg/ml = 0.00026 g/ml.

   The concentration of fd phage = 0.00026 g/ml × $6.022 \times 10^{23}/2.17 \times 10^7$ = ~7.2×10^{12} virions/ml.
8. The mechanism of the activation of the carboxylic group in PPa by EDC can be found in Fig. 2a.
9. 2-mercaptoethanol is added to quench the unreacted EDC molecules.
10. This protocol is for preparing ~50% partially modified PPPa. If fully modified PPPa is wanted, just add 45 μl of the EDC-activated PPa solution instead of 22.5 μl and keep other steps the same.
11. Phage fd may aggregate when stored at 4 °C. Warm up the stock phage solution and vortex for 5 min before using.
12. The same method can be extended to the modification of other types of viruses to generate novel virus-PPa complexes.
13. The conjugation of the activated carboxylic group in PPa and the amino group on SKBR3 phage is shown in Fig. 2b.
14. PPPa is much more straightened and rigid in morphology than unmodified phage due to the strong hydrophobic interactions between PPa macrocycles [14] (*see* Fig. 3).
15. The UV-Vis absorption spectrum of PPPa should show the similar pattern as that of PPa. In the fluorescence spectrum of

PPPa, an energy hopping can be observed due to the tryptophan-based fluorescence resonance energy transfer (FRET). The FRET is caused by the successful conjugation of PPa with phage particles [14].

16. DPBF can be quenched by singlet oxygen in an organic solvent. Therefore, the decay of the DPBF absorption at 410 nm is evidence for the generation of singlet oxygen by PS [14]. Before tests, make sure the solvent is saturated with oxygen gas, which is also one of the reactants.
17. Other time points (3, and 5 h of culturing) are parallel performed to optimize the conditions for PDT.
18. The best exposure time for different cell types and conditions may vary.
19. Other similar cell viability kits can also be used here.

Acknowledgment

We would like to thank the financial support from National Institutes of Health (EB021339, CA200504 and CA195607) and Oklahoma Center for Adult Stem Cell Research (434003).

References

1. Lucky SS, Soo KC, Zhang Y (2015) Nanoparticles in photodynamic therapy. Chem Rev 115:1990–2042
2. Dougherty TJ (1984) Photodynamic therapy for treatment of cancer. J Opt Soc Am B 1:555–555
3. Marcus SL (1992) Photodynamic therapy of human cancer. Proc IEEE 80:869–889
4. Shirasu N, Nam SO, Kuroki M (2013) Tumor-targeted photodynamic therapy. Anticancer Res 33:2823–2831
5. Cao BR, Yang MY, Zhu Y, Qu XW, Mao CB (2014) Stem cells loaded with nanoparticles as a drug carrier for in vivo breast cancer therapy. Adv Mater 26:4627–4631
6. Roberts DJH, Cairnduff F (1995) Photodynamic therapy of primary skin-cancer–a review. Br J Plast Surg 48:360–370
7. Cao BR, Mao CB (2009) Identification of microtubule-binding domains on microtubule-associated proteins by major coat phage display technique. Biomacromolecules 10:555–564
8. Smith GP, Petrenko VA (1997) Phage display. Chem Rev 97:391–410
9. Murugesan M, Abbineni G, Nimmo SL, Cao BR, Mao CB (2013) Virus-based photo-responsive nanowires formed by linking site-directed mutagenesis and chemical reaction. Sci Rep 3:1820
10. Abbineni G, Modali S, Safiejko-Mroczka B, Petrenko VA, Mao CB (2010) Evolutionary selection of new breast cancer cell-targeting peptides and phages with the cell-targeting peptides fully displayed on the major coat and their effects on actin dynamics during cell internalization. Mol Pharm 7:1629–1642
11. Li X, Mao CB (2014) Using phage as a platform to select cancer cell-targeting peptides. In: Lin B, Ratna B (eds) Virus hybrids as nanomaterials. Humana Press, New York, pp 57–68
12. Savellano MD, Pogue BW, Hoopes PJ, Vitetta ES, Paulsen KD (2005) Multiepitope HER2 targeting enhances photoimmunotherapy of HER2-overexpressing cancer cells with pyropheophorbide-a immunoconjugates. Cancer Res 65:6371–6379
13. Erogbogbo F, Tien CA, Chang CW, Yong KT, Law WC, Ding H, Roy I, Swihart MT, Prasad PN (2011) Bioconjugation of luminescent silicon quantum dots for selective uptake by cancer cells. Bioconjug Chem 22:1081–1088
14. Gandra N, Abbineni G, Qu XW, Huai YY, Wang L, Mao CB (2013) Bacteriophage

bionanowire as a carrier for both cancer-targeting peptides and photosensitizers and its use in selective cancer cell killing by photodynamic therapy. Small 9:215–221

15. Cao BR, Xu H, Mao CB (2014) Phage as a template to grow bone mineral nanocrystals. In: Lin B, Ratna B (eds) Virus hybrids as nanomaterials. Humana Press, New York, pp 123–135
16. Zhu HB, Cao BR, Zhen ZP, Laxmi AA, Li D, Liu SR, Mao CB (2011) Controlled growth and differentiation of MSCs on grooved films assembled from monodisperse biological nanofibers with genetically tunable surface chemistries. Biomaterials 32:4744–4752
17. Grieco SHH, Wong AYK, Dunbar WS, MacGillivray RTA, Curtis SB (2012) Optimization of fermentation parameters in phage production using response surface methodology. J Ind Microbiol Biotechnol 39:1515–1522
18. Unwin RR, Cabanas RA, Yanagishima T, Blower TR, Takahashi H, Salmond GPC, Edwardson JM, Fraden S, Eiser E (2015) DNA driven self-assembly of micron-sized rods using DNA-grafted bacteriophage fd virions. Phys Chem Chem Phys 17:8194–8202

"Chapter 42" - an Afterword

Making VLPs in Plants: The Answer to Life, the Universe and Everything? Facts, Trends, and Prospects

Christina Wege and George P. Lomonossoff

Abstract

As transmitted to humans by Douglas Adams, one has to await the revelation of the critical importance of the number 42 to sneak a peek into the fundamental answers to those questions one has always wanted to ask—but never did. ... Hence, this Chapter 42 provides exceptionally brief and personally biased insights into the (state of the) art of virus nanoparticle (VNP) tailoring and utilization, leaving it to you to speculate on future prospects, and, finally—as this is a methods book—describes our recommended protocol on how to acquire and promote both understanding and application of evolutionarily optimized viral particles. The essential outcome is meant to be advantageous for both the discipline of virus-based nanotechnology and future trends that might be followed in science and society, and might serve as an appetizer for a follow-up course, palatable for students, researchers, politicians, commercially interested readers, and further stakeholders.

Key words Virus nanoparticles (VNPs), Virus-like particles (VLPs), Douglas Adams, Farming, Trends, Applications, Industry, Politics, Prospects

1 Introduction

The international scientific "viral nanoparticle community" is small but tenacious, a bit crazy about the fascinating viral properties, and with numerous explorers enjoying exchange with distinct fields of research beyond their own nose, inspiring literature from science to fiction, and comfortable evenings. This has collectively resulted in creative approaches to overcome the many obstacles that arose during the initial stages of virus nanoparticle (VNP) adaptation to unprecedented uses. These included developing methods for the genetic and chemical modification of VNPs without destroying their fundamental properties, overcoming problems of scaling-up production of the VNPs and last, but not least, overcoming the skepticism of the scientific community. As we endeavor to prove in this final Chapter 42,

Christina Wege and George P. Lomonossoff (eds.), *Virus-Derived Nanoparticles for Advanced Technologies: Methods and Protocols*, Methods in Molecular Biology, vol. 1776, https://doi.org/10.1007/978-1-4939-7808-3,
© Springer Science+Business Media, LLC, part of Springer Nature 2018

acknowledging Douglas Adams who foresaw that this chapter might be offering previously unexpected solutions for life (*see* **Note 1**), there are now solutions to many of these issues. We have to confess, however, that we doubt it will provide the answer to the universe and everything—but one never knows ... ;-)

Here, we will therefore provide you with a final protocol that may help virologists and nanotechnologists to establish and consolidate profound new developments, may interconnect you with respective fields of your virus-focused interest, and might, in the long run, enhance the discipline of virology and the study of virus structures. Hence, if you are a student, young or established or senior researcher, politician, investor, grant proposal reviewer, philosopher, or committed writer, please make use not only of this Chapter 42 (in the sense of Douglas Adams, *see* **Note 1** and [1]), but of the whole of this book, to spread and critically discuss its ideas and concepts.

2 Materials

All components essential for the success of your experiment have to be prepared and/or consumed and digested with commitment and perseverance. The starting point is your idea on "it": your most coveted virus-enabled tool.

2.1 Good Virology and Molecular Biology Literature

Depending on "it", i.e., the intended area of VNP application, take into account basic literature on viruses and host molecular biology before going into details of your experiment or commercial development. We recommend a thorough study of fundamental textbooks and review articles to become an expert on the molecular biology of your virus, expression system of choice, and VNP processing options. Do not miss looking into the following:

1. Hull, R. (2014): (Matthews') Plant Virology. 5th Edition. [2].
2. King, A.M.Q. et al. (2012): Virus Taxonomy. Classification and Nomenclature of Viruses. Ninth Report of the International Committee on Taxonomy of Viruses, and Davison, A.J. (2017): Introduction to 'ICTV Virus Taxonomy Profiles'. The ICTV Online (10th) Report on Virus Taxonomy [3].
3. Khudyakov, Y. and Pumpens, P. (2016). Viral Nanotechnology. [4]
4. Recent textbooks on virus or VNP properties and handling, reviews on and, finally, original publications in your preferred areas of research and development. An impressive number of comprehensive review articles on the progress made with VNPs and virus-assisted products farmed in plants, bacterial, yeast, or animal cell systems, or with viral building blocks assembled

in vitro, has been published within the last decade. A very incomplete choice of respective articles and books is suggested in the references list of this chapter [5–32].

5. Your favorite chapters in THIS book [33].

2.2 A Group of Like-Minded People ...

... in your own research team and collaborating institutions and companies, not only fond of interdisciplinary work but also experts in complementary areas. And never underestimate the importance of enjoying their company!

2.3 All Materials and Equipment You Will Need for Your Purpose

1. Pencil, paper, laptop, or desktop.
2. A glass of real ale, rich wine, tasty tea, or something completely different meeting your preferences for inspiring work (*see* **Note 2**).
3. Viruses or virus components, to be prepared from suitable hosts.
4. Everything else, from lab equipment to 3D printer, soldering iron, and camera.

3 Methods

3.1 Design and Realization of "It" Toward Your Proof-of-Concept Project

Since these steps represent standard good scientific practice, they are described only briefly.

1. Develop your goal and experimental strategy (from Subheading "Good Virology and Molecular Biology Literature", **items 1–5**).
2. Perform the experiment with the office and labs at ambient temperature for several months: proceed up to small-scale fabrication.
3. Harvest the virus-based products and verify both their identity and functionality most critically.
4. Try not to be too disheartened when things do not go entirely according to plan!

3.2 Up-Scaling and Production of "It" in Your Favourite Host

Expression technology by use of agroinfiltration-based methodology may be appropriate for plant-based expression. Alternatively, when working with bacteriophages, think of fermenter production in bacterial cells.

1. Decide on up-scaling methodology.
2. Identify companies suitable for supporting the development of your invention.
3. Become famous for "it" and enjoy your achievements (*see* **Note 3**).

4 Notes

1. Douglas Noel Adams (Cambridge/England 11 March 1952–Montecito/California 11 May 2001) foresaw the importance of Chapter 42 as follows:

   "“Shhh,” said Loonquawl with a slight gesture, “I think Deep Thought is preparing to speak!”

   There was a moment’s expectant pause while panels slowly came to life on the front of the console. Lights flashed on and off experimentally and settled down into a businesslike pattern. A soft low hum came from the communication channel.

   “Good morning,” said Deep Thought at last.

   “Er … good morning, O Deep Thought,” said Loonquawl nervously, “do you have … er, that is …”

   “An answer for you?” interrupted Deep Thought majestically. “Yes. I have.”

   The two men shivered with expectancy. Their waiting had not been in vain.

   “There really is one?” breathed Phouchg.

   “There really is one,” confirmed Deep Thought.

   “To Everything? To the great Question of Life, the Universe and Everything?”

   “Yes.”

   Both of the men had been trained for this moment, their lives had been a preparation for it, they had been selected at birth as those who would witness the answer, but even so they found themselves gasping and squirming like excited children.

   “And you’re ready to give it to us?” urged Loonquawl.

   “I am.”

   “Now?”

   “Now,” said Deep Thought.

   They both licked their dry lips.

   “Though I don’t think,” added Deep Thought, “that you’re going to like it.”

   “Doesn’t matter!” said Phouchg. “We must know it! Now!”

   “Now?” inquired Deep Thought.

   “Yes” Now …”

   “All right,” said the computer, and settled into silence again. The two men fidgeted. The tension was unbearable.

   “You’re really not going to like it,” observed Deep Thought.

"Tell us!"

"All right," said Deep Thought. "The Answer to the Great Question ..."

"Yes ...!"

"Of Life, the Universe and Everything ..." said Deep Thought.

"Yes ...!"

"Is ..." said Deep Thought, and paused.

"Yes ...!"

"Is ..."

"Yes...!!! ...?"

"Forty-two," said Deep Thought, with infinite majesty and calm.'

(*Douglas Adams, The Hitchhiker's Guide to the Galaxy; © 1979 Serious Productions Ltd. [1], there: end of Chapter 27*)

2. Alcoholic drinks are hazardous and are toxic compounds and thus not recommended for consumption in excess.
3. If you have followed the instructions of this chapter up to this note and are now the head of a promising VNP company or the minister of science of your country, it's time for a baton change—why not write a follow-up Methods and Protocols book to ensure long-term success of Molecular and Applied Virology with a focus on phages and plants?

References

1. Adams DN (1979) The Hitchhiker's guide to the galaxy. Pan Books, Macmillan Publishers Ltd., London
2. Hull R (2014) Plant virology, 5th edn. Academic, Boston. https://doi.org/10.1016/B978-0-12-384871-0.00023-6
3. King AMQ, Adams MJ, Carstens EB, Lefkowitz EJ (2012) Virus taxonomy. Classification and nomenclature of viruses. Ninth report of the International Committee on Taxonomy of Viruses (ICTV). Elsevier, Amsterdam; and: The ICTV Online (10th) Report, see https://talk.ictvonline.org/ictv-reports/ as introduced by Davison AJ (2017) Introduction to 'ICTV Virus Taxonomy Profiles'. J Gen Virol. 98:1–1. https://doi.org/10.1099/jgv.0.000686
4. Khudyakov Y, Pumpens P (2016) Viral nanotechnology. CRC Press, Boca Raton. https://doi.org/10.1006/viro.1998.9104
5. Bittner AM, Alonso JM, Górzny ML, Wege C (2013) Nanoscale science and technology with plant viruses and bacteriophages. In: Mateu MG (ed) Structure and physics of viruses: an integrated textbook, Subcellular biochemistry (ISSN: 0306-0225), vol 68. Springer Science+Business Media, Dordrecht, pp 667–702. https://doi.org/10.1007/978-94-007-6552-8_22
6. Crisci E, Bárcena J, Montoya M (2012) Virus-like particles: the new frontier of vaccines for animal viral infections. Vet Immunol Immunopathol 148:211–225. https://doi.org/10.1016/j.vetimm.2012.04.026
7. Culver JN, Brown AD, Zang F, Gnerlich M, Gerasopoulos K, Ghodssi R (2015) Plant virus directed fabrication of nanoscale materials and devices. Virology 479–480:200–212. https://doi.org/10.1016/j.virol.2015.03.008
8. Czapar AE, Steinmetz NF (2017) Plant viruses and bacteriophages for drug delivery in medicine and biotechnology. Curr Opin Chem Biol 38:108–116. https://doi.org/10.1016/j.cbpa.2017.03.013
9. Dragnea B (2017) Virus-based devices: prospects for allopoiesis. ACS Nano 11:3433–3437. https://doi.org/10.1021/acsnano.7b01761
10. Fan XZ, Pomerantseva E, Gnerlich M, Brown A, Gerasopoulos K, McCarthy M, Culver J, Ghodssi R (2013) Tobacco mosaic virus: a bio-

logical building block for micro/nano/biosystems. J Vac Sci Technol A 31:050815. https://doi.org/10.1116/1.4816584

11. Glasgow J, Tullman-Ercek D (2014) Production and applications of engineered viral capsids. Appl Microbiol Biotechnol 98:5847–5858. https://doi.org/10.1007/s00253-014-5787-3
12. Koch C, Eber FJ, Azucena C, Förste A, Walheim S, Schimmel T, Bittner A, Jeske H, Gliemann H, Eiben S, Geiger F, Wege C (2016) Novel roles for well-known players: from tobacco mosaic virus pests to enzymatically active assemblies. Beilstein J Nanotechnol 7:613–629. https://doi.org/10.3762/bjnano.7.54
13. Kolotilin I, Topp E, Cox E, Devriendt B, Conrad U, Joensuu J, Stoger E, Warzecha H, McAllister T, Potter A, McLean MD, Hall JC, Menassa R (2014) Plant-based solutions for veterinary immunotherapeutics and prophylactics. Vet Res 45:117. https://doi.org/10.1186/s13567-014-0117-4
14. Koudelka KJ, Pitek AS, Manchester M, Steinmetz NF (2015) Virus-based nanoparticles as versatile nanomachines. Annu Rev Virol 2:379–401. https://doi.org/10.1146/annurev-virology-100114-055141
15. Lee KL, Twyman RM, Fiering S, Steinmetz NF (2016) Virus-based nanoparticles as platform technologies for modern vaccines. Wiley Interdiscip Rev Nanomed Nanobiotechnol 8:554–578. https://doi.org/10.1002/wnan.1383
16. Lee SY, Lim JS, Harris MT (2012) Synthesis and application of virus-based hybrid nanomaterials. Biotechnol Bioeng 109:16–30. https://doi.org/10.1002/bit.23328
17. Li F, Wang Q (2014) Fabrication of nanoarchitectures templated by virus-based nanoparticles: strategies and applications. Small 10:230–245. https://doi.org/10.1002/smll.201301393
18. Lin B, Ratna B (2014) Virus hybrids as nanomaterials: methods and protocols, Methods in Molecular Biology, vol 1108. Humana Press - Springer, New York
19. Liu Z, Qiao J, Niu Z, Wang Q (2012) Natural supramolecular building blocks: from virus coat proteins to viral nanoparticles. Chem Soc Rev 41:6178–6194. https://doi.org/10.1039/C2CS35108K
20. Mao C, Liu A, Cao B (2009) Virus-based chemical and biological sensing. Angew Chem Int Ed 48:6790–6810. https://doi.org/10.1002/anie.200900231
21. Marsian J, Lomonossoff GP (2016) Molecular pharming—VLPs made in plants. Curr Opin Biotechnol 37:201–206. https://doi.org/10.1016/j.copbio.2015.12.007
22. Paul M (2015) International society for plant molecular farming. Transgenic Res 24:375–380. https://doi.org/10.1007/s11248-014-9861-5
23. Rosales-Mendoza S, Angulo C, Meza B (2016) Food-grade organisms as vaccine biofactories and oral delivery vehicles. Trends Biotechnol 34:124–136. https://doi.org/10.1016/j.tibtech.2015.11.007
24. Sack M, Hofbauer A, Fischer R, Stoger E (2015) The increasing value of plant-made proteins. Curr Opin Biotechnol 32:163–170. https://doi.org/10.1016/j.copbio.2014.12.008
25. Saunders K, Lomonossoff GP (2013) Exploiting plant virus-derived components to achieve in planta expression and for templates for synthetic biology applications. New Phytol 200:16–26. https://doi.org/10.1111/nph.12204
26. Scotti N, Rybicki EP (2013) Virus-like particles produced in plants as potential vaccines. Expert Rev Vaccines 12:211–224. https://doi.org/10.1586/erv.12.147
27. Shirbaghaee Z, Bolhassani A (2016) Different applications of virus-like particles in biology and medicine: vaccination and delivery systems. Biopolymers 105:113–132. https://doi.org/10.1002/bip.22759
28. Soto CM, Ratna BR (2010) Virus hybrids as nanomaterials for biotechnology. Curr Opin Biotechnol 21:426–438. https://doi.org/10.1016/j.copbio.2010.07.004
29. Steele JFC, Peyret H, Saunders K, Castells-Graells R, Marsian J, Meshcheriakova Y, Lomonossoff GP (2017) Synthetic plant virology for nanobiotechnology and nanomedicine. Wiley Interdiscip Rev Nanomed Nanobiotechnol 9:e1447. https://doi.org/10.1002/wnan.1447
30. Wen AM, Steinmetz NF (2016) Design of virus-based nanomaterials for medicine, biotechnology, and energy. Chem Soc Rev 45:4074–4126. https://doi.org/10.1039/C5CS00287G
31. Young M, Willits D, Uchida M, Douglas T (2008) Plant viruses as biotemplates for materials and their use in nanotechnology. Annu Rev Phytopathol 46:361–384. https://doi.org/10.1146/annurev.phyto.032508.131939
32. Zeltins A (2013) Construction and characterization of virus-like particles: a review. Mol Biotechnol 53:92–107. https://doi.org/10.1007/s12033-012-9598-4
33. Wege C, Lomonossoff GP (2018) Virus-derived nanoparticles for advanced technologies, Methods in Molecular Biology, vol 1776. Humana Press, Springer Science+Business Media, LLC, part of Springer Nature, New York

Index

A

B

Christina Wege and George P. Lomonossoff (eds.), *Virus-Derived Nanoparticles for Advanced Technologies: Methods and Protocols*, Methods in Molecular Biology, vol. 1776, https://doi.org/10.1007/978-1-4939-7808-3,
© Springer Science+Business Media, LLC, part of Springer Nature 2018

C

D

E

F

Printed by Printforce, the Netherlands